DATE DUE

DEMCO 38-296

Implementation of Solar Thermal Technology

Solar Heat Technologies: Fundamentals and Applications
Charles A. Bankston, editor-in-chief

Implementation of Solar Thermal Technology

edited by Ronal W. Larson and Ronald E. West

The MIT Press
Cambridge, Massachusetts
London, England

y be reproduced in any form by any electronic or
recording, or information storage and retrieval)
isher.

Trade Typesetting Ltd., Hong Kong.
America.

Implementation of solar
thermal technology

on Data

_, / edited by Ronal W. Larson
and Ronald E. West.
    p.   cm. — (Solar heat technologies; 10)
ISBN 0-262-12187-5 (hc)
1. Solar energy. 2. Solar heating. I. Larson, Ronal. II. West,
Ronald E. (Ronald Emmett), 1933– . III. Series.
TJ809.95.S68   1996 vol. 10
[TJ810]
697'.78 s—dc20
[333.792'315'0973]   95-50765
                                       CIP

Contents

Series Foreword

Charles A. Bankston

This series of twelve volumes summarizes research, development, and implementation of solar thermal energy conversion technologies carried out under federal sponsorship during the last eleven years of the National Solar Energy Program. During the period from 1975 to 1986, the U.S. Department of Energy's Office of Solar Heat Technologies spent more than $1.1 billion on research, development, demonstration, and technology support projects, and the National Technical Information Center added more than 30,000 titles on solar heat technologies to its holdings. So much work was done in such a short period of time that little attention could be paid to the orderly review, evaluation, and archival reporting of the significant results.

It was in response to the concern that the results of the national program might be lost that this documentation project was conceived. It was initiated in 1982 by Frederick H. Morse, director of the Office of Solar Heat Technologies, Department of Energy, who had served as technical coordinator of the 1972 NSF/NASA study Solar Energy as a National Resource that helped start the National Solar Energy Program. The purpose of the project has been to conduct a thorough, objective technical assessment of the findings of the federal program using leading experts from both the public and private sectors, and to document the most significant advances and findings. The resulting volumes are neither handbooks nor textbooks, but benchmark assessments of the state of technology and compendia of important results. There is a historical flavor to many of the chapters, and volume 1 of the series will offer a comprehensive overview of the programs, but the emphasis throughout is on results rather than history.

The goal of the series is to provide both a starting point for the new researcher and a reference tool for the experienced worker. It should also serve the needs of government and private-sector officials who want to see what programs have already been tried and what impact they have had. And it should be a resource for entrepreneurs whose talents lie in translating research results into practical products.

The scope of the series is broad but not universal. It is limited to solar technologies that convert sunlight to heat in order to provide energy for application in the building, industrial, and power sectors. Thus it explicitly excludes photovoltaic and biological energy conversion and such thermally driven processes as wind, hydro, and ocean thermal power. Even with this limitation, though, the series assembles a daunting amount of information. It represents the collective efforts of more than 200 authors and editors. The volumes are logically divided into those dealing with general topics such as the availability, collection, storage, and economic analysis of solar energy and those dealing with applications.

The present volume focuses on the programmatic aspects of the government's involvement in solar energy rather than on the results of research, and it is the only volume to deal with the part of the government solar energy program known as commercialization. When the series was conceived, in 1982, the word "commercialization" was banned from the lexicon of the Reagan Department of Energy, and all programs that were considered "commercialization" were being hastily terminated without any reporting or evaluation of their success or failures. This made the volume controversial and made the goal of producing an objective assessment of the results of the programs difficult for the volume's authors, who did not generally have the resources or access to data necessary to conduct such evaluations. Fortunately, most of the authors were involved in the program themselves and had personal files and records as well as their own recollections and personal contacts to rely upon. The flavor of the book is definitely more historical and programmatic than others of this series, but it still provides the objectivity and comprehensiveness that has characterized the series.

In spite of the initial difficulty in obtaining approval for this volume, or perhaps because of it, this volume is potentially among the most useful of the series. Much of the material found here will not be found in other references. The authors and editors have performed, in part, the function that the new managers of DOE in 1982 did not value, that is, reporting and evaluating programs that had fallen from favor, and in so doing have provided valuable history, insights, assessments, and lessons from some very large and important parts of the government's effort to promote solar energy. Future policy makers should benefit from reading the descriptions and the pros and the cons of the government's attempt to commercialize solar technologies and should learn from the program's mistakes and successes.

Since volume 10 addresses different issues and different parts of the solar program than other books in the series, the reader with an interest in the technologies themselves will want to refer to volumes 2 through 9 for more technical information. Those who want a broader historical perspective of the entire government solar energy program will be interested in volume 1.

Acknowledgments

Dedication to solar energy by many people made this volume possible. Frederick Morse conceived this series and guided its preparation. Charles Bankston was the ever-patient series editor and our adviser. Paul Notari, Lynda McGovern-Orr, Annie Harris, Nancy Reece, Nancy Greer, Barbara Glenn, and Al Berger all provided valuable assistance at times during the life of this endeavor. Oscar Hillig was indispensable in his role as manager of the mechanics of the work. Zaida Burkholder and Debbie Saylor helped to keep us on track and handle manuscript processing. Sheldon Butt and Jackson Gouraud provided useful information and insight.

An important ingredient in the preparation of this volume and this series was review of all the chapters by knowledgeable peers. The comments, suggestions, and information supplied by some eighty reviewers have made this a better volume. It is a pleasure to hereby acknowledge the following persons for their review (we give affiliations as of the time of the review):

John Anderson, National Renewable Energy Laboratory (NREL); Bruce Baccei, Public Service Co. of Colorado; J. Douglas Balcomb, NREL; Charles Bankston, CBY Associates; Donald Beatty, consultant; William Bergman, W. S. Bergman Associates; Roger Bezdek, U.S. Treasury Department; Jon R. Biemer, Bonneville Power Administration: Kenneth Bordner, CAREIRS; Gerald W. Braun, Pacific Gas & Electric; Chris Cameron, Sandia National Laboratory (SNL); Millard E. Carr, U.S. Department of Defense (DOD); Nancy Carson, Office of Technology Assessment, U.S. Congress; Craig Christensen, NREL; John A. Clark, University of Michigan; Terry Clausen, NREL; Robert Cole, Boeing; Carl Conner, U.S. Department of Energy (DOE); Jerry E. Cook, Grundfos Pumps Corp.; William T. Cox, retired; John Dunlop, Minnesota State Energy Office; Kenneth Felton, retired; Freemen Ford, FAFCO Inc.; K. Scott Foster, Land O'Lakes Corp.; Mary Fowler, DOE; William Freeborne, U.S. Department of Housing and Urban Development (HUD); Karen George, University of Colorado; Richard Grossman, consultant; Dan Halacy, consultant; Charles E. Hansen, consultant; David E. Holmes, Naval Civil Engineering Laboratory; Jay Holmes, DOE; J. R. Howell, University of Texas at Austin; Gus Hutchison, Solar Kinetics, Inc.; Mary-Margaret Jenior, DOE; Alec Jenkins, California Energy Commission; Robert Jones, Arizona State University; Ronald Judkoff, NREL; Bea Kenney, consultant; Gene Kimmelman, Consumer

Federation of America; Alex Kotch, University of North Dakota; Frank Kreith, consultant; Charles Kutscher, NREL; Jan Laitos, University of Denver; W. Henry Lambright, Syracuse Research Corp.; Salvatore Lazzari, The Library of Congress; George O. G. Löf, Colorado State University; Frank P. Mancini, Arizona Solar Energy Office; Walter Matson, University of Oregon; E. Kenneth May, Industrial Solar Technology; Joe McCarty, U.S. Department of the Army; David C. Moore, HUD; Liz Moore, retired; Dana Moran, NREL; Frederick Morse, Morse Associates, Inc.; Paul Notari, NREL; Charles Ogburn, Auburn University; Marvin E. Olsen, Michigan State University; John Ortman, DOE; Mike Platt, Sunmaster Corp.; Peter Pollock, City of Boulder; Walt Preysnar, HUD; Richard Rademaker, Rademaker Consulting Corp.; Thomas Ratchford, American Association for the Advancement of Science; Steve Rubin, NREL; Beth Sachs, Vermont Energy Investment Corp.; Stephen L. Sargent, Western Area Power Administration; Stephen W. Sawyer, Bates College; David Schaller, U.S. Environmental Protection Agency (EPA); Robert Shibley, State University of New York at Buffalo; A. C. Skinrood, SNL; Scott Sklar, Solar Energy Industries Association; Phil Stern, consultant; Nancy Tate, Irving Burton Associates; William A. Thomas, Law & Science Associates; Lorin Vant-Hull, University of Houston; George Warfield, consultant; Patricia Weis Taylor, consultant; Ron D. White, NREL; Roland Winston, University of Chicago.

The first four figures, photographs of solar technologies, were supplied by the National Renewable Energy Laboratory.

Preparation of this volume took much longer than it should have; it was done over a period when solar energy implementation efforts virtually disappeared and nearly all of us were moving on to other things. A belief that the work reported in this volume is important kept the many persons named above, and most especially the authors, pursuing its completion. We hope others find the result worthwhile. Despite our efforts, surely errors and discrepancies remain in the final text. These are the sole responsibility of the editors.

Finally, the editors acknowledge the patience and support of their wives, Gretchen Larson and Marlies West, whom we left on too many Saturdays for too many years.

I SOLAR THERMAL TECHNOLOGIES IMPLEMENTATION

1 Introduction

Ronal W. Larson and Ronald E. West

1.1 Overview

1.1.1 Background

The energy shock of 1973, a petroleum supply crisis caused by a reduction of crude oil supplies by the Organization of Petroleum Exporting Countries (OPEC), caused waves that extended well into the 1980s and are still felt in the 1990s. Many people in the United States, both in the general public and in the government, concluded in the early 1970s that a national energy policy was essential for security and economic stability. Renewable energy, and especially solar energy, became one focus for national energy policy. Government-supported research into solar and other renewable energy sources greatly expanded. Indeed, many people perceived that some solar thermal technologies were sufficiently well developed that, in addition to continued research and development, a government program was needed to encourage the movement of these technologies into the private energy supply sector. So began government efforts to help commercialize solar energy.

Commercialization assistance involves support for the creation of profitable businesses. Although this is not a typical activity of the U.S. government, there are many precedents for such support, in fields ranging from agriculture to nuclear energy. Thus a government effort to stimulate solar energy commercialization, while somewhat unusual, was certainly not unique, even in the energy arena.

The government has many methods to help stimulate commercialization, including research, development, and demonstrations; purchases; information dissemination; technology transfer; grants, subsidies, and tax policy; cooperation with trade, technical, and promotional organizations; requirements and regulations (and exceptions therefrom); market analysis; and technical and financial consulting. Through legislation and administrative policy all of these methods were used for solar commercialization, beginning in the mid-1970s. Research and development on solar technologies are the topics of seven other volumes in this series, and market analysis is covered in volume 3 (*Economic Analysis of Solar Energy Systems*). The remainder of the available commercialization methods are covered in this volume.

In 1981 the federal government began to withdraw from solar commercialization in order to concentrate on strictly R&D issues; commercialization, for better or worse, was left to the private sector. This volume thus has a special responsibility to ensure that lessons learned from the federal solar thermal commercialization activities are not lost.

1.1.2   Purposes and Limitations of This Volume

In recording and evaluating federal efforts to commercialize solar thermal energy technologies, mostly in the 1970s and 1980s, we offer explanations of the various successes and failures. This work is intended for two types of readers—those interested in solar energy policy and those who hope to learn from the solar policy experience for transfer to other areas of federal activity.

All of our authors were asked to emphasize the historical events that shaped the many parts of solar thermal commercialization. We have tried to limit this historical review to commercialization. A broader historical perspective emphasizing R&D topics is the subject of volume 1 in this series (*History and Overview of Solar Heat Technologies*).

1.1.3   Description of Solar Thermal Technologies

Solar thermal technologies include all means of turning incoming solar radiation into thermal energy, either for use directly as heat or for conversion to electricity. These technologies have been divided by the federal solar thermal program into four categories—active solar heating and cooling, passive solar heating and cooling, industrial process heat, and solar thermal electricity—which are covered in part II of this volume. Excluded from the volume are four other solar technologies in which thermal energy is either not involved or is indirectly involved: photovoltaics, wind, biomass, and ocean energy. Most of federal government's solar commercialization efforts through the early 1980s were directed at the four solar thermal areas because they were perceived to be nearest to commercial viability.

Readers interested in solar total energy or cogeneration programs will find that it is barely covered in part II; total energy has been an orphan application in the federal solar program structure. There was an early program in this hybrid application area at Sandia National Laboratories; interest was also expressed in this topic at Congress's Office of Technology Assessment, beginning in 1974. As far as we could determine, there

Figure 1.1
An active solar thermal hot water system.

was only a single total energy demonstration project, at a Shenandoah, Georgia, hosiery plant (described briefly in chapter 8).

Before considering whether the correct technical systems were selected for commercialization support, let us first look briefly at the technologies included in this volume, relying in large part on photographs. Much more detail can be found in volumes 4 through 9 of this series.

Figure 1.1 shows a residential active solar thermal hot water system. It is characterized by distinctive south-facing, flat panels on the roof, typically totaling about 40 square feet. The term *active* implies a pump that circulates a fluid (usually water or water plus antifreeze) when the panel is warm enough. Another integral part of this active system is a heat storage tank. The tank is usually designed for overnight storage. The term *active system* also applies to larger heating systems used for residential space heating, but these have almost disappeared from the market, as their economics are rarely attractive. The economics are poorer for space heating because solar availability and system performance are usually poorest when heat is needed most and because the systems sit idle for part of the year. The term *active* also applies to much larger commercial

Figure 1.2
A passive solar thermal space heating system.

building systems and to mechanically driven systems used for cooling, or for both heating and cooling. Active components and systems are covered in detail in volumes 5 and 6 of this series (*Solar Collectors, Energy Storage, and Materials* and *Active Solar Systems*). The federal commercialization program for active solar systems is covered in chapter 4, with many (if not most) of the later chapters covering other parts of the federal efforts to assist the growth of this part of the solar thermal program.

Figure 1.2 shows a passive solar thermal system. The term *passive* implies no (or at least little) use of mechanical or electrical energy to move fluids. The system shown differs from the active system of figure 1.1 in two other obvious regards: it is better integrated into the structure, and it is larger. Here, the application is space heating; water heating is sometimes accomplished with completely passive (no pump) methods as well. These heating systems often involve glazed vertical walls and thermal storage systems integrated into the walls or floors, rather than as a separate tank. Other passive design options include sunspaces and Trombe walls. All these passive design techniques are described in volumes 7

Figure 1.3
An industrial process heat system.

through 9. The federal commercialization program for passive solar buildings is covered in chapters 5, 6, and 23 of this volume.

Figure 1.3 shows a typical solar thermal industrial process heat (IPH) system located on the roof of a factory. The hot water being generated is used in the factory to reduce the amount of fuel needed. These systems are larger and generate higher temperatures than those designed for space heating and domestic hot water heating. However, they also have the difficult commercialization problem of competing most often against natural gas, which is presently a very low cost alternative in the United States.

Industrial process heat systems may include the flat-plate type shown in figure 1.1, if the required process temperature is not too high. The trough of figure 1.3 tracks the sun along only one axis and is termed a medium-temperature approach. Still higher temperature, two-axis tracking approaches that could be used in IPH applications are shown below.

Figure 1.4
A heliostat array and power tower, Barstow, California.

Commercialization activities related to IPH applications are covered in chapter 7.

The final category of solar thermal system covered in this book includes all those approaches that can be used to generate electricity. The collectors include the flat-plate and single-axis tracking approaches shown in figures 1.1 and 1.3 above, but also systems that track the sun completely (with a two-axis tracker). One early two-axis tracking example used in a federal solar thermal electricity demonstration project is shown in figure 1.4. The mirrors in this large array are known as "heliostats"; they re-
direct and concentrate the sun's rays back to a central receiver, known as a "power tower." Paraboloidal "dishes" are also used, in sizes up to 30 or more feet in diameter, with a thermal collector at the focal point. Although these two-axis trackers can achieve higher temperatures and therefore a theoretically higher conversion efficiency, the dominant approach for installed commercial systems involves single-axis trough systems as shown in figure 1.3. The federal commercialization activities that were conducted to promote these electricity-generating applications are discussed in chapter 8.

1.1.4   Chronology of Solar Thermal Commercialization

Solar thermal energy research and development programs were initiated within the National Science Foundation (NSF) around 1970, but with no thought of a federal program to commercialize these technologies. However, in 1972, newly elected Representative Mike McCormack of the State of Washington was appointed to the House Committee on Science and Astronautics. A chemist, McCormack was given the assignment of chairing a new Energy Task Force for this Committee, which was then headed by Representative Olin Teague of Texas. Because of the nature of committee assignments, McCormack only had responsibility for conservation and solar energy; the major conventional energy sources were the responsibilities of other committees. Fortunately for him, though, the federal activities in solar energy were being conducted under a new NSF program called "RANN" (Research Applied to National Needs). Because NSF was also one of the principal oversight responsibilities of the House Science Committee, there were few jurisdictional hurdles for Congressman McCormack to overcome. Few perceived the solar area as essential for national security, but the RANN program had identified the solar area as one to which they could apply as much development as research in the usual federal lumping together of R&D. Despite the *R* in its name, RANN was intended to be on the development end of the R&D spectrum.

Hearings were held by the Energy Task Force throughout 1973, and even prior to the oil embargo of that fall, a determination had been made to submit legislation that would accelerate the commercialization of solar residential hot water and heating systems through a federal demonstration program. Demonstrations are a natural third step for an increase in federal support for a technology. Representative McCormack predicted that the demonstration program would launch industry and the American homeowner "into a new exciting and economically healthy arena" (Senate 1974). In chapter 2 and in chapters 9–13, there is still greater detail on this first solar commercialization legislation and the subsequent demonstration programs that developed out of it.

We will see later that it is important to understand how this first commercialization legislation was viewed at the time of its introduction. Most witnesses stated that solar hot water heaters were simple, had been in use for decades, and could be economical if produced in large quantities;

according to them, there was clear support for the legislation. However, NSF said that the legislation was premature—that the RANN program already had a demonstration character, and that it would be best to let this rapidly growing program continue longer before beginning to commercialize through a larger demonstration program. To put this in context, the original McCormack legislation called for a five-year, $50 million demonstration effort, which dwarfed the total NSF solar budget of $4 million in 1973. Counteracting this NSF testimony were numerous NSF contractors and other solar experts who attested to the technical readiness of solar hot water heaters. The National Aeronautics and Space Administration (NASA) expressed subdued support for the demonstration program as well. Because NASA also received its oversight from the House Committee on Science and Astronautics (now the House Committee on Science and Technology), it was natural to select NASA as a principal in the demonstration program. The oil embargo of October 1973 was followed by a great increase in congressional interest in this solar demonstration legislation. Subsequently it passed in subcommittee, committee, and the full House without great discussion or controversy; in fact, there were only three votes against the legislation in the House.

In the Senate the situation was a great deal more complicated because of jurisdictional issues. The House version was sent to five different Senate committees by their own request; this was said to be a record number of referrals. The reasons for this complexity and the committee deliberations are further discussed in the next chapter. The Senate worked relatively rapidly, given this logistical burden. As the idea traversed the five Senate committees, another strong element was added: the Department for Housing and Urban Development (HUD). HUD was to share the demonstration responsibility with NASA, while NSF retained the research side of solar thermal development. The new demonstration and other programs were now hindered by the need for cooperation among several agencies. The differences in the House and Senate versions were resolved in conference, and the resulting legislation was signed by President Ford as PL 93-409 on 11 September 1974.

Given the oil embargo and the strong enthusiasm for solar legislation by almost everyone, Representative McCormack and the House Science Committee introduced a second piece of solar commercialization legislation even before PL 93-409 had passed. This new legislation increased the national emphasis on solar energy by calling for an integrated federal solar energy R&D effort, but it also called for the formation of a Solar Energy Research Institute (SERI), a national solar energy data bank, and other solar energy commercialization adjuncts. This legislation also became law in 1974 as PL 93-473.

Also in 1974, through PL 93-438, again in response to increasing national concern over the nation's energy future, a new federal energy R&D agency was formed, mostly out of the old Atomic Energy Commission (AEC). This new agency, the Energy Research and Development Administration (ERDA), was not charged with commercialization responsibilities. It took over only some of the demonstration responsibilities of PL 93-409 (with most remaining with HUD and NASA). Because there was internal federal dissension on where the solar program should best reside, eventually Congress mandated that the solar program be transferred out of NSF, but this was not a smooth transition either.

Given the ERDA mandate only in R&D areas, the few solar thermal commercialization efforts were continued under the Federal Energy Administration (FEA). This agency did not have a solar office at the time that PL 93-409 and PL 93-473 were initiated, but a small FEA solar office was in operation by 1974, beginning to look at issues of tax credits and so on.

The years 1975 and 1976 showed continued increases in the solar budget (see section 1.2), primarily for R&D but also for the production of solar information for homeowners. The first solar demonstration projects were selected and begun in 1976. In late 1976, newly elected President Carter committed himself to forming a Department of Energy. This new cabinet-level office was intended to elevate energy concerns in the national dialogue and to centralize efforts that were then scattered. During this period, enthusiasm also developed for a series of regional solar energy centers (RSECs), which would be involved primarily in solar commercialization through information programs. This development was perhaps influenced most by representatives whose districts were not selected for the SERI site, which was announced as going to Golden, Colorado.

In July 1977 SERI opened its doors, reporting to ERDA for a short time, but then was transferred in October to the new Department of Energy (DOE), formed out of ERDA, FEA, and several other organizations. The four RSECs opened their doors in May 1978. The historical splitting of commercialization responsibilities among several agencies continued, however, within the Department of Energy. By law, the

Department of Energy was limited in the number of assistant secretaries that it could establish. Although still relatively small in funding, the emerging solar program was deemed politically important enough to deserve its own assistant secretary, whose title and responsibility was for "Conservation and Solar Energy" (CSE). However, the solar commercialization program and the RSECs were assigned to the deputy assistant secretary for conservation rather than the deputy assistant secretary for solar energy, who assumed responsibility for SERI. This was justified in part because the conservation technologies were more ready for commercialization and because the staff from FEA were assigned under the assistant secretary for conservation. However, because SERI had a one-and-a-half-year head start in commercialization programs and because SERI performed R&D work for both deputy assistant secretaries, there was always some controversy about lead roles in solar commercialization.

Nevertheless, virtually all of the state solar funding in these years was through the RSECs. This state funding topic is covered in chapter 29, from the perspective of two coauthors who saw the state and local action from both SERI (DeAngelis) and RSEC (Wrenn) perspectives. The four RSECs approached the state responsibilities somewhat differently, as described in chapter 20. Because of these organizational hurdles, the solar commercialization specialists at SERI had a roundabout path to communicate with their RSEC counterparts. Neither, moreover, had a direct line of communication to solar experts at HUD and NASA. If this sounds complicated to the reader, rest assured that it was also complicated for the solar employees in each of the above agencies.

Despite these problems, through the last years of the Carter presidency the solar commercialization programs matured and expanded, mostly for the solar thermal technologies, but also for wind, biomass, and photovoltaics. A tax credit was established (after a one-and-a-half-year delay that was reported to seriously hurt the industry) through PL 95-618, which was passed late in the 95th Congress (with its members up for reelection in November 1978). This credit greatly changed the nature of the solar commercialization effort, which was also aided by a second oil shortage in 1979, again with long lines and increasing prices at the gas pump.

The Carter view of commercialization was expressed by then Assistant Secretary for Conservation and Solar Energy Thomas Stelson in this way: "I have a staff Office of Commercialization reporting directly to me. Its specific functions include strengthening the commercialization and market planning capabilities in the program offices, ensuring that commercialization efforts are given proper emphasis by the programs" (Senate 1980).

During the 1980 presidential campaign, Ronald Reagan vowed to close down the Department of Energy, but the new administration was blocked in this effort by Congress. However, based on the view that commercialization is the job of private industry, not the government, the new administration did end most of the federal solar commercialization programs. Because of the oil and gas glut, with rapidly declining energy prices, the Democratic-controlled Congress fought less hard against these changes than it might have under other circumstances. Beginning in 1981, the RSECs were phased out, the commercialization staff at SERI was mostly eliminated, and the tax credit programs were first reduced in scope and then terminated at the end of 1985. This volume therefore has much greater detail up to 1981 than thereafter. Although the solar thermal R&D program was significantly reduced, it was not phased out in its entirety.

The national energy situation in 1981 is critical to understanding the decision to turn to the private sector alone for commercializing solar thermal technologies. That situation is seen in figure 1.5, which tracks over a forty-year time span composite fuel price, and both electric and nonelectric per capita energy consumption. In 1981 the U.S. composite fuel price was approximately $3.00 per million Btu, more than three times larger than ten years earlier and twice as large as today (all in 1990 dollars). Although there has been a precipitous price decline since 1981, the new administration could not know that this would occur. The price decline is likely to have been driven in large part by the worldwide success of energy conservation. Since 1983, per capita U.S. energy consumption has remained nearly constant at pre-1965 values, even though the average price of energy is only 58% as much as in 1983 in constant dollars. We think the plentiful supply and consequent low costs can be attributed in part to a growing awareness of the economic value of energy efficiency measures. We shall return to this point in section 1.4 as we discuss the appropriateness of curtailing the solar thermal commercialization program.

Under President George Bush, the federal government slowly revived a few of its solar thermal commercialization functions. There was, for

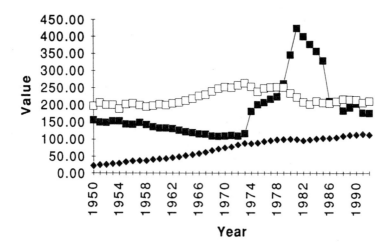

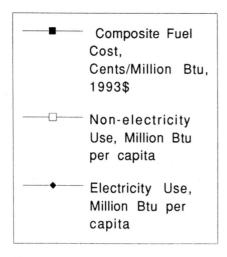

Figure 1.5
U.S. energy consumption and average fuel price since 1950. Source: EIA (1993).

instance, a growing emphasis on technology transfer from the national laboratories. This put the national laboratory staff in the position of having to prove to their DOE program managers that their work was both cutting-edge research that would not duplicate work done in industry and of direct importance to industry. The resulting balancing was not always convincing to all parties. This particular commercialization effort is discussed in chapters 21–24.

As this chapter is being concluded, the administration of President Bill Clinton is getting under way. The solar industry hopes for a change in direction, largely based on Clinton's emphasis on the environment in his campaign, exemplified by his selection of a major solar spokesperson, Senator Al Gore, as his vice presidential running mate. However, an even greater concern about the federal budget deficit may prevent much growth of solar commercialization activity over that proposed by the Bush administration.

Environmental issues have grown in importance over the lifetime of the solar thermal program. The nuclear program in the United States suffered from the accident at the Three Mile Island nuclear power plant on 28 March 1979, and then again in 1986 after the Chernobyl nuclear reactor meltdown in the then Soviet Union. Both of these events increased interest in solar energy, as did a growing concern in the late 1980s about acid rain, global warming, and similar intercontinental consequences of conventional energy use.

A more prominent public event than the first Earth Day (held in 1970), Earth Day 1990 focused public attention for several months on environmental issues. Renewable energy played an important role in both Earth Day observances, but more so in 1990 because of the large growth and subsequent decline in the federal solar program. Also, both were organized by one of the most active of solar advocates—Denis Hayes, who had been director of SERI during the latter part of the Carter administration.

There were both successes and failures in the solar thermal commercialization efforts that developed during the periods described above. Whether the existing commercialization efforts might have had a greater success with a longer lifetime is now a moot question. Nevertheless, we attempt throughout this chapter and volume to learn from the limited existence of those commercialization efforts.

1.1.5   Relationship to Other Volumes and Documents

This volume can stand alone, with no need to refer to other volumes in the series to understand what we wish to convey. However, other volumes contain interesting commercialization information, and several will be of interest from a technical point of view; because of them, we do not go into technical details.

Volumes 1 (Beattie 1996) and 3 (West and Kreith 1988) are specifically related to our volume. To those who are hoping to transfer information learned to another specific federal program, we especially recommend volume 1 (*History and Overview of Solar Heat Technologies*), which provides an overall historical review. Volume 3 (*Economic Analysis of Solar Thermal Energy Systems*) deals much more with economic methodology than with cost competitiveness, but the latter is also there. In the present volume, we do not try to assess the economic viability of any technology, and therefore the reader may find volume 3 to be a useful resource on this fundamental issue behind commercialization.

In chapter 4 of volume 3, Gerald Bennington shows how federal planners justified the measures that are discussed in the present volume using econometric models. Peter Spewak, in chapter 5, explores the rationale for federal action and the appropriateness of various types of action. Finally, volume 3 includes three chapters on the cost requirements for the solar thermal technologies: active solar heating and cooling (Mashuri Warren); passive solar heating and cooling (Charles Hauer); and solar thermal electric and industrial process heat (Ronald Edelstein).

1.1.6   Organization of This Volume and Chapter

This volume is organized into eight broad areas according to six major commercialization categories. Sections 1.2.1 and 1.2.2 justify this taxonomy. An especially important part of understanding any federal program is found in its funding history. Solar commercialization was never the major part of the solar program funding, although at one time it was a much larger fraction than at present. National funding for solar thermal commercialization is summarized in section 1.2.3, with still greater funding and budget detail in most of the subsequent chapters. Section 1.3 provides an introduction to each of the chapters that follow, showing how each fits within an overall policy framework and then summarizing what seems to us to be its main conclusions. Section 1.4 evaluates the federal solar thermal commercialization program, the second purpose of this volume, addressing both overall and specific questions of how well the solar thermal commercialization programs performed. Readers looking for guidance on what may work in very different future federal programs will find, we believe, that this volume indeed contains some answers. Finally, section 1.5 offers a set of learned lessons, which follow from the evaluation questions in section 1.4 but are more global in character. Most authors have also included evaluations and lessons learned; these also are essential to those looking in detail for justification of specific future commercialization policy actions.

1.2   Federal Commercialization Policy Options

In this section we discuss the full list of federal commercialization policy options, beginning with a short discussion of the reasons why these measures were perceived as necessary. Next we show how the dozens of different federal efforts can be sorted logically into six distinct categories and briefly describe several commercialization options, which are covered elsewhere in this series. After introducing the six categories, we discuss their and other funding levels over time.

1.2.1   Rationales for Federal Intervention

Perceived benefits to the nation were the basic reason for government interest in the stimulation of solar energy development and commercialization. In chapter 5 of volume 3 of this series, Spewak presents these national benefits of solar energy use: fuel savings, solar industry economic growth, environmental and health improvements, fewer economic dislocations, as well as the improved national and international security that results from reduced dependence on imported fuels (West and Kreith 1988). Certainly many in Congress and the administration welcomed some of these benefits, especially the last two, in the 1970s.

   Parity—that is, a "level playing field" vis-à-vis other energy sources— is an argument favoring encouragement of solar commercialization that also was advanced by solar advocates. Federal government support for energy technology development is not new, dating at least to tax incentives for petroleum in the 1920s. One estimate is $338 billion (in 1985 dollars) in federal incentives and subsidies for energy through 1977 (Cone, Brenchley, and Brix 1980), about half for oil and the rest for hydropower, coal, natural gas, and nuclear. In contrast, all solar (other than hydro) had received about $2 billion in support through 1978 and $6.6 billion (also in 1985 dollars) through 1984 (Spewak 1988). An independent, detailed analysis by the Alliance to Save Energy (ASE 1993) estimated federal support for (nonhydro) solar energy as being less than 1% of the total energy subsidies of $36 billion (1989 dollars) they found in the year 1989.

The word "barriers" is also often used in discussing the rationale for federal commercialization efforts. A good summary of these barriers to solar thermal introduction is also given by Spewak (1988) and so is not repeated here. Most authors in the following chapters also allude to specific barriers that their particular commercialization programs were designed to overcome. The most important commercialization barriers relate to the economics of the solar energy systems; many of the commercialization programs described in this volume were intended either to help drive the cost down (demonstration programs, technology transfer, etc.) or to find ways to reduce the first cost (through grants, tax credits, utility programs, etc.). Other programs provided information, training, reduced risk, or helped other organizations (such as states, schools, unions, banks, etc.) to do so.

1.2.2 Taxonomy of Commercialization Approaches

The U.S. government support for research, development, and commercialization of solar energy was not unprecedented. Nuclear energy technology was initiated, developed, and supported by the government. Moreover, the government took many steps beyond research and development in an effort to encourage commercialization of nuclear energy, most notably, undertaking fuel preparation, sponsoring demonstration projects, enacting liability limitations, establishing the Nuclear Regulatory and Atomic Energy Commissions, setting standards, and providing for waste disposal. One estimate is that nuclear energy had received $32 billion (1985 dollars) in incentives between 1950 and 1977 (Cone, Brenchley, and Brix 1980). The Alliance to Save Energy (ASE 1993) estimated that there was almost $11 billion in subsidies for the nuclear industry in 1989 alone.

The government can select from a wide range of possible actions to achieve a goal such as the commercialization of solar energy. This volume is organized according to the type of federal commercialization activities that were initiated. The first three (background) chapters are followed by a set of five program area chapters and then two to five chapters each in the six commercialization categories of (1) demonstration and construction, (2) quality assurance, (3) information, (4) technology transfer, (5) incentives, and (6) organizational support.

The management of most of the government's solar thermal activities was divided into four programs, according to technologies, as described in section 1.1.3. Management was mainly concerned with research and development, but each also oversaw commercialization activities. R&D activities are covered in volumes 4–9; the program activities are summarized as well in chapters 4–8 of this volume.

1.2.2.1 Demonstration and Construction

Technology demonstrations and construction of federal solar facilities constituted a significant portion of the federal commercialization effort. The various types of demonstration programs that were adopted were residential demonstrations, commercial demonstrations (including the Schools and Hospitals Program), the Solar in Federal Buildings Program, agricultural demonstrations, and a military construction program. These are described in chapters 9–13.

1.2.2.2 Quality Assurance

Consumers were protected from shoddy production and unethical practices by the establishment of equipment and system performance standards; these are discussed in chapters 14 and 15.

1.2.2.3 Information

Providing information to anyone who wanted it—the public, potential purchasers, manufacturers, technologists, managers, and planners—was an important ingredient in the federal plan to introduce and promote the potential of solar technology. The four adopted measures in this broad category were (1) targeted consumer information dissemination, (2) general public information centers, (3) technical information dissemination, and (4) training and education. These are discussed in chapters 16–19. Because they engaged in several categories of commercialization activity, the regional solar energy centers (RSECs), which played an important role in supplying financial information, are discussed separately in chapter 20.

1.2.2.4 Technology Transfer

The movement of new technology from federal R&D laboratories to the private manufacturing sector is an important step in the commercialization process; with the drastic cutbacks in the five other commercialization programs in recent years, it has become the dominant form of federal commercialization support. Technology transfer is discussed in chapters 21–24.

1.2.2.5 Incentives

Financial incentives in the form of taxation policy, loan guarantees, and grants can provide a powerful market stimulus; these are discussed in chapters 25–27. By comparing manufacturing data in chapter 3 with the taxation expenditures in chapter 25, one can gain an appreciation of the importance of the tax credits to the active solar industry, which can most clearly be seen in the change from 1985 to 1986, following the end of tax credits. The removal of the tax credit led directly to the collapse of the solar collector industry. The much smaller loan and grant programs are discussed in chapters 26 and 27.

Other federal incentive programs that were proposed but *not* adopted were (1) passive system tax credits, (2) tax deductions, (3) tax incentives for resale, (4) accelerated depreciation, (5) removal of tax benefits for conventional fuels, (6) producer payments, (7) producer loans, and (8) priority treatment for solar in government grant, loan, and mortgage insurance programs.

Another category of federal incentive that was adopted, business investment tax credits, is discussed slightly in chapters 8 and 30. Data on the magnitude of these business solar tax credits are provided in section 1.2.3; other incentive actions, such as a tax or tariff on imported oil, or a Btu or carbon tax, are briefly considered in section 1.5.

1.2.2.6 Organizational Support

The federal government provided direct and indirect support for international organizations, state and local levels of government, utility companies, and professional and labor organizations concerned with energy. These are discussed in chapters 28–31. The much smaller federal activities for other important professional communities, such as bankers, realtors, and insurance providers, are included in earlier chapters such as those for architects (chapter 6), information providers (chapter 16), and builders (chapter 23), and on the RSECs (chapter 20). The roles of the federal power marketing authorities, such as TVA (Tennessee Valley Authority), BPA (Bonneville Power Authority), and WAPA (Western Area Power Administration), and of PURPA (the Public Utility Regulatory Policies Act) are likewise discussed in other chapters, such as those on information programs (chapter 16) and utilities (chapter 30).

There are at least five additional commercialization options in the category of organizational support that were discussed by Congress or the executive branch but were never implemented: (1) establishing an industrial energy productivity corporation, (2) mandating utility marginal cost pricing, (3) granting environmental nonattainment exemptions, (4) authorizing generic environmental impact statements, and (5) expediting licensing, permitting, and zoning. Somewhat similar, and possibly to play a future role in the commercialization of solar thermal systems, are the federal mandates for certain utilities to implement integrated resource planning (IRP). This is an approach primarily designed to ensure that electric utilities pay as much attention to demand-side management (DSM) as to supply-side options. Many of the low-temperature solar thermal approaches (such as passive building design), from the perspective of a utility, are indistinguishable from conservation measures and can thus be viewed as DSM measures. Also, IRP rules call for giving consideration to the solar electric options.

Most of the organizational activities listed above were initiated specifically to enhance commercialization of solar thermal energy. There are, however, actions taken with other objectives, such as the Public Utility Regulatory Policies Act (PURPA) and the business investment tax credit, which are included because they had, or potentially had, significant impacts on solar commercialization.

1.2.2.7 Other Commercialization Activities

In addition to the six broad commercialization activities above, there are three other important commercialization activities *not* included in this volume. The first important area of federal commercialization support is monitoring and making available information on the solar resource, the subject of volume 2 in this series (*Solar Resources*; Hulstrum 1989). The second area is market analysis and the establishment of national goals, described in chapter 4 of volume 3 of this series (*Economic Analysis of Solar Thermal Energy Systems*). Through computer simulations, the impacts of such commercialization approaches as various levels of tax credits can be simulated for the use of the program managers in developing their commercialization strategies. National solar thermal goals are also touched on in sections 1.4 and 1.5. The third area of federal commercialization support not included in this volume is funding for research and development. R&D is especially important when driven by observations from the various commercialization programs; Federal R&D activities are covered in volumes 4–9 of this series.

1.2.3 Budget and Forgone Tax Costs for Solar Thermal Commercialization

1.2.3.1 Introduction
The annual budgets for the topics of this volume show quantitatively how the topics were perceived at different times and with respect to each other and to other budget categories. Unfortunately, it has proven difficult to reconstruct the expenditures and forgone revenues. Solar commercialization activity funding generally was imbedded in funding for other activities such as R&D or technology transfer. During some years, the solar thermal budgets were "taxed" to cover the information programs at SERI and the RSECs. Budget rescissions and deferrals also make the tracking job difficult; what was initially appropriated may have been later reduced. Other difficulties arise from the need to track several agencies and sometimes changes within a fiscal year. Also, some of the commercialization budgets were not the responsibility of the main energy agencies but were parts of such organizations as the Department of Housing and Urban Development (HUD), Agriculture (USDA), or Defense (DOD), as congressional authorizing committees sought ways to involve their respective responsible agencies in new energy areas.

The main source of our budget data is a recent Energy Information Administration (EIA) document on energy subsidies (EIA 1992) that covers the period 1978–1992. The earlier history is taken from a SERI (1978) report, while the later years 1993 and 1994 (projected) are from a solar trade publication, the *International Solar Energy Intelligence Report* (*ISEIR* 1993). Only the SERI report provides separate data on IPH, and none do on the passive program. The passive and IPH data were obtained mainly from chapters 5 and 7, respectively, of this volume and subtracted from "buildings" and "solar thermal electricity" totals to obtain the results shown. The taxation incentive values were largely obtained from the 1992 EIA report, supplemented by values from chapter 25 and personal communications (Bezdek 1993; Lagace 1993).

1.2.3.2 Four Solar Thermal Program Budgets

Figure 1.6 shows, in 1993 constant dollars, the total program budgets (R&D, administration, commercialization, etc.) over the twenty-four years from 1971 through 1994 for the four program areas that are covered in chapters 4–8. The peak appropriations (as opposed to "outlays" or "authorizations") for the active, passive, solar electric, and IPH programs were approximately 115, 32, 138, and 13 million dollars (current dollars), respectively. All program peaks occurred in 1980, except for the active program, which peaked in 1978. The cumulative total from 1971 through 1994, expressed in 1993 dollars, was approximately $2.9 billion.

The dominant features of figure 1.6 are

1. The rapid funding rise that followed the oil price rise of 1973.

2. The drop in funding around 1980, following the change from the Carter to Reagan administrations.

3. The budget years for 1974 through 1979 saw the largest budgets for the active systems program. The funding decline beginning in 1979 was a consequence of industry's strong support for a tax credit; a tax credit implies commercial readiness and therefore little need for a federal program.

4. Over the five-year period from 1978 to 1982, the passive component moved from essentially nothing to being at least half of the buildings program. However, because funding for the buildings program was dropping during much of this period, the passive budget has never been huge. The low funding for passive programs in the early years was perhaps mostly due to the relative fragmentation of the passive industry. As the active industry switched its emphasis to tax credits, the passive program began to grow.

5. Since 1980, there has been a steady growth in the percentage for the solar electric area, with about 90% of the recent solar thermal program budget being directed here. Virtually all of this recent total solar thermal budget is in the R&D category, with only minimal amounts for the commercialization topics of this book.

6. There has been small and discontinuous funding for the industrial process heat (IPH) program. These were the most difficult data to obtain because the IPH program has been folded in and out of the solar thermal

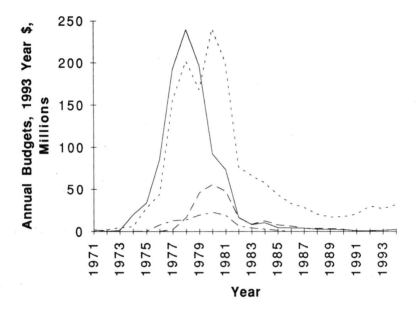

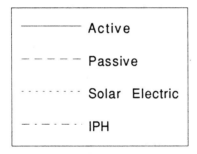

Figure 1.6
Four components of the federal solar thermal program budget.

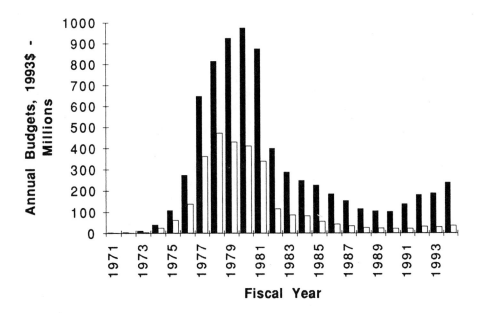

Figure 1.7
Federal solar thermal and total solar program budgets (excluding tax credits).

program. The data shown here were taken from chapter 7 of this volume, largely based on the portion of the program expended on trough collectors versus dish and heliostat approaches. The IPH program never reached 20% of the solar concentrator funding and dropped out of existence for several years.

1.2.3.3 Solar Thermal and Total Solar Budgets

Figure 1.7 compares the budgets for the total solar thermal program (the sum of the four programs shown in figure 1.6) and the total solar program; this plot is also in constant 1993 dollars. The solar thermal programs were at one time the dominant solar program; between 1974 and 1978, the four solar thermal technology programs comprised between

50% and 60% of the total solar budget. Since 1978, there has been a slow and steady decline in the relative funding of the four solar thermal areas; today, the combined solar thermal portion is less than 20% of the total. This decline can be traced to the growing importance of the tax credit for the active solar area, to the switch in federal support in 1981 from both R&D and commercialization to R&D alone, and to growing demands by the other solar programs for increased funding. For FY 1994, the photo-voltaics and biomass fuels programs were both larger than the solar thermal electric program, while the wind program was also only slightly smaller than the solar thermal electric program (*ISEIR* 1993). Cumulative funding for the four solar thermal programs was about $2.9 billion—about 40% of the cumulative total solar funding of about $7.3 billion (1993 dollars).

1.2.3.4   Solar Thermal Commercialization Totals
In figures 1.8 and 1.9, we show solar thermal commercialization budget information from this volume, along with data from Spewak (1988), and the EIA (1992) by our six categories. Readers are encouraged to look at individual chapters for additional commercialization budget details.

1.2.3.4.1   Programmatic Commercialization Budgets   Comparing all the data shows that solar thermal program commercialization accounted for about one-third of the solar thermal total program cost. Figure 1.8 shows annual values for the six commercialization categories. Because it so dominates the total, we discuss the tax credit part of the subsidization category separately in the next subsection. Interesting aspects of figure 1.8 are that the demonstration category was largest, followed closely by the information category, while other categories were either very small or almost nonexistent.

   These data were obtained only from the budget figures shown in the following chapters. Although we believe that they properly reflect trends and relative magnitudes, individual values are at best approximate.

1.2.3.4.2   Tax Credit Forgone Income   The two solar thermal tax cred-its, expressed as forgone revenues, amounted to about 150% of the total solar thermal program budget. Tax credits, therefore, were more than four times larger than the programmatic commercialization expenditures (R&D not included). Figure 1.9 shows the residential and business tax credits by year. In current and 1993 dollars, this tax credit category

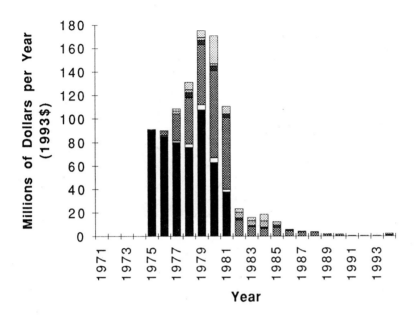

Organizations

Incentives (without tax credits)

Technology Transfer

Information

Quality Assurance

Demonstrations

Figure 1.8
Estimated solar thermal commercialization budgets for six approaches.

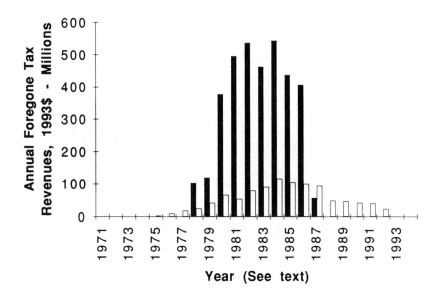

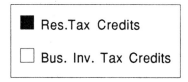

Figure 1.9
Residential and business investment solar tax credits.

reached peaks of about $470 million and $640 million in 1984 and totaled more than $3.1 and $4.3 billion dollars, respectively. Other interesting features of this figure are the large magnitude of both incentives, the factor of roughly 5 in residential over business credits, and the short duration of the residential credits.

1.2.3.5 Solar Thermal Commercialization Budgets for Specific Years

1.2.3.5.1 FY 1980

Commercialization Budgets The 1980 DOE budget request is especially useful in understanding the solar commercialization programs, because it was the largest and because it was the last that wasn't modified by rescissions and deferrals (DOE 1979, vol. 6). There were both decreases and increases in that year as the residential demonstration program was winding down and changing character. That document showed these commercialization programmatic changes between FY 1979 and FY 1980, expressed in the budget categories used in that year (in millions of dollars):

| | Budgets | | Comments |
	1979	1980	
A. Demonstrations			(stated totals in section
Demonstrations—buildings	27.1	14.5	1.2.3.5.2 are slightly
Demonstrations—data			higher than the sum of
collection	12.9	12.8	these subtotals)
IPH demonstrations	7.0	10.3	
B. Quality assurance			
Standards, codes	5.0	4.8	Perhaps half for solar thermal
C. Information	1.4	6.7	Large increase
Market development	13.8	11.5	
Market development-training	.8	8.7	
RSECs	13.6	21.7	See below for detail
D. Technology Transfer			
Technical support			
Buildings	9.9	3.5	
IPH	4.0	3.7	
E. Incentives			None in DOE budget
F. Organizations			Included in other areas above

Other: This document showed these rapidly increasing program totals that presumably included some commercialization also:

Passive	3.9	16.2
IPH	3.4	15.2

RSECs The Senate Committee on Energy and Natural Resources, in hearings on 18 March 1980 had interesting commercialization data on the role of the RSECs, showing a total in FY 1979 of about \$13.6 million, and in FY 1980 of \$21.7 million, approximately equally split between the four RSECs (Senate 1980, 14). The FY 1980 total was split into four categories (in millions of dollars), all of which we believe could be categorized as "commercialization":

Market analysis	3.3
Systems development support	3.2
Market tests and applications	4.2
Market development and training	11.0

There were small allocations in these totals for technology areas other than solar thermal, but the vast majority was for the solar thermal area, which dominated commercial sales in that year. This FY 1980 funding request also showed a projected RSEC total employment of 235 persons—almost all involved in solar thermal commercialization.

1.2.3.5.2   FY 1981

Commercialization Component   In a Senate budget hearing, Assistant Secretary Stelson noted, concerning the $652 million 1981 solar budget request (an increase of about $56 million): "Approximately $465.9 million will be directed to longer-term RD&D [research, development, and demonstration] activities and $186.3 million will be spent for near term development, marketing and commercial applications" (Senate 1980, 35). In percentage terms, this implied that near-term, nondemonstration commercialization efforts were consuming 28% of the proposed total solar (of $652 million) budget—mostly for the solar buildings budget category. Because at least $38 million (of the 1981 RD&D) category was to be spent on solar thermal demonstrations, the total solar thermal programmatic commercialization activities were to be at least one-third of the proposed 1981 total solar budget.

Demonstrations   From the 1981 budget document (DOE 1980, 27) we have these totals (in millions of dollars) for demonstrations alone over a three-year period:

		Budgets	
	1979	1980	1981
Buildings	55.0	36.7	18.0
IPH	11.0	14.0	20.8

This three-year comparison shows a phasing down of the buildings demonstrations, and a growing emphasis on the industrial process heat sector.

DOE Commercialization Staff Stelson's DOE staff request for FY 1981 was for 911 persons, with 652 at headquarters and 259 in the regions. This staff was to support a total (Conservation and Solar Energy) budget of almost $1.8 billion, of which more than one-third ($652 million) was for the solar thermal area. Above, we have estimated that about one-third of the solar thermal funds were used for the subjects of this volume. Thus, if we assume that staff allocations correspond to budget categories, about 100 DOE employees were involved in solar thermal commercialization in this peak year.

1.2.3.5.3 FY 1982 The differences between the Carter and Reagan administrations' solar thermal commercialization philosophies can be seen clearly in their 1982 budget documents. The Carter version was sent to Congress on 15 January 1981, only days before the change in administrations. Although this Carter 1982 budget was required by law, it had little impact; the Reagan 1982 budget that Congress discussed six weeks later bore no relation to the Carter budget. In response to a question from the House Committee on Science and Technology, the new administration stated that the solar energy R&D program would be unchanged: "The major change in direct DOE spending is the elimination of the application, or commercialization, part of our previous program" (House 1981).

The net result was a proposed 67% reduction in the solar 1982 budget from $589 million to $193.3 million (same hearing record). Because this was almost twice the percentage decrease that would be identified with the Carter budget categories, it is clear that the Carter and Reagan administrations' definitions of commercialization were vastly different.

1.2.3.5.4 FY 1983 The Reagan 1983 budget, prepared after one year in office, went even further in deemphasizing the energy area, as it stated: "In view of the limited responsibilities and role of the Federal Government in energy, it is wasteful and inappropriate to maintain a Cabinet-level Department of Energy" (U.S. Budget 1982 [for 1983], 5-42).

The FY 1983 solar thermal budget became a part of a proposed (never realized) new organization termed the "Energy Research and Technology Administration" (ERTA) within the Department of Commerce. Affirming its continued avoidance of federally supported commercialization, the Reagan administration stated in the appendices to that same FY 1983 budget: "In 1983, the solar energy program will conduct *generic and technology base research* in the areas of photovoltaic, solar thermal, bio-mass, and wind energy" (U.S. Budget 1982 [for 1983], I-F44; emphasis added). The same document showed the total solar budget data (in millions of dollars) in the first row of the following table:

	1981	1982	1983
1983 Budget estimates	574.8	454.1	159.2
EIA, 1992 actual	552.0	268.3	201.7

The second row above shows the actual total appropriations obtained from the EIA subsidy report (EIA 1992). The major differences for the 1982 column apparently can be attributed to later successful rescissions and deferrals. It is apparent from this three-year budget history that huge changes were occurring; these were especially felt in solar thermal com-mercialization—as noted in almost all of the following chapters.

1.2.3.5.5 FY 1992 and 1993 After 1982, there were few returns to the commercialization budgets of the peak years. The word "commer-cialization" never reappeared, as the program budgets and the program structure continually changed. The following table shows recent program budget categories (which have recently been presented by end use, rather than technology) as proposed for 1992 and as modified by Congress (in millions of dollars).

Category	1992 budget request (DOE 1991, 4-598)	1992 budget, actual (DOE 1992, App. 1-463)	1993 request
Solar utilities	157	195.1	193
Solar buildings	1.2	2.2	2.0
Solar industry	9.8	10.0	7.1
Solar transportation	29.6	30.0	38.8
Technical and financial assist.	4.0	7.6	6.0

In all cases, Congress raised the Bush administration proposals (the sec-ond column is larger than the first) and (except for transportation, where there was a large increase) the Bush administration proposed a small decrease again for the 1993 (third column) budgets. We clearly see the present emphasis on the solar electrical end-use sector; the solar buildings program is now only about 1% as large. Again, we emphasize that the four solar thermal programs of this volume cover all but the solar transportation (fuels) category. The only commercialization categories in the above are believed to be the technology transfer and information categories in the final line.

1.3   Commercialization Assistance Methods Covered in This Volume

1.3.1   Political and Market Environments and Actions

1.3.1.1   Role of Congress

In chapter 2 J. Glen Moore describes the many actions of Congress in the area of solar thermal commercialization. Moore has been at the very center of this activity as an employee of the Library of Congress's Congressional Research Service (CRS); his CRS experience began well before the first solar legislation and continues to this day.

A key evaluative question for this chapter is, Did Congress play its role properly? One part of the role is timing. Congress in the early years was clearly more aggressive than the administration in pushing for rapid solar commercialization. In hindsight, much of that effort may seem to have been misplaced; federal commercialization of the solar thermal technologies has not been a success. On the other hand, in the early 1970s there was strong corporate and citizen support for action. There was also a shortage of imported oil and a concern for future large price rises. It is certainly not surprising that Congress would choose to act. We concur with Moore in his concluding sentence: "Whatever the future may hold, much of the credit for any future success in the nation's use of solar resources must go to the farsighted efforts of Congress to build and sustain a viable federal program." We believe that many readers will enjoy this very detailed look at those "farsighted efforts."

1.3.1.2   Market Development

Chapter 3 prepared by Carlo LaPorta, provides a history of the production by the solar thermal energy industry. As a staff member and later consultant to the Solar Energy Industries Association (SEIA), which actively promoted solar commercialization and incentive activities, La Porta has an excellent background for this chapter.

LaPorta first describes the history of the various types of federal support or nonsupport. He then describes the market history of each of the solar thermal approaches, concluding on passive solar energy: "The building industry remains to be convinced that no-cost design changes are available that will improve the product and modest cost additions will offer even more superior buildings." Clearly, there is a long way to go in solar thermal commercialization, and this chapter helps to quantitatively document that distance. Although his chapter concentrates on total quantities installed, LaPorta also provides some interesting detail on specific solar energy system costs. Another chapter on costs that will be of interest to some readers is chapter 11 in volume 3 of this series (*Economic Analysis of Solar Thermal Energy Systems*).

1.3.2   Solar Thermal Program Areas

Chapters 4, 5, 7, and 8 give overviews from the perspectives of the four federal offices responsible for solar heat commercialization. These four offices were organized by technology. The organizational structures, as well as the differing perspectives on commercialization dictated by the technologies themselves, are made more clear in these four chapters. Chapter 6 gives a more detailed look at the evolution of one aspect of the passive program—commercial buildings.

1.3.2.1   Active Solar Heating

In chapter 4, from his perspective as a support contractor in the active solar thermal program area with positions at InterTechnology Solar Corporation, Planning Research Corporation, Planning and Management Associates, and Science Applications International Corporation, William Scholten shows how many of the succeeding chapters fit into an historical view of the federal solar heat commercialization strategy. Concluding that the demonstration programs were based on initial perceptions that were overly optimistic, Scholten urges greater early preparation for repair and refurbishment of faulty designs, while avoiding risky designs—procedures essential in the introduction of products in the home heating market. Nevertheless, he also notes that "many systems, particularly those based on sound design experience, proved the practicality of solar heating."

1.3.2.2   Passive Solar Energy Commercialization Program

The passive solar program, described in chapters 5 and 6, began somewhat later than the other three; it developed a strong national constituency (among architects and builders) with a relatively small total federal effort and has a continuing commercialization component. In recognition of these differences, we have described the passive solar program in two parts. We start with an up-to-date account (chapter 5) coauthored by the current DOE passive program manager, Mary Margaret Jenior, and an employee of one of the Department of Energy's support contractors, Robert T. Lorand. Because of the impact of dramatically reduced federal spending over the 1982–1992 time period, which they report in detail, Jenior and Lorand conclude that much of the promise of federal commercialization efforts in passive solar systems is presently impossible to evaluate. Although clearly proud of the program's accomplishments, they see that much more remains to be done: "A sustained commitment over a longer time horizon is needed . . . ."

We next provide an in-depth view (chapter 6) of one subcomponent of the solar heat programs, passive commercial buildings, written by Robert G. Shibley. From his present position as professor of architecture and director of urban design at the State University of New York (SUNY) at Buffalo, Shibley describes the scope and methods of program design, emphasizing the important role of the architect in promoting passive design, and discussing the results of this particular program. During the 1980–1982 period described, he was a project manager and later branch chief of the DOE passive program. "The premature interruption of the Passive and Hybrid Commercial Buildings Program in 1982," Shibley concludes, "has significantly set back attempts to position professional societies, industry groups, and educational institutions to address the structural issues inherent in building energy consumption."

1.3.2.3 Industrial Process Heat

Chapter 7 on the national program for industrial process heat (IPH) was prepared by David W. Kearney, who managed this program at SERI and subsequently served as a vice president at the Luz Corporation. Kearney describes the types of industries that need thermal energy at various temperature levels, and how the national IPH program established a series of demonstration programs to meet those needs. Noting that "the displacement of industrial process heating by conventional energy sources . . . will . . . be very difficult . . . until conventional fuel prices increase," Kearney concludes that "the major success of the DOE program has been to provide a comprehensive set of guidelines and lessons as a basis for reliable and cost-effective . . . systems. While some progress has been made, . . .

large scale ... IPH implementation has still to be achieved." The editors agree with these conclusions. We add that the temporary cancellation of this particular federal program after 1985 was particularly unfortunate, because the potential market is so large, because the system economics are as promising here as for any other solar thermal application, and because the prices of natural gas and oil must certainly increase.

Under a new name ("solar process heat" or SPH), this program was revived in 1990 by the DOE solar program's Office of Industrial Technology, with an emphasis on applications rather than technologies. In mid-1993 there were only two persons working in this revived program area at NREL (National Renewable Energy Laboratory) and three at Sandia in a small program that had then grown to about $2 million.

1.3.2.4   High-Temperature Solar

In chapter 8, J. C. Grosskreutz discusses the national high-temperature solar thermal program. Grosskreutz started approximately twenty years ago as a contractor to the national solar thermal program, when he was at Black & Veatch in Kansas City. Later recruited to the Solar Energy Research Institute (SERI), he continued and expanded upon this early understanding of the program, eventually becoming SERI's manager responsible for all technical programs and then returning to Black & Veatch until his retirement a few years ago.

Observing that: "smaller, modular systems ... were able to proceed through successful field development tests at a fraction of the cost of one central receiver plant," Grosskreutz notes: "The federal planning process overestimated the rate of technology development and component cost reduction"; he summarizes the problems of budget reductions this way: "[I]n a time of lean budgets, it is better to have one well-advanced option than three emerging ones."

1.3.3   Demonstration Programs

Five chapters (9–13) cover the solar thermal demonstration and construction projects that were among the first and keystone aspects of the federal commercialization plan.

1.3.3.1   Residential Demonstrations

Chapter 9 on the federal residential solar thermal demonstration program was prepared by Murrey D. Goldberg, who headed the International Program at SERI, then later edited and produced the final report on the residential demonstration program. This, the first, federal solar commercialization program, began as a congressional initiative. Goldberg discusses the shared responsibility between HUD and DOE (and several predecessors), emphasizing the technical deficiencies of the many installed systems that failed to live up to expectations.

In his final section Goldberg raises a number of evaluative questions. For example: "Might an early-purchase effort aimed, for example, at solar heating of military housing, have had a salutary push of the industry along the learning curve?" Goldberg's response: "Questions such as these can only possibly be answered when comprehensive programs are undertaken and sufficient follow-up data are collected and analyzed. The history of the residential demonstration program leaves most of them unanswered and unanswerable." Unfortunately, this conclusion is one with which we must agree.

1.3.3.2 Commercial Buildings

Chapter 10 on demonstrations of solar energy systems in commercial buildings is written by Myron L. Myers, from the perspective of NASA's Marshall Space Flight Center (MSFC) in Alabama, which played a leading federal role in this effort. Now retired after a career in engineering, Myers was one of the NASA program managers at the time of these demonstrations.

Myers describes the procedural elements of getting the demonstrations underway, the agency interactions, business responses, system technical performance, and evaluation efforts. The program was cut back in 1980, following a somewhat critical General Accounting Office (GAO) report, a report Myers discusses in detail. Among the lessons learned, the author favored one of the later components of the program: "[H]ad the hotel/motel hot water initiative idea come first, things would have been different." Myers closes with a number of success stories that may be helpful to those looking for clues on how to run a demonstration program.

1.3.3.3 Federal Buildings

Oscar Hillig has served two roles in the preparation of this volume. On the one hand, he has summarized in chapter 11 the experiences of the Solar in Federal Buildings Program (SFBP), demonstrating solar thermal technologies in federally owned buildings. Recently retired, Hillig brings to this task an excellent background as one of the principal managers of SFBP while at the Energy Technology Engineering Center of Rockwell

International's Rocketdyne Division; his group provided the program's day-to-day management for the responsible federal agency, Department of Energy. On the other, in the position of Rockwell's manager, Hillig served as the monitor for the Department of Energy for the production of this and all the other volumes in this series. As editors, we would like to acknowledge again his contribution for a job well done in that necessary and unsung part of this volume's production.

The federal buildings solar thermal demonstration program began under the National Energy Conservation Policy Act (NECPA; PL 95-619), with final rulemaking established in late 1979. The results of the program, which eventually covered 700 solar energy systems, were "disappointing," concludes Hillig; there was a need for "better site and application selection criteria." He believes that "the most significant contribution of the SFBP is in the lessons learned, which are in the three manuals .... " The manuals contain extremely valuable information on design, installation, and operation and maintenance of active solar heating and cooling systems. In retrospect, we think that there probably would have been greater national advantage to an earlier start in using federal facilities as a test bed. The federal government could more readily absorb the problems that invariably occur with a new technology than could the users in the private sector.

1.3.3.4 Agricultural Demonstrations

Chapter 12 covers demonstrations done through the Department of Agriculture (USDA)—an agency strongly oriented to information, demonstration, and service, although not directly responsible for energy—and is written by Robert G. Yeck and Marvin D. Hall, who were both active participants in the conduct of this demonstration program, at the national and state level, respectively.

Yeck and Hall make a strong case for the future involvement of USDA's Extension Service in solar commercialization. The agricultural demonstration projects were low-cost, and relatively cost-effective. The authors point out that, because of the seasonal nature of the need, solar crop-drying equipment should be used for other purposes as well; under the rules of the demonstration program, other uses were not permitted.

Evaluation questionnaires indicate favorable response both to the technology and the USDA commercialization program. Nonetheless, the authors conclude that a better-instrumented stage of development should precede demonstrations. One possibly transferable observation, obtained in hindsight, was that "the farmers receiving the most cost sharing were the least practical and economical in solar energy use." The editors concur with the authors' conclusion that future national renewable energy commercialization programs should take greater advantage of the strengths of USDA and its demonstration and information dissemination capabilities. This nation's farmers also have the know-how to make a system work and the longer time horizon and greater willingness to invest that will be needed to commercialize solar technologies.

1.3.3.5 Military Demonstrations

The last in this series on federal demonstrations, chapter 13 was authored by William A. Tolbert, who as an Air Force officer and part of an interservice coordinating group was involved in monitoring the Air Force solar construction programs in the late 1970s. Tolbert later joined the staff of SERI for a time; he has since left SERI to join a large engineering firm.

Tolbert devotes most of his chapter to how the different military services were organized for solar purchases and the types and numbers of projects that each installed. Like other users, observes Tolbert, "the military services were not at all prepared to apply solar thermal technology solutions to the energy problem." These problems were slowly overcome, with the result that for large-scale solar systems "a substantial level of acceptance is now visible." Tolbert concludes: "Demonstration programs have driven the learning curve to the level where solar thermal technologies will continue to be a part of the [energy self-sufficiency] future. After all, in addition to strengthening military missions, every barrel of oil the military saves through the application of solar thermal technology is a barrel of foreign oil they won't someday have to fight for." This advantage has been one of the important reasons used by solar enthusiasts to promote renewables.

1.3.4 Quality Assurance

The next two chapters (14–15) cover government support for assuring the integrity of the product—quality assurance.

1.3.4.1 Codes, Standards, and Certification

The author of chapter 14, Gene A. Zerlaut, has been deeply involved in the subject of solar energy codes, standards, and certification for more than twenty years. As a former owner of a testing laboratory, he is eminently qualified to describe the history of federal and private involvement in this important aspect of commercialization. After recently selling his testing laboratory, Zerlaut formed a consulting business, from which he continues his active involvement in energy matters.

Zerlaut recounts the tortuous path of arriving at consensus standards and testing and certification procedures. He suggests that the federal government's role may have been prematurely pushed and complicated the development of standards at a time when the technology was still developing.

1.3.4.2 Consumer Assurance
The second side of quality assurance is protecting the consumer from unethical practices. Chapter 15 covers consumer assurance from the perspective of Roberta W. Walsh, who was active in consumer assurance for several of the critical years. Walsh has since carried on her interest in the field as a professor at the University of Vermont, specializing in consumer studies.

Much of this chapter reports on three activities of SOLCAN (the Solar Consumer Assurance Network): (1) the activities at the four regional solar energy centers (RSECs); (2) the activities of an interagency working group, which brought in the Federal Trade Commission (FTC); and (3) warranty insurance analysis, which never materialized due to budget reductions. Consumer assurance activities at the state level and by various trade and consumer organizations are also discussed.

Walsh finds that "the situation which prompted a government concern for solar consumer assurance in the first place was re-created: a government-sanctioned market incentive was available with no built-in mechanism to assure minimization of risk. Consumers were left to free-market forces for any protections they may require. This view was held to be desirable in the laissez-faire environment that guided policy and program decisions following the 1980 elections." She concludes: "Whether or not there is a federal solar program, the challenge to the solar industry remains what it has always been: to develop and maintain consumer confidence by providing products and services that do the job well, that do not break down often, and that can be quickly and inexpensively repaired when they do."

1.3.5   Information Programs

There are five chapters (16–20) on federally sponsored information programs.

1.3.5.1   Consumer Information

Solar consumer information generation is discussed in chapter 16, written by Rebecca Vories, who was engaged in this exact activity during its peak period as a staff person at SERI. After SERI eliminated its consumer information role in 1981, she formed her own consulting firm and now specializes in similar areas.

Vories describes the pertinent federal consumer information efforts, beginning with the National Science Foundation (NSF), the Federal Energy Administration (FEA), the Energy Research and Development Administration (ERDA), the Energy Extension Service (EES, under ERDA and DOE), the National Center for Appropriate Technology (NCAT), SERI, American Council to Improve Our Neighborhoods (ACTION), USDA, HEW, and HUD, and two power marketing authorities (TVA and BPA). For each of these groups, she identifies their target audiences and evaluative material on their successes.

Although Vories feels that "there is little doubt that the plethora of information provided through federal government programs had a significant effect on the awareness of the general public regarding the availability of solar thermal technologies," she concludes that "the impact created more awareness than action." Vories ends her chapter with an excellent and extensive set of reasons why this was so and proposes five specific suggestions for a federal information program regarding "appropriate federal role," "specific strategy," "market research," "market segments," and "incentives ... performance."

1.3.5.2   Public Information Dissemination

Chapter 17 on dissemination of solar information directly to the public was coauthored by Kenneth Bordner and Gerald Mara. Mara is a dean at Georgetown University in Washington, D.C.; Bordner was close to the operations of NSHCIC/CAREIRS from his position at the Franklin Research Center in Philadelphia.

Mara and Bordner's chapter is primarily a discussion of the federal solar information program known successively as NSHCIC (National Solar Heating and Cooling Information Center) and CAREIRS (Conservation and Renewable Energy Inquiry and Referral Service). After discussing the legislative background behind this center, the authors show how the various federal parts of this network contributed to getting information to a large inquiring public. In its first year, this organization handled 2,400 phone inquiries in one peak week. Mara and Bordner also report on the drastic reduction in the network's budget, concluding that "the bulk of the inquiries now concern energy conservation and passive solar energy applications. In this curtailed form, information dissemination remains a part of the federal program on conservation and renewable energy."

1.3.5.3 Technical Information

Chapter 18 on technical information was prepared by Paul Notari, who, during much of the period he describes, was responsible for the preparation and production of many of the technical reports coming out of the Solar Energy Research Institute/National Renewable Energy Laboratory (SERI/NREL). Notari also served several terms as president of the American Solar Energy Society (ASES). He is now self-employed as a consultant on the topics of this chapter.

Chapter 18 differs from the preceding chapters by emphasizing material intended to advance the scientific and commercial (rather than the consumer) side of solar energy. This material was generated and disseminated by many more groups than SERI/NREL and ASES; Notari describes all of their efforts. As the federal solar program diminished in size during the Reagan administration, this became the most prominent part of the federal solar information effort. Notari believes that "many of the advances made in solar energy over the last decades are directly attributable to the quality and quantity of information dispensed." "Technical information programs ... character will change," he predicts. "Their importance, however, ... should increase considerably."

1.3.5.4 Training and Education

Kevin O'Connor authored chapter 19 on training and education programs in the field of solar energy. O'Connor was intimately involved in these activities while at the University of Delaware. After coming to SERI, he was responsible for many of the SERI training and education programs described in this chapter. After SERI training and education activity was terminated in the early 1980s, O'Connor was employed elsewhere for a few years but now is back at NREL, active in the national data aspects of alcohol fuels.

According to O'Connor, the federal government provided "both direct program support and stimulation of other programs." A solar education database, with annual publications, was even produced for a few years when solar energy became a "hot" educational area. O'Connor describes the role of many federal agencies in meeting this short-lived demand for federal support. Unfortunately, these were "fragmented efforts with no central focus." Readers interested in employment aspects of solar energy will also find some good data and references in this chapter. O'Connor concludes that "what was and would be needed if renewable energy education has a resurgence is an awareness of what is available, rather than continued funding toward the development of materials and curricula that already exist."

1.3.5.5 RSECs

The regional solar energy centers (RSECs) were created in 1977, at the same time as the announcement that SERI would be located in Golden, Colorado. Some felt that the RSECs were a form of political compensation to the many losers in the intense competition for a site for SERI. However, the RSECs also were clearly formed to serve a different purpose—a commercialization purpose, with close ties to state governments. It took more than a year for the states to organize and to choose from among themselves a single winner in each of the four regions of the country. This shared solar commercialization responsibility then lasted from late 1978 to late 1981. The four regions and the winning headquarters cities were Northeast Solar Energy Center (NESEC) in Boston, the Southern Solar Energy Center (SSEC) in Atlanta, Western Sun in Portland, Oregon, and the Mid-American Solar Energy Complex (MASEC) in Bloomington, Minnesota.

Chapter 20 on the RSECs was written by Donald E. Anderson, the former director of MASEC and now president of Innoventors, Inc. Although it primarily discusses MASEC, Anderson's chapter includes information common to all four RSECs. Noting that "the federal involvement in the advanced research and development stages of new technology is quite different from that in the later steps leading to commercialization," Anderson explains: "In the later steps, many parties involved in the decision-making processes in governmental, regulatory, financial, business, and consumer affairs *must* be participants. The regional centers were thus challenged to perform the anatomically difficult task of standing with both feet firmly in all three (federal, state, and private) sectors." He concludes that the RSECs "did provide the potential" for "accelerating the process of commercialization."

1.3.6 Technology Transfer

The next four chapters (21–24) cover the important commercialization activity of technology transfer, intended to facilitate movement of R&D results from the various national laboratories to the marketplace. Each of the chapters is written by an active participant in the technology transfer process.

1.3.6.1 Liaison with Industry
Chapter 21, by Daniel Halacy, covers the general subject of federal liaison with industry. Halacy has himself written many solar books; he served in the SERI administrative offices as an assistant to SERI's top management during the critical period covered in this book. Now retired, Halacy remains active as a solar enthusiast and author.

Chapter 21 is an introduction to technology transfer as a national laboratory function and to the next three chapters, which were written from the perspective of technology transfer at three national laboratories. Among other topics covered in this chapter are the changing nature of federal patent policy and the role of small business incubators. After reporting on some statistical evidence of technology transfer success, Halacy concludes that: "the lessons learned in the government program will be of great benefit when we begin to make major use of solar energy."

1.3.6.2 Case Study 1: SERI / NREL
Chapter 22 was prepared by Barry L. Butler, a former solar thermal technical administrator at SERI who specialized in solar materials at Sandia National Laboratory prior to joining SERI in its first year. Butler continues solar material activities at Science Applications International Corporation (SAIC) as well as through his own small solar business; he is also active as the vice-president of the Solar Energy Industries Association (SEIA).

Butler's chapter gives examples of cooperative commercialization activities with private sector companies in the three main areas of a solar thermal technology: reflectors, absorbers, and concentrators. Butler cites six major lessons learned, among which: "Industry-driven needs for research results must be present during program planning," "Competition has kept the research performed at SERI from being too esoteric," and "It is sound industrial policy to have industries and national labs work together."

1.3.6.3 Case Study 2: Los Alamos National Laboratory

Chapter 23 has been coauthored by J. Douglas Balcomb and W. Henry Lambright. Much of the Los Alamos National Laboratory (LANL) effort described in this chapter was carried out under Balcomb's direction; for more than a decade, LANL (and former LANL) personnel have continued to refine passive R&D results into useful design guidelines. Now at NREL, Balcomb concludes: "Design guidelines based on targeting a single user constituency and a single locality have proved to be effective."

A university researcher, Lambright made one of the very few evaluative studies of all the federal passive commercialization activities. We include here a condensation of the portion dealing with the Los Alamos National Laboratory passive solar technology transfer activities, together with Lambright's theoretical overview of technology transfer, which is relevant to this entire volume. Among his many interesting conclusions are these: "[T]echnology transfer in passive solar energy is indeed possible. But it is also very difficult. What it takes are linkages ... "; "What is clear is innovation takes time"; and "The Los Alamos case is an example of successful push and pull."

1.3.6.4 Case Study 3: Argonne National Laboratory

In chapter 24 William W. Schertz describes the technology transfer activities of the Argonne National Laboratory. A researcher at Argonne, Schertz worked closely with Roland Winston, a professor of physics at the University of Chicago, to help develop and then to commercialize a novel solar concentrating system called "nonimaging optics." Schertz closes his chapter with a set of seven important lessons, of which his sixth is typical: "The process is an on-going activity that requires the dedication of significant resources to make it happen."

1.3.7 Incentives

This section covers three types of federal incentives: tax credits, financing, and grants. As readers review these three chapters (25–27), some of the evaluative questions they should keep in mind are, Would the solar industry have been healthier without tax credits? Should the incentives have been tied to performance? Should there have been a slower phase out? These and similar questions are discussed in section 1.4.

1.3.7.1 Tax Credits

Chapter 25 on tax credits was coauthored by Daniel Rich, at the University of Delaware, and J. David Roessner, at the Georgia Institute of Technology (previously at SERI). Both have been active researchers on the impact of federal (and state) tax credits during the main period of tax credit availability; both were especially interested in how the tax credits, the most important federal commercialization incentive, affected consumer purchase decisions.

Much of Rich and Roessner's chapter documents the history and turmoil surrounding the beginning and termination of the solar tax credit. The authors provide generally positive evidence of its impact, noting that tax credits "can be a viable part of government commercialization strategies." They conclude: "The elimination of the residential tax credits has contributed to the contraction of the solar industry, and no other policy has been adopted by the federal government to facilitate solar commercialization."

1.3.7.2 Financing

Chapter 26 has been prepared by Steven Ferrey, an early solar financing analyst and solar activist. The history of federal financing of solar thermal systems is not a long one. This and chapter 27 give the only examples of direct financing that we believe exist in the federal solar program. There was an early realization that tax deductions and tax credits were strong stimulatives for wealthy buyers but that some lower-income families would need an attractive loan arrangement to also participate.

Ferrey describes in detail a loan program managed by HUD, the Solar Energy and Energy Conservation Program, that was applied in part to the solar thermal technologies program. As a member of one of two advisory committees for the loan program, which was "neither a bank nor a regulatory program," he was in a unique position to observe and report on its operation and eventual slow demise. Ferrey was among a handful of individuals who sued to release money that Congress had appropriated; much of the chapter reports these legal battles. His final conclusion: "For a thinly funded program, operating at a shadow of its originally designed scope, the Bank demonstrates that 'who' participates in utilizing solar energy and energy conservation technologies, at least with regard to a large segment of the population, is a function of how those technologies are financed."

1.3.7.3   Grants

In writing chapter 27, Seymour Warkov of the department of sociology of the University of Connecticut, and an earlier evaluator of the grants program for Connecticut, was assisted by an associate at the same university, T. P. Schwartz. Warkov and Schwartz conclude: "It is ironic that the grants incentives and other solar programs were dismantled just as solid data began to be produced in 1980–1981." The authors also have a significant commentary on local, bottom-up participation: "Linking solar grants to comprehensive residential energy conservation could have achieved greater energy savings in the household sector ...." We concur on the need for this type of coordinated planning, which many utilities and regulatory commissions are now beginning to adopt under the name "integrated resource planning" (IRP).

1.3.8   Organizational Support

Our final group of chapters (28–31) deals with activities that can be generically described as "organizational." Covering the areas labeled "international," "state and local," "utilities," and "law, environment, and labor," these last four chapters have much to say about barriers to and opportunities for the introduction of any technology.

1.3.8.1   International Programs

Chapter 28 was prepared by Murrey D. Goldberg, the first and only head of the International Program at SERI—between that program's inauguration in 1977 and shutdown in 1981. A physicist and one of the earliest full-time solar experts at a national laboratory, Goldberg headed an energy modeling and data effort at Brookhaven National Laboratory; his main background for the international area was serving as the secretary for an early interagency solar energy coordinating committee.

There was a time in the Carter years, Goldberg begins, when the U.S. international solar program was growing rapidly, as other nations realized the lead position of the United States, and used diplomatic channels to establish a solar relationship with us. With the disinclination of the Reagan administration to utilize this form of diplomacy, the international solar program at SERI was disbanded. Except for this short period of

SERI responsibility, the small U.S cooperative international effort has been managed out of the Department of Energy, where it resides today.

Goldberg points out that the U.S. international program was always primarily of an R&D character and covered many solar technologies: it therefore stretches the definition of solar thermal commercialization a bit to include a more complete discussion of the U.S international solar effort. Several other countries now have a much more extensive solar thermal (especially hot water system) commercialization program than exists in the United States. These commercially successful programs, Goldberg suggests, are probably mostly due to the lack of a natural gas alternative in those other countries. It is likely that future solar thermal commercialization efforts in this country will refer to those successes in other countries.

1.3.8.2 State and Local Programs

Chapter 29 was prepared by two persons with early and long back-grounds in the state and local areas of solar commercialization support. Michael DeAngelis was an early employee at SERI with an interest in the solar commercialization area who later left to do similar work for the state of California. Peggy Wrenn was the first solar staff person for Boulder, Colorado, a city that was an early leader in promoting solar usage and she was active in the programs sponsored by WSUN (Western Solar Utilization Network), one of the RSECs. Both have continued with their solar interests, so this chapter is particularly current.

DeAngelis and Wrenn discuss the various forms of support that state and local governments were able to receive and in turn to supply to pro-spective solar thermal system users. Some parts of this support were from the federal government, but other parts were more locally generated. The wide range of support options provided around the United States can serve in the future for a test of the efficacy of these various options. To our knowledge, this has not yet been done.

1.3.8.3 Utilities

Chapter 30 covers the role of the utilities in solar thermal commercial implementation. Because of their relatively high level of consumer con-fidence and ability to handle large initial investment, the electric utilities have a special potential role in solar implementation.

The chapter was initially written by Stephen L. Feldman, one of the early energy policy analysts with a strong interest in both solar energy and electric utilities. While at the University of Pennsylvania, he was an early recipient of study grants from the National Science Foundation (NSF) covering the possible future roles of electric utilities. Feldman became seriously ill about the time he completed the first draft of this chapter and died before he could revise it. His effort is greatly appreciated. Patricia Weis Taylor, an early employee of SERI with a specialty in technical writing in this exact area and an expertise in the solar commercialization activities of the Electric Power Research Institute (EPRI), was able to complete work on this chapter. We are grateful to her for addressing the recommendations of reviewers and for adding recent material on the commercialization activities of electric utilities.

After discussing early research efforts to understand the potential impacts on and from utilities, Feldman and Weis Taylor concentrate on utility experience with decentralized or dispersed solar thermal technologies. The authors reach two important conclusions on utility involvement: first, that demonstration programs "must include the utility interface as a specific consideration"; and second, that "any solar tax incentive program must also recognize the importance of utilities . . . ."

1.3.8.4   Law, Environment, and Labor

Chapter 31 was written by Alan S. Miller, formerly a professor of law and now the executive director of the Center for Global Change at the University of Maryland. Miller served as a contractor to various federal agencies in both the legal and environmental areas and earlier at the World Resources Institute, the Natural Resources Defense Council, the American Bar Association, and the Environmental Law Institute. Because this chapter covers three significantly different disciplines—law, labor, and environment—Miller's task was especially difficult.

One curious aspect of the early phases of solar enthusiasm is that environmental arguments were not the driving focus that they have since become for the renewable energy technologies. Although "much of the federal effort might be characterized as a search for problems," Miller points out: "Most legal analyses noted the absence of problems for many areas. . . ." Among his conclusions: "Labor issues . . . were never handled well," but "[F]ederal efforts properly emphasized outreach"; "An impressive network was built in a very short period of time"; and "[F]ederally supported efforts did not discourage parallel efforts." "[S]ome important social issues . . . cannot be divorced from policy questions," Miller observes in closing. "This unwelcome fact will no doubt continue to be a source of discomfort for governments to come."

1.4 Evaluation of Solar Thermal Commercialization Activities

In each chapter of this volume, the authors have described how one specific solar thermal commercialization activity was begun and considered how well it was carried out. Evaluation is never easy and the authors' views come from their particular expert perspectives. All authors were closer to their specific programs than we or most expert readers. We do not try to summarize their evaluations in this section but urge readers interested in evaluation to read carefully the final parts of each chapter.

The evaluative questions posed here, being cross-cutting, are more difficult to answer, even in hindsight, than the evaluative questions of the following chapters, which deal with a small subset of commercialization activity in each case. The opinions below, although often finding their origin in specific chapters, are our own; no one else should be held responsible for the following evaluative answers.

1.4.1 General

1.4.1.1 Rationale
Did the right reasons exist for a federal solar thermal commercialization program?

In late 1973 the principal motivations for an accelerated solar thermal commercialization program were the fear of fuel shortages and rising fuel prices. There was talk, especially in New England, of rationing oil and whether to shut schools or industry first. This never came to pass, and oil shortages can now be seen to have been overemphasized. Although the editors believe in a growing shortage of conventional energy, the shortage is not large enough to justify a crash effort for supply reasons alone. However, because so much of our national energy is derived from oil and because such a large percentage of oil is imported, there are certainly sound reasons to reduce that dependence. The validity of these supply concerns has been borne out by the Middle East war in 1990. However, natural gas and not oil is now the principal competitor for solar thermal energy; there is currently a glut of domestic natural gas, and its price is therefore exceedingly low.

The newest national concern, at least for many solar thermal advocates, is environmental—global warming. This concern has also revived interest in nuclear energy, an interest shared by few solar thermal advocates. Although it has not yet resulted in significantly increased sales of solar thermal systems or even in a return to a larger federal commercialization program, this recent environmental concern is apparently genuine and seems to be growing.

1.4.1.2   Timing

Was the attempted commercialization acceleration justified at the time it was initiated?

1973 seemed appropriate to most because of long gas lines and rapidly rising prices. The usual question was why this country had not started earlier. There was substantial public enthusiasm for a large federal role, including the activities covered in this volume. It is our perception that the private sector, for the most part, did encourage federal efforts; Congress was rarely told that its commercialization efforts were misguided (although there certainly were such voices).

However, in hindsight, it now appears to us that the commercialization effort was premature—the solar thermal technologies were just not ready for widespread implementation. In defense of those in charge of commercialization planning, especially Congress, many experts testified in 1973 that the technology was ready and that the only need was to demonstrate this ready technology to a willing public. We conclude that we should be especially cautious in accepting statements from those most involved in a technology. Those most knowledgeable about a technology are often among its strongest proponents and naturally biased in their views.

But hindsight is too easy. The energy situation was very serious in 1973 and perhaps our elected representatives chose exactly the correct course. Perhaps it was the implementation of the commercialization process, and not its timing, that was flawed.

As we discuss tax credits below, it will become obvious that timing is critical to an incentive, and delays can be counterproductive. This question of timing should be kept in mind by readers wishing to understand the solar thermal commercialization experience and to extrapolate from it.

Some long-term observers of the national solar thermal program attribute its failure to a conscious decision to promote other technologies.

They believe that the first demonstration programs were intentionally premature and carried out in such a way that solar thermal systems would receive a bad name. We have seen no evidence to support this theory. There were many credible supporters of the program, including a credible industrial base of support. Readers will find that the authors of the following chapters are generally enthusiastic about the commercialization programs they discuss. If there was a plot, it was never detected by those who carried out the programs.

1.4.1.3 Program Cutbacks

Should the solar thermal commercialization program have been ended so soon?

Among the solar thermal community, the dismantling of solar thermal commercialization—or alternatively of limiting the federal role to R&D—was not a popular decision. However, the new Reagan administration had sent clear signals prior to the election that it favored the private sector over the public; after the election, it carried out the private sector mandate it claimed it had received—especially in the solar area. Clearly, the Reagan administration saw few problems with supply or environment, at least few that could not be handled by the marketplace. The solar thermal commercialization programs were not cut back for budgetary reasons.

The Bush administration apparently saw some environmental issues that justified reversing the Reagan swing of the pendulum. However, by this time, greater concern was also being expressed about the growing budget deficits.

If one accepts a government role in commercialization, then it is clear that most of the commercialization efforts were ended before they had a chance to prove their efficacy and certainly before their performance could be seriously evaluated.

1.4.2 Program Areas

1.4.2.1 Technology Choices

Were the right technologies selected? The four main solar thermal technologies are described in section 1.1.3. Was any technology further advanced than the others? Should any have spent more time in an R&D phase before being promoted in commercialization? To answer these questions, we must go back to each technology separately.

In the active heating and cooling area, there was an early assumption, supported in congressional hearings (by most observers except the NSF program managers), that residential space heating was one of the most ready solar thermal technologies. Many of the first demonstrations therefore focused on home heating. In retrospect, this now seems unwise because the economics for space heating were much poorer than for hot water heating, where the collected energy could be used every day of the year. The home heating market was larger in potential energy savings, but predicted cost reductions from active systems still have not materialized. In addition, builders and homeowners found that big energy-use reductions were possible at lower cost through both conservation and passive solar energy techniques. Today, essentially no active solar thermal residential home heating systems are being marketed.

The reasoning of early federal commercialization planners was sound, but unfortunately not their assumptions. The potential market for solar heating has been greatly reduced by the expansion of natural gas pipelines and by the large drop in natural gas prices. As a result, few homes are being built with oil or electric heating systems, against which active solar home heating systems might reasonably compete.

Early federal R&D planners also assumed that active solar systems would achieve better annual cost effectiveness through combining the heating and cooling functions. So far, these combined systems have not proven cost-competitive. Perhaps too much optimism was shown in 1973, when the energy supply picture looked so bleak. But we should not be too harsh in retrospect; the energy supply picture was a number one citizen concern in the United States, and the active solar approach looked well worth trying as a possible winning solution.

The experiences of the years 1974–1985 established that active solar heating is not cost-competitive with natural gas at current prices. The same systems are, at best, marginally competitive with electricity. The accumulated results provided great improvements in solar heating system design, installation, and operation and maintenance but had little impact on cost effectiveness. Even if a solar energy system could be shown to pay back in three or four years, this would probably not be good enough for the vast majority of U.S. consumers, who are perhaps uncertain about their or their country's future, future energy prices, or the likelihood of technological advance. This point is also made in the chapters on consumer information and tax credits. We still have not, as a nation, solved this fundamental hurdle for high first-cost solar products.

The early NSF and NASA solar R&D programs did not seriously include passive solar thermal design. Accordingly, many of the early federal solar commercialization programs—especially the demonstration programs and the tax credits—ignored the passive option. Passive solar heating and cooling and daylighting eventually became one of the most popular and successful solar technologies. Today, many architects and builders include passive solar options as a standard feature in their design. Because of the federal R&D and later information campaigns, many home buyers now know that they can achieve almost any level of energy consumption; the lowered energy consumption mandated by city and state building codes can be achieved through a combination of energy conservation and passive solar design. But the passive solar option is usually not sold on the basis of its economic competitiveness. Rather, it is sold as adding a comfort amenity, improving reliability of energy supply, and adding to resale value. Features that trade off an increased mortgage payment for a reduced energy bill can also be sold by the architect and builder, provided the lender is aware of the energy-saving benefits.

Because architect, builder, and homeowner cared little in 1975–1985 about whether the annual energy reduction was achieved with passive solar or conservation techniques, the lowest cost approach could be employed. However, within the federal program the integration was not as easily obtained. For instance, in the early days of SERI (now NREL), energy conservation was not a part of either its R&D or commercialization efforts. The reason was not that the benefits of integration were not realized, but rather that different parts of the Office of the Assistant Secretary for Conservation and Renewable Energy were responsible for the conservation and renewable programs. Also, there was substantial competition among several national laboratories, most of which had begun work in both the solar and conservation areas prior to the 1977 start of SERI. The interlaboratory competition seems to have carried over to a competition between approaches that should have been treated cooperatively.

Today, we recognize that solar and conservation measures must be considered simultaneously to achieve a least-cost design. In the early days of the federal solar program, there was a less enlightened attitude for the reasons given above. The passive designs were barely included in the early demonstration programs and not at all in the federal tax credit programs. The tax credits for conservation were always different from those for solar energy systems, and it was sometimes difficult to tell conservation and passive efforts apart. There was also a philosophical refusal to provide a credit for a sunspace or greenhouse, where it was difficult to distinguish between the value for energy saving or the value as an amenity. Within DOE, consideration should therefore be given to further tightening the relationships between conservation and renewable energy to ensure that the two approaches benefit from each other and are only competing for funds based on valid program evaluation measures.

The industrial process heat program had a somewhat similar history. There were many opportunities for low-cost energy conservation in industry, and at least in the early days, it was not seen to be the responsibility of the solar IPH program to design an optimum overall system, combining the best conservation and solar features. The federal IPH program has concentrated on troughs, and the small commercial marketplace has done the same. The competition in this sector was almost always natural gas, which has continued to decline in price—despite the dire predictions of the 1970s. Here again, today's industrial solar heating systems are not cost-competitive at the 1990s' cost of natural gas.

In the solar thermal electric program, there is more of a question. Much of the early R&D funding was for the "power tower" concept, whereas the successful private sector Luz program used troughs. The federal argument for the power tower, expounded in chapter 8, is (1) it is cheaper to move energy around as reflected solar rays rather than through pipes, (2) the efficiency of collection is better with the smaller collection area of a single focal point rather than a focal line, and (3) the theoretical efficiency of conversion from thermal to electrical energy is higher at the higher temperatures possible with the power tower. Despite the logic of these positions, Luz was able to make a success of their trough system, constructing over 350 MW of solar trough collectors before it went out of business in 1990. Luz's explanation was that the business failure was due to the uncertainty of tax credits and environmental requirements as well as the low cost of natural gas (Becker 1992).

Why did Luz make this different choice of solar thermal electricity technologies? Presumably, there were fewer technological hurdles with the lower-temperature troughs. Certainly the capital and time required to field an operational system was less; also, the modularity and smaller scale of the trough systems allowed a greater opportunity for techno-logical change as new systems were built.

Recently, there has been joint public-private funding for dish electric technology development, with a substantial private investment by a major producer of small remote power systems (Cummins Engines). Cummins believes that this third approach will prove to be the most cost-effective for small-scale (and perhaps large-scale, also) electrical systems.

Because there are still advocates for the power tower, and because Luz has folded and no one is yet producing solar dishes, it is premature to conclude which solar electric technology will eventually prove best. In summary, the fact that the solar thermal commercialization program has had only limited commercialization successes cannot be laid to having made early incorrect technological choices. Rather, all of the choices seemed appropriate at the time and it is only with new data that we can see that better choices might have been made.

1.4.2.2 Programmatic Funding Levels
Were sufficient funds allocated?

The total solar thermal commercialization effort, including tax credits exceeded $7 billion (in 1993 dollars). There have been several studies to put this total into perspective by comparing it with federal subsidies in existence for other energy sources, such as for uranium enrichment, waste disposal, drilling, and so on. (See chapter 5 in volume 3 for a review of some of these comparisons.) It is beyond the scope of this chapter and this volume to analyze the appropriateness of these nonsolar expenditures.

Only by looking closely at each chapter's detailed history and findings can one conclude whether the federal financial commitment for solar thermal commercialization was sufficiently large. In retrospect, we con-clude from this chapter-by-chapter review that proponents of the solar thermal technologies underestimated the difficulty of convincing people to purchase solar thermal systems. In this error, they were assisted by those who were convinced that there was plenty of room for cost reductions by manufacturers. Mostly, these cost reductions had not materialized by the time the demonstration programs were terminated. The success of Luz in reducing its costs with large-scale manufacture and installation proves that large-scale reductions were possible.

Most of all, however, the solar proponents believed too easily in the importance of the early rise in prices of the conventional sources of energy. In 1993 natural gas is at an all-time low, whereas many with a responsibility to predict such things twenty years earlier were saying it would be some 2–4 times higher than it presently is. In discussing figure 1.5, we noted the importance of energy conservation in creating the energy surplus that has kept prices in check. If the price of conventional energy sources had not come down, we believe that the decision to stop funding solar thermal commercialization programs would have been seen to be a mistake. Perhaps future generations more concerned about environmental impacts will also lament the decision to stop the commercialization effort. However, given the actual declines in natural gas prices, it now appears that the decision to stop commercialization funding had little effect on actual sales. In summary, a correct decision may have been made—but for the wrong reasons.

1.4.2.3   Federal Organizations
Were the correct institutions involved?

Considering the difficulties associated with the start-up of the Department of Energy, perhaps Congress should have paid more attention to alternative institutions at an early date. However, no other obvious institutions come to mind. One cannot prove that a separate, early solar energy agency would have fared any better. Early federal solar staff and budgets were small, there were few commercial solar activities, and there were no obvious reasons to move programs out of other agencies (such as HUD) that had begun programs prior to the formation of DOE. It does now seem probable to us that there was too much fragmentation of programs, redundancy, and lack of accountability in the multiplicity of agencies, but again we can understand the reasonableness of staying with agencies that had initiated programs that seemed to be popular and were not (to our knowledge) being criticized.

1.4.3   Demonstration Programs

1.4.3.1   General Issues
Was the demonstration component of the solar thermal commercialization program worthwhile?

The main reason for demonstrations of a technology is to convince industry and consumers that the technology is market-ready. To a large extent, the solar thermal demonstrations had the opposite effect—convincing many that the technology was not market-ready. The accomplishments of the demonstration programs, the improvements in the systems were more those of a development phase than a demonstration phase. Therefore, it can be argued that the demonstrations were not successful as demonstrations.

Because not enough was known about failure modes, too many supposedly market-ready systems failed to live up to their billing. However, learning about failure is another reason for having demonstrations when the development phase is intentionally shortened. The program fulfilled that limited mission.

The many changes in solar program administration were not helpful, although they probably could not have been avoided. For those considering demonstration programs for other technologies, we certainly encourage you to utilize one strong existing organization rather than to believe that a series of changing agencies can carry out the demonstration correctly. There must be a single, continuing focal point for accountability, although multiple agencies with separate accountabilities are acceptable to the extent that competition can foster a better program.

It seems reasonable to review the decisions to discontinue the program of demonstrations. If restarted, there should be as large a private sector financial component as can be achieved; it is not reasonable to assume that the private sector can do the demonstration job alone at this time.

1.4.3.2 Purchase Program
Was enough attention paid to government purchases?

The answer would seem to be no. The solar thermal technologies certainly did not follow the path of federal purchases of computers and other electronic hardware. The successors to many early costly, failure-prone electronic units later successfully entered the commercial marketplace—much earlier than if the government had not been involved. Early government purchases of solar thermal hardware would certainly have showed up technical failures well before the consuming public got involved.

Again, however, we must emphasize that the message received by Congress was that these technologies were simple and ready and that the urgency was great. We cannot fault Congress too greatly for not having pursued the purchase approach.

The traditional approach of government purchases to assist a growth industry with national value appears to us to deserve current increased attention. The same rules for life-cycle costing (or integrated resource planning, IRP) that are being urged by DOE for utility planning should apply to all federal purchases. There is little information available on the cost effectiveness of solar thermal products; federal purchases can help to make such information more readily available and believable.

1.4.4   Quality Assurance

1.4.4.1   General Issues

Was the quality assurance component of the solar thermal commercialization program needed, well carried out, worthwhile, and deserving of being continued in the future?

Quality assurance was most definitely needed. The development of test methods, codes, and standards may have been prematurely started and thereby ultimately delayed by the federal government's intervention. On the other hand, the United States is now in a better position to update those codes and standards because of continued government support for participation in international developments.

1.4.5   Information Programs

Was the information component of the solar thermal commercialization program needed, well carried out, worthwhile, and deserving of being continued in the future?

Our answer to all questions is yes, without endorsing any specific future approach. As recommended by Vories in chapter 16, there must be careful attention to targeted, rather than general, consumer information. Technical information accessibility remains good, if not well supported. The RSECs served a useful purpose when there was intense national interest in the solar thermal area; until that interest returns, however, there is little need for them.

1.4.6   Technology Transfer

Was the technology transfer component of the solar thermal commercialization program needed, well carried out, worthwhile, and deserving of being continued in the future?

Based on the data in our chapters, the answer to all questions is yes. However, we have heard negative reactions from some reviewers of this set of chapters to the whole question of national laboratory involvement in areas where commercial firms can compete. Some believe that public versus private competition is not a healthy situation. A delicate balance must be maintained between such competing traditional standards as

- making national laboratory expertise available to all equitably
- using patent protection to allow a corporation to accept the challenges of introducing a new product
- providing the taxpayers a return on their financial support of R&D

Given the continued likelihood of national laboratory involvement in solar R&D, technology transfer must remain a high priority for solar program planners. An evaluation would be helpful to identify strengths and weaknesses of this area.

1.4.7 Incentives

Were the tax credits, loan guarantees, and grants programs successful?

This is a fundamental question for this volume. We have seen in section 1.2 that almost all possible financial incentives were employed. However, only the tax credits were large; the bank and grants programs were quite small. Chapters 25, 26, and 27 have all generally claimed a successful program, except for being hampered by too early a termination. Given that early termination and the lack of any definitive evaluative program, we find it impossible to state anything about the correctness of the proportions.

Also we can find no evidence that any specific program was grossly inappropriate. The largest solar thermal commercialization impact on the federal budget was revenue loss due to tax credits. These credits certainly had a positive impact on sales, right up to their termination. Such small amounts of funds were disbursed by the Solar Bank and the HUD Hot Water Initiative grants that their impact was negligible.

The tax credits worked in the sense that they supported development of a solar collector industry. The individual residential tax credits were not continued in 1985 because the prevailing political climate did not favor them and because decreases in prices for conventional fuels had by then removed any sense of urgency for solar development. However, while in force, these tax credits had an enormous impact on the solar collector industry, leading to dramatic growth from 1978 to 1985. Since 1985 there have been few solar incentives from the federal government and conventional fuel prices have been exceptionally low; after its precipitous decline (described in detail in chapter 3), commercial development has been at a virtual standstill. The collapse of the industry soon after the end of the tax credits leads us to conclude that the tax credits did not achieve their chief purpose of promoting a viable, market-sustained industry.

Another school of thought (including many solar enthusiasts) believes that the solar tax credits were catastrophic in addition to being expensive. Their argument is that the tax credits inflated the costs of solar systems, kept unscrupulous businesses afloat, and thereby generated a negative public image. In this regard, we conclude that the tax credits should have been dependent on system performance standards.

Solar heating of swimming pools and passive home heating received virtually no tax credits, but both continue as viable businesses. This suggests, in hindsight, that cost competitiveness was not adequately weighed in the initial selections of solar technologies to receive commercialization support from tax credits. Similarly, the tax credits should perhaps have been more closely linked to the amount of energy saved rather than to the cost of the system.

More dollars of tax revenues were forgone than all the dollars budgeted for solar thermal programs (by a factor of about 1.5). The tax credits were about five times larger than the programmatic federal commercialization efforts. It now seems that if equivalent expenditures had been made for planned programs, such as federal purchases and monitoring, that greater benefits would have been achieved.

Given this past history, the editors would not recommend a return to a solar tax credit. We believe that alternative support mechanisms are available and more likely to receive congressional approval; these are described below in section 1.5. In contemplating a tax credit for another technology, consideration should be given to a policing mechanism to ensure high-quality, cost-effective products.

1.4.8 Organizational Support

Was the organizational support component of the solar thermal commercialization program needed, well carried out, worthwhile, and deserving of being continued in the future?

Our answer is again yes on all counts. Many of the organizational barriers that necessitated commercialization programs beginning in 1973 still exist. When this nation decides again that solar thermal technologies are important in the future and worthy of a national increase in emphasis, organizational support must be an area that receives attention. Readers are referred to the evaluative comments of the four authors in this category (chapters 28–31).

Certainly, the present, relatively dormant international cooperation activities in the solar thermal area could be expanded. This would help improve U.S. credibility relative to international environmental commitments and would also improve our nation's image as a world citizen. International markets are often more attractive in less developed countries where an electrical grid does not exist, and there is an advantage to smaller, decentralized solar units.

The utilities continue to be appropriate partners for the federal government in solar thermal commercialization. Evaluation should be conducted to understand the reasons for large differences between different U.S. utilities in their involvement in solar thermal energy introduction.

The state and local solar commercialization programs were effective means for implementing many commercialization programs. They might have to be reinvigorated if national goals are reestablished.

The advantages to this country in such areas as labor and the environment have not been subjected to recent evaluative review. We hope that any growth in federally funded solar studies will include this important auxiliary area.

1.4.9 Other Issues

1.4.9.1 Federal versus Private
Is commercialization better done by the private sector?

Certainly we and most authors of this text would agree that the answer is yes—the private sector is more appropriate. No one is asserting anywhere in this volume that the federal government can commercialize anything. A more relevant question rather is, Did the private sector encourage federal intervention for the correct reasons? We think the answer is yes. It now appears to us that it would have been virtually impossible for the private sector to have made any advance without federal involvement. The hurdles, especially those of first cost, and consumer reluctance to be early adopters, were and are too large for an entirely private sector response.

Solar technology showed great promise for a while, but is now mostly in eclipse—and the fault does not lie with the solar industry. We think the reasons for a dormant industry lie with (1) low energy prices, especially for natural gas, (2) larger incentives to competing fuels, and (3) high initial costs of solar energy systems. If we expect an early resurgence of the solar thermal industry, we are again going to need a partnership between industry and government. Solar thermal commercialization cannot be handled by the solar thermal private sector alone, unless we are counting on another crisis to bring that about.

1.4.9.2 Relative Performance
Why did some commercialization programs work better than others?

The reasons some programs enjoyed greater success than others can be traced to three areas:

1. *Experience and organization* (length of time in an endeavor, past history of successes and failures, size, other missions, etc.) There seems little doubt that some of the failures in the commercialization program can be attributed to the lack of organizational experience of the Department of Energy (DOE). New in 1977, DOE had essentially no experience with the private sector nor with commercialization of consumer products. Moreover, its bureaucrats had rarely interacted with other departments, in this case, with NSF, NASA, and HUD, to name only a few. As the Department of Energy matured, these organizational problems disappeared and we have no reason to question present competence.

2. *Application* The passive program seems to have been more successful partly because criteria in addition to energy were involved (comfort, added space, resale, etc.). First-cost problems were solved in part by virtue of folding the systems into the mortgage. Solar thermal electrical systems are also more likely to be accepted because they fit well into existing utility practice.

3. *Timing* Programs that were started later do seem to have been more successful; moreover, success grew in all programs over time. The urgency perceived in 1973, with the rush to get systems in place, undoubtedly led to many of the implementation difficulties. As urgency became less of a factor, success became more likely.

1.4.9.3 Push versus Pull
Should there have been greater or lesser emphasis on either market push or market pull?

We see no easy answer to this question. The tax credits were an attempt to provide a market pull. The demonstration and information programs provided a push. We wonder whether a larger and earlier series of purchases by the federal government would not have been appropriate to provide an additional pull. We also note a general congressional reluctance to work with manufacturers rather than consumers. Possibly, more assistance to manufacturers would have enabled more of them to stay the course.

1.4.9.4 Evaluation

Was there sufficient program evaluation?

Clearly, no. The lack of evaluation is a recurring theme in every section of this volume. Some of the few program evaluations have been performed at the request of Congress by the General Accounting Office (GAO). Recently, some of the best energy evaluations have been by Congress's Office of Technology Assessment (OTA), but these do not seem to have covered the solar thermal program as well as they might have. Moreover, we do not see enough early attention by Congress to its oversight role, which could perhaps have assisted in a more orderly program growth and a less precipitous program decline.

1.4.9.5 Role of Congress

Did Congress micromanage? During the strongest growth periods, Congress was pushing for larger programs; in retrospect, this seems to have been premature. Yet renewable programs were popular and there were indications that conventional supplies were insecure. We do not fault Congress for attention to detail on small but popular programs. However, solar thermal program managers have also reported that congressional offices made an excessive number of inquiries and insisted on determining the location and sizes of demonstration projects, decisions better left to the designated program managers. Congress and congressional staff should readily recognize that federal program managers can exercise good judgment for the right reasons.

Should Congress be blamed for not fighting to retain a stronger program in 1981–1992? In addition to the difficulty of fighting administration officials armed with higher priorities, there was talk of continuing energy glut and energy prices were declining. We understand congressional behavior during this period, while still believing that it should have been otherwise.

Were there excessive efforts to generate projects in home congressional districts (i.e., pork barreling)? The solar funding area was no different from others, although we think many elected representatives would agree that this federal program area was especially easy to find worthy home district projects. With a rising budget in many of the early years, and major constituent concerns over energy supply in virtually every district, we do not find it remarkable that concern should be raised over pork barrel solar thermal projects. The solution to such congressional efforts on behalf of constituents lies outside the solar program arena or this volume.

1.5 Lessons Learned

The following are the editors' personal observations, based both on editing this volume and on a longer history of involvement in the national solar thermal program. These views are exclusively our own and should not be attributed to any of the other authors of this volume.

1.5.1 Programmatic Lessons

1.5.1.1 Crash Efforts
In hindsight, there seems no doubt that more R&D should have preceded the commercialization programs of this volume. There were too many changes in the lead organizations as well as in the substructure within the main responsible organizations. Programs were put in place prematurely because a sense of urgency prevailed. If we could rewrite history, the federal solar thermal R&D programs would have started earlier than in 1971, so that the commercialization programs perceived as so critical in 1973 would have had a stronger base.

1.5.1.2 Program Contraction
The obverse is equally true; we believe that the solar thermal commercialization program was on a substantially correct course when drastic contraction began in 1981. It will be much more difficult and more costly to restart than to have continued a program that had good momentum, but that was terminated before it achieved success and before it could be properly evaluated.

1.5.1.3 Governmental Goals
The national solar thermal program (both R&D and commercialization) presently does not have a well-defined measurable commercialization

objective. In the absence of a measurable objective, we can expect little
interest by the U.S. public in promoting the widespread use of solar thermal technologies. We see value in the reestablishment of realistic solar
thermal program commercialization goals for the years 2000, 2010, and
2020.

1.5.1.4 Budgets

Despite the drastic reduction in solar thermal commercialization program
support from its peak in 1981 and despite low natural gas prices, cost
reductions in solar energy systems and private solar commercial activities
have continued. We believe this indicates that the federal commercialization programs have been and could again be useful; therefore, we believe
they should be expanded at a moderate rate. Our primary justifications
are to level the playing field and to provide environmental insurance.

1.5.1.5 Technology Choices

Although relative priorities should always be reviewed, we see no reason
to drastically modify existing program priorities as exemplified in the
budget discussion of section 1.2. The passive solar program budget, however, seems remarkably low.

1.5.1.6 Governmental Organizations

The present lead role of the Department of Energy seems appropriate;
mandates and goals for many other departments are also appropriate but
not yet well delineated.

1.5.2 General Commercialization Lessons

1.5.2.1 Federal Commercialization Program Evaluation

There has been insufficient attention paid to the evaluation of virtually
all solar thermal commercialization programs. This volume contains only
a few evaluations; many more should have been performed. There is
much to be learned from the evaluation of both existing and terminated
programs.

1.5.2.2 Governmental Design Competitions and Performance Awards

In the present budget mood, solar proponents must look for low-cost
means of promoting commercialization. One of the least costly is to provide regular and expanded federal agency recognition of companies and
individuals contributing significantly to renewable energy commercialization. The Environmental Protection Agency's (EPA's) "green lights"

energy conservation awards program deserves replication in the renewable energy area.

1.5.2.3 Taxation Policy

We are convinced by estimates of environmental damage caused by conventional energy sources and therefore favor a Btu or carbon tax as an important means of accelerating renewable energy commercialization.

1.5.2.4 Commercialization Data

More data should be collected on the performance of existing installations and on the rates of installation of the solar thermal technologies—especially passive solar. These data should emphasize first costs and savings, which we believe are better than many analysts suppose.

1.5.2.5 Removal of Other R&D and Taxation Subsidies

We believe that an appropriate policy is the removal of existing subsidies to the conventional energy sources. The Alliance to Save Energy shows that the subsidy ratio is more than 20 to 1 in favor of conventional sources (ASE 1993). In a time of budget cutbacks, it seems unlikely that any increase in solar commercialization funding would be as successful in reaching national solar goals as a reduction in the subsidies offered to conventional energy forms. Although this would not overcome the head start advantage, it would at least be a step in the right direction. Other mechanisms will have to be developed from the commercialization options discussed throughout this chapter and volume to further assist in establishing a level playing field.

References

ASE (Alliance to Save Energy). 1993. *Federal Energy Subsidies: Energy, Environmental, and Fiscal Impacts*. Washington, DC.

Beattie, D. A., ed. 1996. *History and Overview of Solar Heat Technologies*. Cambridge: MIT Press.

Becker, N. D. 1992. "The Demise of Luz: A Case Study." *Solar Today* January–February: 24–26.

Bezdek, R. 1993. Personal communication, U.S. Department of the Treasury, October.

Cone, B., D. L. Brenchley, and V. L. Brix. 1980. *An Analysis of Federal Incentives Used to Stimulate Energy Production*. IPNL-2410 (Rev. 2). Richland, WA: Battelle Pacific Northwest Laboratories.

DOE (U.S. Department of Energy). 1979. 1980 Budget Request. Washington, DC.

DOE. 1991. 1992 Budget Request. Washington, DC.

DOE. 1992. 1993 Budget Request. Washington, DC.

EIA (Energy Information Administration). 1992. *Federal Energy Subsidies: Direct and Indirect Intervention in Energy Markets.* SR/EMEU-92-02. Washington, DC. November.

EIA. 1993. *Annual Energy Review, 1992.* DOE/EIA-0384(92). Washington, DC, June.

ISEIR (International Solar Energy Intelligence Report). 1993. 4 October, p. 165.

Hulstrum, R. L., ed. 1989. *Solar Resources.* Cambridge: MIT Press.

Lagace, G. 1993. Personal communication, Energy Information Administration, U.S. Department of Energy, November.

SERI (Solar Energy Research Institute). 1978 *Annual Review of Solar Energy.* SERI-TR-64-066. Golden, CO.

Spewak, P. 1988. "Analyzing the Effect of Economic Policy on Solar Markets." In R. E. West and F. Kreith, eds., *Economic Analysis of Solar Thermal Energy Systems.* Cambridge: MIT Press, 1988.

U.S. Budget. 1982. *U.S. Federal Budget for FY 1983.* Washington, DC.

U.S. House. 1981. House Committee on Science and Technology. Budget hearing record, 24 February, p. 1188.

U.S. Senate. 1974. Committee on Aeronautical and Space Sciences. *Hearings on Solar Heating and Cooling.* 93d Cong., 2d sess., 25 February, 50. Testimony of Michael McCormack.

U.S. Senate. 1980. Committee on Energy and Natural Resources. *Hearings.* 96th Congress, 2nd session. 18 March, 14.

West, R. E., and F. Kreith, eds. 1988. *Economic Analysis of Solar Thermal Energy Systems.* Cambridge: MIT Press.

2 The Role of Congress

J. Glen Moore

2.1 Theses

Throughout the solar policy formation process, Congress has consistently championed an expanded federal role in solar resource development and commercialization, while the executive branch has generally favored a more conservative approach. With few possible exceptions, the substance of every solar research, regulatory reform, and economic incentive bill to be enacted into law originated in the Congress. The net result of this advocacy was a larger, more diverse, better-funded research effort, with a more visible commercialization component, than would have been expected from the executive branch alone. Congressional prodding put the solar option on the national agenda in the 1970s; tenacious congressional support kept the federal program, and the solar option, alive in the 1980s.

2.2 Purpose and Scope

Today's federal solar policy is the product of an adversarial, sometimes contentious, relationship between the Congress and the executive branch. It was born out of political ambition, compromise, misinformation, good intentions, impatience, and frustration. In short, it may be both the best and the worst of what our legislative process has to offer.

Federal solar policy evolved within, and is a part of, the larger effort to reach a consensus on a comprehensive national energy strategy. To be fully appreciated, solar policy must be examined in the context of national energy policy. For example, the laws that deregulated oil and gas pricing may be the most important and effective policy instrument yet enacted to accelerate the development of solar technologies. And yet, solar energy is only a footnote in the legislative histories of these laws. No reference to solar energy appears in the text of any oil or gas deregulation bill.

Although not a comprehensive review of national energy policy, this chapter will attempt to show that current federal solar policy has evolved out of an interactive process between the Congress and the executive branch, played out primarily within the larger struggle to reach a consensus on national energy policy. It will attempt to show that policy formation has been, and continues to be, a piecemeal process and that the resulting policy instrument is a patchwork of laws, plans, studies, and policy statements rather than a single, comprehensive law or set of laws.

2.3 Overview

For the purpose of review, it is convenient to divide the congressional role in solar policymaking into three sequential but somewhat overlapping phases: (1) a statutory framework phase; (2) a market development phase; and (3) a maintenance phase. The budget process—an important cross-cutting policymaking tool—will be discussed within the context of these three policymaking phases.

2.3.1 Statutory Framework

The statutory framework phase was carried out over roughly a nine-year period from 1972 through 1980 and took place within the larger effort to reach a consensus on a national energy strategy. The objectives for solar energy and national energy policy were basically the same: first, to reorganize and unify the federal program and, second, to provide a comprehensive program strategy to guide the federal effort.

Even before the oil embargo of 1973, policymakers were beginning to recognize that the existing resource-specific approach to energy development and management was inadequate for dealing with disruptions in energy supply. A new approach was needed—one that would enable the federal government to address short-term energy problems while preparing for longer-range needs in a coordinated, unified manner. The goal was a single energy agency with broad responsibility over energy matters and with direct access to the president. It was further recognized that the energy mix needed to be more diversified so that, in time, the nation could reduce its dependence on oil. New energy research and development (R&D) initiatives would be needed to search out new energy sources to meet future energy needs.

Due to the size and complexity of the task, the reorganization of the federal energy effort would be carried out in two stages. Stage one began with the Nixon administration in the early 1970s and was given impetus by the oil embargo of November 1973. This administration viewed solar as a resource for the longer term, however, and accorded it a low level of attention while it concentrated on immediate and short-term energy problems. The low priority accorded solar by the executive branch set the stage for Congress to take the lead in shaping federal solar policy.

Worsening conditions in the nation's energy supply led to the initiation of stage two in the reorganization process by President Jimmy Carter in 1977. When this administration ended in 1980, the framework for a comprehensive, unified federal energy program was essentially completed. The federal program was in the hands of a single cabinet-level agency, and detailed program guidance was contained in the complex set of laws and plans which constituted the national energy program. Due largely to congressional efforts during this period, the necessary statutory framework was also in place for carrying out a comprehensive solar energy effort.

2.3.2 Market Development

At the start of the 94th Congress in 1975, congressional interest in solar energy shifted from program authority to economic and regulatory initiatives. With energy markets being far from "free," substantial economic incentives would be needed to make solar technologies even marginally competitive with fossil fuel and nuclear sources. A variety of incentives aimed at different parts of the solar market were enthusiastically proffered. With cooperation from the Carter administration in this second phase of solar policymaking, Congress established an extensive network of economic and regulatory support mechanisms intended to "level the playing field" for the fledgling industry. The expanded federal role in solar commercialization was wholly consistent with the then-accepted policy of federal involvement in energy markets.

2.3.3 Maintenance

A third, distinct phase in solar policy formation began in 1981 with the election of Ronald Reagan and the subsequent removal of energy from the national agenda. Strongly committed to the principles of federalism, the new administration was ideologically opposed both to a federal role in solar commercialization and to an aggressive federal solar R&D program. Solar supporters tried to hold on to gains in solar research and economic incentives achieved in the late 1970s but were largely unsuccessful. By the end of President Reagan's first term in 1984, solar commercialization programs had withered under the strain of tight federal budgets and administration opposition; solar research gave way to nuclear and other programs that met the administration's criteria for long-range R&D.

2.3.4 The Budget Process

Throughout the period of solar policy formation, the annual budget review has been an important policymaking tool. In the 1970s Congress used the budget process to influence the size and details of the solar program, and to establish policies favorable to a federal role in commercialization. In the 1980s the Reagan administration, supported by growing congressional concerns over deficit spending, used the budget process to reduce every solar development program (some to near extinction), to eliminate virtually all federal solar demonstration and commercialization programs, and to break the back of the Solar Energy and Energy Conservation Bank. Clearly, the budget process was a major factor in shaping federal solar policy during the 1970s and 1980s.

2.4 Ninety-third Congress: Creating a Statutory Framework

The establishment of a comprehensive national energy policy took place in two stages. The first stage occurred in 1971–1974, in the years immediately preceding and following the October 1973 oil embargo. The second took place in 1977–1980 during the Carter administration. In the intervening years, 1975–1976, Congress and the executive branch jointly pursued a policy of increased federal involvement in domestic energy markets.

Well before the embargo was imposed, policymakers had begun to recognize that the structure of the existing policy organizations and processes that managed energy were obsolete and unable to deal with problems of fuel scarcity (Kash and Rycroft 1984). The existing approach was resource-specific. Over the years separate agencies guided by separate policies had been set up to deal with coal, oil, natural gas, nuclear energy, and electric power. No single agency was charged with formulating a unified policy or making preparations for meeting future energy requirements. This highly fragmented approach to energy management was inappropriate for dealing with oil embargoes or other major shocks to the energy market.

It was recognized in this initial period that much of the country's energy problem was due to an overdependence on foreign oil. Part of the solution was increased domestic oil production, the other part was a more diversified energy mix. Furthermore, nuclear power, which had dominated the federal energy program for twenty years, was not contributing to the mix as had been expected. A more balanced federal research program was needed. Congress and the Nixon administration began a search for new sources of energy and, in particular, sources that would be supplied by new technologies. Various "renewable energy" sources were pursued, including geothermal energy, organic wastes, and solar energy and its derivatives. Energy conservation and technologies designed to use conventional fuels more efficiently were also pursued. Later, the quest would expand to oil shale, tar sands, and heavy oil.

The October 1973 oil embargo focused national attention on the "energy crisis" and the inadequacies of existing energy policies and programs. Neither Congress nor the federal energy program was structured to deal effectively with the resulting shortages, or with larger, underlying organizational problems. Administration proposals for a comprehensive restructuring of the federal energy organization proved unworkable. In the Congress, committee jurisdiction and member leadership over energy matters were not well defined. Energy bills and committee hearings on the crisis flooded the legislative agenda. Overlapping committee jurisdictions and member interest impeded progress toward comprehensive energy legislation; political compromises were negotiated to allow competing factions to participate. In this first stage of consensus building, a piecemeal approach to energy legislation was a matter of political necessity.

At the time of the embargo, executive branch responsibility for energy matters was spread among several agencies, including the Atomic Energy Commission (AEC), the Interior Department, the Federal Power Commission (FPC), and the Environmental Protection Agency (EPA). AEC research on fission and fusion power dominated the federal energy R&D effort, while support for solar and other new sources of energy was negligible. A modest solar research effort was underway in the National Science Foundation (NSF) in photovoltaic and basic research needs in materials. This program, however, was inadequate to deal with the applied engineering and marketing problems that would have to be addressed to move solar technologies from the laboratory to the marketplace.

When the oil embargo hit, Congress was already well into the process of restructuring the federal energy bureaucracy and creating a policy framework to deal with the problem of foreign oil dependency. With supply diversification being one element of the emerging policy, solar was quickly identified as a candidate resource for further research and development. Several hundred energy bills were introduced in the 93d Congress (1973–1974). Of these, approximately twenty-five were solar-related (House 1976b). Of the energy bills enacted, two constituted the first legislative step toward a restructured energy program. These were the Federal Energy Administration Act of 1974 (PL 93-275) and the Energy Reorganization Act of 1974 (PL 93-438). Three others sought to provide detailed program guidance and special emphasis to the nonnuclear portion of the federal energy program, particularly solar. These were the Non-Nuclear Energy Research and Development Act of 1974 (PL 93-577), the Solar Heating and Cooling Demonstration Act of 1974 (PL 93-409), and the Solar Energy Research, Development, and Demonstration (RD&D) Act of 1974 (PL 93-473).

2.4.1 Federal Energy Program Organization

A comprehensive reorganization of the federal energy bureaucracy was first proposed by President Nixon in January 1971 at the beginning of the 92d Congress (House 1971). Seven major departments would be consolidated into four, with one of these an expanded Interior Department. It would be renamed the Department of Natural Resources (DNR) and be responsible for civil nuclear power and most nonnuclear energy programs. DNR enabling legislation was introduced, hearings were held on this and other executive reorganization proposals, but no further legislative action took place.

The energy situation was becoming more difficult when the 93d Congress convened in January 1973. In an energy message to Congress on 18 April 1973, President Nixon announced legislation to establish a Department of Energy and Natural Resources (DENR) based on the 1971 DNR proposal "with heightened emphasis on energy programs." The president's message referred to reorganization as a key to dealing more effectively with energy problems: "If we are to meet the energy challenge, the current fragmented organization of energy-related activities in the executive branch of the Government must be overhauled" (Public Papers of the Presidents of the United States. Richard Nixon, 1973. Special Message to Congress on Energy Policy, April 18, 1973).

Congress received a two part energy reorganization measure on 29 June 1973—one part to establish the DENR, the other to establish an independent Energy Research and Development Administration (ERDA). The DENR component would emphasize data collection, conservation, resource management, policy formation, and other administrative or operating concerns in the energy field, while the ERDA component would emphasize energy R&D.

A significant difference between the 1971 and 1973 proposals was the relationship between the proposed new units and the AEC. In 1971 it was proposed that policy and funding for the R&D functions of the AEC would be transferred to the DNR. In all other respects, however, the AEC would remain unchanged; it would retain its operating functions and identity. In the 1973 proposal the AEC would be merged into the proposed ERDA and would lose its separate identity. Added to this would be transfers to ERDA of most major federal nonnuclear R&D from Interior. In the 1973 proposal the licensing and regulatory functions of the AEC would be separate from the R&D functions and would be carried out by an independent regulatory commission, to be called the "Nuclear Energy Commission" (NEC). The integration of nuclear and nonnuclear functions into the ERDA was a clear indication that the concern for a unified response to energy R&D needs had moved to the forefront of the administration's reorganization efforts.

Legislation was introduced to implement the DENR-ERDA/NEC proposal (H.R. 9090 and S. 2135); hearings were held in the summer of 1973. As the energy crisis intensified, particularly with the October embargo, the need for action on energy reorganization grew more urgent. It became clear, however, that a single bill involving both DENR and ERDA/NEC would be difficult to move through the Congress despite the crisis atmosphere. The need for expediency led President Nixon on 8 November 1973, to ask Congress to proceed with the creation of DENR and ERDA/NEC separately:

Because of the critical role which energy research and development will play in meeting our future energy needs, I am requesting the Congress to give priority attention to the creation of an Energy Research and Development Administration separate from my proposal to create a Department of Energy and Natural Resources. This new administration would direct the $10 billion program aimed at achieving a national capacity for energy self-sufficiency by 1980. (Public Papers of the Presidents of the United States. Richard Nixon, 1973)

Legislation was introduced soon thereafter to create ERDA, as called for by the president. The bill, the Energy Reorganization Act (H.R.

11510 and S. 2744), would not substitute for, nor was it an alternative to, proposed legislation in specific energy fields then under consideration, such as nuclear plant siting, construction of deep-water ports (for oil and liquefied natural gas transport), regulation of strip mining, use of petroleum reserves, or emergency conservation.

Congress proceeded with the ERDA bill, ceasing all work on the comprehensive DENR-ERDA/NEC proposal. Still, there was a need for DENR-type authority. To provide such authority on a temporary basis, Congress gave priority consideration to a separate administration proposal for a Federal Energy Administration (FEA). The FEA was proposed as a two-year agency to deal with emergency fuel shortages, energy conservation, and energy policy. The two proposed agencies—ERDA and the FEA—when established, were expected to work in complementary fashion. In practice, however, conflicts made the need for a single energy agency all the more apparent. Interest continued in a unified DENR, but the merger of energy R&D and policy functions into a single agency would not happen until the 95th Congress with the establishment of the Department of Energy.

2.4.1.1 Energy Reorganization Act

After almost a year and a half of debate, beginning with a presidential statement and draft energy policy legislation being sent to Congress on 29 June 1973, the Energy Reorganization Act of 1974 (PL 93-438) was enacted on 11 October 1974. The debate represented the first attempt ever to examine and revise the federal organization with the specific objective of carrying out comprehensive national energy policies. The act began the process of consolidating and reorganizing the federal energy R&D bureaucracy. It abolished the AEC and established two major new federal entities: the Energy Research and Development Administration (ERDA) and the Nuclear Regulatory Commission (NRC).

ERDA reflected the perceived need for a unified federal approach to energy R&D, as well as a widely held belief that research on energy sources other than nuclear power had to be emphasized. The new agency became the focal point for federal energy R&D. It had broad authority to develop new energy sources consistent with sound environmental and safety practices. Pursuant to the act, NSF programs for solar heating and cooling development and geothermal power development were transferred to ERDA (see section 2.4.2.2 for subsequent transfers of solar authority to ERDA).

The Energy Reorganization Act abolished the AEC and established a new, five-member NRC with responsibility for the licensing, regulatory, and related functions of the former AEC. The transfer of atomic energy R&D functions to ERDA freed the commission from potential charges of conflict of interest associated with advocacy and regulatory responsibilities for atomic energy resting in the same agency. Some members were concerned, however, that with 84% of the personnel and 90% of the funding for the new ERDA transferring from the AEC, nuclear programs would dominate the new agency's agenda. In effect, they were concerned that ERDA would be the AEC by another name.

Concern over program balance was further heightened by the administration's proposed budget for fiscal 1975, which was heavily skewed toward nuclear programs. Several safeguards to ensure program balance were built into the act, the most important of which was the administrative structure given the agency. The act provided an administrator, a deputy administrator, and six assistant administrators, each to be appointed by the president with the advice and consent of the Senate, and each responsible for a separate part of the energy R&D effort. One assistant administrator was responsible for nuclear energy, another for fossil fuels and another for solar, geothermal, and advanced energy systems. On an organization level, the renewable energy program was on par with the fossil fuel and nuclear programs. In subsequent years, Congress would apply the budget process to establish a policy of funding parity for renewables, coal, and fusion energy.

2.4.1.2 Federal Energy Administration Act
When it appeared that quick action on legislation to create the DENR would not be possible, President Nixon asked Congress, on 24 January 1974, to establish the Federal Energy Administration (FEA) as a temporary DENR-type agency (Public Papers of the Presidents of the United States. Richard Nixon, 1974).

Bills to establish FEA had been introduced in December 1973 (S. 2776 and H.R. 11793), and on 7 May 1974, the Federal Energy Administration Act (PL 93-275), was enacted. ERDA and the FEA were independent agencies with separate and distinct responsibilities; ERDA was responsible for R&D, while FEA was responsible for short-term energy resource allocation problems and energy planning.

For dealing with short-term problems, FEA could establish allocation, rationing, price control, and conservation programs to help balance supply and demand for fuels. To help carry out its resource allocation function, FEA collected, analyzed, and assessed data on energy supplies and consumption. Over the longer term, FEA was responsible for expanding the energy supply using available technologies; this included expediting energy resource projects such as the Alaskan pipeline and outer continental shelf leasing, as well as the greater use of coal, oil shale, and other energy sources. FEA would also study the role of government in assuring that adequate economic incentives existed for industry development of domestic energy resources, including solar sources. Thus the policy of a federal role in solar commercialization, which would become so prominent later in the 1970s and so controversial in the 1980s, was rooted in FEA's 1974 legislative mandate.

2.4.2   Solar Program Direction

Solar energy began to attract wide congressional interest in 1973 at the start of the 93d Congress. Statements and materials favorable to solar energy appeared in the *Congressional Record* in increasing numbers. Although, at the time, there were only a few active solar energy systems in operation, the perception in the Congress was that many solar technologies, and especially the technology for solar heating and cooling of buildings, were already developed. All that was needed to get solar heating off the shelf and into the marketplace was a coherent, well-funded national program. Thus, in the initial phase of congressional interest, launching a solar industry was as considered more a political than an engineering problem.

Enthusiasm for solar was fueled by numerous congressional, executive branch, and outside studies. Among those frequently cited were the NSF/ NASA Solar Energy Panel study of 1972, which predicted that by early in the twenty-first century solar energy could economically provide as much as 35% of the nation's gaseous fuel needs, 10% of its liquid fuel needs and over 20% of its total electric requirements (NSF/NASA 1972, 85). The solar panel report prepared in conjunction with a 1973 AEC study entitled *The Nation's Energy Future* was even more optimistic (Technical Review Panel 1973). Although the projections for solar were largely ignored in the full study, the panel report predicted that collectively the various solar technologies could meet 10% to 30% of the nation's required Btu input by the year 2000 and as much as 50% by the year 2020.

The administration increased NSF's solar budget from $4 million in fiscal 1973 to $12 million in 1974. Rather than applauding the increase as a good faith effort by the administration to boost solar research, solar enthusiasts in Congress scorned it as an admission that solar energy had been grossly underfunded in the past. To a growing number of solar supporters, the administration was not doing enough to get solar energy into the national energy mix. Even with the budget increase, solar was receiving less than 2% of the federal energy R&D budget. Furthermore, some members were displeased that solar was either neglected entirely or relegated to the long-term energy future in administration policy statements. For example, in a special energy message to the Congress delivered on 18 April 1973, President Nixon put solar energy and nuclear fusion in the same long-term category:

In the longer run, from 1985 to the beginning of the next century, we will have more sophisticated development of our fossil fuel resources and on the full development of the Liquid Metal Fast Breeder Reactor. Our efforts for the distant future center on the development of technologies—such as nuclear fusion and solar power—that can provide us with a virtually limitless supply of clean energy. (Public Papers of the Presidents of the United States. Richard Nixon, 1973)

In another energy message delivered nine months later the president was more explicit about the timetable for introducing new energy technologies:

For the near term—the period before 1985—we must develop advanced technologies in mining and environmental control that will permit greater direct use of our coal reserves. We must speed the widespread introduction of nuclear power. And we must work to develop more efficient energy-consuming devices for use in both home and industry.... Beyond 1985, we can expect considerable payoffs from our programs in nuclear breeder reactors and in advanced technologies for the production of clean synthetic fuels from coal. By this time, we should also have explored the potential of other resources such as solar and geothermal energy. (Public Papers of the Presidents of the United States. Richard Nixon, 1974)

Some viewed the low priority accorded solar as being inconsistent with the president's goal of energy independence by 1980, inconsistent with solar' s potential as indicated by various forecasts, and an affront to the emerging environmental and antinuclear sentiments of the time. The perception of neglect opened the door to the 93d Congress to take the initiative in formulating federal solar policy. Congress would retain the initiative for the remainder of the 1970s, finally relinquishing it to the Reagan administration in the 1980s, but only after energy was removed from the national agenda.

2.4.2.1 Non-Nuclear Energy Research and Development Act

Despite efforts to bring the nuclear and nonnuclear portions of the ERDA program into balance, it was generally recognized that the two programs being inherited by ERDA were grossly mismatched. The nuclear program was the product of a sustained, twenty-year federal development effort. It was highly organized, well funded, and backed by a network of dedicated national laboratories. Highly detailed program guidance and policy direction for nuclear R&D had been set out in the Atomic Energy Act of 1954, and in subsequent amendments. There was no question that the nuclear portion of the ERDA program would be fully functioning at the time of the transfer.

The need for more emphasis on nonnuclear research was recognized even as the ERDA bill was making its way through the legislative process. During the Senate consideration of the bill on 14 August 1974, Senator Metcalf noted that "by transferring the nuclear development functions into the new Energy Research and Development Administration and splitting off the nuclear regulatory functions, however laudable, we might well be creating a nuclear dominance in the development of an overall energy R&D policy and program" (*Cong. Rec.* 1974b: S14750).

With the exception of coal research, nonnuclear energy sources had received little federal attention. The program for nonnuclear sources was fragmented, underfunded by comparison, and almost devoid of statutory program guidance. The Energy Reorganization Act gave ERDA the organizational structure to carry out nonnuclear research but was short on program guidance. Nuclear research would continue to be guided by the Atomic Energy Act, but no similar guidance existed for nonnuclear sources. Furthermore, the nonnuclear program lacked a sense of urgency, and its priority within an overall energy strategy had not been established. To ensure a balanced federal energy R&D effort, the nonnuclear portion of the program would need special legislative attention.

The Non-Nuclear Energy Research and Development Act of 1974 (PL 93-577), sought to do for nonnuclear energy what the Atomic Energy Act did for nuclear. It would provide program guidance in the form of congressionally defined policies and provisions, and it would provide authorization authority for a federal program in nonnuclear R&D within the newly established ERDA. Outlining specific policies and principles for nonnuclear research, the act (1) assigned emphasis to energy conservation, (2) required attention to the social and environmental consequences of energy technologies, and (3) set forth the principle that, to the extent possible, federal assistance should be confined to research that would not otherwise be carried out by private industry. In addition, it defined innovative forms of federal assistance that could be used by the administrator to carry out the purposes of the act, including establishment of joint federal-industry corporations as well as federal price guarantees for products from commercial demonstrations of new energy technologies. There was no explicit authority in existing law for some of these forms of assistance.

Another important feature of the act was the provision for the annual authorization of the nonnuclear R&D budget. The authorization process established by the act paralleled procedures established by the Joint Committee on Atomic Energy (JCAE) for nuclear program. This was important because, based on the JCAE precedent, oversight committees were given access to highly detailed, internal ERDA budget reviews of the nonnuclear program, which in some years became available even before the budget was formally submitted to the Congress. Congress came to devote close attention to the ERDA budgets for nonnuclear research. The annual budget review process would prove to be an important tool for Congress to work its will on the direction and content of the federal nonnuclear effort.

2.4.2.2 Solar Energy Research, Development, and Demonstration Act
The Solar Energy Research, Development, and Demonstration Act of 1974 (PL 93-473) provided specific, comprehensive policy guidance and the basic administrative tools necessary for a high-visibility, AEC-type solar research program. It made it the policy of the federal government to

• pursue a vigorous and viable program of research and resource assessment of solar energy as a major source of energy for our national needs; and

• provide for the development and demonstration of practicable means to employ solar energy on a commercial scale.

PL 93-473 was entirely a congressional initiative. A Senate version had been introduced by Senator Hubert Humphrey in March 1974 (S. 3234).

Another version, closer to the act's final form, was introduced in the House by Representative McCormack in June (H.R. 15612). The measure went well beyond anything that the administration was proposing at the time. The administration's proposal for comprehensive energy legislation provided no specific program guidance for solar and fell short of the goal of a unified solar program. Indeed, the ERDA bill made the federal solar research effort even more fragmented because it transferred solar heating and cooling to ERDA, but left all other solar research in NSF. PL 93-473 consolidated program authority by transferring all remaining NSF solar programs to a new interim agency. The legislation stipulated that when ERDA was established, this authority would transfer to ERDA. Thus, PL 93-473 unified federal solar research under a single management authority.

The genesis for a unified federal solar research, development, and demonstration effort was contained in an energy report from the Task Force on Energy, chaired by Representative McCormack and issued by the House Committee on Science and Astronautics in 1972. The report noted:

There are many R&D approaches to large scale terrestrial solar energy use, all of them underfunded. Though the National Science Foundation supports some solar energy research, there is no national program to assess these approaches and develop the most promising ones. It is essential that we make a national commitment to bring one or more techniques for large scale terrestrial solar energy collection to commercial demonstration. (House 1972)

The task force and its successor, the Subcommittee on Energy, conducted an extensive investigation of the potential of solar energy technologies. A total of thirteen days of hearings were held, with oral testimony received from about 50 witnesses and written statements from some 100 others. Additional hearings on the need for omnibus solar legislation were held in the Senate by the Committee on Interior and Insular Affairs and the Joint Committee on Atomic Energy. The breadth of hearing activity suggests the level of importance Congress had attached to solar energy and to the federal role in developing it.

2.4.2.3 Solar Heating and Cooling Demonstration Act
PL 93-409 was the first solar-specific bill enacted into law. It evolved from the work of the Task Force on Energy in 1972, from two days of hearings on the status of solar heating and cooling technology before the Subcommittee on Energy of the House Committee on Science and Astro-

nautics in June 1973, and from three additional days of hearings before the subcommittee in November 1973 on a draft bill.

Convinced by industry and research experts that the technology for solar heating and cooling "is approaching economic feasibility" (House 1973) Representative McCormack, who was chairman of the Subcommittee on Energy, determined that legislative action was in order. Encouraged by the witnesses, the subcommittee decided that a federally funded demonstration program was the most efficient and economical way to bring this technology to the marketplace. To this end, Representative McCormack introduced H.R. 11056, the Solar Heating and Cooling Demonstration Act, on 23 October 1973 (a revised bill, H.R. 11864, was reported from the House Committee on Science and Astronautics 7 December 1973); a companion bill, S. 2650, was introduced by Senator Cranston on 2 November.

A demonstration program seemed warranted because of the popular view that solar heating and cooling was very near commercial readiness. In addition, the demonstration approach was seen to have potential advantages over a research-oriented effort in that (1) delays normally incurred in transferring the results of federal R&D programs to the private sector might be avoided; (2) the federal purchase of some 4,000 solar units, and the increased public awareness and acceptance likely to flow from a government-backed program, might stimulate private sector R&D, thereby saving the federal government R&D costs; and (3) a market-oriented program (more so than a research-oriented one) would be more likely to yield cost-effective systems acceptable to consumers.

If successful, the program would break the chicken-and-egg situation perceived in the solar heating and cooling industry, and "promptly launch all related elements of the heating and cooling equipment manufacturing industry, the mortgage industry and the American homeowner into a new exciting and economically healthy arena" (Senate 1974a, 50).

The administration opposed the bill on the grounds that a large-scale demonstration was premature and ran the risk of "set[ting] back solar development for decades" (Senate 1974b, 124). The administration urged a slower, more deliberate approach in order to give the responsible agencies a chance to learn from experimental prototypes before committing to large numbers of units, which might simply replicate mistakes.

The administration's arguments went unheeded, however, and the bill met almost no opposition as it moved through the legislative process.

Representative Symms, one of the few opposing voices heard, raised a philosophical argument anticipating the position the Reagan administration would later take to reduce federal involvement in solar commercialization. He argued that because the basic technology existed and a clear market need was established, the American free enterprise system should be allowed to take its course (*Cong. Rec.* 1974d: H8787–88).

The politics, if not the technology, were favorably disposed for action. The prolonged national debate over the energy crisis had produced few results to report to anxious constituents. The solar demonstration bill with its clear objectives and relatively low cost would be popular in the home districts, demonstrating that Congress could act decisively to relieve the energy problem even if the administration would not. Congress accepted the legislation as a positive "step to alleviate the impending crisis" (*Cong. Rec.* 1974a: H739). The bill became law on 11 September 1974, after receiving overwhelming bipartisan support in both chambers.

2.4.2.4 Introduction of Solar Incentive Legislation

The 93d Congress was aware of the importance of the marketplace in the nation's solar future, but its top priority was statutory authority for an effective R&D program. Consequently, there is relatively little in the legislative record to indicate congressional goals or expectations with respect to solar commercialization. Senator Mike Gravel—a leading critic of nuclear power—was among the first members to articulate the basic cost-benefit rationale for direct federal involvement in solar commercialization. He proposed that

it would be a better bargain to subsidize solar energy during the process of its cost reduction, than to ship our wealth and our jobs out of the country to pay for foreign oil.... By subsidize, I mean tax credits, or tax-free profits, cost-plus contracts, or a combination of these and other temporary incentives.... Even with subsidies, a multibillion dollar solar construction program might turn out to be "free," if free means no extra cost to the American public. Either the program might pay for itself by gradually causing a drop in the prices we pay for whatever foreign oil we still need, or ... by making it unnecessary for our Navy to spend billions on ships to protect our foreign fuel supplies. And by creating jobs for the unemployed, it could save billions. (*Cong. Rec.* 1974c: S17880)

The specific incentive bills introduced in the 93d Congress were offered primarily for discussion purposes, or perhaps to establish member or committee interest in a particular approach. Most were offered without explanation; none was enacted or considered in hearings. Bills were offered calling for tax credits, tax deductions, the use of federal housing programs to demonstrate solar technology, and grants to assist industry in the development of commercial systems. Solar heating and cooling was the only solar technology specifically singled out for legislative action.

2.5 Ninety-fourth Congress: Building a Consensus for an Expanded Federal Role in Energy Markets

A continuing heavy dependence on imported oil coupled with declining domestic production led the 94th Congress to focus on standby energy emergency authorities and fuel switching strategies (Senate 1976, 1). As a result, little progress was made in the 94th Congress toward the further unification of the federal energy program.

The 94th Congress recognized that to a large extent the nation's solar future would be determined by how quickly solar technologies could establish markets. It was recognized, too, that these technologies were at a distinct economic and regulatory disadvantage in competition with conventional fuels and grid power. Therefore, for solar interests in the 94th Congress, legislation to improve the competitive position of solar technologies was not only in order, it was a top priority.

Because the solar industry was in its start-up phase and markets for solar technologies were not well defined, it was not clear what kind or level of support was needed. A number of incentive measures were offered for consideration, but congressional interest centered on tax credits for residential and business investments in solar space conditioning systems. This technology was seen as being closest to market readiness, and having the greatest potential for a major impact on the energy mix.

Tax credits were an attractive option on at least two counts: (1) no annual appropriation was required, although there would be an annual loss in tax revenues (solar interest claimed the losses could be partly or fully offset by employment and other indirect tax revenue gains); and (2) the administrative structure for disbursing the credits (the Internal Revenue Service) was already in place. Shortcomings, such as the credits being a possible giveaway for the wealthy, were offset by the desire to demonstrate assertive action in response to the energy crisis.

Although it showed a great deal of support and enthusiasm for solar incentives, the 94th Congress failed to take a single measure to conclusion. Extensive hearings were held on different tax credit bills with several measures receiving floor action. However, in the end, each fell short, more the victim of the political-legislative process than of a lack of majority support.

Other incentive options were offered for consideration during the 94th Congress but received relatively little attention. Among these were direct, low-interest loans, loan guarantees to finance high-risk solar and wind demonstration facilities, and measures to require the federal government to buy and use solar energy systems in federal buildings and facilities. Some of these, as well as new options, would continue to be the focus of congressional attention during the next two congresses.

2.5.1   Solar Incentive Legislation in the 94th Congress

Solar "incentive" legislation is broadly defined to included legislation that promotes the commercial introduction of solar technologies by reducing the effects of perceived legal, institutional, or economic barriers. Included under this broad definition is federal procurement legislation designed to make the federal government a first market for promising new technologies and legislation designed to discourage the use of other energy sources, thereby making solar more attractive. R&D bills are not included, even though in most cases their ultimate purpose is to foster the development of products attractive to consumers.

The legislative record shows that as early as 1973–1974 Congress was moving toward a policy that the nation's solar future should be determined as much by the marketplace as by government-supported research efforts. Congress came to adopt a view that the federal program should address both research and market needs in a somewhat balanced effort. Overemphasis of one at the expense of the other could risk delaying the full realization of the solar potential.

In its pursuit of a balanced solar program, Congress pushed for an accelerated R&D effort and an increased federal role in commercialization. Administration resistance was encountered on both fronts, with the stiffest opposition on the market side. Unwilling to move at the pace Congress wanted, the executive branch was generally viewed as an obstacle to progress. During the Carter era, however, when Congress and the administration were hammering out national energy policy, some administration resistance to popular solar initiatives was almost certainly politically motivated. It was to the administration's advantage to withhold support until it had extracted congressional support for its initiatives, some of which were far more costly and less popular, such as "synfuels." Both sides used solar commercialization initiatives as cannon fodder in the consensus-building process. Gains in solar commercialization seldom came easily and were frequently marked by long delays, as Congress and the administration worked to resolve differences over more contentious matters.

Because of their bargaining chip role, many solar incentive successes are tucked away in seemingly unrelated bills pertaining to agriculture, military construction, foreign aid, and so on. But as solar interests in and out of Congress quickly learned, legislative success was no guarantee that an incentive would achieve its desired effect. Once enacted, legislation is subject to interpretation and implementation by the administration, and then to reinterpretation by succeeding administrations. If an administration decides that an incentive is inconsistent with its policies, it can use various executive branch prerogatives to neutralize or reduce the impact of the incentive. For example, it can interpret a law narrowly, thereby diminishing the incentive's potential effectiveness. Or, acting within its authority under the budget process, it can delay program implementation by not requesting funds or hiring staff. The administration is bound by the letter of the law but not necessarily by its spirit.

Over the course of carrying out the federal solar program, the executive branch has been more inclined to support the research side of the solar scale. For political and other reasons, it has not embraced solar market incentives as enthusiastically as Congress. Some solar incentives achieved through congressional initiative have therefore not been as effective as expected; some have, in effect, withered and died. It is likely, too, that persistent executive branch resistance to solar economic incentives has had a chilling effect on congressional efforts to do more in this area.

2.5.1.1 Direct Solar Incentive Legislation (Tax Credits)

Congressional interest in solar technology development reached new heights during the 94th Congress with over fifty solar-related bills introduced (House 1976b). Because of the success of the 93d Congress in legislating program guidance for the solar effort, the 94th Congress centered its attention almost exclusively on market incentives. Although none of this attention paid off in legislation, extensive hearings and floor debates provided a forum for consensus building on tax credit and other incentive options. By the close of the 94th Congress, the House and Senate had demonstrated strong bipartisan support for residential and business tax credits as a means of stimulating the solar industry. The absence of an appropriate legislative vehicle appears to have been the main obstacle to a tax credit bill in the 94th Congress. The substance of solar incentive legislation hammered out in congressional conference committees during the 94th Congress would be repackaged in legislation offered by the Carter administration in the 95th Congress. This legislation would eventually be enacted, but for the solar industry, the 94th Congress was an agonizing period of delay and uncertainty.

Of the different incentive options considered in the 94th Congress, tax credits for residential and business applications of solar space heating and cooling dominated the agenda. Solar tax credits were included in three tax bills under active consideration in the final days of the 94th Congress. These were H.R. 2166, H.R. 6860, and H.R. 10612.

H.R. 2166   The Senate became the first chamber to consider and approve a major economic incentive for solar energy when, on 21 March 1975, it accepted a Domenici amendment (64 to 32) to H.R. 2166 providing a tax credit for qualified conservation and solar equipment expenditures. The bill (the Tax Reduction Act of 1975) had passed the House earlier with no solar tax provisions. Senator Domenici's amendment allowed a tax credit for qualified solar energy equipment expenditures for new and existing homes and commercial buildings beginning 1 January 1975 and ending 31 December 1979. The residential credit was for 40% of the first $1,000 spent and 20% of any excess up to $2,000. During the floor debate on the amendment, Senator Domenici argued that "we ought right now to put in the one aspect that everyone agrees upon, which is that insulation and solar technology ought to be stimulated in the American market for their energy-saving capacity and employment potential; and certainly the tax incentive approach will cause that to happen" (*Cong. Rec.* 1975a: S4653).

The conference substitute for H.R. 2166 did not include the Senate-passed energy tax credit provision. The conference decided to defer consideration of this provision because similar incentives were being considered in H.R. 6860, a tax bill then under consideration by the House Ways and Means Committee.

H.R. 6860   The Energy Conservation and Conversion Act was reported (H. Rept. 94-221) from the House Committee on Ways and Means on 15 May 1975, with a solar tax credit provision similar to the one approved by the Senate but dropped from the conference version of H.R. 2166. During the floor debate on the bill, the House agreed to a Wylie amendment (244 to 132) increasing the credit to 25% of the first $8,000 spent, for a maximum credit of $2,000. The debate on the amendment brought out opposing views on the use of tax credits for stimulating solar markets. Proponents of the higher credit level argued that a push from the federal government was essential to achieving a meaningful energy impact in the near to mid term. Opponents labeled it a "government giveaway." Arguing for the amendment, Representative Gude noted the importance of the solar energy market to the nation's long-range energy strategy:

Trying to promote solar heating on the cheap is going to have little meaning for individuals interested in retrofitting their homes or installing solar in their new houses, and it is going to have little effect on our total energy picture.... The Wylie amendment ... would clearly provide a far greater stimulus to solar energy conversion and construction than would the present language because it would more nearly meet the actual costs incurred through the purchase and installation of this equipment. (*Cong. Rec.* 1975b: H5431)

Representative Frenzel opposed the amendment, claiming that it would do little more than subsidize investments by the wealthy which they would have made anyway: "What is going to happen is that people of significant means who can afford to put a $7,000 heating installation in their homes are going to be the beneficiaries of a large tax credit for doing something they would have done anyway" (*Cong. Rec.* 1975b: H5431).

The bill was further amended to extend the business investment credit to solar energy equipment installed before 1 January 1981. Still another amendment provided for a five-year amortization (rather than regular depreciation) for certain equipment, including solar energy equipment, where the investment credit was not claimed.

H.R. 6860 passed the House, as amended, and was referred to the Senate Finance Committee. Hearings were held by that committee in June and July 1975. Before the bill could be reported, however, the Senate amended and passed H.R. 10612, the Tax Reform Act of 1976, with a solar tax package similar to the one included in H.R. 6860. Thus, at this point, both houses of Congress had passed legislation with comparable solar tax credit provisions—H.R. 6860 in the House, and H.R. 10612 in the Senate. Despite a majority support for a solar tax credit in both houses, however, both bills were far from completing the legislative process, and time for action was slipping away.

H.R. 10612   This bill was farther along in the legislative process than H.R. 6860, having been reported and passed by both the House and Senate, even though the Senate substitute differed radically from the House version. H.R. 6860, on the other hand, had not been reported from Senate Finance, putting it two steps behind H.R. 10612 in the legislative process. When H.R. 10612 went to conference, the House insisted that the bill was not the proper vehicle for energy matters because the House version did not contain energy provisions of any kind (House 1976a). During the conference, the House informed the Senate that it retained an interest in H.R. 6860 and hoped that the bill could be enacted. The Senate conferees subsequently agreed to drop all energy provisions from H.R. 10612.

Aware of the action in the conference on H.R. 10612, the Senate Finance Committee reported H.R. 6860 on 27 August 1976, with a substitute amendment that incorporated and added to the energy provisions of H.R. 10612 as passed by the Senate. This effort proved futile, however, as H.R. 6860 failed to reach the floor of the Senate prior to the adjournment of the 94th Congress on 1 October 1976.

2.5.1.2 Indirect Solar Incentive Legislation (Program Authority)
Although no solar-specific legislation passed during the 94th Congress, two landmark energy bills were enacted which would benefit solar efforts indirectly. These were the Energy Policy and Conservation Act (EPCA) of 1975 (PL 94-163) and the Energy Conservation and Production Act (ECPA) of 1976 (PL 94-385). EPCA gave the FEA certain pricing and regulatory authority for use in promoting conservation and fuel switching; ECPA extended the FEA and expanded its role in solar commercialization.

ECPA   Some members of the 94th Congress were dissatisfied with the job ERDA was doing in translating solar R&D efforts into commercial ventures. Representative Ottinger criticized ERDA for being neither inclined nor equipped to see that solar technology already developed got out into the field (*Cong. Rec.* 1976a: H5065). To remedy this situation, Congress approved an amendment to the FEA reauthorization bill (ECPA) that reinforced the FEA's authority to carry out policy and planning functions associated with promoting the commercialization of solar energy. Under a $3 million program established by a Senate amendment, FEA was to prepare a national plan for the accelerated commercialization of solar energy, including workable options for achieving on the order of one million barrels per day of oil equivalency in energy savings by 1985 from a combined total of all solar technology. That FEA would have the lead role in solar commercialization was readily acknowledged by ERDA and the administration (ERDA 1976). The amendment would provide for additional personnel to enable the FEA to carry out its responsibilities in this area. Senator Gary Hart, principal sponsor, noted the importance of the amendment in maintaining a balance between commercialization and R&D activities in the federal solar program:

This research commitment is vitally important to the development of solar energy, but equally important is an aggressive program of commercialization. The amendment I am sponsoring provides FEA with funding and staff necessary to establish this program.... The program is not new. It has been talked about for several years, but somehow has always been left out of administration priorities.... At present there are only two professionals in FEA's solar division, which is barely enough manpower to answer the mail and prepare congressional testimony. My amendment would allow for the addition of 10 staff positions and would inject funds into projects vital to an accelerated solar effort (*Cong. Rec.* 1976b: S9439).

A second major provision, also put into the act by Senate amendment, authorized FEA and HUD to spend $200 million for a national demonstration program to test the feasibility and effectiveness of various conservation and renewable energy incentives. Under the provision, HUD was authorized to make financial assistance available in the form of grants, low-interest loans, interest subsidies, loan guarantees, and other appropriate forms of assistance. With respect to approved renewable energy resource measures, the amount of any grant was limited to $2,000, or 25% of the cost of installing the measure, whichever was less.

Both the FEA solar commercialization effort and the HUD-FEA grant program were enacted into law as part of ECPA, but neither initiative was implemented. PL 94-385 was an authorization bill; an FEA appropriations bill (PL 94-373) had been approved earlier without the necessary funds for either program. Implementation would have required either a supplemental appropriation or a reprogramming of funds, which neither Congress or the administration elected to do.

ECPA was the only major enactment of the 94th Congress to deal directly with energy reorganization. The debate on the bill made it clear that Congress considered FEA temporary and would extend it only for as long as it took to consider a more comprehensive reorganization for energy. While its solar-related provisions were largely symbolic, ECPA did call for an overview study of the existing energy organization, its problems and policies, and options for reorganization (OFEF 1977, 78). Completed during the Ford administration, the substance of the study recommendations was subsequently adopted by the Carter administration and offered in the 95th Congress as part of the comprehensive Department of Energy Organization Act. Thus comprehensive action on the federal energy bureaucracy was deferred to the 95th Congress.

2.6 Ninety-fifth Congress: Completing a Comprehensive National Energy Policy

On 29 April 1977, less than four months after taking office, President Carter proposed his comprehensive National Energy Plan (WCPD 1977). The plan was a framework for a comprehensive national energy strategy, with new energy sources figuring prominently in the overall strategy. It called for a cabinet-level Department of Energy (DOE) to complete the consolidation of the federal energy bureaucracy.

2.6.1 National Energy Plan

As originally proposed, the plan included a series of specific quantitative production and conservation goals to be achieved by 1985. Solar energy—the only nonconventional energy source assigned a goal—was to be in use in 2.5 million homes by 1985. The need for urgent action was predicated on a judgment that the world's demand for oil and natural gas, if not significantly moderated, would exceed world productive capacity as early as 1985, threatening "catastrophe." President Carter called for a national response that would be "the moral equivalent of war."

The plan included a list of ten principles intended to provide "a framework not only for present policies, but also for development of future policies." According to the last principle:

Finally, the use of nonconventional sources of energy must be vigorously expanded. Relatively clean and inexhaustible sources of energy offer a hopeful prospect of supplementing conventional energy sources in this century and becoming major sources of energy in the next.... Traditional forecasts of energy use assume that nonconventional resources, such as solar and geothermal energy, will play only a minor role in the United States energy future. Unless positive and creative actions are taken by Government and the private sector, these forecasts will become self-fulfilling prophecies. . . . Because nonconventional energy sources have great promise, the Government should take all reasonable steps to foster and develop them (National Energy Plan, 1977, 31).

Such rhetoric encouraged the 95th Congress to be even more aggressive with efforts to promote the commercialization of nonconventional sources of energy. Tax credits and other economic incentives that failed to pass in the 94th Congress were quickly reintroduced. Altogether, well over 100 solar or solar-related bills were offered for consideration. Many of these initiatives were incorporated into the National Energy Act (NEA)—the administration's principal legislative vehicle for implementing the National Energy Plan. Other measures were introduced and marshaled through Congress by a loosely organized but effective solar coalition—a group of about 100 House and Senate members strongly supportive of solar energy. After a long and arduous struggle over matters generally not related to solar energy, the National Energy Act passed, with provisions for residential tax credits and other important incentives for solar energy. In addition, the coalition was successful in pushing through other legislative objectives. The 95th proved to be a banner Congress for solar energy, with the enactment of NEA and more than a dozen solar or solar-related incentive measures (Senate 1979, 192).

To implement the energy plan, the administration offered two major legislative proposals: the Department of Energy Organization Act and the National Energy Act. This legislation became the focal point of congressional energy efforts for the 95th Congress and the major vehicle for solar legislative initiatives (Senate 1979).

2.6.2 Department of Energy Organization Act

The federal response to the energy crisis was a major issue in the 1976 elections. President Carter made energy reorganization a priority during his campaign and, on 1 March 1977, presented the 95th Congress with a proposal that was the basis for the Department of Energy Organization Act (H.R. 4263 and S. 826). In transmitting his proposal for a new Department of Energy (DOE), President Carter told the Congress:

Nowhere is the need for reorganization and consolidation greater than in energy policy. All but two of the Executive Branch's Cabinet Departments now have some responsibility for energy policy, but no agency, anywhere in the federal government, has the broad authority needed to deal with our energy problems in a

comprehensive way.... This reorganization can help us bring currently fragmented policies into a structure capable of both developing and implementing an
overall national energy plan. (President 1977)

Congress quickly responded, and on 4 August 1977, President Carter
signed the Department of Energy Organization Act (PL 95-91), making
DOE the twelfth cabinet department in the executive branch. DOE began
operations in October 1977.

As foreseen in earlier reorganization attempts, this comprehensive
measure encompassed both the FEA and ERDA, as well as the Federal
Power Commission (FPC). The Bureau of Reclamation and the power-
marketing agencies, which previously reported to the secretary of the
interior, now reported to DOE. The agency also assumed certain energy
responsibilities previously exercised by the Departments of Housing and
Urban Development, Transportation, Commerce, and Defense, and by
the Interstate Commerce Commission (ICC).

With DOE's creation, the federal government assumed an even larger
role in energy markets. The expanded role raised concerns in both houses
of Congress and would later lead to calls for the dismantlement of the
agency. Other concerns raised during the debate over the bill were that
too much authority was being vested in one agency and that the legislation was moving too fast to allow sufficient time for discussion and
debate. Some members were also concerned that long-term energy R&D
would be neglected, given DOE's organization and focus on short-term
energy problems. This concern prompted a successful floor amendment to
establish an Office of Energy Research to protect and preserve the long-
range R&D functions of the ERDA, particularly those functions relating
to new energy sources. Proposed amendments to allocate specific authorities to each of the eight assistant secretaries and to create a small business
office and a consumer protection office failed; the prevailing feeling was
that the agency should not be unduly restrained in its organization.

2.6.3 National Energy Act

The National Energy Act was a watershed in U.S. energy policy, affecting
virtually every facet of energy supply and demand. As originally introduced, the measure sought to achieve specific reductions in oil and gas use
by a target date of 1985. It would do this through higher energy prices
and a variety of tax and other incentives and disincentives to encourage
fuel switching and voluntary energy conservation. Primary emphasis was

on conservation rather than expanded supply. Specific goals and target dates were later dropped, but the final measure gave the administration unprecedented pricing and fuel allocation authorities. In combination with powers granted by energy legislation enacted during the 93d and 94th Congresses, the National Energy Act substantially increased the federal presence in energy markets.

With respect to renewable energy, the measure provided economic and regulatory incentives to encourage the industry both directly and indirectly: directly through tax credits and other favorable policies; indirectly through energy pricing and regulatory policies aimed at moving energy users away from oil and gas.

Action on the five bills comprising the NEA was completed in the closing days of the second session, after nearly one and a half years of deliberation. This was landmark legislation for solar energy in many respects. In addition to establishing direct and indirect incentives, the bill also gave the solar industry a degree of credibility that was lacking before. The solar bias of the NEA was a signal to the public that the solar market was legitimate and that solar technologies were available to help meet existing as well as future national energy needs. Very briefly, the five acts comprising the NEA did the following for solar energy:

The Public Utilities Regulatory Policy Act (PURPA; PL 95-617) helped establish cogenerators and small power producers by requiring utilities to buy and sell power from such producers at fair rates. Wind and solar thermal power were the solar technologies to benefit most from PURPA.

The Energy Tax Act (PL 95-618) allowed homeowners a tax credit of up to 30% of the first $2,000 and 20% of the next $8,000, for a maximum credit of $2,200. Businesses were allowed to claim an additional 10% investment credit. Both credits were to expire 31 December 1982.

The National Energy Conservation Policy Act (PL 95-619) required utilities to inform customers about energy conservation devices and devices utilizing solar energy or wind power for residential purposes, and to provide a list of businesses in their area that would finance, supply, and install such devices. The act also included provisions to expand lending authority of certain federal housing authorities to make it easier for home buyers to purchase solar-equipped homes.

The Powerplant and Industrial Fuel Use Act (*PL 95-620*) was designed to force major energy users to switch from oil and gas into coal or other energy alternatives, including solar.

The Natural Gas Policy Act (*PL 95-621*) accelerated the deregulation of gas prices, thereby making energy alternatives, including solar, more attractive economically.

2.6.4   Other Solar Commercialization Initiatives

Congress saw the high first cost of solar space and water heating systems and long-standing institutional practices favoring conventional fuels over new sources as the major constraints to solar market penetration. A major objective of the 95th Congress was, therefore, to facilitate solar commercialization by enacting measures to ease first costs and lower institutional barriers.

Solar commercialization was addressed on a wide front. Legislation as diverse as the Youth Employment and Demonstration Projects Act and the GI Bill Improvement Act were passed with congressionally initiated provisions designed to stimulate the market for solar products. Although the National Energy Act was the centerpiece of congressional solar commercialization efforts, its collective parts were only part of the picture. In all, the 95th Congress initiated and enacted some fifteen separate solar incentive measures (OTA 1980). Some of the major enactments are identified below.

2.6.4.1   Food and Agricultural Act of 1977

Congress had long recognized agriculture as an area of the economy where solar energy conversion technology could have an important impact on the use of conventional fuels. The Food and Agricultural Act of 1977 (PL 95-113) was amended to incorporate a comprehensive package of solar-related initiatives first introduced by Representative Brown of California. The Brown amendment provided for the demonstration of solar energy projects on the farm, and encouraged farmers to adopt and install solar power systems, machinery, and production techniques. Most importantly, however, it authorized the establishment of from three to five regional solar energy research and development centers for the performance of agricultural research, extension work, and demonstration projects relating to the use of solar energy on farms. It was under this authority that DOE established four regional solar energy centers, which it located in the Northeast, Southeast, Midwest, and West. These centers carried out R&D and information outreach functions until they were dismantled by the Reagan Administration in 1981–1982.

2.6.4.2. Small Business Energy Loan Act

Small business firms have made a lasting and important contribution to the field of solar energy conversion. The Small Business Energy Loan Act of 1978 (PL 95-315) authorized $30 million in direct loans and $45 million in guaranteed loans for use by small firms for solar energy and energy conservation ventures, thereby helping to assure that these businesses would continue to participate in the development and expansion of the solar energy industry. The financial assistance programs established by the act were administered by the Small Business Administration (SBA). Since the SBA already had authority to help meet the financial needs of small solar business, the Act was, in a sense, a message from Congress that the SBA had not been responsive to the needs of the solar energy industry which was then dominated by small firms.

2.6.4.3 International Cooperation/Foreign Markets

International cooperation in solar energy development has significant market possibilities for the U.S. solar industry, as well as foreign policy implications for the U.S. government. Recognizing this, the 95th Congress acted on a number of bills designed to foster foreign markets for U.S. solar products and to encourage U.S.-foreign cooperation in the development of solar technologies. Several of its enactments added to an existing framework of federal programs that served to promote international cooperation in solar energy; enacted were:

The International Development and Food Assistance Act of 1977 (PL 95-88) authorized $18 million for cooperative programs with developing countries in renewable and nonconventional energy production and conservation;

The Foreign Assistance and Related Programs Appropriation Act of 1978 (PL 95-148) established authority for the United States to participate with other nations in a wide range of conservation and renewable energy activities;

The Nuclear Non-Proliferation Act of 1978 (PL 95-242) created additional authority for the United States to cooperate with developing countries for expanding the development and use of solar energy technologies;

The International Development and Food Assistance Act of 1978 (PL 95-424) provided that in issuing guaranties for housing projects in less-developed countries, preference be given to projects using solar energy where feasible; and

The Foreign Relations Authorization Act, FY 1979 (PL 95-426) provided for the demonstration of solar technologies in U.S. foreign mission buildings, and called on the United Nations to hold a World Alternate Energy Conference in 1981.

2.6.4.4 Federal Buildings and Facilities

In 1975 the federal government owned and operated over 400,000 buildings containing over 2.5 billion square feet of space. In addition, the Postal Service had 36,000 buildings. Congress saw the stock of new and existing federal buildings as both a major market and a proving ground for solar and conservation technologies. Congressional efforts to utilize these facilities for solar marketing purposes began in the 94th Congress with such bills as the Conservation and Solar Energy Federal Buildings Act of 1975 (H.R. 8711), and the Energy Conservation in Buildings Act of 1976 (H.R. 14290). The FEA was strongly supportive of this approach urging, "Where appropriate, substantial numbers of solar heating and cooling systems should be purchased and installed on new and existing Government buildings" (House 1975, 63).

Three congressionally initiated enactments in the 95th Congress relate to the use of solar/conservation techniques in federal buildings and facilities: (1) the National Energy Conservation Policy Act of 1978 (PL 95-619) which was part of the National Energy Act; (2) the Military Construction Authorization Act, FY 1979 (PL 95-356); and (3) H.R. 322, which led to a feasibility study and ultimate funding of solar energy equipment for the House Annex II and the Rayburn House Office Building. The first two of these bills are of particular significance in that they established government policy with respect to the use of solar energy systems in federal buildings.

PL 95-619   Title V of the National Energy Conservation Policy Act of 1978 included a number of initiatives first proposed in the 94th Congress for putting solar/conservation techniques to work in federal facilities. The title included a $100 million demonstration program for solar heating and cooling in federal buildings; a mandate for the retrofit of all federal buildings by 1990 with energy conservation and solar energy systems where such systems are shown to be cost-effective on a life-cycle basis; and a three-year, $98 million federal photovoltaic purchase and utilization program.

PL 95-356   Solar supporters believed that with proper follow-through by the administration, the FY 1979 Military Construction Authorization Act would be one of the most important solar initiatives passed by the 95th Congress (*Cong. Rec.* 1978: S13627–28). Over DOD objection, the bill required that all new military housing use solar energy equipment, if cost-effective, and that at least 25% of all other military construction do the same, on structures started after 8 December 1978. Supporters estimated that the bill would generate $50 million in solar architect-engineering business in 1979, and $100 million annually for the industry over a three- to five-year period (*Cong. Rec.* 1978: S13627–28).

2.6.5   Market Promotion Proposals Not Enacted

While the 95th Congress was remarkably successful for solar interests, three major solar market initiatives were not acted upon. These were in the areas of antitrust legislation, sun rights, and a solar energy development bank.

2.6.5.1   Antitrust Legislation

The feeling on the part of some members that the oil industry had designs on, and could eventually dominate, solar energy development prompted legislation to preserve and promote competition in the energy industry as early as the 93d Congress. Legislation proposed to protect solar interest typically would make it unlawful for integrated oil companies to acquire assets in competing energy technologies or to own or control assets in competing energy technologies within some specified time after enactment of legislation. In December 1977 the Federal Trade Commission's (FTC's) Bureau of Competition sponsored a two-day Solar Energy Symposium to examine the competitive aspects of the solar industry, including oil company involvement in the industry (FTC 1978). The results of the symposium with respect to oil company involvement were inconclusive but were of value in defining the issues.

2.6.5.2   Sun Rights

Under U.S. law, a landowner has the right to receive sunlight directly above his property, but has no right to receive it across the property of a neighbor. Thus, one property owner could legally block the sunlight needed by another property owner to operate a solar device. While the potential seriousness of the sun rights problem was not fully understood, the uncertainty regarding guaranteed access to sunlight was perceived to be a barrier to the widespread use of solar devices. Typical legislation proposed in 95th Congress would mandate solar zoning activities by state and local governments to ensure that one person could not unilaterally cause the sunlight to a preexisting solar device to be blocked in any way. Several bills were introduced to provide legal assurances to users of solar energy equipment that solar energy would always be available.

2.6.5.3 Solar Bank Legislation

The high cost, and in some instances unavailability, of commercial loans for new construction and for home and business improvements was perceived to be a major barrier to solar energy use. To overcome this barrier, several bills were introduced in the 95th Congress to establish a solar energy development bank. In general, the bank would provide low-interest, long-term loans for residential and commercial solar improvements. Most proposals made the bank a new instrumentality of the federal government, reflecting a concern that agencies already administering loan programs were either too incompetent, too biased, or too disinterested in the special problems of solar finance to be effective in these matters. Representative Neal's H.R. 7800 was not the first bill in the 95th Congress to propose a separate solar loan entity, but it drew the strongest support. The Carter administration opposed the measure, however, objecting to the formation of a new federal entity and the large sum ($5 billion) involved in the initial financing. The bill received three days of hearings by a House banking subcommittee (House 1978). Although after the hearing the bill was amended to meet administration objections, it was never reported from committee due in part to the lateness of the session. The 96th Congress would see the administration act on the recommendations of the solar Domestic Policy Review (DPR) and propose its own solar bank bill modeled after H.R. 7800 as amended, the Solar Energy Development Bank Act of 1979 (President, 1979b).

2.6.6 Carter Administration and the Politicization of Solar Development

The hopes of solar advocates had been raised by the election of President Carter in 1976. He was sworn in on a solar-heated reviewing stand; his

National Energy Plan, unveiled soon after he took office, promised to "vigorously expand the use of nonconventional energy"; and the president's version of the National Energy Act included the solar tax credits that the Congress had long been pushing. However, disappointment quickly set in (Frankel 1983, 28).

President Carter's first solar budget request (FY 1978) was uninspired in terms of new initiatives, and identical in dollar amount to the Ford request. The FY 1979 request called for a cut in the level appropriated by the Congress the previous year. For his first secretary of energy, President Carter nominated James Schlesinger, former head of the Atomic Energy Commission and the Department of Defense, raising concerns that the perceived pronuclear bias of ERDA would not be changed. To make matters worse for solar advocates, the president successfully held the National Energy Act solar tax credits hostage, pending action on natural gas pricing and other controversial provisions of the act. This tactic delayed the enactment of the credits by one year, during which time solar sales declined as consumers deferred their solar purchases awaiting congressional action.

Frustrated by their inability to get a meaningful commitment on solar energy from the Carter administration, solar advocates attempted to mobilize public support for solar by holding a national Sun Day on 3 May 1978. This action received strong bipartisan support from Congress. The national attention put President Carter on the defensive, leading him to announce a $100 million supplemental budget request for the DOE solar program in FY 1979 and the formation of a Domestic Policy Review (DPR) for solar energy.

2.7 Ninety-sixth Congress: A Final Period of Policy Expansion

With the passage of the National Energy Act and the Department of Energy Organization Act in the 95th Congress, the feeling was widespread in the Congress that the necessary policy and organizational framework was in place to deal with the energy problem. There were still energy issues to be resolved, as well as some lingering doubts about the efficacy of some of the compromises, but the basic elements of an effective program had been addressed. Building on past accomplishments, the 96th Congress pushed for an even larger federal role in energy matters, including renewable energy. Two major energy bills, the Energy Security

Act and the Crude Oil Windfall Profits Tax Act, increased federal involvement in energy markets to unprecedented levels. The expansion would be short-lived, however, as the Reagan administration would move quickly in the next Congress to roll back the market interventionist policies of the 1970s.

The 96th Congress was in all respects the end of a legislative era that saw Congress lead the federal government toward an ever-expanding role in solar commercialization. The focus of this era was on solar heating and cooling, but the untapped potential of other solar technologies was recognized as well. Also benefiting from the congressional largesse with specific, market-oriented enactments were wind energy, ocean thermal energy, and photovoltaics. Bioconversion benefited as well with alcohol fuels and related legislation. The only solar technology not singled out in legislation was high-temperature solar thermal, which however, was advanced through aggressive congressional support for research and development and through PURPA incentives.

The Carter administration's support of solar commercialization in the 95th Congress did little to ease congressional pressure for expanding the federal role even more. Solar commercialization initiatives remained high on the legislative agenda of the 96th Congress with well over 100 solar or solar-related bills being offered (CRS 1980). At the start of the 96th Congress, Representative Ottinger, a leading solar advocate, criticized the Administration for its weak support of solar energy:

I would like to say that I am very concerned about the priority which the Department of Energy is placing on the solar programs, the amount of money that is being provided, the way regulations are being implemented, and the staffing that is being devoted. . . . There is federal help which can make a significant difference in how quickly solar energy becomes a reality in our country, and insufficient accent is being placed within the Department on making these potentials come true. (House 1979, 3)

Two major policy instruments of the Carter administration account for virtually all of the new legislative authority with respect to solar commercialization in the 96th Congress. The most important of these was the administration's biennial National Energy Plan. Legislation submitted by the administration in conjunction with the plan ultimately led to a substantial increase in the residential and business solar tax credit, and to the establishment of the solar bank that advocates had long sought. The other policy instrument was the Domestic Policy Review (DPR) of solar energy which identified policy options available to the federal government for promoting solar energy. This document was used by solar advocates to justify new initiatives and as a yardstick for measuring administration progress in solar commercialization.

2.7.1   NEP II Legislation

President Carter's second National Energy Plan was delivered to the Congress on 7 May 1979 (President 1979a). Legislation to achieve the policies outlined in the plan were subsequently sent to the Congress. Of these proposals, the Windfall Profits Tax Act and the synthetic fuel bill would have a major impact on the federal solar commercialization effort. In addition, the administration began a phased decontrol of domestic oil prices that made alternate fuels, including solar, more competitive with oil products.

2.7.1.1   Windfall Profits Tax Act of 1980

PL 96-223 was designed to capture and then redistribute a portion of the "windfall" profits oil companies were expected to reap when oil prices were deregulated. With respect to solar energy, the act increased both the residential and business solar tax credits provided under the Energy Tax Act. The conference on the act was particularly difficult because the Senate version specified how the tax revenues would be spent, while the House version did not. Consequently, the House version contained no renewable energy resource provisions.

The act increased the residential solar tax credit to a flat 40% on the first $10,000 of expenditures for qualifying equipment; the termination date of 31 December 1985, was not changed. Under the Senate-approved version, the credit would have been increased to 50% and extended through 1999. Also, a Senate-approved provision for extending tax credits to passive solar equipment was dropped in conference. The act increased the business energy tax credit to 15%, and broadened the list of eligible equipment to include systems using solar or wind energy to generate electricity or to provide heating, cooling, or hot water in a structure; ocean thermal energy equipment for two experimental sites; and equipment using solar energy to provide industrial, agricultural, or commercial process heat. The Senate version would have increased the business credit to 20% and extended the termination date through 1990.

2.7.1.2 Energy Security Act ("Synfuels" Bill)

The Carter administration originally wanted a bill to promote a synthetic fuels industry. However, PL 96-294 acquired major amendments unrelated to synthetic fuels as it was merged with House and Senate bills and was otherwise amended during the legislative process. The Senate's price for supporting the administration's multibillion dollar synfuels corporation was a series of amendments to promote conservation and the use of renewable energy sources, as well as other energy policies. When finally enacted on 30 June 1980, the measure had been transformed into an omnibus energy bill, approaching the National Energy Act in scope (Senate 1982).

Title IV, which clarified and broadened certain existing solar commercialization measures, and Title V, which established the Solar Energy and Energy Conservation Bank, were added largely to garner support from members who were not inclined to support synfuels (Senate 1982, 25):

Title IV, Renewable Resource Initiatives, contained an assortment of provisions promoting the use of renewable energy resources and conservation. Among other actions, sections in this title (1) required DOE to coordinate its solar and conservation outreach activities and report annually to the Congress; (2) created a three-year pilot program within DOE to demonstrate energy self-sufficiency in one or more states through the use of renewable resources; (3) clarified the eligibility of federal facilities which could participate in the Federal Photovoltaic Utilization Act; and (4) relaxed the rules for qualifying facilities under the Public Utility Regulatory Policy Act (PURPA).

Title V, Solar/Conservation Bank, established a combined solar and conservation bank to provide loans for solar and energy conservation installations. President Carter proposed a solar bank on 20 June 1979, to be funded at $100 million annually out of revenues generated by the Windfall Profits Tax. However, the president's proposal was based upon congressional initiatives first introduced in the 94th Congress (H.R. 3849 by Representative Gude and S. 875 by Senator Hart), and was increased in size and scope by committee and floor amendments in both the House and Senate.

2.7.2 Solar Energy Domestic Policy Review

The solar DPR (DOE 1979) had its genesis in a speech made by President Carter in observance of the Sun Day movement on 3 May 1978. The high visibility which the administration accorded the review raised public and congressional expectations for a major new federal initiative in solar development and commercialization. For most solar supporters, however, the final report and follow-up actions by the administration were a major disappointment.

To its credit, the DPR recommended a national goal of 20% solar energy use by the year 2000. This important, albeit symbolic, gesture drew attention to the solar movement and helped establish a sense of purpose and direction for both the federal program and private sector efforts. Solar supporters were disappointed, however, that the DPR did not put forward a coherent, long-range strategy for achieving the goal. The modest set of targeted initiatives recommended by the DPR (which included a solar bank, an increase in the business energy tax credit, and a 20% tax credit for passive solar homes) was basically a mix of pending congressional proposals, which, under the most optimistic assumptions, added up to a solar penetration of 15% by the year 2000. The president's endorsement of a national goal for solar energy use placed the administration in the awkward position of setting a goal without having a program to achieve it.

The DPR lost much of its public relations value when the administration failed to act quickly and enthusiastically upon receiving it from DOE. President Carter endorsed the 20% solar goal and most of the recommended economic initiatives on 20 June 1979, some six months after the report had reached his desk (*WCPD* 1979, 1097–1107). By this time, he had announced a much larger energy initiative that included the decontrol of oil prices and the imposition of a crude oil windfall profits tax. President Carter angered solar proponents by tying the tax credit increase for solar investments to oil decontrol, and he further angered them in July 1979 when he proposed the $100 billion Synthetic Fuels Corporation, dwarfing the solar DPR effort. The way in which the DPR was handled left the solar community with the feeling that the administration was more interested in gaining support for synfuels than it was in promoting a solar future.

In the final analysis, the much-heralded solar DPR proved to be a largely political, largely symbolic, gesture. Some solar advocates denounced the goal and the program to carry it out as the very minimum that the president could have done politically, claiming that the announced program would increase solar use by only two percentage points over what would be expected with no federal program (see, for example, the *Wall Street Journal*, 20 June 1979, or the *Washington Post*, 21 June 1979). Others criticized the goal as being too low, and the administration for not providing anywhere near the level of support needed to achieve a 20% penetration level. Paul Rappaport, the executive director of DOE's Solar Energy Research Institute (SERI), said in a SERI news release that the goal was achievable but would cost an estimated $5 billion per year over the next twenty years. This was far more than the administration was willing to spend, however. In FY 1981, under an austere budget, the administration's solar request plus off-budget costs for market incentives amounted to about $1 billion.

Perhaps most disappointing for solar interests, however, was that after all of the buildup, there was still no long-range federal strategy for solar development and use. Funding and policy decisions were still being made piecemeal. With the federal program operating on a year-to-year basis, industry could not make the long-term capital decisions necessary for the private sector to meet its obligations under the DPR.

Within one year of the president's endorsement of the 20% solar goal, two congressional studies and other information available to Congress left little doubt that the administration was either unable or unwilling to keep its DPR commitment.

• A General Accounting Office (GAO) report found that the administration still had no comprehensive plan for attaining the goal almost one year after its announcement (GAO 1980). GAO also found that DOE had failed to implement several of the administrative measures announced by the president, including the formation of an interagency solar policy coordinating committee.

• A report by the Office of Technology Assessment (OTA) was highly critical of DOE's management of the solar and conservation programs (OTA 1980). OTA found these programs to be understaffed, lacking in direction and suffering from a "pervasive belief within and outside DOE that senior DOE management does not really care" about them.

• A demonstrable commitment to the 20% solar goal would imply significantly enhanced federal funding of solar and conservation programs. However, in late May 1980, the House Science and Technology Committee learned of an internal DOE memorandum from Secretary Duncan that set multiyear funding levels for solar and conservation substantially below their FY 1980–FY 1981 share of the DOE budget (*Cong. Rec.* 1980: E3430). DOE was planning a significant expansion of its coal and nuclear programs in FY 1982–FY 1986, while holding the solar and conservation budgets at the business-as-usual level.

2.8 Solar Policy in the Reagan Era, and the Congressional Response

In the mid- and late 1970s with energy a priority issue on the national agenda, the legislative and executive branches joined in an effort to develop a unified federal energy policy. As a result of this effort, federal control of energy markets grew increasingly tighter. Under the Reagan administration, however, this policy course took a radical turn. The Reagan administration abandoned the market intervention policies of the 1970s in favor of a "free market" course, which holds that energy decisions are best made in the marketplace. As stated by President Reagan, the administration wants to remove government intrusion in energy policy so "native American genius—not arbitrary federal policy—will be free to provide for our energy future" (President 1981).

The administration's free-market approach to energy development had a profound impact on the entire federal energy effort. In the extreme, it precluded the federal government from any activity that might influence private sector decisions, such as applied research, hardware demonstrations, technology-specific market incentives, and even mid- and long-range planning exercises. A free-market policy limits the federal energy effort to long-term, high-risk research projects, which ideally are not technology-specific, and in which the private sector would be unlikely to invest.

The solar program of the 1970s was anathema to Reagan administration policy. Congress had become convinced that a strong federal presence in the marketplace was the key to the nation's solar future. Spurred on by congressionally initiated R&D and incentive legislation, the entire program had become heavily oriented toward applied research, hardware demonstrations, and commercialization incentives. The program was an obvious target for reform under the new policies of the Reagan administration. Indeed, the solar program would have been virtually eliminated if the principles of a free-market energy policy had been carried to the extreme. To the extent that they were applied, the free-market principles of energy development radially changed the federal solar program as it existed in the 1970s. Pressure was brought to bear almost immediately to move federal energy programs away from near-term market development and commercialization activities. The programs most affected were those that supported near-term technologies, such as a solar thermal applications for buildings and small-scale wind systems.

Although the federal solar program changed radically under the Reagan administration, the program's legislative underpinnings were little affected. With the exception of solar-specific tax incentives that expired at the end of 1985, the 1970's legislative authority for an aggressive federal role in solar R&D and commercialization has not expired or been repealed. The seemingly impossible task of turning the federal solar program completely around without repealing basic legislative authority was achieved by the administration, using prerogatives available to it through the budget process. In effect, the Reagan administration used the budget process to engineer a de facto repeal of basic solar program authority; this allowed the administration to introduce change gradually, while avoiding direct up-or-down vote confrontations with solar advocates in the Congress.

As part of the budget process, votes affecting solar policy were imbedded in, and sometimes obscured by, questions of larger public policy concern such as the budget deficit or the DOE role in national defense. In addition, constant pressure from the administration and from within Congress to reduce federal spending and reduce federal involvement in energy markets kept solar proponents divided and on the defensive, making the budget approach to solar program reform all the more effective.

Although brought on by the administration's free-market policies, the radical changes that occurred in the federal solar program required the tacit support of Congress. Throughout this period of policy reform, Congress was a reluctant partner of the administration, drawn in by fiscal pressures. Congress conceded cuts in the federal solar program in the larger struggle to bring federal spending under control. Members generally viewed cuts to renewable energy programs as "drastic, perhaps damaging, but hopefully not fatal, to [the national] strategy for energy independence" (*Cong. Rec.* 1982b: S3207).

The administration wanted to reduce or eliminate as much of the program as possible as quickly as possible, but Congress sought a slower, less drastic approach. Its objective was program reduction without a total sacrifice of program effectiveness. Congress also saw a lack of balance in the way the administration was reducing funding for competing energy technologies, with solar and conservation losing out to nuclear programs. Throughout the 1970s, a major energy policy objective of the Congress had been to bring the nuclear and nonnuclear portions of the federal energy effort into balance. It saw this work being undone by Reagan administration budget proposals. With energy off the national agenda, solar steadily lost ground to nuclear and advanced energy programs, despite congressional support of a balanced federal energy R&D program and despite objections to administration tactics. During the debate on the FY 1982 DOE budget request, the first for which the Reagan administration was fully responsible, Senator Hart objected to the administration's cavalier treatment of energy matters, its favored treatment of nuclear programs, and its apparent disregard for conservation and renewable energy sources (*Cong. Rec.* 1981a: S6848–49). Speaking in support of an amendment to add $200 million each to the solar and conservation budgets, Hart observed

This amendment is not a solar amendment.... It is a priorities amendment. This Congress has established, for the last 8 years, that the priorities of this country dictate that we shall develop all our energy supplies, even those that are not presently economically competitive. The Administration has come in overnight, through the budget process, without an energy statement by the President of the United States, and reversed that course. (*Cong. Rec.* 1981a: S6848–49)

2.8.1 Solar Budget Cuts and the Congressional Response

When the Reagan administration came into office in January 1980, solar spending comprised 11% of the FY 1981 DOE research, development, and demonstration (RD&D) budget. Solar was at funding parity with the fusion program and, under Carter administration projections, would be close to achieving parity with the fission program by FY 1983 (U.S. Budget 1980 [for 1981], 133, 144). Congress appropriated $598 million for DOE solar programs in FY 1981, which was about the same as the FY 1980 level ($595 million in direct program support), suggesting that the legislative and executive branches were in general agreement on an optimum funding level for the market-oriented program then in place.

While RD&D spending appeared to be leveling off, total federal costs for solar were rising in 1980, thanks to the indirect cost of the residential and business solar energy tax credits and solar spending in agencies other than DOE. The cost of the credits was estimated at $208 million in

FY 1980 and $307 million in FY 1981 (U.S. Budget 1980 [for 1981], 133, 144); other-agency solar costs were estimated at $239 million and $463 million (including an estimated $150 million for the Solar Bank) in FY 1980 and FY 1981, respectively (Moore 1980, 18). Thus, when the Reagan administration took over, direct and indirect federal costs for solar applications were estimated at $1.35 billion and rising.

Starting with a budget of $598 million carried over from the Carter administration, the Reagan administration began ramping down direct solar spending almost immediately. It requested a $93 million rescission in the FY 1981 DOE solar budget and the rescission of all HUD funding for the Solar Energy and Energy Conservation Bank. Acceding in part to the request, Congress approved $47 million of the DOE rescission and deferred $49 million, thereby reducing the FY 1981 appropriation to about $502 million. The Solar Bank rescission was approved in full, resulting in the termination of all bank activities for FY 1981.

2.8.1.1 FY 1982 Request

The FY 1982 budget submittal was the first for which the Reagan administration was fully responsible. In it the administration proposed to cut the DOE solar budget to $193 million—about a third of the of the FY 1981 level before the rescission. Secretary Edwards told Congress that this and future budget savings anticipated in the solar RD&D program would not affect the market for solar technologies: "These budget changes will have little effect on solar energy use, which will continue a healthy rate of increase over time as rising conventional energy prices and solar tax incentives stimulate the demand for solar products" (House 1981, 11).

Congress accepted the move away from federally sponsored near-term R&D and commercialization programs, but it could not accept the pace. In this first test of congressional resolve on solar funding, Congress approved the Omnibus Budget Reconciliation Act (PL 97-35) with an authorization of $303 million for DOE solar programs in FY 1982. This was $110 million above the administration's request, but still just half the FY 1981 level before the rescission. The authorization saved several solar programs from termination in FY 1982, including the ocean thermal energy conversion program, large wind systems development, and major elements of the photovoltaics development program. In addition, the Solar Bank was authorized at $50 million, saving it from probable extinction in FY 1982.

The administration's solar budget request received partisan support in the House and Senate, with the Democratic House authorizing $321 million and the Republican Senate accepting the $193 million level as requested. However, even the partisan Senate found it difficult to accept the cut in full; it provided a measure of flexibility in the solar funding level by distributing over several programs a sum of $568 million "left over" after moving the strategic petroleum reserve off budget. It designated $61 million of this amount to energy supply research programs, including solar energy programs, with ocean thermal energy conversion singled out for possible funding in the Appropriations Committee, even though the administration wanted this program terminated. The conference agreement was closer to the House position than the Senate's, with $303 million being authorized for the DOE solar program. The conference also restored the wind, photovoltaic, and ocean thermal programs that the administration wanted to substantially reduce or eliminate.

Senate Democrats supported a Bumpers amendment to the authorization bill that would have added $200 million each to the solar and conservation budgets and doubled the Solar Bank authorization. Although the amendment failed, proponents took the opportunity to object to the pace, process, and lack of balance in the administration's energy policy shift. The amendment was easily defeated (35-64) with opponents arguing that it would add a total of $450 million to the DOE budget cap without offsetting reductions, and would reduce the flexibility of the conference committee in negotiations with the House of Representatives (*Cong. Rec.* 1981b: S6850).

Congress approved FY 1982 appropriations totalling $275 million for DOE solar programs (Energy and Water Development Appropriations Act, PL 97-88). Funding was included for solar programs as authorized in the Budget Reconciliation Act, including those which the administration wanted terminated. The HUD appropriations act (PL 97-101) provided $23 million for the Solar Bank, a sum barely sufficient to keep the bank option alive.

In its report on the FY 1982 appropriations bill, the Senate Committee on Appropriations agreed with the direction in which the administration was moving the federal solar program, but it did not agree with the pace:

The Administration has proposed substantial reductions in solar energy activities. These reductions signal a fundamental change in the level and nature of federal support.... While the Committee agrees with the basic thrusts of this modification

... it is believed that there should be a more gradual transition from federal to private sector support than that proposed by the Administration. (Senate 1981, 89)

In a bipartisan effort, the Congress added about $85 million to the administration's FY 1982 solar budget request. Even so, the net budget was still 58% below the previous year's—hardly a stunning defeat for the administration. Indeed, the administration may have interpreted it as a tacit endorsement of a shutdown of the federal solar program. When Congress approved a $61 million across-the-board cut in FY 1982 DOE funding, the administration took one-third, or $20 million, off the solar program even though solar funding represented just 8% of the agency's total RD&D budget. The absence of a serious congressional challenge to this disproportionate cut may have reinforced a possible administration perception that congressional support for the solar program was weak and that Congress could be squeezed for even greater "savings." In FY 1983 the administration would request a "close-out" solar budget of just $73 million.

2.8.1.2 FY 1983 and FY 1984 Budget Requests
The administration's FY 1983 and FY 1984 budget requests were its most forceful attempts to bring the DOE solar program into conformance with its free-market ideology. In both submittals it sought the wholesale elimination of near-term development and commercialization programs, requesting just $72 million and $86 million, respectively.

The $73 million solar budget request for FY 1983 reflected an administration proposal to eliminate the DOE and reorganize the federal government's role in energy and commerce. The "close-out" budget, as it was called, was a transitional budget leading to the establishment of a new agency, the Energy Research and Technology Administration (ERTA), within the Department of Commerce (DOE 1982, 3, 20). The administration's rationale for ending all but long-range, basic research in solar energy was summed up in its FY 1983 budget justification document:

By relying on the marketplace and private industry, it is no longer necessary for the federal government to support the development, demonstration or commercialization of solar energy, including R&D undertaken for the purpose of accelerating the introduction of new solar technologies. The free market will determine the development and introduction rates of solar technologies consistent with their economic potential. (DOE 1982, 3, 20)

The close-out solar budget served three objectives for the administration: (1) it brought the solar program into conformance with the administration's free-market philosophy; (2) it helped to alleviate the federal deficit; and (3) it prepared the solar program for the eventual termination of the DOE and the transfer of program authority to another agency.

Congress was not persuaded by the administration's rationale. Its bipartisan reaction to the FY 1983 budget proposal was quick, decisive, and largely negative. In the House, Democrats and conservative Republicans attacked the administration's proposal for failing to take nonnuclear energy, and especially energy conservation, seriously, and to present a true free-market budget (see for example, the statement of Representative Vin Weber, a Republican Member of Congress, in House 1983, 10–11). Representative Ottinger, with a bipartisan group of 169 original cosponsors, introduced a resolution to restore balance in the federal energy budget (*Cong. Rec.* 1982a: E1256–57). Ottinger called the Administration's energy proposal "unbalanced, shortsighted, and ill-conceived," and accused the administration of ignoring "repeated congressional action last year that restores diversity and balance to the energy budget." He said further that

this Nation cannot afford to abandon the balanced energy policy crafted over the past decade with the hard work of both political parties. We cannot afford to be lulled into complacency by the current oil glut. I think this resolution and its broad list of cosponsors demonstrates that once again Congress will reject the administration's budget proposals and will maintain its commitment to vital energy conservation and renewable energy programs. (*Cong. Rec.* 1982a: E1256–57)

In the Senate, Republicans joined Democrats in supporting a DeConcini resolution urging that FY 1983 solar and conservation funding be kept at the FY 1982 level (*Cong. Rec.* 1982b: S3207). Senator DeConcini urged members to "dissent, dissent strongly, and persist in our dissension" to administration efforts to eliminate government participation and support for all alternative energy programs (*Cong. Rec.* 1982b: S3207). He attributed the good news on energy market conditions to "worldwide economic stagnation and a severe recession in this country." And he said that for the administration to expect the private sector to pick up the slack and bear the entire burden of research, development and commercialization of emerging solar and conservation technologies was "unrealistic and not in the long-term, or even short-term, public good" (*Cong. Rec.* 1982b:

S3207). Senator Heinz, a Republican and principal cosponsor of the resolution, called the administration's budget submission for conservation
and renewable energy "shortsighted and inequitable" (*Cong. Rec.* 1982c:
S3207–08).

The administration's FY 1983 close-out budget for solar R&D was
rejected by the appropriations committees of both the House and Senate,
handing the administration a clear defeat. In their respective reports on
the FY 1983 DOE appropriations bill, neither committee agreed with the
administration that most solar and renewables research had been developed to the point where the private sector alone could be expected to
carry forward with its development (Energy and Water Development
Appropriation Bill 1983). The Senate Appropriations Committee recommended $188.9 million for solar energy, which was $7.5 million more
than the amount recommended by the Democratic-controlled House
Committee. The committees were in agreement over the funding levels of
most subprograms, including those for active and passive solar building
applications. The administration wanted these programs terminated in
FY 1983, but the committees provided $6.65 million and $5 million,
respectively, continuing them at about half the FY 1982 level.

When Congress was unable to complete action on an FY 1983 appropriation bill for DOE (and several other agencies), it passed an omnibus
budget resolution (PL 97-377) which provided for the continuation of
solar funding in FY 1983 at the FY 1982 level. This was perceived as a
boon for the solar program because the FY 1982 appropriation ($266
million) was substantially more than the amounts recommended by the
appropriation committees of either house. While the resolution gave the
administration wide latitude in setting the final appropriation—anywhere
from its own request to the FY 1982 level—it settled on $202 million.
This level was generally accepted as reasonable, given the tight fiscal
restraints the entire federal budget was under; it would also become a
benchmark by which future solar budget requests would be measured.

A pattern developed in the annual give-and-take over the budget. The
administration would submit a sharply reduced budget request; Congress
would boost it but not enough to meet the level of the previous year. The
steady decline in solar funding meant that Congress was allowing
the administration to use the budget process to dismantle the program.
The pace was slower than the administration wanted, but effective
nonetheless.

2.9   Summary and Outlook

Solar was immensely popular with Congress in the 1970s as a resource that could help to alleviate U.S. dependence on imported oil. Although the program suffered severe budget cuts and political downgrading in the 1980s, Congress would not allow it to die. Today, in light of growing concerns over the effects of fossil fuels on the environment, the solar option is being reexamined for its potential as an environmentally benign alternative. Thanks in large part to the tenacious efforts of many members of Congress over many Congresses, there exits today a federal program capable of exploring this important aspect of solar energy. Whatever the future may hold, much of the credit for any future success in the nation's use of solar resources must go to the farsighted efforts of Congress to build and sustain a viable federal program.

References

Congressional Record. 1974a. Statement by Representative Winn during the House consideration of H.R. 11864. 13 February: H739.

Congressional Record. 1974b. Statement by Senator Metcalf. 14 August: S14750.

Congressional Record. 1974c. Statement by Senator Mike Gravel. 1 October: S17880.

Congressional Record. 1974d. Statement by Representative Symms during the House debate on the conference report on H.R. 11864. 21 August: H8787–88.

Congressional Record. 1975a. Statement by Senator Domenici. 21 March: S4653.

Congressional Record. 1975b. Statements by Representatives Gude and Frenzel. 13 June: H5431.

Congressional Record. 1976a. Statement by Representative Ottinger during the House debate on H.R. 12169, the FEA Extension Act. 1 June: H5065.

Congressional Record. 1976b. Statement by Senator Gary Hart. 15 June: S9439.

Congressional Record. 1978. Statement by Senator Gary Hart in Military Construction Authorization, 1979, Conference Report. 17 August: S13627–28.

Congressional Record. 1980. Statement by Representative Richard Ottinger on DOE's solar and conservation failures. 2 July: E3430.

Congressional Record. 1981a. Statement by Senator Gary Hart. 24 June: S6848–49.

Congressional Record. 1981b. Statement by Senator McClure. 24 June: S6850.

Congressional Record. 1982a. Statement by Representative Ottinger on H.R. 409, resolution to restore balance in the federal energy budget. 24 March: E1256–57.

Congressional Record. 1982b. Senator DeConcini speaking in support of S.R. 355, relating to funding for renewable energy and energy conservation. 31 March: S3207

Congressional Record. 1982c. Statement by Senator Heintz on S.R. 355, introduced by Senator DeConcini with 23 Democratic and 13 Republican original cosponsors. 31 March: S3207–08.

CRS (Congressional Research Service). 1980. *Major Legislation of the [96th] Congress.* Issue no. 14. Washington, DC, September.

DOE (U.S. Department of Energy). 1979. *Domestic Policy Review of Solar Energy.* TD-22834. Washington, DC.

DOE. 1982. *Federal Energy Programs, FY 1983: Budget Highlights.* DOE/MA-0062. Washington, DC, February.

Energy and Water Development Appropriation Bill. 1983. S. Rept. 97-673, to accompany S. 3079. 6 December 1982: H. Rept. 97-850, to accompany H.R. 7145.

ERDA (U.S. Energy Research and Development Agency). 1976. *ERDA 76-1.*

Frankel, G. 1983. "Technology, Politics, and Ideology: The Vicissitudes of Federal Solar Energy Policy, 1973-1983." Draft revision copy. 7 December.

FTC (Federal Trade Commission). Bureau of Competition. 1978. *The Solar Market: Proceedings of the Symposium on Competition in the Solar Energy Industry,* December 1977. Washington, DC.

GAO (U.S. General Accounting Office). 1980. *Twenty Percent Solar Energy Goal: Is There a Plan to Attain It?* EMD-80-64. Washington, DC: 31 March.

Kash, D. E., and R. W. Rycroft. 1984. *U.S. Energy Policy: Crisis and Complacency.* Norman: University of Oklahoma Press.

Moore, J. G. 1980. *Solar Power.* Issue Brief no. 74059. Washington, DC: Congressional Research Service, 6 August.

NSF/NASA (National Science Foundation and National Aeronautics and Space Administration). Solar Energy Panel. 1972. *An Assessment of Solar Energy as a National Energy Resource.* Washington, DC, December.

OFEF (Organization of Federal Energy Functions). 1977. A report. from the President to the Congress. Prepared in accordance with Sec. 162 (b) of PL 94-385. Washington, DC, January.

OTA (Office of Technology Assessment) 1980. *Conservation and Solar Energy Programs of the Department of Energy: A Critique.* Washington, DC.

President. 1977. Message to Congress on proposing Department of Energy (H.R. 4263 and S. 826). H. Doc. 95-91. 1 March.

President. 1979a. "Second National Energy Plan." Message. H. Doc. 96-121. 7 May.

President. 1979b. "Solar Energy Development Bank Act." Communication. H. Doc. 96-170. 27 July.

President. 1981. "Nuclear Policy Statement." 8 October.

Technical Review Panel. Sub-Panel IX. 1973. *Solar Energy Program Report.* Prepared for Dixie Lee Ray, the President's energy R&D Advisor. December.

U.S. Budget. 1980. *Budget of the United States Government, 1981.* S/N 041-001-00185-9. Washington, DC.

U.S. House. 1971. H. Doc. 92-1. 22 January.

U.S. House. Committee on Science and Astronautics. 1972. *Energy Research and Development.* Report of the Task Force on Energy. 92d Cong. 2d sess. Washington, DC, December.

U.S. House. Committee on Science and Astronautics. Subcommittee on Energy. 1973. *Hearings on Solar Energy for Heating and Cooling.* 93d Cong., 1st sess. 7 and 12 June. Testimony by Dr. George Löf.

U.S. House. Committee on Science and Technology. Subcommittee on Energy Research, Development, and Demonstration. 1975. *Oversight Hearings on the Solar Heating and Cooling Demonstration Act of 1974.* 94th Cong., 1st sess. 13-15 May. Testimony by FEA Administrator Frank Zarb.

U.S. House. 1976a. H. Rept. 94-1515 on H.R. 10612. 13 September.

U.S. House. Committee on Science and Technology. Subcommittee on Energy Research, Development, and Demonstration. 1976b [Committee print.] *Solar Energy Legislation through the 94th Congress.* Prepared by the Science Policy Research Division of the Congressional Research Service. Washington, DC December.

U.S. House. Committee on Banking, Finance and Urban Affairs. Subcommittee on Domestic Monetary Policy. 1978. H.R. 7800, The Solar Energy Development Bank Act. Hearings, 95th Cong., 2d sess. 25–27 April. Washington, DC.

U.S. House. Committee on Interstate and Foreign Commerce. Subcommittee on Energy and Power. 1979. *Hearings on Solar Commercialization.* 96th Cong., 1st sess. 10–11 January. Statement by Representative Ottinger, p. 3.

U.S. House. Committee on Appropriations. Subcommittee on Energy and Water Development. 1981. *Hearings on Energy and Water Development Appropriations for 1982.* 25 February. Testimony of Secretary of Energy James Edwards.

U.S. House. 1983. Committee on Science and Technology. *FY 1983 Department of Energy Budget Review Hearings.* 97th Cong., 1st sess. 4 March. Statement of Representative Vin Weber.

U.S. Senate. Committee on Aeronautical and Space Sciences. 1974a. *Hearings on Solar Heating and Cooling.* 93d Cong., 2d sess. 25 February. Testimony by Representative Mike McCormack.

U.S. Senate. Committee on Aeronautical and Space Sciences. 1974b. *Hearings on Solar Heating and Cooling.* 93d Cong., 2d sess. 25 February. Testimony by Michael H. Moskow, Assistant Secretary for Policy Development and Research, Department of Housing and Urban Development.

U.S. Senate. Committee on Interior and Insular Affairs. 1976. [Committee print.] *Energy Legislation in the 94th Congress.* A summary report prepared by the Congressional Research Service. 94th Cong., 2d sess. Washington, DC.

U.S. Senate. Committee on Energy and Natural Resources. 1979. *Energy Initiatives of the 95th Congress.* Prepared by the Congressional Research Service. Washington, DC, May.

U.S. Senate. Committee on Appropriations. 1981. Energy and Water Development Appropriation Bill, 1982. S. Rept. 97-256. 28 October.

U.S. Senate. Committee on Energy and Natural Resources. 1982. *Energy Initiatives of the 96th Congress.* [Committee print.] Prepared by the Congressional Research Service. June.

Public Papers of the Presidents of the United States. Richard M. Nixon, 1973, Special Message to Congress on Energy Policy. 18 April.

WCPD (Weekly Compilation of Presidential Documents). 1973a. 23 April.

WCPD. 1973b, 8 Nov., 1973.

WCPD. Public Papers, R. M. Nixon 1974. 28 January.

WCPD. 1977. *The National Energy Plan.* Executive Office of the President, Energy Policy and Planning. 2 May.

WCPD. 1979. 25 June.

3 Market Development

Carlo LaPorta

3.1 A New Solar Industry

This chapter reviews the market history of solar thermal energy products and the industry that developed to deliver them. While it concentrates on the active solar heating segment of the industry, it also reviews high-temperature solar concentrator and passive solar technologies. ("Passive" here refers to building design and products that maximize use of natural energy flows.)

In 1974, when Congress passed the Solar Energy Research and Development and Demonstration Act, essentially no solar energy market or industry existed in the United States. Indeed, a directory of solar heated homes published at that time listed just 200 installations nationwide. Only "old timers" could easily recall how systems were designed in the 1930s in a commercial market that had existed in Florida for residential solar water heaters. (It was curtailed by the advent of cheap natural gas.)

When the federal and state governments created energy policies in the 1970s to cope with expected energy supply shortages, they strongly supported commercialization of solar energy. The new solar thermal market included three industry segments: (1) active solar heating and cooling; (2) passive solar design and passive design products; and (3) high-temperature solar thermal for process heat and electric power. The active heating industry became the largest and most easily identifiable industry segment.

The passive solar industry remained less well known because it involved design firms and use of common building products. Late in the 1980s, when companies introduced more advanced window technologies and thermal storage techniques, these would be classified as energy efficiency products, not passive solar ones. The higher-temperature solar thermal industry was the smallest segment, numbering less than fifty firms in the early 1980s, nearly all of them primarily engaged in government-sponsored R&D. Only a few of these companies succeeded in making commercial sales, and when conventional energy prices began falling in the mid-1980s, all of these firms but one stopped marketing. Many left the solar energy field.

From 1975 through 1980, a completely new industry formed to market solar heating and cooling systems. Over these six years, it installed about 250,000 systems, mostly water heaters for single-family residences and residential swimming pool heaters. By 1986 the total number of installed systems was close to one million. The core of the new industry was formed by 200 companies that built flat-plate solar collectors. Manufacturing and installation of these systems ultimately involved several thousand companies and peak annual sales near $1 billion.

Tracking passive solar industry development is more difficult. A variety of products that use or control sunlight are available to the building industry. Systematic data collection, however, has failed to establish the exact extent of this segment of the solar industry. It has been estimated that by 1985, 200,000 passive solar homes existed in the United States, but it is nearly impossible to estimate annual gross revenue for the "products" and relate it to an "industry."

The development of a commercial market for high-temperature solar thermal products essentially involved one firm, Luz International. This company successfully built a series of solar thermal electric power plants to sell power independently to a California utility company. By the end of 1990, the solar thermal electric capacity installed commercially totaled approximately 354 MWe. Government investment incentives and third-party capital were essential to the financing of these projects. No commercial market for large solar thermal industrial process heat systems has ever been established. Although a few such systems were built in the early and mid-1980s in the United States, sales were nonexistent until 1990, when a single large parabolic trough system was installed at a California prison.

3.2   1975: Solar Market Development Begins

During the administration of President Richard Nixon, the National Science Foundation's RANN program (Research Applied to National Needs) recommended that solar energy become a meaningful contributor to the national energy supply. RANN funded the earliest solar "proof of concept experiments," essentially solar heating and hot water systems. The combination of these new projects, the impact of the 1973–1974 oil embargo and price hike and studies that predicted resource limits to future economic growth stimulated private sector interest in the solar energy field. When the federal government published projections that U.S. energy use would increase from 75 quads per year to anywhere from 115 to 160 quads by the year 2000 (ERDA 1975), interest intensified, swept along by concern for the nation's energy security, visions of vast new markets, and expectation that government programs in alternative energy represented significant business. Widespread public interest and enthusiasm for solar energy added further impetus to the growth of fledgling solar energy companies.

The first solar companies out of the gate sold pumped flat-plate collector systems for heating swimming pools, water, and building space. Companies entered from every direction. Architects who designed systems established new collector manufacturing firms, and scientists and engineers left existing employers to start their own businesses. Fortune 500 companies established solar divisions, and many existing small and medium-sized firms adapted existing products (e.g., refrigeration heat exchangers) or their manufacturing expertise (e.g., storm doors, automobile radiators) to produce solar collectors.

These companies faced a formidable marketing challenge. All the parts for a solar water heater cost a dealer about $1,500 in 1978 dollars. Adding installation, marketing, and overhead would raise the installed price of a system to $3,000–$4,500. At this time, a conventional gas or electric water heater cost a few hundred dollars. Solar energy thus had to overcome the perceived disadvantage of a high initial cost that is offset over time by very low operating costs. The solar sales force worked hard to convince consumers that energy prices would inflate over time and that investing in a solar energy system was a sound idea. Solar systems to heat swimming pools were easier to sell because they were more competitive and achieved simple payback in a only a few years, especially in climates with fairly long swimming seasons.

Lowering the cost of solar energy technologies was a high priority for government, which approached the problem from a wide technical perspective that extended from materials to performance. In 1975 the federal government's energy research budget for active solar was $12.5 million. A small amount of this was earmarked for analysis of solar thermal concentrator technologies at Sandia National Laboratories. Passive solar would not receive funding until 1977. From this beginning, the active, passive, and solar thermal concentrator budgets would rise to exceed $100 million per year.

With federal funding for solar energy already an important signal to companies, the role of the federal government became even more important when the administration of President Jimmy Carter took office.

Officials at the Federal Energy Administration (FEA) even discussed how the solar industry should be created and what shape it should take. Indeed, the first comprehensive solar product catalogue produced in the United States was prepared by J. Glen Moore of the Congressional Research Service for a congressional committee. The Carter administration also took an interventionist stance when it came to energy markets. Allocations and price controls would be enacted for conventional fuels, and alternative energy resources were subsidized with tax and regulatory incentives.

It was in 1975 that the Federal Energy Administration began gathering solar collector production figures. Initially, the data collection was haphazard. Many companies were missing, others would not reveal their production, and still others probably gave higher than actual figures to make themselves look better. For all its shortcomings, however, the FEA survey stood alone as the only source of such information, and it improved in time. The Department of Energy Information Administration has continued to gather these data.

Solar collector production reported for 1974, the first year tabulated, was 1,137,000 square feet of low-temperature (unglazed) collectors for pool heating, and 137,000 square feet of medium-temperature collectors. The report for pool heating should have been fairly accurate because a single firm, FAFCO, Inc., of Menlo Park, California, dominated this market. The square footage reported probably represented about 3,000 residential installations. The next survey, for 1975, reported production of 3,026,000 square feet of low-temperature collectors and 717,000 square feet of medium-temperature collectors. These figures represented 300% and 500% growth, respectively, and the new solar heating industry was well started on a rapid commercialization path.

3.3 Carter Administration

Before President Carter entered office, Congress had already enacted the key solar research, development, and demonstration legislation that would further stimulate the solar heating and cooling industry. President Carter believed strongly in alternative energy, and besides supporting more R&D, his administration emphasized technology demonstrations to hasten commercialization.

In 1977 the federal budget for active solar rose to $97.6 million from $52.7 million in 1976, and passive solar received its first R&D funding, $1.4 million. A large percent of the federal funds went for purchase of equipment for demonstrations of residential solar systems, through the Department of Housing and Urban Development (HUD), and of commercial systems, through the newly formed Department of Energy (DOE).

At this time, the companies manufacturing active solar heating equipment (cooling was part of the R&D program but never advanced significantly in the market) were searching for the best distribution strategy for their products. Local installers were very important because of the skill needed to install solar systems and because installation represented a significant portion of a systems' final cost. Because solar heating systems cost several times more than conventional water heaters, many collector manufacturers worked directly with dealer installers to eliminate middleman markups. Other firms decided that existing heating, ventilating, and air-conditioning industry distribution channels would be their preferred way to deliver products to consumers. Over time, in the largest markets, distribution companies became an integral part of the product delivery process for systems sold by jobbers, solar specialists, and plumbing companies. Still, a significant number of collector manufacturers built tremendous volume through direct sales to consumers.

In the market delivery process, packaging of all the components into a complete systems became an important approach for the industry. Market experiences indicated that reliability would be enhanced if systems were designed and delivered as a package to a single installer/dealer.

In 1977 it already appeared that the solar collector manufacturers, who dominated the Solar Energy Industries Association (SEIA), a national trade association organized in 1974, would form the political backbone of the solar energy industry. Consumers looked to the collector manufacturers for information, and the federal government treated them as the main players in the industry. Other important participants were the suppliers of key components (absorbers, tanks, pumps, controllers, and collector glazing material); in recognition of their role, SEIA ultimately formed a Suppliers Council in 1983. At the state level, dealers and installers specialized in solar energy also organized as a political force; they, not the collector manufacturers, would be most affected by government policies designed to protect consumers. Collector manufacturers joined the dealers to cooperate in the state-level industry associations.

In 1976 Congress began giving serious consideration to enacting federal solar tax credits as a subsidy to lower the cost of solar energy systems. New Mexico, followed by California, had already passed state tax credit incentives to aid the budding industry. The debate over and timing of the national incentives, however, had a severe impact on the solar industry. When it appeared Congress might pass such legislation, consumers stopped buying. This pause in market development deepened in 1977 and lasted over twelve months.

The solar industry complained strongly about the stalled market. After a major energy message President Carter gave in April 1977, the federal government responded with a grant to consumers of $400 per system for residential water heaters purchased in Florida and ten northeastern states. Only single-family residences were eligible for the grants, which were distributed through HUD. In addition, HUD created a standard that companies would need to meet if their systems were to be eligible for the incentive. As of January 1978, seventeen companies had produced systems that met HUD requirements.

This federal program ultimately allocated funds for 10,000 solar water heaters to the eleven states involved. At an average of 50 square feet per system, this may have accounted for 500,000 square feet of collector production, or about 10% of reported production in 1978.

In anticipation of a federal tax credit for solar energy systems, medium-temperature collector production actually declined in 1978. Finally, in April 1978, Congress did pass tax credits for solar energy in the National Energy Act. With subsequent adjustments to the law, consumers eventually received a credit equal to 40% of the cost of their systems. Swimming pool systems, however, were not eligible. Now, the active solar industry had in place the incentive that would build and sustain sales for the next seven years. At the time the Carter administration took as its goal the installation of 2.5 million active solar systems by 1987, but the industry would not reach that figure.

3.4 Reagan Administration

When President Ronald Reagan took office, it quickly became evident that his senior energy advisers did not believe solar energy should have priority. In addition, the president's own view of the proper role of government did not include support for the commercialization of new tech-

Table 3.1
Budget history, Office of Solar Heat Technologies, Department of Energy (millions of current year dollars)

	Active solar	Passive solar	Total
1975	12.5	0	12.5
1976	52.7	0	52.7
1977	97.6	1.4	99.0
1978	91.3	7.8	99.1
1979	90.4	23.9	114.3
1980	65.1	32.9	98.0
1981	41.4	30.2	71.6
1982	11.5	10.6	22.1
1983	6.6	5.0	11.6
1984	8.2	8.3	16.5
1985	4.8	5.1	9.9
1986			7.9
1987			6.0
1988			5.4
1989			6.1
1990			4.2
1991			2.0

Sources: U.S. House 1983; REI 1985; Sissine 1990.

nologies. Consequently, his administration reoriented the Department of Energy away from commercialization programs and toward the support of high-risk, long-range research and development. The new administration also tried several times (unsuccessfully) to repeal the solar tax incentives before they expired, as planned, in 1985. The solar industry did succeed in getting Congress to extend the solar tax incentive for commercial systems for three years, but Congress lowered the percent of system cost eligible for the credits each year until it reached 10%.

The Reagan administration cut the federal R&D and demonstration funding for active solar from $65.1 million in 1980 to $41.4 million in 1981, and then $11.5 million in 1982. The DOE passive solar budget went from $32.9 million in 1980 to $10.6 million in 1982 (table 3.1). By 1985 funding for both dropped to a total of $9.9 million, reflecting the new policy as well as the perceived commercial status of the technology. The amounts would fall still further, until the two programs were combined into "solar buildings" and transferred to the energy conservation branch of the Department of Energy. By this time, funding had dropped to about $2 million per year. To repeat, the Reagan administration opposed government actively commercializing renewable energy technologies.

3.5 Bush Administration

The administration of President George Bush had an outlook on energy similar to that of President Reagan. In global terms, the administration favored development of the oil and gas industry, continued reliance on coal, and a revival of the nuclear power industry. At the Department of Energy, the new administration revised priorities in the renewable energy program, placing even less emphasis on active solar and passive solar R&D. The programs were combined into "solar buildings" and moved to the energy conservation branch. Solar concentrator technology research and development, however, was raised in priority, and substantial cuts in the budget for central receiver and parabolic dish technology ended after 1989.

After a time, the Bush administration came to support extension of the commercial solar energy tax credit. This credit equaled 10%, and each year from 1988 to 1992 Congress extended it for a one-year period.

How the active solar heating segment of the industry benefited from government R&D programs is described in volumes 5 and 6 of this series. It should be noted that—aside from the R&D—industry also had extensive contact with the federal government through participation in demonstration programs and the government's interest in setting standards for solar equipment and installations. Clearly, for the active solar industry, it was the government's tax incentives that had a major impact on market development.

3.6 Active Solar Thermal Market Development

3.6.1 Low-Temperature Systems for Pool Heating

When the active solar industry began to form in the 1970s, one of the very first markets to develop was for swimming pool heaters. Pool heating was one of the most efficient applications of flat-plate collectors made with unglazed plastic, rubber, and metal flat-plates because the operating temperature was below 100°F. Furthermore, pools provide energy storage

Table 3.2
Solar collector manufacturing (thousands of square feet)

	Low-temperature	Medium-temperature
1974	1,137	137
1975	3,026	717
1976	3,876	1,925
1977	4,743	5,569
1978	5,872	4,988
1979	8,395	5,857
1980	12,233	7,165
1981	8,677	11,456
1982	7,476	11,145
1983	4,853	11,975
1984	4,479	11,939
1985	NA	NA
1986	3,751	1,111
1987	3,157	957
1988	3,326	732
1989	4,283	1,989

Source: U.S. Department of Energy, Energy Information Administration, *Solar Collector Manufacturing Activity*, annual reports.
Note: Because importers were underrepresented in early EIA surveys, the total U.S. market was somewhat larger.

and a pump at no cost. For this application, collector manufacturing companies established a solid business base founded on direct sales to consumers and to dealers (see table 3.2).

As the U.S. solar industry started up, the pool heater manufacturers produced more collector area than medium-temperature manufacturers for seven years, but they found themselves on an equal footing in terms of gross sales volume fairly quickly. The reason the relative gross sales revenue for these two products changed is that pool heating systems are much larger than domestic hot water DHW systems but much less expensive per square foot installed. The factory sales price of a glazed collector was at least three to four times higher than most unglazed collectors. Consequently, the dollar sales volume of medium-temperature systems quickly overtook low-temperature systems. Thus, while the square footage of medium-temperature collectors produced did not exceed low-temperature production until 1981, gross sales revenue attributed to medium-temperature systems was probably equal to that generated by

Table 3.3
Average sizes of active solar installations by application and building type (square feet of collector area)

Building type	Pool heating	Water heating only	Space heating only	Water and space heating
Single-family	376	61	287	224
Multifamily	820	280	312	390
Commercial	1,212	517	340	1,154

Source: EIA 1982.

pool system sales as early as 1975, and by 1980 it exceeded income from pool system sales by a factor of 5. From an industry standpoint, especially for new entrants, it now made more sense to be in the residential water and space heating business.

A national survey of solar installers conducted for DOE collected data that established the average size of active solar installations in 1981. In colder climates, a typical solar water heater would use between 55 and 74 square feet of collectors. In the South, this figure would range from 40 to 65 square feet. Table 3.3 lists the average size of solar energy systems in the residential and commercial sectors at that time.

Another contributing factor to the sales revenue differential between low-temperature and medium-temperature collectors came from the market pull for solar water heaters created by the federal tax incentives. While California state tax credits applied to pool systems, the federal tax credits did not. The federal R&D and demonstration programs also emphasized solar systems for residential energy needs. Finally, sales campaigns in the largest market, California, were diverted toward domestic water heating by a state-mandated utility program. This California initiative stimulated installation of 200,000 solar water heaters, and many solar companies abandoned pool heating technology to capture these more lucrative sales.

This market relationship changed drastically in 1986. The 40% residential tax credit available from the federal government expired at the end of 1985, and sales of medium temperature systems fell by as much as 90%. Since swimming pool systems were unsubsidized, this segment of the industry stabilized at a production level between 3 and 4 million square feet per year from 1986 to 1988, then grew in 1989 to more than 4 million square feet. Although one-third the level of the best year—1981 production of low-temperature panels was more than 12 million square feet—the swimming pool heating market has provided this segment of the solar industry with steady annual growth. In the late 1980s, SEIA surveys for participating manufacturers showed it to be near 20% per year.

3.6.2   Air versus Hydronic Collector Production

Installation of solar energy systems that use a liquid heat transfer medium (a hydronic system) has always exceeded air-type systems. One reason for the disparity was a perceived inefficiency of using air-based collector systems solely to heat water. In addition, two geographic factors were at work. In the northeastern United States, where energy costs were high, most homes used hydronic rather than forced-air heating systems. In California and in warmer southern climates where space heating is less needed, the solar industry concentrated on selling water heaters. As a result, space heating dropped from 40% of the solar market in 1978 to 24% in 1982.

Then, in 1984 and 1985, air collector production increased dramatically. A new concept arose in the South and Midwest: the hot-air, no-storage-space heating system. Solar heat from rooftop panels was directed during daytime hours to the most lived-in areas of a residence. Intense in-the-home selling techniques used by the companies that opened this market tripled production of air collectors from 1983 to 1984. Air collectors rose to 16% of medium-temperature collectors produced, compared to only 11% and 7% the two previous years. At the same time, medium-temperature liquid collector production actually declined (see table 3.4.)

When the active solar heating market shrank in 1986 after expiration of the federal residential solar tax credits, production of both air and hydronic collectors fell by more than 90%. When the market began to revive in 1989, the proportion of air to liquid collectors produced was similar to that in 1982; that is, about 10% of flat-plate collectors produced were air-type.

3.6.3   Thermal Applications

The main market for solar energy systems has been in the single-family residential sector. More than 94% of active solar installations in 1981 were for single-family residences, mainly for water heating. Before the tax credits expired, the vast majority of systems were installed on existing buildings. For example, in 1981, DOE determined that 86.4% of solar water heating installations were retrofit (EIA 1982). Little data collected

Table 3.4
Annual shipments by collector type (thousands of square feet)

	Medium-temperature air	Medium-temperature liquid[a]
1977	726	3,080
1978	723	3,668
1979	846	4,462
1980	649	6,022
1981	436	10,572
1982	1,134	9,357
1983	774	10,584
1984	1,885	9,743
1985	NA	NA
1986	91	1,014
1987		
1988	60	670
1989	202	1,785

Source: U.S. Department of Energy, Energy Information Administration, *Solar Collector Manufacturing Activity*, annual reports.
[a] Medium-temperature liquid includes thermosiphon, integral collector storage, and evacuated tubes in addition to flat-plate collectors for pumped systems.

since 1985 would indicate that the relationship between new construction and retrofit installations has changed very much.

The market for commercial or industrial solar thermal systems requiring higher temperatures has never developed fully. At 15%, federal tax incentives for commercial systems were not high enough to overcome return on investment barriers in the commercial and industrial sector. Solar development companies turned to leasing and energy sales contracts to reach commercial markets using third-party investors that could benefit from the tax incentives. Often, medium-temperature flat-plate collectors could furnish the temperatures needed. When natural gas prices began to decline with oil prices, the commercial and industrial market became even more difficult to open. As a result, industry production of higher-performance concentrating or evacuated tube collectors for building, commercial, or industrial applications never reached a significant level and, as shown in table 3.5, actually declined somewhat before the residential tax credits expired. Some manufacturers did offer evacuated tube or concentrating collectors for small residential water heating and space heating systems, but these never achieved very much market share.

Table 3.5
Annual shipments by collector type (thousands of square feet)

	Concentrators[a]	Evacuated tubes	Total collectors
1978	225	141	10,635
1979	269	229	14,251
1980	325	145	19,398
1981	272	172	20,133
1982	552	84	18,621
1983	435	78	16,828
1984	168	98	16,419
1985	NA	NA	
1986			
1987			
1988	W	4	4,059
1989	W	W	6,271[b]

Source: U.S. Department of Energy, Energy Information Administration, *Solar Collector Manufacturing Activity*, annual reports.
[a] Figures exclude Luz International trough concentrators used for utility solar thermal electric projects.
[b] Figure excludes concentrators, evacuated tubes, and high-temperature concentrators. DOE withholds data to avoid disclosure of individual manufacturer totals.
W = Withheld to avoid disclosure of individual company data.

3.7   Solar Market Location

The location of collector manufacturers in business in the 1970s revealed the three geographic areas where the new industry concentrated: (1) The industrial corridor from Boston to Baltimore, (2) Florida, and (3) California. Very few manufacturers were located in the Midwest. A scattering of companies existed in southern states, with most of these in Texas.

It appears the solar manufacturing industry first located in those regions where industry traditionally existed and where markets were expected to grow fastest. In time, a dozen or so companies successfully established national marketing networks, but for the most part, collector manufacturers and system packagers tended to be located in the region where they generated most of their business. Transportation costs for solar collectors often affected manufacturer participation in geographic markets. State boundaries have also influenced market presence for companies. For example, out-of-state firms have had a difficult time establishing themselves in the Florida market.

In 1984 companies located in California, New York, Florida, New Jersey, and New Mexico produced 97% of low-temperature collectors. For medium temperature collectors produced in 1984, the top five companies were located in California, Florida, Colorado, Arizona, and Ohio. Companies headquartered in these five states accounted for 71% of medium-temperature collector production.

In 1984 California alone accounted for 40% of the low-temperature and 33% of the medium-temperature production. New York held 25% of the low-temperature market, but this was primarily because a leading producer of low-temperature collectors was located there. Florida accounted for approximately 18% of each market; this level indicates the overall production of Florida manufacturers, who held tightly to the market in their own state. Table 3.6 summarizes low- and medium-temperature collector production by the location of company headquarters.

After the residential solar tax credits expired, the solar market remained strongest in states with good solar insolation, high electricity costs, and strong demand for solar swimming pool heaters. Table 3.7 shows the reported destination by state of solar collectors produced in the United States in 1989. These data represent a change in the Energy Information Agency questionnaire from that used in the early 1980s, which reported the company headquarters only. The top four states receiving solar collectors were California, Florida, Arizona, and Oregon. Puerto Rico ranked between Florida and Arizona.

With regard to 1989 production, solar thermal collectors were manufactured in seventeen states. California was the largest producer, with 7,360,000 square feet, which includes about 5 million square feet parabolic troughs produced by Luz International for its solar energy generating station (SEGS) electric power plants. The next three states—New York, Tennessee, and Florida—shipped a total of 2.26 million square feet.

3.8 Sellers

3.8.1 Types of Companies

The active solar energy industry consists of a wide variety of companies, including manufacturers, research firms, distributors, manufacturers representatives, and dealer/installers or maintenance service companies. The

Table 3.6
Annual shipments of solar thermal collectors by company headquarters and collector type, 1982–1984 (thousands of square feet)

	Low-temperature			Medium-temperature[a]		
	1982	1983	1984	1982	1983	1984
Alabama	200	0	W	24	130	W
Arizona	151	15	20	707	875	747
California	3,196	3,071	1,798	2,880	3,508	3,927
Colorado	0	W	9	1,019	W	809
Connecticut	0	W	2	35	27	198
Florida	1,802	96	811	2,074	2,790	2,282
Georgia	0	14	0	27	11	W
Hawaii	0	0	0	26	19	104
Illinois	15	78	W	17	5	W
Indiana	W	W	0	W	W	39
Iowa	0	24	24	57	69	184
Maine	W	W	0	W	W	45
Massachusetts	0	0	0	163	W	49
Michigan	0	0	3	64	W	476
Minnesota	180	W	W	122	41	W
New Hampshire	0	0	W	27	W	W
New Jersey	696	536	550	443	155	90
New Mexico	264	W	71	319	237	269
New York	810	891	1,132	861	1,077	323
North Carolina	0	0	0	250	W	50
Ohio	5	7	13	99	245	667
Oregon	0	0	W	83	65	W
Pennsylvania	1	W	0	173	123	94
South Carolina	0	W	0	20	16	W
Tennessee	33	11	0	332	144	W
Texas	18	W	5	403	594	471
Virginia	11	W	30	372	389	337
Wisconsin	0	0	0	28	W	9
U.S. total[b]	7,476	4,853	4,479	11,145	11,975	11,939

Source: U.S. Department of Energy, Energy Information Administration, *Solar Collector Manufacturing Activity*, annual reports.
Note: Company headquarters may not be site of manufacturing facility.
[a] Includes medium-temperature, special, and other thermal collector types.
[b] Totals include shipments for states in which information was withheld to avoid disclosure of individual company data.
W = Withheld to avoid disclosure of individual company data.

Table 3.7
Destination of solar thermal collectors by state or territory, 1989 (square feet of total collector production, including parabolic troughs for electric power projects)

Alaska	0	Nebraska	0
Alabama	4,372	Nevada	27,216
Arizona	199,016	New Hampshire	387
Arkansas	0	New Jersey	47,653
California	6,907,148	New Mexico	41,076
Colorado	12,891	New York	49,366
Connecticut	58	North Carolina	1,465
Delaware	0	North Dakota	0
District of Columbia	0	Ohio	7,286
Florida	3,117,225	Oklahoma	256
Georgia	2,270	Oregon	87,525
Hawaii	82,192	Pennsylvania	10,418
Idaho	0	Puerto Rico	231,463
Illinois	25,541	Rhode Island	35
Indiana	0	South Carolina	310
Iowa	192	South Dakota	0
Kansas	0	Tennessee	7,286
Kentucky	0	Texas	38,195
Louisiana	0	Utah	699
Maine	24,598	Vermont	67
Maryland	37	Virgin Islands	0
Massachusetts	8,833	Virginia	26,937
Michigan	19,306	Washington	309
Minnesota	0	West Virginia	310
Mississippi	0	Wisconsin	9,780
Missouri	25,320	Wyoming	4,372
Montana	0	Total	11,020,910

Source: EIA 1991.

bulk of the companies involved are engaged in distributing and installing complete systems. Some companies design, manufacture, sell, *and* install their systems. Another key group are the suppliers of major components and parts to the system manufacturers, packagers, and installers; in 1985, forty of these firms formed the Solar Energy Industries Association Suppliers Council.

In the 1970s the new solar industry tried several distribution paths to the market. The most successful firms were collector and system manufacturers that offered their products directly to dealer/installers, who sold and delivered the technology to the end users. These companies packaged complete systems, buying what they did not make themselves from original equipment suppliers; they also benefited from nearly direct customer feedback. There were other companies that sought to establish classical distribution channels typical in the heating and ventilating industry and that expected a stocking wholesale distributor to inventory their products and deal with the dealers and installers. A few companies attempted to work closely with homebuilders, establishing solar water heating as an option for new construction by including it in the home mortgage. In general, however, the industry has never really been organized along the lines of the mature HVAC (heating, ventilating and air-conditioning) distribution chain.

In the early 1980s direct sales marketing to homeowners became a key to marketing success, and firms that trained salesmen and aggressively marketed to end users gained prominence in the industry. Selling relied heavily on government tax incentives, and the system prices for a water heater often ranged between $4,000 and $5,000, which included an adequate amount for sales commissions. The largest marketing companies would manufacture part of their systems and purchase the rest from original equipment manufacturers. When residential tax credits expired in 1985, the system prices had to fall substantially for sales to still occur. Margins tightened substantially, and sales commissions shrank. One of the major problems for the industry after the tax credits expired was retention of a sales force, and this part of the technology distribution system nearly vanished.

Dealers and installers remaining in the market after 1986 found that servicing systems already in place became a very important part of their business. They witnessed a dramatic drop-off in sales; the whole industry shrank to less than 20% of its size in 1985. Many sales and service

Table 3.8
Companies engaged in activities related to solar collector manufactuing

Activity besides collector production	1978	1984	1989
Wholesale distribution	188	186	26
Retail distribution	149	135	24
System design/consulting	173	130	30
Installation	135	115	21
Manufacturing of other system components	107	95	11
Prototype development	238	103	11
Total companies reporting to DOE	340	224	44

Source: U.S. Department of Energy, Energy Information Administration, *Solar Collector Manufacturing Activity*, annual reports.

companies specializing in solar work realized their survival would depend on expanding their activities to include other energy conservation products. Across the country, most traditional plumbing companies that had entered the solar market were quick to leave it when sales fell off dramatically.

As noted above, despite the crucial role of the dealer, the installer, and the service company, collector manufacturers have been regarded as the core of the active solar industry. In 1979 the Department of Energy identified 349 solar collector manufacturers. This number fell to 224 by 1984; some firms fell by the wayside because they were not as skilled as others at selling or had products that did not remain competitive. DOE reporting methods have varied over time, and it became clear at one point that the Energy Information Administration (EIA) was missing the segment of the industry installing integral collector storage and/or thermosiphon systems. Consequently, some of the data are not reliable indicators of what was happening in the industry, especially when comparing 1983–1984 with the 1970s. By 1989, only 44 manufacturers reported collector production to DOE.

Table 3.8 reveals that solar collector manufacturers have been engaged in a wide range of activities besides the manufacture of collectors, from design to actual installation of their equipment in systems. The major difference that exists between 1984 and 1989 is that in 1989 a much smaller percentage of the manufacturers are involved in prototype development, and a larger percentage are involved in system design. That some manufacturers still perform wholesale and retail distribution tasks is a constant from the beginning of the industry.

3.8.2 Size of Companies

Collector manufacturing remains a small business. The large integrated firms that created solar collector manufacturing divisions all quit the industry after 1985. In 1989 the 44 remaining manufacturers reported to DOE that they employed 900 person-years in their solar-related activities, 36 of these manufacturers employed 10 or fewer people, and 27 companies said they had 2 or fewer employees in solar-related activities.

The largest collector manufacturing companies rarely had gross sales over $20 million unless they were also directly involved in the downstream business of delivering the technology to the end users. The largest manufacturing company to ever exist in the active solar industry peaked with annual gross sales between $70 and $80 million; using a large captive sales force, it built its sales on the direct marketing concept. More typical companies involved were likely to have annual sales of less than $10 million. These sales fell dramatically, as much as 80% to 90% for the industry as a whole, after federal residential tax credits expired. Indeed, the largest medium-temperature collector manufacturing and sales company in the industry went out of business in 1986.

The collector manufacturing industry has also exhibited a high degree of market concentration. In 1984, ten companies manufactured 48% of the collectors produced in the United States. Table 3.9 indicates market shares from 1980 to 1984 for the ten largest solar collector manufacturers prior to expiration of federal solar tax credits at the end of 1985.

Table 3.9 includes all solar collector manufacturers. The low-temperature market was significantly more concentrated than the medium-temperature market. In 1984, five companies produced 86.1% of the total market for low-temperature collectors. In the same year twelve medium-temperature manufacturers had 48.5% of the market.

In the late 1980s a heavy concentration in flat-plate collector manufacturing remains. According to DOE figures for 1989, the top four companies would have been responsible for producing 71% of low-temperature and medium-temperature collectors. Counting Luz International production of parabolic troughs, the top ten companies shipped 96.4% of collectors in 1989.

Sales revenue per firm in the early 1980s can be estimated from 1984 data. If 3 million of the 7.925 million square feet produced by the leading ten manufacturers were low-temperature collectors that retailed for

Table 3.9
Annual shipments and market shares of the ten largest manufacturers of active and passive solar collectors

	Rank	Collectors shipped (1,000 ft^2)	Percentage of industry[a]
1980	1–5	9,811	50.5
	6–10	2,650	13.7
1981	1–5	7,129	35.4
	6–10	2,771	13.8
1982	1–5	6,320	33.9
	6–10	2,560	13.7
1983	1–5	5,919	35.2
	6–10	2,752	16.4
1984	1–5	5,320	32.4
	6–10	2,605	15.9

Source: U.S. Department of Energy, Energy Information Administration, *Solar Collector Manufacturing Activity*, annual reports.
[a] Total percentages may not equal sum of components due to independent rounding.

$10 per square foot, and if medium-temperature collectors retailed for $16 per square foot (1984 dollars), and if manufacturers received 30% of the list price for low-temperature collectors and 60% medium-temperature collectors, then the total sales in 1984 for the top ten collector manufacturers would have equaled $57 million, an average of $5.7 million per firm. Actual revenue would have been much higher for any of these companies involved in delivering and installing complete systems to the end user because costs for installed pool heaters ranged between $20 and $30 per square foot and for installed domestic water heaters between $40 and $75 per square foot.

By 1989, the production picture had changed dramatically for manufacturers of medium-temperature collectors. These firms produced about 2 million square feet of collectors, half the amount of metallic and nonmetallic low-temperature collectors. The value of the medium-temperature collector shipments reported to DOE was $14,857,975; the value for low-temperature collectors was $11,157,280, a figure relatively close to sales in 1984 for the top companies. DOE statistics for 1989 indicated that four companies were responsible for 71% of production, meaning that annual sales per company for the top flat plate manufacturers in 1989 averaged $4.62 million. (1989 dollars).

In general, although the leading companies involved in the active solar industry achieved a profitable position, those companies largely dependent on solar sales for their income made the most successful solar businesses. The solar divisions and subsidiaries of larger corporations never made enough money for senior managers to justify remaining in the solar business, and these firms disappeared from the industry.

The companies involved downstream in the distribution network, the system installers and maintainers, were very small; in the early 1980s, many dealer and service firms would install fewer than 25 systems a year. In 1981, of the installation companies surveyed by the Energy Information Administration (EIA), 40.8% reported no installations for the year, 41.6% reported between 1 and 25 installations, 17% reported between 25 and 1,000 installations while only 0.5% reported more than 1,000 installations. At that time, 74.4% of the installers derived 25% of their income or less from their solar business. Growth of solar marketing companies with their own sales outlets in the years 1980–1984, however, probably meant that the number of installers with several hundred to several thousand systems installed per year increased, although there is no way to support or document such a statement.

3.9 Sales

3.9.1 Federal Government Sales

Sales of solar heating systems to government accounted for a modest amount of the sales volume of the industry. An Electric Power Research Institute (EPRI) study found that systems sold for residential solar energy demonstrations supported by the Department of Housing and Urban Development (HUD) in its first two cycles and to the Energy Research and Development Administration (ERDA) for the first cycle of the commercial demonstration program used 262,608 square feet of medium-temperature collectors (DHR, Inc. 1980). Related to medium-temperature collector production reported by the FEA/EIA for 1977, HUD- and ERDA-supported sales amounted to about 7% of total production for that one year. In sum, data suggest that the federal government bought between 313,000 and 424,000 square feet of collectors a year between 1978 and 1981, a figure less than 3% of the total volume of collectors shipped.

Table 3.10
Solar collector manufacturing dollar sales volume (millions of 1984 dollars)

	Low-temperature[a]	Medium-temperature[b]	Total
1974	12.5	10.3	22.8
1975	33.3	53.8	87.1
1976	42.6	144.4	187.0
1977	52.2	417.7	469.9
1978	64.6	374.1	438.7
1979	92.3	439.3	531.6
1980	134.6	537.4	672.0
1981	95.4	859.2	954.6
1982	82.2	835.9	918.1
1983	53.4	898.1	951.5
1984	49.3	895.4	944.7

Source: SEIA estimates based on U.S. Department of Energy, Energy Information Administration, *Solar Collector Manufacturing Activity*, annual reports.
[a] Assuming average cost of \$3,400–\$3,500 and average area of 320ft^2 per system, or $11/\text{ft}^2$.
[b] Assuming average cost of \$4,500 and average area of 60ft^2 per system, or $75/\text{ft}^2$.

The impact of government-supported demonstrations exceeded the small level of production absorbed. These demonstrations helped solar companies establish credibility in the market, and government support of solar energy encouraged consumers to consider buying solar energy systems. Later, critics would contend that too many failed systems from new companies in these demonstrations damaged the technical credibility of the technology, particularly for larger, commercial-scale systems.

3.9.2 Private Sector Sales

As total sales of solar collectors grew steadily between 1974 and 1981, low-temperature collectors, primarily for pool heating, led the way in the early years. Then the market for pool heaters fell off and increasing sales of medium-temperature systems for domestic hot water (DHW) overtook low-temperature system sales.

Table 3.10 indicates market growth in the 1970s and early 1980s of U.S. manufacturers in 1984 dollars. Three factors should be taken into consideration in interpreting these tables. First, until 1984, the Energy Information Administration only reported on the activities of companies that manufactured equipment in the United States and did not include companies that imported equipment. Second, many independent entre-

Table 3.11
Active solar installations (number of systems)

	Pool heating	Hot water	Space heating	Space cooling
1978	21,500	40,300	6,800	700
1979	22,200	47,400	6,700	1,000
1980	31,900	76,800	6,600	600
1981	25,400	115,500	7,900	1,700
1982	18,300	119,400	9,200	300
1983	12,600	149,500	8,100	100
1984	11,500	143,200	9,200	50
Total	142,800	692,300	54,400	4,500

Source: Derived from Energy Information Administration data.

preneurs producing only small numbers of collectors were most likely not included in the EIA data. And third, the 1984 survey made an extensive effort to identify missing manufacturers, resulting in the addition of 71 companies, which makes it difficult to compare figures for collector shipments and industry trends.

The assumptions in table 3.10 came from data compiled in surveys of installations and manufacturer reports (for example, in 1981, the average collector area nationally for solar water heaters was 61 square feet); they reflect systems typical of many installations and were also used to derive the number of active solar installations by application. These estimates are provided in table 3.11.

Figures on the total number of active solar energy installations in the United States vary. The Solar Energy Industry Association points out that more than one million systems have been installed since the mid-1970s. A gap in data exists because DOE only conducted its survey of 1985 data in 1987. Because many firms had quit the business after 1985, it could only estimate production for 1985, which it set at a total of 11.1 million square feet. This figure equals two-thirds the total production for 1984, which suggests that the market had begun to fall off even before the federal residential tax credits expired. Nonetheless 1985 installations and those made in subsequent years should bring the total number of active systems deployed to more than one million.

After the market dropped off, it became much more difficult to track active solar installations by application. With a small number of

Table 3.12
Applications of active solar collector panels produced in 1989 (percentages)

	Low-temperature panels	Medium-temperature panels
Residential pool heating	91	20[a]
Commercial pool heating	9	
Water heating		69
Space heating		10

Source: EIA 1991.
[a] Residential and commercial combined.

manufacturers reporting data, the Department of Energy withheld information that might disclose company specific facts. The Solar Energy Industries Association (SEIA) began collecting quarterly data from its members in 1987, however, not all the companies included in the federal government survey participated. Data available (see table 3.12) indicate the following about the market in 1989:

• Residential solar pool heating continued to be a major application, accounting for 91% of low-temperature flat-plate collector production. The remainder are primarily for commercial pool systems, and in a few cases, low-temperature process heat systems.

• Pool heating also absorbed 20% of the liquid flat-plate medium-temperature panels produced.

• Solar water heating accounted for the bulk of the remaining market, absorbing 69% of medium-temperature collectors produced. About 16% of collector panels used for water heating were manufactured in thermosiphon-type systems.

• Air-type panels, probably used for space heating systems, accounted for the remaining 10% of medium-temperature collectors produced.

The residential sector has always been the primary market for the active and passive solar energy industry. Residential systems accounted for between 70% and 85% of the active solar collector market from 1978 to 1984 (see table 3.13). After 1985, this sector's share actually rose to almost 93%. With oil and natural gas prices declining in the 1980s, the industrial and commercial sectors were simply not interested in solar energy systems.

Table 3.13
Annual shipments by market sector (thousands of square feet)

	Residential	Commercial	Industrial	Agricultural	Other	Total
1978–1984	90,821	17,760	4,997	787	1,687	116,142

Source: U.S. Department of Energy, Energy Information Administration, *Solar Collector Manufacturing Activity*, annual reports.

Table 3.14
Solar collector imports and exports (square feet)

	Low-temperature imports	Low-temperature exports	Medium-temperature imports	Medium-temperature exports	Total imports	Total exports
1978	227,577	761,074	168,017	78,805	395,594	839,879
1979	62,480	711,100	227,925	143,790	290,405	854,890
1980	35,805	899,249	197,788	215,809	235,019	1,115,058
1981	30,662	607,000	165,200	165,000	196,000	771,000
1982	W	318,000	W	137,000	455,000	455,000
1983	W	W	W	W	511,000	159,000
1984	W	230,000	W	118,000	621,000	348,000

Source: U.S. Department of Energy, Energy Information Administration, *Solar Collector Manufacturing Activity*, annual reports.
W = Withheld to avoid disclosure of individual company data.

3.10 Imports and Exports

Before 1985, imported solar collectors accounted for less than 4% of the U.S. market; exports in the early 1980s absorbed less than 10% of U.S. collector production. Although the Energy Information Administration compiled export and import figures in their survey, it failed to survey the companies that imported but did not manufacture collectors in the United States. As a result the figures in table 3.14 understate the import situation, especially because many integral collector storage and thermosiphon system manufacturers were left out of the survey. Israeli and Australian firms, in particular, established important market shares in the United States for these products.

DOE data for 1989 indicate that solar thermal collector imports totaled 1.23 million square feet, which amounted to 11% of total collector shipments in the United States (EIA 1991); these came from four companies in four countries: United Kingdom, Australia, Japan, and Israel. All but 33,000 square feet of the imports were low-temperature collectors.

Export data for 1989 show that, of 461,000 square feet sent abroad, 379,000 square feet were medium-temperature collectors. The balance— 82,000 square feet—were low-temperature panels. Six companies reported exports to fourteen different countries. Foreign markets were not nearly as important for solar thermal collector industry as they were for the photovoltaic industry, where, for some companies, exports represented an imporant source of income.

3.11 Review of Active Solar Industry

3.11.1 Market Size

Annual sales to end users in the market for active solar energy systems reached a peak in 1984, between $700 million and $1 billion. The cumulative number of active solar energy systems installed in the United States probably exceeded 1 million, dominated by pool heating and residential domestic water heating systems. Commercial systems never accounted for more than 10% to 15% of industry sales. The most rapid market growth occurred in the early 1980s, then sales leveled off until 1985; after the residential solar tax credits expired, sales of residential water heaters and space heating systems fell dramatically.

Department of Energy collector production data for 1989 suggest that gross sales for the active solar collector manufacturers amounted to $26 million. If low-temperature systems were installed at $20 per square foot of array, and medium-temperature systems at $50 per square foot of array, then the gross 1989 sales for installed active solar systems would total approximately $140 million.

In a few areas in the United States, market penetration levels in excess of 10% have occurred; in others practically no market for active solar systems exists. With somewhat over 90 million residential dwelling units in the United States, the potential market size is large. Three key limiting factors are lack of consumer knowledge, the still high costs of solar energy compared to conventional energy systems, and the declining prices for conventional energy through the late 1980s, with a resultant decline in interest in conserving energy.

3.11.2   Technology

When the solar industry hit its peak in 1984, production of flat-plate glazed collectors, labeled "medium-temperature," for DHW systems dominated in the market. Production of nonmetallic low-temperature collectors peaked in 1980, and then fell annually through 1987. Production started to rise modestly after that, showing increases of 15% to 20% per year. By the late 1980s, the total area of low-temperature panels produced was double that of glazed, medium-temperature flat plates.

In the 1970s some manufacturers developed components so that solar collectors could be assembled at the installation site. Experience proved that field labor was too expensive and quality control inadequate. Although site-built collectors are low-cost and appeal to the do-it-yourself market, they are also lower in efficiency. By 1981 well over 90% of all collectors installed were shipped assembled from the factory.

What are described as passive systems (integral collector storage and thermosiphon) but are counted as part of the active solar industry rose to a significant level of market penetration in those areas where freezing was not a concern. Small air collector–based space heating systems installed without any thermal storage also began to create a market, principally in the Midwest and Southwest. The growth of this market segment appears to be driven by marketing, rather than by technology or economics.

Sales of water heating systems for residences supported by federal, and often state, tax incentives or utility subsidies were the key to the industry's rapid rise. According to undocumented reports from California in the early 1980s, thermosiphon or integral collector storage (ICS or batch-type) water heaters accounted for 40% to 50% of solar installations; the remainder were the pump-equipped systems from which the name "active solar" came.

3.11.3   Market Location

In the early 1980s California and Florida were the two states with the largest markets for active solar energy systems. Other substantial markets existed in New England, where energy costs and dependence on imported oil were high, and in the Southwest, where high insolation levels increased the economic output of solar energy systems. DOE collector shipment destination data for 1989 show that, as given in table 3.7, California appeared to be the largest market for active solar systems; since 5 million of 6.9 million square feet of collectors delivered to California that year were parabolic troughs, the active solar market in Florida was larger. Shipments of collectors to Florida totaled 3.1 million square feet. The next three largest markets, Arizona, Oregon, and Hawaii, received only 199,000, 88,000, and 82,000 square feet of collectors, respectively (EIA 1991, 7).

3.11.4 Government Role

As noted, direct government purchases supported early industry growth by helping new companies demonstrate their technology and gain market. Additionally, the field experience uncovered faulty installation and design practices for some versions of active solar technology, which, when corrected, led to better system performance and durability. Counted against total industry production, the level of government direct purchases was quite small.

The major impact government had on active solar market development was implementation of federal and state tax incentives. These provided a major marketing tool for the industry, affecting the delivered price of the systems and the means by which active solar energy technology was sold to the public. Much of the growth in sales in 1979 through 1981 can be attributed to the federal tax credits. The slim growth in 1978 can be attributed in part to the uncertainty surrounding the tax credits, as purchases were put off until the situation was resolved in 1979. Later, falling energy prices and lack of public concern about energy prices leveled collector production. Once tax credits expired at the end of 1985, industry sales plummeted.

In the early 1980s, when energy crisis, economic stagnation, and high inflation had consumers worried about the future, much of the solar industry used door-to-door direct sales, telemarketing, and other such labor-intensive, high-pressure, direct marketing approaches to reach consumers. Walk-in sales of solar energy systems have always been limited, Sears and other well-known stores carrying solar water heaters have never created a land rush business; public awareness of the technology has remained limited. As a consequence, the active solar industry had to actively solicit sales prospects; this marketing approach was expensive. Unfortunately, the prospect of sizable sales commissions attracted many high-pressure, fly-by-night outfits interested in quick profits, firms that failed to provide customers long-term support and that approved substandard installations. This dampened public confidence in the active solar industry and put an added heavey burden on firms committed to the business that were trying to introduce a new product with no decided economic advantage.

To generalize, the market development of the active solar industry from the mid-1970s through the mid-1980s was characterized by industry push rather than consumer pull. This accounted for the predominance of single-stage marketing over two-stage marketing, which requires consistent consumer demand. Once the federal tax credit incentives for consumers of solar energy systems ended, the solar industry went through a major adjustment, with sales declining industrywide by 80% or more. Companies in the pool heating segment of the industry, however, held market position well and would see sales continue to rise gradually. The domestic water heating companies were hard hit, and about three-quarters of the collector manufacturing firms active through 1985 disappeared from the business. Only in the late 1980s, as public concern about the environmental impact of using fossil fuels began to seriously affect energy decision making could the solar industry anticipate a more receptive market. Other volumes in this series provide the detail on the technologies, their cost and performance, and on various government solar energy programs.

3.12 Solar Thermal Concentrator Markets

3.12.1 Introduction

Concentrating collector technologies operate at much higher temperatures than active solar heating systems, which typically employ flat-plate collectors. The collectors in high-temperature solar thermal systems track the sun with mirrored (or possibly lens-covered) surfaces to concentrate solar radiation onto a receiver. In the solar flux at the receiver, temperatures from 150° to more than 2,700°F can be achieved, depending on the type of concentrator. This thermal energy can be used directly, converted to steam or another working fluid to produce electricity in a turbine, or used to power chemical reactions, produce products, or eliminate waste materials.

Three basic types of concentrating solar thermal systems exist: central receiver, parabolic trough, and parabolic dish. A central receiver system is labeled "central" because a field of heliostats (two-axis tracking mirrors) focuses its energy on a single, tower-mounted receiver. Such systems are inherently larger in scale, suitable for utility or large industrial applications. Trough and dish systems are referred to as "distributed systems" because the collector field consists of a large number of individual units, each with its own receiver, and the applications are likely to be in smaller, distributed systems, rather than in centralized utility power stations.

High-temperature solar thermal (referred to hereafter as "solar thermal") technology has not yet succeeded in the market like the consumer-oriented flat-plate active solar technology. Solar thermal systems are more suitable for commercial, industrial, and utility applications, where competitiveness with conventional energy systems has been more difficult to achieve, whereas flat-plate active solar technology, given the higher energy prices residential consumers pay, has been able to compete more readily. For a period in the 1980s small parabolic trough systems were used for residential water heating, both single- and multi-family, but they have largely disappeared from the residential market.

3.12.2 Industry Profile

The high-temperature solar thermal industry comprises a very small number of companies. In the mid-1980s approximately thirty companies formed its nucleus, although more played supporting roles as suppliers and furnishers of engineering and construction services; the core participants were evenly divided among companies producing whole systems, engineering companies, system integrators, and component suppliers. The industry was also evenly divided between small companies completely dedicated to solar thermal energy, and large diversified conglomerates in which solar formed only a minor portion of overall business.

Large aerospace manufacturers, such as McDonnell Douglas, Rockwell International, Martin Marietta, Boeing, and Ford Aerospace, as well as large engineering firms and systems integrators such as Babcock & Wilcox, Bechtel, Black & Veatch, and General Atomic were involved extensively in the solar thermal industry from its beginnings. Most of these companies were contractors of Sandia National Laboratory and the Jet Propulsion Laboratory, the two lead research centers in the mid-1970s for development of prototype solar thermal systems. The large companies were attracted by government dollars for R&D, and an energy crisis vision of substantial new markets for large-scale alternative energy technologies.

When the Reagan administration reoriented the Department of Energy's renewable energy program toward longer-range R&D, cutting funds available for solar thermal substantially, the solar industry reacted. Within the large aerospace firms, senior managers authorized discretionary R&D funds only in solar thermal technologies that showed promise of significant upside potential for deployment in large installations. When the federal support for commercialization activities ended, the energy crisis atmosphere waned, and oil and gas prices stabilized and then began to drop, the vision of near-term markets for many utility-scale systems seemed a mistaken one. The only substantial solar thermal market to develop was for solar thermal electric power plants using parabolic troughs, a market created and captured by a nonaerospace small business newly incorporated in 1979. Many large companies decided to end their solar thermal work. Boeing, Ford, Honeywell, General Atomic, Martin Marietta, Jacobs-Del, Nielsen Engineering, and Westinghouse ceased solar thermal research in the early 1980s. Companies remaining in solar thermal reoriented their activities away from market development and competed for the more limited R&D funds for solar thermal research still available from the Department of Energy.

The large aerospace and engineering companies were only part of the solar thermal industry. Also involved were a number of smaller companies, such as Acurex and BDM, firms that were developing parabolic trough technology. In addition, new businesses were created to develop and market solar thermal technology; these included Entech, Industrial Solar Technology, La Jet Energy Company, Luz International, Power Kinetics, Inc., Solar Kinetics, Inc., Solar Steam, Sunsteam, and Suntec. These smaller companies tended to be involved in the distributed solar technology field and were at the forefront of the development of complete systems. When the federal government backed away from supporting market development, many of these companies also reoriented their business to compete for R&D funds available for central receiver and parabolic dish technology; several closed their operations.

Table 3.15 lists companies involved as of 1985 in solar thermal research and development. Some omissions surely exist, but the list is fairly complete. The five utilities that devoted the most resources to central receiver development were Southern California Edison, Pacific Gas & Electric,

Table 3.15
Solar thermal companies, 1985

Parabolic dish	Parabolic trough	Central receiver
Acurex	Acurex	Arco Solar
Advanco	BDM	Babcock & Wilcox
Babcock & Wilcox	Babcock & Wilcox	Bechtel National
Barber-Nichols Eng.	Bechtel National	Black & Veatch
Bechtel National	Entech	Foster Wheeler Dev.
Entech	Industrial Solar Tech.	McDonnell Douglas
Garrett Turbines	Luz International	MTI Solar
LaJet Energy	MTI Solar	Olin
McDonnell Douglas	Olin	Rockwell International
Mechanical Technology	Science Applications	Science Applications
MTI Solar	Solar Kinetics	Solar Kinetics
Olin	Stearns-Catalytic Eng.	Solar Power Eng.
Power Kinetics	3-M	SPECO
Rockwell International		Stearns-Catalytic Eng.
Science Applications		3-M
Sanders Associates		
Solar Kinetics		
Solar Steam		
Stearns-Catalytic Eng.		
Stirling Thermal Motors		
3-M		
United Stirling		
No. of companies = 22	No. of companies = 13	No. of companies = 15

Source: C. LaPorta, R & C Enterprises, Santa Cruz, California.

Arizona Public Service, New Mexico Public Service, and the Los Angeles Department of Water and Power. Their role as potential customers was critical for both the Department of Energy and the solar thermal companies. Southern California Edison and San Diego Gas & Electric signed energy purchase agreements for solar thermal electricity projects, parabolic trough and parabolic dish systems respectively, and Georgia Power managed a multiple-dish total energy demonstration system.

3.12.3   Technology and Commercial Market Developments

As noted above, high-temperature or concentrator solar thermal technology development in the modern era began in the mid-1970s and was funded substantially by the solar thermal program at the Department of Energy. After five to seven years of R&D, several technologies were just getting ready to be demonstrated commercially when conventional energy prices began a slide in the early 1980s, and energy lost its status as an nationally important issue. Consequently, market development of solar thermal technologies stalled badly. The major exception was Luz International, Ltd., a California corporation with a corporate entity in Israel as well, that since 1984, installed 350 megawatts of solar thermal electric power plants in southern California as independent energy producers. Then, in 1989, two firms—Industrial Solar Technology with its development partner, United Solar Technologies, and Sunsteam—also began to revive marketing efforts for parabolic troughs for thermal systems. The three following sections will briefly review the market status of central receiver, parabolic dish, and parabolic trough technologies.

3.12.3.1   Central Receiver

No commercial central receiver systems have been installed. There has been only one complete central receiver electric power plant installed in the United States: Solar One, a 10 MWe system in Barstow, California, using 765,666 ft^2 (71,133 m^2) of heliostats. This prototype experiment, the largest such installation in the world, was designed to prove the concept in a utility setting. Funded by the Department of Energy, with support from the Southern California Edison Company and Los Angeles Department of Water and Power, the plant began operating in 1982; it has been a highly successful experiment, providing many years of utility operating experience. It ceased operation in the late 1980s but has been maintained in anticipation of refitting it with more advanced thermal conversion technology. ARCO Solar, which was developing heliostat technology in addition to photovoltaics, also installed a small central receiver system in Taft, California, in 1983, to demonstrate its potential as an enhanced oil recovery technology; the system generated one megawatt of thermal energy during peak operating conditions, but was never replicated.

When it came to deploying a next-generation central receiver system, the Department of Energy under the Reagan administration had changed emphasis, curtailing support for commercial demonstration of new technologies. Those in industry and government hoped that a combination of federal and state tax incentives and the start given it by Solar One would be enough to carry central receiver technology forward to an initial commercial prototype. Several large aerospace companies, McDonnell Douglas, Martin Marietta, Rockwell International, and Atlantic Richfield's ARCO Solar, together with a few other companies and utilities, attempted to advance to a next generation, but they were unsuccessful.

The failure to develop a near-commercial central receiver installation in the mid-1980s can be attributed to many factors and to no small amount of complexity. The next-generation plant would have employed a more sophisticated thermal energy transfer subsystem based on liquid sodium or molten salt as the working fluid, but the risk was too high for private investors to underwrite any of several projects industry put forth. Nearly everyone agreed that an economically viable central receiver system needed to be near 100 MWe in size, and one group of companies proposed building the first commercial plants at this scale. However, leaping from a 10-MWe pilot plant to a commercial facility 100 MWe in size with a price tag near $500 million could not be done in the 1984–1985 environment. Another group attempted to proceed to the 100-megawatt size through intermediate steps. In this case, a totally unproven receiver system based on liquid sodium would have been employed for the system; however, the partners attempting to develop a 30-megawatt project at Carissa Plains in California were never able to agree on how to share the risks involved. Without one partner standing behind the project, no investors were forthcoming to carry this commercialization attempt forward.

Central receiver technology for process heat systems also stalled in the marketplace because competition with low-priced natural gas was too difficult. ARCO Solar did build a small central receiver system, using 17,000 ft^2 (1,580 m^2) of heliostats, to produce steam for enhanced oil recovery near Taft, California. A proof of concept for ARCO, the plant demonstrated one-man operation of a central receiver system and field-tested a new generation of ARCO heliostats. The project was not economic and never duplicated. Higher value electricity projects were needed to make the economics for central receiver systems work.

After the larger aerospace firms abandoned central receiver commercialization efforts, the technology was carried forward in small steps through the federal government R&D program. Sandia National Laboratory and the Solar Energy Research Institute continued work on full-scale pump and valve tests for a molten salt heat transfer system, developed a new generation of heliostats using stressed membranes to replace metal glass reflectors (with contractors Science Applications International and Solar Kinetics), and conducted research on a new direct absorption receiver that exposed uncontained molten salt to concentrated solar flux. The pump and valve tests would prove that a more efficient heat transfer system was feasible, and the heliostat development would help lower costs for the single most expensive component of a central receiver plant. Several comparative systems studies also helped decide which technology configurations would be most economical for introductory commercial plants. Through the late 1980s, however, no one expected rapid commercial progress to occur until conventional energy prices rose again or until utilities experienced a critical need for new generating capacity that could be installed in modular fashion or that was much cleaner environmentally.

3.12.3.2   Distributed Receivers: Parabolic Troughs

Although an American company built a 50-horsepower parabolic trough–driven pumping system in Egypt in 1913, modern development of this technology in the United States did not really begin until 1973. Sandia National Laboratory and its contractors constructed several prototypes; over time, a fairly uniform design evolved that tracked curved mirrored surfaces and a fixed receiver. The developers anticipated that long-range market potential would develop in the form of total energy systems for industry; but in the near-term search for market applications, government and industry developed industrial process heat and agricultural pumping systems. One such system, using the largest array ever built (23,000 ft^2 or 2,140 m^2) and rated at 150kW, came on line in 1979 in Coolidge, Arizona, and provided thermal energy to an organic Rankine cycle heat engine to make electricity and drive pumps. In FY 1981 DOE launched an ambitious industrial process heat commercialization program that would build a series of demonstration systems across the United States. These systems allowed the companies involved to design, build, repair, and perfect another generation of trough technology, and thereby position themselves for market entry. This demonstration program used all types of distributed collector systems, including parabolic dishes. Many felt that parabolic troughs had arrived technically and economically and could be marketed in regions with good insolation levels and high conventional energy costs.

At the same time the industrial demonstrations were being built, DOE organized its last major project in the development of parabolic trough technology, the Modular Industrial Solar Retrofit (MISR) program. Five manufacturers were selected to design packaged parabolic trough systems with mass production and modularity as guiding principles. Unfortunately, while the MISR program was preparing another generation of technology with the latest improvements, many of the initial demonstrations were experiencing major technical and performance problems. The solar industry would learn that premature demonstration can create customer concerns about a new technology that are difficult to overcome. Unfortunately also, oil and natural gas prices had started to fall about this time. As a result, after companies had solved their technical issues and were able to offer a reliable product, it became nearly impossible to sell solar industrial process heat systems.

One firm, Solar Kinetics of Dallas, Texas, did manage to sell two systems in 1982. The largest was 60,000 square feet, providing process hot water for a metal-film-plating operation in Arizona. The industrial owner, using electricity to heat water, was attracted by the favorable return on investment for the project, given state and federal tax credits available to the purchaser. Solar Kinetics also sold a 20,000-square-foot system in Arizona as a third-party financed deal, but this ended its trough marketing.

Labeling its marketing experience "bloody hell out there," another key U.S. trough manufacturer who survived to 1983 also stopped marketing solar systems. The industry had created a product, but market conditions could not support sales, not when life-cycle costs were still over $10 per million Btu of delivered energy. It was very difficult to approach American industrial managers with projects with economic paybacks any longer than two or three years; the expected industrial and commercial market for parabolic troughs never materialized.

After Solar Kinetics's Arizona projects, the next commercial trough installations occurred in Colorado in 1985 and 1986. Industrial Solar Technology, Inc., of Denver, installed its first and second systems at a recreation center in Aurora and at a county jail in Brighton, Colorado; the company employed metalized polymer reflective film technology rather than metal-glass for its concentrators, which substantially lowered installed costs. The jail system used a 6,000-square-foot array to furnish heat to an organic Rankine cycle engine-generator to provide service hot water and electricity. State and federal tax credits plus a state grant from oil-overcharge funds made the project, a third-party energy sales deal, economically feasible; it has operated reliably since 1986.

Other than another small trough system that also used a lightweight concentrator and reflective film, installed by Sunsteam in California for a metal-plating company, no further commercial parabolic trough projects would be installed through the balance of the 1980s. Marketing experience showed that other than a few projects where site and economic conditions converged nicely, no market development would occur for parabolic trough systems in the United States, with one principal exception.

Luz International Regarding market development of trough technology, the single exception is Luz International, which would build five independent electricity generating stations in California between 1983 and 1987. Acurex and Solar Kinetics Incorporated, the two largest producers of troughs in the United States, had each attempted to finance a project for a 12-megawatt electric power plant to be interconnected to Southern California Edison company as an independent energy producer under the terms of the Public Utility Regulatory Policies Act of 1978 (PURPA). Although neither firm could convince enough investors to support their project, Luz International did succeed with its initial solar thermal electric project.

Incorporated in Israel and in California in 1979, Luz corporate entities received an Israeli government loan almost equal to half of the $4 million it received from private investors. Using a parabolic trough concentrator it engineered with results from government-sponsored programs and its own resources, Luz first began to pursue industrial process heat applications with small (5,000-square-foot) installations. These never succeeded as economic projects, however, and Luz shifted its attention to the electricity market. Able to negotiate a contract with Southern California Edison Company to sell it electricity at 61% above the current avoided costs, Luz doggedly went after investor capital to realize its first project. Completed in 1984, Solar Energy Generating Station 1 (SEGS 1) was located right next to Solar One, the central receiver pilot plant, in Daggett, California. The SEGS 1 collector field exceeded 700,000 square feet, and the plant was rated under peak conditions at 14.7 megawatts electric. Total plant cost to investors was $61 million. Most of the plant was manufactured in Germany (mirrored glass reflectors) and Israel (receivers).

Because solar heat is boosted in temperature with a natural gas boiler at the steam turbine generator, SEGS 1 is considered a hybrid system.

Luz quickly decided it should expand the plant size to lower power costs, and its next six plants were 30 megawatts in size. One was completed in 1985, also at Daggett, two in 1986 at Kramer Junction, and three more, also at Kramer Junction, in 1987 and 1988. Then, in 1989, a change in federal utility regulations allowed Luz to build an 80-megawatt plant in Harper Lake, and start another 80-megawatt facility there in 1990. By the end of 1990, Luz would have 354 megawatts of capacity on line.

All the plants have been hybrid power producers, using natural gas to supplement the solar power. This ensures that Luz is able to be on line whenever Southern California Edison requires peak power, and maximizes the income stream from the project. The SEGS 2, 3, and 4 plants cost nearly $100 million each, more than doubling energy output for less than a 50% gain in price. The collector fields in each of these systems equaled approximately 1.7 million square feet. In an 80 megawatt plant, the collector array approached 5 million square feet.

All the Luz projects were developed as third-party independent energy production facilities under provisions of PURPA. State and federal tax credits and accelerated depreciation figured prominently in the ability of Luz to finance the plants among private and institutional investors. Later plants combined equity and debt finance, often with utility companies participating. No other firm has equaled Luz's ability to market solar energy projects of this scale; as perhaps the most significant market development gain, Luz demonstrated that investors can realize attractive profits from solar thermal energy projects. Luz also demonstrated that economies of scale have a powerful role in bringing down the cost of solar thermal technology as production increases. The thirty-year levelized cost of electricity for SEGS 1 was near 25 cents per kilowatt-hour. By the completion of the third facility, this had fallen to near 15 cents, and the larger 80 megawatt plants would further reduce this figure to 10 cents or less. [Editors' note: In late 1991 the Luz parent company went into bankruptcy, but the operating company continues to run those plants already on line. See Becker 1992.]

3.12.3.3 Distributed Receivers: Parabolic Dishes
Parabolic dishes are point focus systems that track in two axes to maximize solar energy capture, minimize cosine losses, and achieve high concentra-

tion ratios, permitting design of very high temperature systems. Therefore, they have advantages akin to central receivers, but like troughs, the basic module is small, allowing substantial flexibility in meeting energy needs.

Technical development of parabolic dishes progressed more slowly than trough technology, not for any discernible reason other than internal planning at the Department of Energy. When the Reagan administration changed the direction of the R&D programs at DOE, being further behind in the development path had both advantages and disadvantages. Unlike troughs, dishes were not proclaimed by DOE to be "fully developed," therefore funding was not eliminated. Lack of demonstrations, however, meant that after 1982, the development process would be slow and proceed in small increments. Except for one project completed in 1982, there would not be any substantial government buys of large systems that would allow manufacturers to tool and gain valuable production experience.

Basic research on dishes began in the mid-1970s at the Jet Propulsion Laboratory (JPL) in California, with federal funding through the Energy Research and Development Administration (ERDA), and later through DOE. At the outset, two material approaches were taken, one using glassmetal reflectors and the second using metalized polymer film reflectors. Later, stressed membrane construction techniques would also be employed that used plastic film reflectors. In addition, two energy conversion options were developed: one employing an energy collection and transport system that delivered thermal energy to a ground mounted conversion system or load; the second a dish-mounted heat engine generator group to make electricity at the dish.

The first large-scale demonstration of dish technology was a total energy system at Shenandoah, Georgia. With participation of Georgia Power Company, this 1982 installation at a knitwear factory produced steam for industrial process heat, for cooling, and to make electricity. With 114 collectors, each 23 feet in diameter, this was the largest dish installation in existence at that time. Solar Kinetics Inc. produced the collectors. This system successfully demonstrated the concept of collecting energy from each dish and delivering it to a central energy conversion unit.

At the same time, JPL and its industry subcontractors were developing technology for dish-mounted energy conversion systems. Three engine concepts were tested: an organic Rankine cycle, a Brayton cycle, and a Stirling cycle. Ultimately, JPL let a contract to Advanco, a spin-off company from Fairchild, for a metal-glass concentrator with a 25-kilowatt United Stirling AB engine-generator. Fabrication was completed in early 1984, and the unit, named "Vanguard," set and still holds several world records for conversion of sunlight to energy.

Advanco, dependent on the DOE program, did not survive, however. Long interested in central receiver technology, McDonnell Douglas began to hedge its position, as development of a 100-megawatt central receiver plant for Southern California Edison Company floundered; it entered an exclusive joint venture with United Stirling, securing rights to the engine Advanco had used. Failing to win a DOE procurement for a next system, Advanco ceased operations. Meanwhile, McDonnell Douglas lay marketing plans and ultimately built eight prototype dish-Stirling systems, a few of which were sold to utility companies. A corporate decision to leave the energy business saw McDonnell Douglas ultimately sell its dish-Stirling technology to Southern California Edison and abandon the solar energy field. Meanwhile, several other companies have continued developing Stirling engines, often with support from NASA for space station application.

After 1985–1986, commercialization initiatives to deploy dish-Stirling technology practically halted, but the technology development process continued. Stirling Thermal Motors refined its kinematic engine and began testing it at Sandia National Laboratories in New Mexico. In addition, DOE and NASA each supported Stirling engine technology development, and several companies began work on free piston Stirling engines, and several more are positioning themselves with this option. Although a few other firms continued with organic Rankine cycle technology and one retained some momentum with a Brayton cycle system, the Department of Energy decided to uniquely support Stirling engines for dish applications. By 1990 Cummins Engine Company emerged as a leading system integrator with plans to deploy a prototype 4-kilowatt free piston system with a LaJet Energy Company dish.

LaJet Energy Company   The only large-scale commercial deployment of dish technology in the United States was a project (Solar Plant 1) built at Warner Springs, California. Louisiana Jet Petroleum Company formed a subsidiary to develop a lightweight, low-cost dish by using space frame concepts and aluminized mylar film in a unique vacuum-assisted focusing design. The LaJet Energy Company, Abilene, Texas, worked with Merrill Lynch White Weld to organize a third-party finance deal (based on federal and state tax credits) for an independent energy project interconnected to San Diego Gas & Electric Company. LaJet started construction even before financing was closed, expecting the enthusiasm evident in the alternative energy, PURPA-driven environment in California to transfer easily to its project. Ultimately, the parent firm had to invest millions of additional dollars in the project to fully finance it, which restricted the amount of capital available to LaJet Energy Company for further development.

The Solar Plant 1 system employed 700 dishes (each 460 square feet in area) divided into a preheater section and a superheater section. Two ground-mounted turbines, one 3.68-megawatts, the second 1.24, make up the 4.92-megawatt system. LaJet experienced problems with its system and finally modified it through a joint venture with Cummins Engine to a solar/diesel hybrid unit. No subsequent commercial projects from LaJet ever developed. The company later became directly engaged in the DOE dish R&D program, and received a contract to build a large version of its original collector as an innovative concentrator.

PKI, Inc., and Acurex also received such contracts and assembled their innovative dish concentrators at Sandia for testing. Then the DOE program turned to a new design, employing a stressed-membrane technology developed for lower-cost heliostats; it awarded contracts to build stressed-membrane dishes to the two companies that had developed prototype heliostats—Science Applications International and Solar Kinetics. By 1990 LaJet activity in this field had faded from view.

Both LaJet and McDonnell Douglas fell victim to the downturn in interest in solar energy as a result of low oil prices and lack of public concern about energy. They and other companies have either adjusted to emphasize R&D or quit the industry. Meanwhile, some private companies continue to develop dish technology on their own (e.g., Solar Steam of Fox Island, Washington, built a 40-foot geodesic frame concentrator using metal-glass reflectors for thermal loads) and others participate in the DOE program as it continues to support dish research. Other than Cummins's plans to deploy small electric power units in the mid-1990s,

Table 3.16
Solar thermal program funding (millions of current year dollars)

	Central receiver	Distributed receiver	Total
1975	7.6	7.9	15.5
1976	19.0	20.3	39.3
1977	39.6	36.0	75.6
1978	64.8	49.6	114.4
1979	56.0	53.4	109.4
1980	64.1	79.1	143.2
1981	52.0	86.3	138.3
1982	24.1	30.9	55.0
1983	25.3	19.2	44.4
1984			38.7
1985			33.0
1986			25.5
1987			23.0
1988			17.0
1989			15.0
1990			15.4

Sources: For 1975–1983, see Radosevich 1982; for 1984–1990, see Sissine 1990.

no near-term commercial projects were on the horizon as the 1980s ended.

3.12.4   Review of Solar Concentrator Market Development

3.12.4.1   Government Funding
After 1981, federal solar R&D and demonstration funding fell dramatically (see table 3.16). A round of demonstrations had been constructed, and no new major hardware purchases would be forthcoming. Support to keep demonstrations running was also cut; over time, the basic R&D program was reduced as well. For example, in 1982 dollars, the fiscal year 1990 appropriation for concentrator solar thermal R&D equaled $11.8 million, which is only a quarter of the real resources devoted to solar thermal in 1982. The effect of these cuts, lower oil prices, and negative signals regarding the role of solar energy in the nation's future discouraged many companies. The number of firms involved in solar thermal technology development dropped by more than 50% from 1981 to 1985. The large aerospace companies, mainly engaged in central receiver development, were notable departees.

3.12.4.2 Competitive Costs

Ultimately, any new energy technology must prove its worth in the competitive market by offering lower energy costs or a reliable supply at a price acceptable to the market. The Solar Energy Industries Association Five-Year Research and Development Plan for 1985–1989 stated that current costs for solar thermal electricity production were approximately $0.16/kWh, levelized in constant 1984 dollars. Costs for industrial process heat were given as $20/MBtu. Cost projections for commercial solar thermal energy systems were seldom attractive enough to convince potential users to purchase the technology. A few cases existed where high-value applications combined with incentives would return a competitive return on investment to financial participants, but widespread sales of solar thermal energy systems simply did not occur.

Central Receiver Systems In 1986, based on the experience of Solar One, the Department of Energy calculated the cost of central receiver systems with present technology and present costs. Assuming a 100-MWe plant, with a 50% capacity factor, and an annual efficiency of 17%, glass-metal heliostats costing $18.60/ft^2 ($200/m^2), a molten salt cavity receiver at $7.45/ft^2 ($80/m^2), and with storage of 2,600 thermal megawatt-hours, the total overnight installed system cost was estimated to be $3,100/kWe peak (DOE 1986, 10).

Table 3.17 indicates that in the mid-1980s, DOE expected the capital cost for midterm deployment of central receivers to fall to $1,800 per kilowatt, and that levelized electricity costs would be near 8 cents per kilowatt-hour. For a technology with practically no air emissions, such an energy cost is close to the competitive range. It will take a number of installations, however, before the $1,800 per kilowatt installed figure can be met.

Parabolic Dishes The LaJet Solar Plant 1 provided the only cost data for an actual commercial dish system. LaJet's system used 321,800 ft^2 (29,900 m^2) of parabolic dishes and a central generator to produce 5 MWe. The installation was operating at a 29% capacity factor with an annual efficiency of 13%. The system cost was $3,400/kWe peak, concentrator cost was $14.60/ft^2 ($157/m^2), receiver cost was $3.60/ft^2 ($39/m^2), and transport cost was $6.50/ft^2 ($70/m^2) (DOE 1986, 13). Table 3.17 projected an energy cost for dishes of 7 cents per kilowatt-hour when

Table 3.17
DOE solar thermal five-year (1986–1990) and long-term goals

	Electricity central receivers	Dishes	Heat troughs
System annual efficiency	20%/22%	17%/28%	36%/56%
System capital cost[a] (1984$/kW peak)	$1,800/$1,000	$2,100/$1,300	$590/$370
Capacity factor	0.5/0.5	0.26/0.26	0.24/0.24
System energy cost[b] (1984$)	$.08/$.04 per kWhe	$.07/$.05 per kWhe	$23/9 per MBtu

Source: DOE 1986.
Note: Five-year goal/long-term goal.
[a] Normalized to turbine or process capable of handling peak field thermal output; includes indirect costs.
[b] System goals levelized in real dollars; long-term values levelized in nominal dollars (assuming 7% inflation) are $.11/kWhe, $14/MBtu. The $9/MBtu (1984$) industrial process heat long-term target is the levelized cost of delivered energy in the 1990s; it is derived from current fossil fuel costs of $5/MBtu (1984$).

systems reached $2,100 per kilowatt, about two-thirds the cost of the Solar Plant 1 installation, which was a prototype commercial installation.

Parabolic Trough Electricity Systems   Luz International SEGS power plants in California offer the most current cost data for commercial line focus electricity systems. SEGS 1, a 1984, 13.8-MWe net plant, cost (1984 dollars) $4,300 per kilowatt of electric power capacity. In 1985, SEGS 2, a 30-MWe net plant cost $3,200/kWe. SEGS 8, an 80-megawatt plant constructed in 1989 had a total price of $230 million (1989 dollars), or $2,875 per kilowatt. Luz reported that for its 1989 installation, its thirty-year levelized electricity cost had dropped to under 10 cents per kilowatt-hour.

Parabolic Trough Industrial Process Heat Systems   According to Department of Energy estimates in 1986, IPH line focus technology had achieved a 15% capacity factor and an annual solar conversion efficiency of 32%. At that time, estimated total system costs were $760 per thermal kilowatt, with concentrator costs at $18.60/ft^2 ($200/m^2), receiver cost of $3.70/ft^2 ($40/m^2), and transport cost of $3.70/ft^2 ($40/m^2) (DOE 1986, 10, 34). Besides Luz International, which developed a utility scale trough that has an installed cost closer to $16.70 per square foot ($180 per square meter) for just the collector field, two small businesses were preparing a light-weight trough for the market that promised to have an installed cost for the complete system that was between $18.60 and $23.20 per square foot ($200 and $250 per square meter). Based on performance of an existing system, Industrial Solar Technology maintained that even with minimal production levels it could install an industrial process heat system for close to $520 per thermal kilowatt and a complete system cost of $21.85 per square foot ($235 per square meter). On a twenty-year levelized cost basis, such a system would offer thermal energy at a cost below $10 per million Btu and could begin to reopen interest in solar thermal energy systems for industrial process heat.

In summary, aside from one company's success marketing parabolic trough electric systems, solar thermal concentrator technology had not yet reached commercial status by the end of the 1980s. With federal government support, central receiver and parabolic dish technology was continuing to advance technically. A few companies were still planning for future commercialization; a few small businesses were also reintroducing parabolic trough technology for industrial process and commercial heating applications, but their progress was slow.

3.13   Passive Solar Energy Market Development

3.13.1   Introduction

Passive solar energy is sometimes hard to define. Essentially, it is building design and use of materials that take advantage of or mitigate natural energy flows to increase occupant comfort and reduce conventional energy consumption. Normally, describing market development of a product will include compilation of statistics on an industry, but what is the passive solar industry? A narrow definition would include only commercial builders that construct a thoroughly designed passive solar building, plus those supporting companies that produce unique materials and components used in such strictly designed buildings. A loose definition would include any builder and all window companies that furnish south-facing, double- or triple-glazed windows in any structure. Tracking "industry" development is therefore a function of which definition is chosen. For this review, the passive solar industry will be loosely defined.

3.13.2   Market Development

Solar architecture is an old science, dating from man's first attempt to build comfortable shelter. Since the Middle Ages, whenever energy has been expensive or difficult to obtain, building designers have paid more attention to available natural energy—sunlight, shade, and air currents; whenever energy has been available and cheap, builders have generally cut costs and ignored energy-conserving options. In the United States, several waves of enthusiasm for solar heating have risen and fallen. Thus far, they have generally receded without the building industry firmly adopting and incorporating solar energy heating systems in their building designs.

The most recent wave of enthusiasm began in the mid-1970s energy crisis. The computer and tens of millions of research dollars spent over an eight-year period advanced the scientific understanding of solar energy impacts on buildings to a new level of sophistication. Where only a hundred passive solar buildings had existed in the United States in the early 1970s, several hundred thousand were built from 1977 to 1987. But this wave has receded as well, and neither consumers in general nor the building industry have embraced passive solar except for isolated pockets of the market. The building sector remains firmly first-cost-conscious, and only a small minority of consumers and builders are willing to invest more in building construction in order to significantly reduce operating costs.

Acceptance remains limited despite the dramatic gains of passive solar technology and design since the early 1970s. Advanced storage and glazing materials have been developed and marketed; a wide variety of passive solar buildings has been monitored over several years to provide data on passive solar features; and design software and worksheets have been developed to help architects and builders design and construct efficient passive solar homes. Nevertheless, even with these important advances, the development of the passive solar energy market has been very difficult to quantify, and no reliable statistics on it exist.

3.13.3   Mid-1970s Technology Diffusion

At the outset of the 1970s, what passive solar homes existed belonged mostly to solar energy pioneers, who, except for a handful of architects, were not involved at all in the commercial building business. When the nation's energy consciousness was raised, a few individuals recognized the massive energy conservation potential of passive solar architecture for buildings, and a landmark event for passive solar was publication of the very popular *Solar Home Book* (Anderson 1976), which featured a number of passive solar designs, discussed basic principles of design, and included essential tables to help readers prepare their own designs. A popular technology diffusion process was started, and many consumer-oriented publications on solar building design appeared.

After first placing greater emphasis on active solar heating and cooling, the federal solar energy research and development program adopted passive solar energy as a separate discipline and began to fund research on direct-gain systems and heat transfer in buildings. As scientists, architects, and professional engineers became more knowledgeable in the disciplines of passive solar design and construction, a core of knowledge was developed. Many organizations became involved in the education process, including the American Institute of Architects, the National Concrete Masonry Association, the Solar Energy Research Institute (SERI) and some of the national laboratories, as well as several regional research centers and state energy offices.

The Carter administration's emphasis on support for commercialization of alternative energy technologies also played a large role in disseminating information on passive solar energy. Besides the millions of dollars in research and demonstration, the federal government sponsored many activities such as design contests, projects aimed at incorporating passive solar design features in manufactured housing, conferences, and large-scale printing of passive solar energy publications to augment dissemination of the technology. Such federal support for passive solar was slower in coming, however, than for active solar technologies. In the Department of Housing and Urban Development (HUD) solar demonstration program, a minuscule percentage of the buildings receiving subsidies for solar energy systems were passive solar; at DOE, budgets for passive solar were slower to rise than those for active solar technologies.

On the technical front significant progress was also made. Many of the initial solar home designs encountered an overheating problem in fall and spring. Research on the role of mass in sun-heated structures was a critical advance. In addition, support for design studies and design contests helped the public appreciate that passive solar homes would not be "freaks" in appearance.

Congress's reluctance to support market development initiatives for passive solar, as it had for other new technologies, stemmed, in part, from the difficulty of defining what constitutes a passive solar building and, in part, from the feeling that because many passive solar design measures added marginal cost to a structure, consumers should adopt them without much incentive. As a result, federal tax credits for solar energy were restricted to exclude any building components that also had a structural role, for example a masonry wall for thermal storage could not be claimed if it supported a roof. Legislation to enact a passive solar tax credit bill did pass the House of Representatives in 1980, but not the Senate. The tax incentives that did exist expired in 1985, although a few states passed tax incentives to help stimulate consumer acceptance of passive solar buildings.

Dissemination of passive solar technology to home builders was not an easy process; indeed, for the mainstream home building industry in the United States, it has just not happened. Those builders who did adopt passive solar approaches were most often independents, outside the mainstream industry; in states like New Mexico, such builders would only do one, two, or three homes a year—just enough to keep busy and make a living. Some would not even categorize themselves as part of the home building industry and steadfastly maintained their separateness.

The federal government and some state governments did undertake specific initiatives to interest home builders directly in passive solar, and in some cases these were successful, at least in part. The driving forces for commercial builders, however, are costs and consumer preferences. When home buyers demanded passive solar, builders would offer passive solar homes for the extra cost; few were willing to offer more expensive but more energy-efficient homes as a matter of course.

3.13.4   Passive Solar Products

Several different passive solar design tools have been developed, including design books, descriptions of and data on successful designs, computer programs, and worksheets, to help builders and architects design efficient passive solar homes. Builders can also buy a variety of products for passive solar applications, including phase-change materials; brick, concrete, tile, and other heat storage materials; glazing, skylight, greenhouse or sunspace, and other aperture materials; and miscellaneous supplementary materials including vents, paint, and insulation.

Design software, phase-change materials, sunspaces, low-emissivity films and glazings, and some new photochromic and electrochromic glazing materials are the products that can be claimed by the industry as passive solar products. The federal government research program has continued to support development of materials, especially advanced glazings. Although some large integrated manufacturing companies with passive solar departments recognize the large potential of passive solar heating in the building sector and continue to invest modestly in product development, there has been nothing like a groundswell movement.

3.13.5 Industry Profile

Passive solar energy could be described more in terms of design than of product. Establishing the value of the market is difficult because the product is generally a home and the costs of the solar features are often hard to pinpoint. The layout, thermal mass, window area, and ventilation design determine whether a home is passively heated by the sun, rather than the presence of specific products manufactured and sold by an identifiable industry. Although there are some true passive solar products and some builders and architects dedicated to passive designs, they do not form an industry according to normal definitions. Even where passive solar products exist, such as greenhouses or sunspaces and phase-change materials, consensus opinions on the size of the market do not exist. Companies are generally unwilling to divulge their own production or to give estimates of the size of the total market.

As a consequence, the passive solar "industry" is composed of a wide variety of professionals, including architects, engineers, and builders; manufacturers of materials, including glass, concrete, brick, and tile; manufacturers of components, including greenhouses and windows; manufacturers of sensible heat and phase-change storage materials; and research institutions, including universities, several federally supported research centers, the National Association of Home Builders, the National Concrete Masonry Association, and the American Institute of Architects. There is even a business association, Passive Solar Industries Council.

In general, the companies that form the industry are either small or derive only part of their income from passive solar work; few depend entirely on passive solar energy products or buildings. The large home building companies are not actively pushing passive solar, although some builders in small regions, particularly Santa Fe, New Mexico, Boulder, Colorado, and Richmond, Virginia, have established healthy businesses building passive solar homes. Some individual architects and engineers are very dedicated to the field, but, as with passive solar builders, their impact is small compared to the entire building industry. Dow Chemical produces a phase-change material (PCM), but its commercial success with PCM modules has been limited. Other PCM manufacturers are generally small firms. Greenhouse manufacturers with a passive solar product, that is, one designed to provide heat, also tend to be relatively small businesses, and the major greenhouse manufacturers generally do not put a heavy emphasis on the energy aspects of their products. Some glass manufacturers have quite large businesses, but only a fraction of their products go to passive solar applications.

3.13.6 Sales

It was not until the 1930s and 1940s that the first modern passive solar homes were built in the United States. Several innovative architects and builders pushed passive solar in the wartime and immediate postwar period when conservation was in vogue; however, with the prosperity and abundant energy of the 1950s, passive solar was largely forgotten. A few innovators kept experimenting over the years, but very few passive solar homes were built. Not until energy prices began to rise in the early 1970s did the passive solar concept begin to gain attention among consumers.

The first passive solar buildings in the United States were generally individual houses, sometimes in small developments created by innovative architects and developers. Passive solar caught on in certain areas, particularly Boulder, Sante Fe, and Richmond, and examples of such market pockets exist across the country. As stated above, passive solar heating has never caught on with the major builders and has yet to become the standard in most areas.

In the 1980s gains in energy efficiency, especially improved insulation measures and more fuel efficient heaters, restrained market penetration of passive solar heating. Restrictions on tax incentives for passive solar measures posed another barrier; many passive design features were not eligible for the solar energy tax credits. As the energy crisis abated, passive solar was seen as a luxury and experienced competition from other luxury add-ons to home building: customers were faced with choosing between spending extra money on passive features or on an expanded bathroom, nicer kitchen fixtures, or other more luxurious trappings.

The National Association of Home Builders (NAHB) estimates that about 200,000 passive solar homes had been built in the United States by the end of 1984. Subsequent estimates prepared for the Environmental Protection Agency (EPA) raised the estimate to 270,000 buildings by 1989 (ICF Inc. 1990). Almost all passive solar buildings are detached single family residences. Attached townhouses are the next most common application, followed by multifamily and commercial buildings (Finneran 1986). Over 50% of passive solar homes are custom-designed, whereas only about 10% of all homes are custom-designed. Only a handful of commercial buildings with daylighting and other passive solar features have been built. Glassed atriums are quite popular in commercial buildings, but these are built primarily for aesthetic value, not for their lighting and energy savings.

Sales have been mostly to socially conscious, upper-middle-class families in their thirties and forties, the same group that leads the way in most purchases of innovative products. A review of the market indicates value, aesthetics, and prestige have become more important factors than energy or social consciousness in choosing passive solar homes, although in some areas, builders have successfully sold passive solar homes to first-time home buyers and others in the low end of the housing market.

Growth of the solar heating market has depended on growth in the building sector and on the perception of the importance of energy prices. With energy prices stable or falling, concern for energy savings has diminished. Energy is still an important criterion, but it usually falls below many other factors in importance for the home buyer, including location, style, price, and bathroom, kitchen, and bedroom design.

3.13.7   Sunspaces

The use of a greenhouse, or sunspace, for solar heating began in the early to mid-1970s. The first installations were by do-it-yourselfers; it was not until 1978 that the first mail-order sunspace kit was offered. The do-it-yourself market was highest in the beginning, but in time, wealthier customers began paying professionals to install their sunspaces. In 1985 manufacturers' estimates of total greenhouse sales for residences ran from $50 million to as high as $300 million. While a consensus opinion has not been developed on the size of the market, many agree that it has experienced rapid growth as a desirable feature in a comfortable, stylish home. According to Kevin Finneran (1986), "Attached sunspaces, sold as kits, are the most popular passive solar product of the 1980s." Sunspace additions to restaurants have become very popular, although these are usually considered economic for their ambiance, rather than their energy value; many, if not most, such additions are actual energy losers because of the extensive overhead glass.

3.13.8   Glazing

Glazing straddles the defining line between energy conservation and passive solar. Double-, triple-, and quadruple-pane glass windows have been developed to decrease heat loss through windows—a conservation measure—but these same windows have also been modified to increase the proportion of sunlight admitted and thus the heat value of windows—a passive solar measure. Low-iron glass, coatings, and films are available to increase the transmissivity, to lower emissivity, and to decrease heat loss through windows. Sales have not been large, but these products have made great strides, and expectations for future growth are high. Also under development for passive solar applications are switchable glazings, which respond to an electric current or sunlight to become increasingly opaque (as photosensitive sunglasses do). Although not widely available, these have the potential to capture the public's imagination in the near future.

3.13.9   Thermal Mass Heat Storage

Heat storage materials have had a modicum of success in the market. Natural solid substances used in construction to increase thermal mass have been quite successful, including rock, brick, cement, and tile. Approximately two dozen companies produce phase-change materials (PCM), a more sophisticated material that absorbs and can store a large quantity of heat when it changes from a solid to a liquid state. Knowledge of their sales is very limited, but it appears that no company experienced significant success in the 1980s. Cost, the durability of the container, and separation of the solutions have been the major problems for most PCM materials, preventing them from gaining widespread acceptance. Solid-to-solid PCM materials are also under development, and their incorporation in traditional building materials may prove to be a very marketable concept. Using water for heat storage is a basically simple approach, but homeowners have been reluctant to try it.

3.13.10   New Mexico Sales by Technology

As noted above, few statistics exist to plot the market development of passive solar. New Mexico, a state with a fairly large number of passive solar builders, proves an exception, and some data exist there. The New Mexico climate favors such an approach to buildings, locally abundant

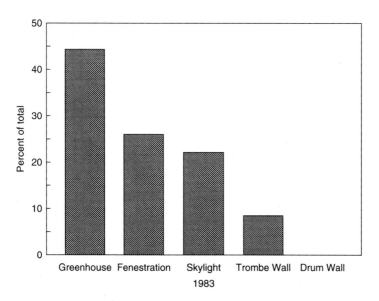

Figure 3.1
Percentage breakdown of passive solar installations in New Mexico, 1983.

adobe has a long history of building applications, and key work on development of passive solar design tools was carried on at the Los Alamos National Laboatory. Consequently, passive solar gained wider acceptance in New Mexico, and a group of highly skilled professionals was busy using and disseminating passive solar technology there.

The legislature in New Mexico passed state tax credits for consumers of passive solar products. The state fiscal authority has compiled statistics on the number of credits claimed by technology. Figure 3.1 shows the percentage of claims by technology for one tax year, 1983 (LaPorta n.d.). With a 44% "market share," greenhouses were the passive solar feature most often selected by consumers. Fenestration and skylights were the next most popular items. These figures support the finding that passive solar features are most popular when they add to the quality of life aesthetically for the upscale consumers responsible for the demand.

Figure 3.2 traces the number of claims for passive solar installations in New Mexico by year from 1977 to 1983; data were not available for 1981. This figure illustrates that, through the late 1970s, the technology was not well known, and market penetration minimal; after the oil price rise in 1979 and educational initiatives took hold in the state, the number of

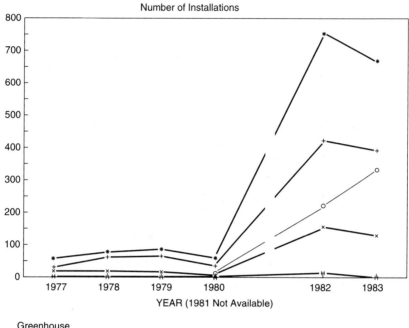

Number of Installations

YEAR (1981 Not Available)

Greenhouse
—— * ——
Fenestration
—— + ——
Skylight
—— o ——
Trombe Wall
—— x ——
Drum Wall
—— H ——

Figure 3.2
Passive solar installations by type, New Mexico, 1977–1983.

installations shot up significantly. A combination of falling conventional energy prices, a relaxed attitude among the public about energy, and a recession in the state's economy led to a later decline in market, as indicated by the reduced number of tax credits claimed for passive solar installations. These trends track closely with the market development path of the active solar industry across the country.

3.13.11 Summary Review: Industry and Market

Passive solar designs and products have made significant progress since the early 1970s, and although their market success has not been dramatic, progress has been made. The lack of dramatic market success in passive solar can be attributed to several factors: no one has been promoting the technology; there are few passive solar products for manufacturers to promote; builders sell buildings, not energy; and the government has backed away from educating the public and promoting the technology. Because few are promoting passive solar, public awareness and market demand are the primary forces behind its acceptance. The end of the energy crisis caused energy costs to drop to near the bottom of the list of factors home owners and potential home buyers consider. There is still interest in the market; although demand is growing, mostly from the acceptance of informed consumers and the sales efforts of technically advanced builders and architects, it remains weak. Thus passive solar heating makes advances, but it does so slowly.

By the end of the 1980s solar building design had not yet entered the mainstream residential construction industry and was even more rarely found in the commercial building sector. The majority of systems in place have been direct gain systems. The building industry remains to be convinced that no-cost design changes are available that will improve their product and that modest cost additions will offer even more superior buildings.

References

Anderson, B. 1976. *The Solar Home Book*. Harrisville, NH: Cheshire Books.

Becker, N. 1992. "The Demise of Luz: A Case Study." *Solar Today* January/February.

DHR, Inc. 1980. *Supply Problems in the Solar Heating and Cooling Industry*. Palo Alto, CA: Electric Power Research Institute.

DOE (U.S. Department of Energy). 1986. *Draft National Solar Thermal Technology Program: Five Year Research and Development Plan, 1986–1990*. Washington, DC.

EIA (Energy Information Administration). 1982. *1981 Active Solar Installations Survey*. Washington, DC.

EIA. 1991. *Solar Collector Manufacturing Activity 1989*. Washington, DC.

ERDA (U.S. Energy Research and Development Agency). 1975. *A National Plan for Energy Research, Development, and Demonstration: Creating Energy Choices for the Future*. ERDA-48. Washington, DC.

Finneran, K. 1986. *Status of the Renewable Energy Industry: Passive Solar Heating and Cooling*. Washington, DC: Solar Energy Industries Association.

ICF Inc. 1990. "Renewable Energy Market and Pollutant Reduction Potential for the United States. Unpublished study for the Office of Air and Radiation, U.S. Environmental Protection Agency, March.

LaPorta, C. n.d. Data compiled for New Mexico Energy Research and Development Institute. R&C Enterprises, Santa Cruz, CA.

Radosevich, L. G. 1982. "Federal Plan Changes." Presentation to the EPRI Solar Thermal Evaluation and Program Development Workshop, Palo Alto, California, 31 March.

REI (Renewable Energy Institute). 1985. *Annual Renewable Energy Review: Progress through 1984*. Washington, DC.

Sissine, F. 1990. *Renewable Energy: Federal Programs*. Washington, DC: Congressional Research Service, Library of Congress, 12 February.

U.S. House. 1983. *A Multi-Year Framework for Federal Solar Energy Research and Development*. Staff Report to the Committee on Science and Technology. Washington, DC, July.

II SOLAR THERMAL PROGRAMMATIC PERSPECTIVES

4 Active Heating and Cooling

William Scholten

One of the earliest solar technologies to receive the interest of the federal government was active heating and cooling for buildings. In the context of active solar heating and cooling, the word "active" describes systems that require some external (generally, electric) power to operate the system. The typical active feature is an electric pump to force circulation of a heat collection fluid through a loop containing a solar collector and a storage subsystem. Passive systems, on the other hand, will function with solar energy being the only energy received by the system.

4.1 Historical Background

Experimental and commercial applications of active solar heating in the United States date back to the early 1900s. Thousands of solar hot water heating units were installed in the Southeast in the 1930s and 1940s. By and large, these systems functioned adequately, but nearly all were displaced by systems using inexpensive electricity or natural gas when these became available to virtually all building owners and operators. The Massachusetts Institute of Technology built the first solar-heated house in 1938, followed by about a dozen similar demonstrations in the ensuing thirty years. There was a flurry of interest in solar energy research at a few universities in the mid-1950s, but it died away in the face of the abundance of "cheap" energy—coal- and nuclear-generated electricity, oil, and natural gas—that prevailed at that time.

Starting in 1971, the National Science Foundation (NSF), under its Research Applied to National Needs (RANN) program, maintained a low-level research program in the area of solar energy conversion (DOE 1978b, I-1). The program was subdivided into six major technology areas (FEA 1975, 7): solar heating and cooling of buildings, wind energy conversion, solar thermal conversion, ocean thermal conversion, photovoltaic electric power systems, and bioconversion to fuels.

The oil shortages of 1973 focused national attention on the possibilities for reducing dependence on nonrenewable fossil fuel energy resources through energy conservation practices and the exploitation of renewable energy resources. Beginning in FY 1973, federal funding for solar energy increased rapidly. The NSF solar energy budget for FY 1971 through FY

1975 was $1.2 million, $1.66 million, $3.96 million, $14.8 million, and approximately $50 million, respectively (FEA 1975, 2).

Of the various solar energy technologies, solar space heating and water heating were perceived to be the most developed; the general perception was that effective active solar heating systems could be constructed from available "off-the-shelf" components, with the possible exception of collectors, which could be easily constructed from appropriate metal plates and tubes. In addition, using the collected solar heat to run a (heat-driven) absorption chiller for cooling in the summer time was seen as a practical extension of a solar heating system's capabilities that could be accomplished with a minimum amount of modification to existing equipment.

These perceptions were exploited around the end of 1973, when the National Science Foundation undertook an effort to demonstrate the feasibility of active solar heating and cooling by contracting for "proof of concept" systems to be installed in four schools around the country: in Boston; Timonium, Maryland; Warrenton, Virginia; and Brooklyn Center, Minnesota.

All of these active solar systems provided space heating and one (Timonium) provided space cooling as well. In order to demonstrate solar heating in a timely manner (i.e., before the onset of the 1974–75 heating season), projects were initiated on a crash basis to provide operational systems before the end of the 1973–74 heating season. A typical schedule was that for the Fauquier High School (Warrenton) system, for which the contract was let on 22 January 1974 and the system was declared operational on 19 March 1974 (NSF 1974, 1). These projects later became part of the Energy Research and Development Administration's (ERDA) Commercial Solar Heating and Cooling Demonstration program. In August 1976 total estimated federal funding for these projects was $2.75 million (ERDA 1976a, 31, 32, 36, 54).

Writings on solar energy in the early 1970s supposed that development needs would be quite modest. George Löf indicated in 1971 that "an optimized system of residential solar heating and water heating, with a suitable portion of auxiliary heat, can now be competitive with electric heating in most areas of the United States, and possibly with gas or oil heat in a few special situations" (ITC 1971, app. G). Löf then concluded:

The foregoing discussion of the current status of the technology and the economics of residential uses of solar energy for comfort heating, water heating, and

comfort cooling indicates that all three of these uses for solar energy are technically feasible but that there are numerous problems associated with types of
system, their design, their optimization, their control, their combination with
conventional energy sources, and many other factors. It is also evident that the
economic potential of these solar energy uses is good, given incentives to produce
substantial numbers of units for public use—first in areas of comparatively high
energy costs and later fairly widely distributed over the United States. (ITC 1971
app. G)

"As to the total budget for a program to place solar heating and cooling in position for commercial sale and practical use within 10 years", Löf
observed, "a high degree of uncertainty must be recognized. On the basis
of experience to date, coupled with appraisal of the task outlined below,
a total budget of $10 to $20 million should be adequate". (ITC 1971
app. G)

4.2 Legislative Mandates

In 1974 the 93rd Congress adopted four acts that established administrative procedures and guidelines for a federal program to develop solar
energy technologies (see chapter 2 in this volume for details). Of all the
solar energy technologies, only heating and cooling was treated with separate legislation to provide for an immediate demonstration program,
reflecting the perception noted above that this technology was at a more
advanced stage of development than the others.

The stated purpose of the Solar Heating and Cooling Demonstration
Act of 1974 (PL 93-409) was "to provide for the demonstration within a
three-year period of the practical use of solar heating technology, and to
provide for the development and demonstration within a five-year period
of the practical use of combined heating and cooling technology." The
foundations of the residential and commercial solar heating and solar
heating and cooling demonstration programs were laid in this legislation.
In addition, PL 93-409 directed the secretary of defense to arrange for the
installation of systems "in a substantial number of residential dwellings
which are located on Federal or federally administered property where
the performance and operation of such systems can be regularly and effectively observed and monitored by designated Federal personnel." (This
program is discussed further in chapter 13.) The National Aeronautics
and Space Administration (NASA), the Department of Housing and

Urban Development (HUD), the National Bureau of Standards (NBS), and the National Science Foundation (NSF) were named in the legislation as the agencies responsible for implementation of the act, with the additional stipulation that functions vested in NASA and NSF could be transferred to ERDA, upon its formation, by the Office of Management and Budget (OMB). The act included FY 1975 authorizations of $5 million each to NASA and HUD and an aggregate authorization of $50 million for FY 1976 through FY 1979.

The Energy Reorganization Act of 1974 established ERDA, subsequently the major part of the Department of Energy (DOE), which was established in 1977 by the Department of Energy Organization Act (PL 95-91). The secretary of energy was also directed by the Energy Conservation and Production Act (PL 94-385) "to provide overall coordination of Federal solar energy commercialization and to carry out a program to promote solar commercialization" (DOE 1978a, 3). The six solar energy technology areas established by NSF were largely maintained by the ERDA and DOE program organizations. However, the solar program within ERDA organized into the three main technology areas of direct thermal applications, solar electric conversion, and fuels from biomass. Solar heating and cooling of buildings was included under direct thermal applications.

An additional solar heating and cooling program that became known as the "Federal Buildings Program"(not to be confused with the federal buildings portion of the demonstration program) was initiated by Congress with the passage of the National Energy Conservation Policy Act of 1978.

4.3   National Program for Solar Heating and Cooling of Buildings

ERDA's Division of Solar Energy established the National Program for Solar Heating and Cooling of Buildings as the mechanism for implementing the Solar Heating and Cooling Demonstration Act of 1974. In conducting this program, the assistance of HUD, NASA, DOD, NBS and other federal agencies was utilized (ERDA 1976b, 1). The primary goal of the program was "to work with industry in the development and early introduction of economically acceptable solar energy systems to help meet National energy requirements" (ERDA 1976b, 1). The goal was later refined further to "to assist in the early establishment of a viable solar industry for the design, manufacture, distribution, sales, installation, and maintenance of the solar heating and cooling systems" (DOE 1978a, 6). It was envisioned at that time that solar energy could provide at least 10% of the heating and cooling requirements of commercial and residential buildings in the year 2000 (DOE 1978a, 6). The program was divided into three basic elements: residential and commercial demonstrations (see chapters 9 and 10); development in support of demonstrations; and research and development. In addition, the program sponsored studies to identify potential barriers to the widespread use of solar energy (e.g., building codes, taxation, legislation, and financing). (Legal issues are discussed further in chapter 31.) The program established a Technical Information Center for storage and dissemination of program-generated information and the National Solar Heating and Cooling Information Center for dissemination of information from various sources to the general public. (Further details on information programs are discussed in chapters 16, 17, and 18.) The Commercial and Residential Demonstration Program was further divided into federal and private segments. Responsibility for management and coordination of the residential demonstration program was shared by ERDA with HUD for the private sector segment and with DOD for the federal segment. NASA was given responsibility for the demonstration support element of the program. Responsibility for the research and development element remained in ERDA, under the research and development branch established for that purpose. NBS received primary responsibility for the development of standards and performance criteria for solar heating and cooling systems, subsystems and components, and for establishing a program for the accreditation of laboratories qualified to test solar heating and cooling subsystems and components. (The development of standards is discussed further in chapter 14.)

4.3.1 Residential and Commercial Demonstrations

The demonstration program was implemented through a series of cycles of awards utilizing a variety of systems, building types, and demonstration regions. (The program is discussed further in chapters 9 and 10.) The mechanism of succeeding cycles allowed advances in the state of the art to be incorporated in subsequent projects. Building types were divided into residential and commercial, and further subdivided into federal and non-federal within each of these categories. The federal projects are referred to as the "Federal Buildings Program" in some of the early documentation (e.g, ERDA 1976b); however, these projects were distinct from those in the later Federal Buildings Program mandated by Congress under provisions of the National Energy Conservation Policy Act of 1978. Awards under this demonstration program were made to owners of existing and planned buildings in amounts usually equal to the projected cost of purchasing and installing the solar equipment. Manufacturers and suppliers of the specified solar systems usually collaborated with builders, architects, and owners in the preparation of proposals and in designing the solar facilities.

At the inception of DOE, the demonstration program was placed under the Demonstration Branch of the Division of Solar Applications, headed by Ron Scott. The Demonstration Branch was headed by Bill Corcoran, and included Carl Conner, Walt Preysner, Dave Pellish, Roger Bezdek, Gene Doering, Bill Lemeshewsky, H. Jackson Hale, Frank DeSerio, and Noel Kiley. Jack Hale headed the National Solar Data Network (NSDN) Program, and Bill Lemeshewsky headed the Solar in Federal Buildings Program, mandated by the National Energy Conservation and Policy Act of 1978. (This federal part of the demonstration program is discussed in detail in chapter 11.)

In the spring of 1977 active solar hot water systems were perceived as being market-ready; because of this perception, both the residential and commercial demonstration programs included a hot water initiative. The hot water initiative represented an increase in scope of the demonstration program that was undertaken to encourage the development of a market demand for solar hot water systems. (The hot water initiative is discussed further in chapter 27.)

4.3.1.1 Residential Demonstrations

The nonfederal residential demonstration program was directed by HUD. The program consisted of a series of five cycles of Request for Grant Application (RFGA) solicitations issued in the fall of 1975, July 1976, January 1977, November 1977, May 1978, and October–November 1979. The solicitations of November 1977 and May 1978 were both part of cycle 4, and were designated as "Cycle 4" and "Cycle 4A," respectively (HUD 1979, i). The number of projects and amount of grant awards for each cycle are shown in table 4.1 (see also chapter 9).

The U.S. Department of Defense (DOD) administered a federal residential demonstration program that met with minimal results: "Bids

Table 4.1
Projects and expenditures for residential demonstration program

Cycle	Number of systems	Total grant amount (millions of dollars)
1	55	1.0
2	102	3.9
3	169	6.0
4 and 4A	144	8.5
5	316	1.5
Total	786	20.9

Source: DOE (1978a) and HUD (1979).

obtained for 50 individual residential projects on several military bases were too high to allow for a cost-effective demonstration. One project was implemented in 1977" (DOE 1978a, 20). Ultimately, only seven projects were reported under this program (DOE 1979a, III-ii; see also chapter 13).

4.3.1.2 Commercial Demonstrations

The bulk of projects under the nonfederal buildings portion of the commercial demonstration program were selected from proposals solicited in three cycles of Program Opportunity Notices (PONs) by ERDA and, later, DOE. The program also included eighteen projects inherited from NSF (which included the four proof-of-concept school experiments), eight "Phase 1" projects, and fifty projects resulting from a "Hot Water Initiative for Motel/Hotel Installations." Table 4.2 presents the distribution of projects by source and building type. (The commercial demonstration program is discussed further in chapters 10 and 13.)

In order to ensure an early start of demonstration projects with a representative mix of locations and building types, ERDA initiated a "Phase 1" program in which the study contractors (General Electric and Inter-Technology) were required to identify specific demonstration project opportunities and submit them for consideration. Eight projects were selected from these recommendations.

The three PONs calling for proposals to incorporate solar energy in new or existing commercial buildings were released in the fall seasons of 1975, 1976, and 1977, respectively. The first emphasized space heating and hot water systems. The second accommodated a wider range of solar

Table 4.2
Number of DOE commercial nonfederal demonstration projects

Group	Early NSF	Phase I	PON I	PON II	PON III	Hot water initiative	Total
Building type and solar application mix							
Schools/universities							
Htg/clg only	1			6	4		11
DHW only	1						1
Htg/clg and DHW	1		5	12	8		26
Htg and clg	3		1	2	2		8
Htg, clg, and DHW	3		1	2	3		9
	9	0	7	22	17	0	55
Office buildings							
Htg/clg only		3	3	8	8		22
DHW only			1	1			2
Htg/clg and DHW		1	3	9	14		27
Htg and clg	2		1	3			6
Htg, clg, and DHW		1	3	3	8		15
	2	5	11	24	30	0	72
Public buildings							
Htg/clg only	1		2	2			5
DHW only	1		1				2
Htg/clg and DHW	2		3	14	4		23
Htg and clg			2	2			4
Htg, clg, and DHW	1	1		6	3		11
	5	1	8	24	7	0	45

Hotels/Motels							
Htg/clg only		1					1
DHW only	2		2			50	54
Htg/clg and DHW		2		2	1		5
Htg and clg							
Htg, clg, and DHW			1				1
	2	3	3	2	1	50	61
Commercial/industrial bldgs							
Htg/clg only			1	1	4		6
DHW only			1				1
Htg/clg and DHW				4	9		13
Htg and clg							0
Htg, clg, and DHW			1		3		4
	0	0	3	5	16	0	24
All building types							
Htg/clg only	2	3	7	17	16		45
DHW only	4	2	3	1	1	50	61
Htg/clg and DHW	3	1	13	41	35		93
Htg and clg	5		4	7	2		18
Htg, clg, and DHW	4	2	5	12	17		40
	18	8	32	78	71	50	257

Source: (DOE 1978a), augmented with data for the PON III projects from (DOE 1979a).

applications and required specific attention to energy conservation techniques. The third emphasized combined heating and cooling and required cost sharing for the incremental costs of the solar energy system. Projects under the hot water initiative were selected from responses to a PON issued in May 1977 for active solar hot water systems in hotels or motels.

In addition to the demonstrations in nonfederal buildings, thirty-eight commercial demonstration projects in federal buildings are reported for the initial commercial demonstration conducted under the Solar Heating and Cooling Demonstration Act of 1974 (DOE 1979a, II-ii,II-iii). Subsequent awards for projects in federal buildings were made under the Federal Buildings Program that was set up under the National Energy Act legislation.

4.3.2   Development in Support of Demonstrations

The Development in Support of Demonstration Program was run by the Marshall Space Flight Center (MSFC) of the National Aeronautics and Space Administration (NASA) under an interagency agreement with ERDA and, later, DOE. The purpose of the Development in Support of Demonstration Program was to provide a base for stimulating an industrial capability for mass production and marketing of solar heating and cooling equipment. The basic methodology adopted for the program was to test subsystems and systems in a laboratory environment, followed by a period of field-testing in "operational test sites," in which the systems were installed, operated, and monitored in a "real building" environment. The program was divided into four distinct tasks (DOE 1978a, 32–34):

1. Systems integration of marketable subsystems—to purchase marketable subsystems and integrate them into prototype systems.

2. Existing subsystems that require additional development—to allow continuation of development work on promising subsystems (collectors, pumps, thermal energy storage, heat transfer fluids, microprocessor controls, and thermosyphon heat exchangers) designs.

3. Existing systems that require additional development—to continue development testing work on existing active heating and cooling system designs, and

4. Systems design and development—aimed at the development of economically packaged systems to reduce costs and minimize technical risk in the demonstration program.

Table 4.3
Budget authority for development in support of demonstration (millions of dollars)

Prior to FY 1976	FY 1977	FY 1978	FY 1979	FY 1980	FY 1981	Total
12.6	19.8	11.3	9.4	5.7	1.9	60.7

Source: Internal DOE budget document.

All of the subsystems and systems were eventually scheduled to be installed in one of forty-five planned operational test sites. These test sites were all operational by the end of FY 1979. Monitoring of some of the last sites to become operational continued into FY 1980. The approximate total ERDA/DOE budget authority for this program was 60.7 million dollars, annually distributed as shown in table 4.3.

4.3.3   Research and Development Program

In addition to funding demonstrations of solar energy systems, DOE sponsored a research and development (R&D) program to "assist creating a viable solar energy industry by improving the performance and reliability of solar energy components and systems, and by reducing their costs" (DOE 1978b, I-1). The origins of the R&D program date back to 1971, when NSF designated $0.54 million for R&D in the solar heating and cooling of buildings under its Research Applied to National Needs (RANN) program (ERDA 1976c, 1):

From 1971 to 1973, the funding level for solar heating and cooling R&D remained essentially constant, with all awards being made in response to unsolicited proposals submitted to NSF. By September 1973, the solar heating and cooling R&D program consisted of eight projects, at a total funding level of $1.2 million.

In the fall of 1973, NSF issued a Request for Proposals (RFP) for solar heating and cooling R&D in seven areas:

• Solar collectors

• Thermal energy storage subsystems

• Solar space cooling subsystems

• Solar-assisted heat pumps

• Advanced solar collectors for heating and cooling systems

• Alternative approaches to solar energy systems

• Evaluation of buildings having solar energy systems

As a result of this procurement, 33 new projects were initiated, at a total funding level of approximately $3 million.

The NSF R&D projects were transferred to ERDA and then to DOE as those organizations were formed. By the fall of 1975 the solar heating and cooling R&D program had grown to fifty-eight projects in five categories: collectors, thermal energy storage, air-conditioning and heat pumps, systems and controls, and program support (ERDA 1976c, 1).

In addition to the Demonstration Branch, the Division of Solar Applications also included the Research and Development Branch, headed by Fred Morse. This branch was responsible for the Solar Heating and Cooling Research and Development Program. It had three program managers in the active solar area: Steve Sargent, collectors; Mike Davis, heat pumps; and Bob LeChevalier, cooling; in addition, the branch included the passive solar R&D and, for a part of the time, agricultural and industrial process heat.

In the winter of 1975–1976, a series of meetings were held to "assess the state of the art of solar technologies, identify impediments to improved performance and reduced costs, and recommend R&D tasks to resolve these problems" (DOE 1979f, I-1). As a result of these meetings, an applications-oriented research and development program was initiated (DOE 1979f, I-1). In FY 1977, a series of solicitations resulted in 1,200 proposals, from which 157 projects were selected (DOE 1979f, I-2). In FY 1978 a total of 257 projects received funding from the research and development program, of which 42 were for passive systems, 26 were for systems analysis, and 15 were for program support (DOE 1979f, I-2). The proportion between active and passive of the latter two categories cannot be determined.

4.3.4 Data Collection, Evaluation, and Dissemination

In 1977, DOE established the National Solar Data Network (NSDN) to collect and process performance data from selected commercial and residential demonstration sites and to disseminate results to a wide range of potential users. The objectives of the NSDN were (Hale 1978):

1. Evaluate and publish thermal performance of representative solar demonstration systems.

2. Evaluate and publish operational performance and cost/benefits.

3. Publish comparative analysis reports leading to improved solar systems.

   The Federal Systems Division of IBM Corporation was awarded the first contract for instrumentation, collection, and analysis of data from solar demonstration projects. IBM was also responsible for the design, installation, and maintenance of instrumentation on commercial sites, while Boeing, the support contractor for the residential demonstration program, had this responsibility for the residential sites. Data were collected on-site using an automatically recording site data acquisition system, and were transmitted daily over voice-grade telephone lines to a central data processing system. Standard primary performance factors defined by NBS were calculated by a host computer using site-specific equations.

   Three basic steps were required to prepare each site for connection to the network (Murphy 1978):

1. Analyze each site and its instrumentation requirements.

2. Select sensors and personalize standard data acquisition equipment to meet specific site requirements.

3. Install and check out site instrumentation.

   Results from NSDN activities were reported in a series of reports that included:

1. *Solar Energy Project Description* for each site, describing the system, the site, and the required instrumentation and also presenting the predicted system performance.

2. *Monthly Performance Report* for each site, containing system status and a tabulation of monthly performance data generated by the host computer.

3. *Solar Energy System Evaluation Report*, published periodically for each site, containing a more comprehensive analysis of system performance.

4. *Comparative Analysis Reports*, using data from various sites to evaluate performance of a given class of systems (e.g., solar space heating systems).

5. *Solar Bulletins*, generated in a timely manner when information of use to the newly developing solar industry became evident.

6. Reports pertaining to special studies defined for the demonstration program.

In January 1979, 148 systems were included in the NSDN (DOE 1979b, iii). After two years, the original NSDN contract period was over, and a new request for proposals was issued. In September 1979 Vitro Laboratories, a division of Automation Industry, Inc., was awarded the contract for continuation and expansion of the NSDN operation.

4.3.5   Codes, Standards, and Performance Criteria

At the start of the demonstration program, because of the newness of the technology and its rapid growth, a lot of the basic skills found throughout industry were not evident for the solar industry. There was no single accepted method of estimating the effectiveness of the solar application, no standard building codes, no "cookbook" designs for the heating, ventilating, and air-conditioning (HVAC) contractors, no skilled installation personnel, and few skilled design personnel.

The need for model codes and standards to help alleviate this situation was readily apparent. In an attempt to accelerate the development of codes and standards for active solar heating and cooling, DOE initiated a program in concert with the major trade and professional organizations of the building and heating, ventilating, and air-conditioning industries to establish standardized methods by which solar heating and cooling equipment could be tested, rated, and evaluated. The major elements to be established were a model solar code, test methods and performance criteria, and procedures for evaluating and rating systems' performance.

More than fifty associations, led by the Council of American Building Officials (CABO) were involved in drafting a consensus model building code (DOE 1979c) of requirements for solar installations. This code was adopted in whole or in part by many state and local agencies. This effort also involved a pilot training program for code officials to help translate the document's message into practical experience. (The development of codes and standards is discussed in more detail in chapter 14.)

NBS served as DOE's major laboratory in the development of test methods. Trade organizations such as the American Society of Heating, Refrigerating, and Air-Conditioning Engineers (ASHRAE), the Air-Conditioning and Refrigeration Institute (ARI), and the American Society for Testing of Materials (ASTM) incorporated the results of NBS research into industry test methods or equipment standards. Closely tied to the development of standard test methods was the establishment of uniform rating methods. DOE worked closely with industry to establish the Solar Rating and Certification Corporation (SRCC). Typical of the results of these efforts are five ASHRAE-published standards listed below:

ASHRAE 93: "Method of Testing to Determine the Thermal Performance of Solar Collectors"

ASHRAE 94.1: "Method of Testing Active Latent Heat Storage Devices Based on Thermal Performance"

ASHRAE 94.3: "Metering and Testing Active Sensible Thermal Energy Storage Devices Based on Thermal Performance"

ASHRAE 95: "Methods of Testing to Determine the Thermal Performance of Solar Domestic Water Heating Systems"

ASHRAE 96: "Methods of Testing to Determine the Thermal Performance of Unglazed Flat-Plate Liquid-Type Solar Collectors"

4.3.6 Market Development

An early report prepared for the Federal Energy Administration (FEA) in 1976 found that the solar industry infrastructure was quite fragmented and that a market of about one billion square feet of collectors would be required to drive the cost down to the "should cost" of $1.20/square foot (ITSC 1976). The report also pointed out that the various supporting parts of the industry were not in place and that the best hope for low costs would be complete vertical integration of the industry.

As the demonstration program unfolded it soon became obvious that there were problems associated with the installation of solar systems; many did not work at all and many created more problems than they solved. A system of design reviews eliminated some of the more obvious flaws in the designs, but the costs did not decrease as anticipated. This prompted a look by the DOE program at methods used throughout industry to implement commercialization of conventional heating and cooling systems serving the same market envisioned for the active solar systems in the demonstration program. It was found that it takes on the order of ten to twenty years for the typical industry-developed cooling system to become fully commercial after its initial conception. During this time, the system goes through a "product development sequence" that is typified by the following eight phases: (1) basic and materials research, (2) market research and product justification, (3) component development and testing, (4) prototype systems development and test, (5) engineering field test, (6) market test, (7) production, and (8) marketing and sales. In the engineering field test systems are set up and operated in real installations for the purpose of identifying and correcting problems. This is then followed by a "market test," in which systems are sold on a test basis to confirm their market readiness. In this sequence, demonstration, if it occurs at all, does not occur until the product is fully tested and operational, and on the market. It was thus seen that the demonstrations of the DOE program did not really constitute any portion of the normal development sequence of conventional products.

In FY 1979 emphasis on demonstration was reduced; in response to the National Energy Act of 1978, a comprehensive role supporting the growth of a viable solar industry in all phases of product development was adopted. Market development received increasing emphasis, as did planning directed toward accelerated commercialization. By this time it had become clear that the cost reductions needed to effect a significant market penetration were not going to occur unless production rates were greatly increased. The Federal Buildings Program was mandated to augment the other demonstration programs as one method of increasing the demand for production. Congress also passed a series of partial tax credits for owners of installed solar energy and other energy conserving equipment. Federal credits included a 40% residential energy credit, the regular investment tax credit (6% to 10%), a business energy tax credit (9% to 15%), and an R&D credit (25%). Many states, most notably California, passed additional tax credit measures where a substantial active solar energy industry existed during the term of the tax credit programs. In many cases, state credits could be combined with federal credits to offer total benefits as high as 75% of the cost of a typical solar energy system (Mihlmester Zabek, and Friedman n.d.). The Individual Return Analysis Section of the Internal Revenue Service reports the federal tax credits for residential renewable energy for the years 1978 through 1982 shown in table 4.4. Although these include all renewable energy systems (active, passive, wind, geothermal, and photovoltaic), it has been estimated that solar heat energy represents about 90% of the total (Mihlmester, Zabek, and Friedman n.d.). (Tax credits are discussed further in chapter 25.)

Table 4.4
Federal residential energy tax credits, 1978–1985 (millions of dollars)

	1978	1979	1980	1981	1982	1983	1984	1985	Total
Credit	32	44	166	263	322	325	325	330	1,807

Source: U.S. Department of Treasury, Internal Revenue Service, Individual Return Analysis Section through 1982; estimates by Office of Tax Analysis for 1983–1985.

Another result of the passage of the National Energy Act was the establishment of four regional solar energy centers (RSECs) to conduct major outreach activities of local importance in areas of market research and evaluation, education and training, model codes and standards development, and consumer assurance. These regional centers were the Northeast Solar Energy Center (NESEC) in Boston, the Mid-America Solar Energy Complex (MASEC) in Minneapolis, the Western Solar Utilization Network (WSUN) in Portland, Oregon, and the Southern Solar Energy Center (SSEC) in Atlanta. The regional solar energy centers were slated to play an ever-increasing role in technology transfer and other marketing activities, with a correspondingly shrinking role for the centrally directed program. For example, in 1978 DOE sponsored four regional conferences highlighting the objectives, plans, and experience of the National Commercial Solar Heating and Cooling Demonstration Program and the National Solar Data Program (DOE 1978c). In 1979 these conferences were repeated, and hosted by the regional centers (DOE 1979d, 1979e). (The RSECs are discussed further in chapter 20.)

Around the summer of 1979 Fred Morse replaced Ron Scott, who had left DOE, as director of the Division of Conservation and Solar Applications of DOE. One of Morse's first tasks was to reorganize the Division of Conservation and Solar Applications into the Office of Solar Heating and Cooling, as a result of the shifting emphasis away from demonstration and toward commercialization and market development. In addition to active heating and cooling, the passive solar and agricultural and industrial process heat technologies were included in the office. Two divisions, the Systems Development Division and the Market Development Division, were formed along functional lines. The R&D activities were focused more toward systems integration activities in the Systems Development Division, headed by Mike Davis. The Market Development Division was headed by Bob Jordan. The FY 1980 budgets for the two divisions were $71 million for systems development and $69 million for

Table 4.5
Federal budgets for active solar heating and cooling (millions of dollars)

Area	1974	1975	1976	1977	1978	1979	1980	1981	1982	1983	Total
Research and development	6.0	4.4	18.8	25.5	21.1	27.5	20.1	20.0	11.3	6.5	161.2
Demonstrations		6.5	28.3	62.0	58.7	60.2	34.0	11.0	0.0	0.0	260.7
Commercialization and technology transfer				1.6	1.4	2.7	11.0	10.4	0.2	0.2	27.5
Total	6.0	10.9	47.1	89.1	81.2	90.4	65.1	41.4	11.5	6.7	449.4

Source: R. L. Lorand, Science Applications International Corporation.

market development, reflecting an essentially equal emphasis on each activity.

In 1984, under a support services contract to DOE, Science Applications International Corporation (SAIC) contacted the relevant offices of DOE and compiled expenditures for three active solar heating and cooling activities: research and development, demonstrations, and commercialization and technology transfer. The compilation excluded funding for the agricultural and process heat segment of the overall program. The results are shown in table 4.5. The budget for commercialization and technology transfer activities increased substantially in FY 1980 and FY 1981, while the program emphasis shifted abruptly from commercialization and market development activities to "long-range, high-risk" research and development. The responsibility for commercialization of solar energy was shifted entirely to the solar industry. The DOE budget for renewable energy was drastically reduced, and a major reorganization took place, with a significant reduction in personnel. Active solar research activities were placed in the Active Heating and Cooling Division of the Office of Solar Heat Technologies under the assistant secretary for conservation and renewable energy.

4.4 Lessons Learned

DOE never formally evaluated the demonstration program against its original objectives. However, in 1980, The General Accounting Office (GAO) reviewed DOE's Solar Heating and Cooling Demonstration Program and found that:

- most projects funded under the program have not demonstrated that solar heating and cooling are practical,
- data dissemination has not been very successful, and
- the extent the program has aided in developing a viable solar industry is unknown. (GAO 1980, p. i)

Morse (1980, 1) responded to these conclusions in an internal memorandum:

Our disagreement with the findings and conclusions of the report is based on a number of different factors. These include (1) a basic disagreement on the interpretation of the intent of Congress in promulgating the original Act; (2) a belief that GAO neither understood, appreciated nor took into consideration the primitive state of the solar art at the outset of the program; (3) the fact that GAO reviewed only the early stages of the program (i.e., predominantly the NSF and PON I projects) and did not consider the amount of time required to design and construct fully operational projects (advanced conventional and solar) and to collect sufficient data to conduct a meaningful analysis and publish the results.

The original projections of many solar enthusiasts, typified by Löf's earlier quote, proved to be optimistic. This optimism carried with it the perception that solar energy systems could be readily installed, that they would perform as predicted in their design, and that they would operate in a virtually maintenance-free manner. Based on these implicit assumptions, the demonstration program did not at first include any accommodation for repair or refurbishment of the installed systems, nor was the NSDN designed to be easily used for system maintenance and improvement. Another problem in the demonstration program was the awarding of funds for installation of many systems based on unproven concepts, risky designs, and poor materials. The program thus served as a vehicle for testing ideas rather than demonstrating the practicality of well-proven systems; disappointments and failures were natural outcomes. On the positive side, however, many systems, particularly those based on sound design experience, proved the practicality of solar heating.

In retrospect, it would have been more appropriate to structure the solar development path along lines followed by the development of a typical commercial heating and/or cooling system, with engineering field test and market test phases leading to fully reliable functioning systems before public exposure. The demonstration program can be viewed at best as a "hybrid" of these two phases. In contrast, the NASA Development in Support of Demonstration Program was structured more in line with the engineering field test concept; it utilized a data collection program essentially the same as NSDN, but with the designed purpose of identifying problems that could be subsequently corrected, resulting in upgraded system performance. Undoubtedly the demonstration program would have gained a more favorable reaction from the public if it had been based on a more realistic perception at the start.

References

DOE (U.S. Department of Energy). 1978a. *National Program for Solar Heating and Cooling of Buildings, Annual Report.* DOE/CS-0007. Washington, DC.

DOE. 1978b. *Solar Heating and Cooling Research and Development Project Summaries.* DOE/CS-0010. Washington, DC.

DOE. 1978c. *Revised Proceedings of The Department of Energy's Solar Update.* CONF-780701. Washington, DC.

DOE. 1979a. *Solar Heating and Cooling Demonstration Project Summaries.* DOE/CS-0038-2. Washington, DC.

DOE. 1979b. *Instrumented Solar Demonstration Systems: Project Description Summaries.* SOLAR/0017-79/34. Washington, DC.

DOE. 1979c. *Model Document for Code Officials on Solar Heating and Cooling of Buildings.* DOE/CS/4281-1. Washington, DC.

DOE. 1979d. *Proceedings of The U.S. Department of Energy's Regional Solar Updates.* Vol. 1, *Federal Program Presentations and National Solar Data Program.* CONF-790758-Vol I. Washington, DC.

DOE. 1979e. *Proceedings of The U.S. Department of Energy's Regional Solar Updates.* Vol. 2: *Invited Papers and Appendices.* CONF-790758-Vol II. Washington, DC.

DOE. 1979f. *Solar Heating and Cooling Research and Development Project Summaries.* DOE/CS-0010. Washington, DC.

ERDA (U.S. Energy Research and Development Administration). 1976a. *National Program for Solar Heating and Cooling of Buildings Project Data Summaries.* Vol. 1 *Commercial and Residential Demonstrations.* ERDA 76–127. Washington, DC.

ERDA. 1976b. *National Program for Solar Heating and Cooling of Buildings.* ERDA 76-6. Washington, DC.

ERDA. 1976c. *National Program Plan for Solar Heating and Cooling of Buildings Project Summaries.* Vol. III Research and Development. ERDA 76-145. Washington, DC.

FEA (Federal Energy Adminstration). 1975. *Solar Energy Projects of the Federal Government.* FEA/C-75/247. Washington, DC.

GAO (U.S. General Accounting Office). 1980. *Report to Congress: Federal Demonstrations of Solar Heating and Cooling on Commercial Buildings Have Not Been Very Effective.* EMD-80-41. Washington, DC.

Hale, J. H. 1978. "The National Solar Data Program" *Revised Proceedings of the Department of Energy's Solar Update,* CONF-780701. Washington, DC: U.S. Department of Energy.

HUD (U.S. Department of Housing and Urban Development). 1979. *Solar Heating and Cooling Demonstration Program: A Descriptive Summary of HUD Cycle 4 and 4A Solar Residential Projects.* HUD-PDR-455. Washington, DC.

ITSC (InterTechnology Solar Corporation). 1976. *Industry-Market Infrastructure Analysis.* FEA order no. P-05-76-2382-0. Washington, DC: U.S. Federal Energy Administration.

ITC (InterTechnology Corporation). 1971. *The U.S. Energy Problem.* Vol. II, *Appendices,* Pt. A. Warrenton, VA.

Mihlmester, P. E., E. Zabek, and D. M. Friedman. N.d. *The Effects of Tax Credits on Solar Energy Markets.* Silver Spring, MD: Applied Management Sciences.

Morse, F. M. 1980. *SA Response to GAO Report on Commercial Solar Demonstration Program.* DOE memorandum. Washington, DC: U.S. Department of Energy.

Murphy, L. J. 1978. "The Solar National Data Network." *Revised Proceedings of the Department of Energy's Solar Update.* CONF-780701. Washington, DC: U.S. Department of Energy.

NSF (National Science Foundation). 1974. *Solar Energy Heating Augmentation Experiment Design Construction and Initial Operation.* ITC report no. 090974, NSF-RA-N-74-019. Washington, DC.

Public Law 93-409 1974. Solar Heating and Cooling Demonstration Act of 1974, H.R. 11684. 3 September.

5 Passive Technologies

Mary-Margaret Jenior and Robert T. Lorand

During the late 1970s and continuing into the early 1980s, the U.S. Department of Energy (DOE) greatly expanded its role in promoting the use of passive solar technology in buildings (see figure 5.1). Programs ranged from research and development to demonstrations, along with a variety of market development activities. Organizations such as the Solar Energy Research Institute (SERI; now the National Renewable Energy Laboratory or NREL), the Los Alamos National Laboratory (LANL), and regional solar energy centers (RSECs) were given broad mandates to work with the research community and the building industry in order to transfer the results of the federal passive solar program to the private sector and overcome market barriers to widespread commercialization.

More than a decade after these initiatives were launched, this ambitious objective has yet to be achieved. However, significant strides have been made and valuable lessons learned. This chapter examines five of the more significant of these earlier programs, with a view toward better focusing future efforts. The five programs examined are

• Denver Metro Home Builders Program

• Passive Solar Industrialized Buildings Program

• Nonresidential Experimental Buildings Program

• Passive Solar Design Tool Development Program

• Monitoring and Evaluation Program

In addition to these programs, the Solar in Federal Buildings Program (SFBP) and the residential portion of the Solar Heating and Cooling Demonstration Program administered by the U.S. Department of Housing and Urban Development (HUD) had commercialization aspects. The SFBP in particular, intended to use federal purchasing power to expand the market for active and passive technologies. However, both of these programs were geared primarily to active rather than passive systems, and are therefore not covered.

The five programs selected are briefly described below, with much of the background information coming from the document entitled *Improving the Transfer of Passive Solar Energy from DOE National Laboratories: Linkages and Decision Processes* (Lambright and Sheehan 1985). Following the overview of the programs is a discussion of the lessons

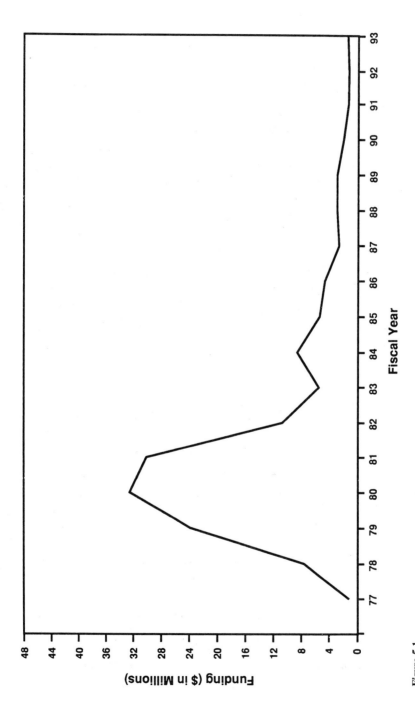

Figure 5.1
Passive solar residential program funding history.

learned and some observations regarding the adoption or commercialization of passive solar technologies.

5.1   Denver Metro Home Builders Program

In 1980 the Solar Energy Research Institute or SERI (now the National Renewable Energy Laboratory or NREL), in conjunction with the Metropolitan Denver Home Builders Association (HBA), the Western Solar Utilization Network (Western SUN or WSUN) regional solar energy center, and the Colorado State Energy Office (CSEO), launched a passive solar design competition targeted at low to moderate income housing. The objective of the program was to encourage local home builders to incorporate passive solar design features in their buildings. SERI provided funding for the program, and technical assistance. The Denver metro HBA served as the conduit for program information, workshops, and so on, to its members, and Western SUN and CSEO assisted in publicizing the program.

The program design involved two stages. In the first, a design competition was held to determine which builder submittals should be selected for design funding. Out of forty-seven proposals, twelve were originally selected for funding. The selected builder's designated architect or design consultant was paid directly by SERI. No commitments were required on the part of the selected builders, except to provide designs for critique by SERI reviewers. During this stage, SERI provided technical assistance in the form of workshops for the participants, as well as workshops for interested nonparticipants under the auspices of the Denver metro HBA. The home builders had complete responsibility for construction, with SERI serving simply in an advisory capacity. Construction on all twelve homes was completed in early 1981.

The second stage involved publicizing and marketing the program, and the subsequent monitoring of building performance and occupant attitudes. The builders were paid for the right to monitor the dwellings after occupancy. The centerpiece of the publicity campaign was the Denver metro HBA's "Parade of Homes" event. This two-week event resulted in thirty-one sales contracts and projections of more than twice that many additional sales.

Based on these preliminary results, it appeared that the Denver metro program could serve as a model for additional efforts within Colorado, as well as other parts of the United States. Indeed, several other localities sought to replicate the program. However, a dramatic change in SERI's mission—the elimination of virtually all its market development activities—and the economic downturn made these more ambitious efforts impossible. SERI's passive solar budget was reduced from $3.7 million in 1981 to less than $1 million in 1982 (see DOE 1980, 1982). Locally, the efforts of the Denver metro HBA and the presence of SERI, continued to make the Denver area one of the most favorable for passive solar buildings in the nation.

What was learned from this program was the importance and effectiveness of working with the local building industry, and using their delivery mechanisms to introduce new technologies in the marketplace. In addition, the value of using SERI staff as facilitators to help bring together both builders and architects in developing practical solar design strategies was proven. What was not proven was the effectiveness of the program in spurring the widespread use of passive solar. There is no question that the awareness of solar by the local design and building community was raised. However, the downturn in the economy, and the significant cutbacks in government funding, as well as the redirection of the government program to almost exclusively high-risk, long-term R&D, appears to have deterred the large-scale replication that was envisioned. The abrupt termination of "follow-on" efforts of the Denver metro type by SERI no doubt raised concerns among the building community, which may have substantially reduced the program's long-term impacts.

5.2   Passive Solar Industrialized Buildings Program

The Passive Solar Industrialized Buildings Program (also known as the "Passive Solar Manufactured Buildings Program") was launched as the fourth cycle of demonstrations under the Solar Heating and Cooling Demonstration Program in 1979 (DOE 1979b). Its objective was to support the development of passive and hybrid solar energy design in industrialized (e.g., primarily factory-built) buildings. Under this program, DOE intended to provide funding in three phases: (1) design, (2) prototype fabrication and field tests, and (3) test marketing. Only projects that had successfully met the requirements of each phase would receive funding for the subsequent phase. A total of twenty-six manufacturers were given phase 1 awards. Of these, eight participated in phases 2 and 3.

The significant reduction in numbers from phase 1 was the result of DOE budget reductions (see figure 5.1) and the weakening housing market/economy that occurred in 1981.

Various designs were developed, monitored, and found to perform well, among which were designs for panelized and modular buildings, as well as for mobile homes and metal structures. Passive design elements included direct-gain, indirect-gain (mass wall), and isolated-gain systems (e.g., sunspaces, thermosiphon air panels), with a range of storage media, and distribution and control elements. Solar contributions ranging from 20% to over 60% were documented, along with estimated incremental costs typically from 0% to 10% of a comparable nonsolar building (SERI 1986). However, there was little replication of the designs and no widespread offerings by the participating manufacturers. Only two of the manufacturers continued to offer designs based on experiences under the program. This was due in large measure to the state of the economy at the time, as well as some of the program's "fits and starts" resulting from budget cutbacks at DOE.

5.3   Nonresidential Experimental Buildings Program

Like the Passive Solar Industrialized Buildings Program, the Nonresidential Experimental Buildings Program (formerly called the "Passive Solar Commercial Buildings Program"—see chapter 6 for detailed discussion), was an attempt by DOE to focus the later stages of the Solar Heating and Cooling Demonstration Program on passive solar technologies (DOE 1979a). Under this program, launched in 1979, funds were provided for demonstrating the effectiveness of passive solar design strategies for "light commercial buildings." These were perceived to be most amenable to scale-up of residential passive solar technologies, which represented a large segment of the nonresidential building population.

Funds were to be made available in three phases: (1) design, (2) installation, and (3) evaluation and commercialization. Information dissemination and documentation of the design process were also covered under the latter two phases. In addition, participants were required to agree to varying degrees of data collection for up to five years after project construction. In all, thirty-five projects were selected for phase 1 funding from more than 300 proposals. Twenty-two of these actually completed their designs; a total of nineteen received phase 2 and phase 3 funding.

From the outset, this program involved close coordination between DOE staff and contractors and the design teams. An advisory team was established to review the designs and to ensure that only sound designs would proceed to construction. The result was a set of projects that represented a broad array of building types and solar strategies and that, by and large, worked well. Program results were documented through case studies and disseminated through the American Institute of Architects (AIA), magazine articles, presentations at workshops/conferences, and so forth. *Commercial Building Design: Integrating Climate, Comfort, and Cost*, which won an award for excellence from *Progressive Architecture* magazine, documented the design experiences, as well as the performance and cost attributes of the various buildings (Burt Hill and Min Kantrowitz 1987). The results indicated that nonresidential buildings incorporating passive solar design features could reduce conventional energy requirements by an average of 46%, and do so at costs that were comparable to conventional construction. About half the buildings compared were at or below the average cost for comparable buildings based on R. S. Means and F. W. Dodge data; most were within the normal range for similar nonsolar buildings. The program also pointed out the need for improved design tools for these applications.

Again, what was not accomplished was any wide-scale replication of the projects. This can be attributed in part to lack of sufficient DOE funding support for a sustained information dissemination effort and in part to the variation and uniqueness of the buildings designed, as well as to the nature of passive solar design. The intent was not to create "cookie-cutter" buildings, but to explore the effectiveness of different design strategies, and to obtain sufficient information about them—performance, cost, occupant attitudes, and so on—so that the successful strategies could be adapted to other environments.

5.4 Passive Solar Design Tool Development Program

Passive solar design tool development was originally initiated by DOE's predecessor agencies as an adjunct to the R&D program—not as a commercialization or market development activity. During the program's earliest stage, researchers at the Los Alamos National Laboratory (LANL) conducted experiments in simple test cells to determine how various solar concepts worked and to quantify their results. The empirical data were used to validate performance models, and these models in turn were used to develop correlations, which provided a relatively straightforward method of predicting the impacts of specific solar features on the solar contribution of the design. Soon after its development in the mid-1970s, this "solar load ratio" (SLR) approach became widely accepted as the performance prediction method of choice for residential passive solar-heated buildings.

Through DOE sponsorship, the LANL researchers converted the methodology into a series of passive solar design handbooks (see chapter 23 of this volume), which covered solar design principles, methods of sizing various components to obtain desired solar savings levels (without sacrificing required comfort levels), and passive solar economics. While the value of this series to the design community was immediately apparent, DOE budget cutbacks and an emphasis on long-term, high-risk R&D slowed its evolution. Despite this, improvements were made with the intent of widely distributing the results. Refined versions of the handbooks were published under the auspices of the American Society of Heating, Refrigeration, and Air-Conditioning Engineers (ASHRAE)—to lend the added credibility of ASHRAE to the handbooks and to increase their visibility among the professional design community.

At about the same time, the utility of the design methods as a means of helping home builders design passive solar homes meet specific performance levels was demonstrated through the New Mexico Showcase of Solar Homes project. The showcase was a large, privately funded, passive solar development in Eldorado, New Mexico, that sought to emulate the results of the Denver metro program. It was intended to provide homes that not only made substantial use of solar energy to reduce utility bills, but that were also relatively conventional-looking in their design. The design tool methodology was used to develop a series of worksheets or guidelines that would enable typical home builders—not just architects, engineers, or solar design professionals—to determine the performance of various building designs incorporating passive solar and energy conservation features.

The simplified guidelines proved very successful. Public Service Company of New Mexico, one of the advisers to the New Mexico showcase project, and a key advocate of the guidelines approach, worked with the New Mexico Solar Energy Research Institute and the Edison Electric Institute to promote the use of its guidelines throughout its service territory, and to encourage other utilities to do the same. It also sponsored academic training in passive solar design using the guidelines, and helped to put in place low-interest loans, through the Farmers Home Administration (FmHA), for passive solar designs based on the guidelines. Within the Eldorado subdivision itself, more than a thousand passive solar homes and a school were designed using the guidelines. To this day, the development continues to follow the guidelines to ensure that new passive solar homes meet minimum requirements.

Building on this success and the recognition that home builders held the key to broader use of passive solar designs in residences, DOE called on the Passive Solar Industries Council (PSIC) and its members, especially the National Association of Home Builders (NAHB), and the National Renewable Energy Laboratory (NREL) to develop a building guidelines design tool. Essentially, this was a further refinement of the methodology and materials developed for the New Mexico Showcase of Solar Homes. A methodology for developing specific worksheets and guidelines for more than 2,400 locations throughout the United States was developed; this was called *Passive Solar Design Strategies: Guidelines for Home Builders*, or *Guidelines* for short. A computerized version of the guidelines, *BuilderGuide*, was later modified to aid in designing and evaluating retrofit/renovation passive solar projects. Through PSIC and local cosponsors such as local home builders associations (HBAs), workshops were and continue to be held throughout the United States. Over 3,000 builders and developers have been educated to use the guidelines through June 1994. A comparable set of guidelines for small nonresidential buildings has been developed. Energy-10, as the software is called, is the first simplified tool to combine passive solar heating, cooling, daylighting, and energy efficiency strategies in one package.

5.5 Monitoring and Evaluation Program

From the outset of the early passive solar programs, it was recognized that monitoring and evaluation efforts would be central to determining the effectiveness of the various technologies and to building confidence on the part of the designers, builders, and consumers. In order to validate building performance predictions and associated analytical and design tools, a program of modeling, code development, and validation work was initiated.

Among the most successful of the monitoring and evaluation efforts was the design tool comparative assessment work implemented under an International Energy Agency (IEA) Solar Heating and Cooling Program task, and the development of a short-term energy-monitoring method (STEM). Both were used to determine the relative accuracy of various passive solar design and analysis tools in predicting thermal performance. A number of tools had been developed in the United States and abroad, but there was little available information documenting their accuracy. Under IEA task 8, more than 200 different tools were identified and evaluated. The evaluation involved comparisons of actual versus predicted data and model to model performance. The result was the identification of model deficiencies and the subsequent development of improved algorithms and improved models. Sophisticated models such as the SERI-RES (SUNCODE) model are now used as benchmarks against which to compare other models or tools.

In a related development, the use of simplified performance monitoring methods has led to an increased capability to accurately forecast the thermal performance of passive solar buildings. The need for simplified, less costly, yet accurate methods became obvious as a result of the monitoring experiences under the various passive solar demonstration programs. Performance monitoring proved to be time-consuming and expensive. "Class B" monitoring efforts cost on the order of $50,000–$100,000 per building (Burch 1992). In order to enable a larger number of projects to be monitored, it was determined that a simplified methodology would be needed. Accordingly, DOE sponsored the development of the short-term energy-monitoring (STEM) method for passive buildings (Balcomb, Burch, and Subbarao 1993). The method has been refined to the point that a series of measurements taken over several days can now be used to accurately predict annual energy performance.

The advent of a model validation methodology and a low-cost short-term performance test method has helped aiding in the establishment of Home Energy-Rating Systems (HERS). For example, the state of Colorado is using the DOE-developed methodology as part of its HERS program. The national HERS program, which is currently under development, will also use elements of the methodology to establish the credibility of HERS calculation tools and their application. Ultimately, the HERS program will enable the development of performance-based building energy standards that can be effectively monitored. It will also increase the confidence of utilities in promoting energy conserving and passive solar homes, and of lending institutions in providing loans that reflect the reduced operating costs of passive solar homes.

5.6 Lessons Learned

The foremost result of the various programs has been the demonstration that passive solar buildings work well—typically providing 30%–50% of a building's requirements for heating, cooling, or lighting—and need cost little or no more than conventional buildings. These savings have been achieved through building designs that integrate passive solar and energy conserving features to enhance both performance and comfort. Furthermore, occupant surveys indicate that the buildings are well liked, and appear to contribute to increased productivity. For example, some of the owners/occupants of the buildings in the Nonresidential Experimental Buildings Program had the following observations (ASES 1990):

"We have enjoyed a pleasant, comfortable atmosphere in the addition since 1982 because of our passive solar system"—Monsignor Frank J. Hendrick, St. Mary's Parish, referring to the St. Mary's School Gymnasium addition, whose design resulted in a projected two-thirds reduction in energy use as compared to a comparable gymnasium.

"Without question, our passive solar library is a source of pride and will serve our needs for years to come"—Mayor W. Maynard Beamer, Mt. Airy, North Carolina, commenting on the Mt. Airy Public Library, which uses less than one-half the energy of a comparable building of the same size.

"The use of the Visitor Center has been extended beyond our imagination"—Professor Walter Kroner, Director of Center for Architecture Research, RPI, commenting on the Rensselaer Polytechnic Institute Visitor Center, which uses approximately one-third the energy of a comparable nonsolar building.

"The Center is wonderful and offers an excellent opportunity to teach Scouts about architecture and energy conservation"—Judy Borie, Executive Director, Girl Scouts of Greater Philadelphia, commenting on the Shelly Ridge Girl Scout Center, which provides significant savings in heating and lighting energy costs and which has received two design awards.

A review of the residential buildings data indicates that passive solar homes may actually sell more quickly and command a higher sales price over comparable nonsolar homes. The review, entitled *Consumer Acceptance, Marketability, and Financing of Passive Solar Homes* (NAHB 1990), documents these findings, drawing on information obtained from owners and builders of twenty-one homes monitored by the DOE passive solar program. It should also be noted that, while the programs showed significant success in demonstrating the effectiveness of passive solar heating and daylighting, there was insufficient focus on passive solar cooling and load avoidance, where most problems (and missed opportunities) with passive were encountered.

Integrated design solutions—carefully thought-through designs that properly account for the interactions of the various building elements, including mechanical systems—are essential to successful system design. The *BuilderGuide* design tool reflects this philosophy and has been refined to enable home builders to make the necessary performance and design trade-offs without creating comfort problems. A key lesson is that simple design solutions work best and are most readily accepted by the design and building industries and by building owners. Passive solar design strategies produce buildings that are more reliable, are more easily maintained, and retain their resale value better than those produced by more complex approaches.

Cost sharing is a desirable feature whenever possible, but may limit participation. This is particularly true for smaller companies or for many companies during periods of economic slowdown. This appears to have reduced participation in the Passive Solar Industrialized Buildings Program, and possibly the Nonresidential Experimental Buildings Program.

The effectiveness of DOE and its laboratories in establishing programs hinges on its ability to understand the workings of the building industry and to work closely with local and national representatives at key stages of each project. These programs underscored the need for a partnership with industry and for teamwork from beginning to end; DOE or laboratory staff served most effectively as technical advisers and facilitators.

Lack of sustained funding and continuity in program thrusts can severely limit program effectiveness, and reduce its potential. For example, the Denver metro program's impacts were significantly reduced by lack of support. Similarly, the change in focus from broad-based R&D, demonstrations, and commercialization activities to primarily long-term R&D eliminated much of the market development aspect of the Passive Solar Industrialized Buildings Program and the Nonresidential Experimental Buildings Program. Unfortunately, it was this aspect of the programs that was expected to generate the most interest in replicating program results.

Given the nature of the building industry, it is overly ambitious to assume that demonstrating the success of passive solar design strategies in a relatively small number of buildings will lead to widespread replication. However, if a sufficient number of small-scale projects can be established, working from successful models, and if there is *sustained* support, widespread replication may be possible. This requires a greater commitment of funds over a longer time frame and greater emphasis on training and information dissemination.

The diversity of the industry, its conservative nature, and the large number of small firms make diffusion of new technologies a relatively slow process. Incremental steps, such as providing the necessary design tools can yield substantial benefits, as can working to incorporate more rigorous performance criteria in local codes and standards.

5.7   Observations on the Adoption and Commercialization of Passive Solar Technologies

Design competitions and demonstrations, if properly managed, can be an effective way to raise local awareness and obtain necessary information on projects. However, there is no evidence that these demonstrations must be performed on a large scale; from a marketing/replication viewpoint, most of the results are regional. Consequently, all that is required is a relatively limited number of well-designed projects, targeted at markets that show the greatest potential.

New technologies and strategies can be successfully introduced, but until there is sufficient experience with them, their widespread adoption is unlikely. Programs geared to increase market penetration need to develop a phased approach that builds on local successes, using the most effective delivery mechanisms. Education and training through the appropriate channels appear particularly effective and could be enhanced if accompanied by broader publicity and widespread information dissemination.

However, since this latter tactic—broad publicity—was never fully employed, there is insufficient data to draw firm conclusions.

A "whole-building" approach to design is a necessity for passive solar technologies. It is not productive to think of a passive solar building as a collection of components—conservation versus passive elements—or to consider mechanical equipment in isolation. Successful, marketable passive solar designs integrate various elements to minimize conventional energy requirements economically, without sacrificing comfort requirements. Indeed, work to date has pointed to the need for smaller, integrated auxiliary heating, ventilating, and air-conditioning systems (HVAC), to take advantage of the performance impacts of passive solar design features. This could be a promising area of research. Along the same lines, a "conservation" versus "renewable" mentality has served to fragment research efforts and to blur what should be the focus of the research—the whole building.

A sustained commitment over a longer time horizon is needed on the part of the government to give its commercialization efforts a chance for broad success. Through the 1980s the prevailing philosophy of long-term, high-risk R&D, to the exclusion of virtually all other types of activities, made it extremely difficult to achieve the results envisioned when the initial passive solar market development programs were launched. Indeed, these efforts needed to account for the prevailing philosophy before programs could be completed (albeit in scaled-back form). As a result of this, the United States is currently in the position of seeing other countries—most notably, those in the IEA—adopt some of our know-how and of having to look to them for solar building advance. For example, the United Kingdom has developed an improved version of the SERI-RES tool, and the IEA has developed generalized design guidance for atria. The IEA countries, Germany in particular, have put considerable effort into advancing the use of transparent insulation materials for glazings, opaque surfaces, and active solar collectors for both new and retrofit building markets. The value of developing U.S. know-how in these areas, as a potential export commodity, is another argument in favor of sustained passive solar program support.

References

ASES (American Solar Energy Society). 1990. *Savings from the Sun: Passive Solar Design for Institutional Buildings.* Boulder, CO, June.

Balcomb, J. D., J. D. Burch, and K. Subbarao. 1993. *Short-Term Energy Monitoring of Residences*. Atlanta: American Society of Heating, Refrigerating and Air-Conditioning Engineers.

Burch, J. 1992. Private communication from Jay Burch of National Renewable Energy Laboratory (NREL), December.

Burt Hill Kosar Rittelmann Associates and Min Kantrowitz Associates. 1987. *Commercial Building Design: Integrating Climate, Comfort, and Cost*. New York: Van Nostrand Reinhold.

DOE (U.S. Department of Energy). 1979a. *Passive Solar Commercial Buildings Design Assistance and Demonstration, Program, Opportunity Notice*. Chicago: DOE Operations and Regional Office, July.

DOE. 1979b. *Passive and Hybrid Solar Manufactured Housing and Buildings, Program Opportunity Notice*. Chicago: DOE Operations and Regional Office.

DOE. 1980. *FY 1981 Annual Operating Plan for the Office of Solar Heat Technologies*. Washington, DC, December.

DOE. 1982. *FY 1982 Annual Operating Plan for the Office of Solar Heat Technologies*. Washington, DC, February.

Lambright, W. H., and S. E. Sheehan. 1985. *Improving the Transfer of Passive Solar Energy from DOE National Laboratories: Linkages and Decision Processes*. San Francisco: DOE Operations Office.

NAHB (National Association of Home Builders). 1990. *Consumer Acceptance, Marketability, and Financing of Passive Solar Homes*. Rockville, MD: NAHB Research Foundation.

SERI (Solar Energy Research Institute). 1986. *Passive Solar Manufactured Buildings: Design, Construction, and Class B Results*. 2d ed., Golden, CO, June.

6 Passive Commercial Buildings Activities

Robert G. Shibley

6.1 Program Origins and Goals

The explicit legislated goals for the Passive and Hybrid Solar Commercial Buildings Program (now the Nonresidential Experimental Buildings Program) called for the development and commercialization of passive and hybrid solar energy systems in commercial buildings. These goals were initially quantified as part of the overall goals for solar heat technologies by specifying the percentage of annual building starts that were to incorporate solar energy systems. This interest in commercialization was transferred from the old Atomic Energy Commission and the National Science Foundation in 1975 to the Energy Research and Development Administration. In October 1977 there was a large increase in support for solar as the responsibility was again shifted through legislation to the U.S. Department of Energy (DOE). Some of the most modest efforts of the program from 1977 forward were couched in terms of facilitation or acceleration of energy-producing or -conserving strategies already under development in industry or in other government agencies, while other efforts ambitiously set out to establish new passive and hybrid systems or whole new approaches to the production of buildings. By 1978 DOE was describing "ideal" goals for new construction and retrofitting in existing buildings as follows:

• By 1980: Incorporate solar energy systems in at least 10% of the annual residential and commercial building starts and install retrofit systems annually on 2,500 residential and 200 commercial buildings.

• By 1985: Incorporate solar energy systems in at least 10% of annual residential and commercial building starts and install retrofit systems annually on 25,000 residential and 1,000 commercial buildings. (DOE 1978, vii)

By 1980 DOE was projecting that all solar technology programs could produce 18.5 quads annually by the year 2000 (DOE 1980). The portion of that goal to be produced by programs under the Passive and Hybrid Solar Division totaled 1.5 quads distributed across the division's residential, commercial, solar energy products, solar cities and towns, and agricultural branches (Maybaum 1980). In particular, the Commercial

Buildings Branch of the division was assigned .25 quads of annual energy production as its initial share of the goal (Shibley 1980).

Measuring progress against the .25 quad goal is extremely difficult because of the fragmentation of the building industry, the wide variation of possible ways to measure energy use in buildings, and the termination of most of the program well before it had a chance to take full effect. All of the proposed measures are dependent on assumptions and speculations about what people would have done under different circumstances (in the absence of the Commercial Buildings Program). Disaggregating consumption savings according to conservation strategies, passive strategies, or active solar system strategies is also extremely difficult when hybrid systems are involved that employ combinations of all such strategies in an integrated manner. This difficulty is further exacerbated when the program is specifically intended to encourage development in some areas while it actually performs basic research, exploratory development, and test of concept work in others. Finally, further adding to these difficulties, the program was drastically cut in 1982 before many of its efforts had time to mature.

The background to the legislation authorizing work on commercial buildings reveals a very complex set of objectives addressing virtually every aspect of the production and use of buildings as well as the delivery of energy to them (FEA 1974; ERDA 1975, 1976, 1977). It was precisely becaue they were much more sophisticated than simple conservation or production target goals that these objectives attracted hundreds of individuals to the program, individuals who continue to commit much of their lives to the exploration and implementation of passive and hybrid solar building practices. Such objectives and the real accomplishments in implementing them transcend the narrow range of concerns about the energy consumption of buildings and contribute significantly to the methods of inquiry used to investigate the social and physical production as well as the habitation of buildings.

6.1.1   Commercial Buildings: A Tough Problem

In 1979 buildings accounted for an estimated 35% of the nation's consumption of energy. *Project Independence* (FEA 1974) had predicted five years earlier that transportation and industrial energy consumption would be reduced at rates faster than the building sector of the economy because the building sector of the economy is technically a more difficult area in which to employ conservation and renewable energy strategies, and the industrial and transportation sectors of the economy could achieve more dramatic effects with first measures. Furthermore, the decision to conserve is held by a relatively few key people in transportation and industry, while the building sector of the economy is very disaggregated, requiring education and decision by literally thousands of people (Shibley 1983). For technical, economic, and organizational reasons, serious efforts to develop and employ energy conservation strategies and renewable energy sources in the building sector of the economy were seen as a very difficult proposition.

The goals for the Commercial Buildings Program which emerged in the face of such difficulties were comprehensive in their scope. The hope was to establish commercialization of passive and hybrid solar as a matter of routine in the business of building rather than the radical experiment it had been in the mid-1970s. The program is based on the simple desire to increase the efficiency of commercial building energy production and utilization through the employment of passive and hybrid solar energy systems, building components, and design strategies. This was accomplished through a series of activities that built on the residential experience with passive and hybrid solar in establishing commercial systems and strategies; embraced the environmentalist vision of a sustainable ecology; considered the full range of social issues involved when one intervenes in urban environments, human organizations, and within the building industry; improved the efficiency of the existing building stock through renovation and adaptive reuse; and facilitated the education of all those who had to participate in the realization of program goals and objectives.

6.1.2 Building on the Residential Experience: Establishing Commercial Passive Systems and Strategies

Almost in spite of *Project Independence* predictions, by 1979 the residential sector of the building industry had already made serious progress in residential heating and was beginning to establish a quantified basis for residential cooling and ventilation as well (see chapter 8 of *Passive Solar Buildings*, volume 7 of this series for more details). The assumption in the establishment of the Commercial Buildings Program was that this success in relatively simple building systems could generalize to larger and more complex functional and multizonal facilities characteristic of commercial buildings.

The research and applications effort borrowed extensively from the terminology, systems, and problem-solving approaches of the residential precedent. Table 6.1 summarizes the full array of the projects' work efforts, each of which has its residential counterpart. Detailed laboratory studies of multizone convection and ventilation, dehumidification, computer code construction and validation, and parametric studies were well represented. As will become clear, the objectives for the program fully embraced the contributions of building physics and went well beyond them.

Borrowing on the success of the residential experience had the advantage of helping to capture support for the investigation into commercial buildings, but it later became a real limitation. The radically different motives, construction practices, industry organization, building functions, and design processes employed in commercial construction were to require a fundamentally different understanding of the problem of building energy use from the one needed for successful passive and hybrid solar residential construction.

6.1.3   Embracing the Environmentalist Vision

The history of passive solar architecture is clearly thousands of years old, dating back to the time when primitive and preindustrial cultures built vernacular architecture in accordance with regional climatic conditions. The early success with the modern version of passive solar in residential buildings was coupled with a kind of religious zeal; passive solar architecture was referred to as a "movement" even within the Department of Energy. Early disciples of the passive religion were out to revolutionize building design; their goal was to create environments where a balance with nature could be achieved. The mission of individuals engaging passive solar was seldom defined in the narrow terms of reducing fossil fuel consumption. The long-term time perspective went well beyond the year 2000, to a time when people would no longer feel the need to overpower nature and would be inclined to live in harmony with it. Those individuals within or near the "movement" found many similarities between the passive solar approach to energy in buildings and the issues environmentalists marched for in the 1960s. The energy "problem" was (and is still) a very important point of departure for a broad range of ecological inquiry.

A significant part of the environmentalist vision included the recognition that solar and ecological strategies had to be made to appeal to the

Table 6.1
Functions and activities of the Passive and Hybrid Solar Commercial Buildings Program: Developing heating systems, cooling systems, and combined/integrated systems

Basic physical studies — **Accomplishment** Measure passive phenomena	Assessment and analysis — **Accomplishment** Forecast system performance	Design development and testing — **Accomplishment** Predict system performance in buildings	Promote use — **Accomplishment** Real performance results	Sponsor education
■ Develop research methods ■ Detailed computer simulation of physical phenomena ■ Laboratory scale testing ■ Detailed basic physical phenomena studies: • Heat transfer • Environmental resources • Assessment • Interior and site environment studies ■ Characterization of basic system, material, and component concepts ■ Develop new materials through analytical and lab experiments	■ Test procedures and performance criteria development ■ Exploration of specific system and component concepts ■ System, subsystem, and component concept descriptions ■ Systems integration (or combination) potential analysis ■ Technology and analysis tools needs identification (user requirements) ■ Large-scale computer tool development ■ Prototype systems data acquisition (including test cell data) ■ Verification of bench model performance ■ Preliminary evaluation of manufacturing and economic feasibility of systems and materials	■ Proof-of-concept and feasibility: • Parametric studies of prototype systems and components ■ Large-scale analysis tool integration ■ Determination of cost-to-performance targets ■ Interior and microclimatic data acquisition ■ Target audience characterization	■ System design ■ Construction documents and cost estimates ■ Simplified design tool development ■ Prototype and proof-of-concept tests ■ Design tool verification ■ Seasonal and multiclimate data collection ■ Analysis of prototype system performance	■ Handbooks ■ Technical manuals ■ Design guidelines ■ Curriculum development ■ Industry and R&D workshops, conferences, and trade shows

full range of economic and business interests that shaped the U.S. marketplace. As such, although the vision was an ideal one (perhaps even radical), it was also practical and firmly grounded in the day-to-day reality of building economics.

6.1.4 Considering the Social Implications of Building and Building Processes

As part of this practical idealism, people within the passive solar movement and certainly people within the Commercial Buildings Program saw the program objectives as having a strong social component. For example, the work of the Solar Cities and Towns Program within DOE's Commercial Buildings Branch (1979 through 1981) was instrumental in addressing the social structure of the urban/suburban environments of U.S. cities; it sought to relate the urban design of places to issues of equity, economic development, and the cooperation needed to establish a sound (sustainable) urban ecology.

Early in the residential program it became clear that while we could predict the energy consumption of uninhabited test cells within plus or minus 5% or better, variations in occupant behavior in similar dwellings would account for as much as 50% difference in the actual consumption of energy (Seligman, Darley, and Becker 1978). The efforts in marketing both residential and commercial passive and hybrid solar had therefore to address the attitudes and values of the human consumer, not just the physics of energy transfer in the artifacts we call buildings (Kantrowitz 1983).

Still another social concern the program had to face was the social construction of the building industry and the professionals that made it work. A large part of the success of any significant commercial passive solar building was seen to be rooted in the intense collaboration required between building owners, occupants, architects, engineers, contractors, developers, financiers, public utility companies, and even building manufacturers. The fragmentation of the commercial building industry and a relatively weak tradition of close collaboration were seen as major impediments to developing a workable passive and hybrid solar energy strategy for commercial buildings.

Perhaps one of the most important social impacts of passive solar relates to the potential development of new industry. By 1980 program briefings were citing goals of the Passive and Hybrid Solar Division that included 100,000 new jobs by the year 2000, over 1 million construction industry workers involved in passive solar building construction, and over $10 billion in annual construction cost in the year 2000. The argument supporting such industry stimulation was that passive solar buildings required more and better materials and a higher level of craft than conventional construction. DOE projected that this would stimulate growth in existing material supply companies, create new products, facilitate an overall improvement in construction quality, and increase U.S. competitiveness abroad with new export products and skills (DOE 1980). The program sustained massive reductions in funding in 1982, making the goals unreachable. Even so, the Passive Solar Industries Council (PSIC), created in 1980, continues (with modest DOE support) to provide the building industry with practical information about passive solar, advancing its use in both commercial and residential construction (PSIC 1992).

6.1.5   Improving the Efficiency of the Existing Building Stock

An emerging objective of the program in the late 1970s established a strong focus on the existing building stock. Given that an estimated 80% of all the buildings that would exist in the year 2000 already existed, it was believed that retrofit or renovation strategies were an essential ingredient in reaching significant energy efficiency in buildings. The logic of this goal was further reinforced by Richard Stein (1977) as he developed the argument for what he called the "embodied energy of buildings." Stein's argument was that one had to consider the energy it took to make a building, as well as the energy to demolish and replace it, in order to understand its energy profile. Through this argument Stein introduced a different kind of building economics in the decision to build new. This new economics required one to include the full energy life cycle of the building in returns projected for new construction versus rehabilitation.

6.1.6   Facilitating the Education of Those Who Contribute to the Energy Consumption of Commercial Buildings

The education of all participants in the commercial building process was also seen as an important item on the program agenda. Architects and engineers had to reconsider the problems and opportunities of more climate-responsive design, including the possibility of what was then referred to as an energy aesthetic. Financiers had to reconceptualize economic return based on the potential future costs of energy. Public utilities had to examine plant capacities in light of peak demand periods, rate incentives, and the costs of increasing plant capacities. Building owners and managers had to consider the behavior of occupants as a key variable in their ability to run profitable and healthy building environments. Building regulators had to relate energy concerns to health, safety, and comfort standards in order to fully address the complexity of building systems and to increase the efficiency of such systems. The education efforts were parallel to the full scope of the research and systems development efforts.

In summary, even though buildings were seen as more difficult than transportation to impact, significant opportunities for savings and increased independence were forecast by DOE experts in both the residential and commercial sectors. The background to legislation supporting the Passive and Hybrid Solar Commercial Buildings Program included allusions to many of the concerns discussed above, but the explicit agenda was to help the building industry and building inhabitants become more energy-efficient.

6.2 Scope and Formal Structure of the Commercial Buildings Program

Like many of the programs in the Department of Energy of the late 1970s and early 1980s, the Passive and Hybrid Solar Commercial Buildings Program used the Department of Defense program budgeting process, which dates back to Robert McNamara's service as Secretary of Defense. The Commercial Buildings Program was funded in six categories, ranging from the most basic research, to prototype and engineering field testing, to demonstration and commercialization activities. The program was initially designed to support various systems, which presumably would move through the stages of development to successful commercialization at different and somewhat independent rates (DOE 1980).

The Commercial Buildings Program was formalized in FY 1980 by "(1) aggregating several projects (totaling $7.5 million) in the Passive and Hybrid Solar Division directed at the commercial buildings market sector; (2) preparing a five-year development plan; and (3) developing the administrative frameworks at DOE headquarters and in the field to support research, development and marketing" (Shibley 1980, 3-1). The organization to do the work of the program was segregated in much the same way as the program budget with different DOE field offices, regional solar energy centers (no longer part of the program), national laboratories

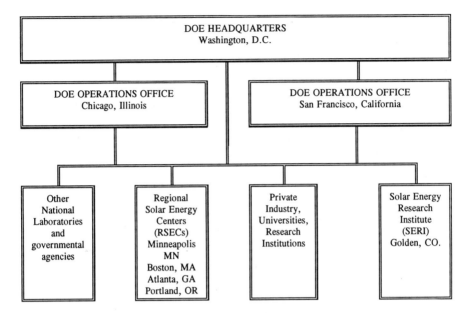

Figure 6.1
Organization of the passive and hybrid solar commercial buildings program (DOE, 1982).

and the Solar Energy Research Institute (now the National Renewable Energy Laboratory). For the most part the national laboratories were tasked with the basic research and initial engineering design of systems, with the Los Alamos and Lawrence Berkeley Laboratories taking much of the lead. Los Alamos tended to have greater responsibility for passive heating systems, while Lawrence Berkeley tended to focus more on passive cooling and daylighting. Assistance in the basic work of systems development was provided by the Brookhaven and Argonne Laboratories, and other laboratories and university research centers (through contracts from the DOE field offices).

Engineering field tests and demonstration programs were managed largely through the Chicago Field Office of the Department of Energy for the Commercial Buildings Program with a large effort in cooling supported by the San Francisco Field Office. The DOE field offices worked through the issuance of Requests for Proposals (RFPs) or through Program Opportunity Notices (PONs) keyed to specific program and project objectives. Responses to the RFPs came from university research centers, private contractors, independent solar energy centers, and consultants.

The responsibility for what came to be known as "commercialization activities" was largely delegated to the four regional solar energy centers (see chapter 20 for additional background on the RSECs). The Solar Energy Research Institute (SERI) played a central role in helping to coordinate and monitor the activities of the participants in the program, documenting work in support of a technology transfer function as well as participating in systems development in several different stages, depending on the talent available to the institute in a given fiscal year. In reality, however, all of the various organizational elements contributed to each stage of program development.

As one would expect, several aspects of the program tended to be driven more by skillful advocacy and talent than by the rationalized program structure. Skilled researchers were not asked to simply turn over their work to the field offices for "further development testing and demonstration," nor did the program managers and contractors lose interest in the work when it came time to commercialize systems. In fact, an informal and often-sanctioned crossover of responsibilities served the interests of the Commercial Buildings Program very well. Some of the most interesting basic research objectives and most provocative research happened in what were billed as "demonstration programs." Some of the best outreach and commercialization happened through the technology transfer programs at the national labs and SERI. The regional centers gave us some of the best field data for basic research by revealing the problems in implementation and commercialization.

Figure 6.2 outlines the topics initially explored by the Passive and Hybrid Solar Commercial Buildings Program in its systems categories. What occurred over time was the realization that, while the categories seemed to fit the requirements of program management and budgeting, they were dysfunctional when related to the complex dynamics involved with the production of buildings. In many ways, building programming, design, construction, financing, regulation, and management require an in-depth understanding and integration of interdependent systems. One building component often must serve several functions, only some of which relate specifically to the production or conservation of energy. Systems or approaches shown to work well in engineering field tests were often not practical in actual use for reasons that had no relation to energy consumption. In many ways, the emerging need to study the whole building and to understand individual systems as part of an

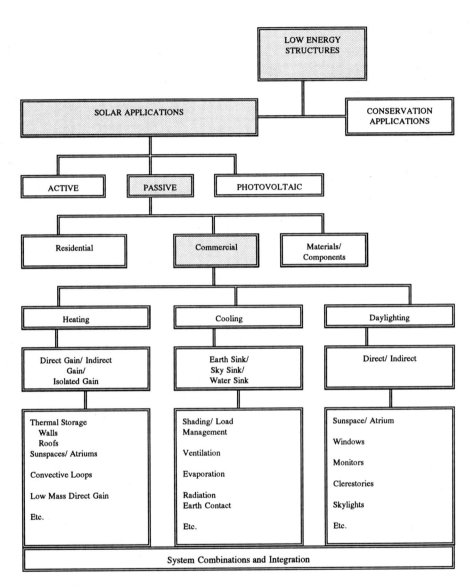

Figure 6.2
Initial systems classification for the Commercial Buildings Program (Shibley, 1982).

interdependent network of systems, components, and human behaviors contributed greatly to the emerging literature and interest in what is called "whole buildings research." More recent efforts by the National Institute of Building Sciences (BTECC 1988) and efforts by the American Society of Testing Materials, both seeking to establish a performance evaluation standard for whole buildings, are partially attributable to those who struggled with the development of systems in the Commercial Buildings Program.

Passive solar cooling was specifically problematic in the development of the Commercial Buildings Program. Solar heat technologies were well established, and cooling was seen as a kind of stepchild, not yet mature enough to hold its own against the more fully established area of inquiry in the competition for scarce resources. By initially focusing a category of inquiry on cooling, the program gave the topic a chance to mature so that it could become part of the emerging discussion on whole buildings research. In effect, DOE established an incubator for a new area of inquiry so it could take its rightful place in the building research establishment.

The subject of whole buildings research that emerged during the Commercial Buildings Program reveals some limitations in the fit between the dominant modes of inquiry employed in the 1970s and those employed in the 1980s. For the purposes of this interpretation of the Commercial Buildings Program, it is helpful to make the oversimplified distinction between science and design as primary modes of inquiry. Both ways of thinking and knowing have contributed greatly to the evolution of passive and hybrid solar architecture. The Commercial Buildings Program afforded some powerful opportunities to relate the insights of basic science to the practice of design and to bring to bear the full insights of design on basic science. In essence, the Commercial Buildings Program has afforded several chances to rethink the processes and artifacts of design and design research—the question of fit between problem and method.

6.3 Methods of Inquiry

The question of fit between problems to be addressed by the program and methods of inquiry has as its backdrop the political, economic, social, and ecological objectives outlined in the first section of the chapter. As such, there is an almost paralyzing complexity. No single methodology can address the full range of issues. An overreliance on laboratory science versus field research, for example, cannot establish the climate for acceptance in the field by commercial interests in passive solar. Conversely, the marketplace is often as interested in demonstration as it is in scientific proof, whether in the laboratory or the field. Finally, design investigation is especially well suited to explore relationships leading to such demonstrations.

Single-variable empirical testing has been the cornerstone of inquiry in the area of energy consumption in buildings. Heat transfer has been explored in the Passive and Hybrid Solar Commercial Buildings Program through the laboratory tests, test cells, and test building experiments. As a result, single-zone heat transfer through convection, conduction, and radiation is understood well enough for tentative transfer functions to be derived which are applicable to most heating and several cooling strategies. The bulk of two-variable testing in energy research has been related to single-zone heat transfer, most specifically useful in understanding what occurs in residential construction. There is a good match between empirical science and the problems of heat transfer.

More complex variable interaction work is done through rational extrapolation from the single- and dual-variable testing in both laboratory and field tests. Performance prediction based on statistical probabilities and the assumption of variable dependence is the foundation of computer codes like the "Building Loads and System Thermodynamics" program, developed by the U.S. Army, or "DOE #2," developed by the Lawrence Berkeley Laboratory. The codes extrapolate from relatively simple experiments, attempting to offer rational models of the actual complexity of commercial buildings; they focus on multizone configurations, mechanical system variations, and differing occupancy conditions, in combination with different conservation and solar strategies. The strength of the codes is their ability to address many complex variables. The weakness is that code validation through empirical testing is extremely difficult, precisely because the codes deal with so many complex conditions. Even with greatly improved laboratory validation tests, the actual functioning of a building is subject to far more interdependent variables than the most sophisticated codes are able to address.

The simulations of building performance based on complex computer codes are very useful in informing the architect of trade-offs during design. Although the rational method of inquiry used to develop the codes appeared to deal with relationships, in reality the programs were developed through a series of parametric studies that modified one variable at a time while holding all others constant, using, for example, the solar load correlations method developed by the Los Alamos National Laboratory. Because of the power of computing, this was done many times, yielding trends or patterns that form the basis for assumptions about multivariable interactions. Unfortunately, the social production and habitation of buildings are not processes that resemble holding several variables constant while one is optimized to fit. The trends are useful as hypotheses to test in design but they do not simulate the reality of making and dwelling in buildings.

Computer codes, correlation methods, and assumptions all helped enhance the predictability of performance, but it seems doubtful that they would ever be able to address the complex relationships between all the variables involved in energy use in buildings. Metaphorically, one can understand that modeling the process of building design and construction would be like modeling the results of several pool players each striking balls on the same table, sometimes at the same time, sometimes in sequence, with widely divergent ranges of talent, and with some of the players randomly removing balls from the table and refusing to play.

In the metaphor the pool players are not capricious. Their behaviors are motivated by both internal and external variables that all have an interdependent effect on the movement of balls on the table. This is the essence of the problem of organized complexity present in architectural design; each variable interacts with all of the others as it also influences the specific problem under study. The fact remains that although the complexity of energy use in buildings is sometimes overwhelming, it is trivial when compared with the full range of issues that must be addressed in building design. The largest single issue for architecture/energy inquiry in the Commercial Buildings Program was how to place the description of energy consumption in buildings in the organized and complex set of relationships that idiosyncratically come together on every construction project and during any building's life cycle.

The Commercial Buildings Program addressed empirical test and prediction models specifically related to energy use through science. The categories of investigation (see figure 6.2) were derived from the first principles of physics and from the residential experience that attempted to apply them to various building components and materials. It was precisely in such application that the limitations of science-based methods were constraining. The end goal in the Commercial Buildings Program came to be less about application than about the integration of energy considerations with the economic, functional/behavioral, symbolic, social, ecological, and engineering frames of reference which come to bear on the art and science of building. Integration found design to be a powerful method of inquiry and architecture to be the good host discipline. (See also Schneekloth and Shibley 1987 for further discussion on the relationship between research and practice in architecture.)

6.3.1 Design as a Fundamental Method of Inquiry

This concern for integration in the context of real building experience became part of the core of the program in passive commercial buildings. There was an opportunity to use the design experience to acquire an appreciation of the unique confluence of concerns that occur whenever people build. It did not result in the ability to generalize static facts but rather provided a kind of "procedural knowledge" (Belenky et al. 1986; Schneekloth and Shibley 1986) as well as insights that informed the scientists in the program which problems can be disaggregated and better understood through simple or disorganized complex problem solving. The result was the acquisition of an understanding of energy in context, in the same way the Harvard Business School case studies improve a manager's appreciation of the unique confluence of issues involved in the complexities of business development.

Geoffrey Vickers, in *The Art of Judgment* (1965), develops the case for what he calls "appreciation," comparable to the understandings developed during the program. These include judgments made about objective reality, judgments about the value attributed to such reality, and judgments about what action should occur. In forming judgments about what is real, the empirical and rational methods of the physical sciences served well in the Commercial Buildings Program. Forming judgments about how such reality is valued and what action should be taken required both social science and design.

Of the many projects in the total program, two (which each spanned five years) will serve to illustrate this effort to augment basic building

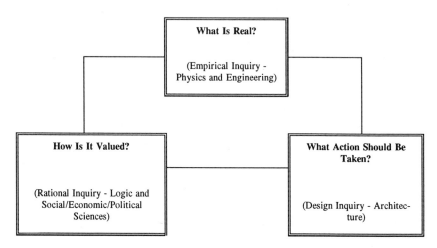

Figure 6.3
Appreciative system that addresses all modes of inquiry employed in the Commercial Buildings Program (Vickers, 1965; Shibley, 1981).

physics and logic by design. One is the Passive Solar Commercial Buildings Program Opportunity Notice, which came to be known as the Commercial Building Design Assistance Program, and the second was the Passive Solar Curriculum Project. One involved the design of commercial facilities and the other involved the design of curriculum modules for collegiate schools of architecture.

6.4   The Passive Solar Commercial Buildings Program Opportunity Notice

The Passive Solar Commercial Buildings Program Opportunity Notice (PON) was issued by the Chicago Operations Office of DOE in 1979 as part of what was then described as a demonstration program on the application of passive solar principles to light commercial buildings. It is one of several examples of how the program used design as a fundamental method of inquiry. As part of the PON program, resources were focused on design assistance to participants who planned to build light commercial structures. There were also allocations provided to offset the capital costs of building in accordance with sound climate-responsive principles employing passive and hybrid systems. In a second program similar to the commercial PON, building manufacturers were solicited and a search begun for premanufactured components that could be integrated with the preengineered and prefabricated building industry. Both the Commercial and the Manufactured Buildings Program Opportunity Notices emphasized real design and construction experience through three phases of development: design, construction, and performance evaluation. Essentially these projects defined areas for further development by the professions and/or the construction industry as well as a research agenda.

In the commercial PON forty design teams with varying levels of passive solar experience were initially selected to participate in the three-phase program, with twenty-three projects surviving after design development. The experimental building prototypes included a wide variety of building categories, ranging in size from 600 to about 64,000 square feet. In order to capture information from the process of design, an interactive review process was established, which allowed design teams an opportunity to consult with nationally recognized technical experts. Each team participated in a series of reviews that monitored project progress and provided design assistance:

• *Group schematic reviews* became forums where designers informally exchanged ideas among themselves and technical experts who specialized in particular aspects of passive solar technology. Morning sessions consisted of project presentations to the panel of technical experts and other design teams. Afternoon sessions focused on architectural integration, daylighting, analytical techniques, and heating. Each team met with the technical experts on an individual basis.

• *Interim reviews* provided additional design assistance at the request of the design teams. These meetings were coordinated by technical monitors who provided technical and program assistance for each project on an ongoing basis.

• *Final design reviews* assessed project and process documentation.

All reviews, particularly the group schematic reviews, established an educational rapport that alleviated initial hesitancy on the part of the design professionals. Several conditions contributed to making the review climate a positive and constructive one:

1. The review teams did not interfere between the client and designer; they worked with the design team in response to the client criteria.

2. Both the reviewers and design teams were intent on learning; as a result, the review was not so much about finding error as it was collaboratively defining what constituted error and why.

3. The reviews were conducted in a manner consistent with the tenets of peer review rather than a superior/subordinate exchange.

4. The physical settings for the reviews were conducive to the review tasks.

5. Technical monitors followed the projects throughout the design process. They served as project liaisons between DOE and each project to assure that technical and documentation assistance was available for the evolving designs.

6. Finally, comments regarding passive solar integration were offered at a time when the project was still in schematic design. Thus input from recognized authorities could be considered before too much time and expense had been invested in project planning.

This series of reviews did much to shape the designs that were finally constructed. In the absence of passive commercial building intuitive knowledge, a framework was established for design professionals to learn from one another. The process used in the commercial PON is a model for incremental improvement in the design professions that can be applied with or without federal subsidy. (See Shibley 1983 for further description of architectural inquiry about energy.)

The method of work in the PON is perhaps best described as "discuss at length and take good notes." In the discourse between researchers, passive solar experts, building owner/developers, senior architectural critics, and project designers, rich insights were added to the appreciation of the integration of energy concerns in design. The products of the PON, including volumes of design review and performance appraisal technical reports, are perhaps best summarized in *Commercial Building Design* (Burt Hill Kosar Rittleman Associates and Min Kantrowitz Associates 1987), which in partial validation of the PON's methods of inquiry, received a *Progressive Architecture* award for research in 1988. As less tangible and perhaps even more valuable product of the effort, senior research observers and consultants left the reviews with a clearly more sophisticated insight into the nature of the social production and habitation of buildings and of the problems faced in designing and developing passive and hybrid solar commercial buildings.

6.5   Passive Solar Curriculum Project

The University of Pennsylvania was awarded a contract to develop curriculum modules for collegiate schools of architecture through the Commercial Buildings Program in 1979. The intent was to infuse the insights of passive solar into the academic foundations of the profession. The University of Pennsylvania issued a Request for Proposals under the contract to all the collegiate schools of architecture, selecting eleven to develop and test individual curriculum modules. In the course of the development and testing process the work was reviewed by all participants; the review sessions were analogous to the process used in the Passive Solar Commercial Buildings Program Opportunity Notice. Upon completion of the University of Pennsylvania contract, eleven curriculum books were published and distributed to all the collegiate schools of architecture. Figures 6.4 and 6.5, based on the journal published by the project, illustrate the types of exercises developed in each curricular module and their relationship to the design process.

A follow-up contract was then issued to the Association of Collegiate Schools of Architecture (ACSA) to further test the eleven modules by schools other than those that developed them. Nine schools were selected to run the tests based on their responses to the ACSA Request for Proposals. Again there was a process of group review and discussion over the period of a year and a half. A journal assessing the experience of the schools with the curriculum modules, *Architecture, Energy, and Education* (Shibley 1984a), was then issued and distributed to all the collegiate schools of architecture. The journal identified major impediments to the teaching of passive solar architecture directly related to the organized complexity of both architecture and pedagogy and recommended strategies to moderate such impediments.

It is useful to note that the curriculum project was awarded a *Progressive Architecture* "citation for research" in 1985 by a completely different jury than the one citing the Passive Solar Commercial Buildings Program Opportunity Notice. Since the initial distribution of the modules, over forty collegiate schools of architecture have directly employed the

	Studio		Seminar		Lecture		Lab	
	Intro.	Adv.	Intro.	Adv.	Intro.	Adv.	Intro.	Adv.
Carnegie-Mellon University	●		●		○			
Georgia Institute of Technology			○	○			●	●
Kent State University	●	○						
Massachusetts Institute of Technology	▢		■		○			
North Carolina State University			●		○			
New Jersey Institute of Technology	▢				▢			
University of Oregon	○				●	○		
Rice University			○	○	○	○		
Rensselaer Polytechnic Institute		○	○	○	○	○		
University of California at Los Angeles	○		●				○	○
Yale University			○	●			○	○

●   Particularly useful in this setting
○   Parts or aspects can be useful in this setting
■   Particularly useful in this setting and aimed primarily at faculty
▢   Parts or aspects can be useful in this setting and primarily aimed at faculty

Figure 6.4
Passive solar curriculum module academic context analysis (Fraker and Prowler, 1981).

DESIGN PROCESS

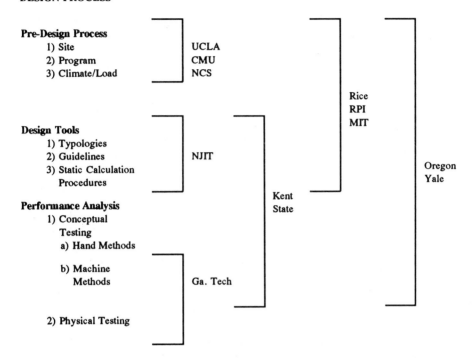

Pre-Design Process
   1) Site
   2) Program
   3) Climate/Load

UCLA
CMU
NCS

Rice
RPI
MIT

Design Tools
   1) Typologies
   2) Guidelines
   3) Static Calculation
      Procedures

NJIT

Oregon
Yale

Performance Analysis
   1) Conceptual
      Testing
      a) Hand Methods

Kent
State

      b) Machine
         Methods

Ga. Tech

   2) Physical Testing

Figure 6.5
Design process topics engaged by universities selected to develop passive and hybrid solar curriculum modules.

Table 6.2
Major barriers to the teaching of passive solar in collegiate schools of architecture

Methodological barriers	deal with the differing processes most designers feel they are employing when they "design" as opposed to when they are addressing energy concerns.
Structural barriers	deal with the academic division between studio and technical courses and the seemingly omnipresent feeling that there is not enough time to "get technical" in the design studio.
Attitudinal barriers	deal with both student and faculty beliefs that energy concerns are unimportant, too complex or too difficult to address, and too limiting to the designer.
Informational barriers	deal with the lack of knowledge or appropriate access to knowledge about what constitutes energy efficiency in buildings.

Source: Shibley 1984a, 1984b.

texts in course work; in addition, the curriculum modules are catalogued in every collegiate school of architecture in North America. The authors of the curricular modules used the publishing and presentation opportunities to help position themselves in tenured rank positions or positions of leadership in collegiate schools of architecture where their influence on the topics related to passive solar design continue to be felt. Since 1982, three of the project participants went on to be department chairs and two went on to become deans in collegiate schools of architecture.

Reviews of these two specific programs within the Commercial Buildings Program reveal the nature of insights derived through design and design instruction in addition to more rational or empirical methods. For example, the empirical and rational insights derived through science were augmented by:

• seeking to redefine what constitutes error in addition to reducing the incidence of predefined error.

• giving as much credence to the assumption of irrationality as to the assumption of rationality.

• utilizing inductive reasoning to complement deduction.

• utilizing the client/designer team values, motivations, and philosophies as the major component of inquiry, while, at times, attempting to objectively separate inquirer from inquiry.

• assuming that reliability depends more on the participation of key persons than on the application of objective measures. Low reliability can therefore be expected when key people are ambivalent or internally inconsistent.

• attributing as much importance to the uniqueness of inquiry results as to their generalizability. (Shibley 1984a, 1984b)

6.6   Results

A report issued by DOE in February 1982, entitled Passive Solar Commercial Buildings Program Summary: *1980–82*, (Gialanella and Rockwell International 1982) addressed the program results in terms of, among other things, industry involvement, research impacts, and professional practice. The following sections cite abstracts of the conclusions of that report and offer commentary on current conditions related to the 1982 assessment of progress.

6.6.1   Industry Involvement

1982

Although much work remains to be done, indications are that many segments of the industry are becoming motivated to incorporate passive solar design concepts and continue the work begun by the government.

Preliminary indications are from members of the Passive Solar Industries Council (PSIC) who represent just part of the diversified building industry. Among the members are the American Institute of Architects (AIA), American Wood Council, Architecture Aluminum Manufacturers Association, Brick Institute of America (BIA), International Masonry Institute (IMI), National Association of Home Builders (NAHB), National Concrete Masonry Association (NCMA), Passive Solar Products Association, Tile Council of America, United Brotherhood of Carpenters and Joiners, and the Northeast Retail Lumbermen's Association (NRLA).

• AIA is sponsoring a five-year energy program.

• IMI is emphasizing passive solar efforts in research and design, education, market analysis, communications, and legislation.

• BIA has an active program which includes demonstration projects in several states, research on thermal performance of passive buildings at Princeton University, and publications for consumers, tradesmen, and professionals including a design manual for architects and engineers.

• NCMA has made passive solar a top program priority, primarily through publication, seminars, and research on thermal mass.

• PSIC publishes a monthly newsletter, *Passive on the Move*, distributed to approximately 4,000 industry members, which contains current technology and R&D activities. (Gialanella and Rockwell International 1982, 14)

1992

Sharp funding reductions in 1982 led to significant drops in industry involvement. Although the materials developed in the late 1970s and early 1980s are still available for use, the loss of commercialization support to the regional solar energy centers (RSECs) significantly hindered publication and distribution activities. The Passive Solar Industries Council (PSIC) and its members continue to publish materials for use by member organizations (PSIC 1992), but the vast majority of their guidelines are concerned with residential construction rather than commercial buildings, and are based on work done in passive residential in the late 1970s through the late 1980s largely supported by federal funds.

Commercialization activities in the nonresidential sector had further to go and less time to get there. A renewed government emphasis on developing materials and on involving the nonresidential building industry would take advantage of the willingness of the PSIC member organizations to distribute such materials. The industry was clearly positioned to move but was dependent on government support for the development of the primary materials for training and (in many cases) for the development of technology and products. (See chapter 5 in this volume.)

6.6.2   Research Impacts

1982

The research in cooling and daylighting design concepts for commercial buildings is being performed at university test facilities. These developments will be reported to designers at state-organized workshops as they become available.

• Major indications are that daylighting is more critical to the design of larger, more common use buildings such as high-rise office buildings. The research to date emphasizes that greater energy efficiency is achievable in these buildings.

• It appears that significant savings will be realized from the 23 Commercial Buildings Design Assistance Projects [Passive Solar Commercial Buildings Program Opportunity Notice]. The savings resulting from the passive solar components may be as high as 86% over conventional commercial buildings of the same size. In all 23 projects, energy use is projected to be substantially lower than in conventional buildings.

• Of total annual sales nationally for all metal buildings of $1.1 billion, contractors involved in passive solar design projects account for $328 million in sales, which represents 29% of total sales. The metal buildings industry is also benefiting from the information network established by Butler Manufacturing Company for transferring research developments to the builders. Over 2,000 people, including designers, builders and potential clients have toured Butler's research facility during the testing of their passive concepts funded (in 1980 and 1981) by the Commercial Buildings Program. Butler's prominence in the industry, associated with 750 builders nationally, cannot be underestimated. (Gialanella and Rockwell International 1982, 15)

1992

In 1982 the description of results projected potential impact based on how the industry was positioned and on the continuation of the federal initiatives in basic research, field testing, demonstration, and commercialization activities. Continued research on daylighting, for example, is needed for the full savings potential to be realized.

The Commercial Program completed an analysis of the twenty-one buildings that remained in the program after 1982 (down two from 1982 and down nineteen from the original forty projects). Energy cost savings analysis from the project reveals that an average improvement of 51% was achieved for a sample of twelve of the projects. A 46% savings was reported on a larger sample by Burt Hill Kosar Rittleman Associates and Min Kantrowitz Associates (1987).

What came to be known as the Industrialized (or Manufactured) Buildings Program led to eight manufacturers completing participation out of the original twenty-six. DOE currently reports that there has been little replication of the designs or broad product distribution by participating manufacturers. (See chapter 5 in this volume.)

6.6.3   Professional Practice

1982

It is impossible to quantify all the impacts that have occurred throughout the past three years as a result of conference attendance, information dissemination, and subsequent awareness and utilization of passive solar design tools. The following indications and responses to questionnaires done by the industry lead us to believe that the extent of impact, especially on professional practice, is far-reaching. For example:

• 80% of the attenders at the Passive Studio, developed by the AIA Research Corporation, indicated that they used passive solar design in commissions following their attendance.

• The series of project reviews for the DOE Commercial Buildings Design Assistance effort [Passive Solar Commercial Buldings Program Opportunity Notice] established a framework for design professionals to learn from one another. Project participants assert that the process used is a model for incremental improvement in the design professions which is applicable with or without federal subsidy.

• The refinement of the solar design tools developed with DOE support has enabled designers to make reliable predictions regarding building energy performance. Architects, engineers, consultants, and owners of the 23 project teams who participated in the DOE Passive Solar Commercial Buildings Assistance Program indicate a positive impact on their businesses and an obvious intent to continue and expand their work in this area. (Gialanella and Rockwell International 1982, 15)

1992

While the profession at large is largely market-driven and the market for energy efficiency is weaker than it was in the 1970s and early 1980s because of relatively low energy costs, the American Institute of Architects (AIA) continues to have an interest in passive solar design strategies as expressed through its activities with the new American Institute of

Architects/Association of Collegiate Schools of Architecture Research Council (Shibley 1987). The environmental theme of the 1993 annual AIA convention includes a strong awareness of energy issues as part of the larger scope of ecological design.

The Passive Solar Industries Council (PSIC) continues to offer educational programs on passive solar energy, but the demand is significantly less than it was in 1982. Part of this can be attributed to the relative saturation of the profession over the 1979–1982 time period, but it is mostly a direct reflection of the drop in client demand for passive commercial building design.

6.7 Conclusions

The premature interruption of the Passive and Hybrid Solar Commercial Buildings Program in 1982 has significantly set back its attempts to position professional societies, industry groups, and educational institutions to best address the structural issues inherent in building energy consumption. This interruption has also substantially reduced the basic research and engineering development necessary for well-positioned professional, industrial, and educational associations to successfully influence the commercialization of passive and hybrid solar approaches to building design. There is no reduction in the need for such influence, however, because gross energy consumption by buildings is increasing. For example, the percentage of total energy consumption by the building sector of the economy was 1.35% more in 1986 than in 1985, and about 11% more in 1986 than in 1978 (BTECC 1988). Even though much of this increase is attributable to the predicted faster rate of improvement in the transportation and industry sectors of the economy than in the building sector, the statistic still indicates significant room for improvement in the building sector.

Another measure of overall energy use effects on the economy presents a more positive picture. In 1988 the Building Thermal Envelope Coordinating Council (BTECC) reported: "In 1973 it took 27,100 Btu to generate each 1972 constant dollar of GNP. In 1986 it was down to 20,000 Btu, a reduction of 26%. The 1986 figure also represents about a 1.5% reduction from 1985" (BTECC 1988, 7).

The same BTECC report, however, summarizes several forecasts of our national oil demand through 1995 by asserting that the demand will increase at a rate of .05% to 1% per year, resulting ultimately in U.S. imports exceeding the 1977 peak of 8.6 million barrels per day. As a consequence of this the United States continues to face threats to its economic well-being and ability to compete in the world market. Continued U.S. dependence on imported energy sources could mean poorly heated and cooled buildings, families in economic crisis because of high energy costs, national economic instability related to a continuing unfavorable balance of trade, obsolescence of large inventories of building stock due to high energy costs, and continued construction of energy-inefficient buildings because of inadequate technical guidance, testing procedures, standards, and regulations (BTECC 1988, 9).

The structure and intentions of the Passive and Hybrid Solar Commercial Buildings Program were well suited to address the range of concerns expressed by the National Institute of Building Sciences through its Building Thermal Envelope Coordinating Council. These same kinds of threats are also articulated in several other publications, including Vice President Al Gore's *Earth in the Balance* (1992). Gore proposes governmental leadership through a Strategic Environment Initiative (SEI) to address these and other threats not just to the United States but to the entire planet. His SEI includes calls for action comparable to those of the solar heat technologies programs, and of the Passive and Hybrid Solar Commercial Buildings Program in particular. For example, Gore calls for government purchasing programs to develop early marketable versions of new technology products, for the establishment of rigorous and sophisticated technological assessment procedures, and for the founding of a network of training centers.

The Department of Energy continues its active participation in the International Energy Agency, but resource constraints have precluded real leadership in this area. We must become more competitive in the marketplace of passive and hybrid solar product and building design approaches development. Concurrently, we must be cooperative with international efforts to reduce worldwide energy demands.

Nicholas Lenssen, writing for the United Nations–sponsored Worldwatch Paper series, sums up the need for "northern" leadership in the development of a "new energy equation" applicable worldwide by suggesting: "It is critical that support from the North include not only the billions of dollars annually provided as foreign assistance but also the power of example" (Lenssen 1992, 37). The Passive and Hybrid Solar

Commercial Buildings Program, though prematurely concluded, was one part of a still very much needed international set of model programs that can assist both national and foreign aid policy on support for passive and hybrid solar.

Acknowledgments

This analysis was partially sponsored by a grant to IDEA, Inc., Syracuse, New York, from the National Endowment for the Arts, Design Arts Program on the topic of "Design as a Method of Inquiry," with Robert G. Shibley, AIA, AICP as principal investigator. Additional support was provided through a contract with Rockwell International from the U.S. Department of Energy, through the ETEC Solar Documentation Project, contract no. 85-356373.

References

Belenky, M. F., B. M. Clinchy, N. R. Goldberger, and J. M. Tarule. 1986. *Women's Ways of Knowing: The Development of Self, Voice, and Mind.* New York: Basic Books.

BTECC (Building Thermal Envelope Coordinating Council). 1988. *National Program Plan: Building Thermal Envelope.* ed. E. Stamper. Washington, DC: National Institute of Building Sciences.

Burt Hill Kosar Rittelmann Associates and Min Kantrowitz Associates. 1987. *Commercial Buildings: Integrating Climate, Comfort, and Cost.* New York: Van Nostrand Reinhold.

DOE (U.S. Department of Energy). 1978. *Fundamentals of Solar Heating.* Washington, DC.

DOE. Passive and Hybrid Solar Division. 1980. *DOE Passive and Hybrid Solar Program Briefing.* Washington DC, December.

DOE and Booz Allen & Hamilton, Inc. 1982. *Design and Performance Trends in Energy-Efficient Commercial Buildings.* Washington, DC: U.S. Department of Energy.

ERDA (Energy Research and Development Administration). 1975. *A National Plan for Energy Research, Development and Demonstration: Creating Energy Choices for the Future.* Vol. 1, *The Plan.* ERDA-48. Washington, DC.

ERDA. Division of Solar Energy, 1976. *National Program for Solar Heating and Cooling of Buildings.* Washington, DC.

ERDA. Division of Solar Energy. 1977. *Solar Energy in America's Future: A Preliminary Assessment.* DSE-115/1. Washington, DC.

FEA (Federal Energy Administration). 1974. *Project Independence: Final Report of the Solar Energy Panel.* Washington, DC.

Fraker, H., and D. Prowler, 1981. *Project Journal: Teaching Passive Design in Architecture.* Philadelphia: University of Pennsylvania.

Gialanella, L., and the Rockwell International Energy Technology Engineering Center. 1982. *Passive Solar Commercial Buildings Program Summary: 1980–82.* Governmental reports abstract NTIS-PR-360. Washington, DC: U.S. Department of Energy.

Gore, A. 1992. *Earth in the Balance: Ecology and the Human Spirit.* Boston: Houghton Mifflin.

Jacobs, J. 1961. *The Life and Death of Great American Cities.* New York: Random House.

Kantrowitz, M. 1983. "Occupant Effects and Interactions in Passive Solar Commercial Buildings: Preliminary Findings from the U.S. DOE Passive Solar Commercial Buildings Program." *8th National Passive Solar Conference*, ed. J. Hayes and D, Andrejko. New York: American Solar Energy Society.

Lenssen, N. 1992. "Empowering Development: The New Energy Equation." In *Worldwatch Paper 111.* Washington DC: Worldwatch Institute.

Maybaum, M. W. 1980. "Overview of the Passive and Hybrid Solar Energy Program." *Proceedings of the Annual DOE Passive and Hybrid Solar Energy Program Update Meeting.* Washington DC: U.S. Department of Energy.

PSIC (Passive Solar Industries Council). 1992. *The Passive Solar Industries Council.* Promotional brochure. Washington, DC.

Schneekloth, L., and R. Shibley. 1986. "Toward Process Knowledge: The Case for Learning by Doing with the Basic Assumption of Human Competence." In R. Westrum, ed., *Organizations, Designs, and the Future.* Ypsilanti: Eastern Michigan University and the National Science Foundation.

Schneekloth, L., and R. Shibley. 1987. "Research/Practice: Thoughts on an Interactive Paradigm." In R. Shibley, ed., *Research and Architecture: Scope, Methods, and Institutional Traditions.* Washington DC: American Institute of Architects/Association of Collegiate Schools of Architecture Research Council.

Seligman, C., J. Darley, and L. Becker. 1978. "Behavioral Approaches to Residential Energy Conservation." In R. Socolow, ed., *Saving Energy in the Home.* New York: Ballinger.

Shibley, R. 1980. "Toward Passive and Hybrid Solar Commercial Buildings." *Proceedings of the Annual DOE Passive and Hybrid Solar Energy Program Update Meeting*, Washington DC: U.S. Department of Energy.

Shibley, R. 1981. "Passive and Hybrid Solar Commercial Buildings." *Passive and Hybrid Solar Update.* Washington, DC: U.S. Department of Energy.

Shibley, R. 1983. "Quads and Quality: Architectural Inquiry about Energy." In J. Snyder, ed., *Architectural Research.* New York: Van Nostrand Reinhold.

Shibley, R. 1984a. *Architecture, Energy, and Education.* Washington DC: Association of Collegiate Schools of Architecture.

Shibley, R. 1984b. "Barriers to Successful Integration of Energy in Architecture." *Proceedings of the Ninth National Passive Solar Conference*, ed. J. Hayes and A. Wilson. Boulder, CO: American Solar Energy Society.

Shibley, R., ed. 1987. *Research and Architecture: Scope, Methods, and Institutional Traditions.* Washington, DC: American Institute of Architects/Association of Collegiate Schools of Architecture.

Stein, R. G. 1977. Architecture and Energy. Garden City, NY: Anchor Press/Doubleday.

Vickers, G. 1965. *The Art of Judgment.* New York: Basic Books.

7 Industrial Process Heat

David W. Kearney

7.1 Solar Industrial Process Heat Applications and Program

Solar thermal energy has significant potential as a valuable energy resource for process heat in industry. The industrial sector is a large energy user—approximately 27 quads or 37% of total U.S. demand; an early focus of the federal program to implement solar thermal technology was therefore the substitution of solar thermal energy for fossil fuel–derived thermal energy in the industrial sector. Historically, this effort included process heat uses in the agricultural sector as well. These two sectors appeared to be particularly well suited to solar technology implementation: not only is their thermal energy use high, but the wide temperature range of that energy use matches the capability of solar thermal systems (DOE 1978a, 1978b; SERI 1980; Kearney and Lewandowski 1983).

A large portion of process heat energy is required for diverse but relatively straightforward lower-temperature applications such as process water heating, hot air drying, and industrial plant use of low to intermediate-pressure steam. Industrial applications offer the possibility of more effective and economic uses of solar energy because annual loads are often constant and not seasonal, capable maintenance personnel are likely to be available, and large collector arrays offer the potential for economies of scale. On the other hand, the ready acceptance of solar energy by industrial users must overcome several obvious barriers, such as land availability, diurnal matching of the resource to the demand, and economics.

Industrial requirements for the commercialization of solar technology demand the satisfaction of relatively difficult and sophisticated economic and technical criteria, compared to residential applications of solar energy. Low system costs are needed for delivered thermal energy to be competitive with fossil fuels and to meet specified corporate criteria, such as return on investment or payback period (typically 2–3 years or less). Coincidentally, industrial users and decision makers also typically demand equipment with a proven record of reliability, availability, performance, and suitability for the end-use application. In the early stages of introducing a new technology such as solar industrial process heat (IPH), the decision to adopt a system in a particular application is complicated, more often than not, by the need for a series of company approvals in a chain from individual plant through central corporate managements.

7.1.1 Federal Program Structure

The federal solar program, recognizing that development and stimulation were required in several areas to further its overall objectives, formed an initiative to encourage both solar technology developers and potential industrial users to apply solar energy to the agricultural and industrial sectors. The implementation program depended on complementary efforts to improve and develop the solar technology needed for agricultural and industrial applications through separate federal technology development activities and the burgeoning solar industry. Consequently, the Agricultural and Industrial Process Heat (AIPH) Program focused on demonstration (or field test) projects to promote system development, to provide an early market, and to demonstrate successful operation and reliability.

The early concept and steps to promote solar implementation in AIPH had its beginnings in 1975 in ERDA (the Energy Research and Development Administration, established in 1974) with a series of demonstration projects led by W. Cherry. The projects in the agricultural sector were carried out in cooperation with the Department of Agriculture. Funding levels were typically from $25,000–75,000 per project. The agricultural applications included food processing (9 projects), grain drying (11), crop drying (10), heating of livestock shelters (11), and the heating/cooling of greenhouses (16). (See chapter 12 regarding agricultural demonstrations.) The first industrial projects were also funded in the 1975–1976 period and were chosen to demonstrate the use of solar energy for process hot water (3 projects), process hot air (4), and process steam (4). The IPH projects were typically on the order of $750,000 for design and construction, followed by $150–200,000 per year for operation. Key project managers at Department of Energy (the successor of ERDA) from this period on were J. Greyerbiehl and W. Auer; in late 1977, as the AIPH program gained in size and importance, J. Dollard assumed leadership to manage its development and implementation. Outside of the federal DOE management, the solar IPH program was largely implemented by Sandia National Laboratories, the Solar Energy Research Institute (SERI), the Energy Technology Engineering Center (ETEC) of Rockwell International, and the San Francisco DOE Field Office. There were a large number of contributors from these organizations involved in all aspects of the program. The formal IPH program at DOE ended about 1985, although some effort in this area was renewed in 1991.

The subsequent discussion in this chapter is restricted to the industrial portion of the DOE AIPH program. The federal funding level for solar industrial process heat increased rapidly as the size and number of projects grew. Figure 7.1 shows the approximate federal budget for the projects (not including separate technology development and commercialization activities) from 1975 through 1985.

The federal program broadened somewhat beyond the design and field-testing of specific solar IPH systems and sought to encourage the early commercialization of solar energy in the IPH sector through endeavors including market assessment, evaluation of recommendation of federal and state incentives, and technology development with an IPH focus. Market development activities, centered at SERI for the IPH program, sought to identify and provide information to key potential participants in the development and use of solar IPH, including manufacturers, users, and engineering and construction firms (see Brown 1988 for an extensive discussion of market issues).

7.1.2   Industrial Applications

Process heat uses vary widely in the industrial manufacturing sector. Individual process systems are often large, consuming substantial amounts of energy. Industrial process loads are, in general, predictable and relatively constant, allowing for effective solar system design. Thermal storage may or may not be desirable or necessary; for very large loads the solar system may simply supplement fossil fuel use when solar radiation is available or, in other applications such as liquid heating, large storage tanks may be an integral feature of the industrial system.

Energy use in industry is found over a broad range of heat transport or process fluids (air, water, steam, oils) and temperature ranges. Only a small percentage (less than 5%) of the energy is used for low-temperature applications (below 212°F), usually for washing with hot water in food processing or parts cleaning. About 30% of the energy satifies intermediate applications (212°–350°F), with the remaining use at higher temperatures. Intermediate-temperature applications include a wide variety of low to medium-pressure steam requirements for heating and drying in the food processing, textile, and paper and pulp industries. Higher-temperature applications include high-temperature steam in the chemical, metal, and petroleum industries. Appropriate solar thermal technologies must be matched to these applications, largely depending on temperature

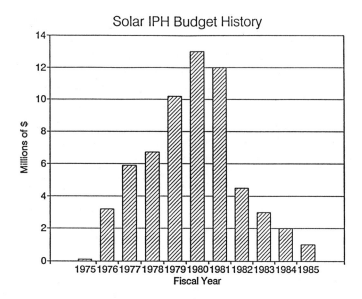

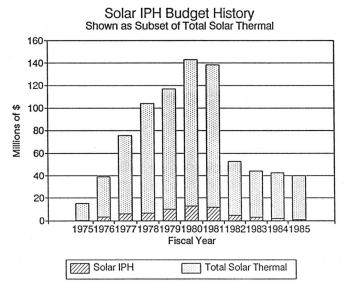

Figure 7.1
Federal budget history for solar IPH (based on personal communications with D. Kumar, Meridian Corp., and J. Anderson, NREL). Solar IPH refers here to system field tests and IPH-related system design and commercialization activities; a portion of the remaining solar thermal budget included technology development that directly benefited the IPH program (e.g., concentrators and reflector materials).

requirements. Figure 7.2 summarizes these points through an illustration of the industrial process medium, relative proportion of energy use, and appropriate solar collector technology as a function of temperature for industrial process heat applications.

A number of considerations—positive and negative—enter into the match of solar energy to an industrial process heat application:

• Large applications offer the possibility for large solar collector production runs, bringing cost reductions associated with the economies of scale. Similarly, system engineering costs and construction become a smaller fraction of total system cost.

• The wide temperature range of industrial applications offers opportunities for diverse utilization of solar technologies, from flat-plate to concentrating collectors.

• Large solar energy systems require large land or rooftop areas. Many industrial locations may offer excellent settings in this regard, but land availability can be a crucial barrier, particularly in an urban setting, and is a key issue. This need may also dictate an energy supply system located far from the application.

• Industrial applications suggest availability of good operation and maintenance (O&M) practices because skilled labor may be on site. On the other hand, industry favors low O&M requirements as a matter of cost and convenience.

• Most industrial applications differ in engineering requirements due to layout, load schedule, and end use, possibly necessitating costly plant-specific system design. Standardized systems for recurring applications can minimize such costs.

• Industrial effluents that degrade collector surfaces may be present, and this can strongly influence material selection and/or economics of the application, although tighter environmental controls may be lessening this issue.

• Energy efficiency measures in industry can reduce or eliminate the need for lower-temperature solar collector systems and consequently should be utilized prior to solar IPH system implementation.

• Care must be taken to match nonconstant solar energy delivery with the industrial demand, possibly through the use of thermal energy storage.

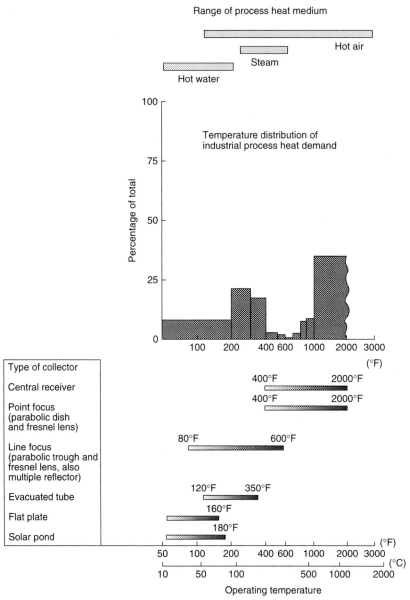

Figure 7.2
Temperature dependence of industrial needs and solar technology selection (Kearney and Lewandowski, 1983).

7.2 Solar IPH Field Test Program

7.2.1 Field Test Characteristics

From 1975 through 1981 a series of solar IPH field tests were funded for design or construction to provide opportunities for industry, solar collector manufacturers, and solar energy system developers to gain experience in the IPH marketplace. These projects, listed in table 7.1, initially consisted of a group of fifteen medium-sized systems with collector areas on the order of 10,000 ft^2, followed by a second group of eight large systems designed with collector areas on the order of 50,000 ft^2. At the inception of the IPH projects it was planned that, upon completion of the project operating phase (normally 1–2 years), the ownership and operation of the systems would be transferred to the plant owners to provide a regular source of thermal energy and serve as examples of solar IPH applications for other prospective users.

7.2.2 Field Experience and Performance

It is clear from table 7.1 that the field test program covered a broad spectrum of industrial applications, locations, and solar energy system types. Numerous lessons were learned over the life of the program related to engineering design, technical performance, industry issues, and market opportunities. Many of the projects did not measure up to initial expectations, leading some observers to conclude that the success of the projects was quite limited. As a foundation for future guidance to the use of solar energy in industrial processes, however, this series of projects provided a valuable and solid base for the proper selection and design of the solar energy system, industrial user, and overall system characteristics.

The projects were funded in cycles, or stages, with little or limited feedback from one cycle to the next because of project timing. For example, the design phase of the low-pressure steam projects commenced before the operational phase of the hot water and hot air projects could provide useful experience. It was not recognized at first that reliable maintenance of the systems would be an important issue in the field tests, but this aspect became a serious matter for numerous systems. Maintenance problems were related not only to collector and system design features but also to the dedication of the industrial user to keeping the systems operational. For example, if the solar energy system required

Table 7.1
Solar IPH field test projects

Low-temperature hot water/hot air IPH field tests (140°–212°F, 60°–160°C)

Location	Process	Industrial partner/contractor	Collectors	Array size (ft²)	Collection temperature	Designed annual energy delivery (1,000 Btu/ft²/yr)
		HOT WATER				
Sacramento, CA	Soup can washing	Campbell Soup Co./Acurex Corp.	Flat-plate and parabolic trough	7,335	150°F (66°C)	300
La France, SC	Textile dyeing	Riegel Textile Corp./General Electric Co.	Evacuated tube	6,680	270°F (132°C)	210
Harrisburg, PA	Concrete block curing	York Building Products, Inc./ AAI Corp.	Multiple reflector	9,216	135°F (57°C)	163
		HOT AIR				
Gilroy, CA	Onion drying	Gilroy Foods, Inc./Trident Engr. Assoc.	Evacuated tube	5,950	210°F (99°C)	393
Decatur, AL	Soybean drying	Gold Kist, Inc./Teledyne-Brown Engr.	Flat-plate	13,104	140°F (60°C)	282
Canton, MS	Kiln drying of lumber	J. A. LaCour Kiln Services, Inc./Lockheed Missiles and Space Co.	Flat-plate	2,520[a]	142°F (61°C)	357
Fresno, CA	Fruit drying	Lamanuzzi & Pantaleo Foods/Calif. Polytechnic State U., San Luis Obispo	Flat-plate	21,000	145°F (63°C)	110

[a] Plus 2,400 ft² reflectors.

Intermediate-temperature steam IPH field tests (350°–550°F, 177°–288°C)

Location	Process	Industrial partner/contractor	Collectors	Array size (ft²)	Steam conditions	Designed annual energy delivery (1,000 Btu/ft²/yr)
Dalton, GA	Latex production	Dow Chemical/Foster-Wheeler Devlp. Corp.	Parabolic trough	9,930	366°F (184°C) 165 psia (1.14 MPa)	252
San Antonio, TX	Brewery	Lone Star Brewing Co./Southwest Research Inst.	Parabolic trough	9,450	351°F (177°C) 140 psia (0.96 MPa)	338[a]
Ontario, OR	Potato processing	Ore-Ida Co./TRW	Parabolic trough	9,520	417°F (214°C) 300 psia (2.07 MPa)	200
Hobbs, NM	Oil refinery	Southern Union Co./Monument Solar Corp.	Parabolic trough	10,080	380°F (193°C) 185 psia (1.27 MPa)	351

[a]Collector output.

Low-temperature steam IPH field tests (212°–350°F, 100°–177°C)

Location	Process	Industrial partner/contractor	Collectors	Array size (ft²)	Steam conditions	Designed annual energy delivery (1,000 Btu/ft²/yr)
Topeka, KS	Concrete curing	Capitol Concrete Products/Applied Concepts Corp–JPL	Parabolic dish	860	316°F (158°C) 73 psia (0.5 MPa)	100[a]
Pasadena, CA	Commercial laundry	Home Cleaning and Laundry/Jacobs-Del Solar Systems	Parabolic trough	6,496	340°F (171°C) 125 psia (0.86 MPa)	249
Sherman, TX	Gauze bleaching	Johnson & Johnson/Acurex Corp.	Parabolic trough	11,520	345°F (174°C) 125 psia (0.86 MPa)	122
Bradenton, FL	Orange juice pasteurizing	Tropicana Pro, Inc./Genral Electric Co.	Evacuated tube	10,000	350°F (177°C) 150 psia (1.03 MPa)	270
Fairfax, AL	Fabric drying	West Point Pepperall/Honeywell, Inc.	Parabolic trough	8,313	317°F (158°C) 85 psia (0.58 MPa)	132

[a]Sponsored by the JPL parabolic dish program in the early 1980s; entry supplied by author.

Table 7.1 (continued)

Intermediate-temperature cost-shared IPH field tests (212°–550°F, 100°–288°C)

Location	Process	Contracting team	Collectors	Array size (ft^2)	Fluid conditions	Designed annual energy delivery (1,000 Btu/ft^2/yr)
Fort Worth, TX	Corrugated board production	Bates Container/BDM	Parabolic trough	34,720	Steam 440°F (227°C) 150 psi (10.1 MPa)	220[a]
San Leandro, CA	Pressurized hot water for washing	Caterpillar Tractor/SWRI	Parabolic trough	50,400	Water 235°F (113°C) 30 psi (2.0 MPa)	269
Pepaekeo, HI	Cane processing and electricity	Hilo Coast Processing Co./ Team, Inc.	Parabolic trough	44,100	Steam 400°F (204°C) 165 psi (11.1 MPa)	305
Haverhill, OH	Chemical plant: Polystyrene	U.S.S. Chemicals/Columbia Gas/H. A. Williams and Assoc.	Parabolic trough	50,400	Steam 440°F (227°C) 150 psi (10.1 MPa)	143[a]

Source: Compiled by R. Davenport, Solar Energy Research Institute, 1980.
[a]These projects did not reach the construction stage.

more than normal maintenance attention, but supplied only a small fraction of the plant process heat needs, management could (and often did) choose to fall back to full reliance on their conventional thermal energy source.

The operating experience with the projects varied considerably (SNLA and Stine 1989). Several projects went into normal operation and successfully served the plant beyond the normal DOE-sponsored operational phase. Others were shut down soon after the operational phase ended because of problems with the solar equipment, high operational cost, or plant shutdowns. These systems tended to be those operating at higher temperatures and utilizing concentrating collectors (parabolic troughs). Several projects never reached the operational stage because of collector and/or system problems. The scarcity of projects with high performance and high reliability led to some concern and criticism of the effectiveness of the entire program (see, for example, Lumsdaine 1981). It has been observed that a longer project operating phase, perhaps as long as five years, may have resulted in systems that were truly operational and much more reliable.

Both Sandia National Laboratories, Albuquerque (SNLA), and the Solar Energy Research Institute (SERI) participated in technical evaluations of the projects and identified areas where improvements were possible (SNLA and Stine 1989; Lewandowski, Gee, and May 1984; Kutscher and Davenport 1980). Design features to improve collector reliability and performance included tracking and drive systems as well as flexible hoses on parabolic troughs. It was found that excessive heat losses from long piping requirements could be reduced by locating the solar field close to boilers and heat exchangers, by paying adequate attention to heat loss through supports and other connections, by using sufficient insulation and by eliminating oversized piping to decrease thermal capacitance of both the piping mass and the contained process fluid. Several publications evolved from the IPH program that provided design guidelines for solar industrial applications (Harrigan 1981; Kutscher et al. 1982; Kutscher 1985).

Deficiencies in data acquisition, IPH system reliability, and industrial process limitations made consistent data sparse and careful tracking and prediction of long-term performance difficult. The monthly thermal efficiency of the systems based on the delivered energy normalized by the incident direct radiation in the aperture plane of the collector generally

varied from 20% to 30%, with some data as high as 40% and others as low
as 10%. Lewandowski, Gee, and May (1984) projected annual energy
delivery efficiencies from 23%–52% for six of the later systems, somewhat
higher than extrapolated annual measurements in the 17%–43% range for
those systems. As reported below, early commercial systems installed subsequent to the field tests improved markedly on this experience, reaching
annual thermal delivery efficiencies in the 50%–60% range.

7.2.3 MISR Program

The industrial solar IPH program could not have proceeded without
parallel activities in collector and system development across the full
range of solar thermal technology. One program that specifically supported the need to reduce the engineering costs and improve the reliability
of low- and moderate-pressure steam systems for industrial use was the
Modular Industrial Solar Retrofit (MISR) program, conceived and managed by Sandia National Laboratories, Albuquerque (Cameron, Dudley
and Lewandowski 1987). The primary objective of this effort was to design and test standard system modules utilizing parabolic troughs to produce saturated steam up to 250 psig. Five modular systems were designed
and installed at Sandia and SERI for testing (four at Sandia; one
at SERI). Collector field sizes were simulated at sizes on the order of
25,000 ft^2, made up of 1,920–5,040 ft^2 of collectors per system and supplementary propane-fired heaters to simulate the remainder of the energy
input. Optical efficiencies of the collectors approached 80% with glass
reflectors, giving module thermal efficiencies in the 60%–70% range. The
MISR program emphasized system performance, including balance-of-
plant equipment and controls in the test system.

7.2.4 User Response to IPH Field Test Projects

A study was funded by Sandia in 1982 to obtain feedback from the industrial users at the IPH field test sites (Martin 1983). With respect to
technology issues, almost half the firms (out of eighteen field test projects)
felt that the installed systems lacked sufficient testing and validation prior
to use at an industrial site. Assessments of system value varied with the
industrial users, ranging from confidence that the system contribution to
IPH needs was useful and economic (albeit small—from 2% to 15% of
total load), to concern that the energy contribution was too low compared
to expectations, or to an inability to reach a conclusion because the data

acquisition system was inaccurate or unreliable. As a general rule, thermal performance of the solar IPH systems was less than had been predicted, partially because initial estimates of expected solar radiation levels may have been optimistic. Operation and maintenance experience was generally positive, though this observation appears to be in conflict with the lack of system reliability. The IPH system interface with the industrial process was reported to be good, although the solar IPH system contribution was usually a small part of the energy use. Data acquisition reliability appeared to be a widespread problem.

Despite these drawbacks, industry's attitude, based on interviews with management, appears to have been generally favorable, with the use of solar energy to supply IPH considered an acceptable alternative to conventional energy sources. The experience strongly suggests that markedly fewer problems would have been encountered if the installed systems had been more mature, decreasing or eliminating issues with solar system availability, performance prediction, and measurement of energy delivery. These, of course, were major goals of the MISR program.

7.3 Commercialization Issues

7.3.1 Additional IPH Studies and Projects

A number of assessments of IPH applications were made under the DOE program that were not implemented in actual projects. Table 7.2 lists some of these studies, which included utilization of solar collectors from flat plates through central receiver systems (see Brown 1988 for additional information on end-use assessments and preliminary evaluations of solar IPH potential).

Moreover, several solar IPH projects were constructed in the middle to late 1980s that were not part of the DOE-sponsored series of field test projects and that constituted the first real steps to commercialization of this application of solar energy. The selection and design of these projects benefited significantly from the earlier federal projects. Table 7.2 also provides data on these applications.

7.3.2 Lessons Learned

The process of selection of the DOE projects was not designed to carefully target applications nor collector types prior to the solicitation. Rather, it

Table 7.2
Other solar IPH studies and projects

Low-temperature cost-shared IPH studies (to 212°F, 100°C)

Location	Process	Contracting team	Array size (ft^2)	Delivery temperature	Designed annual energy delivery (1,000 Btu/ft^2/yr)
Santa Isabel, PR	Fruit juice and nectar pasteurization	Nestle Enerprises/General Electric/ Center for Energy and Environmental Research/ O'Kelly, Mendez, and Brunner A/E	50,000	210°F (99°C)	180
Santa Cruz, CA	Leather tanning and finishing	Salz Leathers/Pacific Sun/Chilton Engr.	35,200	85°–140°F (25°–60°C)	315
Oxnard, CA	Sodium alginate processing	Stauffer Chemical/Wormster Scientific/ Desert Research Inst.	38,000[a]	125°F (52°C)	321
Des Moines, IA	Meat processing	Team, Inc. of Va./Oscar Mayer/ Univ. of Wisconsin	40,320	180°F (82°C)	217
Shelbyville, TN	Poultry processing	Tyson Foods/Lockheed	50,400	140°F (60°C)	292

[a] Plus 43,780 ft^2 of reflectors

"Repowering" industrial retrofit studies (central receivers)

Location	Process	Contracting team	Receiver type	Number of 540 ft² (50 m²) heliostats	Delivery temperature	Solar size (MW$_t$)
Bakersfield, CA	Natural gas processing	Arco Oil & Gas Co./Northrup, Inc.	Oil external	437	572°F (300°C)	5.7
Bakersfield, CA	Enhanced oil recovery	Exxon/Martin Marietta/Foster Wheeler	Water/steam cavity	537	518°–545°F (270°–285°C)	29.3
San Mateo, NM	Uranium ore processing	Gulf R&D/McDonnell Douglas/Foster-Wheeler/Univ. of Houston	Water/steam external	413	366°F (186°C)	13.9
Mobile, AZ	Oil distillation	Provident Energy Corp. Foster-Wheeler/McDonnell Douglas	Water/steam external	1,154	500°F (260°C)	43
Sweetwater, TX	Gypsum board drying	U.S. Gypsum/Boeing/Inst. Gas Technology	Air cavity	418	900°F (482°C)	13.9
El Centro, CA	Reforming ammonia	Valley Nitrogen Producers/PFR Engr. Systems/McDonnell Douglas	Gas cavity	1,153	1,472°F (800°C)	15

Solar enhanced oil recovery studies

Location	Contracting team	Collectors	Approximate array size (ft²)	Delivery temperature	Designed annual energy delivery (MBtu)
Bakersfield, CA	Exxon/Martin Marietta/Foster-Wheeler	N-S parabolic trough	245,000	250°–554°F (121°–282°C)	70,000
Bakersfield, CA	General Atomic/Petro-Lewis Oil	E-W parabolic trough	340,000	500°F (260°C)	70,000

Table 7.2 (continued)

First commercial solar IPH projects

Location/year	Process	User/developer	Array type and size ft² (m²)	Delivery temperature °F (°C)	Design annual energy delivery (1,000 Btu/ft²/yr)
Chandler, AZ 1982	Hot water for electrochemical processing	Gould Foil Division/Solar Kinetics, Inc.	Parabolic trough 60,470 (5,620)	500 (260)	—
Shderot, Israel 1984	Steam for potato processing	Shaar Hanegev Industrial Pk./Luz Industries Israel	Parabolic trough 5,270 (490)	473 (245) 365 psia (2.5 MPa)	—
Aurora, CO 1985	Commercial hot water: Pool heating and domestic	Paul Beck Recreation Center/Industrial Solar Technology	Parabolic trough 2,400 (223)	160 (71)	242
Brighton, CO 1986	Commercial hot water: Prison domestic water	Adams Co. Detention Fac./Industrial Solar Technology	Parabolic trough 7,800 (725)	260 (127)	220
Tehachapi, CA 1990	Commercial hot water: Prison domestic water	California Correctional Inst./United Solar Technologies/Industrial Solar Technology	Parabolic trough 28,800 (2,677)	260 (127)	233

was left to solar developers and industry to team up and propose a project in response to a broad solicitation. The best of the responses received as a result of programmatic Requests for Proposals (RFPs) were evaluated and selected.

Hindsight suggests that targeting entry-level applications would have offered a better selection of projects and a higher chance of success. Experience from the IPH projects has led to the identification of seven factors that have characterized the most successful systems:

1. Solar system developer expertise

2. Location with a high solar resource

3. Well-designed, reliable, and proven solar collectors, with strong manufacturer support

4. Good industrial user O&M practices

5. Excellent match between solar resource and industrial application with respect to system integration and load cycle

6. Adequate inexpensive land area proximate to thermal energy load

7. Firm industrial user commitment

A significant benefit of the DOE program was to highlight these factors and the realization that industrial solar IPH systems could be successful if also coupled with reduced costs, higher performance, and a competitive marketplace. The leading example of the application of these factors to industrial applications has been the installation of several hot water solar IPH systems (May and Gee 1992) by a Denver company that was an early spinoff of the DOE IPH activities. These systems, which have demonstrated a high measure of reliability, achieved improvements in cost reductions by a factor of 2 or more in cost per unit area and performance gains of a factor of 4 or higher in annual thermal energy delivered per unit area (Anderson 1992). Recent projections (Williams and Hale 1993) suggest that improved solar industrial solar process heat systems can offer a viable alternative to conventional natural gas systems under certain conditions.

The evidence suggests that success can be achieved in solar IPH applications, given markets where solar energy can be competitive with alternative sources. In niche applications such as the displacement of electricity or diesel oil as an energy source, or in specialized applications such as heating large oil storage tanks at low temperatures, the economics are more favorable, but the fact remains that the displacement of industrial process heating by conventional energy sources such as oil or gas will continue to be a very difficult hurdle until conventional fuel prices increase beyond post–energy crisis levels.

7.4 Summary

In summary, the major success of the DOE program has been to provide a comprehensive set of guidelines and lessons as a basis for reliable and cost-effective solar industrial process heat systems. While some progress has been made in commercialization of solar IPH subsequent to the DOE program described here, large-scale industrial solar IPH implementation is still to be achieved. To encourage and accelerate this effort, a new initiative (Hewitt 1992) to systematically target specific promising applications has been undertaken by DOE's Solar Industrial Program managed by the National Renewable Energy Laboratory with support from Sandia National Laboratories. Although still its early stages, this program encompasses and builds upon the important knowledge gained from earlier efforts, focusing on good fits in the marketplace, low-temperature applications, high conventional energy costs, mature solar system hardware, and strong user commitment.

References

Anderson, J. 1992. "Overview of Solar Process Heat Program." *SOLTECH '92 Proceedings: Solar Process Heat Program*. Albuquerque NM: SOLTECH Conference.

Brown, K. C. 1988. "End-Use Matching and Applications Analysis." In R. E. West and F. Kreith, eds., *Economic Analysis of Solar Thermal Energy Systems*. Cambridge: MIT Press.

Cameron, C., Dudley, V., and Lewandowski, A. 1987. *MISR Qualification Test Reports*. SAND-85-2316–85-2320. Albuquerque, NM: Sandia National Laboratories.

DOE (U.S. Department of Energy). Division of Solar Energy. 1978a. *Solar Domestic Policy Review: Final Report*. Washington, DC.

DOE. Division of Solar Energy. 1978b. *Solar Energy of Agricultural and Industrial Process Heat: Program Summary*. Washington, DC.

Harrigan, R. W. 1981. *Handbook for the Conceptual Design of Parabolic Trough Solar Energy Systems for Process Heat Applications*. SAND-81-0763. Albuquerque, NM: Sandia National Laboratories.

Hewitt, R. 1992. "Solar Process Heat Prefeasibility Studies." *SOLTECH '92 Proceedings: Solar Process Heat Program*. Albuquerque, NM: SOLTECH Conference.

Kearney, D., and A. Lewandowski. 1983. "Solar Energy Industrial Applications." In *Mechanical Engineering*. New York: American Society of Mechanical Engineers.

Kutscher, C. F. 1985. "Design Considerations for Solar Industrial Process Heat Systems." *Proceedings of the IEA Conference on the Design and Performance of Large Solar Collector Arrays*. SERI/SP-291-2664. Golden, CO: Solar Energy Research Institute.

Kutscher, C. F., and R. L. Davenport. 1980. *Preliminary Operational Results of the Low-Temperature Solar Industrial Process Heat Field Tests*. SERI/TR-632-385. Golden CO: Solar Energy Research Institute.

Kutscher, C. F., R. L. Davenport, D. A. Dougherty, R. C. Gee, P. M. Masterson, and E. K. May. 1982. *Design Approaches for Solar Industrial Process Heat Systems*. SERI/TR-253-1356. Golden, CO: Solar Energy Research Institute.

Lewandowski, A., R. Gee, and K. May. 1984. *Industrial Process Heat Data Analysis and Evaluation*. SERI/TR-253-2161. Golden, CO: Solar Energy Research Institute.

Lumsdaine, E. 1981. "Solar Industrial Process Heat." Paper presented at the Solar Technology Assessment Conference, Orlando, FL.

Martin, D. 1983. *Management Response to Their Solar Industrial Process Heat Field Experiments*. Lawrence: University of Kansas Center for Research.

May, K., and R. Gee. 1992. "IST's Colorado Trough Systems: Performance Update and Review." *SOLTECH '92 Proceedings: Solar Process Heat Program*. Albuquerque, NM: SOLTECH Conference.

SNLA (Sandia National Laboratories, Albuquerque) and W. Stine. 1989. *Solar Industrial Process Heat Project: Final Report*. SAND-89-1968. Albuquerque, NM.

SERI (Solar Energy Research Institute). 1980. *Putting the Sun to Work in Industry*. SERI/SP-34-175R. Golden, CO.

Williams, T., and M. J. Hale. 1993. "Economic Status and Prospects of Solar Thermal Industrial Heat." *Proceedings of the Solar '93 Conference*. Washington, DC: American Solar Energy Society.

8 High-Temperature Technologies

J. C. Grosskreutz

This chapter describes the federal program for development and demonstration of high-temperature solar thermal technology on a commercial scale and includes an evaluation of the implementation program. The scope is limited to concentrator technologies that produce heat at temperatures greater than 600°F (315°C) by means of central receivers, parabolic troughs, and parabolic dishes. The heat so generated can be used directly to drive industrial processes including the production of fuels and chemicals or for conversion to electricity by means of Rankine, Brayton, or Stirling thermal cycles.

8.1 Initial Mandate

Solar thermal technologies have a long development history, dating back to the 1870s and the work of Augustin Mouchot (Butti and Perlin 1980), who used cone-shaped, solar concentrators to drive simple heat engines. In 1913, an array of large parabolic troughs was constructed in Egypt which produced steam to drive a 55-horsepower irrigation pump (Butti and Perlin 1980). Thus it was natural that high-temperature, more efficient solar concentrator systems were considered as alternative energy sources in the early 1970s when the oil price shocks and energy shortages occurred in the United States.

The actual mandate for development and demonstration came with passage of the Solar Energy Research, Development, and Demonstration Act of 1974 (PL 93-473). That law states that "it is the policy of the Federal government to provide for the development and demonstration of practical means to employ solar energy on a commercial scale (PL 93-473). Further, the law established a Solar Energy Coordination and Management Project whose chairman was authorized to "initiate a program to design and construct, in specific solar energy technologies (including ... thermal energy conversion ... for the generation of electricity and the production of chemical fuels), facilities or power plants of sufficient size to demonstrate the technical and economic feasibility of utilizing the various forms of solar energy" (PL 93-473). Included were pilot plant construction and operation and other activities "which may be necessary to show commercial viability" of solar thermal technology (PL 93-473).

The first comprehensive assessment of solar thermal energy as a national energy resource occurred, however, two years before enactment of this legislation. In January 1972 the National Science Foundation (NSF) and the National Aeronautics and Space Administration (NASA) established a solar energy panel, which prepared a long-range solar thermal development plan, *Solar Energy as a National Energy Resource* (NSF/NASA 1972, 54). This plan formed the basis for defining the objectives and utilization goals of the federal high-temperature solar thermal program that were contained in the Federal Energy Administration's *Project Independence* task force report (FEA 1974). This report stated the major goals of the program as "(1) to provide a full system capability for the widespread production of supplementary electric and thermal power in the 1980s to meet electric utility requirements .., and (2) to provide a full system capability for total energy systems for urban and rural communities, industrial load centers and military bases." These goals have remained intact during the ensuing years.

8.2   Organizational History

From the beginnings of the federal solar thermal program, the lead Washington energy agency has maintained overall program direction and planning authority and has been responsible for yearly budgets and funding of the various program elements. As the program matured, the details of program implementation and management of industrial and university contractors were delegated to the federal field operations offices and certain national laboratories.

8.2.1   Federal Program Direction

Solar thermal programs began at the National Science Foundation (NSF) in 1971 under the RANN (Research Applied to National Needs) program. Lloyd Herwig, Dwain Spencer, and George Kaplan provided leadership during the period 1972–1974 and were responsible for the first comprehensive program plan for solar thermal (FEA 1974). With the formation of the Energy Research and Development Administration (ERDA) in January 1975, responsibility was shifted from NSF to the new agency and the solar programs and personnel active at both NSF and the Atomic Energy Commission (AEC) were transferred as well. High-temperature solar thermal was made part of the ERDA Division of Solar Energy and came under the solar electric technologies branch, which was managed by Dick Blieden.

In September 1975 Hank Marvin became director of the Division of Solar Energy and took a strong role in solar thermal programs. The arrival of Marvin, who had formerly been with General Electric, brought an industrial management point of view that was to have an important effect in solar thermal implementation, especially the central receiver system. Gerry Braun was recruited in 1977 from the Southern California Edison Company to head a now rapidly expanding program. In 1978, as part of the emerging national energy policy, the Department of Energy (DOE) was established; the solar thermal program found itself in yet another new home but with its cadre of experienced personnel intact.

The creation of DOE gave a much-needed stability to the program after five years of ever-changing lead agencies and personnel and the inability of Congress and the administration to define and agree on a national energy policy. Serving as director of the Solar Thermal Division until late 1983 during its most active years, Braun solidified planning, budgeting, and program management activities. He was succeeded by Howard Coleman, who had earlier served as deputy to Marvin at ERDA and in 1978 was also in charge of centralized solar systems, including solar thermal.

8.2.2   Program Management

8.2.2.1   DOE Operations Offices and National Laboratories
The old AEC regional operations offices and the national laboratories began the job of solar thermal contract procurement and management as early as 1975, making good use of the time-tested AEC experience. The San Francisco (SAN) and the Albuquerque (ALO) DOE operations offices have had continuous and strong roles. Bob Hughey of SAN and George Pappas and Joe Weisiger of ALO have played influential roles in the thrust toward bringing the technology to the point of commercial application. These two offices have been ably supported in program management by Sandia Laboratories, both at Livermore (SNLL) and at Albuquerque (SNLA). (Very early in the program, 1973–1974, NASA–Lewis Research Center had a seminal role in managing the evaluation and design of central receivers for bulk electricity production under George Kaplan.) Key individuals at Sandia have been Al Skinrood (SNLL) and John Otts (SNLA) continuously since 1974–1975. Other strong contributors were Cliff Selvage, Bill Wilson, Rick Wayne, and Jim

Wright, all of SNLL, and Glen Brandvoldt, Don Schueler, and Jim Leonard of SNLA. The Jet Propulsion Laboratory (JPL) became involved in developing the parabolic dish electric program to commercial readiness in the late 1970s, with Vince Truscello as the program leader under Mickey Alper. (JPL's involvement terminated in 1984 as all program activities were consolidated.) The Aerospace Corporation also played a strong role in the first few years of the central receiver program. Under contract to NSF, they were engaged in program and project evaluation and generated the mission analysis for the central receiver electric program in 1973–1974. Primary participants were Mason Watson and Piet Bos.

The management style of the national laboratories played a key role in moving high-temperature solar thermal technology toward the goals of commercial readiness during 1975–1983. Able and experienced project managers who were conversant with the technology were assigned to contracts placed with the private sector. In addition to their interaction with the laboratories, operations offices, and Washington, contractors attended semiannual review meetings, where they gave oral progress reports on their individual projects and heard programmatic and budget overview reports by federal personnel. Over a period of years, these review meetings became a powerful tool for building an interested and involved technical community and for management review of program progress.

8.2.2.2 Management Stability

The program management organization described in section 8.2.2.1 proved to be very effective. The operations offices and national laboratories provided the institutional and personnel stability that buffered the contractor community from the ever-changing winds of national energy policy in Washington (which of necessity required ever-increasing attention from the DOE headquarters staff). Moreover, Sandia and JPL became important reservoirs of technical knowledge and expertise in high-temperature solar thermal energy. They have also served as a knowledgeable interface in maintaining the interest and commitment of the private sector, the ultimate implementers of commercial solar thermal systems.

8.2.3 Role of EPRI

The high-temperature solar thermal program at the Electric Power Research Institute (EPRI) was formulated by Dwain Spencer, who left NSF in April 1974 to join the new institute. The EPRI program strove to complement the much larger federal program and emphasized the Brayton hybrid cycle for southwestern U.S. applications of the central receiver. EPRI also supported the development of the lightweight, enclosed (bubble) heliostat concept. John Bigger initially managed EPRI's solar thermal program, followed by Ed DeMeo.

Cooperation with DOE programs that hold a strong interest for utilities has been an EPRI objective. Sandia, JPL, and ERDA/DOE solar personnel have been included in EPRI project reviews and vice versa. A milestone in cooperation was achieved with the Molten Salt Electric Experiment (MSEE), a proof-of-concept demonstration of a 0.75-MWe central receiver system utilizing molten nitrate salt as the primary heat exchange and transport fluid. The experiment was conducted at the Central Receiver Test Facility at Sandia Albuquerque and was cost-shared by DOE (50%), EPRI (25%), and private industrial and utility companies (25%). The MSEE completed a year of testing and operation in July 1985 (Bergan 1986)

In 1986 EPRI discontinued its funding of solar thermal electric technology and concentrated its resources on photovoltaic device development. As a result, utility coordination of solar thermal programmatic priorities was shifted to individual utilities and EPRI's previous focus on solar thermal technology was no longer available to DOE.

8.3 Annual Budgets

Funding of the high-temperature solar thermal program grew rapidly in the 1970s, a period of escalating oil prices and fuel shortages in the United States. This growth made it possible to plan, construct, and operate a 10-MWe central receiver pilot plant and a 400-kWe parabolic dish total energy demonstration. Annual budgets through FY 1986 are shown graphically in figure 8.1 Strong support for these budgets has been given by the solar thermal industry, both by the Solar Energy Industries Association (SEIA) and by individual companies and utilities. SEIA has, through the years, prepared an industry-initiated "shadow budget" for the solar thermal program, which it makes available to DOE and the Congress during the budgeting process each year.

Solar thermal federal budgets fell on hard times in the 1980s as oil prices declined and federal involvement in renewable energy technologies

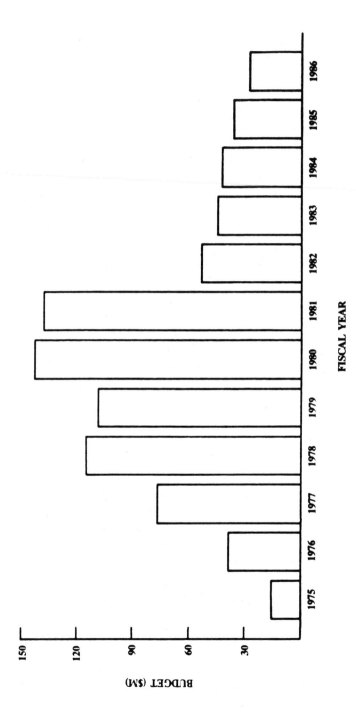

Figure 8.1
Annual federal budgets for the high-temperature solar thermal program in current dollars.

was sharply reduced. Compounding these effects have been the slackening of electricity demand and uncertainty over the future national economic outlook. Planning for program phaseout and zero budgets began as early as 1981, but continued support from Congress and industry has allowed a more orderly phasedown in activity. Industry participants in the solar thermal program provided sensible guidance to DOE through the SEIA's shadow budgets and five-year plans. Future budget trends for all program activities except research are highly uncertain.

8.4 Program Activities

At the heart of the program implementation effort were the projects carried out by private contractors and managed by the national laboratories; of even more importance was the strategy devised for implementation.

8.4.1 Strategy for Implementation

The initial mandate identified the major end-use products of solar thermal energy as the "generation of electricity and the production of chemical fuels" (PL 93-473). High-grade process heat was added later as a natural programmatic objective. It was recognized in 1973–1974 that the ability to produce these products on a commercial scale would depend critically on the development and support of an industry base for component design, development, and manufacture and for system design and construction. Of equal importance, especially in the earlier phases of the program, was the development of in-house R&D programs at Sandia and JPL to support and guide their management of the implementation effort. To support both the laboratories and the industrial participants, the program made limited use of university research (University of Houston and Georgia Institute of Technology) until the late 1970s, when new, more innovative ways of utilizing high-temperature solar thermal energy were sought as an important factor in meeting the longer-term national needs. The Solar Thermal Users' Association was formed in 1976 and served as a channel for evaluation of university proposals and for DOE funding for university research.

In almost all previous federal efforts to develop a new technology (e.g., jet aircraft, nuclear energy), the government was also the prime customer; but in the case of solar energy, an uninitiated and largely skeptical private, commercial set of customers was involved. Federal program

planners, directors and managers therefore set about to include electric
utilities and industrial users of electricity and high-temperature process
heat in their development and demonstration contracts with private
industry. These contracts began with feasibility studies and conceptual
designs to support projected system performance and costs. Preferred
systems and component designs emerged, and competitive cost targets
and goals were set for eventual commercial deployment.

Federal procurement and testing of heliostats, parabolic dish and
trough concentrators, and receivers provided designers and manufacturers
the opportunity to build toward the future production capability essential
to any commercial deployment. The first large federal purchase of heliostats for the Central Receiver Test Facility (CRTF) at Sandia Albuquerque in 1976 and of parabolic dishes for the total energy facility at
Shenandoah, Georgia, in 1980 provided the much-needed, initial data
points on the unit cost learning curves for concentrators.

8.4.2 Program to Implement the Commercialization Strategy

The federal programs to implement central receiver and distributed receiver system demonstrations are similar in nature but different enough in
detail to merit separate descriptions.

8.4.2.1 Central Receiver Systems

This program was geared toward a nominal 100-MWe commercial demonstration plant, first planned in 1974 to be operational in the 1985–1986
time frame (FEA 1974). It was successful in reaching the pilot plant stage.
A 10-MWe water/steam central receiver system was completed at Barstow, California, in 1982, and testing was concluded in 1984. A power
production phase, conducted by Southern California Edison, was completed in 1987. The steps and programs that led to this point are described
briefly below.

8.4.2.1.1 Pilot Plant System Designs and Subsystem Experiments In
mid-1975 industrial teams composed of aerospace system integrators,
architect-engineers, boiler manufacturers, and universities developed three
competing conceptual system designs utilizing a water/steam Rankine
cycle. Two-year contracts were awarded, each of which called for a prototype heliostat to be tested and evaluated and for a 5-MWt receiver to be
constructed and tested under operating conditions. At the conclusion of
the contracts in mid-1977, the designs were evaluated by Sandia National

Laboratories, Livermore (SNLL), and many of the features of the Mc-Donnell Douglas design were selected for the pilot plant.

Concurrently with these system designs, and prior to completion of the Central Receiver Test Facility, a 1-MWt water/steam cavity receiver was designed and built by Martin Marietta under an ERDA contract. Successful testing of this first U.S.-built receiver with concentrated sunlight in a central receiver configuration, complete with heliostats, was carried out at the solar furnace located in the Pyrenees mountains at Odeillo, France.

8.4.2.1.2 Central Receiver Test Facility To support the central receiver pilot plant effort, it was necessary to construct a large, outdoor test facility to evaluate and qualify the heliostat and receiver designs. One of the first actions by Hank Marvin after his arrival at ERDA was to place this facility on the critical path for central receiver demonstration and to give its construction funding first priority in the FY 1976 budget.

Conceptual design for the 5-MWt Central Receiver Test Facility (CRTF), located at Sandia Albuquerque Laboratories, began in mid-1975 by an industrial/university team led by Black & Veatch Consulting Engineers. The $24 million facility was completed in late 1977 on budget and on schedule, and the first receiver tests were conducted in mid-1978. Since its completion, CRTF has been utilized continuously to test advanced receiver designs; second-, third-, and fourth-generation heliostat designs (some privately developed); thermal storage subsystems; molten salt steam generations, and a full-system, molten salt experiment.

8.4.2.1.3 10-MWe Pilot Plant The site for this plant was selected in a competitive, cost-sharing competition among U.S. electric utilities. Southern California Edison (SCE) won the award by proposing an SCE-owned site near Barstow, California, and offering to cost-share 20% of the estimated capital cost of the pilot plant. Later, a team led by McDonnell Douglas won the design contract, and Townsend & Bottum was awarded the construction contract. Turbine roll occurred in April 1982. The cost of construction for this first-of-a-kind pilot plant was S140 million, which included an extensive data acquisition and analysis system as well as extensive documentation and review of the construction.

8.4.2.1.4 Utility Repowering Designs An important step beyond the pilot plant stage in the program was implemented in 1979 by a competitive procurement. The strategy, conceived by Public Service of New Mexico, was to "repower" existing oil and gas fired utility units with solar thermal central receiver steam source.* The existing steam turbine generator and all balance-of-plant facilities could be utilized without additional capital investment. Successful proposers offered designs that ranged in size from 30 to 120 MWe for the solar-powered generating capacity. Eventually, there were eight participating major utilities involved in these design and economic evaluation studies.

The economics of solar repowering rests on the value of the fossil fuel displaced by solar energy. Given the estimated capital costs of the solar plant additions (which were based on heliostat costs for delivery in the mid-1980s), only 25%–30% of the required costs could be recovered by fuel savings; the remaining 70%–75% required federal funding. By 1983 it was clear to all but the most devoted participants that federal cost-sharing for any demonstration of any renewable energy technology would not be forthcoming.

8.4.2.1.5 Advanced Central Receiver Systems   The results and experience from the system design studies, receiver subsystem tests, and the utility repowering projects showed that using water as the sole heat transfer medium in central receivers had major operational disadvantages, particularly with respect to thermal efficiency and energy storage capacity. To achieve a competitive position in commercial utility markets, a modified approach was implemented, and a competitive procurement to develop advanced central receiver designs was made. These studies showed that molten nitrate salts or liquid sodium as the primary receiver heat transfer and thermal storage fluid offer significant advantages in operational flexibility and thermal efficiency. The principal advantage is the availability of 6–8 hours of thermal energy storage which allows operators to shift plant operation to the early evening hours when electrical demand is high and the sun is not available. A steam generator between the hot, molten storage medium and the electric power generation subsystem enables the turbine generator to operate continuously and smoothly through transient cloud conditions, another significant advantage. The early morning startup time for a molten salt or sodium receiver is also much shorter than for water/steam receivers.

---

*Although Public Service of New Mexico (PNM) submitted the first proposal to DOE containing this strategy and a plan for repowering one of their own units, DOE elected not to fund PNM, but used the idea for a competitive procurement in which PNM did not participate.

Successful testing and demonstration of molten salt and sodium receivers, molten salt thermal storage and a molten salt steam generator at CRTF in the early 1980s provided the motivation for the industrial community's decision to recommend that these new systems be the basis for demonstrating central receiver technology on a commercial scale (Arizona Public Service 1988).

8.4.2.1.6 Molten Salt Electric Experiment A proof-of-concept experiment (which also served as a technical demonstration for utilities) to verify complete system operation of a 750-kWe molten salt central receiver system was successfully implemented at CRTF during the period 1983–1985 and confirmed the flexibility of operations (from thermal storage) during cloud transients and solar outages. Using and adapting major subsystems already built and tested individually at CRTF, the Molten Salt Electric Experiment (MSEE) was accomplished by the cooperation and evenly divided cost-sharing between the federal government and the private sector, including EPRI, three utilities, and nine industrial participants.

The positive results of the MSEE (Bergan 1986), the subsequent molten salt pump and valve test program, and the advanced molten salt receiver test at CRTF provided the basis in the mid-1980s for the technology of choice for future demonstration of central receiver technology. Although nitrate salts have some disadvantages (e.g., periodic removal of impurities) and molten salt valve design is still evolving, this technology was the unanimous choice for commercialization, in the 1990s, of the utilities and engineers still involved in the program in 1988 (Arizona Public Service 1988).

8.4.2.2 Distributed Receiver Systems

Once the central receiver program was well launched, ERDA turned its attention to the second goal of the program: total energy systems for small communities, industrial load centers and military bases (FEA 1974). The result of this effort was two programs for the development of high-temperature, parabolic dish systems, first as a total energy system and later as small, self-contained electricity producing modules. These later modules, called "dish-electric systems," incorporate a receiver, a heat engine, and an electric generator all located at the focal point of the parabolic dish.

Parabolic trough concentrators were also included in the ERDA solar thermal program, but the temperatures reached with the technology were nearly always below 600°F and therefore fall outside the scope of this chapter on high-temperature systems. However, the parabolic trough solar electric systems developed privately by Luz International in 1987–1988 were capable of heating an oil-base heat transfer fluid to 700°F by solar energy. The Luz systems are described in section 8.5.3 of this chapter.

8.4.2.2.1 Total Energy Systems This effort was begun and completed as a single project. A competitive procurement was made in early 1978 for a system design and site with no restrictions on the technology to be selected. This procurement resulted in a cooperative, cost-shared agreement between DOE and General Electric/Georgia Power Company to design, construct, and operate a total energy system located at a knitwear factory near Shenandoah, Georgia. Using parabolic dishes, the facility was designed to produce 3 megawatts of thermal energy at 750°F (400°C), to be collected at a central location and converted to 400 kW of electrical energy, to 1,385 lb (630 kg)/hr of 350°F (175°C) steam for clothes pressing, and to 1,430 MJ/hr of chilled water for air-conditioning. Construction of the system was completed in early 1982, and after resolution of start-up problems, turbine-generator operation began in January 1983. The system was formally transferred to Georgia Power in late 1984 and continues to operate and generate useful energy and operational data. Sandia Albuquerque (SNLA) was the responsible program manager for DOE. Ed Ney of Georgia Power and Bob Hunke of SNLA were key personnel in the implementation of this project.

A distributed receiver test bed was built at Sandia National Laboratories, Albuquerque, in 1974–1975 to support the development effort for parabolic trough and dish technology for total energy applications.

8.4.2.2.2 Dish Electric System Module This program to develop a stand-alone, dish-electric system module was originally conceived and carried out by JPL for DOE. While there was little technical interchange between it and the Shenandoah project due to considerably different designs, the experience gained on the total energy project was influential in the programmatic decisions by DOE to pursue dish-electric technology. Beginning in FY78, the program was structured similarly to the central receiver program. Program elements have included development of concentrator, receiver, and heat engine subsystem designs, test and evaluation of these designs at a Parabolic Dish Test Site (PDTS) at JPL, field testing of a complete dish electric module at 25 kWe, and planned deployment of a small community experiment.

The module development program goals were to develop a commercial prototype module (25 kWe) as well as the industrial manufacturing capability to support future deployment of these modules. Several options were pursued for the receiver/heat engine subsystems: organic Rankine cycle, Brayton cycle, and Stirling cycle. In each case, the receiver and heat engine designs were fully integrated, including a heat rejection capability, so as to form a compact, reliable package for focal point mounting. These three designs were carried forward in a parallel effort through FY84 when budget reductions forced cutbacks and elimination of the Brayton cycle.

A Parabolic Dish Test Site was established by JPL and located at Edwards AFB, California, in 1982 for testing and evaluating the modules and subsystems being produced under the module development program. A special 11 m diameter test bed dish concentrator was constructed at the site and operated successfully until late 1984, when management of the dish program was transferred to Sandia Albuquerque (SNLA). The test bed concentrator was shipped to Albuquerque and formed the nucleus of a new Distributed Receiver Test Facility (DRTF) adjacent to the Central Receiver Test Facility (CRTF) that included both troughs and dishes.

In a competitive procurement by JPL in FY 1983, Advanco Corporation was awarded a contract to construct and demonstrate their 25-kWe dish-Stirling commercial prototype module design at the Southern California Edison's Santa Rosa substation located in Rancho Mirage. This 34-foot (11-meter)-diameter module, called "Vanguard," was successfully demonstrated in early 1984, achieving a record design point efficiency of 29% for electricity production and a projected net annual average efficiency of 18%, the highest achieved by any solar electric system.

Since this successful experiment, the dish-Stirling concept was developed by McDonnell Douglas and United Stirling as a private venture to market the technology commercially. (McDonnell Douglas dropped out of this venture when the company made a corporate decision in 1987 to leave the field of solar energy commercialization.) One of their dishes was tested near the Barstow 10-MWe pilot plant by Southern California Edison.

8.4.2.2.3 Small Community Experiments Two demonstrations of dish-electric technology at Osage City, Kansas, and Molokai, Hawaii, at about 50 kWe were awarded to Power Kinetics Inc. (PKI) on the basis of competitive procurements in 1983 and 1984. PKI utilizes a concentrator design and engine module not part of the mainline DOE module development program, although an early version of the PKI moving-slat concentrator was tested briefly at a concrete block company in Topeka in the early 1980s. In addition, a small solar desalination plant utilizing eighteen of the PKI collectors has been built in Yanbu, Saudi Arabia, under the joint Saudi-American solar agreement, SOLERAS (see chapter 28). The concentrator and other components for the Osage City and Molokai demonstrations are being readied (1986) for test and evaluation at the DRTF in Albuquerque. The rather unusual ability of this dish design to capture the majority of the DOE dish-electric demonstrations is due in large measure to the low production costs reported for the Yanbu installation by the manufacturing company, a spin-off of Rensselaer Polytechnic Institute (RPI).

8.4.2.2.4 Industrial Process Heat Systems The application of solar thermal heat at temperatures above 600°F for industrial process heat (IPH) has been limited, primarily because of thermal storage capacity limitations, and is not described here. High-temperature IPH processes, such as those used to produce fuels and chemicals, require continuous 24-hour operation for both technical and economic reasons (Black & Veatch and Ralph M. Parsons 1982). High-temperature continuous processes require extended startup and shutdown times because of the need to retain process stability and to minimize damage to containment vessels, piping, and valves from thermal stress. Therefore, shutdown is always avoided except for scheduled maintenance. Furthermore, continuous high process temperatures are required to maintain high process reaction efficiencies. Both high efficiency and maximum utilization of the capital-intensive process equipment are necessary for a competitive, low-cost product.

As of 1988, high-temperature, economic batch processes for the production of fuels and chemicals had not been successfully identified.

8.4.3 Execution of the Programs

Section 8.4.1 described the program implementation strategy as consisting of three parts: (1) development of an industrial base for design and manufacture of the technology, (2) development of in-house R&D capability at DOE laboratories, and (3) development of a user and customer base for high-temperature solar thermal systems. In effect, the three parts amount to the formation of an industry-government partnership for the purpose of bringing a new, civilian technology to the point of commercial application, a feat without parallel in the past.

8.4.3.1 Industry-Government Partnership
The federal government had established a national policy, and hence a commitment, to the development of practical means to employ solar energy on a commercial scale (PL 93-473). In so doing the government had assumed the financial risk of this development. On the other hand, a private sector industry had to be created and nurtured to successfully implement the national policy on the scale envisioned in the 1970s. This industry would bring design, manufacturing, and construction expertise, and, most important, a sense of what was required for private sector user acceptance of the new technology.

8.4.3.2 Building the Industrial Contractor Base
Industrial interest and capabilities were built by the simple process of public solicitation of competitive bids to deliver certain services to the federal government. This mechanism was helped along by two very important circumstances: (1) solar energy was an attractive technology with nearly universal appeal; and (2) the aerospace industry was looking for something to take up the slack in a flat Department of Defense market in the mid-1970s. The earliest respondents to the RFPs issued by NASA, SAN, ALO and Sandia National Laboratories were the aerospace systems integrators, who in turn organized industrial teams to provide all of the capabilities in power engineering, component design, and manufacture required for project execution. Universities were also added to these teams where special research expertise was required—in the traditional manner of industry-university relationships. More importantly, electric utility companies were part of many of these teams, especially in the central receiver program, to bring the user point of view to bear. A partial listing of the major participants in the program is given in table 8.1.

Table 8.1
Major industrial and university participants in the solar thermal program

Aerospace	Boeing
	Honeywell
	Martin Marietta Corporation
	McDonnell Douglas Astronautics Co.
	Rockwell International
Engineers-architects (traditional system designers/integrators of utility and industrial power plants)	Bechtel
	Black & Veatch
	Gibbs & Hill
	Stearns Roger
Component and system manufacturers	Acurex
	Advanco
	Babcock & Wilcox
	Foster Wheeler
	Power Kinetics, Inc.
	Sanders Associates
	Solar Kinetics, Inc.
Utility users	Arizona Public Service
	El Paso Electric
	Georgia Power
	Pacific Gas & Electric
	Public Service of New Mexico
	Public Service of Oklahoma
	Sierra Pacific
	Southern California Edison
	Western Power Division of Centel
Industrial users	American Gypsum
	Amfac Sugar Mills
	Capital Concrete
	Valley Nitrogen
Universities	Georgia Institute of Technology
	University of Houston
	University of Minnesota

Table 8.2
Technical accomplishments of the federal central receiver program (1988)

Elements	Accomplishments
Test facilities	
• Central Receiver Test Facility (CRTF)	5-MW$_t$ facility built and successfully operated for eleven years at Sandia National Laboratories, Albuquerque, NM.
Components/subsystems	
• Heliostats (glass/metal)	Three generations built and tested. Over 2,100 heliostats deployed.
• Receivers	Three generations of water/steam receivers built and tested.
	One sodium receiver built and tested.
	Two generations of molten nitrate salt receivers built, tested and operated.
• Thermal storage	Oil/rock system built, tested and operated.
	Molten salt system built, tested and operated.
• Steam generators	Hot molten nitrate salt-to-steam heat exchanger built, tested and operated.
	Hot sodium-to-steam heat exchanger built, tested and operated. International Energy Agency (IEA).
Systems	
• Proof-of-concept experiments	0.5-MWe sodium/steam experiment built and operated at Almeria, Spain (IEA).
	750-kWe molten salt experiment built and successfully operated at the Central Receiver Test Facility, Albuquerque, NM.
• Pilot plant	10 MWe water/steam plant built and successfully operated at Barstow, CA.

8.5   Evaluation of the Implementation Program

The evaluation of program implementation is made in terms of the initial mandate for successful "development and demonstration of practical means to employ solar thermal energy on a commercial scale" (PL 93-473)

8.5.1   Technical Accomplishments

To answer the question of whether technically practical means were developed, the accomplishments of both central and distributed receiver programs are listed in tables 8.2 and 8.3. These tables represent the most visible proof of program implementation, and the accomplishments must be judged successful in view of the end products at the close of 1988.

Table 8.3
Technical accomplishments of the federal distributed receiver program (1988)

Elements	Status
Test facilities	
• Test bed concentrators	Two dishes built and operated at JPL, Pasadena, CA.
• Distributed Receiver Test Facility (DRTF)	Facility built and operation initiated in 1984 at Sandia National Laboratories, Albuquerque, NM.
Components/subsystems	
• Concentrators	Three generations built and tested. 136 parabolic dishes deployed.
• Receivers	Prototype organic liquid, air, and helium receivers built, tested, and operated.
• Heat engines	Prototype organic Rankine cycle and Stirling cycle engines built, tested, and operated.
Systems[a]	
• Total Energy	3-MW_t facility built and successfully operated at Shenandoah, GA.
• Dish Electric	25-kWe Vanguard module built and successfully operated at Rancho Mirage, CA.
• Desalination	53,000 gallons ($200m^3$) per day fresh water facility built and operated at Yanbu, Saudi Arabia.

[a] Privately funded projects such as the LaJet system at Warner Springs, CA, and the Luz International SEGS units near Daggett and Kramer Junction, CA. are included in section 8.5.3.

The Barstow, California, pilot plant, the Shenandoah, Georgia, total energy facility, and the Vanguard dish-electric module are the appropriate flagships of the central receiver and distributed receiver (precommercial) implementation programs. They stop considerably short of actual demonstration of these technologies in the usual accepted sense, that is, providing evidence of technical maturity and economic feasibility sufficient to attract private investment capital in the first commercial plants. Nevertheless, these projects provide the evidence that the two principal, high-temperature solar thermal technologies of this century can be built and operated in the utility/industrial user environment that they are intended to serve. The plants were not intended to be the ultimate in efficiency, reliability, and cost of operation but rather to give the country experience and confidence in high-temperature solar thermal systems sufficient to provide direction for further development and commercial deployment. They have achieved those objectives. How well the implementation program enabled the private industrial suppliers and users to build economically competitive commercial plants is discussed in section 8.5.3.

8.5.2   Industrial Supplier and User Base

The ability to employ solar thermal energy on a commercial scale requires a supplier industry and interested customers. Table 8.1 lists the major industrial and user participants in the federal solar thermal program during 1973–1985. The effort to develop multiple private sources of design, manufacture, construction, and operational expertise was very successful, particularly when measured by the willingness of industrial and user participants to cost-share with the government to keep development activities alive during 1982–1985. However, rapidly shrinking federal budgets largely eroded this base during the last half of the decade.

Southwestern electric utilities have been and remain committed to the use of central receiver technology for bulk power generation when it becomes economically competitive. However, a user/customer base for parabolic dish technology has been slow to develop, and its application role is still not clearly defined.

8.5.3   Competitive Position and Private Ventures

The drive to develop economically competitive high-temperature solar thermal energy systems was sustained and dominant throughout the history of the federal solar program. Consequently, considerable effort has been expended toward reducing component and system costs, and the trends in cost reduction provide the evidence for evaluation of the implementation program. Both central receiver and distributed system implementation programs have set component and system cost targets that will yield commercially competitive energy from such systems (SNLL 1984).

For central receiver systems, heliostats represent the largest single cost component. Figure 8.2 depicts the historical data for costs per unit area of reflecting surface: the federal purchase of 222 heliostats for the Central Receiver Test Facility at Albuquerque in 1976–1977; the federal purchase of 1,818 heliostats for the 10-MWe Barstow, California, pilot plant in 1980; and the ARCO Solar firm price quote to PG&E for the 30-MWe prototype commercial plant at Carissa Plains, California, in 1983 (not built). The cost trend is very encouraging and the reduction still required may possibly be achieved with the new stressed-membrane heliostat now under development.

Total system costs for commercial-sized central receivers have been estimated many times based largely on scale-up of preliminary design

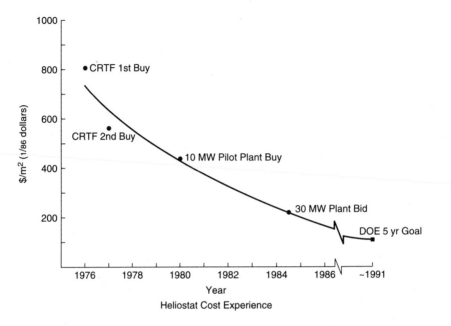

Figure 8.2
Heliostat cost experience curve.

information (see table 8.4 and figure 8.3); but these are not reliable figures for evaluation of competitive position. Serious attempts to develop private financing in 1982–1984 for a 30-MWe and 100-MWe plant to be located at Carissa Plains and Lucerne, California, respectively, are more instructive. Despite liberal federal and state tax incentives then available, the cost and performance risks of central receiver plants were still too high to attract the needed private capital ($150–500 million) and the plants had not been built as of early 1990.

The federal development of parabolic dish systems and components has not progressed sufficiently to provide hard, public cost numbers for trend analysis. On the other hand, a private venture by LaJet Energy Company in 1984 succeeded in financing, building, and operating a 4.4-MWe parabolic dish, distributed collector (but centralized generating facility) at Warner Springs, California, for the sale of electricity to San Diego Gas & Electric Company. The plants consists of 700 stretched-membrane dish collectors, each consisting of 24 stretched polymer membrane facets each 5 feet in diameter. Superheated steam at 750°F is generated in the collec-

Table 8.4
Solar thermal central receiver direct capital costs historical trends: 1981–1987

Plant/data source	Size MWe (net)	Solar multiple	Cost basis	Direct capital cost (1/1/87: $/kWe)	Heliostat cost (1/1/87: $/ft² ($/m²))	Direct capital cost less heliostat cost 1/1/87 ($/kWe)
Solar One (water/steam) SAN86-8002	10	1.0	As constructed (1981)	10,800	59.60 (642)	6,228
Solar 100 (salt)			As bid (1982)			
MMC	100	1.25	MMC	4,216	31.40 (338)	2,602
MDAC	100	1.23	MDAC	4,110	25.60 (276)	2,813
Saguaro stand-alone (salt) APS/B&V	58	1.1	Preliminary design (1983)	3,963	36.20 (390)	1,054
EPRI TAG (water/steam hybrid) AP-3982	152	1.0	Bechtel study (1984)	2,769	19.60 (211)	1,824
APS utility study (salt)	100	1.8	Conceptual design (1987)	2,516	9.30 (100)	1,638
	200	1.8		1,959	7 (75)	1,283

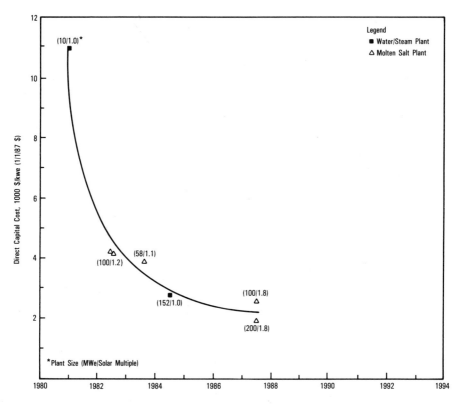

Figure 8.3
Direct capital costs of solar thermal central receiver plants, 1981–1987 (based on table 8.4).

tor/receivers and piped to a central Rankine cycle turbine generator. In 1985 McDonnell Douglas formed a wholly owned subsidiary, Energy Systems Ventures (ESV), to manufacture, market, and deploy parabolic dish–Stirling engine modules with a capacity of 25 kWe each but abandoned it in 1987. Without question, the federal program, especially the Shenandoah, Georgia, and Vanguard system experiments, has provided the technical basis for these ventures. (A more extensive review of component and system costs and their trends under the federal program is given by Hansen and Tennant (1988).

Electric power generation systems based on parabolic trough collectors have enjoyed excellent success in a private commercial venture by Luz International, a solar thermal electric development company based in Los Angeles. The trough collector designs drew heavily on those developed by the U.S. federal solar thermal program. Luz constructed seven plants totaling 194 MWe of installed capacity near Daggett and Kramer Junction, California. The plants are owned by investor-owned limited partnerships and operated by Luz, and the power is sold to the Southern California Edison Company (SCE) under power purchase agreements based on SCE's avoided costs under the Public Utility Regulation Policies Act (PURPA) legislation. The 194 MW of capacity is comprised of an initial 14-MW module and six succeeding 30-MW modules called "Solar Energy Generating Systems" (SEGS). All of these SEGS modules are hybrid, solar/fossil fuel energy systems in which both solar energy and natural gas are used to generate superheated steam to achieve high electrical conversion efficiency. The steam conditions so obtained range from 555 psia/780°F in the initial unit to 1,450 psia/950°F in the final two units constructed (Kearney, Price, and Jensen 1989). This hybrid solar/fossil fuel energy arrangement also allows Luz to deliver the guaranteed electrical production to SCE throughout the year. A final 400 MWe of capacity, made up of five 80-MWe SEGS modules, began construction near Harper Lake, California, 1988.

These investor-owned SEGS systems have been able to provide economic electrical energy by virtue of a combination of favorable circumstances that were recognized and vigorously pursued by Luz:

1. The inherent modulal nature of distributed, solar trough collector systems makes it possible to proceed with a commercial venture in small steps that can be financed more easily. Equally important, modularity also allows improvements in design and operation as experience is gained with succeeding modules.

2. The federal distributed receiver systems program had already provided a technology development history and foundation on which Luz could base its parabolic trough collector/receiver and system designs.

3. The PURPA legislation and the active encouragement of the California Energy Commission in the early 1980s made it possible for Luz to sell its electric power to the Southern California Edison Company.

4. Southern California Edison was willing to sign a series of five present and future power purchase contracts with Luz using California's Standard Offer no. 4, which, designed to encourage the development of renewable energy sources, was very attractive financially to independent power producers.

5. The existence of federal tax credits for this type of venture.

6. The availability of natural gas at low, regulated rates. Circumstances 1 through 5 were the most crucial for success. Other major factors were the willingness of Luz to take high risks, and the creative financing methods that were developed to reduce investor risk.

These circumstances notwithstanding, the SEGS plants represent the high point of the 1980s in the implementation of commercially viable solar thermal electric power plants in the world.

[Editors' note: In late 1991 the Luz parent company went into bankruptcy, but the operating company continues to run the already-built plants. See Becker 1992.]

8.5.4 Summary Evaluation

The high-temperature solar thermal program has at least met the spirit of the two goals stated in the 1974 Project Independence report to provide a full-system capability for electrical power and for total energy systems. But the wider goal of "widespread production in the 1980s to meet utility requirements" was clearly not met. The investor-owned Luz SEGS units, delivering over 300 MWe, represent the only significant contribution to this goal. Under federal funding, central receiver systems have been brought to a higher degree of technical maturity than distributed, parabolic dish systems. This situation is due mainly to much earlier emphasis on central receivers at both the programmatic and funding levels. A coordinated and centralized component and subsystem test facility at Sandia Albuquerque (CRTF) also played an important role in widening this disparity. Similar facilities for parabolic dish systems did not become operational under 3–4 years later. Ironically, parabolic trough technology for electric power production was never given priority under the federal program.

Development of an industrial design and manufacturing base for both central and distributed receiver systems (including parabolic troughs) was quite successful. End users of central receivers were much more visible and involved in the technology development than end users of the dish systems, thanks primarily to the successful early efforts to involve the well-defined utility industry, which is oriented toward large centralized power plants.

Both the Barstow, California, and Shenandoah, Georgia, plants have successfully achieved most of their original objectives and have provided valuable operating data and experience for continued technology development toward commercially viable systems. Unfortunately, Barstow, by virtue of its size, cost and visibility, has been measured by some cost and performance objectives that it was never designed or intended to satisfy.

By satisfying the overall objective of getting high-temperature solar thermal technology into the private sector, the parabolic trough and dish technology must be judged the more successful, although the customer/ user base for the dish is still untefined. This success is due principally to the modular character of distributed systems and the much lower, overall capital cost of deploying a reasonably sized, commercial system. By any measure, however, both central and distributed high-temperature solar thermal programs must be judged successful in view of their ability to field operating systems of sufficient size to prove their ultimate potential for satisfying a portion of the country's future energy requirements. Significant cost reductions by factors of 2–3 and equipment performance guarantees are still required to achieve commercial feasibility, in the absence of tax incentives and favorable power purchase contracts, but no one doubts that the goals are achievable with continued, sustained, federal funding of development and demonstration programs. Unfortunately, because of a drastically changed funding philosophy regarding renewable energy sources, such support all but disappeared in the mid-1980s.

8.6 Lessons Learned

Fifteen years of federal effort to develop and demonstrate high-temperature solar thermal technology have not been sufficient to attract private investors for commercial development on a wide scale, although the 360 MWe of investor-owned Luz SEGS units constitutes a significant beginning. The failure of wide-scale development lies less with the technology itself than with a combination of external factors, principally the fall in oil prices and electricity demand at a time when solar thermal development was hitting its stride in the early 1980s. However, much can still be learned from the experience gained in this first-of-a-kind attempt by the federal government to develop an essentially untried technology that depended on a private customer base largely unfamiliar with the technology and to bring it to the point of commercial feasibility.

Specific Lessons

• The development of a government-industry partnership for implementing a "civilian" energy technology like solar thermal proved to be a very effective strategy. It was important to define carefully the role of each of the participants and to develop a mutual respect for what each contributed.

• Central receiver systems as originally envisioned for utility applications were too large and expensive to allow development through both pilot and demonstration states without substantial federal budget outlays, which, though recognized by many as necessary, were never forthcoming. As a result, only the Barstow, California, 10-MWe pilot plant was built. The administration and Congress were unwilling to commit funds to larger utility demonstrations and developers were unable to attract sufficient private money.

Meanwhile, smaller, modular systems like the parabolic trough and dish systems were able to proceed through successive field development tests at a fraction of the cost of a single central receiver plant and to attract private investment. Recognition of this advantage came late in the central receiver program and essentially stymied central receiver development until smaller cost-effective modules were designed in 1987–1988. Unfortunately, the federal funding of this program had by then shrunk to such a low level that even the design studies could not be continued.

In fact, most technologies that are now commercial began on a small scale and grew larger as economies of scale made continued technology improvements feasible.

• The federal planning process was too optimistic in estimating achievable rates of technology development and of component cost reduction, while the federal budgeting process underestimated the costs involved in reaching and achieving successful demonstration. In projecting commercialization activity, there was often confusion between component and system cost goals and achievable costs with current and developing designs.

• Mass production of heliostats was often invoked as a means of reaching commercial cost goals. Such a strategy resulted in a chicken-and-egg conceptual dilemma wherein large, cost-competitive central receiver systems could not be built until such systems were actually ordered and deployed—a scenario not likely to attract investors. However, the correlation between mass-production and collector costs was largely disproved when it was, in fact, demonstrated by ARCO Solar in the production of 16,145,600 ft² (150,000 m²) of heliostat-type photovoltaic trackers for the Carissa Plains, California, installation, that the savings projected by tooling up for a production capacity of ten times this amount was zero, if not negative (Caldwell 1986). This is not to say, however, that larger production rates and continued R&D will not lead to lower costs over time.

• The issues of system reliability, availability, and annual energy production—bottom-line attributes for utility investors whose return on investment depends on them—were not given sufficient attention by the national laboratories and the aerospace system integrators until very late in the program. They were not alone in this oversight, however, as attractive power purchase contracts under the Public Utility Regulatory Practices Act (PURPA) and tax incentives became powerful motivators to rush unproven systems into commercial operation before they were technically ready—for example, many early wind machines in California in the early 1980s.

• Budget allocations among technology options, especially in the dish-electric program, became increasingly difficult as budgets declined, but program managers and contractors were reluctant to eliminate any of several parallel development paths (organic Rankine, air-Brayton, and Stirling cycles). As a consequence, no one technology emerged as the clear choice in a time of drastically reduced budgets and possible program phaseout. The lesson is that in a time of lean budgets, it is better to have one well-advanced option than three emerging ones.

General Lessons

• A federally planned and funded technology development and demonstration program that includes the involvement and commitment of private sector designers, manufacturers, and users is basic to the goal of commercial deployment of a new energy resource.

• Sustaining the federal commitment is crucial, when such programs are deemed by the Congress to be in the national interest and are enacted and signed into law. Later withdrawal of that commitment without due process undermines private sector confidence in the government as a reliable partner.

• The difficult transition from technical readiness to commercialization is a process in which technical and economic risks are reduced in a series of successful demonstrations. Private technology development and cost reductions are a continuing process even after commercial deployment begins.

• Significant cost, performance, and/or regulatory compliance advantages over conventional, state-of-the-art competing technologies must be evident before private sector deployment of a new energy technology can be realized on a significant scale. Moving from the conventional to the new always requires economic motivation above and beyond a simple, long-term affection for a new technology.

Acknowledgments

Grateful acknowledgment is given to the following individuals who, through their significant involvement in the solar thermal program, were willing to share their experiences and memories of the early days: Gerry Braun, Howard Coleman, Lloyd Herwig, Jim Leonard, Hank Marvin, Jim Rannels, and Dwain Spencer. Any errors found here are the author's, not theirs.

References

Arizona Public Service Company. 1988. *Utility Solar Central Receiver Study*. DOE/AL/38741-1. Washington, DC: U.S. Department of Energy, November.

Becker, N. 1992. "The Demise of Luz: A Case Study." *Solar Today* January/February.

Bergan, N. E. 1986. *Testing of the Molten Salt Electric ExPeriment Solar Central Receiver in an External Configuration*. SAND-86-8010. Albuquerque, NM: Sandia National Laboratories.

Black & Veatch and Ralph M. Parsons Co. 1982. *Assessment of Fuels and Chemicals Production Using Solar Thermal Energy: Final Report*. DOE contract DE-AC03-81SF11496. January.

Butti, K., and J. Perlin. 1980. *A Golden Thread*. New York: Van Nostrand Reinhold.

Caldwell, J. H., Jr. 1984. Private communication, 15 February.

FEA (Federal Energy Administration). 1974. *Project Independence: FEA Final Task Force Report on Solar Energy*. Sec. 3. Washington, DC.

Hansen, C. E. and W. L. Tennant. 1988. "Historical Cost Review." In *Solar Heat Technologies: Fundamentals and Applications*. Vol. 3, *Economic Analysis of Solar Thermal Energy Systems*, ed. R. E. West and F. Kreith. Cambridge: MIT Press.

Kearney, D. W., H. Price, and C. Jensen. 1989. "Design and Operation of the Luz Parabolic Trough Solar Electric Generating Plants." Paper presented at the American Solar Energy Society Annual Meeting, June.

NSF/NASA (National Science Foundation/National Aeronautics and Space Administration). 1972. Solar Energy Panel. *Solar Energy as a National Resource.* Washington, DC, October.

Public Law 93-473. Solar Energy Research, Development, and Demonstration Act of 1974. 26 October.

SNLL (Sandia National Laboratories, Livermore). 1984. *National Solar Thermal Technology Program: 5-year R&D Development Plan, 1985–89.* Prepared for U.S. Department of Energy. Livermore, CA, December.

III SOLAR THERMAL DEMONSTRATIONS AND CONSTRUCTION

9 Residential Buildings

Murrey D. Goldberg

9.1 Residential Demonstration Program

In 1973, in response to the OPEC oil embargo and the resultant price escalations and shortages, the nation began to explore strategies to reduce its dependence on nonrenewable energy resources, particularly imported oil. The theme was energy independence, the eventual replacement of imported fuels with resources indigenous to the country.

One approach explored by the Congress was a national program to encourage increased use of solar energy. To implement this program, Congress passed several acts, including the Solar Heating and Cooling Demonstration Act of 1974 (PL 93-409; see chapter 2 for more information on the passage of this legislation). PL 93-409 provided for a number of research, development, and demonstration activities. Overall program authority was eventually given to the Energy Research and Development Administration (ERDA) which eventually became part of the Department of Energy (DOE). The Department of Housing and Urban Development (HUD) was legislatively mandated to plan and conduct the residential demonstration called for in the law.

During the debate that preceded passage of PL 93-409, proponents of a commercial and residential demonstration maintained that the solar industry was ready for a major marketplace demonstration. They claimed (1) that the technology was known and available to the solar energy industry, (2) that the necessary products were already being manufactured, and (3) that a significant demonstration program was all the industry needed to provide economically viable and technically reliable solar systems on a large scale.

Other testimony at the congressional hearings, including the official government position as approved by the Office of Management and Budget (OMB), questioned both the state of the art and the state of the industry, particularly its lack of a reliable design, production, installation, and service infrastructure. As a principal spokesman for this point of view, HUD proposed instead additional research and development and a smaller, more controlled demonstration.

Few current manufacturers, HUD argued, provided a system delivery approach to the marketplace or were involved in the design or installation of systems containing their components. Their only formal ties to installers were loosely drawn distributership arrangements, mostly with established local heating, ventilating, and air-conditioning (HVAC) contractors. Although reliable for HVAC work, these contractors lacked the understanding and expertise needed to service solar energy systems. In addition, neither the compatibility of the various components packaged into solar energy systems nor the reliability of the systems themselves had been tested or evaluated in different environments. No industry standards, not even interim ones, existed to determine or compare system or component suitability for particular applications. A final point of concern regarding a large-scale marketplace demonstration was the diverse and fragmented nature of the industry and the problems this raised with respect to reasonable and enforceable purchaser warranties, including the availability of knowledgeable maintenance and repair organizations to provide services both during and after warranty periods.

In view of the above concerns, HUD concluded there was a significant risk to unsuspecting purchasers of homes with demonstration systems, who might take government involvement in and sponsorship of a major demonstration as assuring reliability of the systems. This would place HUD and the government in an unwarranted position of moral liability if such systems proved faulty.

Despite these concerns, Congress chose to proceed with a large-scale demonstration of both residential and commercial systems through PL 93-409, although it did include concurrent research and development activities. Interim performance criteria were to be determined and disseminated, testing procedures to be developed, and a contingency fund to be created for system maintenance and repair. Other activities included data collection and analysis, design and installation training, and performance monitoring.

9.1.1 Goals and Objectives

The primary goal of the residential demonstration program was to induce the growth of residential use of solar energy in both new and retrofit construction. Upon completion of the program, a viable and competitive solar industry was to be in place that could respond to increased market demand for reliable solar products.

The main program objectives aimed at reaching this goal were to

• identify and demonstrate solar equipment available for use in new and existing dwelling units;

• provide a database of hardware characteristics, performance, and acceptability;

• identify institutional barriers to the widespread use of solar heating and cooling in residential applications and recommend ways to remove these barriers; and

• provide industry and regulatory bodies with experience necessary to continue use of solar energy in residential buildings after the program ended.

Development of a functioning solar industry, with established marketing ties to the builder/developer community, was considered an absolute necessity for ready acceptance of solar products in the marketplace. The demonstration program was structured to stimulate direct interactions between industry and community, both highly diverse and fractionated, with government encouragement and overview. HUD elected to use a grant approach, rather than contract directly for project or solar system construction, to avoid direct involvement in the market process. Designers, builders, and solar equipment contractors, therefore, functioned in a normal, free-market atmosphere. HUD could provide oversight as needed and technical assistance when requested but on a generally hands-off basis. Developing market forces were left to work as would be required in a future self-sustaining industry.

9.1.2 Methodology

Successful builder applicants received grants for all or part of the costs for design, purchase, and installation of the solar system. Funds were provided through 943 grants for either purchase and installation of solar heating and cooling equipment or design of such systems, or both. When adjustment is made for design-only and terminated grants, 497 grants resulted in construction. In all, 10,098 living units (individual family house or apartment) were constructed, using 1,255 solar energy systems.

There were three categories of grants included in the program (each described in greater detail below):

• *Site-system projects* involved applicants from specific locations, determined by HUD, who were willing to build a project using a HUD-prescribed system that would be integrated into the design.

• *Design-only projects* involved applicants who wished to produce a design for a passive residence and allow HUD to publish the design for use by others.

• *Integrated projects* involved applicants from any locale who wished to incorporate a particular system of their choice into their proposed construction project.

Awarded in cycles through eight formal national competitive solicitations, grants were given to builder/developers, housing authorities, universities, local government agencies, and similar organizations throughout the United States and its territories. First awards were made in early 1976; final awards were made in late 1979. Management support for the program was provided by contract with Boeing Aerospace Company (now BE&C Engineers) and by subcontracts with Dubin Bloome Associates, the Real Estate Research Corporation (RERC), and the American Institute of Architects Research Corporation.

9.1.3 Data Collection

Data collection, analysis, and dissemination were considered important program elements. Three types of data were compiled: nontechnical, technical, and instrumented. Nontechnical data, which were collected through interviews and market surveys, covered marketing activities, public acceptance, financing, utility costs, and repair and maintenance. Market acceptance analyses were conducted to determine, among other issues, to what extent solar made a difference in how a home was perceived or treated by various marketplace participants. Questionnaires were tailored to such specific participants as builders, purchasers, mortgage lenders, and code officials. Questionnaire results were compared with results from similar surveys conducted in the nonsolar marketplace to determine policy and attitude differences.

Utility consumption data were gathered from more than 11,000 utility bills. Energy use in solar homes was compared with that in nonsolar homes in the same area. Other energy consumption comparisons were made after adjusting for climate or for building size.

Building and solar system design characteristics constituted the bulk of the technical data collected. Data for active solar energy systems were organized to make possible computer or hand calculations of performance by means of the F-Chart method (see *Active Solar Systems*, volume 6 of this series). These data included collector tilt and azimuth, absorber material and coatings, storage size and type, kind of freeze protection, and expected operating parameters. The solar load ratio (SLR) method (see *Passive Solar Buildings*, volume 7 of this series) was used to calculate performance for passive systems. Predicted performance was calculated for each system, using the appropriate method, as a basis for comparison with actual performance data.

Instrumented data were collected as part of a cooperative effort between HUD and DOE through the National Solar Data Network (NSDN). HUD established each instrumentation design and procured, calibrated, and installed the sensors. Thermal energy flow data from onsite instruments were sent to a central computer and processed into user-oriented performance factors. More than 100 solar energy systems from this program were instrumented, and performance results were included as part of the NSDN analysis effort.

9.2   Program Grants

9.2.1   Site-System Projects

For these projects, HUD preselected a set of site locations from among standard metropolitan statistical areas (SMSAs) and a set of system types. A matrix plan matched systems to sites so that comparisons could be made of system suitability in various parts of the country (ADL 1976). Builders were to integrate the selected solar systems into residences of their design, with the program paying all costs of the solar addition. Five annual cycles of ten matched site-system pairs each were initially planned.

The response to a solicitation of interest from builders for the first set of ten locations in 1976 was very disappointing. All a builder had to do to apply was indicate an interest in a single-family or multifamily project that the builder intended to build and provide a statement of qualifications. No technical effort was required at this stage. In spite of the minimal effort required, only eleven applications were received by the closing date for the solicitation.

The program plan called for a successful applicant to receive a fixed-price grant to integrate the solar equipment into the project design and to prepare a cost estimate for solar construction costs. Following this, the applicant would negotiate with HUD for a lump-sum addition to the grant to cover solar costs. If a project were selected for instrumentation, funds would be added to cover design and installation. A phased, negotiated grant, it was thought, would negate applicant concerns about risk.

A poll of those receiving the solicitation who did not respond revealed many reasons for their decision not to reply. The most consistently stated reasons included (1) an unwillingness to accept responsibility for a system type and manufacture not of their choosing, (2) a feeling that, given their inexperience, a fixed-price design grant exposed them to unknown risks, and (3) the judgment that the significance of domestic hot water systems (the majority of the selected system types) did not justify the effort. In July 1976 the Grant Application Review Panel (GARP) awarded twelve grants in seven of the first set of ten locations: Boston, Atlanta, Columbus, Denver, Tucson, Los Angeles, and Honolulu; no responses were received from Richmond, Des Moines, and Albany.

In addition to the limited response, there were many other disappointments with the site-system approach. Builder design integration efforts were crude and complicated, and communications between builders and assigned manufacturers were sometimes strained and contentious. The supplier-installer-builder interaction clearly showed the lack of choice. If a major program goal was to create a free-standing industry, the forced relationships of this approach were not the way. Most builders had difficulty coping with the design involvement required of them; they were most comfortable with developing a complete house design package, then proceeding promptly to field construction. Their heavy involvement in field activities made design coordination difficult; concurrent design and construction phases interrupted normal building practices. Intrusion of needs for instrumentation substantially compounded builder problems.

The original twelve grants were awarded for forty-four solar energy systems, all for single-family detached houses. Nine of the twelve grantees made it through design review without withdrawing. One of the nine withdrew after award of construction funds but before construction began; six of the remaining eight grantees chose to proceed with fewer units than originally intended. The eight finalists completed construction of twenty-two systems on twenty-two houses. Only four grantees went through with the instrumentation of seven systems. Progress on all projects was very slow; all were finally completed in the first quarter of 1980.

In December 1976 HUD decided to cancel all further site-system solicitations. It was clear that builders were much more interested in the integrated projects approach.

9.2.2 Integrated Projects

There were six cycles of integrated project awards, the first in December 1975 and the last in October–November 1979. As successive cycles over the four years were reviewed by HUD and its consultants, previous proposal deficiencies and field problems were translated into updated instructions and requirements for the next cycle. In this way, HUD sought to improve both proposal quality and the understanding of solar requirements by designers, developers, and contractors.

Progressive development of better-designed and more reliable solar products was not limited to technical information and design calculations. A warranty to the purchaser was required in cycle 1, but, by cycle 4, all solar systems had to be designed, manufactured, and installed in accordance with HUD's *Intermediate Minimum Property Standards* (*IMPS*) (HUD 1977). Builders were required to offer an acceptable warranty of not less than five years for collectors and one year for system installation. In sum, an award winner in cycle 2 could not have been one in cycle 4 without having become better qualified and without having shown more understanding of the residential solar application process.

All project awards in each cycle were judged finally by the GARP after evaluation by a panel of experts for applicant qualification, project development and opportunity, and technical acceptability. The complex review process allowed numerous opportunities for review of expert rejections. Following the full review process, applications were graded and presented to the GARP in rank order for award consideration. Each presentation was made by a reviewer team that acted as a project advocate, detailing the project and answering any technical or nontechnical questions raised. GARP-approved scores were then used to regrade and rerank the projects. Costs were tallied and an award cutoff established to match available funds in the cycle. Table 9.1 recapitulates integrated project awards by cycle; figure 9.1 shows the geographical dispersion, by state, of the awards.

9.2.3 Design-Only Projects

Design-only project grants were awarded at three points in the program. There were 108 awards during a passive design competition, 61 during cycle 5, and two during the site-system cycle 1. The passive design competition and cycle 5 grants were planned from the start. The site-system

Table 9.1
Summary of integrated project grants

Cycle	Awarded	Rejected	Cancelled	Completed
1	55	6	9	40
2	102	7	25	70
3	169	2	22	145
4	48	0	12	36
4A	96	1	20	75
5	105	5	31	69
Total	575	21	119	435

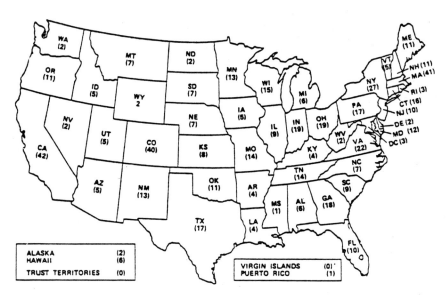

Figure 9.1
Location of grants, Integrated Projects Series.

designs resulted simply from two grantees unable or unwilling to proceed with construction. They are included to clarify the numbers.

9.2.4   Passive Design Competition

This competition was announced in May 1978. Qualified parties were invited to apply for a design award for either a new construction (category A) or a retrofit (category B). Both categories were limited to single-family housing, attached or detached. New construction awards were set at a lump sum of $5,000; retrofit awards were set at $2,000. In addition, new construction projects planned for speculative, open-market sale were eligible for construction grants covering one to five units. These grants were fixed at $7,000 for the first unit and $2,000 each for up to four additional units.

Each project application required the designer to establish his or her eligibility and to provide sufficient information to determine the appropriate grant category. Designers had to provide a set of required design considerations and calculations for a proper passive solar design using tables and formulas supplied by HUD. This provided data that were common to all designs. For technically acceptable applications, the awards jury considered project marketability and repeatability.

The GARP evaluated all technical concerns that turned up in the review process and passed them on to grantees at the time of award. In some cases, applicants were invited to a design workshop before award. Hands-on technical consultation was provided for resolving design deficiencies and clarifying design needs. Table 9.2 summarizes the passive design competition; figure 9.2 shows the geographical dispersion, by state, of the awards.

9.2.5   Cycle 5 Projects

Cycle 5 was advertised in March 1977. In a two-step process, applicants had to apply for a project in one of two categories. Category 1 covered retrofits for low- to moderate-income, urban, multifamily buildings sponsored by neighborhood associations. Category 2 included new single-family houses built for sale on the open market. Projects in either category had to include significant energy conservation features and reasonable passive solar elements. Active solar elements could be included if they complemented the conservation and passive solar features.

Table 9.2
Summary of passive design competition grants

	Grants Awarded	Grants Completed
Design Retrofit	17	17 (17)*
Design New Construction	145	145 (91)
Construction	80	54 (0)
Total	242	216 (108)

*( ) portion of grants that were for design only

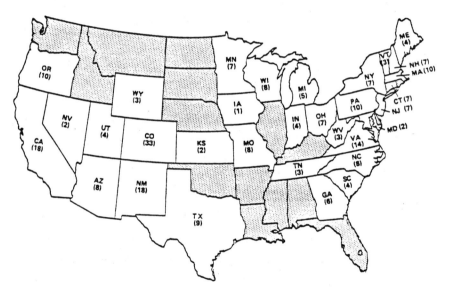

Figure 9.2
Location of grants, Passive Design Competition.

Application requirements for a step one design assistance award were modest, consisting of a one-page form and four attachments. Three attachments described the qualifications of the applicant, the project designer, and the solar system designer. The fourth provided a general description of the project, proposed energy conservation and solar features, a project schedule, and information on funding sources other than the grant.

Applicants had to be selected for a step one design award to be eligible for a step two construction award. Each had to complete a step one

Table 9.3
Summary of Cycle 5, Step One grants

Category	Awarded	Rejected	Annulled	Completed
1	25	0	1	24
2	114	3	5	106
Total	139	3	6	130

statement of work before receiving payment—$5,000 for a category 1 project and $2,000 for a category 2 project. A request for a step two construction grant was included in the step one work statement. All who satisfied step one requirements were paid whether or not they received a step two award.

An evaluation panel reviewed the 880 step one applications received by the closing date. Following consideration by the GARP, HUD announced the award of 25 category 1 and 114 category 2 grants on 22 May 1979. Category 1 winners were all neighborhood or community development groups, experienced in carrying out revitalization projects for low- to moderate-income housing. Category 2 winners were all established single-family home builders.

A midcourse design workshop was held in July 1979 to provide grantees with hands-on architectural and engineering consultation. Design details and calculations were reviewed, after which final designs were submitted. HUD received 130 step one work statements, including requests for step two funds. As listed in table 9.1, 105 construction awards were made. Table 9.3 recapitulates the step one (design-only) cycle 5 awards; figure 9.3 shows the geographical dispersion, by state, of the awards.

9.3 Buildings, Systems, Participants

Analysis of data collected in the program revealed many interesting details on the buildings constructed, solar energy systems used, and various participant groups. Some findings were highly quantitative and provide good benchmarks for measuring progress in solar energy application; others were qualitative and reflect participant expectations and reactions.

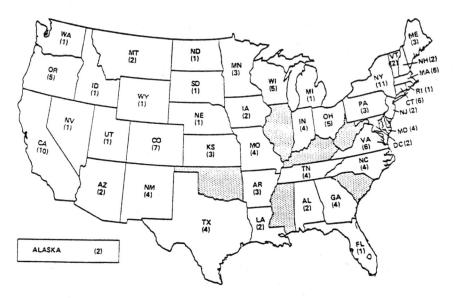

Figure 9.3
Location of grants, Cycle 5, Step One.

9.3.1 Buildings

The demonstration program included a full range of residential dwellings (one family occupies one dwelling unit). Most dwelling units (90%) were in multiple unit buildings, that is, apartments. Retrofits constituted 70% of the total units. Use of large, centralized solar systems for multifamily dwellings was common, with each of these larger systems serving an average of more than 42 units, primarily to supply domestic hot water. Single-family dwellings, served by systems on a nearly one-to-one basis, accounted for a majority (83%) of the units. Sometimes two systems, one active and one passive, served one dwelling.

Demonstration houses were similar in style to new conventionally heated homes in the same area and offered similar amenities and features. The average active solar house was a three-bedroom, two-bath, single-family detached unit. Typically, the house was a two-story or split-level of contemporary design with a wood exterior, a garage, and a patio or balcony. Standard amenities and appliances usually included central air-conditioning, dishwasher, garbage disposal, range/oven, partial carpeting, and a fireplace. Overall, active solar dwelling units were smaller than other homes on the market, with a median living space of 1,650 ft^2 (153 m^2) compared with 1,850 ft^2 (172 m^2) for a conventional home. Active solar unit prices were concentrated in homes priced in the $40,000 to $80,000 range (1976–1979 dollars).

Marketing analysis indicated that solar builders could anticipate most success with contemporary single-family detached units, particularly split-level and one-story. Active solar energy systems were integrated with conventional space and water heating systems, so occupants did not have to perform any unfamiliar functions to operate the solar systems. Typical active systems were south-facing and had collector areas between 250 and 500 ft^2 (23 and 46 m^2).

A typical passive solar home was a contemporary, two-story house with wood exterior, two or three bedrooms, two baths, and a wood stove or fireplace, with a finished living area of 1,800 ft^2 (167 m^2). Solar features were substantially different from those of active solar homes. Many different passive heating and cooling techniques were used (figures 9.4 and 9.5). Occupants had to operate the systems, but they indicated that this requirement was reasonable. The mean selling price for these single-family detached passive solar homes was $68,000 (figure 9.6). For single-family attached solar homes, the mean price was $63,471. Mortgages obtained by purchasers of single-family solar dwellings averaged about $50,000 for 30 years at 9% to 10% interest. Interest rates rose toward the end of the program.

Residences built under the program were all for sale or rent on the open market. This policy was established to examine public acceptance of solar buildings on the speculative market, for which the majority of homes are built. Home sales began early in 1976 and continued into 1982; the demand for housing varied substantially during that time. Some homes sold rapidly; others remained unsold for long periods. Important factors were availability of financing; home location, features, and layout; and price.

9.3.2 System Characteristics

Most systems provided either space heating, domestic hot water, or both (table 9.4). A few systems provided space cooling. Most active systems used a liquid for heat transport; the passive systems all used radiant heat or a combination of radiant heat and air. However, the 569 active heating/domestic hot water systems for single-family dwellings were almost equally divided between use of liquid (289) and air (280) collectors. In

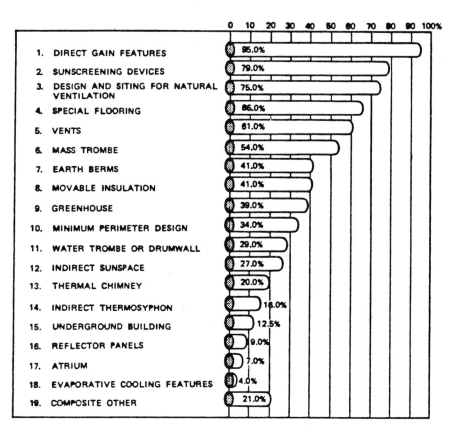

Figure 9.4
Frequency of passive characteristics (speculative houses).

approximately two-thirds of the systems, collector area was between 100 and 500 ft^2 (9 and 46 m^2) (figure 9.7).

Active systems that used liquid heat transfer had the following average sizes and capacities:

• Domestic hot water systems—50 ft^2 (5 m^2) of collector and 100 gallons (0.4 m^3) of storage.

• Space and water heating systems—290 ft^2 (27 m^2) of collector, for an average of 160 ft^2 (15 m^2) per 1,000 ft^2 (93 m^2) of heated floor space, and 80 ft^3 (2.3 m^3) of storage material.

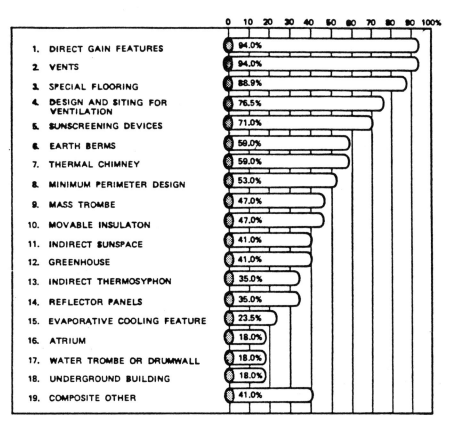

Figure 9.5
Frequency of passive characteristics (custom houses).

Passive solar systems fell into one of three classifications: direct, indirect, and isolated (figure 9.8). Direct-gain systems were most common and were used in more than 80% of the houses. Indirect-gain systems were included in nearly half; isolated, in more than a third. More than half used combinations of the three types, for example, combining direct-gain windows and indirect-gain Trombe walls; some homes had all three types. Statistically, houses with combinations were counted as having one system. On that basis, there were 266 passive systems in the demonstration.

9.3.3 System Costs and Savings

Median incremental active solar cost was $6.66/ft^2 ($71.69/m^2) of living space, while median passive solar cost was nearly half this, at $3.47/ft^2

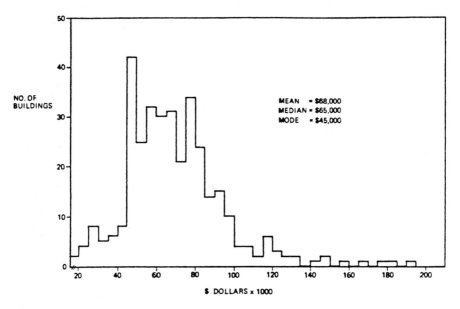

Figure 9.6
Sales price for single-family projects.

Table 9.4
Number of Solar Systems Constructed

	Single Family		Multifamily		
System Use	Active	Passive	Active	Passive	Total
Heating Only	41	110	3	13	167
Heating/Domestic Hot Water	569	5	72	0	646
Domestic Hot Water Only	ʼ93	23	102	17	335
Heating/Cooling	7	5	2	0	14
Heating/Cooling/Domestic Hot Water	0	90	1	0	393
Total	810	233	179	33	1255

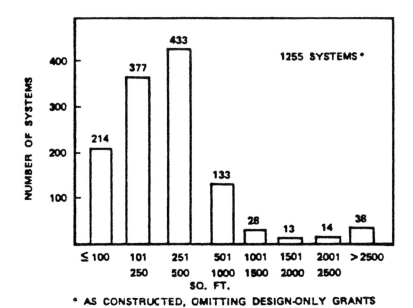

Figure 9.7
Collector area distribution.

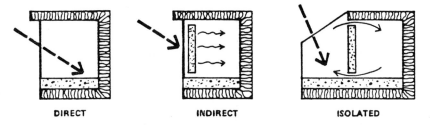

Figure 9.8
Characteristics of passive systems.

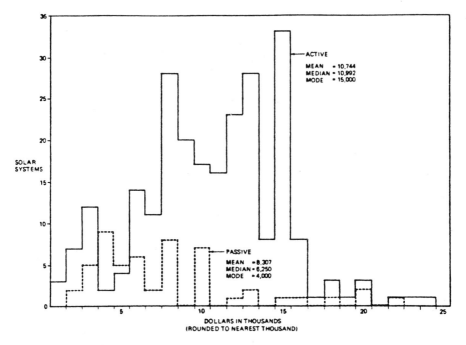

Figure 9.9
Builder-reported solar system cost.

($37.35/m²), or 15% and 6% of house cost, respectively (figure 9.9 shows total, incremental solar system costs). Fuel savings for both types of solar homes were similar (Boeing 1977). Median total costs attributed to passive or active features (above and beyond what would be included in a conventional home) and costs per unit area were derived from information provided by builders. Figure 9.10 is a chart of builder responses to the question: "How much additional cost attributable to passive features do you think the market can bear (based on total selling price)?" Increased costs of 6% to 10% were thought tolerable by 40% of the builders.

Adjusted energy consumption figures, derived from analysis of utility bills, showed that the average single-family solar dwelling used 37% less energy (per unit area per degree-day) during the time monitored than did the average nonsolar house. (See table 9.5 for details. "Comparatives" are comparable nonsolar dwellings, and "all solar" is active and passive dwellings averaged together.) Not all savings were attributable to solar energy systems; energy efficiency (e.g., amount of insulation) and occu-

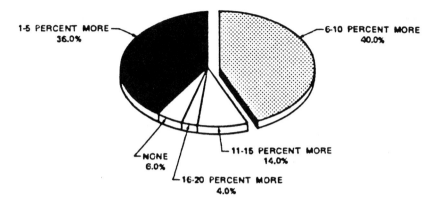

Figure 9.10
How much added cost for solar will the market bear?

pant behavior were also major factors in reducing energy use (NBS/Boeing 1981a).

9.3.4 Predicted System Performance

System performance, as calculated from the collected technical design data, was important in the evaluation of the viability of solar heating and cooling designs. Final technical design details, submitted as a contractual requirement, provided input data for mathematical calculations of anticipated solar system performance.

Figures 9.11 and 9.12 show solar fractions for submitted designs (solar fraction is the ratio of heat provided by the solar system to total building heat loss) as calculated by F-Chart. Active systems used solely for domestic hot water had calculated solar fractions that varied widely, from less than 5% to more than 80%, with a mean of 49% (figure 9.11). Active systems used for both space heating and domestic hot water had calculated solar fractions that also varied widely, from 5% to 85%, with a mean of 43% (figure 9.12). For systems used for space heating only, the mean calculated solar fraction was 40%.

In passive solar energy systems the house itself is generally the collector (or part of the collector). As with an active collector, enclosure thermal efficiency is an important factor in the amount of usable or retained heat. Because heat is both gained and lost through a passive collection system, the difference is the net solar contribution. The solar fraction is the net amount gained compared to the heat lost by the entire house.

Table 9.5
Mean Utility Energy Consumption (Single-Family Dwellings; Solar Contributes to Space Heating)

	Gas			Electric			Total		
	Mean	# #	STDEV	Mean	# #	STDEV	Mean	# #	STDEV
Heated Living Area (sq. ft.)									
Comparatives	1897	45	525	1914	24	42	1913	69	505
All Solar	1989	60	594	1763	70	414	1867	130	515
Active	1985	56	584	1789	57	362	1886	113	500
Passive	2044	4	822	1646	13	535	1740	17	610
Heat Only	1799	16	640	No Cases			1799	16	640
Million Btu/Year									
Comparatives	148.75	45	67.50	78.53	24	31.54	124.33	69	66.48
All Solar	72.58	60	41.36	64.07	70	25.74	67.99	130	33.98
Active	73.55	56	41.97	66.62	57	26.41	70.06	113	35.01
Passive	58.88	4	32.77	52.87	13	19.69	54.28	17	22.34
Heat Only	66.81	16	65.08	No Cases			66.81	16	65.08
Degree Days/Year									
Comparatives	5607	45	1798	5096	24	2103	5429	69	1910
All Solar	4879	60	1845	5394	70	1750	5157	130	1806
Active	4824	56	1843	5288	57	1810	5058	113	1833
Passive	5658	4	1946	5861	13	1423	5813	17	1496
Heat Only	4453	16	1900	No Cases			4453	16	1900
Btu/Sq. Ft./Year									
Comparatives	75746	45	29953	45284	24	24386	65151	69	31548
All Solar	37230	60	20564	38480	70	19235	37903	130	19792
Active	37855	56	21040	39438	57	20362	38654	113	20623
Passive	28470	4	9580	34281	13	13008	32913	17	12270
Heat Only	36321	16	31197	No Cases			35321	16	31197
Btus/Sq. Ft./Degree Day									
Comparatives	14.78	45	6.07	10.27	24	7.02	13.21	69	6.72
All Solar	9.33	60	8.32	7.58	70	3.77	8.38	130	6.33
Active	9.61	56	8.53	7.84	57	3.76	8.72	113	6.60
Passive	5.37	4	3.51	6.40	13	3.72	6.16	17	3.35
Heat Only	8.87	16	9.33	No Cases			8.87	16	9.33

#—Number of cases
STDEV—Standard deviation

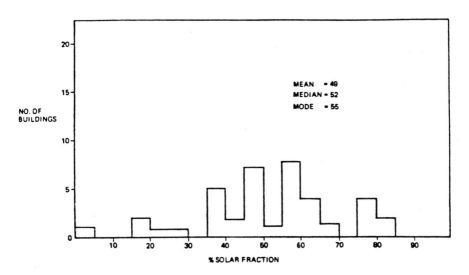

Figure 9.11
Predicted solar fraction, DHW systems.

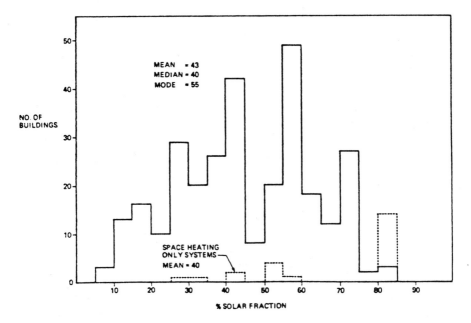

Figure 9.12
Predicted solar fraction, combined space-heating and DHW systems.

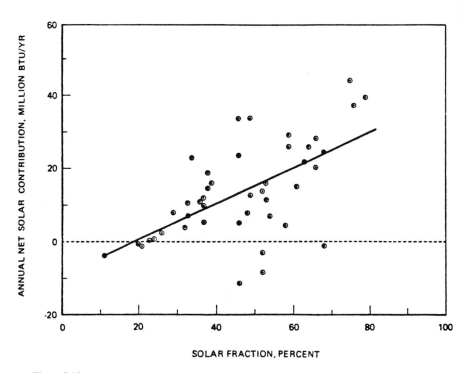

Figure 9.13
Annual net solar contribution vs. solar fraction.

Figure 9.13 shows a plot of net solar contribution versus solar fraction for a large sample of passive systems as calculated by the solar load ratio (SLR) method. This analysis shows that a positive solar fraction is possible with a negative net solar contribution. That is, more heat leaves the collector (glazing) than enters on an annual basis. The loss may be acceptable if glazing is included to provide a view or light; it will not be if glazing is included specifically for its passive contribution.

9.3.5 Builders and Bankers

Builders participating in the demonstration were typical U.S. residential builders; their average experience in residential construction was ten years, compared with an industry average of slightly more than eight. Participating firms built from fewer than 10 to more than 750 homes per year, for an average of 21 housing units per year; more than half had not previously built a solar home. These builders were affected by the industrywide housing slump, and construction rates cited were before the industry's decline. About 85% of the builders, interviewed when their active solar systems neared completion, expressed satisfaction with their experience in the program. Their sales experience on the whole was successful, and their costs were recovered.

Most new single-family homes required a seven-month construction period; retrofitting solar equipment on existing single-family structures took less time. Multifamily projects took longer to complete, with construction time varying greatly. Most builders reported having no problems with equipment breakdown or delivery, labor or materials availability, or installation of the solar systems.

Builders reported few problems in obtaining financing, zoning and code approval, and utility hookups. Interviews with lending officers, municipal and county officials, utility representatives, and insurance agents revealed few existing financial, administrative, or legislative impediments to residential solar applications. Similarly, solar home purchasers apparently were being treated no differently from buyers of conventional homes in dealing with banks, utilities, assessors, and insurance companies. Overall, institutional policies neither encouraged nor discouraged residential solar development.

9.3.6 Purchaser Characteristics and Satisfaction

Results of a market analysis survey indicated that purchasers of active solar systems were in the mainstream of the American home-buying public (NBS/Boeing/RERC 1981). People in many age, income, and occupation groups bought solar homes during the survey period (1976–1979). A typical surveyed household had three members, an annual income of more than $32,000, and a 38-year-old head of the household. Buyers generally had previously owned a home in the same metropolitan area.

The solar energy system played a major role in attracting potential purchasers to a given house. The system was not, however, the most important element in the final purchase decision. The presence of energy-saving materials and the value of the house were equally important. Overall, solar and nonsolar purchasers considered the same factors in deciding to buy a home, although solar buyers, as a group, did appear to be more conscious of saving energy.

9.3.6.1 Active Solar Energy System Purchaser Interviews

Initial interviews with active solar system buyers showed that, among those who expressed an opinion, most (66%) were satisfied with their solar energy system. In follow-up interviews conducted at approximately six-month intervals, 76% expressed continued satisfaction with the system, more than 95% expressed satisfaction with the house and development, and 76% would consider buying solar again. However, some active solar home purchasers were dissatisfied with their solar systems.

In general, five factors influenced consumer satisfaction with active solar energy systems: perceived fuel cost savings, need for major system repairs, attainment of an adequate temperature comfort level in the home, ability of the occupants to provide routine maintenance, and complexity of operating procedures.

9.3.6.2 Passive Solar Energy System Purchaser Interviews

The typical household in a passive solar home built for the speculative market consisted of two well-educated adults in their late thirties employed in professional, technical, or related occupations. Median household income was estimated at $35,900 (based on data collected between 1979 and 1981).

Purchasers of speculative passive homes cited fuel cost savings, energy savings, environmental concerns, and interest in new technology among their reasons for wanting a house with passive solar features (RERC 1982). They first became aware of the home's features through real estate agents, builders, newspaper classified ads, or outdoor signs. "Potential cost savings" were cited most often as the focus of marketing efforts for the solar homes. When asked specifically why they purchased their house, most attributed their purchase to the passive solar features.

Among custom home purchasers who did not design or build their homes, most had initiated the process of designing a passive solar home and all had selected their site. Few imposed constraints on the designer with regard to passive solar features.

Occupants of both speculative and custom homes were equally divided on the question of whether they paid more for a passive solar home than they would for a "well-designed energy-conserving conventional house." A slight majority said yes. Only six purchasers out of forty-five who financed their houses experienced problems with lending institutions.

Initial interviews, conducted shortly after occupancy, revealed that most purchasers were well satisfied with their houses, the performance of their passive solar systems, and the level of comfort. For the most part, positive expectations had been met and negative expectations had not materialized. Some problems were reported, but practically all purchasers agreed that they would consider buying another solar house.

Approximately six months after the initial interviews, follow-up interviews were conducted with some of the households. Expectations regarding fuel savings, workability, comfort, and environmental concerns had been met or exceeded. All respondents reported that their backup systems had performed reliably, and they had experienced no major problems with the passive systems. The follow-up interviews revealed a high level of satisfaction, and all respondents indicated a more positive attitude toward passive solar energy since moving into their homes (RERC 1980).

9.4 Repair Program

9.4.1 Origins

As active solar systems became operational through the various award cycles, problems began to surface. Routine site visits, ongoing purchaser surveys, investigations of instrumentation anomalies, and direct complaints by owners or grantees indicated the existence of serious problems with almost all types of active systems. Most direct complaints arose after failed attempts to get satisfaction from the original system contractors.

Initially, HUD strove to maintain a hands-off posture toward builder-purchaser relations. Field representatives of the management contractor did encourage manufacturers and builders to respond to purchaser complaints, but they intervened directly only where problems occurred with instrumented systems. Most of the earliest problems involved badly leaking air systems, where instrumentation would not work properly.

Reports of troubles increased rapidly during 1978. In January 1979 HUD began responding, on a case-by-case basis, to a few complaints beyond those involving instrumented systems. By this time, reported problems included leaking liquid systems, repeated pump failures, and malfunctioning controls. By summer 1979 it was clear that many problems were recurring and that a formal review and correction process would be needed if purchasers were not to sustain significant financial losses (NBS/Boeing 1981b).

In a report dated 9 October 1979, the Government Accounting Office (GAO 1979) criticized the program for not taking action to ensure that all systems were operating properly. GAO's findings, based on an independent investigation of twenty grants involving 91 dwelling units, revealed a serious, short-term failure rate roughly equivalent to HUD's own statistics. Both sets of data indicated the failure rate was time-related and could be expected to worsen considerably.

At about the same time GAO issued its report, HUD established a review board to conduct a methodical assessment of reported problems and to prescribe corrective actions. The board consisted of the HUD division director, the solar program manager, and members of the solar program technical staff; key Department of Energy and National Bureau of Standards staff were invited to participate as consultants/advisers.

9.4.2 Process Description

The System Operating Problem Review Board (the "board") promptly set up a procedure for problem identification, investigation, board review, and resolution. Due to the size of the complaint backlog and the rate of new problem reports, the board decided not to undertake a full-scale survey of all projects; available, experienced staff were only sufficient to assimilate the available material.

Problem reports came to HUD through direct complaints from purchasers or grantees, through indirect complaints received by field representatives, through direct observation made during routine site visits, and through data received from instrumented systems. A System Operating Problem Report (SOPR) for each complaint provided a summary of actions taken, from first report to final resolution. These SOPRs served as the board's basic working tool.

The board held its first monthly meeting in October 1979, reviewing all current problems and determining actions to be taken by the management contractor and its subcontractors. A technical review of the situation generally resulted in one of several actions:

• The problem was immediately resolved by the grantee, the home owner, or the management subcontractor during a site visit.

• A determination was made that the system was repairable and that the responsible party (grantee, builder, installer, etc.) was capable and willing to make the repair. HUD technical assistance was authorized.

• Repairs beyond the responsibility and capability of the grantee team were authorized by HUD.

• Systems not likely to perform satisfactorily after necessary repairs were removed and replaced by equivalent conventional systems. Where systems were repairable but the owner did not wish to risk future problems, HUD authorized system removal.

Once any repair work was begun at a site, whether by a grantee or under a HUD-funded repair contract, a field representative maintained periodic surveillance of the work. Monthly status reports were reviewed by the board until the required action was completed.

A decision was made by the board to involve original grantees and installers in the repair or removal work wherever possible. Because most problems resulted from a lack of understanding, not from deliberate oversight, sharing in the learning experience provided by system repair or removal improved the ability of grantees and installers to serve future solar customers satisfactorily.

The repair program continued to address only reported problems. No attempts were made to look for further problems through general field surveys; this operating mode provided sufficient work for the available manpower and funding. In May 1980, however, a roof fire caused by an overheated collector on a building in Boulder, Colorado, triggered an investigation into the potential hazards of collector materials and configurations. Thirty-three grant locations, involving 54 systems where hazardous conditions theoretically existed, were surveyed. Although no new hazards were located, most of the surveyed systems entered the repair program for other problems found. This investigation led to a further survey of all houses with solar attics, involving another 16 grants and 23 systems; all were incorporated into the repair program.

By early 1981 field reports of problems had dwindled to two or three per month. This rate suggested that the repair program could continue for years; something else was required to bring about an effective, timely conclusion. It seemed reasonable to assume that many problems remained and that many purchasers no longer knew where to go with complaints. A severe recession in the housing market meant that many original builders were not accessible to the homeowners. A cutback in monitoring funds was also reducing the identification of potential problems.

After assessing the situation, the board reached the following conclusions:

• Available repair funds could maintain significant activity and be used to optimize the number of repairs accomplished before the management support contract ended.

• Funds were not sufficient to correct every problem that might be uncovered by a random survey. An equitable approach would provide for all potential hazards to life, safety, or health first. Remaining funds would be used to attack problems based on their relative severity.

• Problem reports reaching HUD from outside sources would be investigated and incorporated into the repair program as before.

A review of potential hazards led ultimately to removal of all copper-clad plywood absorbers. Corrosion in liquid systems was identified as the most serious problem after potential hazards to life, safety, and health. System configurations in the program were ranked as to degree of potential trouble, and survey priorities were set accordingly. Field surveys were begun in July 1981 and continued through May 1982. All projects surveyed were incorporated into the repair program for action.

9.4.3 Summary

Of the 497 total grants included in the demonstration program, 229 were involved in the repair program. Although the most severe system problems were addressed, there is reason to believe that many other problems were never reported or resolved. It should also be noted that, on a percentage basis, complaints about passive systems were negligible. (Details of the technical nature of the many problems identified in the repair program will be found in other volumes in this series, particularly in volume 6, *Active Solar Systems.*)

9.5 In Retrospect

Awards for construction under the residential demonstration program were made from early 1975 to late 1979, and most construction was completed by early 1981. Data collection was carried out from 1977 to 1981. During 1980 system repairs became the major activity, and this continued to be the case up to program termination in 1983. Many aspects of the full HUD demonstration program represented first-time activities for the federal government in direct promotion of a technology in the consumer marketplace. Looking back, how did the program fare?

A retrospective assessment of the program quickly reveals many more questions than answers. Congress chose, through passage of PL 93-409, to accept the testimony of those who claimed that the technology was sufficiently mature, with a manufacturing capability already in place, to justify a major demonstration effort that would result in an economically viable and technically reliable consumer industry. Although the resulting demonstration program could point to many successful installations (particularly of passive solar systems), the number and types of problems dealt with in the repair program suggest a level of readiness for the consumer market well below that normally considered acceptable. Was the government oversold on the product? Did the government oversell the product to builders, bankers, and purchasers? Were program duration and funding adequate to accomplish the federal objective?

To carry out the congressional mandate, HUD adopted a strategy and an associated methodology that called for minimum government intrusion in normal relations between architects, builders, component suppliers, and purchasers. The use of grants, the timing of award cycles, the procedures adopted for funding competition, and the selection of geographical and solar energy system criteria were all part of that methodology. Were the strategy and methodology correct? Were they effectively implemented and carried through? In a subsequent demonstration program, for hot water solar energy systems, HUD adopted a more market-intrusive strategy (see chapter 27) and was roundly criticized in the process. What then is an acceptable and productive government role in such programs?

The federal government has frequently had a major impact on consumer markets through large-scale purchases of products that have both public sector and private sector potential; aircraft and computers are two obvious examples. Might an early-purchase effort, aimed, for example, at solar heating for military housing, have given the industry a salutary push along the learning curve? Would such an effort have generated the desired opening into the general housing market?

Questions such as these can be answered satisfactorily only when comprehensive programs are undertaken and sufficient follow-up data collected and analyzed. The history of the residential demonstration program leaves most of them unanswered and unanswerable.

References

ADL (Arthur D. Little, Inc.). 1976. *A Location Matrix Plan for the Residential Solar Heating and Cooling Demonstration Program*. Vol. 1, *Findings and Recommendations*. Cambridge, MA.

Boeing Aerospace Company. 1977. *Energy Consumption and Cost Data Acquisition Plan*. FOSD-SH-101. Seattle.

HUD (U.S. Department of Housing and Urban Development). 1977. *Intermediate Minimum Property Standards*. Vol. 5, *Solar Heating and Domestic Hot Water Systems*. Washington, DC.

NBS/Boeing (National Bureau of Standards and Boeing Aerospace Company). 1981a. *Utility Consumption File*. Final computer report. Gaithersburg, MD.

NBS/Boeing. 1981b. *Technical Concerns File*. Final computer report. Gaithersburg, MD.

NBS/Boeing/RERC (National Bureau of Standards, Boeing Aerospace Company, and Real Estate Research Corporation). 1981. *Marketing and Consumer Acceptance File*. 2 vols. Gaithersburg, MD.

RERC (Real Estate Research Corporation). 1980. *Marketing and Market Acceptance Data from the Residential Solar Demonstration*. Vol. 1. Washington, DC: U.S. Department of Housing and Urban Development.

RERC. 1982. *Passive Solar Homes in the Marketplace*. Vol. 2. Washington, DC: U.S. Department of Housing and Urban Development.

10 Commercial Buildings

Myron L. Myers

10.1 Program Authority

The Solar Heating and Cooling Demonstration Act of 1974 (PL 93-409) established a national program to develop solar heating and cooling systems and to demonstrate to the public that such systems were practical and economical. The act, signed by President Gerald Ford on 3 September 1974, proposed, in part, "to provide for the demonstration within a three-year period of the practical use of solar heating technology, and to provide for the development and demonstration, within a five-year period of the practical use of combined heating and cooling technology." The act provided for involvement by the National Science Foundation (NSF) for basic research and demonstration and by the Department of Housing and Urban Development (HUD) for the residential aspects. The act also amended the National Aeronautics and Space Act of 1958 to enable the National Aeronautics and Space Administration (NASA) to perform "research, development, demonstration, and other related activities in solar heating and cooling technologies as provided for by the Act." This was the basis for the program and for the major role assigned to NASA.

10.2 Organizational History

PL 93-438, provided for the creation of the Energy Research and Development Administration (ERDA) staffed primarily with former Atomic Energy Commission (AEC) personnel. Under ERDA, which came into existence on 19 January, 1975, NASA continued to actively support a major portion of the commercial demonstration program. NSF and HUD also continued in their previously assigned roles: NSF with research, design, development, and demonstration of large solar enegry systems for space heating, water heating and cooling; and HUD with design, development, and installation of smaller solar energy systems serving the same purposes in single-family and multifamily residential units.

To implement the NASA role, NASA headquarters assigned the commercial demonstration task to the George C. Marshall Space Flight Center (MSFC) located in Huntsville, Alabama. In response, MSFC established the Solar Heating and Cooling Project Office. To fully implement the assignment in keeping with the intent of the act, including the amendment

to the Aeronautics and Space Act of 1958, MSFC established two subsidiary offices—the Systems Demonstration Office and the Systems Development Office. The first was to manage MSFC's assigned portion of
the Commercial demonstration program and the second was to provide
for the development of systems and subsystems in support of the commercial demonstration program; including prototype solar water heating
systems, solar space heating systems, solar space cooling systems and
combinations of these. These systems involved development of a data
acquisition system as well as collectors, instrumentation, and absorption
air-conditioning subsystems.

10.3 Program Management Activities

MSFC's role in the commercial demonstration program was recognized
as an important assignment by NASA; therefore, both of the above-
named offices were headed by experienced project managers, engineers
who had previously worked on major MSFC space programs such as
Saturn V (Apollo) and Skylab. Commercial demonstration program
activities at MSFC spanned a period of five years (1975–1980), during
which the System Demonstration Office (later named the "Commercial
Demonstration Office") eventually managed 118 commercial demonstration projects (approximately 42% of the total). The Solar Heating and
Cooling Demonstration Act of 1974 also directed HUD and NASA to
prepare a comprehensive plan for conducting the development and demonstration activities called for under the residential portion of the act.
After obtaining plan approval, and in compliance therewith, the Systems
Development Office managed over 30 contracts, which produced a data
acquisition system and 42 solar energy systems installed in various types of
buildings, called "operational test sites," located throughout the United
States. Most of the systems were installed in residences for heating and
hot water; however, larger buildings were equipped with solar-powered
air-conditioning systems as well. Twenty-two subsystems were also procured through the above-mentioned contracts, in the form of marketable
subsystems, available from industry at the time, and subsystems requiring
additional development; these subsystems were integrated into the operational test sites.

Over the five-year period MSFC was actively engaged in the federal
solar energy demonstration program, these two offices expended a total of

Table 10.1
Funding

Resource	System development program	Commercial demonstration program
Funds	$60 million[a]	$9 million[b]
Person-years	340	137

[a] This figure, taken from page C-3 of an MSFC document entitled *Solar Energy Applications*, dated October 1980, includes all costs associated with the Systems Development Office, that is, the cost of the data acquisition system, 42 developmental solar energy systems, 22 subsystems, testing equipment and materials, and all manpower and travel associated with system development.

[b] This figure is based on information found on page 75 of NASA 1980; it includes only those costs incurred for manpower, travel, and equipment used for diagnostic testing by the Site Assessment Group. The cost of the 118 Commercial Demonstration projects was borne directly by the DOE and the owners on a cost-share basis.

approximately $69 million and 477 person-years of labor. Approximately 68% of the person-years were used by MSFC's Science and Engineering (S&E) laboratories. The S&E personnel acted in a direct technical support role by providing on-site consultation, analytical diagnostics, laboratory testing, and verifications. Approximately 75% of the labor expended by the Systems Development Office came from the S&E laboratories and approximately 50% of the Commercial Demonstration Office labor was furnished by these laboratories. The approximate expenditures by the DOE for the services rendered by the MSFC are shown in table 10.1. Subtasks managed by the Commercial Demonstration Office are discussed below; detailed activities of the System Development Office are beyond the scope of this volume.

10.3.1 Program Opportunity Notices (PONs)

A total of four PONs were issued for the commercial demonstration program, two by ERDA and two by its successor, the Department of Energy (DOE). The PONs required cost sharing by respondents. The nominal forty-page PON document was introduced by a one-page letter from the procurement official setting forth the major ground rules and giving the due date for receipt of proposals. Approximately one-third of the PON was devoted to such things as introduction (purpose, requirements, conditions, government assistance and participation, and submission of the proposals), instructions for preparing the proposals, a description of the proposal evaluation and selection procedures, information on small business participation and a short dissertation on the national energy policy and its implementation. The remainder of the document was devoted to appendices with numerous preformatted pages concerning project data, system and subsystem performance/technical data, the preferred proposal format, and a description of data collection anticipated by the government and the methods by which this would be accomplished.

From prior experience, some of ERDA's top management personnel were aware of and understood the value of MSFC's procurement capabilities, as well as the center's unique technical and engineering management skills. ERDA therefore chose to rely heavily on MSFC as an important source of input for preparation of the PONs.

The people in ERDA who had the responsibility to plan management of the commercial demonstration program, also had the foresight to solicit assistance from experts in the heating and air-conditioning field, namely, members of the American Society of Heating, Refrigerating, and Air-Conditioning Engineers (ASHRAE), a highly respected nationwide organization. ASHRAE members had direct contact with the trades and could effectively forestall problems that might thwart the efforts to interest the public in this "new" endeavor. Three ASHRAE standards were cited for compliance in the PONs: *ASHRAE 90-75* (ASHRAE 1975) in PON 1 and *ASHRAE 93-77* (ASHRAE 1977a) and *ASHRAE 94-77* (ASHRAE 1977b) in later PONs. ASHRAE's role in the commercial demonstration program began early and continued throughout the program's five significant years.

In consonance with the intent of the act, the first PON issued by ERDA in September 1975 stressed less sophisticated solar energy systems, systems that would provide space heating and/or hot water. Heading the preparation of this first PON at MSFC was Donald R. Bowden; the depth of Bowden's prior experience and responsibility demonstrates the importance that MSFC placed on the assignment given to the center.

According to the second PON, released by ERDA on 8 October 1976: "Responses concerning space-heating systems will receive primary emphasis at this time." Assisting in the preparation of this PON was John Price, deputy manager of the Solar Heating and Cooling Project Office, who had also held numerous managerial positions at MSFC.

On 1 October 1977, between PON 2 and 3, the Department of Energy (DOE) was created out of ERDA and several other agencies; responsibility for the preparation of the PONs was turned over to DOE's Chicago operations office. In PON 3 (November 1977), cooling systems were given more consideration; in PON 4 (June 1979), the fourth and final solicitation, passive systems were emphasized.

MSFC was assigned the responsibility for the technical management of all 31 PON 1 projects. Management of PON 2 and 3 projects was distributed geographically between MSFC and the DOE operations offices in Chicago and San Francisco; management of PON 4 projects was retained by DOE.

At about this same time, MSFC was assigned responsibility for another DOE solar energy program—the Solar in Federal Buildings Program. From January through March 1980, MSFC was engaged in evaluating 900 federal building program proposals submitted by some 16 government agencies. MSFC's role in this endeavor, following proposal evaluation, was to negotiate interagency agreements for the 843 projects selected by DOE and then to assist the various agencies in the design of the solar energy systems, acquisition of the system components, construction, installation, testing and checkout, and evaluation of system performance. All of this required significant effort from MSFC's limited staff. Appropriately, MSFC's participation in the commercial demonstration program was systematically phased out in 1980, and as the solar projects became fully operational, management responsibility was transferred to DOE. When full operation of a project was achieved by a private owner, there was no reason for MSFC to remain in the management loop; the MSFC phaseout was also consistent with the five-year period established by the act.

10.3.2 Proposal Evaluations

The technical proposal evaluations for PONs 1 and 2 were performed at MSFC by the Technical Advisory Committee (TAC). For both of these evaluations, the TAC chairman was an MSFC employee; the chairman for the PON 1 evaluation was John Price, and the author was chairman for the PON 2 evaluation. ERDA, of course, made the final selection of the projects to be funded. For large procurements, MSFC formally evaluates proposals through the use of a Source Evaluation Board (SEB) and a Source Selection Official (SSO). The evaluation process for PONs 1 and 2 followed closely that of an SEB, with ERDA performing the SSO function. The last two PONs (3 and 4) were evaluated at the Argonne

National Laboratory (ANL) under the authority of DOE's Chicago Operations Office. At the request of that office, MSFC supported the evaluations at ANL with approximately ten employees, most of whom were project managers from the Commercial Demonstration Office—the remainder were from the S&E laboratories.

Because the author was directly involved with the evaluation of PON 2, the mechanics and statistics of that evaluation will be discussed in greater detail as an example of how responses to the PON's were processed. Late in November 1976, 307 proposals were received at MSFC. The technical evaluation was completed 10 February 1977; results were delivered to ERDA the following day. On 16 February 1977, ERDA requested cost proposals from 175 of the original respondents—proposals judged "technically adequate" by the TAC. The cost proposals were to be postmarked not later than midnight, 11 March 1977. A business subcommittee within the TAC took the lead in performing a cost summarization for ERDA; on April 4 the resulting cost summary was presented to the ERDA Source Evaluation Panel (SEP), chaired by Ron Scott. On 28 April 1977, the SEP, with the author as an adviser, presented the facts of the evaluation to the ERDA Source Selection Official (SSO), H. H. Marvin. From the 175, ERDA selected 80 projects for cost negotiations.

The technical evaluation, exclusive of the cost summarization activity, took 81 calendar days and involved approximately 45 people from the MSFC, Battelle Memorial Institute, ERDA's Chicago Operations Office (ANL personnel), and other ERDA personnel. The evaluation was a full-time activity for the typical evaluator. The PON 2 milestone summary is shown in table 10.2.

10.3.3   Contract Negotiations

The MSFC project managers assigned to the Commercial Demonstration Office took an active part in all phases of the program—contract negotiation with the successful proposers was no exception. The project managers were of invaluable assistance to the contracting officer and his negotiators because it was at this time that ambiguities were pursued with the contractor. If required, the project managers would call on scientists and engineers from the MSFC S&E laboratories to assist in obtaining clarifications. The resulting contracts, although funded on a cost-sharing basis with the government, were fixed-price contracts as far as the government's cost incurrence was concerned. So it was imperative from the

Table 10.2
PON 2 milestone summary

10/08/76—PON 2 released
11/19/76—PON 2 closing date
11/21/76—Proposals received at MSFC
11/22/76—Evaluation process started by TAC
02/10/77—TAC completed technical evaluation
02/11/77—Technical evaluation results delivered to ERDA
02/16/77—ERDA requested cost proposals (175)
03/09/77—TAC summary presentation to ERDA SEP
03/11/77—Cost proposals postmarked NLT midnight
03/21/77—TAC began cost summarization
04/04/77—ERDA received cost summarization
04/28/77—SEP presentation to SSO
05/77—ERDA selected 80 projects for negotiation
05–08/77—Negotiations resulted in 70 contracts

standpoint of both parties that what was being proposed could in fact be delivered within the proposed funding. The government wanted projects to become good demonstrations without compromise because of funding; the owners, of course, had an equal interest in cutting energy and maintenance costs. Within a four-month period, 70 of the 80 contract negotiations were successfully concluded at a contract value of $16.86 million; negotiations for the other 10 were terminated for various reasons.

10.3.4 Design, Construction, Installation, Start-up, Checkout, and Acceptance Testing

Of the 283 commercial demonstration projects in existence in mid-1979, 139 (approximately 49%) were either installed or scheduled for installation during the construction of the host structures. Although the solicitations did not require that the host structures be under construction at the same time as the solar energy systems; systems installed during construction of the host facility proved to be among the most successful from the standpoint of trouble-free operation and overall efficiency. Also, adherence to the schedule for construction of a new facility is always driven by the need to occupy the building as soon as possible; this, of course, enhanced the timely completion of the integrated solar system. To illustrate this point, as of 30 April 1980, for PON 2 projects, 80% of the new host structures had fully operational systems; while only 71% of the buildings being retrofitted had fully operational solar energy systems; even more dramatic, for PON 3 projects, 41% of the new structures had fully operational systems, while *none* of the retrofit structures had (NASA 1980).

10.3.4.1 Design

Having no direct knowledge of how DOE Headquarters or the Chicago and San Francisco Operations Offices operated in this or any other phase of a project, the author can only describe how the MSFC Commercial Demonstration Office monitored projects. Each project manager attended a minimum of two reviews for every solar energy project for which he carried responsibility (normally these were the preliminary and the final design reviews). The number of projects assigned to a given manager varied with complexity of the system and extent of the manager's experience. For instance, one manager, who had extensive management experience, carried twenty projects, nine of which were complicated; the balance were, for the most part, simple, similar hot water systems for motels owned by the same chain. The other managers carried from seven to thirteen projects each.

One of the most valuable contributions made by these project managers at the design reviews was the knowledge they were able to transfer between projects—not only from those they managed, but from the others as well. This came from daily dialogue between the managers and with other knowledgable MSFC personnel (S&E laboratories and facilities). Because solar energy projects of these proportions were relatively new, errors were not uncommon. Many solar energy system owners had been only somewhat involved in the construction of their own buildings, but they now found themselves deeply involved in the installation of solar systems in these buildings. Although eager to get into the field, most of the local architects engaged by these owners were not experienced in the design of solar energy systems, nor could they always provide competent solar engineers. In these cases the gap was often bridged by experience brought by the Project Manager.

10.3.4.2 Construction and Installation

During this phase of the project, the owner would sometimes ask for assistance from the government's project manager. The project manager might then travel to the site and offer advice or, in the case of MSFC, call upon people from the S&E laboratories, or sometimes the MSFC Facility Office would be asked to assist. The Facility Office mechanical engineers had gained considerable solar system knowledge from work they had done with the Huntsville-based Corps of Engineers Division Office, which designed and constructed a Solar Heating and Cooling Test Facility (Breadboard) on a 20-acre site at MSFC. The Breadboard facility conducted outdoor testing of both active and passive solar energy system components for use by the Systems Development Office.

The costly delays encountered most often during the design phase of a project dealt with improper installation of the control system. Especially critical was the location of sensors and the contractor's understanding of the control logic. This is where the extensive instrumentation and control experience of MSFC personnel became important. Sometimes, depending upon which of the ten Commercial Demonstration Office project managers was involved, the problem could be solved by the project manager; at other times an engineer from the S&E laboratories would be consulted.

10.3.4.3 Start-up
The proof of the design and installation was with system start-up. For those systems (approximately 26%) that had instrumentation feeding into a Site Data Acquisition System (SDAS), the performance of the new system could be checked with a high degree of confidence. Just getting a new system to operate at all was sometimes an achievement, let alone having it attain optimum performance. The presence of the project manager at this milestone for a system was helpful. For example, when collectors did not fill, it might be a simple matter of a closed valve to a vent; because this problem had arisen before, the project manager would be the first to recognize it again.

10.3.4.4 Checkout and Acceptance Testing
Following start-up, the system was checked out and an acceptance test was conducted. Because NASA/MSFC project managers were accustomed to performing any and all tests in their aerospace work using detailed procedures, this method for checkout and acceptance testing was little more than second nature. Many of the steps in the acceptance tests were common for all active solar systems of the same type (heating, hot water, or cooling); however, since there were differences from one system to another, each MSFC-managed project was subjected to an Acceptance

Test Procedure (ATP), which was carefully prepared by the owner's project manager and reviewed and approved by the MSFC project manager before the test was conducted. The MSFC project manager usually observed this test and assisted in solution of any problems that developed. The ATP checklist for final inspection provided formal documentation of reasonable expectations. The scope and degree of detail of the ATP assured the site owner that system performance met the design requirements. The ATP ensured that

• components and subsystems were of correct material, size, and capacity;

• components and subsystems were installed correctly;

• components and subsystems functioned as designed.

• system was properly coordinated and balanced; and

• system functioned without deleterious effects on the building in which it was installed.

ATP included (where applicable)

A. identity (location) and demonstration of fail-safe controls;

B. identity (location) and demonstration of pressure relief valves and/or stagnation temperature-controlled relief dampers;

C. assurance of no leaks in system at 150% pressure (for liquids; minimum loss in air systems);

D. assurance that there was no growth of algae, fungi, mold, or mildew;

E. assurance that backflow was prevented;

F. assurance that there was no fluttering of valves;

G. demonstration of fill and drain (freeze test) in liquid systems;

H. test for pressure head of pumps;

I. measuring amperage and voltage for pumps and fans;

J. measuring of pressure drops at inlet and outlet of collectors and determining system pressure drop;

K. measuring collector flow (GPM);

L. recording of temperature measurements at

1. collector inlet and outlet headers;

2. heat exchanger inlets and outlets;

3. conditioned space(s);

4. storage tank inlet and outlet;

5. preheat coil inlet and outlet;

M. conduct of operational test of system in following modes:

1. solar energy collection with space heating;

2. solar energy collection without space heating;

3. space heating from storage (no collection);

4. all auxiliary heating (no storage heat; no collection);

N. verification of heat collection performance to determine whether it was within a reasonable margin of predicted performance, e.g. insolation on collectors versus output);

O. verification that all equipment, piping, controls, and so on, were installed in the manner specified;

P. verification that all pumps, controls, dampers, fans, and air distribution system were operating as specified.

10.3.5 System Dedications

Public awareness was of primary concern to the commercial demonstration program. Therefore, the DOE and NASA were highly sensitive to the need to have well-advertised public dedications of the completed solar systems. Invitations went out to the leaders in the communities as well as state and federal representatives. Local high school bands, glee clubs, and the like were sometimes called upon to entertain at the dedications. The project manager was always given a spot on the program to present a brief description of the national program and the significance of the subject project. For PON 1, the average time between contract signing (April 1976) and system dedication was about twenty months, with the first dedication in August 1976.

10.3.6 Performance Data

As mentioned previously, the Systems Development Office developed a Site Data Acquisition System (SDAS) that was installed on approximately 26% of the commercial demonstration projects through June 1979. Data transmitted from an instrumented system on a daily basis were collected at a central location, stored, and entered each month into a "site formula" unique to the particular site. The formula contained all the pertinent physical parameters of the system to enable a computer to process the values of on-site insolation, ambient temperatures, wind speeds and directions, storage temperatures, and various other sensor data into reduced data and performance reports for analysis and dissemination on both a monthly and seasonal basis. Reports were first published by IBM, the designer and fabricator of the SDAS, and later (beginning in 1979) by Vitro Laboratories, a division of Automation Industries. By the end of 1979, reports had been published for 26 of the 41 MSFC commercial demonstration projects designated for instrumentation installation. Reports for 3 of the 26 projects had been published as early as September 1977. For sites where there was no SDAS, the project managers relied upon visits or monthly telephone contact with the responsible person at the site for objective information on how well the system was performing. In the case of a retrofit system, the information requested included monthly records of both past and present utility usage versus heating degree-days. From the design specifications, the project manager could readily determine whether the system was functioning as intended.

10.3.7 Operational Anomalies

Problems with solar energy commercial demonstration systems occurred primarily because these systems were new to the architects, engineers, construction contractors, and tradespeople. As the number of systems came on line, the number of problems also grew; it was recognized that special attention would have to be given the problems. Therefore, with the approval of DOE, MSFC set up a special team, called the "Site Equipment and Assessment Group" (SEAG) and made up of a cadre of four men with experience in instrumentation, mechanical systems, and data analysis. The SEAG would investigate problems either as single members or in teams of two or three; if a problem required special knowledge, then a specialist from the S&E laboratories would join the group. The SEAG worked only as directed by the project manager; that is the project manager, first made aware of a problem from his own analysis or from the concern of on-site personnel, would analyze the problem to determine the most probable cause and then seek the assistance of the SEAG. The group obtained special instrumentation devices in order to properly perform its task. For example, the SEAG was provided with a briefcase-sized computer system into which a tape of the site formula for a particular site could be introduced. Then the system under investigation was equipped with strap-on instrumentation for temperature, as well as inserts for flow and pressure measurements, insolation meters, anemometers, and so forth. The carry-on computer permitted an on-the-spot analysis to determine the cause of unwanted performance. Another unique tool used by the SEAG in its data analysis was an infrared device calibrated to provide temperature gradients. This was invaluable for scanning an array to determine where a flow blockage or air pocket might exist in a collector or series of collectors or to find voids in insulation.

Although many special instruments and devices were made available to this group, the SEAG's most important asset was its collective experience; SEAG members had spent years in testing various components, subsystems, and systems in the laboratories at MSFC. Often, as a result of its investigations, SEAG would conclude that there was a major problem with a particular system that could not be remedied by simply changing a valve or relocating a vent line. DOE, recognizing that major problems do occur, instituted a refurbishment program. Candidates for refurbishment were brought before a committee, which met periodically in DOE Headquarters; facts concerning the poor performance of a given system were provided this committee. Composed of key personnel from the DOE Headquarters, DOE Operations Offices, MSFC, and ASHRAE, the committee was chaired by Carl Conner, the DOE commercial demonstration program manager, who reviewed the recommendations of the investigative body (often MSFC's SEAG). After receiving the committee's advice, Conner would decide whether a system should be refurbished. Of the approximately 30 projects considered in this program, slightly more than half were accommodated.

10.3.8 Monthly Status Reports

Keeping interested parties informed via written reports on the status of the work had always been the cardinal rule at MSFC. For the commercial demonstration program, this was accomplished with the *Solar Heating and Cooling Commercial Demonstration Program Management Report* issued monthly. Each month the project managers would have either visited each project site or would have called the responsible person at the site to obtain the latest project information. Recorded on a carefully formatted data sheet for each project, the information delineated such things as the project title, contract number, date of the status report, name, address, and telephone number of the owner's project manager, name and telephone number of the government's project manager, whether or not the site was instrumented, and, if so, the number of sensors. Also recorded were cost information: owner's and government's share, amount vouchered by the owner to date, and so forth; collector information: area, number of collectors, type (e.g., liquid flat-plate, evacuated tube, air, etc.), and manufacturer's name; and solar energy storage information: volume, location, and any auxiliary energy source. A short, yet detailed description of the system was also provided, giving the anticipated solar contribution for water heating, space heating and/or cooling as appropriate. The type of facility was shown, along with its location, including latitude. The conditioned area was also recorded. In addition to the static information listed above, the one-page status sheet provided information on significant events and problems (if any) and schedules for design, construction, installation, and sensor installation, also showing system start-up, anticipated or actual. A key part of the current status report was a brief notation, a few words preceded by a code number, that told at a glance whether the system was under design, in construction, had been installated, had completed acceptance testing, was partially or fully operational, or was down. The project managers' input to this report was used by numerous levels of management in MSFC, NASA Headquarters and DOE to monitor the overall progress of the demonstration program.

Table 10.3 lists 118 commercial demonstration projects covered in the monthly *Solar Heating and Cooling Commercial Demonstration Program Management Report*, with examples of some static information listed for those projects.

10.3.9 Oral Reports

MSFC project managers and the author were each required to give oral presentations covering his area of responsibility in the commercial demonstration program to MSFC/NASA top management as well as DOE. By keeping everyone informed through the monthly status reports described above, the oral presentations proved to be a useful adjunct; meaningful, updated information was discussed and action assignments made at these meetings. DOE often received as many assignments as MSFC, affording quick response to concerns plaguing the MSFC project managers and the contractors; this proved especially important during the period of refurbishment discussed in Section 10.3.7.

10.3.10 Presentations to Technical Groups

The government's entry into the market with partnership-like support for procurement and installation of large solar energy systems aroused interest among technical societies and the trades. All across the United States where the commercial demonstration program projects were gaining attention, groups of architects, engineers, and tradespeople were anxious to learn, firsthand, from those intimately associated with this program. A conservative estimate is that MSFC project managers spent 6% to 7% of their time in preparation, travel, and presentations to such groups. These presentations were made to a national convention of ASHRAE, to many local chapter meetings of solar groups, to gatherings of architects, mechanical engineers, and so on, at Sun Day observances, and even to a convention of American Homemakers, where the author addressed thousands of housewives. All afforded invaluable opportunities to promote the use of solar energy; each of us was keenly aware of the positive impact we could make through this medium. The author remembers some hostile encounters with people who did not believe that the government had a role in this endeavor, although this was not the majority belief. Most professional people in the technical fields recognized that initial entry into the commercial-sized solar energy market, at a time when manufacture of solar components was in its infancy, could be achieved only through some type of subsidy arrangement, coupled with an organization that could shepherd the program full-time. The energy crisis loomed very real; the government, which possessed the necessary resources to spearhead the effort, was perceived by most interested parties as the most likely advocate for early introduction of renewable energy systems.

10.3.11 Contractors' Reviews

In December 1977 the first DOE-sponsored Solar Heating and Cooling Demonstration Program Contractors' Review was held at New Orleans (DOE 1977). This review of the commercial demonstration program covered contracts in effect at the time, namely, those awarded by NSF, DOE, other federal agencies. During this annual review and the two that followed, the government's project managers took an active part on panels to discuss numerous problems confronting this relatively new venture. The problems ranged from the mundane, such as insurance agents not understanding solar components (one insurance company wanted to treat

Table 10.3
Commercial demonstration projects managed by MSFC

Project/location state/terr.	Fac use	Prog[a]	N/R[b]	Array area (ft²) (× .0929 = m²)	Coll type[c]	Cond'd area (ft²)	Sys des %[d] H	HW	C
1. Alabama Power Co., Montevallo, AL	Ofc Bldg	2	N	2,340	FPL	17,000	57	38	18
2. Huntsville Senior Citizens Center, Huntsville, AL	Sr Ctr	2	N	1,945	FPL	12,922	85	85	0
3. La Quinta Motor Inn, Mobile, AL	Motel	H/M	N	200	FPL	N/A	0	67	0
4. Reynolds Metals Co., Lister Hill, AL	Shower Fac	2	N	2,344	FPL	3,542	79	58	0
5. Shoney's South, Inc., Little Rock, AR	Restaurant	3	N	1,428	FPL	3,542	79	58	0
6. Tempe Union High School, Tempe, AZ	School	1	N	20,621	FPL	264,000	78	89	0
7. Aretex Services, Fresno, CA	Laundry	1	R	2,340	FPL	N/A	0	19	0
8. Environmental Center, San Francisco, CA	Ofc Bldg	1	N	900	FPL	55,000	80	100	0
9. El Camino Elementary School, Irvine, CA	School	1	R	5,000	ET	41,000	50	0	50
10. Iris Images, Mill Valley, CA	Film Lab	1	N	640	FPL	N/A	0	59	0
11. Fire Station, Lake Tahoe, CA	Fire Sta	1	R	352	FPL	3,200	50	50	0
12. Project Sage, El Toro, CA	Apt Bldg	NSF	R	1,008	FPL	N/A	0	70	0
13. Project Sage, Upland, CA	Apt Bldg	NSF	N	800	FPL	N/A	0	70	0
14. Santa Clara Community Center, Santa Clara, CA	Comnty Ctr	NSF	N	7,085	FPL	27,000	85	0	65
15. Hogates Restaurant, Washington, D.C.	Restaurant	NSF	R	6,300	FPL	N/A	0	55	0
16. Lutz-Sotire Partnership, Stamford, CT	Ofc Bldg	1	R	2,561	FPL	25,000	50	0	0
17. Wilmington Swim School, Inc., Newcastle, DE	Swim Sch	3	N	2,700	FPL	12,400	40	30	0
18. Brandon Swimming Association, Brandon, FL	Health Ctr	2	N	5,356	FPL	18,440	90	89	59
19. Dade Co. Elementary School, Coral Gables, FL	School	NSF	N	18,613	FPL	70,000	0	90	70
20. Days Inn, Altamonte Springs, FL	Motel	H/M	R	1,000	FPL	N/A	0	TBD	0
21. Days Inn, Clermont, FL	Motel	H/M	R	1,000	FPL	N/A	0	TBD	0
22. Days Inn, Jacksonville, FL	Motel	H/M	R	900	FPL	N/A	0	65	0

23. Florida Solar Center, Cape Canaveral, FL	Ofc Bldg	2	R	1,781	ET	5,200	95	0	70
24. Florida I-95 Visitors Center, Yulee, FL	Welcome Ctr	1	R	2,720	T/C	3,300	99	0	88
25. Reedy Creek Utilities, Lake Buena Vista, FL	Ofc Bldg	1	N	3,840	CPR	6,100	100	100	80
26. Quality Inn of Key West, Key West, FL	Motel	H/M	R	1,392	FPL	N/A	0	47	0
27. Days Inn, Atlanta, GA	Motel	H/M	R	960	FPL	N/A	0	81	0
28. Days Inn, Savannah, GA	Motel	H/M	R	900	FPL	N/A	0	50	0
29. Georgia Power Co., Atlanta, GA	Ofc Bldg	3	N	23,760	CPR	650,000	65	25	22
30. North Georgia Planning and Development Commission, Dalton, GA	Ofc Bldg	2	N	2,001	FPL	7,500	93	86	55
31. Savannah Science Museum, Savannah, GA	Museum	3	N	3,870	FPL	24,238	99	100	75
32. Shenandoah Community Center, Shenandoah, GA	Comnty Ctr	NSF	N	11,213	FPL	58,000	90	90	60
33. Towns Elementary School, Towson, MD	School	NSF	R	10,360	FPL	32,000	50	80	60
34. Scattergood School, West Branch, IA	Gym	1	N	2,496	FPA	8,000	75	75	0
35. Clarksville Elementary School, Clarksville, IN	Gym	2	R	6,402	FPL	145,200	100	0	0
36. La Quinta Motor Inn, Indianapolis, IN	Motel	H/M	N	2,000	FPL	N/A	0	66	0
37. La Quinta Motor Inn, Merriville, IN	Motel	H/M	N	2,000	FPL	N/A	0	60	0
38. Citizens Mutual Savings & Loan, Leavenworth, KS	S&L Bldg	3	N	2,520	FPL	9,260	53	98	52
39. Ducat Investments, Inc., Kansas City, KS	Warehouse	2	N	7,800	FPA	39,000	40	0	0
40. Kansas Cosmoshpere & Discovery Center, Hutchinson, KS	Museum	3	N	3,398	FPL	30,581	94	100	0
41. Kaw Valley State Bank, Topeka, KS	Bank Bldg	2	N	1,253	ET	5,600	47	95	74
42. Medical Office Center, Wichita, KS	Med Ofc	2	N	1,059	VWA	4,764	85	0	0
43. Unified School District 306, Mentor, KS	School	2	N	5,125	FPL	138,000	52	84	0
44. Rademaker Corporation, Louisville, KY	Ofc & Wrhs	1	R	435	FP/LA	1,080	50	15	0
45. LSU Field House, Baton Rouge, LA	Fld House	1	R	5,700	FPL	101,200	28	25	0
46. Grover Cleveland School, Dorchester, MA	School	NSF	R	4,600	FPL	21,000	65	0	0
47. Baltimore County Jail, Towson, MD	Jail	3	N	11,300	FPL	120,000	33	90	0

Table 10.3 (continued)

Project/location state/terr.	Fac use	Prog[a]	N/R[b]	Array area (tf²) (×.0929 = m²)	Coll type[c]	Cond'd area (ft²)	Sys des %[d] H	HW	C
48. Baltimore Fire Station, Baltimore, MD	Fire Sta	2	N	2,000	FPL	8,500	78	100	0
49. Carrol County Commissioners, Union Mills, MD	6 Bldgs	3	R	4,150	NET	6,100	56	76	0
50. City of Baltimore, Baltimore, MD	Rec Bldg	1	N	3,100	FPL	17,000	64	0	0
51. Ferguson Corporation, Annapolis, MD	Motel	1	R	2,170	FPL	N/A	0	68	0
52. Maryland National Park, Gaithersburg, MD	Maint Ofc	1	N	1,500	NET	3,000	60	0	0
53. Montgomery Community College, Germantown, MD	School	2	N	4,500	FPL	18,000	60	88	0
54. Timonium Elementary School, Towson, MD	School	NSF	R	5,300	FPL	2,500	60	0	90
55. Ihgham County Board of Commissioners, Okemos, MI	Med Care	1	N	9,374	FPL	N/A	0	25	0
56. Raddison Plaza, St. Paul, MN	Hotel	1	N	4,752 / 3,744	FPL / FPA	193,770	26	75	0
57. Telex Communications, Blue Earth, MN	Mfg & Assy	1	R	11,578	FPL	97,000	60	0	0
58. Kansas City Fire Station, Kansas City, MO	Fire Sta	1	N	2,808	FPA	8,800	47	75	0
59. Stevens College, Columbia, MO	Visitor Ctr	2	N	3,168	FPL	13,000	71	0	0
60. William Tao, St. Louis, Mo	Ofc Bldg	2	R	244	NET	1,007	45	50	0
61. World Plan Executive council, Waverly, MO	Ed Fac	3	N	2,400	FPL	30,000	35	0	0
62. New Mexico State University, Las Vegas, NM	Ofc Bldg	NSF	N	6,730	FP/C	25,535	100	100	50
63. First National Bank of Clarksdale, Marks, MS	Bank Bldg	3	N	469	FPA	2,000	60	0	0
64. Security National Bank, Starkville, MS	Br Bank	3	N	312	FPA	768	55	0	0
65. Charlotte Memorial Hospital, Charlotte, NC	Ed Bldg	1	N	3,950	FPL	19,000	40	79	0
66. Montgomery Community College, Germantown, MD	School	3	N	1,320	CPT	37,900	89	95	0

#	Location	Type								
67.	Days Inn #1, Woodland Rd., Charlotte, NC	Motel	H/M	R	3,000	FPL	N/A	0	80	0
68.	Days Inn #2, Sugar Creek Road, Charlotte, NC	Motel	H/M	R	1,000	FPL	N/A	0	50	0
69.	Days Inn #3, Tuckaseegee Rd., Charlotte, NC	Motel	H/M	R	1,000	FPL	N/A	0	54	0
70.	Days Inn #4, Statesville, NC	Motel	H/M	R	1,000	FPL	N/A	0	54	0
71.	Kalwall Corporation, Manchester, NH	Warehouse	1	R	1,700	TFP	8,640	60	0	0
72.	Garden State Racing Association, Cheery Hill, NJ	Motel	2	R	5,759	ET	70,000	32	30	0
73.	RKL Factory, Lumberton, NJ	Factory	1	N	6,000	FPL	40,000	80	0	70
74.	Sea Loft, Long Branch, NJ	Restaurant	2	N	1,436	ET	24,700	43	50	28
75.	La Quinta Motor Inn, Las Vegas, NV	Motel	H/M	N	2,000	FPL	N/A	0	66	0
76.	Troy Miami Public Library, Troy OH	Library	1	R	3,264	ET	23,000	67	0	0
77.	Columbia Gas System Service Group, Columbus, OH	Ofc Bldg	2	R	3,047	CPT	25,000	35	72	24
78.	Columbus Technical Institute, Columbus, OH	Ed & Lab	2	N	3,712	ET	43,715	77	0	38
79.	Oklahoma Department of transportation, Buffalo, OK	Ofc Bldg	3	N	6,000	FPL	TBD	35	80	45
80.	Factory, San Juan, Puerto Rico	Factory	NSF	N	7,308	ET	10,800	0	0	60
81.	Blakedale Professional Center, Greenwood, SC	Ofc Bldg	1	N	954	FPL	4,400	85	100	0
82.	Charleston County School District, Charleston, SC	School	3	R	3,750	ET	22,000	75	0	55
83.	Abbeville Fire Station, Abbeville, SC	Fire Sta	3	N	2,000	AIR	3,000	67	50	0
84.	Days Inn, Anderson, SC	Motel	H/M	R	750	FPL	N/A	0	45	0
85.	Mt. Rushmore Visitor Center, Keystone, SD	Visitor Ctr	1	R	2,000	FPL	7,250	45	0	45
86.	Days Inn, To Be Determind, somewhere in the southeast U.S.	Motel	H/M	N	1,000	FPL	N/A	0	TBD	0
87.	Days Inn, To Be Determined, somewhere in the southwest U.S.	Motel	H/M	N	1,000	FPL	N/A	0	TBD	0
88.	Belz Investment Co., Memphis, TN	Ofc Bldgs	3	N	3,120	FPA	25,720	48	0	0
89.	Coca Cola Bottling Works, Jackson, TN	Btlg Plant	2	R	10,000	ET	70,000	53	54	0

Table 10.3 (continued)

Project/location state/terr.	Fac use	Prog[a]	N/R[b]	Array area (ft²) (×.0929 = m²)	Coll type[c]	Cond'd area (ft²)	Sys des %[d] H	HW	C
90. La Quinta Motor Inn, Nashville, TN	Motel	H/M	N	1,500	FPL	N/A	0	57	0
91. City of Dallas, Dallas, TX	Rec Ctr	NSF	R	3,650	FPL	8,000	80	90	48
92. Days Inn, (Garland), Dallas, TX	Motel	H/M	R	1,000	FPL	N/A	0	65	0
93. Days Inn, (Forest Lane), Dallas, TX	Motel	H/M	R	1,000	FPL	N/A	0	65	0
94. Days Inn, (Valley View), Dallas, TX	Motel	H/M	R	1,000	FPL	N/A	0	65	0
95. Days Inn, Houston, TX	Motel	H/M	R	1,000	FPL	N/A	0	65	0
96. La Quinta Motor Inn, Beaumont, TX	Motel	H/M	N	2,000	FPL	N/A	0	59	0
97. La Quinta Motor Inn, San Antonio, TX	Motel	H/M	R	1,800	FPL	N/A	0	69	0
98. La Quinta Motor Inn, Texas City, TX	Motel	H/M	N	2,000	FPL	N/A	0	63	0
99. North Dallas High School, Dallas, TX	School	3	R	4,800	FPL	126,705	24	32	0
100. Radian Corporation, Austin, TX	Engr Ofc	1	R	350	T/C	850	90	90	80
101. Sky Harbour Elementary School, San Antonio, TX	School	3	N	5,616	CPT	59,864	87	64	67
102. Thompson Motel, Taylor, TX	Motel	H/M	N	1,232	FPL	N/A	0	73	0
103. Travis-Braun & Associates, Dallas, TX	Ofc Bldg	2	N	1,596	FPL	9,338	87	100	0
104. Trinity University, San Antonio TX	Ed/6 Dorms	1	R	16,080	T/C	284,928	66	66	5
105. Woodlands Aquatic Center, Woodlands, TX	Aqtc Ctr	H/M	R	4,080	FPL	N/A	0	56	0
106. Woodlands Inn, Woodlands, TX	Conf Ctr	H/M	R	7,140	FPL	N/A	0	58	0
107. Arlington Racquet Club, arlington, VA	Rec Fac	3	R	3,100	FPL	4,000	94	30	0
108. City of Richmond, Richmond, VA	Library	3	N	1,646	FPL	7,530	89	83	58
109. Fauquier High School, Warrenton, VA	School	NSF	R	1,600	FPL	TBD	77	0	0
110. Loudon county School Board, Leesburg, VA	Tech Ctr	1	N	1,225	FPL	N/A	0	26	0
111. Portsmouth Public Schools, Portsmouth, VA	School	3	N	3,737	FPL	59,000	63	86	0
112. Terrell E. Moseley, Lynchburg, VA	Ofc Bldg	1	R	400	FPL	1,780	73	70	0
113. Wilson Psychiatric Hospital, Charlottesville, VA	Phsyc Hosp	2	R	1,600	FPL	7,218	67	71	0

114. Olympic Engineering Corporation, Richland, WA	Engr Ofc	1	R	1,500	NET	3,000	60	0	0
115. Alderson-Broaddus College, Philippi, WV	Col Bldg	3	N	3,150	FPL	50,000	39	39	0
116. Bethany College, Bethany, WV	Dining/Motl	2	R	4,111	FPL	7,200	33	86	0
117. Page-Jackson School, Charleston, WV	School	NSF	R	10,943	FPL	52,600	85	0	51
118. La Quinta Motor Inn, Casper, WY	Motel	H/M	N	2,000	FPL	N/A	0	58	0

[a] Prog:
1 = PON 1 project
2 = PON 2 project
3 = PON 3 project
H/M = Hotel/Motel Hot Water Initiative project
NSF = National Science Foundation–initiated project
[b] N/R:
N = New host structure for solar energy system
R = Retrofit
[c] Coll type:
AIR = Air collector integral with roof structure
CPR = Concentrating parabolic reflector type
CPT = Concentrating parabolic tracking
ET = Evacuated tube type
FPA = Flat plate, air
FP/C = Flat plate with concentrators
FPL = Flat plate, liquid
FP/LA = Flat plate, 2 types for 2 systems, i.e., liquid and air
NET = Nonevacuated tube
T/C = Tracking with concentrators
TFP = Translucent fiberglass panel
VWA = Vertical wall with air as transport medium
[d] Sys des %: Number shown under H (heating), HW (hot water) and C (cooling) denote the design expectation for the percentage of the calculated facility requirement to be contributed by the solar energy system.

each collector as a boiler!), to the more technical concerns such as controls, their proper application and limitations. The second and third DOE-sponsored national contractors' reviews were held in San Diego in December 1978 and in Norfolk, Virginia, in December 1979 (DOE 1978b, 1979b). In the interim, there were a number of regional solar update conferences, in Atlanta, Chicago, San Francisco, Boston, and Orlando, Florida. At each successive review and update, one could perceive a growing maturity in the participants as everyone became better informed.

10.3.12 Special Studies

On several occasions, the MSFC project managers were called upon to investigate unique problems that developed either during installation, at start-up, or in the first few weeks of system operation. At a small motel with a solar hot water system, for example, the motel owner reported that people who ate in the motel's restaurant, after installation of the system, complained about bad-tasting gravy and other foods. A study was conducted by the project manager, and the source of the ill-tasting food was found to be "free phenol" that had leached out of an epoxy paint used to coat the inside of the solar system's storage tank. Because there was no expansion tank or check valve, the water from the storage tank contaminated the cold water supply used by the restaurant. There were numerous other "ministudies" performed by project managers that obviated the need for extended investigations and involvement of the S&E laboratories and/or other specialists.

10.3.13 MSFC'S Science and Engineering Laboratories' Role

During the five years of active participation by NASA/MSFC in the commercial demonstration program, the MSFC project managers regularly relied on the significant resources of the S&E laboratories. The use of solar energy was not new to a select group of these managers; this energy had been harnessed to power the satellites and to regulate the temperature of spacecraft. Because most of the project managers had been chosen from the laboratories, their knowledge of S&E was invaluable. Each year, when the budget was being formulated for presentation to the DOE, one of the most important components was manpower and funding requirements for S&E support. The nature of the commercial demonstration program made it relatively simple to predetermine which

of the various laboratories in S&E would be needed. By projecting the
number of solar energy projects there would be in any given phase
(design, construction, installation, start-up, acceptance testing, or operation) during the fiscal year under consideration, manpower, operating,
and travel fund requirements could be quantified with a fairly high degree
of accuracy. This kind of planning was necessary from two perspectives: from MSFC's, because at this time the space shuttle program was
demanding much engineering, and of course from DOE's, because it expected sound supporting evidence to accompany funding requests. Having made provision for S&E support, the project managers would call in
for assistance when it was required. A more sophisticated use was made of
the laboratories in support of the Commercial Demonstration Office's
team of special troubleshooters, the SEAG, which was discussed earlier.
The S&E laboratories' technical assistance to the commercial demonstration program began with technical input during the formulation of the
Program Opportunity Notice and continued through to fully operational
systems. Examples that come to mind include many which were handled
by the materials laboratory. One, highlighted by the 11,578-square-foot
collector array at Telex, Blue Earth, Minnesota, had to do with flexible
connections for the collectors' inlets and outlets. Using standard automobile hose was not successful, partly because of ultraviolet deterioration. Hard line connections were susceptible to the stresses of expansion
and contraction with ensuing fatigue failures, while temperature limitations of some materials made high-temperature soldering or brazing
impractical. The S&E materials laboratory conducted extensive tests on
various tube materials and clamping devices and determined that high-
temperature reinforced silicone hose, smooth metal nipples (no machining
burrs), spring-type clamps or stainless steel screw-type clamps with a
smooth inner liner should be used to avoid the problems previously
encountered.

10.4   Hotel/Motel Hot Water Initiative Program

In 1977 DOE's San Francisco Operations Office managed the solicitation
and evaluation of a separate group of proposals for hotel and motel solar
hot water systems. A few MSFC personnel from the S&E laboratories
assisted in the evaluation. Forty-nine contracts were awarded as a result

of this solicitation. In December 1977, 27 of these projects were assigned
MSFC for management within the Commercial Demonstration Office.
In 1980, after deletions and additions, 47 hotel/motel projects remained;
29 were being managed by MSFC; the remaining 18 by the DOE. The
majority of the MSFC projects were Days Inns: 12 owned by the parent
company based in Atlanta and 4 under a Days Inn franchise to Office
Parks of Charlotte, North Carolina. La Quinta Motor Inns accounted for
9 more of the MSFC projects; Quality Inns for 1 located at Key West,
Florida. The remaining 3 motels were locally owned and operated: 2 in
Woodlands, Texas, and 1 in Taylor, Texas.

The Hotel/Motel Hot Water Initiative Program was one of the better
ways to publicize solar energy for commercial use. The systems were
prominently affixed to the structures and were seen by a new group of
people every day. Days Inn management was well satisfied with the performance of their systems and passed this on to the general public by
conducting special events that highlighted the use of solar energy as an
attractive alternative to the use of nonrenewable energy.

10.5   National Science Foundation Projects

Prior to the establishment of ERDA, the National Science Foundation
(NSF) sponsored 24 commercial-sized demonstration projects throughout
the United States—one with located in Puerto Rico. Beginning in 1977
and continuing into 1979, a total of 15 NSF projects had been reassigned
to MSFC for management within the Commercial Demonstration Office;
management of the other 9 was retained by DOE.

By the second quarter of 1980, 13 of the 15 NSF Projects under MSFC
management, had been "turned on." Of the 13, 9 were fully operational,
and 1 was partially operational. At that time, management for these 10
projects was returned to DOE for project monitoring. Of the remaining 3
projects that had been "turned on," 2 (Fauquier High School in Warrenton, Virginia, and Towns Elementary School in Atlanta) were in-
operative. The other project, Page-Jackson School in Charlestown, West
Virginia, was only partially operational and was not considered report-
able because it was delivering less than 50% of the design goal. In mid-
1980, 2 of the solar energy systems originated by the NSF, were still under
construction; these were located at the Dade County Elementary School

in Coral Gables, Florida, and at a factory in San Juan, Puerto Rico. By 1981 all MSFC activity on these remaining 5 projects had been transferred to DOE.

Of all the commercial demonstration program projects managed by MSFC, the NSF projects presented the greatest challenge. A problem with some of these systems was excessively complicated controls. One such system the author vividly recalls was located in Shenandoah, Georgia, about 25 miles southwest of Atlanta. Installed in the Shenandoah Community Center, this system sported a showpiece glassed-in control area with colorful piping. However, it was fraught with operational problems, many of which could be directly linked to the complicated control system. It was further complicated by the fact that it delivered energy for four subsystems; space heating, domestic water heating, swimming pool water heating (in autumn and spring), and space cooling (it provided approximately 60% of the heat required to drive a 100-ton lithium bromide absorption chiller).

From a paper on the Shenandoah Solar Recreational Center (Williams and Craig in DOE 1978b, pp. 349–357) delivered at the Second Solar Heating and Cooling Commercial Demonstration Program Contractors' Review under "Lessons Learned" it reads in part:

There is little that can be done to reduce the system's complexity. In future designs, more use may be made of microprocessor-based controls, but it is not clear at this point that this will inherently improve reliability since the basic mechanical complexity and the number of sensors will likely remain the same. The only obvious remedy is to directly reduce the system's complexity, e.g., the number of modes, valves, etc., and accept less overall performance in return for reduced operating and maintenance cost. Truly, the last joule of solar energy extracted from the solar collectors may often only be obtained at a great cost.

This is a statement that should not be taken lightly.

Beyond what is stated above, it is the author's opinion that many of the problems encountered with the NSF projects were caused by a lack of experience in construction management within the scientific and academic communities. Although their technical capabilities were beyond reproach, these NSF project designers had been thrust into the construction management aspect of the solar energy field, where the majority of them had little or no experience. Designing and constructing an on-campus laboratory type experiment is one thing; designing and constructing a functional facility is quite another.

10.6 Data Acquisition and Analysis

As discussed in section 10.3.6, one of the supporting functions of the
Systems Development Office was to develop a Site Data Acquisition
System (SDAS). The system included sensors necessary to measure insolation, flow rates, temperatures, power consumption, wind speed, wind
direction, and so forth. Once each day (at around 2 A.M.) this information,
which had been collected several times per hour during the previous 24
hours and stored on a cassette tape, was sent to the Central Data Processing System via the standard commercial telephone network. Once
each month these data were entered into custom system performance
equations; the resulting performance information and other reduced data
such as heating degree-days for the period, insolation data, and so forth,
were distributed to interested parties, such as the owner of the solar
energy system, the owner's project manager, the national data program
manager, the government's project manager, and the DOE's Technical
Information Center.

As of mid-1979, 74 out of a total of 283 DOE commercial demonstration projects (26%) were in the instrumentation program; only 17 of
these projects (23% or 6% of the entire program) were fully operational.
Even so, the General Accounting Office's report to the Congress (GAO
1980) states in the "Digest" portion of the report: "The data dissemination program cost for commercial demonstrations, through fiscal year
1979, exceeded $13 million. The benefits from this program thus far have
been limited." Even though the act itself was five years old at that time,
the contracts for PON 1 commercial demonstration program projects had
not been awarded until the middle of FY 1976. As of the third quarter of
FY 1979, 41 of the 74 projects (55%) designated to receive instrumentation were still either in design, construction or checkout; only 5 of the 74
systems (less than 7%) were not operating. The remaining 28 systems
(38%) were partially operational. In summary, the GAO had drawn its
conclusion using a mere 33 instrumented projects having an average
existence of approximately 1.5 years.

Eventually, the SDAS proved its value by collecting and processing
real-time sensor data from instrumented sites. This enabled reviewers to
evaluate design assumptions, determine real energy savings, pinpoint problem areas, compare various system and system components performances,

and disseminate the information in order to enhance the design of future systems and assist in the refurbishment of existing ones.

10.7 Lessons Learned

It has been said that experience is a harsh teacher, it gives the test first and the lesson later. Some of the "tests" encountered in the commercial demonstration program were indeed harsh. Although the greatest number of these "tests" were experienced in the technical areas; many were also found in the marketplace in the area of public acceptance.

10.7.1 Institutional Lessons Learned

The Solar Heating and Cooling Demonstration Act of 1974 was prompted by the energy crisis of 1974 and was immediately popular. However, as time went by, the crisis was found to be something less than critical, and, as has been said elsewhere: "The memory of the American people has a half-life of about two weeks." If one acknowledges that there is some degree of truth in this statement, then it becomes easy to understand why this relatively new technology, with its somewhat intricate design considerations, experienced difficulty when it came to maintaining public interest. This made the task of widespread commercialization extremely challenging. There were other forces at work that were quite detrimental to the program. The GAO's untimely report (GAO 1980) to Congress did little to help the program. The information it contained was some 9 to 10 months old. To illustrate, on page 15 of the report, under a section entitled "Conclusions," the second paragraph reads: "Very few commercial demonstration projects are operating as designed. As of June 1979, only 104 of the 238 projects funded had been constructed and their systems started up. The solar systems on 55 of those projects, or over 53 percent, were either down, partially operational, or being tested.... "

A status review (DOE 1979a) of the 104 MSFC-managed projects at June 1979 (the time cited by GAO 1980) compared these same projects. It is revealing inasmuch as it clearly relates a diametrically opposite message from the one stated by the GAO. Table 10.4 illustrates this point.

The status descriptions shown in table 10.4 were devised jointly in 1978 by Carl Conner, program manager, Demonstration Project Management, DOE, Jackson Hale, program manager, National Solar Data Program,

Table 10.4
Status of 104 MSFC-managed projects, June 1979 and April 1980

	Number of projects	
Status	June 1979	April 1980
Fully operational	25	52
Partially operational, reportable	3	3
Partially operational, nonreportable	0	0
System down	6	6
System in checkout	13	7
System under construction	31	23
System in design	26	13

and the author to provide a uniform method of reporting project status. These status categories, along with definitions for each, appeared in all official reporting thereafter, including the reports used to gather the information recorded in the table. The 1979 data are from a June 1979 DOE publication of Project Data for the Solar Heating and Cooling Commercial Demonstration Program (DOE 1979a) and the 1980 data are from and official MSFC monthly report of the type mentioned earlier.

In its 1980 report, the GAO indicated that approximately 20% of the solar energy demonstration projects were "fully operational" as of June 1979; in table 10.4, MSFC reported a slightly higher percentage, namely 24%. It can also be seen that only 9 to 10 months later, MSFC reported exactly 50% of the 104 projects as fully operational.

The target audience for the GAO report, the Congress, is the one group that can seal the fate of government-funded programs. The author believes that had the GAO made their study 9 or 10 months later, the conclusions drawn by that thorough group would have been entirely different and the effect on the program would have been sufficiently positive to have convinced the Congress that the commercial demonstration program was worth continued support. But what happened? Congress cut off funding.

In retrospect, I believe that had the hotel/motel hot water initiative idea come first, things would have been different. I believe this for three reasons:

1. Such applications use solar energy in its simplest form to satisfy a year-round demand.

2. Good, workable systems, not much different from those used in Florida since the 1930s, could have been replicated quite easily with little more than site adaptation.

3. The vast majority of non–technically oriented taxpayers in the public sector would have been sufficiently impressed by the results to back federal funding of solar energy. Then, when the time was right, the scientists in our midst could have proceeded, with heightened public backing, to the more sophisticated solar energy systems.

I also believe that if the federal solar program were to start up again, it should be initially kept to a simple application—heating water. I further believe that installing solar hot water systems in motels—with collectors, in plain view—would be the best place to start. Hotels, from a practical standpoint, must almost always mount the collectors on the roof of their multistory buildings—out of the public eye.

10.7.2 Technical Lessons Learned

In the technical area of the program numerous lessons were learned; a few stand out in the author's memory.

One of the PON 1 Projects, Scattergood School, a small, parochial school with approximately 65 students, is located in West Branch, Iowa. An air system was chosen for the demonstration. The school's project manager at the time held a doctorate in physics; he was also the headmaster of the school. This combination—a project manager with a vested interest, highly knowledgeable in the physical sciences, and in full-time residence at the site—was made to order for the commercial demonstration program. Perhaps in a subtle way this example also serves to illustrate why commercialization was so difficult, because even this well-nurtured system experienced a failure which, although not catastrophic, was nevertheless, detrimental to the host structure—a gymnasium. The air system provided both space heating and heating for domestic water. In November 1977, during a night when the ambient temperature dropped to $-15°F$ ($-26°C$), the building was being heated from storage. Unfortunately, a motorized damper, designed to keep air from flowing from the collectors, did not seal properly, permitting subzero air to flow across a heat exchanger. The heat exchanger froze and burst; approximately 2,000 gallons (7,570 liters) of water flooded the gymnasium before the leak was discovered. To rectify the problem, the motorized damper was properly adjusted and an additional damper installed between the collectors and the heat exchanger. The costs of the damage were covered by the owner, and eventually the project did overcome this negative beginning. The lesson was that dampers in conventional systems can leak without causing undo concern, but the results of leakage past dampers in a solar air-to-hydronic system can be devastating.

An insolation sensor at the Kaw Valley State Bank project in Topeka was, because of its size and location, bare of ice and snow. On the morning of 8 January 1979, it was bitter cold. Ice and snow covered the evacuated tube collectors, but the sun was bright and as soon as the insolation reached 35 Btu/ft^2/hr (94.9 Kcal/m^2/hr), the system controller reacted to the bare sensor and the system was turned on. Subfreezing ethylene glycol/water solution flowed to the heat exchanger. When this occurred, two things happened: first, the water froze in the heat exchanger, and this burst the walls of the exchanger, and second, the water then mixed with and diluted the glycol solution, which was carried back to the collectors and burst the glass tubes. The majority of the system had to be replaced. To prevent recurrence, a temperature-controlled diverting valve was installed to pass transport fluid to the heat exchanger when the temperature of the fluid rose to 75°F (23.9°C) and to by-pass the heat exchanger when the temperature of the fluid dropped below 65°F (18.3°C). As a further precaution, a fail-safe controller was introduced, just upstream of the heat exchanger, that would cut off the collector pump if it sensed the temperature fall to 40°F (4.4°C). The lesson was that components usually function as intended; but the designer cannot always foresee the entire spectrum of the component's proper functioning; "what ifs" should be investigated to the point of exhaustion.

There were many other lessons learned during the commercial demonstration program, but allotted space does not permit recounting them all. These experiences have been dutifully recorded in numerous publications such as the proceedings of the first, second, and third contractors' reviews, the preliminary issue of the *Experiences Handbook*, and the ASHRAE handbook of experiences in the design and installation of solar energy systems.

Even though there were many problems encountered during the short life of the commercial demonstration program, there is, after all, a bright side. We can profit from the program's errors—this is where recorded history enjoys its greatest value. The errors made in the design, installation, and operation of the program's solar energy systems need never be repeated.

10.8   Success Stories

Contrary to what the previously cited 1980 GAO report might imply, there were many success stories. Let me close with some of these.

The Scattergood School project, mentioned above, was a project that had an abundance of success. The school raised and marketed approximately 500 hogs per year; this required grain (corn), which was also raised by the school. There was a 6,000-bushel grain-drying silo that cost approximately \$1,200/year to operate. Realizing that a lot of heat was available in the early fall, when it was not required for the gymnasium, NASA and, in turn, DOE were approached with a request to divert some of this heat to dry the corn. Permission was given. At no additional cost to the government, heated air from the rock storage bin was diverted to the drying silo and the job was accomplished. (See chapter 12 on similar examples of crop drying.)

During the December 1979 third contractors' review, the project manager reported (DOE 1979b, pp. 413–418):

The space heating requirement of the complex from September, 1978, through April, 1979, including grain drying, was 152×10^6 Btu. The solar heating system supplied 129×10^6 Btu, or 84% of this amount. This percentage is significantly higher than the 75% solar used in the design calculations. The figure is even more surprising because of the severity of the winter—21% more degree-days and 10% less incident solar energy than long-term average [upon which the system design calculations were based].

After the successful venture with the grain storage, the project manager again approached NASA with a request to pipe some of the solar-heated water to a girl's dormitory. As with the silo, this would again be at no cost to the government. And again permission was granted.

Another shining example is the Telex Communications project in Blue Earth, Minnesota. This project involved a solar space heating system for a large (97,000 ft^2 or 9,011 m^2) manufacturing plant. The system employs 11,578 ft^2 (1,075 m^2) of collectors; it was designed to contribute 60% of the heating requirements. In 1979 solar energy met 100% of the heating needs until December 26, even during periods when the temperature dipped to $-9°F$ ($-22.8°C$)! This project, incidentally, had also experienced a few problems after start-up, but they were soon fixed permanently.

In still another case, a heating, cooling, and hot water system had been installed with the construction of a new 7,500-ft^2 (697-m^2) office building in Dalton, Georgia, owned by the North Georgia Planning and Development Commission. On the night of 9 January 1979, the temperature was approximately $10°F$ ($-12.2°C$). The igniter for the conventional furnace failed. At 7 A.M. the temperature in the building had dropped to $40°F$ ($4.4°C$). The storage tank for the solar energy system was found to be $109°F$ ($42.8°C$), so the solar automatic control was manually overridden. For three consecutive days (until the igniter could be replaced), the solar energy system carried the building load; this permitted fifty-six employees to remain on the job.

Although the author has no way of knowing how the above experiences influenced the sale (or commercialization) of solar energy systems; I do know that these success stories and many like them were recounted before various groups throughout the United States. The author personally presented them at Sun Day celebrations, before various technical societies, and even on a local TV version of "Meet the Press" one Sunday afternoon in Dallas.

References

ASHRAE (American Society of Heating, Refrigerating, and Air-Conditioning Engineers). 1975. *ASHRAE 90-75*. "Energy Conservation in New Building Design." August.

ASHRAE. 1977a. *ASHRAE 93-77*. "Methods of Testing to Determine the Thermal Performance of Solar Collectors." February.

ASHRAE. 1977b. *ASHRAE 94-77*. "Methods of Testing Thermal Storage Devices Based on Thermal Performance." February.

DOE (U.S. Department of Energy). Office of the Assistant Secretary for Conservation and Solar Applications. 1977. *Proceedings of the Solar Heating and Cooling Demonstration Program Contractors' Review*, New Orleans, 5–7 December. Prepared under contract EG-C-01-2522 by Planning Research Corporation/Energy Analysis Company, McLean, VA.

DOE. 1978a. *Solar Heating and Cooling Project Experiences Handbook*. Prepared under contract EC-78-C-01-4131. Prepared jointly by: The U.S. Dept. of Energy. Project Management Centers (Chicago Office, San Francisco Office, Marshall Space Flight Center–NASA), the ASHRAE and the Univ. of Alabama in Huntsville (UAH), July.

DOE. Office of the Assistant Secretary for Conservation and Solar Applications. 1978b. *Proceedings of the Second Solar Heating and Cooling Program Contractors' Review*, San Diego, CA, 13–15 December. Prepared under contract EC-73-C-01-4131 by Johnson Environmental and Energy Center, University of Alabama, Huntsville. Sponsored by: U.S. Department of Energy Assistant Secretary for Conservation and Solar Applications, Washington, DC.

DOE. Office of the Assistant Secretary for Conservation and Solar Applications. 1979a. *Project Data: Solar Heating and Cooling Commercial Demonstration Project*. Prepared under contract EG-77-C-01-2522 by Planning Research Corporation/Energy Analysis Company. McLean, VA, June.

DOE. Office of the Assistant Secretary for Conservation and Solar Energy. 1979b. *Proceedings of the Third Solar Heating and Cooling Program Contractors' Review*, Norfolk, VA, 16–19 December. Prepared under contract DE-AC01-CS-30114 by Kaba Associates, Inc. Washington, DC.

GAO (U.S. General Accounting Office). 1980. *Federal Demonstrations of Solar Heating and Cooling on Commercial Buildings Have Not Been Very Effective*. EMD-80-41. Report to U.S. Congress. Washington DC, 15 April.

NASA (National Aeronautics and Space Administration). 1980. *Solar Heating and Cooling Commercial Demonstration Program Management Report*. SHC1004-1. Huntsville, AL: George C. Marshall Space Flight Center, 30 April.

Ward, D. S., and H. S. Oberoi. 1980. *Handbook of Experiences in the Design and Installation of Solar Heating and Cooling Systems*. An ASHRAE document. Fort Collins: Solar Energy Applications Laboratory, Colorado State University, July.

11 Federal Buildings

Oscar Hillig

The Solar in Federal Buildings Program (SFBP) was legislated by Congress, with the final rulemaking published in October 1979 (DOE 1979c). The intent of the program was to demonstrate the feasibility and promote the use of solar energy in federal buildings throughout the United States. The program was managed by the Department of Energy (DOE), with technical and administrative support of the program provided by the National Aeronautics and Space Administration (NASA), and later by the Energy Technology Engineering Center (ETEC) operated by Rockwell international. This chapter has been adapted from the final report on SFBP (ETEC 1992).

A major portion of the program involved the installation of approximately 700 solar energy systems owned by various government agencies and located on their sites throughout the United States. The program was divided into two parts: (1) design and construction; and (2) performance monitoring and documentation. The first part of the program involved site surveys, design reviews, acceptance testing, and general solar technical assistance to the government agencies. The solar systems were to be state of the art, using off-the-shelf components, with the owning government agency responsible for contracting for design and construction and for operation and maintenance of the systems. In the second part of the program, operational and maintenance data were gathered and the thermal performance of some of the larger systems was monitored.

Over the course of the program considerable information was collected, studied, and summarized in a series of reports. With the assistance of experts from DOE laboratories and the solar industry, information from this program pertaining to large active solar heating systems has been integrated and presented in a set of three manuals: *Active Solar Heating Systems Design Manual, Active Solar Heating Systems Installation Manual* and *Guide for Preparing Active Solar Heating Systems Operation and Maintenance Manuals* (ASHRAE 1988, 1990, 1991).

11.1 Legislative/Historical

Under Title V, Part 2 of the National Energy Conservation Policy Act (PL 95-619) the Department of Energy (DOE) was charged by Congress with carrying out a program to demonstrate applications of solar heating and cooling technology to a variety of types of federal buildings, using commercial equipment.

The Solar in Federal Buildings Demonstration Program (SFBP) was a multiyear program wherein DOE provided funds to federal agencies for the design, acquisition, construction, and installation of commercially available solar hot water heating, space heating and cooling, industrial process heating, and passive systems in new and existing federal buildings. The intent of the program was to stimulate the development of the solar industry by encouraging federal agencies to design, purchase, and install commercially available solar energy systems in new and existing federal buildings, and to commit the federal government to begin to use solar technology in all appropriate applications.

The federal government objectives for SFBP (DOE 1979a) were

• to encourage more rapid commercialization of solar energy systems in the public and private sectors;

• to establish a leadership role in supporting solar applications;

• to demonstrate confidence in and support of the solar industry;

• to obtain design, installation, maintenance, and operating experience with solar systems;

• to support efforts to shift from nonrenewable to renewable energy sources;

• to maximize industry/federal agency interface through joint solar efforts;

• to increase consumer confidence in solar energy applications;

• to lower solar system costs by providing education through experience to solar system designers and installers, and by the economies of scale related to large procurements.

The SFBP was implemented by a Notice of Proposed Rulemaking (NOPR) published by DOE on 2 April 1979 (DOE 1979a). These rules set forth requirements and procedures for the submission and content of federal agency proposals, criteria for evaluation and selection of projects to be funded, transfer of funds for approved projects, and periodic reports with respect to the maintenance and operation of active and passive solar heating and solar heating and cooling demonstration projects in federal buildings.

A public hearing was held regarding the NOPR on 17 April 1979, and written comments were accepted until 31 May 1979. Preproposal conferences were held in each of the ten DOE regions, introducing and explaining the proposed program requirements to federal agency personnel. Based on responses to the NOPR, a notice of final rulemaking was drafted and published in the *Federal Register* on 19 October 1979 (DOE 1979c).

Concurrent with the review of the NOPR, the DOE prepared an environmental assessment of the SFBP and concluded that the rules did not have the kind of impacts that call for a regulatory analysis or an environmental impact statement under the National Environmental Policy Act of 1969 (DOE 1979b).

Under the SFBP rules, the life-cycle cost (LCC) method was one factor to be used in evaluating a proposed solar demonstration project. In January 1980 the DOE published a final LCC rule that established the methodology and procedures for computing LCC (DOE 1980; the LCC method is described in Ruegg and Short 1988).

The initial SFBP legislation authorized funding up to $100 million but budgetary revisions reduced the appropriated funds to $57,420,000; an additional $1.8 million was appropriated for FY 1981, bringing the total appropriation for carrying out program activities to $59,220,000.

Many experts in the solar field outside of DOE, including industry representatives, provided significant additional technical assistance throughout the program. The Solar Energy Industries Association (SEIA), which was involved and supported this program from the beginning, helped inform industry so that a best effort could be obtained.

11.2 Management Structure

The DOE retained full responsibility for managing all aspects of the SFBP operation under the assistant secretary for conservation and renewable energy, Active Solar Heating and Cooling Division. Technical and administrative services were provided by the National Aeronautics and Space Administration (NASA) at their Huntsville, Alabama, site through an interagency agreement. The four DOE regional solar energy centers (RSECs) were charged with providing regional knowledge about, and participation in, the SFBP (see chapter 20).

In mid-1981 NASA decided to discontinue participation in all solar energy activities not directed to space applications. Rockwell International's Energy Technology Engineering Center (ETEC) assumed the technical and administrative support functions of the SFBP in August 1981, reporting to DOE/SAN (DOE's San Francisco Operations Office).

11.3 Project Selection

Preproposal meetings were held in April and May of 1979 in each of the ten DOE regions, introducing and explaining the proposed program requirements to concerned federal agency personnel. Official proposal forms A-1, Technical Data Requirements, and A-2, Cost Data, were issued 5 October 1979, to be completed by each agency proposing to participate in the SFBP.

The basic information requested was a conceptual design, site information and cost data, including the amount of cost to be shared by the participating agency. To encourage maximum agency participation, the level of detail required to complete the forms was kept to the minimum needed for feasibility assessment.

DOE sponsored ten technical workshops in 1979 and 1980 for federal agency personnel. These week-long workshops were conducted by the Solar Energy Research Institute (SERI; now the National Renewable Energy Laboratory or NREL). They offered basic information on solar energy system design, economics, installation, and operation to assist federal agencies proposing to participate in the SFBP.

Proposal submission began after the life-cycle cost procedures rule (DOE 1980) was issued. The proposals were first evaluated by NASA in a two-step process: (1) review for completeness and consistency with proposal requirements; and (2) pass/fail screening to determine (a) that adequate nonsolar conservation measures were in place or scheduled, (b) that the solar energy system design was sound, (c) that the cost estimate was realistic, and (d) that the estimated cost to DOE did not exceed program limits. Projects that passed these two steps were graded according to the following criteria: (a) relative cost effectiveness (20 points), (b) extent of cost-sharing (20 points), (c) potential for replication (15 points), (d) proximity to solar market areas (15 points), (e) extent of passive solar measures included (15 points), (f) visibility and accessibility (5 points),

(g) innovative application (5 points) and finally, (h) geographical distribution (5 points).

In all, 912 proposals were evaluated, from which DOE selected 843 in May 1980, awarding $31,100,000 to sixteen different federal agencies, with projects in all fifty states and the District of Columbia. During the course of the program, 142 of these projects were withdrawn or canceled for technical, cost or schedule reasons, for contractor default, or at the request of the proposing agency. Some projects included more than one system; 718 systems were built over the course of the program.

11.4 Project Funding

Following DOE's selection of a project for the SFBP, NASA prepared an Order for Supplies or Services (H-Order), setting forth the cost, schedule, and statement of work for the project. When signed by the agency proposing the project, this H-Order became a contract, and the design funds specified in the order were transferred from DOE to the agency. NASA completed H-Orders for all projects accepted by DOE for the SFBP.

When the final system design and cost estimate had been completed and accepted by NASA (or later, ETEC) the acquisition, construction, and installation (ACI) funds were transferred to the agency. Subsequent requests by the agency for changes in schedule or funding were approved by NASA in a modification to the original order, or by ETEC by letter with DOE/SAN's concurrence; a maximum increase of 25% of the original approved ACI funding was permitted under SFBP rules for such causes as price escalation, need for larger systems, or design changes to improve efficiency. The agency was responsible for distributing the funds to the design and construction contractors.

11.5 Project Review and Control

11.5.1 Site Inspection

Following DOE's selection of a project for the SFBP, a site visit was made by NASA or ETEC staff to determine if the site was suitable for the project. The collector field location was checked for adequate area to allow proper spacing between collector rows, for proper collector orientation, and for potential shading problems. NASA conducted site surveys on all but 60 of the projects selected. ETEC completed the site survey for these projects at the time of the first design review if it was held at the site, or from photographs submitted by the proposing agency. None of the sites selected by DOE was rejected as unsuitable by the site survey team.

11.5.2 Design Review

The design of a project was done by the agency staff when in-house experience existed or, more often, by an architectural/engineering (A/E) firm under contract to the agency. Some agencies chose to award a single contract for design and construction. In any case, a final design review was held when the design was 90% complete, with the designer submitting his work to the agency and NASA (later, ETEC) staff. Consultants specializing in solar energy system engineering were sometimes included in the design review team. The design was reviewed to be sure it complied with the performance requirements of the agency's proposal and could be built for the funding approved. Concerns about technical details of the design (called "Design Review Comments") were listed for resolution with the designer and agency. When NASA or ETEC considered it advisable, a review meeting was held with the designer at this time. For larger systems, a preliminary design review was held when design was 30%–35% complete to assure that the designer was meeting the intent of the agency's specification, and to resolve any questions about the design while changes were easy to make. When the design review report was submitted, the ACI funds were released to the agency, and a construction contract was awarded by competitive bidding (unless a design-and-construct basis had been used). Final decisions rested with the agency; there was no requirement in the SFBP rules that the agency accept the recommendations of the NASA/ETEC design review.

11.5.3 Project Cost

A final cost estimate was prepared as part of the final design package to confirm that the project proposed by the agency and approved by DOE could be built for the approved funding. If the estimated cost was above the funding, the project was redesigned or downsized to fit the available funds at that time. The agency had the option of requesting an increase in ACI funds up to 25% of the original approved amount; such requests were reviewed by NASA/ETEC staff who then recommended approval or disapproval to DOE. Redesign/downsizing/funding increase action was also required when construction bids were received and found to be higher than the approved funding.

11.5.4 Construction

Once construction began, the agency reported progress monthly. The agency was responsible for controlling quality and maintaining schedule of the construction work. Whenever warranted, NASA/ETEC technical staff visited the construction site as an additional quality check. When the project was complete and successfully operated for a short period (usually two weeks), the NASA/ETEC technical staff carried out an acceptance test to verify that the system had been built in accordance with the design and to evaluate the system's thermal performance over one or two days' operation. After the acceptance test, a checklist of discrepancies was given to the agency for correction by the contractor, and an acceptance test report analyzing the system's performance was prepared for the agency.

The SFBP rules required the agency to submit an operation and maintenance (O&M) report to ETEC each quarter for the first year of operation and annually for the next two years. The agency was also required to prepare a final construction report as a record of the as-built system. In spite of DOE's efforts, response to those requirements was sporadic; only half of the projects supplied final construction reports, and fewer than one-third of the large projects supplied O&M reports for one year. The small hot water systems in the National Park Service (NPS) did fulfill the three-year O&M report requirement.

11.6   Construction Summary

The types of systems constructed are shown in table 11.1. Acceptance tests were completed on all but thirteen of the NPS systems; these were not ready when the park was visited, and a return visit was not cost-effective. Verification of subsequent construction and satisfactory operation of these systems was provided by the park ranger.

Acceptance tests were completed on all but seven active and one passive of the larger systems; again, those systems were not ready to be tested in their scheduled sequence, and a return visit was judged not to be cost-effective. Four of those active systems and the one passive system did operate successfully. The other three active systems never were brought

Table 11.1
Type of systems constructed

Small systems (537)	
Small hot water systems for National Park Service (NPS)	537
Large systems (181)	
Other water heating systems	98
Space heating	11
Space heating plus hot water	32
Swimming pool heating	3
Swimming pool heating plus hot water	3
Industrial process hot water heating	8
Industrial process hot water plus space heating	1
Space cooling	2
Space cooling plus space heating	3
Space cooling plus hot water heating	1
Space cooling, space heating, and hot water heating	1
Passive space heating and/or daylighting[a]	16
Hybrid (passive plus active hot water heating)	2
	718

[a] The responsibility for the passive projects was turned over to the Solar Energy Research Institute (SERI) midway in the program because of SERI's extensive passive program.

to successful operational status and were subsequently dismantled at the recommendation of the owning agency. DOE concurred with the recommendations to dismantle.

11.7 Data Reporting

11.7.1 Databases

Following DOE's selection of the 843 projects to be funded for the SFBP, NASA prepared a database that held pertinent details fully describing each project. ETEC revised and expanded this database, adding and updating information as each project progressed through design, construction, and acceptance testing. This database, called "SFBP DATA-BASE," was used to prepare the two key management reports, *The Project Description Directory* (vol. 1), and *The Program Management Report* (vol. 2) that were updated and issued monthly. The database was invaluable for cost and schedule management of the SFBP during the design and construction phases.

The SFBP DATABASE remains the definitive record of SFBP project data. In addition, two other databases were set up of selected files from the first database. For the 181 larger projects (excluding the small household hot water systems installed in the National Parks by the Department of the Interior) performance, operation and maintenance (PO&M) data were extracted from the quarterly PO&M reports, final construction reports, and final acceptance test reports. In 1987 and 1988 project contacts at each site were questioned to determine their system's operating status at that time and to learn of any operational problems with the system. These operational data were used to prepare the dBASE III database named "PROJDAT1.DBF."

Another dBASE III database, called "LL.DBF," included review comments from the final design review of seventy-five of the larger projects. LL.DBF was studied to see if a relationship could be established between design problems and subsequent system performance. The final acceptance test reports for these projects were then reviewed for comments that could relate to design problems and this information was added to the database. This became the basis for the "lessons learned" section of the three ASHRAE manuals, and for other project evaluation studies, such as refurbishment recommendations. The various parameters used to set up the databases are listed in table 11.2.

The three databases were searched and sorted (e.g., by owning agency, site, collector manufacturer, etc.) as required. A partial listing of a typical sort showing cost and performance numbers is included in table 11.3.

Some important generalizations may be made from various database summaries:

• The majority of the large systems included in the PROJDAT1.DBF database had collector areas less than 2,000 ft^2; only ten projects cost more than $500,000.

• The design cost typically was 5% to 15% of the construction cost but ranged up to 30% on some smaller systems.

• Installed collector loop efficiencies were about 20% less than the ASHRAE single-collector test efficiency.

• SFBP projects were built in forty of the fifty states, necessarily placing many (about two-thirds) in locations with insufficient insolation to make

Table 11.2
Project data in the SFBP databases

- Project number SFBP-____
- Funding order number H-____ with effective date
- Owning agency contact with Telephone number
- Owning agency
- Location: city and state
- Approximate insolation available, annual average Btu/ft^2-day
- Building use
- Solar application
- Freeze protection method
- Collector field area, gross ft^2
- Collector type
- Number of glazings
- Collector manufacturer
- Storage tank volume, gallons
- Ratio, storage volume to collector area
- Collector-loop heat exchanger type, with number of walls
- Final design review status
- Construction status
- Acceptance test status
- Final construction report status
- PO&M report status
- ETEC project engineer and technical monitor
- Proposed design cost

- Proposed construction cost
- Proposed DOE cost
- Proposed agency cost share
- Proposed total cost
- Actual design cost
- Actual construction cost
- Actual DOE cost
- Actual agency cost
- Actual total cost
- Ratio, construction cost to collector area, $/ft^2
- Ratio, design cost to construction cost
- Actual collector efficiency, average during acceptance test, and as percentage of manufacturer's *ASHRAE 93* test efficiency
- Collector operating point at actual average efficiency
- Site assessment
- Subjective design assessment
- Quantitative design assessment
- Installation assessment
- Operation assessment
- Maintenance assessment
- Agency cooperation assessment
- Loadside rating number
- Overall rating

the project cost-effective. Space heating systems that could produce useful heat only in winter were common, but all were uneconomical by an LCC standard. Only 4% of the projects were built in very good insolation areas. Freezing caused serious damage to six projects because of faulty design or installation or injudicious choice of freeze protection methods. Positive, simple freeze protection design must be mandatory.

- SFBP included collectors from about forty-five manufacturers; all but a handful were out of business by 1990.

- Of the projects that were contacted for status update, fully half were not operating five to seven years later. Further, for 75% of these nonoperating

Table 11.3
Cost and performance of some SFBP projects

Agency	Location	Construction cost, $	Design cost, $	Total cost, $	Construction cost per collector area, $/ft²	Average collector efficiency (%)
Transportation	Portsmouth, VA	66,000	9,900	75,900	71	57
GSA	Santa Ana, CA	30,000	6,500	36,500	78	42
GSA	Cheyenne, WY	26,000	6,500	32,500	91	45
GSA	Ogden, UT	26,000	6,500	32,500	80	28
GSA	Aberdeen, SD	26,000	6,500	32,500	78	16
GSA	Concord, NH	31,250	6,000	37,250	109	35
GSA	Hartford, CT	38,530	5,000	43,530	80	56
GSA	Auburn, WA	29,041	6,500	35,541	111	54
VA	Togus, ME	121,835	15,944	137,779	148	52
Treasury	Sherwood, ND	5,662	170	5,832	70	69
Defense	Fort Devens, MA	212,800	27,972	240,772	83	45
Energy	Bartlesville, OK	244,984	27,400	272,384	184	51
Interior	Tonalea, AZ	134,763	24,219	158,982	112	37
Treasury	Richford, VT	49,788	2,950	52,738	71	38
Treasury	Churubusco, NY	43,728	2,950	46,678	85	17
Treasury	Maida, ND	34,772	11,819	46,590	54	47

GSA = General Services Administration.
VA = Veterans Administration.
Source: ETEC 1992.

systems, there were no plans by the owning agency to restore them to operation.

• Selective sorting of data from these databases was used to identify the best candidates for refurbishment among nonoperating systems, using potential energy collection per dollar of refurbishment expense as a criterion.

11.7.2 Cost Estimation Procedures

Thirteen SFBP projects, operating successfully and judged to be typical applications of solar energy for service hot water heating, industrial process water heating, or space heating or space cooling, were selected for a detailed cost analysis. Actual costs were not used. To remove any cost effect of government sponsorship, the analysis used national average labor and materials costs from R. S. Means Publications to estimate the cost of building each of the thirteen projects for a commercial customer. As-built drawings, final construction specifications, and (when available) acceptance test reports were furnished by ETEC to the cost analyst. A "takeoff" of detailed components was made from the project drawings, quotations were obtained, and installation labor costs were calculated.

Miscellaneous cost factors such as insurance, sales tax, contingency, and profit were applied. A "national mean" construction cost was then calculated and subsequently adjusted to reflect site-specific costs such as labor, material, taxes. The cost of building the system at its home site and at other climatically different sites was compared. Also, the cost of building a system 50% or 100% larger than the actual SFBP project was calculated.

The following conclusions were drawn from these cost analyses:

• Construction costs were strongly dependent on the adaptability of the building to solar components and the extent of modification required.

• For systems with more than 2,000 ft^2 of collectors, only small reductions in cost/ft^2 were realized as number of collectors increased.

• The cost to build the same system in different cities varies only a maximum of 15% from highest to lowest and only a maximum of 8% from the national average.

Because these cost analyses were made using actual as-built drawings and specifications of large, successful operating solar projects, there was a high confidence in the results. Using these analyses to formulate a series of linear equations relating cost to collector area, ETEC developed a quick, reliable method to estimate construction costs for large active solar energy systems. The method, which can be used by small design firms without access to a detailed estimating procedure and which is fully described in the ASHRAE *Active Solar Heating Systems Design Manual* (ASHRAE 1988), requires only the following information about the solar energy system:

• Site location

• Solar application

• Collector type

• Total collector area

• Freight distance for major component delivery

• Cost data indexes for the site (such as provided by R. S. Means Publications).

11.7.3   Freeze Damage

Freeze protection, in active solar systems using water as the heat transfer medium, is provided by antifreeze or by design. Water was used as coolant in 63 SFBP projects. Designed-in freeze protection was provided by one of three designs: drain-back (47), recirculation (10), and drain-down (6). A freeze incident with resulting damage to collectors and piping occurred at six of these projects. Four of these had recirculation design and two drain-back design. The potential for a freeze incident was discovered at another drain-back site and the system was modified before any damage was done. All seven projects were in southern states: Arkansas, Arizona (2), Georgia, Kentucky, North Carolina, and Texas (Nadae and Balkwill 1990).

The freeze incidents were attributed to design errors (5 sites), installation error (2 sites), component failure (2 sites), operating error (3 sites), or improper system modification (1 site); some sites showed more than one failure mode. Details of events leading to freeze incidents, the damage done, failure analyses, and corrective system modifications recommended are described fully in Nakae and Balkwill (1990).

Failure analysis led to the following conclusions:

• The use of recirculation as a freeze protection method should be restricted to small installations in mild climates.

• Recirculation must have backup protection; either groundwater flush or drain-down will do.

• Groundwater flush can only be used as a backup where adequate flow rate at a high enough pressure to service the whole collector field can be guaranteed.

• Drain-down is acceptable as a primary freeze protection method only in mild climates and where the annual energy loss to the sewer drain-down is not significant. Drain-back is an acceptable primary freeze protection method at any SFBP site, providing certain design requirements can be met, the most important of which is the ability to drain completely whenever the collector pump stops, without the need for an automatic or controlled valve to operate.

• The simplest control system for freeze protection is best.

• Any system design change during construction or after start-up must be reviewed and approved by the system designer, no matter how insignificant it may seem to the change requester.

• System checkout, operator training, and the on-site operation and maintenance manual must cover the system's freeze protection method thoroughly, for both normal and emergency situations.

These conclusions have been included in the ASHRAE *Active Solar Heating Systems Design Manual* (ASHRAE 1988) as recommended design features for freeze protection.

11.7.4   Instrumentation Reliability

Btu meters are used to measure the energy produced by active heating solar systems. Their accuracy is often questioned, so a study was done to check the performance of these meters in the field. System performance of twenty-six selected large SFBP projects was reviewed and analyzed by using data from the PO&M reports submitted monthly by the owner. The data reported were to be from the standard Btu meters required at each project by the NOPR. ETEC staff determined the accuracy of the Btu meters at these projects during site visits. Flow rates were checked by using an ultrasonic flowmeter (calibrated at ETEC); temperatures were checked by probes (also ETEC-calibrated) and digital readouts. The observed Btu readout was compared with the Btu total calculated from the ETEC measurements.

As a result of these checkouts, ETEC found the following:

• Nineteen of the Btu meters (73%) were found to be operating with adequate accuracy to generate data useful for assessing the thermal performance of the solar energy system. On two of the bad systems, the flowmeter was not working; on the other five, either the flowmeter or integrator was providing inaccurate data.

• Although Btu meter manufacturers provide standard repair kits for their instruments, not one site had such a kit. Simple repair at the site is cheaper and quicker than returning the unit to the manufacturer.

• At several sites, manufacturers' standard installation procedures had not been followed. The most common violation was to install the flowmeter in the wrong-sized pipe. Three sites had poor sensor locations. Meters must be checked during design review and again during construction to ensure proper location.

• The average error measured in the 19 working meters was 4.1% in flow rate and 2.9% in integrated Btu. This compares favorably with the checkout of nine Btu meters by an independent testing laboratory, in which the error of the Btu output and flow rate averaged $\pm 5\%$.

• Commercially available Btu meters can be used to monitor performance of solar-heated hot water systems with high confidence that the error will be 5% or less, provided that the meters are sized, installed, and maintained in strict accordance with the manufacturer's recommendations.

11.7.5   Performance Monitoring

As the acquisition, construction, and installation (ACI) phase of the SFBP neared completion, a team was assembled to plan the performance monitoring and reporting phase. This team included representatives from all parts of the solar community—architect/engineers, component manufacturers, research and development personnel of universities and government agencies, and members of professional and technical societies.

The review team met on 5–7 April 1983 and concluded that the successfully operating solar energy systems in the SFBP could be an excellent source of performance and reliability and maintenance (R&M) data. Such data would be very useful in fulfilling the objective of the SFBP and would provide direction to DOE for future research and development work. The team stressed the merits for high-quality data from a few highly instrumented, good sites, rather than minimum quality data collected from all sites. Eight sites were selected and highly instrumented using techniques developed in the National Solar Data Network (NSDN). Remote monitoring and data acquisition were used; all data were transmitted to a central location for recording, analyzing and reporting. This monitoring was carried out by the Vitro Corporation.

The eight systems' design and operating parameters, and some monitoring results, are shown in table 11.4. The complete monitoring activity for each site is reported in a separate document (see Logee 1987). An R&M study was performed for the eight systems at the end of the monitoring activity. Some important observations can be made from these monitoring efforts:

• These systems performed better than previously monitored NSDN projects.

• Operational availability of the systems was better than 90%.

• Systems with simple controls had fewer problems.

• In most cases, the design load was overestimated, leading to poor system utilization.

• Collector efficiency in most systems was 33%–42% and was between 50%–100% of ASHRAE single-collector test efficiency typical for off-the-shelf collectors in 1980.

• In most cases, the measured solar fraction was lower than that predicted by F-Chart.

• In-tank heat exchangers showed very low efficiencies compared with manufacturers' claims.

• Tank and piping thermal losses were higher than predicted.

• The annual performance of systems for space cooling and space heating/cooling was very low—substantially below the commonly understood potential of multifunction systems. These data indicate that service hot water heating and outside air preheat for space heating are the best applications for solar energy systems.

• Design errors were found even in these good systems; for example, control sensor placed in the wrong location, pumps controlled by insolation sensor or timer, oversized collector field.

• Thorough design reviews and acceptance testing are essential to good thermal performance.

• Operator training and monitoring are both necessary to ensure that a solar system is operating and performing properly.

• A site-specific O&M manual is required.

11.7.6 Operational and Performance Experience

Significant changes affecting the solar industry, its customers, and the general community had taken place by the late 1980s. Fuel prices had declined, tax credits were no longer available, and as a consequence the solar industry was greatly reduced in number and size. The budgets of many federal agencies also were reduced. Consequently, most federal agencies lost interest in solar projects, either existing or proposed; sometimes the systems were abandoned.

The rules of the SFBP (DOE 1979c) required that the owner/agency prepare an operation and maintenance report on each system and submit it to DOE quarterly for the first year and annually for the next two years of operation. In spite of the regulations and follow-up action by DOE, few reports were submitted; from those submitted, reliability and maintenance (R&M) data for the small hot water systems were collected by ETEC and supplied to SERI. These data showed a marked improvement in reliability compared to small solar water heating systems in earlier programs.

ETEC conducted several telephone surveys in 1988. One survey showed that the small hot water systems, in general, still were working well. Problems on these systems were mainly non–solar-related, such as failures of heating elements in the auxiliary hot water heater and the pumps. Table 11.5 shows the agency responses to questions related to system operation and maintenance for the larger systems.

Over 50% of the agency contacts answered "don't know," "probably yes" or "probably no" to the question: "Is the system operational?" This lack of current knowledge of their systems, and the lack of response to reporting requirements, indicates a corresponding lack of motivation and interest in solar energy systems by many of the facility managers. DOE had neither the funds nor the legislative mandate to repair or rehabilitate the SFBP systems.

ETEC's engineers involved in the SFBP selected thirty-five projects they felt were better-performing systems and, using their experience from

Table 11.4
ETEC SFBP monitoring results summary

	Monitored buildings							
	Tucson JCC Tucson, AZ	HQ AFF-ES Building Dallas, TX	Gainesville JCC Gainesville, FL	Fort Devon Launderette Boston, MA	Eisenhower Museum Abilene, KS	Homestead Launderette Miami, FL	National Agriculture Library Beltsville, MD	Cherokee Indian Hospital Cherokee, NC
A. Solar energy system use								
• Space heating					×		×	×
• Space cooling		×	×				×	
• Service hot water	×	×	×	×		×		×
B. Number of people for which system designed	200	2,400						200
C. Number of visitors	22,000			25,000	200,000	15,000		2,775
D. Type of solar system freeze protection								
• Drain-back		×				×	×	×
• Ethylene glycol					×			
• Propylene glycol				×				
• Other	×							
E. Type and size of collector								
• Collector Type								
Flat plate, black								
Flat plate, selective		×		×	×			
1 glass				×	×			
2 glass				×				
Evacuated tube			×					×
• Manufacturer	Libby-Owens-Ford	Air Craftman	Owens-Illinois	Sunworks	U.S. Solar	Energy Transfer Systems	Sunmaster	Owens-Illinois
• Size (ft²)	1,659	1,147	6,144	2,563	4,201	1,480	5,314	5,517
• Number of collectors		60	192	108	110		200	320

F. Storage tank								
• Number of tanks	1	2	1	1	2	2	1	1
• Size (gallons)	2,200	1,329	3,000	3,800	1,980	1,000	3,800	6,335
F.1 Average collector efficiency (%)	36	42	46	42	33	37	12	
F.2 Building area (ft^2)			18,000				282,400	
F.3 Array piping loss (Btu $\times 10^6$/yr)							3.32	
F.4 Cost of solar heat, 20 years ($/10^6$ Btu)	16.65	15.27		17.07	31.4	11.55		
G. Yearly performance (Btu/ft^2·yr)	199,517	136,878	147,460	147,460	119,355	173,375	47,450	75,555
H. Savings/costs								
• Fuel saved	Gas	Gas	LPG/electric	Gas/oil		LPG	Gas/electric	Oil
10^6 Btu/yr	552	262	477/52	627	840	485	31.7/30	689
$ saved/yr	3,488	1,389	2,570	3,943	3,915	2,609	200/640	4,200
• Parasitic energy	Electric	Electric		Electric	Electric	Electric	Electric	Electric
$\times 10^6$ Btu/yr	10	8.98		14.5	37	6.63		39
kWh /yr	2,946	2,630		4,246	10,834	1,942	3,852	11,376
$ Cost/yr	199	176		377	807	130		761
I. Solar system availability (%)		98		99	100		98	98
J. Load Btu/10^6/yr								
• SDHW	488		328			363		207
% of design			50			44		11
• Heating								
% of design								
• Cooling								
% of design								

Table 11.4 (continued)

	Monitored buildings							
	Tucson JCC Tucson, AZ	HQ AFF-ES Building Dallas, TX	Gainesville JCC Gainesville, FL	Fort Devon Launderette Boston, MA	Eisenhower Museum Abilene, KS	Homestead Launderette Miami, FL	National Agriculture Library Beltsville, MD	Cherokee Indian Hospital Cherokee, NC
K. Solar delivered (Btu × 10^6/yr)								
• Actual	331	157		377	504	291	254	413
• F-Chart predicted		349		388	916	226	725	463
• % of incident to load	28	28		32	20			17
L. Solar Fraction (%)								
• Actual	68	21	83	36	49	80	9	6
• F-Chart	86	34	29	41	34	75		6
M. Insolation close to average								
• Yes		×	×	×		×	×	×
• No	×				× (Lower)			
N. Losses from storage (Btu × 10^6/yr)	89.8	39.3	206	6.7	148	37	27	122
• Effective R-value	1.3	2.2	1.9	7.7	2.1	4	6.7	3.2
• Design R-value	20	30	33	30		30	14.5	
• Average water temp. (°F)		111	163	76	135	123	166	123
O. Control system								
• Worked well				×	×	×		×
• Had problems							×	
• Remarks							64% available	
P. Collector loop flow rate (GPM)								
• Measured	54		77		128	36	54	68
• Designed	42		75		84	41	54.6	64

		Dormitory	Cafeteria	Laundry	Museum	Laundry	Library	Hospital
Q.	Effectiveness of collector loop heat exchanger							
	• In Tank		0.10					
	• Outside				0.36	0.32		0.07
R.	Collector parameters							
	• F_rT_a	0.66	0.775	0.714	0.76	0.72	0.5	0.391
	• F_rU_l	0.69	0.670	0.709	0.72	0.9	0.19	0.224
	• Area (gross ft²)		19.1				23.3	17.17
	• Tilt	30°		35°			37°	
S.	Collector performance with respect to ASHRAE (%)	100	50–100				9 below	50–100
T.	Ratio of operating collector/incident	0.77	0.87	0.80	0.88	0.88		
U.	System cost (1985$)							
	• Total	99,259	78,855	121,914	212,080	89,762		
	• Cost ($/ft²)	55.83	68.75	47.70	50.50	62.33		
V.	General remarks	70,000 gal. hot water/mo	Average chiller COP = 0.53 monthly average water use 43,760 gal.	10% collector shading, Dec.		Rewire system	North-facing reflectors ineffective, no op. manual	
W.	Building type	Dormitory	Cafeteria	Laundry	Museum	Laundry	Library	Hospital

Source: ETEC 1992.

Table 11.5
System status, 1988

Question	Yes	Probably yes	Don't know	Probably no	No
Is system operational? (136 responses)	23%	21%	5%	32%	19%
Are you planning to fix the system? (65 responses)	14%	0%	9%	0%	77%
Is the system being maintained? (113 responses)	11%	11%	11%	4%	65%

Source: ETEC 1992.

the design reviews and the available operational data, assessed the overall quality of these systems. In every case ETEC staff felt the system design could be improved. Even though these designs had been reviewed by experts (whose recommendations had often not been followed), much had been learned during the course of the program. The ETEC staff felt that the quality of installation and of operation and maintenance also could be improved. As a result of these observations, DOE decided that a set of comprehensive solar manuals should be written to address design, installation, and operation and maintenance of large systems.

11.8   Refurbishment

The ETEC survey of large SFBP projects in 1987–1988 determined that many systems were not operating for lack of repairs to pumps, instruments, or other vital components. A database analysis established the maximum funding available to restore the system to operation with a reasonable payback time. These reviews showed that with the major capital investment of procuring and installing the system already made by SFBP, the refurbishment expense would be moderate and cost-reffective.

One such refurbishment was carried out at Kelly Air Force Base in San Antonio. The recirculation freeze protection system had malfunctioned, with severe freeze damage; the system was redesigned to use glycol coolant in the collector loop and a new control system was installed. The original $500,000-system was restored to successful operation for less than

$80,000. The acceptance test of the redesigned system showed excellent thermal performance; the integrated system efficiency over the one-day, five-hour test was 45%. Further, the acceptance test team studied the use pattern of the solar-heated water and was able to recommend changes in matching hot water use to load demand that further improved the collector array's efficiency.

11.9   SFBP Manuals

It was decided that the best way to accomplish the program objectives to "obtain design, installation, maintenance, and operating experience and data with solar systems in the federal sector; to lower solar system design and installation costs by providing education through experience to solar system designers and installers" (DOE 1979c) was to develop a set of manuals for the design, installation, operation, and maintenance of large active solar heating systems. An expert group of highly experienced solar energy system designers and installers was established by DOE/ETEC; the group recommended developing a set of new manuals, using existing manuals as a resource. The new manuals would draw upon and include the experience of recognized solar experts and would include the knowledge and experience gained from the SFBP and other relevant programs. Industry, through the Solar Energy Industries Association (SEIA) and the American Society of Heating, Refrigerating, and Air-Conditioning Engineers (ASHRAE), played a major role in this effort. DOE directed ETEC to take the lead in coordinating and assembling the manuals. ETEC also authored many sections and introduced information from the lessons learned from the SFBP. Several working group meetings were held for each manual, where the documents were thoroughly reviewed and a consensus reached; after approval by ASHRAE and SEIA, the manuals were sent to ASHRAE for publication (ASHRAE 1988, 1990, 1991).

The design manual (ASHRAE 1988), augmented with lessons learned, details step-by-step design procedures. The O&M guide manual (ASHRAE 1990) shows designers of large active solar heating systems how to prepare a site-specific O&M manual. The installation manual (ASHRAE 1991) provides industry's consensus of the best available installation procedures for large active solar heating systems.

11.10   Findings, Conclusions, and Recommendations for Future Work

The SFBP contributed to the knowledge and experience base for active solar heating systems. Although the legislated directives to the program were implemented, the program's broader intentions were not fully met because of the difficult environment for solar use in the 1980s—falling energy prices, a growing budget deficit, the loss of the tax credit, and the closing of many solar companies. Had the conditions of the late 1970s (oil crises, political atmosphere, solar tax incentives, etc.) persisted, the program as originally defined would have met its objectives and been a success. As these conditions changed, DOE redirected the program first to support research and then to support industry through the development of the three ASHRAE solar manuals.

The ASHRAE manuals are perhaps the most significant result of the program. The greatest value of the SFBP is in the lessons learned, and the manuals include those lessons. The key factors required to design and build a cost-effective solar heating system are identified. As the country begins to deal with growing environmental pressures, the use of solar systems will grow. The manuals from the Solar in Federal Buildings Program, provide a solid foundation for the design, installation, and operation of future solar systems with superior performance. The ingredients of high solar insolation, high replacement fuel costs, low-temperature heating, year-round load, good design and installation, good O&M manual, and strong owner interest are all essential to the success of a solar project. These requirements are stressed in the manuals, along with step-by-step procedures for design, installation, and operation and maintenance.

In many respects the results of the SFBP were disappointing. In retrospect, the SFBP would have produced better systems if, for example, better site and application selection criteria had been applied, more owner involvement and commitment had been shown, and more control over design decisions had been exercised by DOE.

11.10.1   Findings

1. Of the 718 solar energy systems installed, every system was operational at the time of acceptance testing; continuing operation of each system was in the hands of the owner agency. Some systems are working well a decade later, while many others have been shut down by the owners.

2. Not all types of solar systems were found to be cost-effective at the time of this program. Only large water heating systems with year-round demand could be amortized in twenty years (standard system design life) in the economic climate of the late 1980s. The following applications of solar energy systems were found to be recommended and developmental:

Recommended	Developmental
DHW, single-family	Space heating
DHW, multifamily	Space cooling
Service hot water, nonresidential	Combinations of heating, cooling, DHW

Pool heating

3. The following types of facilities were found to be good candidates for the recommended solar energy systems:

- Swimming pool
- Laundry
- Low rise apartment
- Dormitory
- Health clinic
- Nursing home
- Residential
- Cafeteria
- Low-rise office building
- Hospital
- Gymnasium

4. The following were found to be important characteristics for successful applications of active solar energy systems:

- High to good solar insolation
- Expensive alternate energy
- Facility life of twenty years or more
- Year-round load
- Small variations in load
- System size based on measured or calculated load
- Low temperature required ($75°$ to $150°F$)
- Project participants are interested in using solar
- Dependable, proven designers and installers
- Thorough design review and acceptance test
- Adequate space for system components
- Standard designs and proven components

- Component sizes based on economics, not thermal performance
- Distribution lines short and well-insulated
- High effectiveness heat exchangers
- Storage subsystem properly designed
- Adequate instrumentation to assess system operation
- Preventive and remedial maintenance

5. Of the projects where the facility manager was contacted for status update, half were not operating; most of these were five to seven years old. Further, for 75% of these nonoperating systems, there were no plans by the owning agency to restore them to operation.

6. Facility managers' motivation and interest are essential to good continued operations. Cost sharing generates motivation.

7. A good operation and maintenance manual is essential to good continued operations.

8. Design costs were typically 5% to 15% of total cost, and up to 30% on some smaller systems.

9. Collector loop efficiencies were as much as 20% less than ASHRAE single-collector test efficiencies.

10. Freezing caused serious damage to six projects because of faulty design or installation or injudicious choice of freezing protection methods. Positive, simple freeze protection design must be mandatory.

11. SFBP included collectors from about forty-five manufacturers; all but a handful are out of business now.

11.10.2   Conclusions

The following objectives of the SFBP were met:

- The program was carried out in accordance with legislation.
- Development of the solar industry was stimulated in the early 1980s by the building of over 700 solar systems and the development of design, installation, and operation and maintenance manuals.
- The federal government did begin to use solar technology (and would have done more had the incentives continued).
- The objective of obtaining design, installation, and operation and maintenance experience and data was well met.
- Good applications for the use of solar energy were identified.

Because of changes in the political atmosphere, falling energy prices, reduced budgets, and the lack of agency motivation and interest, some objectives were either not met or only marginally met:

• There was a minimal increase in consumer confidence in the use of solar energy.

• Large procurements were not made and therefore the economies of scale were not realized.

• There was no major shift to solar energy use.

• Industry/federal agency interaction through joint solar efforts was stimulated to only a small degree.

11.10.3   Recommendations

Future government-funded solar energy demonstration programs should be limited to applications and projects that can be expected to be amortized over twenty years. Applications that are not cost-effective at the design state may be built because of other consideration; for example, an owner's decision to reduce use of fossil fuels. In such cases the owner should provide enough funding to reduce the government's cost to the cost-effective level.

Control and final approval of project design and funding should remain firmly with the funding manager. Once the design is approved, changes (and funding increases to make the changes) should be discouraged. A strong design review procedure should be used to accomplish this.

System design should be as simple as possible consistent with the purpose of the project. The ASHRAE design manual recommendations should be followed.

Low design costs are vital in keeping overall system costs within the cost-effective limit. In a multiproject program, the designer should attempt to standardize system design as much as possible, customizing only to fit into space available at the site and to match the design load.

A good operation and maintenance manual must be a requirement. The owner must demonstrate that adequate and competent maintenance will be provided to the system. The owner must have a positive and approving attitude on the use of solar energy in the building.

References

ASHRAE (American Society of Heating, Refrigerating and Air-Conditioning Engineers). 1988. *Active Solar Heating Systems Design Manual.*

ASHRAE. 1990. *Guidance for Preparation of Active Solar Heating Systems Operations and Maintenance Manuals.*

ASHRAE. 1991. *Active Solar Heating Systems Installation Manual.*

DOE (U.S. Department of Energy). 1979a. *Federal Register* 44:19328. 2 April.

DOE. 1979b. Environmental assessment. DOE/EA-0083. Washington, DC, June.

DOE. 1979c. "Solar in Federal Buildings Demonstration Program Rules." *Federal Register* 44:60664–60673. 19 October.

DOE. 1980. *Federal Register.* 45:5620. 23 January.

ETEC (Energy Technology Engineering Center). 1992. *Solar in Federal Buildings Program: Draft Final Report.* Rockwell International, 2 November.

Logee, T. L. 1987. *Quality Site Seasonal Report, Fort Devens Launderette, SFBP 1751, December 1984 through June 1985.* ETEC-87-7. Energy Technology Engineering Center, Rockwell International, 15 October.

Nakae, T. S., and J. K. Balkwill, 1990. *Freeze Incidents at Solar in Federal Buildings Program (SFBP) Projects.* SFBP-XT-0129. Energy Technology Engineering Center, Rockwell International, 1 June.

Ruegg, R., and W. Short. 1988. "Economic Methods." In R. E. West and F. Kreith, eds., *Economic Analysis of Solar Thermal Energy Systems.* Cambridge: MIT Press.

12 Agricultural Demonstration Programs

Robert G. Yeck and Marvin D. Hall

12.1 Rationale for Inclusion of Agriculture

Solar energy is the principal driving force for agricultural processes. It energizes photosynthesis in plants, which are not only the origin of our food supply but also the origin of our fossil fuels and the basis for alternative sources of energy such as wood, vegetable oils, and alcohol. Greenhouses are designed to maximize the photosynthetic effect of solar energy. The natural (solar) drying of forages, fruits, and vegetables can be enhanced by solar collectors, although fossil fuels, when readily available, provide a more expeditious method of drying.

Farm animal production is very much affected by adverse weather. The role of solar energy has been actively studied both as a contributor to heat stress and as protection against cold stress. Livestock shelters have been designed to guard against excess solar heat in the summer and to take advantage of solar heat in the winter; these are primarily passive solar structures, using open-sided or windowed southern exposures. Supplemental heat is often needed, particularly in nurturing young animals, for which fossil fuels, when available, are the most expeditious choice.

The U.S. Department of Agriculture (USDA) has had a continuing interest in solar energy and the technologies for enhancing its usefulness. After World War II, as it reinstituted its R&D programs, several solar projects were initiated. At Athens, Georgia, some experimental, cabin-sized houses were fitted with plastic film–covered collectors and rock-filled heat storage systems to investigate the feasibility of using solar technology to reduce heating costs for low-income rural families (Simons and Haynes 1963). In 1949 projects were initiated to demonstrate the benefits of insulated glass windows in improving the solar heat gain in dairy barns in Wisconsin (LaRock and Yeck 1956) and poultry houses in Pennsylvania (Bressler 1953). During this period plastic film–covered greenhouses were put to use in lieu of the more expensive glass-covered structures. Considerable knowledge of solar irradiation potential was gained from studies of environmental stress on animals in Missouri (Yeck and Stewart 1959) and California (Kelly and Ittner 1948); shade design criteria were substantially advanced. Biomass technologies were inherent in crop research. The search for broader uses of agricultural products led to the development of plastics from oil seed crops; although plastics produced by the petrochemical industry are now more economically competitive, agriculture provides a fallback position.

In 1974 the National Science Foundation (NSF) forwarded $1 million in pass-through grants to USDA's Agricultural Research Service for investigations in solar energy applications in agriculture (Kerr 1987). This activity was continued as the U.S. energy programs were eventually gathered under the administration of the Department of Energy (DOE). An inventory of research showed approximately $9.5 million of federal and state agricultural appropriations under the heading of "Substitution of Renewable Energy R&D" for FY 1977. Nearly $2 million was identified as "technology transfer" (NASULGC/USDA 1978).

One of the results of the NSF funding was an assessment of solar energy application in agriculture (Harris 1976). This study placed highest priority for direct solar use in agricultural crop drying, in its strategy recommendations calling attention to the success of the Extension Service (ES), created by the Smith-Lever Act of 1914. The ES system is tripartite consisting of the U.S. Department of Agriculture, land grant universities, and local governments. The Extension Service has a core of federal specialists paralleled by specialists at university levels who work closely with other federal agencies and national organizations in interpreting research, preparing educational materials, and providing technical guidance for a network of area and county agents operating in nearly every county of the United States. The county agents work directly with individuals, families, and groups to show farmers and other users how to apply best practices and new technology. The Harris report stated that the Extension Service had attained a high degree of creditability that could greatly enhance the adoption of new energy technology.

It was, therefore, very appropriate that the DOE demonstration program under PL 93-409 include agriculture. The USDA had the appropriate congressional authorization and had the desired experience in research.

12.2 Organization Agreements

The agricultural demonstration programs were organized following a series of meetings between DOE and USDA representatives. Interagency agreements were developed for On-Farm Demonstration of Solar Heating of Livestock Shelters (DE-A101-78 CS 35149) and On-Farm Demonstration of Solar Drying of Crops and Grains (DE-A101-80 CS 30459). Key agencies were the Extension Service, USDA, and DOE's Division of Solar Thermal Energy Systems.

Major objectives were to determine the technical and economic feasibility of using solar energy for heating livestock shelters and for drying of crops. The program was expected to identify incentives and opportunities for widespread applications of these technologies. Known technologies were applied under farm operating conditions. Initiation of solar heating systems by the farmers, themselves, was considered an integral part of feasibility assessments. With the success of these exemplary systems, acceptance could be expected to spread to other farmers.

The agreements called for DOE to provide $1.2 million for the livestock shelter projects and $1.095 million for the drying projects. These funds were augmented by farmer inputs and by input that was already present in the USDA system. Several ancillary services and, in some cases, travel expenses were furnished from existing resources within USDA without charge to the agreement. These services were from professionals and their staffs within USDA, state universities, and county agent offices. No financial accounting was made of this augmentation effort, but the cost of the program would have been substantially higher without this input. The more significant value of the input, however, was the involvement of key people in the agricultural infrastructure—enhancing the chance of acceptance. Advantageously, the USDA input also included expertise from current research and extension efforts in other energy programs of the USDA.

Participation was arranged through states. Conceptually, each state would propose ten farm cooperators representing different types of systems and/or farm product orientation. Farmers were to pay one-half the cost; none of the DOE funds could be used for new buildings or facilities. The state university extension would provide design criteria, assist with construction, collect and evaluate data, conduct tours, and prepare information materials.

12.3 Selection of Participants: Livestock Program

The livestock shelter program was initiated first. Its operational mode proved sufficiently satisfactory to be used with the solar drying program. The sequence for each program was as follows: (1) notify all states of the nature and availability of the program; and (2) invite each state that was interested to submit a preproposal. The preproposals were evaluated by engineers, architects, and economists.

Livestock shelter panel members were W. T. Cox, Extension Service; R. G. Yeck, Agricultural Research Service; R. Slater, Farmer Home Administration; and D. Van Dyne, Economic Research Service—all of the USDA; and Paul Barnes, who represented DOE. Of 21 states submitting preproposals, 15 were invited to submit complete proposals; the same panel selected 9 states representing 84 farms. Selections were based on geographical distribution with respect to the livestock confinement operations, representation of different kinds of solar energy systems, and chances of success.

State project managers, in cooperation with farmer participants, submitted completed designs for each farm to an Extension Service advisory committee of state university livestock shelter experts: V. P. Mason, Virginia; C. K. Spillman, Kansas; and K. E. Felton, Maryland; with Paul Barnes representing DOE. The committee disapproved, approved, or approved with comments as they saw fit.

12.4   Participant Selection: Grain Drying

Grain and forage storage proposal panel members were W. E. Matson, Extension Service; D. L. Spearman, Price Support and Loan Division; B. R. Beckham, Farmers Home Administration; G. Reisner, Economic Research Service; and G. Klein, Agricultural Research Service—all of the USDA; Paul Barnes of Oak Ridge National Laboratory again represented DOE. Of 14 proposals, 12 were invited and 9 were selected, representing 75 systems and 12 different crops; geographical distribution was among prominent grain-producing areas in humid climates, the dry climates having little need for supplemental heat. An advisory committee operated much the same as the livestock shelter committee. Members were G. H. Foster, Indiana; M. D. Hall, Illinois; and G. Klein, USDA; Steve Kaplan, Oak Ridge National Laboratory, represented DOE.

The recruiting and selection of farm cooperators were conducted by state project managers, all of whom had been active in their state extension programs. In many instances, the project managers had previously been contacted for solar energy advice by farmers or their county agents. At the University of Maryland, the county agents were invited to canvass their counties for potential cooperators. This was particularly successful in the grain-drying project. One of the Maryland recruits had seen an item in the press and contacted federal officials, who in turn referred the inquiry to the project manager. The farmer proved to be a good cooperator and provided the program with a hay-drying demonstration that might not otherwise have been included.

12.5   System Erection and Monitoring

One of the criteria of the program was to have the demonstration effort simulate, as closely as possible, the situation that the average farmer would face when installing the system. In real life, this would mean that the farmer would need to make his own decisions on system selection, procurement, and erection. The need for expediting the program and maintaining control dictated that the choice of system be made in agreement with program managers and the program's advisory committees. Farmers were to have procured or constructed the systems on their own. In order to assure desired quality and timeliness of operations, program managers and their staffs needed to assist some farmers in procurement and provide advice for system erection. For the most part, actual fabrication was done by the farmer. In both programs, however, fabrication of collectors and placement of duct systems and components were done on site—each being tailored to the individual farm situations. Commercial packages of entire systems were not available. Solar collection panels were locally fabricated—usually on site. It should be noted that the choice of on site fabrication was somewhat biased due to relatively low cost of on-farm labor. In some cases, work was performed during off-peak times, when farm laborers might not otherwise have been productively engaged.

Having farmers provide and install instrumentation for monitoring of systems was considered impractical for the typical farm solar thermal energy system; instrumentation usually was provided and installed by the project managers and their staff. Data were automatically recorded where practical. In some cases, farmers participated in data collection, which had the added benefit of maintaining the farmer's interest, awareness, and familiarity with the system—important factors when using the farmer as a promoter for a demonstration project. Generally, new instrumentation was not procured; projects made use of whatever instrumentation was on hand at cooperating institutions. Considerable variability in monitoring procedures among projects therefore occurred; the program had neither the funds nor the time to further improve the monitoring procedures.

For the shelter program, however, all states had sufficient resources to make reasonably sound evaluations. Monitoring was less complete for the drying program, which was prematurely terminated.

12.6 Evaluation and Demonstration Procedures

The universities, through their project managers, were responsible for performance analysis and dissemination of the information from the demonstration project. They arranged for tours of demonstration sites, held field days, placed displays at fairs, presented results before farm groups, distributed press releases, and prepared extension service bulletins for public distribution. Most of the farmers within each of the participating states were made aware of the demonstration projects through one or more publicity measures. Individual state reports indicate that interest was high among farmers. Publications were the major vehicle for out-of-state publicity. The evaluation aspects were complicated by somewhat limited performance data and by the multiplicity of factors contribute to economic feasibility, including projected fossil fuel costs, inflation factors, tax advantages, and appropriate credit for farm labor. Nevertheless, economic assessments were made that provided a frame of reference for disseminated information.

12.7 Livestock Shelter Program Systems and Their Performance

The cooperative agreements with the states for the On-Farm Demonstration of Solar Heating of Livestock Shelters Program were initiated in the spring of 1978. Although many contacts had been made in anticipation of the program, much additional time and effort were required to complete agreements with cooperating farmers and to gain mutual acceptance of design between the program manager and the farmers. This, plus fabrication delays, resulted in the last systems not being in place until the fall of 1980—about one year of slippage. Most monitoring was completed by 1981; the final report of the program was issued in December 1982 (USDA 1982). The following states participated: Illinois, Iowa, Kansas, Minnesota, Missouri, Ohio, Nebraska, Vermont, and Virginia.

The demonstration aspects began with site visits by neighbors during construction and continued several years beyond the completion of the final report, a prime benefit of having the program conducted within the University Extension Service. Additional benefits were gained from parallel solar research programs that were underway at the various state universities. For instance, the University of Maryland had a solar heating project for one of its experimental broiler houses, which provided data helpful in assessing demonstration project performance. A USDA poultry laboratory at State College, Mississippi, had a similar project.

Most applications provided supplemental heat in buildings for baby animals and chicks. Some were multipurpose—including grain drying. Other applications were swine (62), dairy (18), poultry (3), and veal (1).

The collectors' designs varied widely. Air was used for thermal transfer in 65 collectors; the other 19 used water. Only 7 of the collectors were free-standing; the rest were incorporated into the structures. Modified Trombe walls were used for 28; modified roof or attic systems accounted for 24. The remainder could not be simply classified. All emphasized economy of construction and adaptability to the site.

Solar input versus output efficiencies varied widely and tended to be much lower than those obtained with well-designed glass covered commercial collectors. Nevertheless, all alleviated the need for fossil fuel–derived energy. Many had investment recovery periods of six years or less, but ten- to twenty-year paybacks might be a more appropriate average for all systems. The longest payback period was forty-three years—it was not well designed. Systems for heating dairy wash water were not considered economically competitive with heat recovery systems that made use of heat from milk and/or heat from milk refrigeration equipment but would be viable competition for fossil fuel–fired water heaters. Some examples of individual systems, as well as other management details within the livestock shelter program, are discussed in greater detail later in this chapter under the more detailed discussion of the Illinois solar energy projects (section 12.9).

12.8   Solar Drying Program Systems and Their Performance

As with the shelter programs, the On-Farm Demonstration of Solar Drying of Crops and Grains Program was funded by DOE through an interagency agreement with USDA. The Washington office of the USDA

Extension Service entered into cooperative agreement with each participating state extension office. The following states participated in the
drying program: Florida, Illinois, Kansas, Maryland, Missouri, Michigan, Tennessee, and Virginia. Cooperative agreements with each state
were tendered in December 1980; the program allowed one year for design and construction and two years for evaluation. The program, however, was terminated after two years due to changes in federal budgeting
priorities; monitoring aspects were therefore limited to one year. As with
the shelter project, farmer negotiation and construction aspects required
more than one year and some projects did not have any data at the time
of the final program report of March 1983 (USDA 1983). The demonstration aspects of the program followed the same pattern as those of the
shelter program.

The systems that were used in the drying program were quite different
from those in the shelter program, although a few systems served both
programs. Indeed, when selecting cooperators and systems, priority was
given to multipurpose application. Drying grain is done over a very short
season, often less than one month for any one crop. It can be lengthened
to three months or longer when crops with different harvest times are
included. Heating livestock shelters and/or farm homes was considered as
an additional multipurpose application, but the early termination of the
project negated most of the chances for trying such alternate uses. Although some installations relied entirely on solar drying, the risk of not
have supplemental heating available during cloudy days caused many
program managers to look on solar as a complement to fossil fuels rather
than a complete replacement.

The numbers of systems by crop were as follows: hay, 3; fruit, 2; vegetables, 2; peanuts, 9; small cereal grains, 35; corn, 52; soybeans, 29; to-
bacco, 3; and unclassified, 6. Multiple use applications for space heating
were residence, 12; shop, 24; swine, 10; dairy, 3; poultry, 1; and unclassified, 4. Two systems were designed to provide residential hot water.

The designs of systems varied widely: some wrapped around round
grain bins, some used attic or roof systems, and others used some sort of a
free-standing collector. Only 8 of the 76 systems used storage; 74 used air
for heat transfer. Most were fabricated on-site; practically all emphasized
economy of construction rather than collector efficiency. Consequently,
efficiencies were usually well below those expected for commercial glass-
covered collectors. One attic system was only 5% efficient yet had a pay-

back period of five years. Most of the dryer system had efficiencies between 25% and 50%.

As with the shelter program, payback period calculations were complex. The methods of calculation varied among states. Results even indicated differences by year, which was not surprising because climatic condition varied from year to year. Most payback periods fell within six to ten years but ranged from 1.8 to forty years. In all cases, significant amounts of fossil fuel energy were saved. The dissemination of information followed a pattern similar to that for the shelter program. Excellent publications on the subject were distributed nationally. One of the solar drying systems was rated sufficiently successful to warrant national distribution of plans for its designs (USDA 1980).

12.9   Illinois Experiences with Solar Energy Projects

The experience at the University of Illinois provides a more detailed example of the program. The use of solar energy in Illinois agriculture started as early as 1964 because the University of Illinois Extension engineers had already taken an interest in improving production efficiency. Hundreds of bare flat-plate solar collectors were built in swine-farrowing and nursery units. Crop drying became a major effort after 1976; except for engineering help from the University of Illinois Cooperative Extension Service, projects were funded entirely by farmers.

Early solar projects in Illinois stressed low-cost, site-built construction that would show a profitable return in a fairly short time frame, three to five years; this philosophy was and still is being advised in the state. In 1979, although Illinois had many solar farm buildings, documented research reports were not readily available. When DOE made federal funds available through the USDA to set up demonstration projects, Illinois was selected for both crop-processing and livestock systems.

12.9.1   Illinois Program Initiation and Funding

In May 1979 the Department of Agricultural Engineering, University of Illinois Extension Service, signed an agreement with the Science and Education Administration of USDA to demonstrate solar energy use in livestock production systems. In January 1981 a similar agreement was made to demonstrate solar energy potential in crop-drying systems on farms. The USDA then made DOE pass-through funds available to the

University of Illinois to offset part of the costs incurred in designing, building, and monitoring the projects. Although the projects were to extend over a three-year period, the third year of the grain project was canceled. Funding for the livestock project was $132,000. Funding was $110,000 for the grain project and was reduced to $69,850 when the final year's program was canceled. Cost sharing by the cooperating farmers on the livestock project was $33,290, approximately 50% of the extra cost of solar materials. The grain project was cost-shared at $16,700. or approximately 35% of actual costs.

County and state extension service people contacted all farmers who had shown previous interest in using solar energy. This was followed by an evaluation process lasting several months. Many of the farmers who were contacted had several years' experience using solar energy. Of the farms selected, all ten livestock farms and five of the seven crop-drying farms had been working with M. D. Hall of Illinois Extension Service on some aspects of solar application before the project began. Generally, this extent of prior contact was not as prevalent among the other states in the project.

12.9.2   Illinois Program Goals

Following the DOE program criteria, the Illinois goals for the two programs were

1. To demonstrate that simple, low-cost solar collectors could provide significant amounts of heat energy for agricultural applications.

2. To demonstrate that significant amounts of energy from gas or electric systems could be saved by using solar energy.

3. To demonstrate that walls and roofs of buildings could serve as solar collectors.

4. To evaluate the suitability of various building materials for use in solar collectors as part of the roof or walls of the structure.

5. To demonstrate that free-standing, movable collectors could be used to dry grain, heat farm shops, and preheat ventilation air in livestock buildings.

6. To demonstrate that solar collectors attached to grain-drying bins can be used as for low-heat drying.

7. To improve solar designs and develop improved plans for solar construction on farm buildings.

12.9.3 Illinois Program Management

In Illinois the solar heat for the livestock project and the solar heat for the crop-drying project were two separate studies, although many of the systems overlapped in functions. One or more of the University of Illinois Extension Service engineers worked on each project during construction, monitoring and publicizing the results.

The livestock study started first under a full-time project investigator, C. E. Rahn. A. J. Muehling and W. H. Peterson, Extension Service agricultural engineers, University of Illinois, served on a part-time basis as manager and associate manager, respectively. Solar heat for crop drying was carried out by project investigator, David W. Morrison, with W. H. Peterson serving as project manager.

12.9.4 Details for the Illinois Projects

Details for each of the seventeen projects was fully documented in the final project reports (USDA 1982, 1983). Results are summarized in tables 12.1 and 12.2.

Item 1 of table 1 was the Aden project. Known as the "swine farrow-to-finish building," the unit was designed to produce approximately 2,000 market-weight pigs per year. The 44 ft × 148 ft building had an east-west orientation, and the solar system was the south-facing slope of the roof 4/12 (18.4%). The west 88 feet of roof functioned as a covered plated collector; clear, corrugated fiberglass served as the cover. The east 96 feet of the south-facing roof and the entire north-facing roof functioned as a bare plate collector, with colored steel as the collector. Eight hundred feet of $\frac{3}{4}$-inch black iron pipe was laid in the south slope below the fiberglass to heat water that in turn would be used to heat the building floor. The ventilation fans inside the swine rooms pulled air through the solar cavity.

The costs for the covered plate collector were $1.53/ft^2; costs were not calculated on the bare plate because they would be almost zero. The 2,112 ft^2 of covered collector showed a simple payback of approximately four years. The bare plate would have a much faster payback, but because was not monitored, no value is known. The black iron pipe failed to heat water to the minimum 120°F heater requirement and was therefore not feasible. The preheat of the ventilation system also reclaimed heat lost through ceiling insulation as air was brought into the building from every square foot of roof surface. The heating season for this area was approximately October through April with 250,000 Btu/ft^2 of solar energy

Table 12.1
Animal space heating systems

Project type	Collector area, ft^2 (m^2)	Predicted annual energy, 10^6 Btu (GJ)	Cost, $/ft^2 ($/m^2)	Predicted simple payback (yrs)	Special notes
Swine building roof	2,112 (196)	100 (106)	1.50 (16.00)	4	Roof collector of fiberglass
Grain storage processing roof	4,636 (431)	33 (35)	0.50 (5.00)	8.1	Solar grain drying system used for swine building
Swine building roof and wall	6,480 (602)	128 (135)	1.20 (13.00)	7.9	Payback period does not include grain drying
Swine building add-on wall	336 (31)	361 (380)	5.50 (60.00)	6.4	Rock storage included in the system
Swine building roof	2,080 (193)	75 (79)	2.30 (25.00)	7.8	Solar also used to dry grain (not included in study)
Swine building roof	2,304 (214)	12 (13)	2.00 (22.00)	N/A	Annual heat shown is heat transferred to adjoining building
Swine building roof	1,784 (166)	N/A	9.50 (100.00)	N/A	Freeze-up occurred—system had to be down for repairs (no data collected)
Swine building roof	6,480 (602)	369 (390)	1.40 (15.00)	5.6	North-south ridge with collectors facing east-west
Swine building roof	1,050 (98)	129 (137)	9.20 (98.00)	9.1	Hot water allowed zone heating for a longer heating season
Stand-alone collector to dairy parlor	192 (17.8)	2.3 (2.4)	31.00 (335.00)	325	Includes both space heating and water heating

Table 12.2
Drying systems

Project type	Collector area, ft^2 (m^2)	Predicted annual energy, 10^6 Btu (GJ)	Cost, \$/ft^2 (\$/m^2)	Predicted simple payback (yrs)	Special notes
Grain storage and drying bin (solar wall)	760 (70.6)	11 (11)	1.70 (18.00)	4–5	South 2/3 of bin covered with fiberglass to form collector
Grain drying and home heating/portable collector	288 (26.8)	20 (22)	5.00 (59.00)	6.8	Portable collector but not easily moved. Space heating and grain drying
Shop and machine storage shed, roof and walls	4,700 (437)	73 (77)	1.50 (16.00)	4–5	Shop heating values are not included in total annual energy. Grain drying and shop heating
Stand-alone water-type collector	720 (66.9)	N/A	24.00 (250.00)	N/A	Farmer only used system 2 days during drying season—no data for space heating. Grain drying and livestock heat
Hay storage, drying wall and roof collector	9,500 (883)	175 (184)	0.80 (8.50)	5.2	High heat gains due to time of year for hay drying
Grain storage and drying building	2,380 (221)	64 (68)	1.10 (12.00)	6.7	Solar collector as part of roof and south wall
Livestock building for grain drying	3,415 (317)	43 (45)	2.80 (30.00)	8.6	Total energy and payback are for grain only. Does not include space heating for pigs

available on an 18.4% slope. This building, not counting the bare plate collector, produced 100 million Btu for the heating season.

Item 1 of table 2, known as the "Ponder project," demonstrated low-temperature drying of corn. The southern two-thirds of the bin surface was painted black, enclosed with fiberglass, and used as a collector; the fan and motor were also covered. A 3-inch space between fiberglass and bin provided the air channel for solar gain. The cost of adding the solar system to the bin was \$1,300. There are many variables to consider in grain drying, but one concern of most producers is the energy cost per point of moisture removed. Electric energy used in 1981 was 0.14 kWh per point per bushel of corn and 0.20 kWh in 1982. The total Btu collected was 11 million, and the bin was rated 42% efficient as a collector. The 1981 data showed a 4.5-year payback at 6 cents per kiloWatt-hour, and the 1982 data were somewhat less favorable.

12.9.5   Output of Illinois Projects

All Illinois goals were successfully completed, some more successfully than others. Solar building plans were available for distribution; information on materials, labor, skills, and the economics of incorporating solar into an agricultural system was also developed.

The crop-drying study resulted in the following:

1   blueprint plan
1   radio program
2   tours for public (70 people)
3   publications
32 public meetings (1,105 people)

The livestock program resulted in the following:

1   videotape
3   blueprint plans
9   publications
33 public meetings (1,545 people)

Farmers are very receptive to ideas and new methods that will be profitable, even at some risk. To them, providing engineering consultations, plans, and follow-up management may be more important than cost-sharing. Cost-sharing with any type of producer tends to make the producer less concerned about making the system economically sound. In Illinois, it appears that the farmers receiving the most cost-sharing were the least practical and economical in solar energy use.

The Illinois keep-it-simple, low-cost philosophy was somewhat thwarted on the funded projects because some participants thought higher thermal efficiency might be more profitable. The testing made possible by this demonstration showed that, in most cases, the lowest-first-cost system turned out to be the most profitable.

12.9.6   Illinois Design Recommendations

1. Keep cost of collectors low (less than $3/ft^2) and payback period short (less than five years).

2. Make solar collectors part of farm structures (roof, walls, etc.)

3. Lay out buildings to run east-west.

4. Make effective use of insulation, vapor barriers, windbreaks, and summer venting.

5. Try to design collectors so they can be used for more than one job (ventilation preheat, grain drying, space heating, etc.).

6. Do not spend extra money to increase roof angle.

7. Make collector large enough to be effective in getting the job done.

8. Disregard heat storage in most cases, but some systems can have heat storage with very little cost for space heating.

9. Design for simple maintenance and repair and easy access.

10. Design system so that it can be easily enlarged.

11. Allow for expansion and contraction in all parts of the solar system.

12. Allow for management error in system as much as possible.

13. Use 5-ounce weight fiberglass of the best quality you can find—heavier if you are using fiberglass as a solar cover. Be sure some type of ultraviolet shield is designed into or on the material to prevent discoloration.

14. When using corrugated or ribbed fiberglass for cover sheets, caulk all laps.

15. Use screws with neoprene washers to secure fiberglass.

16. Do not store fiberglass outside in stacks without a good protection from the sun and rain.

17. Be sure insulation and structural components of collector can hold up under temperature extremes of the design. Fires are a hazard.

12.9.7   Illinois Construction Precautions

1. Have a plan before starting construction.

2. Roll and tie long fiberglass sheets when moving into place on roofs to avoid damage from wind.

3. Use slip clutches on screw guns to avoid overtightening.

4. Do not walk on fiberglass sheets any more than absolutely necessary.

5. Install cover sheets on solar building last—particularly during summer months.

6. Take extra care on air systems to avoid leaks to ambient air. Reduced efficiency will result from sloppy workmanship, inadequate fasteners, or lack of good caulking.

12.9.8 Illinois Follow-up Survey

A random survey of 24 farmers was conducted in the spring of 1984, and 19 returned the questionnaire. Eight of the 24 who received the questionnaire were included in the DOE-funded field study group. A summary of the questionnaire is listed below; no statistical analysis was attempted.

Number sent out: 24

Number returned: 19 (79%)

Average size of collector: 8,652 ft^2

Function of collector: Corn drying and livestock ventilation preheat

Replies to question: "Do you feel your collector has been a good investment?"
Yes: 95.8%   No: 4.2%

Replies to question: "Knowing what you now know, would you build or buy another solar collector?"
Yes: 84.2%   No: 5.3%   Maybe: 10.5%

Replies to question: "Do you feel the University of Illinois should be doing more or less to promote solar energy?"
More: 73.9%   Less: 5.3%   Same: 15.8%

Average number of years experience with solar from people returning survey: 6

Typical comments by farmers:

"Very cost-effective."

"Keep information in front of the people."

"Collector has paid for itself many times."

"After seven years I think I should replace fiberglass as efficiency seems to be dropping."

"I am happy with collector and it has been maintenance-free."

"We are well satisfied."

"Maintenance higher than anticipated."

"I plan to start building another solar building this year."

"Collector is doing good job."

"Does a good job when sun shines."

"University of Illinois has been a leader in solar technology and should continue in solar research."

"Hired help don't manage system very well."

"Grain drying easier to manage than livestock ventilation."

"I would probably make changes, but I like the principles—it has basically worked well."

"Commercial building suppliers try to talk farmers out of solar buildings as if they were unproven."

12.10   Illinois Projects versus Projects in Other States

Illinois was used as an example because it had both the livestock and the drying programs and it had a relatively large number of participating projects. Prior to the outset of the projects, it had a more active solar experimentation program than many of the other states—but almost all states had some type of solar promotional activity prior to their projects. Despite differences in the design selections and alternate system uses among the states, results were similar for all states. Feasibility was well demonstrated on some farms but not as favorably demonstrated at others. The selection of design and farmer operational procedures impacted on system effectiveness. Information distribution processes were very similar for all states.

12.11   Program Benefits and Information Dissemination

The demonstration programs for both the livestock shelter and the crop-drying programs resulted in a successful demonstration of the state of solar thermal energy technology as applied to these two activities. Systems needed extensive tailoring to each farm situation. Many of the systems had only faint resemblance to residential and industrial technologies, but they worked. A large percentage of the systems had payback periods of ten years or less. The USDA Extension Service was effective as an institutional arrangement for implementing the program, bringing the entire agricultural extension network, federal, state, county, and farmer, into play. Participating farmers were usually pleased with the results and served as proponents of the solar energy systems. Although the energy savings potential of solar in agriculture is not as great as in residential applications, its chance for adoption is high. Economic paybacks showed promise. Solar was effective for farm use and farmers made it work for them. Furthermore, solar technology applied to a sensitive national interest—food supply. During an energy crisis, farmers are somewhat at the end of the priority chain for energy sales, and farming energy needs may not be met at the time of need. Self-sufficiency helps avoid the risk of nonavailability and can help mute the effect of rising energy costs on food supply.

The duration of both programs was relatively short. The full potential of publicity was not realized. The final reports of the programs indicated that many bulletins and technical paper publications were still in the developmental stage. Nonetheless, each state had several presentations before farm groups (Iowa had 93). Plans of designs were made available by some of the states; several states had radio and T.V. programs and disseminated video tapes. State extension service newsletters distributed several times a year were "quick-release" devices that were widely used. Iowa indicated that a tremendous plus for the success of the program is continuing its visibility long after termination. For all practical purposes, the projects were completed in 1982. The final report for the livestock shelter solar heating program was completed in December 1982 (USDA 1982), and the final report for the solar drying of crops and grains was completed March 1983 (USDA 1983). Thanks to the involvement of the Extension Service, particularly at the state specialist level, follow-up evaluations and publicity has continued. For instance, the 1987 Winter International Meeting of the American Society of Agricultural Engineers at Chicago included a session with four presentations that represented a follow-up on the solar grain-drying projects. Despite greatly reduced incentives when fossil fuel prices dropped, many of the systems are still being operated. It was interesting to learn that several installations were no longer in operation due to deterioration of the collector. The short time line given for getting the program underway and tight money with high interest rates of the early 1980s brought about some cost-saving measures that were not cost-effective. The greatest problem was failure to use the more durable pressure-preservation-treated lumber; many systems evidently rotted out. Another major problem was reliance on inexperienced farm labor to build solar projects. Despite oversight of design and preliminary instructions by state specialists, units often were not sturdy enough or were poorly anchored; wind and other weather damage decommissioned some collectors. The most favorable presentation (Stewart and Gird 1987) reported on four of the six active participants in the Maryland project. These four were still operating their systems at the time of the presentation. One was a 1,200 ft^2 wrap-around collector built at \$2.85/ft^2. Two were modified "Purdue Type" portable collectors, 1,152 ft^2 at \$5.42/ft^2 and 576 ft^2 at \$6.08/ft^2. The fourth was also an air exchange plastic-covered collector but built in a two-slope configuration, 45 degrees and 30 degrees from the horizontal, from roof to ground alongside a barn that contained the hay being dried. The farmer had already started construction when he entered the project, and cost data were not available. The first system saved about 342 gallons of liquefied petroleum gas (LPG) equivalent per year; the second, 539 gallons. The third system was not amenable to heat equivalent calculations, and the fourth saved about 723 gallons of LPG equivalent per year. The more important feature was that all four systems were able to depend entirely on solar for grain and hay drying. The major sacrifice over the use of fossil fuels was that solar took longer. For instance, in the first Maryland example, 32,400 bushels of grain were dried from 20% moisture to a safe storage level of 14% in forty days. Fossil fuel drying is usually accomplished in substantially less time.

12.12 Greenhouse and Other USDA Energy Activities

Although more appropriately in the province of technology development than demonstration, the USDA through its Agricultural Research Service also managed greenhouse energy conservation, greenhouse residential heating, and wind energy projects for DOE. Several USDA agencies were also involved in biomass energy research and development projects involving wood fuel, alcohol fuels, vegetable oil fuels, methane from agricultural wastes, and gasifiers using agricultural wastes. Some of these projects were funded directly from USDA appropriations; others were jointly or entirely supported by DOE funds.

Although all of these activities are included in the broader definition of solar technology, greenhouse technologies are most directly identified with solar thermal technology. In the case of greenhouse energy conservation, effort was directed to retaining solar heat gained during the day. Experimental systems included (1) layered systems, with air space between sheets of glass or plastic, or combinations of the two materials; (2) insulating blankets that were drawn under the glass covers at night; and (3) heat sink devices like barrels of water and beds of rock under plant stands. All proved successful.

Greenhouse heating systems, on the other hand, were applications of the greenhouse concept as a solar collector. Similar in principle to solar plenum attics or wall systems except that the space behind the cover is large enough for people movement and, if desired, for use as a indoor garden, greenhouses proved to be effective as heat collectors but, of course, not as thermally efficient as well-designed glass-covered collectors. A greenhouse is, however, more esthetically attractive than the standard glass collector and provides an expanded living space for a home. The garden aspects are of questionable economic benefit, but families seem to enjoy raising their own flowers and produce, which are certainly fresher than those obtained through the market. Families who have tried the greenhouse concept like it. Greenhouses save energy, but part of the cost must be charged to occupant pleasure to justify using them over standard collectors. The solar greenhouse program involved USDA federal laboratories and several state experiment stations.

T. E. Bond, director of the USDA Agricultural Research Service's Rural Housing Laboratory at Clemson, South Carolina, served as program manager for this DOE-USDA effort. The solar attic concept was developed at the Rural Housing Laboratory.

Because the USDA Agricultural Research Service and state experiment stations work closely with the USDA Extension Service, many of the greenhouse projects concurrently served as demonstration projects. The benefits of a piece of research conducted under this institutional arrangement thus have an excellent chance of living on; the system that is developed as a research item often becomes a demonstration item for the Extension Service.

Results of the research projects were published as state bulletins and in professional journals such as *Transactions of ASAE*. Proceedings such as the Fourth Annual Conference on Solar Energy for Heating of Greenhouses and Greenhouse-Residence Combinations (Mears 1979) were another valuable source of information. Some construction plans were made available through the Cooperative Farm Building Plan Exchange of the USDA Extension Service.

The DOE funding support of greenhouse projects demonstration was essential to gain the rapid advance in technology needed for energy conservation in greenhouses and to satisfy public interest in greenhouse-residence systems. Greenhouse applications provided significant reductions in fossil fuel energy usage and have been widely adopted by greenhouse operators.

12.13 Procedural Change Recommendations

A major policy issue with demonstration projects is whether technology is sufficiently developed to enter the commercialization phase that a demonstration project fosters. In the cases of solar livestock and drying technologies, there was a substantial information base relative to the need and method of delivery of solar heat. Several different design concepts for collecting solar heat were also available. Some wedding of the collectors with the delivery needs had also been done. There was, however, considerable uncertainty over the economic feasibility for each of the collector systems. The DOE programs had evaluation written into them, but almost ancillary to the role of demonstration. There should have been an intermediate development stage that stressed measurement of performance under farm practice. Funding would need to be augmented to obtain and operate instrumentation; demonstration systems and a promotional phase with publicity, farm visits, and the like, could occur thereafter. It is possible that sufficient manufacturer interest might then arise to have commercial packages available for turnkey installation. If so, still another phase, involving more farmers could be begun to promote acceptance; perhaps the last phase might involve cost-offsetting incentives, with the manufacturers doing the promoting.

Operational management and implementation procedures had both pluses and minuses. The use of the USDA Extension Service as the overall coordinator worked well. The national extension network was in the best position to blend DOE needs with the ability of the state and local extension offices to follow through. The Extension Service knew the qualifications of personnel within the system and had the credibility to garner the cooperation of potential implementors. Mechanisms were in place (or were quickly put in place) by which financial exchanges could be made—on the surface, a rather complex picture with federal, inter-governmental agency, state, county, and farmer involvement. The solicitation of farmer cooperators was often done through county agents—a promotional device in itself; the interest of county agents and, even more so, the interest of participating farmers brought forth excellent visitation of demonstration projects. The proposal submission procedure, the panel for selecting the state for cooperation, and the final site-by-site design review worked well, although the review process did cause some delays. Finally, the major benefit of the use of the Extension Service system was the continuation of activity long after the formal funding ended—something that often does not occur with private contractors.

Those at the state and farm levels who implemented the agreements and the construction often noted delays during various steps of the process. Of particular importance is the need to move with the minimum of delay once a farmer is selected as cooperator. Farmers' interest can rapidly wane—particularly when, as was the case in the demonstration projects, their economic situation deteriorates. Excessively high interest rates at the time caused farmers to avoid borrowing capital. The farmers had to pay for the system "up front," which eliminated many potential cooperators. This was even more awkward when the economic feasibility was very much in question. The delays in implementation at the DOE level reduced incentives because they extended solicitation of cooperators into a period in which the fossil fuel crises was beginning to abate. An essential need for farmer participation is authorization and funding to assist farmers in financing their systems.

It is hoped that solar energy application technology for agriculture will continue to be improved. In that way, the premature rush for commercialization that will probably occur again during the next fossil fuel energy crisis will have more complete system performance data upon which investment decisions can be made.

A tremendous plus for the success of the program is the continuation of its visibility long after termination. Thanks to the involvement of the Extension Service, particularly at the state specialist level, publicity follow-up, evaluations, and publicity have continued.

References

Bressler, G. O. 1953. "Thermopane Windows for Poultry Houses." *Everybody's Poultry Magazine* 58:10, 24–25 June.

Harris, W. L. 1976. *Solar Energy Applications in Agriculture Potential; Research Needs and Adoption Studies.* NSF/RA-760021. Prepared the for Agricultural Research Service, U.S. Department of Agriculture with grant funds from National Science Foundation Grant PTP 75-10573. College Park: Agricultural Experiment Station, University of Maryland, January.

Kelly, C. F., and N. R. Ittner. 1948. "Artificial Shades for Livestock in Hot Climates." *Agricultural Engineering* 29:239–242. June.

Kerr, N. A. 1987. *The Legacy: A Centennial History of State Agricultural Experiment Stations, 1887–1987.* Columbia: Missouri Agricultural Experiment Station, University of Missouri, March.

LaRock, M. J., and R. G. Yeck. 1956. *Bank Barns Compared with Above-Ground Barns for Housing Dairy Cattle.* Production research report no. 2. Washington, DC: U.S. Department of Agriculture.

Mears, D. R. 1979. *Proceedings of the Fourth Annual Conference on Solar Energy for Heating Greenhouses and Greenhouse-Residence Combinations.* UC-59b. Sponsored by DOE, ASAE, and Rutgers University. New Brunswick, N.J.: Cooks College, Rutgers University.

NASULGC/USDA (National Association of State Universities and Land Grand Colleges and U.S. Department of Agriculture). 1978. *Inventory of Agricultural Energy Research and Development and Technology Transfer for FY 1977.* Washington, DC, May.

Simons, J. W., and B. C. Haynes. 1963. "Solar Radiation for Farm Homes." *Georgia Agricultural Research* 2:9–10.

Stewart, L. E., and J. W. Gird. 1987. "An Evaluation of Solar Drying in Maryland." Paper 87-4538, 1987 International Meeting of American Society of Agricultural Engineers (ASAE), St. Joseph, MI, December.

USDA (U.S. Department of Agriculature). 1980 "Low-Cost Collector for Grain Drier." Plan no. 6341. Cooperative Farm Building Plan Exchange, Extension Service, Washington, DC.

USDA. 1982. *On-Farm Demonstration of Solar Heating of Livestock Shelters.* Final Report for DOE. Interagency Agreement DE-A101-78CS35149. NTIS cat. no. DE-83-004940. Washington, DC: Extension Service, December.

USDA. 1983. *On Farm Demonstration of Solar Drying of Crops and Grains.* Final Report for DOE. Interagency Agreement DE-A101-80CS30459. NTIS cat. no. DE-83-012082. Washington, DC: Extension Service, March.

Yeck, R. G., and R. E. Stewart. 1959. "A Ten-Year Summary of the Psychroenergetic Laboratory Dairy Cattle Research at the University of Missouri." *Transactions of the ASAE* 2:71–77.

13 Military Demonstration Programs

William A. Tolbert

13.1 Program Characteristics

Solar thermal technology demonstration programs within the military have been unique for a variety of reasons. First, military demonstration programs involve a full spectrum of solar thermal technologies, including active and passive space heating, active and passive domestic water heating, solar industrial process heat, and solar thermal electric power. Although the level of activity in each of these areas varies significantly, demonstrations of solar thermal technologies in each area do exist. Second, military demonstration programs encompass the full spectrum of end use, including residential, commercial, and industrial applications. This is because military installations are usually the equivalent of self-contained communities that just happen to have fences around their boundaries. These installations normally include homes, schools, hospitals, office buildings, commercial-type facilities, utility systems, and often, industrial complexes. The Department of Defense (DOD) has over 500 major installations, and there are approximately 2.4 billion square feet of space in some 400,000 DOD buildings.

Military solar thermal technology demonstration programs fall into a variety of classifications. Some are part of major federal demonstration programs (such as the Solar in Federal Buildings Program), while others are developed, funded, and managed totally within the military services. The latter category includes large-scale and small-scale projects funded by each military service through a variety of sources, such as Military Construction Program (MCP) appropriations (under which a project can take five to ten years to be designed, funded, and constructed), a variety of research and development (R&D) funds, or some form of discretionary funding. The nature of these demonstration programs varies by military service, although the programs are often correlated with the lessons learned across the services. (For the purposes of this review, Marine Corps activities will be included within the Navy program.)

Most military solar thermal programs do have at least one thing in common: they are all subject to significant congressional direction and cost guidance. Although solar thermal technologies can enhance military mission accomplishment in a variety of ways (including energy security),

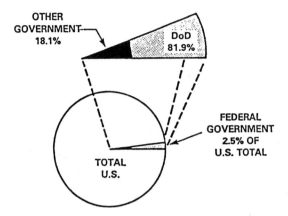

Figure 13.1
Energy consumption, FY 1986 (Office of the Deputy Assistant Secretary of Defense, 1985).
In this year total U.S. consumption was 74.33 Quads, of which the federal government
accounted for 1.82 Quads and the Department of Defense for 1.49 Quads.

the primary driving force has always been, and for at least the short term
will continue to be, economics.

This leads to the final characteristic. The DOD is the largest energy
consumer within the federal government and accounts for over 2% of all
the nation's energy usage (1.49 quadrillion Btu, or 1.5 EJ, in FY 1986).
These relationships are shown in figure 13.1 (DOE 1987). Figure 13.2
shows DOD energy consumption by fuel type, with an additional breakdown of petroleum usage. The cost to DOD of the energy consumed in
FY 1986 was approximately $6.3 billion. DOD energy costs in FY 1984
accounted for 4.4% of total DOD appropriations.

13.2 Organizational Structure

Many organizations may be involved in the development, approval, and
implementation of solar thermal technology demonstration programs
within the military. The organizational structures of energy programs
within the DOD and the individual services vary significantly and are
complicated by the variety of activities involved (policymaking, appropriations, research and development, engineering, construction, and operations). In addition, organizational structure, office designations, and
staffing levels have changed many times over the past several years,
clouding the identification of consistent roles and responsibilities.

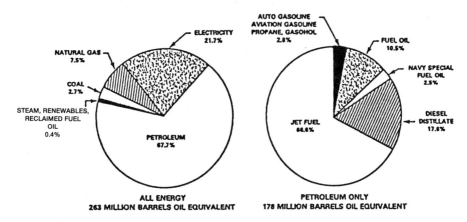

Figure 13.2
Department of Defense energy consumption by fuel type, FY 1984.

13.2.1 Department of Defense Energy Organization

DOD energy organizations primarily provide overall energy policy and guidance to the individual military services (DOD 1985). DOD does not provide direct funding (R&D or MCP) for any energy program. The Defense Energy Organization is shown in figure 13.3. The Director of Energy and Transportation Policy is under the Deputy Assistant Secretary for Defense (Logistics), DASD(L). The Defense Energy Policy Council (DEPC) is the senior-level advisory group that provides DASD(L) the means to coordinate energy policy at the highest levels. It also provides a mechanism to contribute feedback on energy programs and problems. The Defense Energy Action Group (DEAG) is the working-level group that assists the Director of Energy Policy and Transportation in the development and coordination of energy policy. The Defense Energy Data Analysis Panel (DEDAP) is composed of representatives from each DOD component (military service) concerned with the Defense Energy Information System (DEIS) and the DOD Management Information System.

13.2.2 Army Energy Organization

The Army Energy Organization, shown in figure 13.4 (Department of Army 1985), consists of several key elements: A Special Assistant for Energy, the Army Advisory Group on Energy (AAGE), and the Army Energy Office (AEO). The Special Assistant for Energy, a senior position

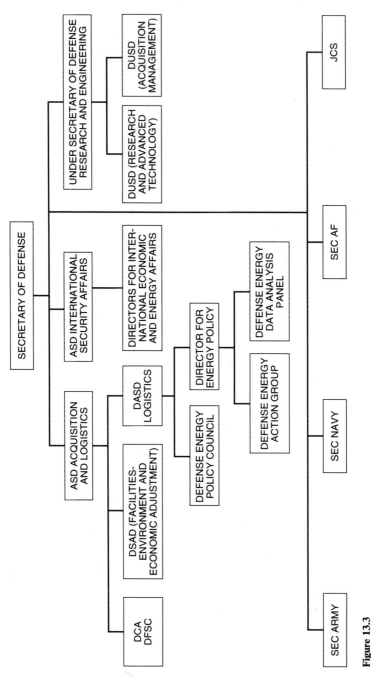

Figure 13.3
Department of Defense energy management.

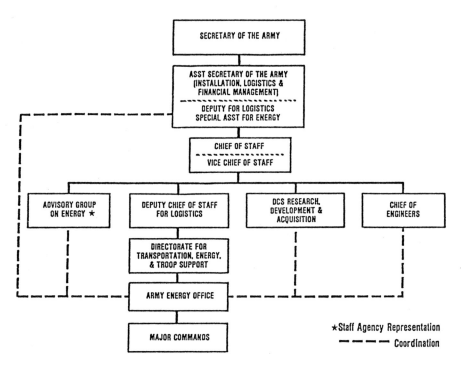

Figure 13.4
Army energy organization, January 1986.

on the staff of the Secretary of the Army, represents the Army on the Defense Energy Policy Council (DEPC), implements tasks and initiatives coming from the DEPC, and monitors the Army Energy Program. The AAGE is a general officer-level organization that continually reviews Army programs, policies, and procedures for their impact on energy. The AAGE also provides a forum for the exchange of information and ideas related to energy conservation and energy self-sufficiency. The Army Energy Office is responsible for supervising the Army Energy Program, recommending allocation and use of Army energy resources, and executing the Army energy conservation program.

13.2.3 Navy Energy Organization

The Navy Energy Organization is shown in figure 13.5 (Department of the Navy 1984). The Deputy Chief of Naval Operations (DCNO) for Logistics (OP-04) is the lead Navy authority for energy matters. The

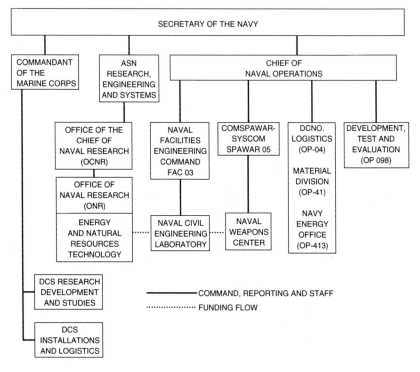

Figure 13.5
Navy energy organization, January 1986.

Director of Material Division (OP-41) provides the principal staff support for energy matters and serves as the Navy Special Assistant for Energy to the Secretary of the Navy. The Navy Energy Office (OP-413) is responsible for planning and monitoring the efficient use of energy throughout the Navy and for implementing DOD energy priorities and policies. The Navy Energy Office also acts as a central point of contact for Navy energy matters.

13.2.4 Air Force Energy Organization

The Air Force Energy Organization is shown in figure 13.6 (Department of the Air Force 1984). Overall management of the energy program is the responsibility of the Director of Maintenance and Supply (USAF/LEY). USAF/LEY reviews and coordinates Air Force energy planning, serves as a member of the Defense Energy Policy Council, and chairs the HQ

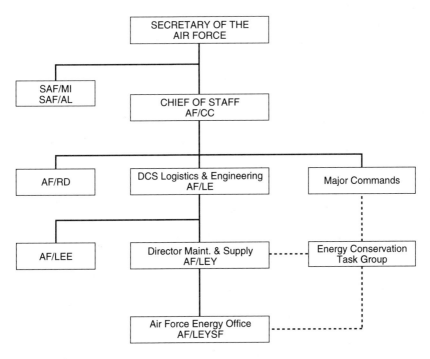

Figure 13.6
Air Force energy organization, January 1986.

USAF Energy Conservation Task Group. USAF/LEY functions are carried out by the Air Force Energy Office, which is the focal point for all energy matters in the Air Force. The office is also responsible for developing and publishing specific Air Force energy goals and policies as well as monitoring progress toward energy goals. In addition, the Assistant Secretary for Research, Development, and Logistics (SAF/AL) reviews mobility energy policies, while the Assistant Secretary for Manpower, Reserve Affairs, and Installations (SAF/MI) reviews energy policies affecting facilities.

13.2.5 Military Energy Research and Development Organizations

Each of the military services has research centers or laboratories that conduct solar thermal technology development and demonstrations. Using appropriated R&D funding, these organizations conduct a wide spectrum of research, development, and demonstration and are often involved in the instrumentation and evaluation of non-R&D solar demonstration programs. The principal R&D organizations by military service are

Army Civil Engineering Research Laboratory (CERL), Champaign, Illinois

Facilities Engineering Support Agency (FESA), Fort Belvoir, Virginia

Cold Regions Research and Engineering Laboratory (CRREL), Hanover, New Hampshire

Navy Naval Civil Engineering Laboratory (NCEL), Port Hueneme, California

Naval Research Laboratory (NRL), Washington, D.C.

Naval Weapons Center (NWC), China Lake, California

Air Force Air Force Engineering and Services Laboratory (AFESC/RD), Tyndall Air Force Base, Florida

Air Force Wright Aeronautical Laboratories (AFWAL), Wright-Patterson Air Force Base, Ohio

13.3   Energy Policy and Guidance

Energy policy and guidance to the DOD and the individual military services comes through a variety of sources and mechanisms.

13.3.1   National Policy

National policy may come in the form of Executive Orders (such as Executive Order 12003), Public Law (such as the National Energy Conservation Policy Act of 1979 or the Military Construction Authorization Act of 1981), or Department of Energy (DOE) regulations and plans (such as the National Energy Plans I and II).

13.3.2   Defense Department Statements

Defense guidance normally comes in one of two forms. The Defense Energy Management Plan (DEMP) is the basic DOD planning document and incorporates planning undertaken within DOD to maintain peacetime readiness through adequate and secure supplies of energy. The

DEMP supplements and interprets other energy policy documents and reports the accomplishments of recent programs, plans for future activity, and progress toward goals established by executive order. In contrast, Defense Energy Policy Program Memoranda (DEPPM) provide direct guidance to the military services by translating national goals into specific military service goals. The first DEPPM issued each year also states the priorities—and the various energy programs that fall under them—required to enhance energy preparedness to meet worldwide military and industrial base requirements at the lowest possible cost.

13.3.3 Individual Service Documents

Each military service incorporates all higher general guidance into either its own energy policy program memorandum (such as AFEPPM) or its official five-year energy planning document (such as the Navy Energy Plan, FY 1984–1990). In addition, the services often establish energy goals, policies, or restrictions more specifically tailored to their requirements and operations.

Figure 13.7 shows how solar energy policy and guidance flow from the top down within the Air Force. The figure also shows the reporting/feedback mechanisms that are used to track progress against goals.

13.3.4 Key Guidance/Goals

In addition to the Solar Heating and Cooling Demonstration Act of 1974, a wide variety of energy policy and guidance documentation identified above directly applied to solar thermal technology demonstration programs.

The national energy goals were established in the National Energy Plan I (NEP I) in 1977 and were incorporated for the executive branch in Executive Order 12003. These goals were further defined by DEPPM 80-6, DEPPM 86-3, and individual service memoranda. In addition to conservation, DEPPM 80-6 included the following goals:

• Provide at least 1% of energy from renewable energy sources by FY 1985, 5% by FY 1990, 10% by FY 1995, and 20% by FY 2000 (all services).

• Achieve by 1985 a 45% reduction of energy consumption per gross square foot in new buildings compared with existing buildings (all services).

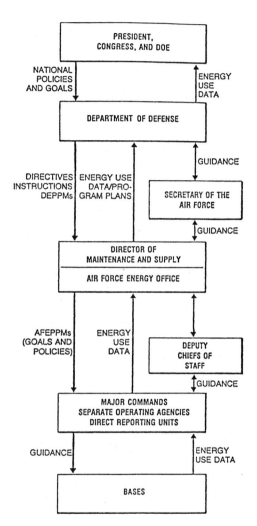

Figure 13.7
Energy Policy and Guidance Flow, U.S. Air Force.

• Reduce dependence on nonrenewable and scarce fuels by the year 2000 (Army).

• Provide energy self-sufficiency at remote sites by 1990 (Air Force).

• Expand development and application of renewable and alternate energy systems where cost-effective in terms of life cycle (DOD).

13.4   Funding

Solar thermal technology demonstration programs within the military have been funded through a variety of appropriated and nonappropriated funding mechanisms.

13.4.1   Appropriated Funds

The majority of funding for demonstration programs has come through congressional appropriations. These appropriations have varied significantly over the years and include funds appropriated by other federal agencies (ERDA, DOE, NASA, etc.) for demonstrations on military installations, and military appropriations for military programs. It is extremely difficult to identify solar thermal technology demonstration program budget levels within these appropriations for a variety of reasons. First, the "solar funds" are not distinct line items within each service's appropriation. Within the Military Construction Program (MCP) appropriations, for instance, project costs include all design and construction costs. Likewise, research and development "solar funds" are often included in appropriations for specific weapons systems or fall within larger program line items (such as environmental programs), where they are obscured and difficult to aggregate. Second, although some military appropriations are "fenced" or "set aside" specifically for energy programs as opposed to construction programs, the funds are usually identified to include all renewable energy applications and often even energy conservation projects, making identification of "solar thermal funds" virtually impossible. Finally, no clear database has been established by any of the services that effectively extracts and consolidates key funding information. Because of these constraints, reasonable estimates of the total military investment in solar thermal are virtually impossible, and solar thermal costs will be included in this chapter only where clear documentation exists for individual projects.

13.4.2 Nonappropriated Funds

At least two mechanisms exist to demonstrate solar thermal technologies within the military using nonappropriated funds. By far the largest
such potential exists within the Air Force and Army Exchange Service
(AAFES) in their solar heating/cooling of Army and Air Force exchange
complexes. These mini-mall/shopping center demonstration programs
have had high visibility within the DOD.

Section 2394 of the Military Construction Codification Act (PL 97-
214), permits the military services to enter into long-term contracts for the
purchase of energy from production facilities on or off military installations (Sklar and Noun 1982). The contract, similar to a standard utility
contract, requires approvals through functional channels up to the secretary of defense and notification of contract terms to the Congress. The
Congress has expressed strong interest in and support for this concept,
which allows the private sector to demonstrate solar thermal technologies
on a large scale on military installations while benefiting from energy and
business investment tax credits and revenues from the energy produced.
The impact of this legislation within military solar programs is described
in more detail in section 13.6.4.

13.5 Demonstration Projects/Programs

Solar thermal technology demonstration projects within the military
began in earnest after the oil embargo of 1973 and the resultant rise in
fuel costs. Partly because of a lack of programmed funds, and partly
because of a lack of familiarity with solar thermal technologies, early
military solar thermal demonstration activities started off as small but
highly visible R&D programs.

13.5.1 U.S. Air Force Academy Demonstration Program

The first full-scale demonstration of solar heating technology within the
DOD began in 1974 at the U.S. Air Force Academy as part of a research
program sponsored by Air Force Systems Command (Nay et al. 1976).
The project involved retrofitting an existing military family housing
(MFH) unit with an active, liquid, flat-plate collector system. The solar
energy system consisted of 546 ft^2 (51 m^2) of collectors (flat-plate, liquid,
nonselective surface, copper absorbing plate, double-glazed) manufac-

tured by Revere Copper Corporation and a reinforced concrete storage tank with 2,500-gallon (10-m^3) capacity filled with water and vented to the atmosphere. The system and the home were completely instrumented to record all environmental and system parameters. Aquisition cost for this project (not including military labor) was approximately $60,000 (FY 1975).

Officially, the original project objectives included developing design criteria for solar thermal systems and obtaining construction, operations, maintenance, and cost data upon which to make future funding decisions. Unofficially, the objectives were to see if military engineers could use published data to design a system and to show the military community that solar heating really did work.

The project provided an opportunity to showcase solar thermal technology to the military, from the chairman of the Joint Chiefs of Staff on down. Its significant lessons included the following:

• Military engineers could design, construct, and operate solar heating systems.

• Solar heating systems really could work (43% average energy savings over four years).

• Much of the design guidance available in 1974 was based on supposition or heavy extrapolation.

• Military operations and maintenance personnel (plumbers, electricians, etc.) were often intimidated by the "new" technology.

• Existing control strategies and equipment were very inadequate.

• Electronic components (sensors, switches, etc.) were unreliable at elevated temperatures.

• Many mechanical devices (pumps, bleed valves, gauges, etc.) were unreliable because of high temperatures, the fluids used, and cycling effects.

• Corrosion/materials problems were significant.

• Complex design or operations schemes resulted in complex problems— simple was better.

• Collector optimization (number of square feet of collector) was not as important as minimizing building modifications (costs).

• Only fractions of the collected energy were actually used in heating the quarters or providing domestic hot water.

• Costs greatly exceeded savings.

In summary, the project proved that solar heating could work but had to be better designed and more cost-effective to be widely used in the military.

13.5.2 Advanced Energy Utilization Test Bed

Built in 1977 at the Naval Civil Engineering Laboratory (NCEL) at Port Hueneme, California, the Advanced Energy Utilization Test Bed (AEUTB) gave the Navy an opportunity to do a full-scale demonstration of solar thermal technologies (NCEL 1985). This was the first solar energy technology demonstration site within the Navy, and also the first effort within the DOD to evaluate the interaction of different conservation techniques and solar thermal systems. The AEUTB looked like other two-story townhouses found in the Navy, but unlike conventional structures, it was built with a strengthened roof (to accommodate any brand of solar collector), a reinforced attic (to allow space for thermosiphon systems), and a 2,000-ft^2 storage vault for either hot rock or water storage systems.

NCEL engineers used the AEUTB to evaluate solar heating system configurations and to develop a detailed database to determine the most effective systems for meeting heating and hot water requirements in typical Navy housing units. The hands-on experience obtained by Navy engineers and technicians working on the AEUTB led to the development of extensive design guides, technical notes, and operation and maintenance (O&M) manuals; it advanced the application of solar thermal technologies within all of the military services.

13.5.3 DOD-ERDA Demonstration Program

The Solar Heating and Cooling Demonstration Act of 1974 was designed to foster demonstrations of the practical use of solar heating in various U.S. geographic and climatic regions within three years from the effective date of the act. ERDA was given overall responsibility for managing and coordinating the demonstration programs. DOD's designated role was to demonstrate the use of solar technology on federal residences; and ERDA provided funding authority for two such projects.

In the first project, DOD was to install solar heating devices on 35 new and 15 existing military family housing units (residential units) at various military installations across the country, for operation during the 1975–76 winter season. The estimated cost was $1.69 million. The second project called for installing a central collector system for heating systems on eighty military family housing units. This project was initiated in August 1976, and the estimated cost was approximately $1.4 million.

Both of these demonstration programs proved to be real eye-openers for both DOD and ERDA. After completing the first two phases (designing the solar energy systems and purchasing the solar collectors from various manufacturers), DOD went out for bids. The results were 200%–600% over government estimates. There were a variety of causes for the excessively high bids, but the dominant one was overdesign of the systems. Instead of trading off performance against construction costs, the design contractor consistently designed complex systems that optimized performance without regard to installation costs. For example, if the roof of the residence could only accommodate 650 ft^2 of collectors but the optimum design required 700 ft^2, the designers called for the roof to be extended. The result was solar energy system costs that often exceeded the replacement cost of the entire residence. For this reason, only one project (quadriplex at Sheppard Air Force Base in Texas) was ever completed. The vast quantity of collectors that had been purchased were distributed throughout the three military services for a variety of applications, and the remaining ERDA program funds were redirected to fund other solar design projects.

The second demonstration project was also cancelled due to the increased costs encountered as a result of impractical designs.

The first DOD demonstration projects thus ended before they really began. The only real results were several GAO reports (e.g., GAO 1978), numerous unsatisfactory designs, and the expenditure of over $700,000 (FY 1977).

13.6   Programs Funded from Military Appropriations

By 1980 the military services had over 61 solar energy systems in operation, 38 under construction, and 166 under design. In response to requests from the House Committee on Armed Services, DOD published a series of consolidated listings of these projects called "DOD Solar Energy

Project Summaries." The last entry in this series was published in April 1981 (DOD 1981). The lack of large-scale dedicated solar thermal demonstration programs within DOD also makes a detailed analysis of individual projects impractical. I will therefore limit myself to identifying in summary format solar thermal projects within each military service. The listings will provide brief systems descriptions and cost information where available.

13.6.1 Army Solar Thermal Programs

Table 13.1 lists thirty active solar heating and cooling projects undertaken by the Army, including residential, industrial, and commercial-type applications. Table 13.2 shows seven passive solar heating and cooling projects, consisting exclusively of military family housing (residential) applications. In these projects housing contractors were paid additional funds to incorporate passive solar energy systems because of calculated energy savings.

13.6.2 Navy Solar Thermal Programs

The Navy has successfully completed over fifty active solar energy system projects. Descriptions of eleven operational solar thermal systems are contained in table 13.3 (these are representative systems for which performance data are available). The projects included flat-plate and line focus concentrating collectors on residential and commercial-type applications. The Navy also completed two of the three solar military family housing projects included in table 13.4. These solar thermal projects incorporate passive solar systems into new Navy housing.

13.6.3 Air Force Solar Thermal Programs

As of FY 1985, the Air Force had completed over fifty demonstrations of solar thermal technology worldwide. These projects involved active solar heating and cooling, solar DHW, and passive solar heating applications on a wide range of residential, industrial, and commercial-type facilities. The eleven projects shown in table 13.5 were among those completed and evaluated.

The Air Force has completed only one passive solar heating project (a child care center) and is only recently pursuing the application of passive solar heating in military family housing. This was due to a policy determination in 1980 that passive solar technologies were generally unproven

Table 13.1
Army active solar thermal energy systems

Location	Building description	Collector area (ft^2)	Collector type	Application (% solar)	System cost ($ in year of construction)
Norfolk, VA	Multipurpose building	11,520	Evac. tube	DHW—90 C—56 H—97	839,000
Greenwood, MI	Reserve Ctr.	6,600	Flat plate	DHW—99 C—86 H—97	387,000
Fort Bliss, TX	Medical ctr.	13,200	Flat plate	DHW—50	577,000
Albuquerque, NM	Reserve ctr.	10,600	Flat plate	DHW—100 C—98 H—100	780,000
Fort Belvoir, VA	Kingman bldg.	10,000	Evac. tube	[a]	[a]
Fort Riley, KS	Bachelor qtrs.	2,700	Flat plate	DHW—30	374,000
Fort Hood, TX	Batallion HQ	5,600	Flat plate	C—83 H—93	
Fort Hood, TX	EM barracks	3,470	Flat plate	DHW—67	116,000
Fort Hood, TX	Dental clinic	4,394	Parabolic concentr.	DHW—55 C—55 H—55	398,000
Fort Hood, TX	Darnell Med. Hospital	4,300	Flat plate	DHW—51	425,000
Fort Polk, LA	Dining facility	4,836	Flat plate	DHW—84	
Fort Polk, LA	Housing, 260 units	10,400	Flat plate	DHW—66	837,000
Fort Polk, LA	Hospital	11,000	Evac tube	DHW—21	1,440,000
Fort Polk, LA	Central solar energy plant	8,416	Evac. tube	C—100 H—21	[a]
Fort Polk, LA	Post exchange	11,700	Evac. tube	DHW, C, H[a]	[a]
Fort Benning, GA	Mech. shop	50	Flat plate	DHW—45	57,000
Fort Benning, GA	Laundry/DHW/central heating make-up	27,000	Solar pond	DHW—50	4,300,000
Yuma Proving Grd	Range ops. ctr.	13,000	Tracking	DHW—100	900,000
Fort Huachuca, AZ	Academic bldg.	6,500	Tracking concentr.	DHW—18 C—75 H—100	1,080,000

Table 13.1 (continued)

Location	Building description	Collector area (ft²)	Collector type	Application (% solar)	System cost ($ in year of construction)
Fort Huachuca, AZ	Field house	900	Drain-back	DHW[a]	[a]
Fort Huachuca, AZ	Field house	2,000	Unglazed flat plate	Pool H—75	53,000
Seagoville, TX	Reserve ctr.	12,050	Flat plate	DHW—98 C—94 H—100	527,000
Fort Stewart, GA	Housing	10,560	Flat plate	DHW—50	[a]
Fort Stewart, GA	Dining hall	3,000	Flat plate	DHW—30	195,000
Fort Bragg, NC	Dining hall	1,950	Flat plate	DHW—30	330,000
Fort Bragg, NC	Dining hall	3,060	Flat plate	DHW—30	293,000
Fort Bragg, NC	Admin. bldg.	37,905	?	DHW, H[a]	[a]
Fort Ord, CA	Dining hall	1,808	Flat plate	DHW—50	[a]
Fort Ord, CA	Security	87	Flat plate	DHW—75	14,000
Fort Ord, CA	Housing	28,000	Flat plate	DHW—55	1,920,000

[a] No data available.
DHW = Domestic hot water, C = Cooling, H = Heating.

Table 13.2
Army passive solar thermal energy systems

FY	Location	Building type	Application
1982	Ft Drum, NY	Family housing, 232 units	Passive heating (direct gain design)
1982	Picatinny Arsenal, NJ	Family housing, 26 units	Passive heating (direct gain design)
1983	Fort Lewis, WA	Family housing, 115 units	Passive heating
1983	Fort Irwin, CA	Family housing, 144 units	Passive cooling (shading, heat rejection)
1984	Fort Stewart, GA	Family housing, 244 units	Passive cooling (shading, heat rejection)
1984	Fort Polk, LA	Family housing, 200 units	Passive cooling (shading, heat rejection)
1984	Aberdeen Proving Ground, MD	Family housing, 208 units	Passive heating (direct gain design)

Table 13.3
Navy active solar thermal energy systems

Location	Building description	Collector area (ft^2)	Type collector	Application (% solar)	System cost ($ in year of construction)
El Toro, CA	Family housing, 216 units	9,245	Flat plate	DHW—49	578,000
Pearl Harbor, HI	Family housing, 190 units	20,000	Flat plate	DHW—90	487,000
Kauai, HI	Cafeteria, qtrs. support bldg.	486	Plat plate	DHW—75	10,100
Roosevelt Roads, PR	Family housing, 300 units	19,500	Flat plate	DHW—90	717,000
Ballston Spa, NY	Family housing, 100 units	14,400	Flat plate	DHW—70 H—43	433,000
Camp Pendleton, CA	Dining hall	2,500	Flat plate	DHW—60 H—38	341,000
Camp Pendleton, CA	Bachelor qtrs.	2,800	Flat plate	DHW—45	276,000
29 Palms CA	Bachelor qtrs.	[144 units]	Con-centr.	DHW—43	238,000
Camp Pendleton, CA	Swimming pool	4,864	Flat plate, unglazed	PH—42	609,000
Pearl Harbor, HI	Family housing, 306 units	29,376	Flat plate	DHW—87	[a]
El Toro, CA	Bachelor qtrs.	8,440	Flat plate	DHW[a]	586,000

[a] No data available.
DHW = Domestic hot water, H = Heating, PH = Process heat.

Table 13.4
Navy passive solar thermal energy systems

FY	Location	Building type	Application
1982	Seal Beach NAS, CA	Family housing, 200 units	Passive heating (prefab Trombe units)
1984	Camp Pendleton, CA	Family housing, 104 units	Passive heating (prefab Trombe units)
1984	Adak, AK	Family housing, 405 units	Passive heating (prefab sunspace)

Table 13.5
Air Force active solar themal energy systems

Location	Building description	Collector area (ft^2)	Collector type	Application (% solar)	System cost ($ in year of construction)
Nellis AFB, NV	Bachelor qtrs.	[a]	Flat plate	DHW—78 H[a]	646,000
Sheppard AFB, TX	Family housing	[a]	Flat plate	DHW—60 H—60	98,000
Robins AFB, GA	Corrosion ctrl.	18,000	Flat plate	DHW[a]	982,000
Edwards AFB, CA	Dormitory	576	Flat plate	DHW—39	39,000
Edwards AFB, CA	Library	262	Flat plate	DHW—65 H—10	21,000
Norton AFB, CA	Audio-visual bldg.	8,000	Flat plate	PW—50	435,000
Griffis AFB, NY	Fire station	1,344	Flat plate	DHW—60	40,000
Mountain Home, ID	[a]	624	Flat plate	DHW[a]	21,000
USAF Academy, CO	Youth ctr.	1,800	Flat plate	PH—49	122,000
Mather AFB, CA	Personnel bldg.	2,400	Flat plate	DHW—73	167,000
Eglin AFB, FL	Dormitory	3,200	Flat plate	DHW—82	93,000

[a] No data available.
PW = Process water (for washing A/C parts), PH = Process heat, DHW = Domestic hot water, H = Heating.

and that "unique" passive solar features would therefore not be allowed in new Air Force construction. The policy was subsequently repealed, but the "pipeline" had been sufficiently disrupted to impact near-term demonstrations of passive solar thermal technology.

13.6.4 Lessons Learned

In 1984 the military services jointly funded through the Los Alamos National Laboratory (LANL) an assessment of many existing solar thermal projects (LANL 1982–1984). The Army completed its assessment of operational experience with the 30 sites included in table 13.1 and identified 86 separate problems from the systems inspected (table 13.6).

An overriding conclusion from the assessment was that the Army had too many one-of-a-kind systems in its inventory. Of the 30 systems

Table 13.6
Summary of problems encountered in DOD solar thermal projects

Problem description	Frequency		
	Army	Navy	Air Force
Improper design	44 (51%)	10 (29%)	15 (39%)
Inadequate specifications	5 (6%)	2 (6%)	5 (13%)
Equipment malfunction	17 (20%)	9 (26%)	12 (30%)
Improper operation	13 (15%)	7 (21%)	6 (15%)
Installation error	7 (8%)	6 (18%)	1 (3%)

inspected, the Army found 28 different system schematics—many with overly complex designs incorporating up to 10 distinct modes of operation. In addition, a common theme of poor performance/payback emerged from the solar air-conditioning applications.

The results have been more encouraging for the seven passive applications in table 13.2. The majority of systems have provided the heating or cooling anticipated with little maintenance or repair required. These projects have also demonstrated the acceptability of passive designs to the military families occupying the homes. As a result, passive features will be incorporated in future Army housing projects where they can be justified on the basis of anticipated energy cost savings.

The assessment completed by the Navy for the eleven projects in table 13.3 identified 34 separate system problems (table 13.6; NCEL 1985). As with the Army, the Navy found its systems to be generally unique; however, the Navy's systems were simpler in design and were therefore operating more effectively. As a result of its findings, the Navy developed guidelines on the most common faults found, troubleshooting guides for field use, and preventive maintenance programs.

Owing to the newness of the three Navy passive solar applications in table 13.4, no operational experience is available at this time. Nevertheless, because proposals were selected which incorporated passive solar features in their designs, more and more contractors are incorporating passive solar features into current submittals.

As a result of its experience with active solar heating, solar DHW, and passive solar heating, the Navy has established a policy of incorporating only solar DHW and passive solar heating systems (where justified on the basis of life-cycle cost) in the design of new Navy family housing.

The assessment completed by the Air Force for the eleven projects in table 13.3 identified 39 separate system problems (table 13.6). Like the other services, the Air Force was surprised at the complexity of its solar energy system designs, the overall lack of "standard" designs, and the lack of operational procedures and guidelines. In response, the Air Force emphasized tighter design reviews and the development of maintenance manuals to accompany all solar system installations.

13.6.5   Air Force Solar IPH Demonstration

In 1981 the Air Force completed an analysis of all Air Force industrial process heat (IPH) requirements (JPL 1981). The study provided some significant findings:

• On-hand Air Force databases on IPH requirements were generally incomplete; sufficient data to select appropriate solar thermal technologies needed to be collected at individual installations.

• The majority of Air Force IPH requirements were low-temperature requirements currently being met by high-temperature systems operating at low efficiencies.

• Solar thermal technologies had the technical and operational capabilities to meet a large variety of Air Force process heat needs.

These results prompted the demonstration of a solar thermal industrial process heat system at Hill Air Force Base in Utah (ACC 1983). Manufactured by Power Kinetics, Inc. (PKI), of Troy, New York, and installed in 1982, the parabolic dish/system provided steam at 360°F (182°C) and 125 psi (858 kPa) directly to the Air Force Worldwide Landing Gear Facility at the Ogden Air Logistics Center for various industrial refurbishing processes. The PKI parabolic dish had an effective reflective surface area of 864 ft^2 (80.3 m^2) and was rated at 100,000 Btu per hour (29 kJ/s) under a full sun. Cost of the installed system, including instrumentation and controls, was $88,000.

The project was constructed in excellent time and experienced only minor tracking problems prior to initial operation. Once operating, the system functioned well and required minimal maintenance (1 hour/week on average).

The plant continued to prove its capability to deliver the desired energy product in an Air Force industrial environment until the concentrator was damaged in a freak windstorm (80 + mph). The system was then deactivated because of the lack of additional funding. The system did operate well enough and long enough to dispel questions about the readiness of solar thermal IPH technology, leaving only cost issues to be addressed.

13.6.6   DOD-DOE MX-RES Project

The MX-RES Project was a joint DOD-DOE effort to use renewable energy sources (RES) to power the MX land-based missile system. Besides providing energy for the remote MX sites, it was hoped that the project would lead the development of commercial RES by supplying the demonstration opportunity and the initial volume markets needed by the industries involved.

The MX program was approved by President Carter on 7 September 1979, with construction scheduled to begin in 1983. The heart of the MX system were 4,600 horizontal shelters, grouped into 200 clusters. Each cluster included 23 shelters (each requiring 20.6 kW continuous electrical power), a cluster maintenance facility (225 kW), and other facilities that would have brought the electrical power requirement of a cluster to approximately .75 MW, or 150 MW for all 200 clusters. Two bases would have also been located in the deployment area to house support personnel and operations, requiring an additional 33 MW of power.

Although the MX-RES Project considered various technologies (solar thermal, photovoltaic, wind, geothermal, and biomass), solar thermal electric power was considered one of the most viable technologies when the first major solicitation was released in early 1981.

The MX-RES Project was terminated when the MX deployment concept was significantly changed by DOD and Congress—and the need for centralized remote power significantly reduced. Had the project gone forward, this would have been the largest federally funded demonstration of solar thermal technology in the country.

13.7   Demonstration Programs Using Nonappropriated Funds

13.7.1   Army Air Force Exchange Service

By far the most extensive demonstration of solar thermal technology using nonappropriated funds was accomplished by the Army and Air Force Exchange Service (AAFES). AAFES constructs and operates the

Table 13.7
Early AAFES solar DHW, heating (H), and cooling (C) projects

Location	Collector size (ft²)/ manufacturer	ERDA cost ($ in year of construction)	AAFES cost ($ in year of construction)	% solar
Randolph AFB, TX	12,500/Raypack	468,613	148,000	31.8
Kirtland AFB, NM	8,000/Raypack	423,397	52,000	42.9
Fort Polk, LA	11,700/GE	532,240	123,601	38.4
Bolling AFB, DC	15,392/GE	372,268	333,109	47.5

retail shopping facilities on military installations. These facilities vary from enclosed malls to individual stores. As an example, the Fort Polk Exchange in Louisiana includes 56,723 ft² (5,270 m²) of retail space and cost approximately $4.4 million to construct.

AAFES demonstration projects fall into two general categories. The early projects (1977–1979), jointly funded with ERDA, focused on the four exchange complexes identified in table 13.7. Projects designed and constructed after the initial four were funded solely by AAFES (table 13.8); these projects generally reflected a significant reduction in the scale of activities.

The initial AAFES solar thermal technology demonstrations projects were largely disappointments, providing more opportunities to learn what not to do than opportunities to save energy. None of the systems performed well when activated, and all required excessive ongoing maintenance and repair downtimes. All four systems were analyzed in detail by NASA (1980–1981), and design deficiencies were consistently reported. The general category of "design deficiencies" included the following:

• Complex operating modes and controls often conflicted with each other.

• Major components were oversized, undersized, or mismatched.

• Dissimilar materials were improperly used, causing corrosion problems.

• Excessive roof penetrations were used.

Under the general category of "workmanship," the following deficiencies were found:

• Pumps and valves were installed backwards or upside-down.

Table 13.8
Follow-on AAFES solar DHW projects

Location	Application	Collector size (ft²)/type/ manufacturer	Designer
Fort Gordon, GA	Post exchange launder-ette, restrooms, beauty shop, barber	1,217/flat plate/ Aircraftsman	Scientific Analysis Inc. Montgomery, AL
Fort Devens, MA	Base launderette	2,556/flat plate	Mountain Associates Princeton, MA
Fort Campbell, TN	Post exchange launder-ette, restrooms, fast food activity	3,454/flat plate/ TVAEC	Tennessee Valley Alter-native Energy Group Johnson City, TN
AAFES Dallas, TX	Office bldg. (5-story), photo lab, cafeteria, restrooms	1,147/flat plate/ Aircraftsman	Scientific Analysis Inc. Montgomery, AL
Langley AFB, VA	Base exchange beauty shop, barber, restrooms	633/flat plate	PNG Conservation Co. Charlotte, NC
Homestead AFB, FL	Base launderette	1,475/flat plate/ Energy Transfer Sys.	Sun Energy Connections Inc. Orlando, FL
Scott AFB, IL	Base launderette	745/flat plate/ Aircraftsman	Scientific Analysis Inc. Montgomery, AL
Robbins AFB, GA	Base exchange beauty shop, restrooms	380/flat plate	PNG Construction Co. Charlotte, NC
Offutt AFB, NE	Base exchange flower shop, restrooms, launderette	1,601/flat plate/ Gulf Thermal	Dressler Energy Con-sulting & Design Corpo-ration Overland Park, KS

- Liquid flows were not properly balanced.
- Nonspec equipment or materials were used.

  Several "materials" problems were also encountered:

- Seals, valves, pumps, and piping were found to leak.
- High rust content was found in working fluids.
- Over 2,700 evacuated collector tubes were broken in operation, mostly due to thermal shock

Correction of the deficiencies in the four original systems required signif-icant redesign and reconfiguration, at an estimated total cost of over $60,000.

The later AAFES projects applied the lessons learned in the earlier projects, resulting in better designs, better construction, and better initial performance.

In summary, AAFES experiences with the demonstration of solar thermal technology led to the following conclusions:

• Solar system designs must be kept as simple as possible.

• Solar system components and controls must be kept simple and to a minimum.

• Contractors, equipment vendors, and their submittals must be carefully screened to assure all project requirements are met.

• Installation engineers need better operations manuals and maintenance training.

13.7.2 Private Sector Financing

The Military Construction Codification Act, FY 1982 (PL 97-217) encouraged the military to solicit the development of energy production facilities on military installations by the private sector. Although many of the projects resulting from this legislation focused on conventional fuels or geothermal energy, at least one substantial solar thermal project was pursued within the military.

During 1981 the U.S. Air Force Academy (USAFA) in Colorado decided to supplement its central heating plant with a solar thermal system. The heating plant was judged a good application with several attractive features. The system used hot water as opposed to steam, and served a daily load as opposed to the weekday load found in many industrial applications. Finally, open land suitable for large solar collector fields was available near the heating plant. In 1982 the academy, with the assistance of the Solar Energy Research Institute, developed a Request for Proposals (RFP) that was later released to the public. After months of negotiation with three qualified firms, a proposal from the Acurex Solar Corporation (ASC) was accepted and an Air Force Utility Service Contract was developed, negotiated, and signed by both parties (ASC 1982; USAFA 1984).

The ASC system consisted of 100,800 ft^2 (9365 m^2) of parabolic trough collectors sited on eight acres of academy land. The system was designed to produce 31 billion Btu (32.7 TJ) of net energy (7% of annual plant hot water energy load). The ASC concept included taking advantage of

the federal investment tax credit, the federal business energy investment credit, the Colorado solar tax credit, and accelerated depreciation
methods.

Although signed, the Utility Service Contract was never implemented
because of problems stemming from the law's requirement that approvals
be obtained up to the level of the secretary of defense. The DOD procurement and legal systems were not accustomed to this type of long-term
"joint venture" contract with the private sector. Although many obstacles
were overcome and much new ground broken, the inertia encountered
precluded activation of the contract and construction of the plant prior to
the expiration of the federal tax credits. Although other types of private
sector–financed energy plants are still being pursued by the military, with
the loss of the solar tax credits, a solar thermal demonstration under this
category does not appear likely.

13.8 Economics

Within the military, the economic justification for demonstrating solar
thermal technology shifted significantly from the mid-1970s to the mid-
1980s. During the mid-1970s the emphasis was on changing from scarce
petroleum to renewable energy technologies wherever possible. In the
early 1980s the emphasis changed from considering solar in all new construction to considering solar where it could be shown to be cost-effective
and soon thereafter, to considering solar only where stringent cost effectiveness could be proven. The impact of this shift from security to economics led to significant attrition in the number of military solar projects
funded and constructed from 1981 onward.

The most significant impacts came as a result of Section 116 of the
Military Construction Appropriations Act of 1981 (PL 96-436), which
directed that (1) a 7% discount factor be used to establish the present
worth of future costs, (2) maintenance costs be included for the solar
energy systems, and (3) "marginal fuel costs as determined by the Secretary of Defense" be used. Previously, life-cycle cost effectiveness had been
determined using (1) undiscounted constant dollars, (2) no additional
maintenance costs related to the solar energy system, and (3) more liberal
projected fuel cost inflation factors.

Application of the more liberal economic criteria contained in various earlier "guidance documents," including Section 804 of the Military

Construction Act of 1981 (PL 96-418), would have resulted in the Navy alone having 36 "economically viable" solar thermal technology demonstration projects in 1981. Application of the more stringent criteria contained in Section 116 of PL 96-436 resulted in only four "viable" projects for 1981—all that in fact were built. The other military services encountered the same type of attrition owing to the more stringent economic criteria. Today all of the military services study the application of solar on essentially all applicable projects but incorporate it into the design only if the stringent life-cycle cost effectiveness requirements are met.

Additional problems were created for the military when it came time to validate the cost effectiveness of solar thermal technology demonstration projects. First, individual building metering was, and to a great extent today still is, almost nonexistent. This made baselining energy consumption at the individual building level almost impossible, and calculating actual solar energy savings speculative at best. Second, solar thermal technology demonstrations that included some sort of data acquistion system in the original design generally lost funding for that part of the package prior to construction. Third, projects whose performance was later evaluated in detail by NASA or DOE organizations generally showed actual performance levels, and therefore energy savings levels, that were much lower than anticipated. These problems undoubtedly contributed to the difficulty that the military services often encountered in even getting construction funds authorized for some of the "viable" projects that remained.

This leads to a last economic consideration, somewhat unique to the military. The military builds its new facilities to meet very specific mission requirements and is subject to significant funding limitations. If a military construction project containing an active solar heating or cooling system runs into funding shortages of any kind, the preceived trade-off becomes square footage (mission requirements) versus amenities. Many times the simplest perceived solution is to delete high first-cost solar energy system additives, ignoring the life-cycle impacts. Although efforts are often made to compensate by allowing for "solar retrofitting" in the design, the loss remains. This natural attrition process is not so common in passive solar applications, which are more difficult to remove from a design package prior to construction.

The military experience with solar thermal technology demonstration programs to date has resulted in a perception that solar energy systems are seldom cost-effective unless they are simple (usually active solar DHW and passive solar heating only) and well matched with regard to end use and geographical location.

13.9   Conclusions

In 1973, when rapidly increasing energy costs during the oil embargo began to shatter military operations budgets, the military services were not at all prepared to apply solar thermal technology solutions to the energy problem. Solar thermal system design, construction, operation, and maintenance experience within the military services was virtually nil. In addition, large-scale technology demonstration programs had seldom been undertaken, with the exception of weapons system prototype development programs. The solar energy industry was also in its infancy, with perhaps more promise than actual experience and discipline. As a result, inexperienced military engineers had a difficult time getting accurate or objective information with which to implement successful projects. It is no wonder that early military solar thermal technology demonstration and application programs, driven by a sense of urgency and generally without regard for cost effectiveness, were characterized by poorly planned, ineptly designed, badly constructed, and inadequately maintained solar systems. As military engineers became more experienced and sophisticated customers, systems were increasingly installed that met necessary standards of reasonable reliability.

In the mid- to late 1970s the military gained its first hands-on experience with solar thermal technology and learned many lessons that it applied to future designs. At the same time, the need to design and construct solar energy systems to meet specific life-cycle cost goals became more pressing, providing an additional driving force toward simpler designs and more efficient performance.

By the early 1980s economic constraints had become much more restrictive, but the military had gained enough in-house expertise in residential, industrial, and commercial-type applications to be "smart" consumers of solar thermal technologies. The establishment of strict project selection criteria, effective construction management capabilities, accurate design and operation manuals, and trained operation and maintenance personnel had increased the success rate of military solar projects significantly. Although large-scale solar application programs are still not commonplace in the military, and are considered economical only in certain applications and locations, a substantial level of acceptance is now visible.

Ahead lie many new challenges for the application of solar thermal technology within the military. The prime driving force to date has been economics—paying growing utility bills—and yet the military has not met its goals for applying renewable energy technologies. This continuing institutional conflict—establishing required renewable energy goals but then making them virtually impossible to meet through congressionally mandated requirements for stringent life-cycle cost effectiveness—has created the largest obstacle to the expanded application of solar thermal technologies in the military. Tomorrow's challenges will continue to include economics but will expand to include operational readiness, mission enhancement, and national energy security. The military is already being called upon to reduce the vulnerability of its military operations through increased energy self-sufficiency and reliability. Demonstration programs have driven the learning curve to the level where solar thermal technologies will continue to be a part of that future. After all, in addition to strengthening military missions, every barrel of oil the military saves through the application of solar thermal technology is a barrel of foreign oil we will not someday have to fight for.

References

ACC (Applied Concepts Corporation). 1983. *Air Force Logistics Command (AFLC) Solar Thermal Plant.* 15 April.

ASC (Acurex Solar Corporation). 1982. *Solar Production of Energy for Space Heating the U.S. Air Force Academy.* October.

Department of the Air Force. 1984. *1984 Air Force Energy Plan.* Washington, DC.

Department of the Army. 1985. *Army Energy Plan.* Washington, DC, 25 February.

Department of the Navy. 1984. *Energy Plan FY 1984–1990.* Washington, DC.

DOD (Department of Defense). 1981. "Solar Energy Project Summaries." Washington, DC, April.

DOD (U.S. Department of Defense). Office of the Deputy Assistant Secretary of Defense. 1985. *Defense Energy Management Plan.* Washington, DC, September.

DOE (U.S. Department of Energy). 1987. *Annual Report of Federal Government Energy Management, Fiscal Year 1986.* Washington, DC, 3 June.

GAO (U.S. General Accounting Office). 1978. *Solar Demonstrations on Federal Residences: Better Planning and Management Control Needed.* Washington, DC, 13 April.

JPL (Jet Propulsion Laboratory). 1981. "USAF Thermal Applications Requirements Definition," 21 October.

LANL (Los Alamos National Laboratory). 1982–1984. "Letter Reports." Los Alamos, NM.

NASA (National Aeronautics and Space Administration). 1980–1981. "Letter Reports." Washington, DC.

Nay, M. W. Jr., J. M. Davis, R. L. Schmiesing, and W. A. Tolbert. 1976. *Solar Heating Retrofit of Military Family Housing.* FJSRL Technical Report 76-0008. September.

NCEL (Naval Civil Engineering Laboratory). 1984. *The Navy's Energy House.* Port Hueneme, CA, March.

NCEL. 1985. *Evaluation of Installed Solar Systems at Navy, Army, and Air Force Bases.* Port Hueneme, CA, December.

Sklar, H., and R. Noun. 1982. *Analysis of Air Force Base Energy Acquisition through Third-Party Contracts.* Golden, CO: Solar Energy Research Institute, July.

USAFA (U.S. Air Force Academy). 1984. *Utility Service Contract* (F05611-84-D-001) *for Solar Generated High-Temperature Hot Water.* Colorado Springs, CO, 6 April.

IV SOLAR THERMAL QUALITY ASSURANCE

14 Testing, Standards, and Certification

Gene A. Zerlaut

An active, free marketplace for solar energy products is fundamental to the development of a healthy national solar energy industry, and voluntary consensus standards are perhaps the single most important requirement for shaping a viable solar industry. (Zerlaut 1981)

The voluntary consensus process traditionally has been the means by which industry and society develop and produce the standards that form the basis for commerce. This is the only process that stands between economic chaos and government over-regulation on the one hand, and the exercise of a truly competitive marketplace on the other. In the absence of voluntary consensus standards, the temptation can be overwhelming for narrow segments of society to erect impediments to market penetration and for government to react with so-called mandatory standards and costly regulations. As a result, both processes tend to become barriers to the free market development of a healthy commerce. (Zerlaut and Garner 1983)

14.1 The Role of Standards

It can safely be said that essentially all parties to the development of a solar industry in the United States believed that it was necessary to put in place as quickly as possible a national program for testing, rating, labeling, and certifying solar collectors and solar energy systems, as well as for establishing solar standards. This perceived need was the result of the emergence, as early as 1974, of a number of products with either or both low thermal performance and inferior reliability. The mushrooming of poor products in the mid- to late 1970s resulted from unscrupulous or from poorly trained, incompetent manufacturers and installers (who otherwise may have been scrupulous).

In discussing the history of the U.S. solar industry, and the extensive standardization activities that marked the course of its development, is it instructive to first review the concepts associated with "standardization." More than sixty years ago, the U.S. Supreme Court stated that standardization can be "beneficial to the industry and to consumers" in a landmark ruling dealing with the *Maple Flooring Manufacturers Association v. the United States* (268 U.S. 563, 566 [1925]). The Voluntary Standards and Accreditation Act of 1977 (S.825, Sec. 5(2)) defines a standard as being "a set of rules, or requirements, applied to products, established as a result of a standards-development activity by a standards-development organization." This act also defines certification, in Sec. 5(18), as being "a process of product evaluation, including testing and analysis, administered and managed by a certification agency, whereby producers or suppliers are authorized to attest that their products or services satisfy the applicable standards" (Hoffman 1978).

The processes of writing meaningful consensus standards may be described as one that involves a delicate balance between a multiplicity of "conflicts of interest." Although this description is an oversimplification, it does indeed speak to the concept that "balanced interests" are a necessary condition of avoiding the legal ramifications of antitrust—particularly restraint-of-trade—actions.

Thus, even though motivation—one of the most critical elements necessary to the preparation and implementation of the standards required for a developing industry—was present in the case of solar technologies, it was quickly evident that establishing these standards would be no easy task.

First, perhaps never in the history of U.S. technology has such a plethora of organizations been utilized and, in many cases, been formed, for the purpose of fostering the growth, and regulation, of an industry as on behalf of solar energy in the period from 1974 through 1986. As shown in table 14.1, more than thirty U.S. and international organizations and institutions were involved either directly or indirectly in U.S. efforts to create a national testing, rating, labeling, and product certification program for solar collectors. Of the thirty or more organizations that were exclusive to the United States, eight were created solely for the purpose of supporting and regulating the solar energy industry. Second, with only one or two exceptions, test methods that were at once acceptable to both the solar industry and to government and public interests were simply not available and had to be developed.

This chapter will elucidate, if not explain and rationalize, the morass of problems and confusions created by the involvement of so many organizations, whose attempts to make the "solar dream" come true were all too often counterproductive and competitive, and often totally at odds with one another.

14.2   National Solar Mandates

14.2.1   Impetus for Federal and State Mandates

Impetus for a national effort to harness solar energy came from the realization during the late 1960s and early 1970s that the United States

Table 14.1
Glossary of acronyms for U.S. and international solar organizations

ANSI	American National Standards Institute
ARI	Air-Conditioning and Refrigeration Institute
ASES	American Solar Energy Society
ASHRAE	American Society of Heating, Refrigeration, and Air-Conditioning Engineers
ASTM	American Society for Testing and Materials
DOE	U.S. Department of Energy
DOC	U.S. Department of Commerce
ERDA	Energy Research and Development Administration (became DOE in October 1977)
FEA	Federal Energy Administration (sunset in 1980)
FSEC	Florida Solar Energy Center
HUD	U.S. Department of Housing and Urban Development
ISCC	Interstate Solar Coordination Council
LRC	Lewis Research Center (NASA)
NASA	National Aeronautics and Space Administration
MSFC	George C. Marshall Space Flight Center (NASA)
NBS	National Bureau of Standards (DOC), now National Institute of Standards and Technology (NIST)
NSF/RANN	National Science Foundation/Resarch Applied to National Needs
NSHCIC	National Solar Heating and Cooling Information Center
RSEC	Regional Solar Energy Center (DOE-funded)
SEIA	Solar Energy Industries Assoication
SEREF	Solar Energy Research and Education Foundation
SERI	Solar Energy Research Institute
SRCC	Solar Rating and Certification Corporation
SSCC	Solar Standards Coordinating Council (ANSI)
TIPSE	California Testing and Inspection Program for Solar Equipment
TVA	Tennessee Valley Authority

International solar organizations (having influence on U.S. solar efforts)

AFNOR	Association Française de Normalisation (Paris)
CEC	Commission of the European Community (Brussels)
DIN	Deutsches Institut für Normung (Berlin)
EC	European Community (same as CEC)
EFTA	European Free Trade Association (Rotating)
IEA	International Energy Agency (Rotating)
IEC	International Electrotechnical Commission (Geneva)
ISO	International Organization for Standardization (Geneva)
SAA	Standards Association of Australia (Sydney)
WMO	World Meteorological Organization (Geneva)
VAMAS	Versailles Agreement on Materials and Standards

needed to work toward energy independence by decreasing the nation's dependence on foreign oil and by relying on coal and nuclear energy as long-term solutions to the country's energy problems. While the latter two concerns related largely to environmental and safety issues, and to the ultimate safety of mankind, there was increasing concern that the United States could not depend on the long-term availability of low-cost crude oil from the oil-exporting nations.

The promise of nuclear energy was beginning to fade as the cost of new plant, improved safety, and waste disposal became more obvious. Legislators began to take note of predictions that increased reliance on coal as an energy source of the future would result in global warming. These issues for the future predisposed Congress toward favorable consideration of the solar bills that were offered during the early 1970s.

14.2.2   Federal Enactments of the 1970s

In 1974 Congress provided funding for initiating research and development activities in solar technologies and for supporting the commercialization of solar energy. This solar legislation is discussed in detail in chapter 2.

The law that directed HUD to design and fund a broad-based solar demonstration program had the most direct influence on solar device testing. It required manufacturers to submit test results from third-party laboratories, in order to have their equipment considered for distribution to qualified homeowners under the HUD Solar Hot Water Initiative. The Energy Conservation and Production Act (PL 94-385), had considerable influence on the future of the nation's solar testing, rating, and certification programs.

The 1978 energy tax act, PL 95-618, provided for residential solar tax credits, business solar tax credits, and residential conservation energy tax credits. This law had the greatest influence of all federal enactments on the ultimate development of a national program for testing, rating, labeling, and certifying solar collectors and systems; furthermore, the tax credits were a major step in the commercialization of solar.

Even through product testing, rating, and certification were never tied to either the federal residential or the federal business income tax credits (the latter passed in 1980), federal income tax credits created a much stronger national market for the solar industry. The failure to tie product performance standards to the federal income tax credits may have had far-reaching consequences for both the consumer and the solar industry, and this concept is explored further in this chapter.

14.2.3 Solar Mandates within the States

With the passage of CH.165 Laws in 1974 exempting solar energy devices from property taxes through 1984, Arizona became the first state to enact solar legislation. This was followed by CH.93 Laws of 1975, amended in 1976 and 1978, which provided tax incentives for solar energy devices used in residential, business, or investment property. These laws permitted the taxpayer either to depreciate the cost of a solar device, amortized over 36 months, or to claim an income tax credit. Arizona subsequently became one of the first states to require both solar collector and solar system product testing, rating and certification as a condition of taking one of the tax incentives.

By 1981 every state except Kentucky, Pennsylvania, West Virginia, and Wyoming had enacted either tax incentives or other solar-related legislation. Further, forty-two of the forty-six states with solar legislation on the books provided some form of tax incentives. Only Minnesota, Florida, Arizona, California, New Mexico, Oregon, and Utah tied the election of tax credits to the testing, rating, certification, and labeling of solar collectors; only Arizona, California, Oregon, and Utah required that domestic hot water solar systems be tested and certified to be eligible for tax credits.

Representatives from the energy offices of these and other states were all participants on the Interstate Solar Coordination Council (ISCC; see section 14.3.2.3.2). Established largely to provide certification reciprocity between the states, the ISCC later helped form the Solar Rating and Certification Corporation (SRCC), a positive mechanism for adopting solar certification in these states.

Florida and California each developed their own product certification programs, as mandated by their legislatures, although both states accepted rating and certification of collectors done by either SRCC or the Air-Conditioning and Refrigeration Institute (ARI), the two national solar product rating and certification programs then in operation. After 1986, only the SRCC and the Florida programs remained operational.

14.3   Organizations: A History of Federal and Private Involvement

14.3.1   Federal Organizations

14.3.1.1   National Science Foundation/Research Applied to National Needs

In the late 1960s the National Science Foundation (NSF) formed an umbrella office called "Research Applied to National Needs" (RANN). In mid-1972, NSF funded an evaluation of various methodologies for testing the thermal performance of solar collectors conducted by the National Bureau of Standards (NBS). Results of this evaluation, published as *NBSIR 74-635*, are presented in section 14.3.1.5 (Hill and Kusuda 1974).

NSF/RANN became the first federal agency to recognize that one of the most significant problems facing the utilization of solar energy was a lack of accurate information on the exterior durability of candidate materials for solar collectors. In mid-1974 it funded a broad-based solar materials test program at the IIT Research Institute (IITRI) to determine the effect of both accelerated and real-time weathering on the materials' optical and mechanical properties. The concept was to prepare a handbook of relevant mechanical and solar optical properties that would aid solar device design engineers in the selection of the proper materials, a program that was completed in 1979 (Gilligan, Brzuskiewicz, and Brzuskiewicz 1978).

14.3.1.2   Department of Housing and Urban Development

In 1974 HUD was directed to demonstrate solar energy applications for the purposes of establishing (1) the viability of solar technologies and (2) product and solar materials testing standards. Third-party performance testing of solar collectors supported the 1977 HUD Hot Water Initiative Program (see chapters 9 and 27).

Funds provided by HUD under PL 93-383 were used by NBS to contract with the Air-Conditioning and Refrigeration Institute (ARI) to identify laboratories capable of performing thermal performance tests of solar collectors (see section 14.3.1.5).

In 1976 HUD also funded the National Bureau of Standards to aid in developing Intermediate Minimum Property Standards (IMPS) for Solar Heating and Domestic Hot Water Systems (HUD 1977). The IMPS provided the technical basis for the design of solar heating and domestic hot water systems. In July 1977 the IMPS became effective for all Federal Housing Administration (FHA), Veterans Administration (VA), and Farmers Home Administration (FmHA) projects utilizing solar energy systems.

Attempts were made to impose the IMPS on the Solar Energy Research and Education Foundation (SEREF) solar collector rating, certification and labeling (RCL) program (see section 14.3.2.2). The acceptance by the RCL program of the tests specified in the solar IMPS and in its companion document *NBSIR-73-1305A* (Waksman, Streed, et al. 1978) would have had the effect of imposing on the solar industry mandatory government testing and performance standards rather than those developed through voluntary consensus by either American Society for Testing and Materials (ASTM) or American National Standards Institute (ANSI). Implementation of the often costly and unnecessary government-developed test procedures and requirements would have had an adverse effect on the solar industry (Butt 1978a, Zerlaut 1978).

14.3.1.3 Federal Energy Administration

Formed in 1974 to deal primarily with the nation's nonnuclear energy needs, the Federal Energy Administration (FEA) began to support solar technologies in 1975, largely within the federal buildings program. Indeed, FEA had been charged (under PL 94-385) with overseeing the installation of $100 million of solar facilities in federal buildings by 1978.

As a result of extensive conversations with the Solar Energy Industries Association (SEIA), FEA issued a Request for Proposals in early 1977 for the development of a "program of testing, certification, and labeling" as a high-priority objective of the Solar in Federal Buildings Program (FEA 1977). A contract was awarded by FEA to SEREF, the educational arm of SEIA. The specific authority to fund this work resulted from PL 94-385, which authorized FEA to "carry out a program to develop the policies, plans, implementation strategies and program definition for promoting accelerated utilization and widespread commercialization of solar energy." FEA undertook this effort in support of the Solar in Federal Buildings Program.

The FEA contract to SEREF was awarded in anticipation of a companion program under the Energy Research and Development Administration (ERDA) to test some 200 solar collectors for thermal performance, and durability and reliability. The award was made on 26 September 1977, five days before the Department of Energy (DOE) was formed by merging ERDA (nonnuclear energy research and development), FEA (nonnuclear energy commercialization), and the U.S. Atomic Energy Commission (AEC) into one super agency having the responsibility for all energy research, development, and commercialization.

14.3.1.4 Energy Research and Development Administration

Formed in 1974 when Congress voted into law the Energy Reorganization Act (PL 93-438), the Energy Research and Development Administration (ERDA) took over most of the solar energy research activities that had been initiated within NSF/RANN. In cooperation with the FEA, ERDA established a national program for testing, rating, labeling, and certifying solar collectors to support the broad-based commercialization efforts directed by the 93d Congress. ERDA then funded the NBS to develop criteria for assessing the capabilities of laboratories to test the thermal performance of solar collectors. The results of this study, published as NBSIR 78-1535, are discussed in detail in section 14.3.1.5 (Neissing 1978).

14.3.1.5 National Bureau of Standards

The National Bureau of Standards (NBS), now the National Institute of Standards and Technology (NIST), first became involved in solar collector test technologies in the early 1970s, funded by NSF/RANN. The NBS solar program focused on (1) research on methods of testing the thermal performance of collectors and systems, (2) development of methods of testing materials used in the manufacture of solar devices, and (3) development of methods to assess the reliability and durability of solar collectors.

14.3.1.5.1 *Thermal Performance Testing of Solar Collectors*

Solar Collector Testing The first "standard" test procedures for thermal performance testing and rating of solar collectors were developed by NBS (Hill and Kusuda 1974). This report was prepared in the general format of an ASHRAE standard method of testing, and covered both liquid and air as the heat transfer fluid. The report also presented minimum requirements for instrumentation and provided schematics of suggested test loops for testing both liquid and air collectors. It became the principal resource document for the American Society of Heating, Refrigeration, and Air-Conditioning Engineers (ASHRAE) Standards Committee 93, which was subsequently charged with the responsibility for developing a national consensus standard for testing the thermal performance of solar collectors. NBS later published Technical Note 899, which included testing thermal storage devices (Hill et al. 1979).

In 1979 the Center for Building Technology published NBS Building Science Series 117, which specified experimental verification procedures for the standard thermal performance testing of solar collectors (Hill, Jenkins, and Jones 1979). In 1981, through its Interim Solar Collector Testing Program, DOE funded the SEREF RCL program in order "(1) [to determine] overall uncertainty in collector thermal performance data reported by commercial test facilities . . . and (2) [to provide] illustrations of how the uncertainty in individual collector test results affects the rating and selection of collectors for system design and for determining degradation with exposure or operating time" (Streed and Waksman 1981). Streed and Waksman concluded that measurement error was most probably the major contributor to lack of intralaboratory repeatability, but that environmental effects contributed to interlaboratory differences by influencing the loss coefficient for the collectors.

Interlaboratory Testing   In May 1975 NBS solicited information from test laboratories interested in testing the thermal performance of solar collectors. With funding from DOE, NBS procured approximately twenty-five solar hot water collectors of two different types from two different manufacturers. NBS then contracted with twenty-three laboratories to perform round-robin tests to determine the comparability of results from a large number of test laboratories from different geographic locations in the United States. Eleven laboratories participated; nine performed the tests using outdoor test loops, one performed only an indoor test employing a solar simulator, and one performed tests using both an indoor solar simulator and an outdoor test loop.

The round-robin tests showed a wide variation in the slope of efficiency versus the temperature parameter. A subsequent analysis of the results from about half of the laboratories revealed that most of the differences arose from systematic variations between laboratories (Streed et al. 1978). Indeed, when data were analyzed from only the six laboratories most closely adhering to the *ASHRAE 93-77* requirements, the scatter observed was within expected limits for the test procedure. The analysis identified erroneous measurements of solar radiation and fluid flow as being most responsible for the poor agreement between the larger sample of laboratories (Streed 1977; Wood 1978). These results were subsequently used in designing test conditions for the SEREF and SRCC solar collector rating, labeling, and certification programs.

Laboratory Approvals Subsequent to the round-robin test program, NBS received funding from ERDA to develop criteria for qualification of laboratories to test solar collectors in accordance with draft standard *ASHRAE 93-P*. Two additional objectives of this program were (1) to identify laboratories possessing the necessary qualifications and (2) to develop a commercially adaptable plan to certify solar collectors on the basis of thermal performance test results. NBS contracted with the Air-Conditioning and Refrigeration Institute's subsidiary, the ARI Foundation, Inc., to perform these two tasks. Identification of laboratories was based on criteria established by the Building Economics and Regulatory Technology Division of NBS's Center for Building Technology (Niessing 1978). The criteria included laboratory quality control and quality assurance procedures, adherence to good laboratory practice in accordance with *ASTM E548* (ASTM 1984), and the ability to test in accordance with *ASHRAE 93-P*.

Although the contract with ARI appeared to duplicate work by SEREF, NBS chose to incorporate ARI's substantial experience in testing and product certification for the HVAC industry. NBS had traditionally insisted on accreditation of multiple-qualified laboratories, but they granted the contract even though ARI had historically utilized only a single laboratory for each of its HVAC programs.

Thirty-five laboratories responded to solicitations in December 1976, but only fifteen offered complete responses to a second questionnaire. Of these, eleven laboratories were approved to test liquid collectors, and eight of these were selected to test air collectors.

Approved laboratories were used by DOE in its Solar Collector Testing Program (in support of the SEREF rating, certification, and labeling program). NBS further intended that this study form the basis for a subsequent laboratory accreditation program under the aegis of the National Voluntary Laboratory Accreditation Program (NAVLAP), established within the Department of Commerce in February 1976. The Department of Energy, however, declined to "find a need" as required by NAVLAP, and alternatives emerged in the private sector (SRCC and ARI) and in the state programs of California and Florida.

14.3.1.5.2   Methods of Testing Solar Device Materials   The Center for Building Technology (CBT) of NBS vigorously pursued development of solar materials testing methods. Responding to an informal suggestion in 1977 by DSET Laboratories, Inc., that it test samples of cover and absorber materials under nonoperational, stagnation conditions (Zerlaut 1977), CBT's materials branch initiated a three-site exposure test program. Results of the test program are presented in NBS Technical Notes 1132 (Clark et al. 1980) and 1196 (Waksman, Thomas, and Streed 1984). These studies led in turn to development of the following ASTM standards on the weathering exposure of transparent solar collector covers and absorptive solar receiver materials under conditions simulating both operational and stagnation modes:

E 744   "Practice for Evaluating Solar Absorptive Materials for Thermal Applications"

E 781   "Practice for Evaluating Absorptive Solar Receiver Materials When Exposed to Conditions Simulating Stagnation in Solar Collectors with Cover Plates"

E 782   "Practice for Exposure of Cover Materials for Solar Collectors to Natural Weathering Under Conditions Simulating Operational Mode"

E 881   "Practice for Exposure of Cover Materials to Natural Weathering Under Conditions Simulating Stagnation Mode"

14.3.1.5.3   Collector Durability and Reliability   One of the more controversial documents published by NBS was *NBSIR 77-1305*, "Provisional Flat-Plate Solar Collector Testing Procedures" and the first revision, *NBSIR 78-1305A* (Waksman, Streed, et al. 1978). The solar industry viewed the testing requirements of *NBSIR 78-1305A* as "complete overkill." The industry was especially concerned when DOE announced that the requirements would be implemented in the DOE Collector Test Program and the SEREF-developed RCL program (Butt 1978a; Zerlaut 1978). SEREF protested that (1) the tests and pass/fail criteria were arbitrarily chosen without sufficient field experience, (2) industry had not been given time to review the document, (3) geographical differences had not been evaluated, and (4) there had been no correlation with consensus standards then in development (Butt 1977). A few of the durability and reliability tests were later accepted for the DOE Interim Collector Performance Testing Program, and the ARI- and SEREF-

developed rating, certification, and labeling programs. In response to
the industry criticisms, CBT developed a comprehensive research plan
(Waksman, Streed, and Seiler 1981; Waksman, Thomas, and Streed 1984).

14.3.2   Trade Associations and Other Private Organizations

14.3.2.1   Solar Energy Industries Association

14.3.2.1.1   Organization   Founded in 1974 as the nation's solar-oriented
manufacturers' trade association, the Solar Energy Industry Association
(SEIA) was created to promote solar energy utilization and to ensure that
solar products were deserving of acceptance by both the consumer and
government regulatory agencies. By early 1978 SEIA had grown to eight
divisions, more than fifteen advisory and standing committees, and four
state chapters. Working closely with the Congress in those years, SEIA
helped provide the impetus for increased government purchases of solar
equipment, the initiation of federally funded demonstration programs,
and the adoption of federal and state solar tax incentives. Through its
voluntary technical committees, SEIA initiated development of a national
testing, certification, rating, and labeling program in the late 1970s.

14.3.2.1.2   SEIA Rating, Labeling, and Certification Program   The SEIA
"Solar Collector Thermal Performance Rating, Labeling, and Certification Program" proceeded through a series of steps from the preparation of an initial proposal to the Federal Energy Administration, to the
successful implementation of the program with representatives of state
governments under the auspices of the Solar Rating and Certification
Corporation (SRCC). The interrelationship of the various organizations
in the development of the successful solar product certification program is
depicted in figure 14.1.

The Solar Energy Research and Education Foundation (SEREF) was
created to manage the DOE-funded development of a national rating,
labeling, and certification (RCL) program. SEIA intended that this program would be implemented either within the trade association itself or
as a SEIA-sponsored, legally separate entity. But because the RCL plan
was federally funded, the information it provided would be public and
therefore could be utilized by other organizations to implement either
national or regional programs for rating, certification, and labeling that
might compete with SEIA's own interests.

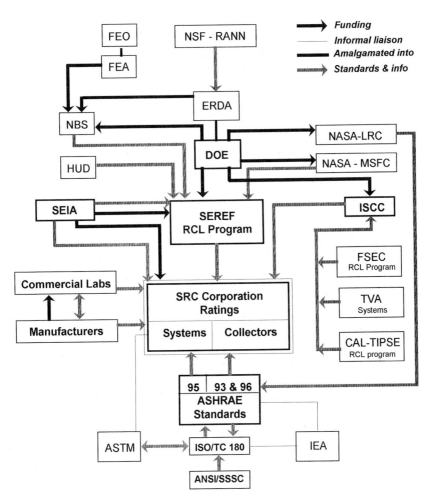

Figure 14.1
Organization interactions with Solar Rating and Certification Corporation (see table 14.1 for
definitions of acronyms).

Accordingly, SEIA established a parallel RCL program in early 1978 in order to quickly implement an industry-tailored version of the SEREF program. Because of the somewhat more focused industry view that would accrue to the parallel effort of SEIA, it was recognized early on that there would be some differences between the two programs. The SEREF program, being federally funded, quite naturally had a broader input in terms of government, industry and consumer interests. Nonetheless, SEIA adopted laboratory accreditation and RCL operational procedures that were based largely on SEREF-developed approaches. Only in the actual performance rating methodology did SEIA differ from the rating methodology proposed by SEREF, and then only in nonsubstantive technical areas.

In January 1978 SEIA formally adopted *DI-78P*, "Recommended Thermal Rating Methodology," which covered

• qualification and intercorrelation of test laboratories to perform *ASHRAE 93-77* thermal performance tests of solar collectors

• generation of thermal performance curves and incident angle modifiers to fit experimentally determined data

• calculation of all-day thermal energy collection values for each of several conditions

• reporting of a computed standard day thermal energy collection value as a "rating number" (Butt 1979a).

By early May 1978 SEIA had adopted a comprehensive product certification standard (*PCS-1-79*) incorporating the laboratory accreditation procedure, the product certification and labeling implementation procedure, and a revised rating methodology (*DI-78P*) into a single document. This new rating methodology permitted rating of a broader range of collectors and incorporated several basic reliability tests adopted from *NBSIR-78-1305A* (to meet California's requirements). By October 1979, twelve collector models from ten manufacturers had been rated, and a total of twenty-six manufacturers were participating in the SEIA RCL programs (SEIA 1979a).

14.3.2.2 Solar Energy Research and Education Foundation

14.3.2.2.1 Organization The Solar Energy Research and Education Foundation (SEREF) was founded by SEIA in August 1977 for the following purposes:

• To educate the public in the uses and benefits of solar energy, and to promote an increase in public interest

• To impose cooperation between the solar industry and other sectors of commerce, and with and among utilities

• To encourage private giving for educational, scientific, and literary purposes related to solar energy

• To acquire, preserve, and disseminate data and information pertaining to solar energy issues

• To solicit, receive and maintain funds to accomplish these purposes (SEREF 1977).

It was under these broad guidelines that SEREF received the FEA contract to develop a national rating, certification and labeling program.

14.3.2.2.2 SEREF Rating, Certification, and Labeling Program   Development of the SEREF Solar Collector Rating, Certification and Labeling Program was funded by the Federal Energy Administration only days after creation of the Department of Energy (FEA 1977). Hence, the SEREF program was subsequently administered by the Office of Solar Applications of the Department of Energy.

SEREF adopted rules of procedure that complied with the requirements of (1) the American National Standards Institute, (2) the Department of Energy, and (3) the Federal Trade Commission. The National Energy Act (PL 93-403) required consultation with the FEA "with regard to any product or material standard which is relied upon in implementing parts of the Act" (Vakerics 1977). SEREF created the Solar Standards Steering and Oversight Committee (SSSOC) to coordinate standards development.

SSSOC members were appointed by the SEREF board or the committee itself; the majority represented organizations that were not members of SEIA, including banking, insurance, and utility companies, professional societies, other trade associations, consumer organizations, and architectural firms. Hence, the committee was considered to be "balanced" (Butt 1978b).

SEREF laboratory accreditation criteria and procedures were based largely on *ASTM E548*, addressing a laboratory's organization, human resources, physical resources, and quality assurance system in detail (ASTM 1984). Particular attention was devoted to calibration frequency and procedures, and to the absence of existing or potential conflicts of interest. Unique to the SEREF accreditation process was the use of a black box apparatus to test the ability of a laboratory to accurately determine the thermal performance of solar collectors. This apparatus, called "LITE" (Laboratory Intercorrelation Test Equipment), was subsequently employed in the laboratory approval stage of the DOE 200 solar collector test program, known as the "Interim Solar Collector Test (ISCT) Program".

The SEREF Rating Methods Committee adopted *ASHRAE 93-77* as the test method for use in determining the thermal performance of flat-plate and concentrating solar collectors. The rating method employed was an all-day thermal performance computation using "standard" conditions based on the *ASHRAE 93-77* test results. The SEREF Labeling and Certification Committee developed procedures for the random selection of collectors, handling of confidential test information, and labeling of collectors.

DOE provided validation of the ISCT ratings and underwrote the costs associated with the collector tests and preparation of a manual of test results. The test program also provided a resource for accurate performance data on solar collectors. In early 1978 SEREF undertook the provisional accreditation of laboratories using the LITE test equipment. Each of eight approved laboratories tested collector models randomly selected from 108 manufacturers. The Solar Energy Research Institute (SERI) checked the data for completeness and compiled a catalog giving performance results of each collector (Kirkpatrick 1983). These data later provided the basis for the development of the SRCC program (see section 14.3.2.3). The data showed that the "actual range of interlaboratory variation was found to be within the predicted limits of accuracy of the procedure." Additionally, testing laboratories with LITE was highly useful in "identifying correctable instrumentation or procedural variance" in most of the laboratories.

14.3.2.3   Solar Rating and Certification Corporation

14.3.2.3.1   Attempts to Amalgamate the SEIA and ARI Programs   SEIA attempted to combine its and the SEREF RCL programs with that of the Air-Conditioning and Refrigeration Institute (ARI), primarily because DOE wanted to replace the private sector RCL programs with a government-run program. The DOE program was modeled after the energy efficiency rating (EER) then being developed for application to conventional heating and cooling systems and also supported by the National Bureau of Standards (May, Hill, and Streed 1979; Hill, May, and Streed 1979). However, the government's E.E.R.-based RCL program did not come to fruition largely because of the substantial progress already made in the SEIA, SEREF, and SRCC RCL programs. Also, the assistant secretary for policy, as well as DOE's Solar Branch, favored an industry program (Butt 1979b).

In early July 1979 SEIA and ARI reached accord on many aspects of a joint RCL program. The two organizations envisioned a program administered from a single office with joint governance, the accreditation of multiple laboratories, the use of the ARI fee structure, random selection, and the use of a label with both ARI and SEIA logos. Rating thermal performance of collectors was the most difficult issue in reaching an acceptable methodology. Each organization had approached the rating from different technical perspectives. While both rating methodologies were based on *ASHRAE 93-1977* thermal performance testing, they differed significantly in how the ratings would be determined and reported. The SEIA/SEREF ratings were based on an all-day energy production computation for a specific reference climate and solar day. The ARI rating approach, on the other hand, was based on the instantaneous efficiency taken from *ASHRAE 93-1977*.

The Air-Conditioning and Refrigeration Institute favored a single, industry-approved laboratory to perform all rating-related performance testing. ARI felt that a single test laboratory was preferable because (1) all collectors would be subjected to the same climatic conditions in thermal performance tests, and (2) the same systematic errors would normalize the comparative rating values of competitive solar collectors and solar energy systems. ARI further wanted to reduce test-to-test variability by confining thermal performance tests to the use of an indoor solar simulator.

The SEIA Rating Subcommittee approved the Joint ARI/SEIA Accords in October 1979 (Mather 1979). The SEIA Executive Committee approved the accords as a technical document in October 1979, subject to reservations pertaining to administrative details, and to ARI's insistence that no more than two laboratories be accredited for testing in support of the joint program (SEIA 1979).

Negotiations between the SEIA RCL program manager and committee chairmen, on the one hand, and the ARI Solar Products Committee, on the other, continued over the next nine months with little progress in resolving details, particularly those involving the rating methodology and the number of laboratories to be accredited. In late 1980 the parties broke off negotiations. ARI concentrated on setting up their own solar product RCL program (see section 14.3.2.4). With ISCC representation on the board of directors, SRCC became the major actor in solar collector and solar energy system rating, certification, and labeling until 31 December 1985, when Congress declined to extend the tax credits for solar technologies, causing the catastrophic downturn of the solar industry that many had predicted (Zerlaut 1985).

14.3.2.3.2   Interstate Solar Coordinating Council   In early 1979 representatives of the Florida and California solar certification programs became concerned about the incompatible features of the various standards and certification programs across the country. On 2 April 1979 the Department of Energy's Office of Solar Applications held a meeting with representatives of a number of states and DOE personnel to discuss the concept of a national program for rating, certifying, and labeling solar collectors. At this meeting, the Office of Solar Applications distributed a "Statement of Cooperative Intent: Proposed Voluntary Framework of Federal, State, and Local Agencies to Provide for Coordination and Reciprocity of Testing and Certification Programs for Solar Energy Equipment" (Butt 1979c). The statement, prepared in February 1979, reviewed the status of both federal and state initiatives to force solar product testing and certification to provide consumer protection, as well as the areas in which federal support had been given for the accelerated development of consensus solar standards for product and materials testing; it provided an initial blueprint for a cooperative enterprise between the states that had, or were intending to develop, state-sponsored product testing and certification programs (DOE 1979).

The Department of Energy endorsed the desire of the states to develop a single consensus process for testing, evaluating, certifying, and rating solar equipment. As a result of this initial meeting between the states and DOE, DOE agreed to support the states' standards development activities. The Florida Solar Energy Center was awarded funding to administer the Solar Public Interest Coordinating Committee, which later became the Interstate Solar Coordination Council (ISCC), and to provide travel funds for state solar coordinators to attend standards development workshops.

ISCC representatives met frequently for the next year developing the procedures for solar collector certification. The final version of the consensus documents reduced the cost of testing and certification from previous procedures by requiring only one thermal performance test, with a teardown and inspection of the collector after testing was concluded. ISCC took this concept to the industry certification organizations, and endorsed the idea of a single RCL program jointly administered by SEIA and ARI. Following the failure of the joint accord by SEIA and ARI, ISCC and SEIA agreed to form the Solar Rating and Certification Corporation (SRCC).

At the time of the demise of the federal solar tax credits in early 1986, forty-two states were represented on the council, along with representatives from Puerto Rico, the Bonneville Power Administration, and the Tennessee Valley Authority (SRCC 1986a). Because the major task of ISCC was to coordinate state-sponsored solar testing, rating, certification, and labeling programs and to work out reciprocity protocols among the states, the council became a major force in the development of the final versions of the rating methods.

14.3.2.3.3 Organization of SRCC The SRCC was incorporated in Washington, D.C., in 1980 as an independent, nonprofit organization to certify and rate the reliability and thermal performance of solar energy collectors, solar energy systems, and other solar equipment. It was formed by the Solar Energy Industries Association and the Interstate Solar Coordination Council, and its initial board of directors was drawn largely from these two organizations. SRCC's board of directors currently comprises technical and solar business experts from both the private and public sectors; the private sector members are elected by the SRCC participants, and the public sector members are elected by ISCC. SRCC is financed partially through fees collected from the manufacturer at the time of application and licensing fees collected annually based on the sales of labeled units, and partially by DOE.

As of 1 December 1986, SRCC had certified and rated 13 unglazed, swimming-pool-type collectors; 150 glazed, flat-plate, liquid-type collectors from 36 manufacturers; 42 air collectors from 26 manufacturers; 3 linear tracking concentrator collectors from a single manufacturer; 1 boiling-liquid collector with an integral heat exchanger; and 38 solar domestic hot water systems (SRCC 1986b). The water heaters included 10 forced-circulation systems, 9 integral collector storage systems, and 19 thermosyphon systems (SRCC 1987). (The numbers provided represent the collectors listed in the 1986 SRCC directory and do not include collectors no longer certified.)

14.3.2.3.4 SRCC Solar Collector Certification and Rating Program

SRCC evaluates and accredits independent laboratories to perform thermal performance tests. The test methods employed by the laboratories are all based on consensus standards: *ASHRAE 93-1977* for glazed, hot-liquid collectors, *ASHRAE 96-1980* for unglazed, swimming-pool-type collectors, and *ASHRAE 95-1981* for solar hot water heating systems (see section 14.4.3). The manufacturer initiates the certification process by selecting an accredited laboratory and applying to SRCC for certification. Upon acceptance of the manufacturer's application, an SRCC representative randomly selects from the production line a unit of each model to be tested and certified.

The manufacturer pays the shipping costs and test fees. Once SRCC is informed by the laboratory that the equipment meets the structural integrity and reliability requirements, SRCC certifies the unit and calculates the rating values. A label containing basic product information and thermal performance ratings is then required to be affixed to each and every unit distributed under that brand name and model number (SRCC 1981b).

The operating guidelines also provide for innovative solar equipment, certification of similar models, qualification of prior testing under either the Florida Solar Energy Center (FSEC) or California TIPSE programs, private branding, procedures for challenging the ratings, and SRCC-invoked revocation of certification (SRCC 1981b).

14.3.2.3.5 SRCC Rating Methodologies for Collectors and Systems

Rating Documents   Solar collectors to be rated are first tested in accordance with *ASHRAE 93-1977* or *96-1980*, for glazed and unglazed collectors, respectively (ASHRAE 1977, 1980). However, they are rated in accordance with the SRCC rating method *RM-1* (SRCC 1981a) for non-tracking, flat-plate collectors, and *RM-2* (SRCC 1982) for solar concentrating collectors. System testing and rating procedures performed in support of the SRCC solar system RCL program are based on *ASHRAE 95-1981*, and ratings are calculated in accordance with *SRCC 200-82* (SRCC 1983).

Flat-Plate Collectors Collectors are first given a pressure test, then mounted outside for exposure to the weather for thirty days of minimum solar conditions. The collector is next tested for tolerance to thermal stresses and again pressure-tested. If no problems occur, the collector is tested for thermal performance, the most expensive step in the test sequence. Finally, the collector is disassembled and inspected for water leakage or other degradation in the solar collector components.

The ASHRAE performance curve is applied to three "standard days" specified by SRCC, representing typical winter and summer solstices and the equinoxes in the United States. In addition, the inlet temperatures to the collectors are simulated to represent various applications of solar heating systems—swimming pools, water heating and space heating. The resulting daily energy production for each of the conditions is presented in a matrix on the solar collector certification label.

With differences dictated by the optical design compared to flat-plate collectors, concentrating, and therefore tracking solar collectors are tested and rated in much the same manner as flat-plate units.

Domestic Hot Water Systems Solar water heating systems are required to be tested as a unit under specified solar irradiance conditions. Consequently, the solar testing must typically be conducted using a solar simulator. However, if the collectors to be used in the system have been tested in accordance with *ASHRAE 93-1977*, the remaining components can be tested indoors using an electric resistance heater to provide the heat that would have been provided by the solar panels. The ratings of solar water heating systems present the amount of heat provided by the system on a "standard" day with a load profile specified by SRCC. The ratings can therefore be used with confidence to compare systems but offer little assistance in predicting long-term (annual) performance of a system under varying load conditions.

14.3.2.3.6 SRCC Labeling System Photographs of the SRCC certification award and the rating table are shown in figures 14.2 and 14.3. The manufacturer must affix the label to each production unit of the certified collector model. Certified equipment must be retested, recertified, and rerated every three years for as long as the manufacturer remains in the program and offers that model for sale (SRCC 1981a). A photograph of the label required for solar hot water systems certification is given in figure 14.4.

COLLECTOR RATING NUMBERS								
METRIC (SI Units)					ENGLISH (Inch-Pound Units)			
Megajoules Per Panel Per Day					Thousands of Btu Per Panel Per Day			
CATE-GORY (Ti - Ta)	CLEAR DAY 23MJ/m².d	MILDLY CLOUDY DAY 17MJ/m².d	CLOUDY DAY 11MJ/m².d		CATE-GORY (Ti - Ta)	CLEAR DAY 2000 Btu/ft².d	MILDLY CLOUDY DAY 1500 Btu/ft².d	CLOUDY DAY 1000 Btu/ft².d
A (-5°C)	83	66	49		A (-9°F)	78	62	46
B (5°C)	53	37	20		B (9°F)	51	35	19
C (20°C)	20	7	0	SRCC Standard 100-81	C (36°F)	19	7	0
D (50°C)	-	-	-		D (90°F)	-	-	-
E (80°C)	-	-	-		E (144°F)	-	-	-

Figure 14.2
SRCC rating table.

SRCC relaxed its retest policy following elimination of the federal tax credits in order to prevent a mass exodus from the program. Nonetheless, there was a significant reduction in testing activity causing major laboratories, such as DSET Laboratories, Inc., and Wyle Laboratories, Inc., to terminate their accreditation. DSET Laboratories, Inc., once the major solar collector test laboratory in the world, officially ceased accepting solar collector tests in late 1989. As of the end of 1990, only two laboratories remain accredited by SRCC to test solar collectors: the Florida Solar Energy Center, Cape Canaveral, Florida, and the National Solar Test Facility (of Canada) operated by ORTEC International near Toronto, Canada.

14.3.2.4 Air-Conditioning and Refrigeration Institute

In January 1978 the Air-Conditioning and Refrigeration Institute (ARI) announced the "initiation of an industry voluntary certification program for thermal performance of solar collectors." This was ARI's twelfth product certification program and was prompted by several large members of ARI who were manufacturers of both air-conditioning equipment and solar collectors. The two major administrative aspects of the ARI program that distinguished it from SRCC were (1) its requirement that all models offered for sale must be certified (to preclude an unethical manufacturer from selling the same collector as a different model to avoid certification costs and licensing fees); and (2) its reliance on a single laboratory for performance testing (to preclude differences in laboratory-

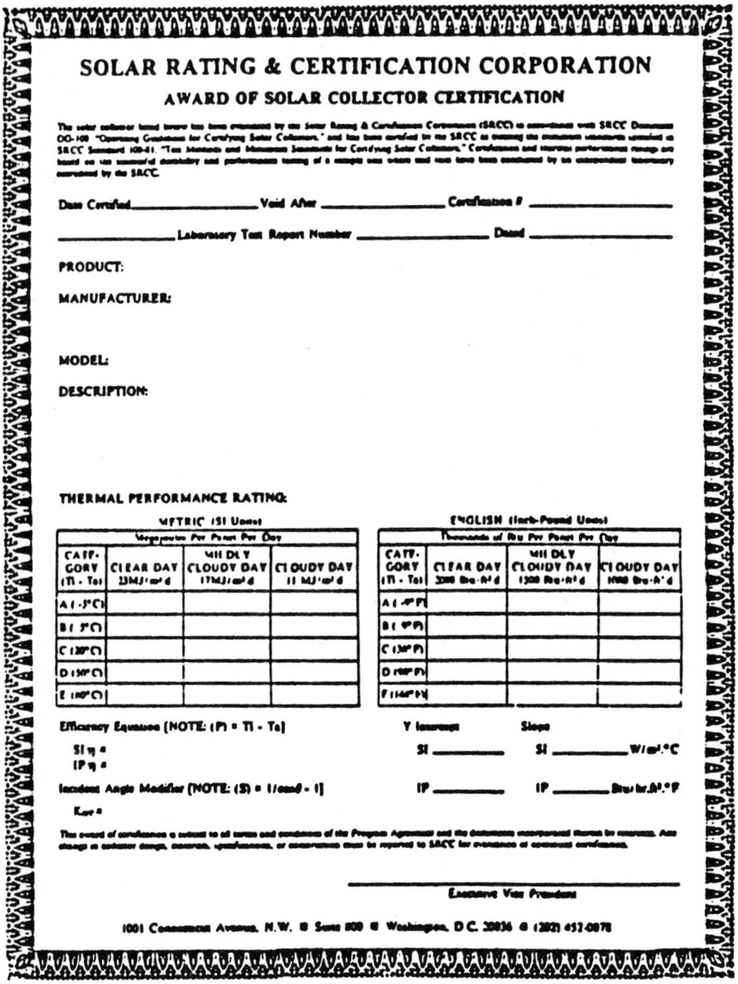

SOLAR RATING & CERTIFICATION CORPORATION

AWARD OF SOLAR COLLECTOR CERTIFICATION

PRODUCT:

MANUFACTURER:

MODEL:

DESCRIPTION:

THERMAL PERFORMANCE RATING:

Figure 14.3
SRCC Certification Award.

	RATING CATEGORY	TERM	METRIC (SI Units)	ENGLISH (IP Units)
	SOLAR ENERGY DELIVERED	Q_{SAV}	24.6 MJ/d	23300 Btu/day *
	FRACTIONAL ENERGY SAVINGS (Electric)	FES(E)	50 %	50 %
	FRACTIONAL ENERGY SAVINGS (Gas)	FES(G)	%	%
	Heat Loss Coefficient	L	5.2 W/°C	9.9 Btu/hr.°F
	Auxiliary Energy Capacity	Q_{CAP}	18.7 MJ	17700 Btu
	Auxiliary Energy Consumption	Q_{AUX}	21.9 MJ/d	20800 Btu/day
SOLAR WATER	Parasitic Energy Consumption	Q_{PAR}	MJ/d	Btu/day
HEATING SYSTEM STANDARD	System Set Temperature	T_{SET}	48.9 °C	120 °F
200-82	Standard Test Load	Q_{DL}	42.3 MJ/d	40119 Btu/day
	Delivered Test Load	Q_{DEL}	43.1 MJ/d	40800 Btu/day

Figure 14.4
SRCC solar water heating system standard.

specific systematic errors and climatic variables). Also, unlike SRCC, ARI required that 30% of all models be retested each year.

While both the ARI and SEIA rating methods were based on the results of testing according to *ASHRAE 93-1977*, there were, in fact, significant differences. ARI selected as the rating numbers specific points on the regression plot of the so-called Hottel-Whillier efficiency equation (Hottel and Whillier 1958; Hottel and Woertz 1942; Whillier and Richards 1961), while SRCC calculated all-day performance for "standard days" under "standard conditions" using the data from the *ASHRAE 93-1977* tests. ARI's reliability tests were, like the SRCC requirements, selected from those presented by NBS in *NBSIR 73-1305A* (Waksman, Streed, et al. 1978). The major difference between the two certification programs in this respect is the thirty-day pretest stagnation exposure in the test sequence that is required by SRCC but not ARI.

In 1981 ARI chose DSET Laboratories, Inc., as its test laboratory; using *ARI 910-82, 920-82*, and *930-82*, ARI began certifying and rating solar collectors and solar hot water systems (ARI 1981a, 1981b, 1982). In the period from initiation through 30 June 1984, ARI certified 39 collectors from 24 manufacturers.

14.3.3   State and Other Federal Programs

14.3.3.1   Tennessee Valley Authority Residential Hot Water Project
The Tennessee Valley Authority (TVA) is a corporate instrumentality of the United States and was organized pursuant to the Tennessee Valley

Authority Act of 1933. In 1982 TVA identified manufacturers of solar domestic hot water systems (SDHW) in its power service area and established a procedure for loans on residential SDHW systems throughout that area.

Through electric utility newsletters and advertising methods, TVA and its service area manufacturers and distributors notified area homeowners of the availability of approved SDHW systems. Interested homeowners were provided a list of manufacturers and contractors participating in the project. A homeowner could select a SDHW system, provide for its installation, and arrange with TVA for an inspection of the system upon installation. The local participating TVA power distributors, upon approval of an installation, provided for payment to the contractor, installer, or the homeowner (for "do-it-yourselfers") the as-installed cost of the SDHW system (for amounts up to $3,400). The homeowner was required to repay the funds by a loan "secured through a purchase money security interest" on the SDHW system.

Manufacturers and distributors were required to submit their proposed SDHW system to a TVA-approved test laboratory for system thermal performance testing. DSET Laboratories, Inc., and Wyle Laboratories, Inc. (Huntsville, Alabama), were selected by TVA for this certification program. Because no laboratory had an approved solar simulator at that time, the test method chosen was the *thermal simulation* method of *ASHRAE 95-1981* (ASHRAE 1981). However, the manufacturer, or distributor, was required to have the solar hot water system collectors tested for thermal performance in accordance with *ASHRAE 93-1977* (ASHRAE 1977) prior to submittal to either DSET or Wyle, and the test laboratory for this phase could have been any one of the laboratories approved for the HUD Hot Water Cycles (HUD 1978). It is noted that DSET had previously published results of testing using its thermal simulation test facility (Geisheker, Putman and Bard, 1981; Bard and Rupp 1981). In testing approximately 65 SDHW systems for the TVA program, DSET and Whyle learned much that became of importance in the subsequent SRCC solar hot water systems rating, certification, and labeling program (Bard 1984). DSET Laboratories went on to construct a large solar simulator for systems testing for the SRCC programs. In fact, once solar simulator testing became available, thermal simulation was seldom used in thermal performance testing of hot water systems.

14.3.3.2 California Energy Commission Inspection Program
The California Testing and Inspection Program for Solar Equipment (TIPSE) was inaugurated in early 1978 to "set testing standards for manufacturers of flat plate glazed solar collectors" (CEC 1980). The TIPSE program was authorized by California Assembly Bill 1512 in September 1977.

Collectors were randomly selected from participating manufacturers by TIPSE representatives and submitted to one of seven laboratories accredited by the State of California for *ASHRAE 93-1977* thermal performance testing. In addition, collectors were evaluated for durability by performing thirty-day stagnation, thermal shock/cold fill, thermal shock/water spray, and static pressure tests. *ASHRAE 93-1977* thermal performance tests were performed before and after the durability tests, and collectors whose thermal efficiency decreased by more than 10% were not certified for the California state tax incentives. The ratings and efficiency were contained on the California Energy Commission (CEC) label (see figure 14.5) that the manufacturer then affixed to each collector sold.

A total of 387 solar collectors, representing 136 different solar manufacturers from across the United States, were certified by the California Energy Commission. The California TIPSE program was terminated on 1 March 1982 as the ARI and SRCC solar collector RCL programs became a reality.

14.3.3.3 Florida Solar Certification and the Florida Solar Energy Center
The Florida Solar Energy Center (FSEC), located at Cape Canaveral, was created by the Florida Legislature in 1975. In October 1976 the Florida

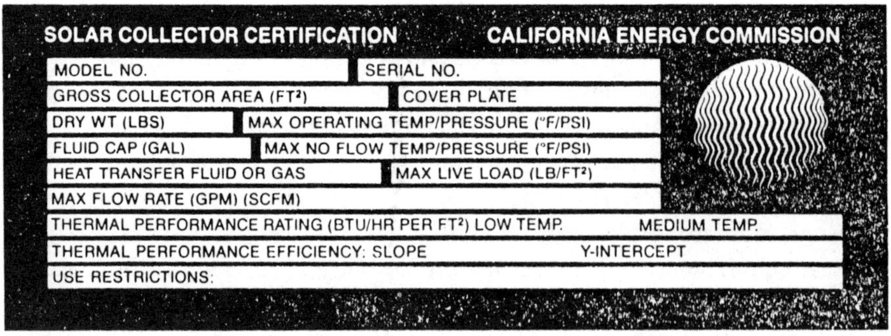
SOLAR COLLECTOR CERTIFICATION CALIFORNIA ENERGY COMMISSION
MODEL NO. SERIAL NO.
GROSS COLLECTOR AREA (FT²) COVER PLATE
DRY WT (LBS) MAX OPERATING TEMP/PRESSURE (°F/PSI)
FLUID CAP (GAL) MAX NO FLOW TEMP/PRESSURE (°F/PSI)
HEAT TRANSFER FLUID OR GAS MAX LIVE LOAD (LB/FT²)
MAX FLOW RATE (GPM) (SCFM)
THERMAL PERFORMANCE RATING (BTU/HR PER FT²) LOW TEMP. MEDIUM TEMP.
THERMAL PERFORMANCE EFFICIENCY: SLOPE Y-INTERCEPT
USE RESTRICTIONS:

Figure 14.5
California Energy Commission label.

Solar Energy Standards Act was passed, and FSEC was directed to (1) develop standards for solar energy collectors and systems sold or manu- factured in Florida, (2) establish criteria for determining the thermal per- formance of such equipment, and (3) establish and maintain a testing facility for evaluating solar collector durability and thermal performance (FSEC 1980b). In 1978 an amendment required that all solar energy sys- tems sold or manufactured in Florida after 1 January 1980 meet the FSEC standards and be so labeled.

The operations document, FSEC 77-6-R80, was based "almost in its entirety" on operating guidelines (*ISCC 80-2*) developed by the Interstate Solar Coordinating Council (ISCC 1980b). The FSEC administrative guidelines provide for levying a testing fee, accreditation of "competitor" test laboratories, and for a testing sequence similar to that of the Cal- ifornia TIPSE program (section 14.3.3.2), except that only a postexposure *ASHRAE 93-1977* thermal performance test is required. The durability and reliability testing required by FSEC are contained in a separate document (FSEC 1980a), also taken largely from ISCC (ISCC 1980a). Collectors rated, certified and labeled by SRCC are automatically certified by FSEC.

As of the end of 1990, FSEC had certified 1,025 collectors, based on 521 thermal performance tests from a total of 381 manufacturers. Of these, 227 tests were performed by laboratories other than FSEC, and 251 manufacturers were from other states (FSEC 1990). Although FSEC has not constructed a solar simulator, they have performed extensive R&D on outdoor test methods for determining the thermal performance of SDHW systems (Chandra and Khattar 1980).

14.4   U.S. Standards Organization

14.4.1   Solar Energy and the Consensus Standards Process

14.4.1.1   Early Problems Confronting the Standards Community

As solar energy utilization gained momentum in the mid-1970s, the need for solar standards and standardization became increasingly apparent to government, consumer groups, the standards writing organizations, and industry itself. But as awareness of the need for standards increased, so did confusion between the definition of *physical* standards and *paper* standards, between performance standards and specifications, between certification and rating, between testing standards and rating standards, and between testing to performance standards and to pass-fail criteria.

One segment of the industry wanted testing and rating standards, the other did not. The latter cited unfairness, barriers to innovation, and anticompetitiveness among their reasons. The more conservative members of the segments opposing standards at times sounded like the more radical of the consumer action groups: citing standards as potential barriers to innovation. And certain consumer advocates voiced support for pass-fail testing to keep undesirable products off the market, a process that others argued would have had the opposite affect by causing industry to design products to just pass, without regard to long-term reliability or adequate performance.

The solar market was not, in the mid- to late 1970s, a naturally evolving market; its growth was being forced by all of the actors noted above, but most notably by government and the industry itself. Thus the normal processes of natural development of selling techniques, test standards, product specifications, warranty policies, and product certification schemes, were all greatly accelerated. In retrospect, it is no longer as perplexing to those of us who have worked in standards writing for many years that this fledgling and unwieldy industry was so confused at first by the many new concepts associated with standardization. Solar manufacturers were told they must participate in the process, but they did not understand why it was important or how they would use it. And I am now amazed at how quickly some of the loudest detractors of standards and certification learned the meaning of "due process" and "balanced committees," that "consensus doesn't mean unanimity," and how to "compromise for the common good."

This section will detail the activities of the more important of these standards organizations, who they represented, how they worked, and the material they produced.

14.4.1.2   Consensus Standards Development

Courts today generally hold that the "rule of reason" shall be applied to the restraint of trade associated with standards development, in which case the "unlawfulness of a particular standard depends solely on whether it *unreasonably* restrains trade" (Hoffman 1978). Hence, it is necessary that a consensus process be employed in the arguments pertaining to the standard, and that due process be applied to the consideration of opposing arguments and viewpoints.

The Antitrust Division of the Department of Justice has recommended procedures to ensure due process that include the following (S.825 Sec.102(b)(1) and (c)(1)):

• Formulation of the proposed standard should include a large number of "balanced" interests among producers (manufacturers), users, consumers, and general interest.

• Before publication, the formulated standard must be given the widest possible circulation among persons and organizations affected by the standard.

• Adequate notice of the proposed standard shall be given to all affected parties.

• All comments must be considered and proper weight given to them, with negatives requiring special care and documentation.

• The right of appeal shall be a critical element of the standards development process, and an independent and competent appeals body shall be convened as required (Hoffman 1978).

It was against these concepts of due process, covering balance, notice, consideration of all opinions, resolution of negatives, and rights of appeal, that the activities in solar standards development began in the mid-1970s. Although often foreign to the large group of experts assembled to prepare solar standards, these concepts represented the core and essence of the standards bodies that formed the framework within which we worked.

14.4.2   American National Standards Institute

14.4.2.1   Organization
The American National Standards Institute (ANSI) was founded in 1918 by five engineering societies and the U.S. Departments of War, Navy, and Commerce to end duplication and conflict in the voluntary standards community at the time, which, seventy years later, remains one of ANSI's chief functions and goals (ANSI 1988). Today, ANSI is a federation of more than 1,000 large and small businesses and more than 180 voluntary organizations, ranging from other standards organizations to trade organizations, organized labor, and consumer groups. ANSI's principal responsibilities are (1) to approve American National Standards, (2) to act as the coordinator for voluntary standards activities in the United States, (3) to

serve as the U.S. member of the International Organization for Standardization (ISO) and the International Electrotechnical Commission (IEC),
(4) to provide forums for government-industry cooperation, and (5) to act
as the clearinghouse and resource for information dealing with national
and international standards.

ANSI has developed procedures to ensure "due process" and criteria
for approval of American National Standards. In this respect, ANSI accredits standards writing organizations and the standards committees of
technical and trade organizations that meet its criteria for standards development, as well as publishing procedures for development of American
National Standards (ANSI 1982). ANSI defines consensus as "substantial
agreement reached by concerned interests according to the judgment of a
duly appointed authority, after a concerted attempt at resolving differences. Consensus implies much more than the concept of a simple majority, but it does not necessarily imply unanimity" (ANSI 1982).

14.4.2.2 ANSI Steering Committee on Solar Energy Standards Development

ANSI formed the Steering Committee on Solar Energy Standards Development (SCSESD) in January 1976. The committee was an outgrowth
of a meeting on 28 April 1975 between the Federal Energy Administration and the American National Standards Institute—with representatives from ASTM, American Society of Mechanical Engineers (ASME),
the Institute of Electrical and Electronic Engineers (IEEE), and the electric power industry—to discuss the need for standards development in the
field of solar energy (Landis 1975). As a direct result of this meeting,
ANSI asked ASTM to convene a conference in the fall of 1975 on solar
energy standards needs. ANSI subsequently formed the Solar Standards
Oversight Committee (SSOC) in response to the needs expressed at the
ASTM conference.

By the summer of 1978 the ANSI SSOC, which was funded by DOE,
had some twenty-two organizational members. These included federal
agencies such as DOE, HUD, NBS, NASA, and the General Services
Administration (GSA); standards-writing organizations such as the
American Society for Testing and Materials (ASTM), the American
Society for Heating, Refrigeration, and Air-Conditioning Engineers
(ASHRAE), the American Society of Mechanical Engineers (ASME);
industry organizations such as the Solar Energy Industries Association

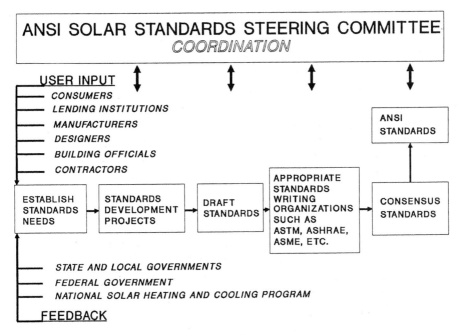

Figure 14.6
Organization of solar standards development in the United States.

(SEIA), the Air-Conditioning and Refrigeration Institute (ARI), the American Gas Association (AGA), the Hydronics Institute, American Institute of Architects (AIA), Underwriters Laboratories, and many other organizations.

The purpose of the ANSI Steering Committee on Solar Standards Development (SCSESD) was to provide information on needs and priorities, the existence of expertise and resources, and progress of standardization within member organizations. The committee defined its role in liaison with the National Bureau of Standards, which had received funding from DOE to develop a national plan for the development of solar energy standards. This role is pictured schematically in figure 14.6, which shows the relationship between the various inputs, standards-writing organizations, and ANSI. These organizational relationships were proposed by NBS in NBSIR 78-1143A (Waksman, Pielert, et al. 1978) but were never fully consummated. The huge matrix of standards and codes ultimately collapsed of its own weight. Moreover, ANSI SCSESD became so embroiled in the controversy surrounding the development of a national program for rating, certifying, and labeling solar collector products that little time was left to devote to other issues.

Nonetheless, the ANSI Steering Committee on Solar Energy Standards Development (1) provided information on the status of standards development among the member organizations; in this manner (2) coordinated standards development within the member organizations; and (3) became the impetus for, and architect of, the development of the formal U.S. technical advisory groups (TAGs) for international standards activities in both solar heating and cooling and photovoltaic technologies. These international standards organizations and their solar activities are discussed in section 14.5.

Unfortunately, with respect to the development of a national solar collector certification and rating program, ANSI's role shifted from coordinator to mediator, and the steering committee failed in its new role. The efforts of the NBS to institute a National Voluntary Laboratory Accreditation Program (NAVLAP) and an Energy Efficiency Rating (EER) RCL program were both brought to the table in the ANSI solar standards steering committee, where industry differences were actually exacerbated by the prolonged attack upon the SEIA/SEREF industry programs. As a result, SEIA and ARI were forced to seek a resolution to the controversy outside of the committee (see section 14.3.2.3.1).

The possibility that an eventual resolution of the RCL imbroglio could be effected within the context of the ANSI solar standards steering committee appeared extremely remote following the committee's June 1979 meeting (ANSI 1979b). Arguing that, if the industry could not reach agreement on a single national RCL program, the federal government would "do it for us," I persuaded Sheldon H. Butt, president of SEIA, and G. R. "Munk" Munger, president of ARI, to allow me an opportunity to work out a compromise rating, certification, and labeling program that did not compromise either the technical or the procedural intents of the SEIA and ARI programs. Both men agreed, and Munger appointed Alvin Newton to represent ARI in these negotiations, which resulted in preliminary agreements (Zerlaut 1979; ANSI 1979c) but which, after nearly two years of effort, were halted as a result of failure to reach an agreement on one or two issues (see section 14.3.2.3.1).

In early 1979 George T. Pytlinski, director of the New Mexico Solar Energy Institute, proposed that ANSI establish international standards activities in solar energy in the United States (Pytlinski 1979). The matter was referred to the ANSI solar standards steering committee for a determination. It was pointed out in this referral that U.S. participation in an International Standards Organization (ISO) activity in solar energy would carry significant financial responsibilities. This issue was referred to the member organizations for further consideration at a future meeting of the SCSESD (ANSI 1979a). SEIA went on record as opposing the U.S. initiation of international activity in solar heating and cooling technologies at that time (Butt 1979d). SEIA pointed out that the United States, without strong support from the solar industry, would be ill advised to initiate ISO activity; indeed, that such activity might have an adverse effect on the development of domestic standards in the United States.

As chairman of ASTM Committee E44 on Solar Energy Conversion (see section 14.4.4.3), a member of the board of directors of SEIA, and an experienced member of the U.S. delegation to ISO Technical Committee ISO/TC61, I presented to the ANSI solar standards steering committee a synopsis of the benefits of international standardization. As a result, an ad hoc committee on international activities was appointed to develop positions for ANSI in solar heating and cooling, and photovoltaic technologies (ANSI 1979b). It was at the first meeting that International Energy Agency (IEA) activities were first brought to the attention of the ANSI Solar Standards Steering Committee. These activities were to later have considerable bearing on U.S. delegates chosen to represent the United States on the ISO/TC180 Committee on Solar Energy (see section 14.5.2.2).

Subsequently, DOE indicated its willingness to amend the ANSI contract that established the Solar Standards Steering Committee to include funding for the development of two U.S. technical advisory groups (TAGs). These groups would represent the United States in any ISO or IEC standards activities. DOE expected the ANSI solar standards steering committee to positively commit to supporting an international solar standards involvement (ANSI 1979c).

In June 1979 the ANSI Solar Standards Steering Committee added photovoltaics to its purview, effectively creating two thrusts, one for solar heating and cooling technologies within ISO, and one in photovoltaic technologies within IEC (ANSI 1979b).

During 1980 the steering committee discontinued discussions of the so-called ARI-SEIA/SEREF accords because SRCC was becoming a reality, and the accords were doomed to failure in any case. The committee concentrated on funding and staffing issues associated with U.S. participation in international standards, particularly ISO/TC 180 on solar heating. I chaired the task group charged with developing the protocols and operational procedures for a U.S. Technical Advisory Group for TC 180, and for selection and preparation of a delegation to the organizational meeting held in Sydney in the spring of 1981.

14.4.3 American Society of Heating, Refrigeration, and Air-Conditioning Engineers

14.4.3.1 Organization

The American Society of Heating, Refrigeration, and Air-Conditioning Engineers (ASHRAE) is a professional and technical engineering society that undertakes the development and publication of voluntary test standards and recommended practices in the fields of heating, ventilation, air-conditioning, and refrigeration. ASHRAE is a nonprofit organization of some 45,000 members from throughout the world and is accredited by ANSI as a standards-writing organization.

ASHRAE publishes the annual volumes of the *ASHRAE Handbooks*, the *ASHRAE Journal*, and the *ASHRAE Composite Indexes of Technical Articles*. Both its handbooks and voluntary consensus standards are developed under the auspices of ASHRAE technical committees with scientific and technical jurisdiction for specific areas of heating, ventilating, refrigerating, and air-conditioning. New and revised standards are approved by the ASHRAE Board of Directors.

14.4.3.2 ASHRAE Solar Standards

*14.4.3.2.1 **ASHRAE 93-1977** (First Performance Test Standard)* In early 1974 ASHRAE formed the Technical Committee on Solar Energy Applications, TC6.7. Then in 1975, in response to a proposal from TC6.7, DOE gave ASHRAE funding to develop a standard method of testing solar collectors based on *NBSIR-74-635* (Hill and Kusuda 1974). The major purposes of the ASHRAE effort were to (1) accelerate the usually slow process of consensus standards development; and (2) to bring to the effort the nation's most experienced practitioners of thermal performance testing of solar collectors. TC6.7 formed Committee 93 and asked John I. Yellott of Arizona State University to chair the committee.

Committee 93 met throughout 1975 to develop a testing standard that dealt with the many diverse opinions as how to make reliably accurate temperature and mass flow measurements of the transfer fluid and of solar irradiance. The use of a solar simulator by NASA's Lewis Research Center (Simon 1975) drew a great deal of controversy, as did the question of determining the area of a collector "for rating purposes." Also, the question of using thermopiles as opposed to resistance thermometry consumed a great deal of time.

During this process, because certain federal procurements required that a so-called standard thermal performance test be performed, a number of laboratories began testing with draft versions of the test standard, designated 93-P. Even though this was contrary to good standards development practice, and officially frowned upon by both ASHRAE and Committee 93, such testing brought the committee added experience in a number of controversial or unresolved areas.

ASHRAE 93 was approved by the ASHRAE Standards Committee in February 1966 and by the ANSI Board of Directors as *ASHRAE 93-1977* in February 1977 (ASHRAE 1977).

In 1985, for the purposes of assuring "acceleration and consensus," ASHRAE reconvened Committee 93 as Committee 93R to revise *93-1977*, under the chairmanship of Byard D. Wood of Arizona State University and with DOE funding. This was done to include new information resulting from several years of experience using *93-1977*, and to provide guidance for the use of *ASHRAE 93* in thermal performance testing using indoor solar simulators (primarily in support of testing for the SRCC rating, certification, and labeling program). *ASHRAE 93-1986* was approved and published in 1986 (ASHRAE 1986).

14.4.3.2.2 ASHRAE 95-1981 (Systems Performance Test Method)
Under the chairmanship of James E. Hill of NBS, ASHRAE convened Committee 95 in February 1977 to develop a test standard for determining the thermal performance of packaged solar hot water heating systems. After more than two years of deliberations, Committee 95 published a draft for public review (*ASHRAE 95-P*). A second draft was published in April 1980. The final draft standard was approved by the ASHRAE Board of Directors as *ASHRAE 95-1981* in June 1981 (ASHRAE 1981). *ASHRAE 95-1981* was also discussed in the section dealing with the SRCC solar energy systems rating procedures (see 14.3.2.3.5.).

14.4.3.2.3 **ASHRAE 96-1980** *(Performance Testing of Swimming Pool Collectors)* As a result of the dissatisfaction with the results of swimming pool solar collector tests in accordance with *ASHRAE 93-1977*, ASHRAE Committee 96 was formed to develop a thermal performance standard pertaining specifically to the outdoor testing of unglazed solar collectors. The work of this committee, chaired by Robert E. Cook of the National Swimming Pool Institute (NSPI), was not funded by DOE.

The revisions in *ASHRAE 93-1977* procedures required for low-temperature collector testing were completed and a draft standard was submitted to the ASHRAE Standards Committee in October 1979. *ASHRAE 96-1980* was approved by the ASHRAE Board of Directors in January 1980. However, it has been subsequently argued by the test laboratories that the stringent requirements for temperature difference and wind speed measurements are unrealistic and that commercial temperature and air velocity measurement devices meeting these requirements are either not available or are prohibitively costly.

14.4.4 American Society for Testing and Materials

14.4.4.1 Organization

The American Society for Testing and Materials (ASTM) was founded in 1898 as a scientific and technical organization devoted to the development of standards relating to the performance of materials, products, and systems. It "is the world's largest source of voluntary consensus standards" (ASTM 1988). Committees and subcommittees operate under procedures that ensure balanced representation between producers, users, general interest, and consumer participants (ASTM 1988).

14.4.4.2 Operation of ASTM Committees

ASTM Committees are usually comprised of a number of subcommittees whose scopes encompass the major technical divisions of work anticipated. Committee chairmen are selected by ballot of all committee members; subcommittee chairmen are usually appointed by the chairman. In most subcommittees, the structure is further defined by forming a number of task groups for the express purpose of writing specific standards.

Standards are developed in accordance with generally accepted procedures, described in figure 14.7. However, regulations governing ASTM technical committees include detailed requirements for meeting consensus, due process, and general operation of technical committees, task

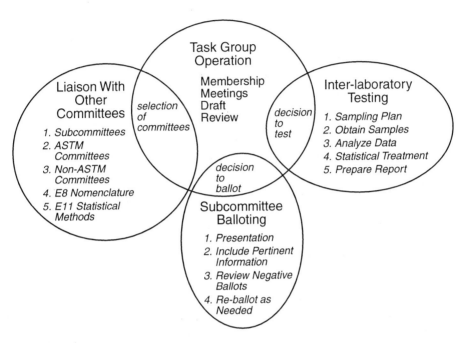

Figure 14.7
Operation of ASTM subcommittee task groups.

groups, interlaboratory testing, meetings, and ballot actions (ASTM 1988).

14.4.4.3 Committee E-44 on Solar Energy Conversion

14.4.4.3.1 Organization and Federal Funding ASTM organized Committee E-44 on Solar Energy Conversion in June 1978. It was organized as a "classified" committee by combining two subcommittees of ASTM Committee E-21 on Space Simulation and Applications of Space Technology. These were E21.09 on Solar Electric, Thermal and Stored Energy Utilization, and E21.10 on Solar Heating and Cooling Applications (Zerlaut 1981). Committee E-44 was classified in the sense that its membership was required to be balanced among producers, users, and general interest members. Later, classified committees were also required to "have balance," with consumer-interest participants. Indeed, Committee E-44 helped to pioneer consumer involvement in standards writing by ensuring consumer participation.

By 1981, Committee E-44 had more than 300 members, 15 active sub-committees, and was one of ASTM's most active technical committees. As a result of the workload, and the high national and international interest in solar standardization, E-44 had to deal with a unique set of problems. These included

• pressure to accelerate the development of materials and related standards,

• involvement of a significant number of consumer advocates on its various technical subcommittees,

• mutual lack of understanding between "consumer participants" and technical experts from and/or sponsored by industry, and

• scarcity of technical experts with experience in voluntary consensus standards development.

As a result of a DOE grant (DOE 1978) from the Office of Solar Applications, the increased use of technical experts helped accelerate the development of ASTM standards. This grant provided for (1) directed funding of technical experts to write urgently needed first-draft standards; (2) reimbursement of the travel expenses of regular committee and sub-committee members selected to attend interim task group meetings, thereby accelerating the process; (3) reimbursement of the travel expenses of from three to five consumer-oriented participants plus expert consultants to attend regular meetings; and (4) low-level funding of verification testing that might be required to prepare a test standard.

A major impediment to the initiation of standards for new technologies is usually a paucity of technical information. In the case of Committee E-44, needed technical and scientific information was accumulated as it became available, and from the DOE-funded materials studies administered by SERI and NBS (Clark et al. 1980; Waksman, Streed, and Seiler, 1981; Waksman, Thomas, and Streed, 1984, Clark and Roberts 1982; Masters et al. 1981; and Zerlaut and Anderson 1989).

14.4.4.3.2 Committee Structure   ASTM Committee E-44 was charged with "the promotion of knowledge, stimulation of research, and development of standards concerned with the technology for conversion of solar energy to directly usable energy forms, and the application of such technology for the public benefit" (ASTM 1978). As late as 1986, fifteen subcommittees were dealing with standardization in solar technologies.

These subcommittees are listed in table 14.2. Of these, sixteen are under the jurisdiction of E44.04 on materials, twelve under E44.09 on photovoltaics, and six under E44.02 on Environmental Parameters.

The importance of the standardization efforts of ASTM Committee E44 may be seen in light of the following:

• The standard on definitions developed by Subcommittee E44.01, *E772*, was the major English language resource document for ISO/TC180/WG1 on nomenclature (See section 14.5.3).

• The four solar radiometer calibration standards developed by Subcommittee E44.02 were resource documents for ISO/TC180/SC1 on climate; *E891* and *E892* are finding use worldwide in many technologies, including ISO/TC180/SC1 and IEC/TC82 on photovoltaics.

• A number of materials standards have been adapted by ISO/TC180/WG2 on materials for international standardization. These include *E903*, *E781*, *E765*, *E881*, and *E744*. *E838* has been adopted as *G90* by ASTM Committee G3 on nonmetallic materials.

• All of the standards developed in E44.09 on photovoltaics have become resource material for international standardization under the aegis of IEC Technical Committee 82. A number have been adopted directly.

Recently, Committee E44 streamlined its committee structure as a result of reduced interest in solar energy in the United States. Indeed, it has now merged with the Committee on Geothermal Systems in order to receive the benefits of greater support from ASTM (the so-called energy glut has affected all ASTM committees associated with alternative energy technologies).

14.5   International Organizations Involved in Standardization

14.5.1   World Meteorological Organization

In December 1951 the General Assembly of the United Nations recognized the World Meteorological Organization (WMO) as a specialized agency of the United Nations (Johnson 1984). A number of technical regulations have been promulgated for use by member countries that deal with meteorological measurement procedures and practices.

The preface to the fifth edition of the *Guide to Meteorological Practices* (WMO 1983) states that it contains "advice on methods required to keep

Table 14.2
ASTM Committee E-44 subcommittees

No.	Title	Scope[a]
E44.01	Nomenclature	Establish definitions for terms used in solar energy
E44.02	Environmental Parameters	Identify environmental parameters and establish measuring procedures for solar and meteorological data
E44.03	Safety	Review test methodologies, installations, and operations for safety
E44.04	Materials Performance	Develop standards related to the reliability and durability of materials
E44.05	Heating and Cooling Subsystems	Standards for evaluation of design, performance, and reliability of solar energy device subsystems
E44.06	Heating and Cooling Systems	Standards relating to methods and applications including active heating of swimming pools, DHW, space heating/cooling
E44.07	Process Heating Systems	Standards related to design and performance analysis of desalinization and process heat systems
E44.08	Thermal Conversion Power	Standards required to evaluate design and performance of solar thermal conversion power systems
E44.09	Photovoltaic Systems	Standards for evaluating design and performance of photovoltaic power systems
E44.10	Wind-Driven Power Systems	Standards for wind energy
E44.11	Ocean Thermal Power Systems	Standards for ocean thermal power conversion systems
E44.12	Biomass Conversion Systems	Standards relating to biomass as a fuel and chemical source
E44.13	Advanced Energy Systems	Standards needed to aid technically feasible but unproven solar energy conversion systems
E44.14	Passive Heating/Cooling	Standards needed to evaluate performance of materials, products, components and system used in passive solar applications
E44.15	Environmental and Societal Impact of Solar Energy Conversion Systems	Criteria and methods for measuring environmental and societal impact of solar applications

[a] Abridged scope statements.

observing stations up to international standards, recommendations for international observing practices for the taking of observations before coding, information about uniform procedures for applying correction with a view to eliminating errors, and, in general, the best methods of obtaining correct observations." As such, the "guide is not intended to be a detailed instruction manual for the use of observers but to form a basis for the preparation of such manuals by each Meteorological Service to meet its own particular needs. It should thus lead to the desired degree of standardization and of uniformity of methods of observation throughout the world."

WMO has two world radiation centers (WRCs), located at Davos, Switzerland, and St. Petersburg, Russia (formerly Leningrad, USSR), and numerous regional centers. Region IV, which includes the United States, has its regional centers in Toronto, Ontario, Canada, and Boulder, Colorado. (The reader is referred to Zerlaut 1989, the author's chapter in *Solar Resources*, Volume 2 of this series for a list of regional centers.)

WMO activities pertinent to solar energy utilization have involved liaison with both the International Energy Agency (IEA) and the International Organization for Standardization (ISO), largely in areas dealing with the measurement of both hemispherical and direct solar radiation, and in the calibration of solar radiometers. WMO's objective of ensuring that member countries adopt uniform measuring and calibration procedures have fallen short of the needs of the worldwide solar energy industry in terms of the precision and accuracy required of solar instrumentation employed in testing for product certification. As a consequence, it became necessary for both the IEA and ISO to undertake further efforts. ISO concentrated on international standards in the form of practices for the measurement of solar radiation and methods for the characterization and calibration of solar radiation measuring instruments, or radiometers.

14.5.2 International Energy Agency

14.5.2.1 Organization
Recognizing the need for a coordinated effort to solve certain energy-related problems facing the petroleum-importing countries, in 1974 the member countries of the Organization for Economic Cooperation and Development (OECD) created the International Energy Agency (IEA). IEA is funded by the member countries on a task-participation basis: that

Table 14.3
Status of tastks and responsibilities within the IEA Solar Heating and Cooling Program

Task	Title (subject)	Status	Operating agent
I	Investigation of the performance of solar heatng and cooling systems	Completed	Denmark
II	Coordination of R&D on solar heating and cooling components	Completed	AIST, Japan
III	Performance testing of solar collectors	Active	
IV	Development of an insolation	Completed	DOE, U.S.
V	Use of existing meteorological information for solar energy application	Completed	Sweden
VI	Performance of solar heating, cooling, and hot water systems using evacuated collectors	Active	DOE, U.S.
VII	Central solar heating with seasonal storage	Active	CBR, Sweden
VIII	Passive and hybrid solar low-energy buildings	Active	DOE, U.S.
IX	Solar radiation and pyranometry studies	Active	AES, Canada
X	Research on materials for solar heating and cooling systems	Active	MITI, Japan
XI	Passive and hybrid solar commercial buildings	Active	SFOE, Switzerland
XII	Building energy analysis tools for solar applications	Active	DOE, U.S.
XIII	Advanced solar low-energy buildings	Active	NIT, Norway
XIV	Active solar energy systems	Active	MEMR, Canada

Source: Blum 1991.
Blum, S., Jan. 1991, 1990 Annual Report of the International Energy Agency Solar Heating and Cooling Program, IEA/SHC/AR/90, International Planning Associates, Inc., Silver Spring, Md. 20901.

is, the greater cost of participation is borne by countries that manage certain tasks and subtasks, compared to countries that participate.

In 1975 IEA established a solar heating and cooling program as one of sixteen technology areas deemed of critical importance to the OECD. Eighteen countries and the European Community support the solar heating and cooling program through agreed-upon collaborative research projects. By 1990, fourteeen tasks had been initiated; Table 14.3 is a list of these tasks, showing the current status and operating agents for each.

14.5.2.2 Solar Heating and Cooling Activities of the IEA
Much of IEA's activities in solar radiation measurements and radiometry was undertaken under the auspices of Task III, Performance Testing of Solar Collectors (see also Zerlaut 1989).

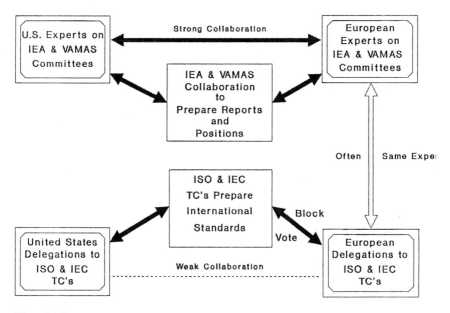

Figure 14.8
Comparison of practices in Europe's EC and EFTA countries with practices in the United States.

A problem, peculiar to the United States in my opinion, results from the significant differences in the relationship between international standards development and treaty-derived collaborative research in the countries of the European Community (the EC) and European Free Trade Association (EFTA), as compared to the United States. This difference is depicted by the chart presented in figure 14.8 (which was prepared by the author to argue for a change in the manner by which NIST deals with technical experts in the VAMAS (Versailles Agreement on Materials and Standards) program, another international collaboration in prestandardization research).

The U.S. delegation to ISO/TC 180 (for a discussion of ISO, see the next section) included experts with experience in domestic standards from such organizations as ASTM, ASHRAE, and so on. These experts were seldom involved in officially sponsored research prompted by international collaboration within the framework of the International Energy Agency. This all too often left the U.S. delegation to ISO/TC 180 facing de facto agreements when they attended the subcommittee meetings that were supposed to be the beneficiaries of the collaborative research in IEA. The U.S. delegation to TC 180 often found that the Europeans were uniformly aligned on a particular standard, the subject of which had been studied within the IEA community with significant input from U.S. experts. The European delegates to ISO/TC 180 and the IEA experts were often the same people, while this was seldom the case for U.S delegates. Hence, the Europeans were frequently puzzled that we on the U.S. delegation (1) were unaware of IEA work and of the U.S. IEA positions, and (2) were unable to simply accept the European consensus! Only in the area of solar radiometry (SC 1) did the U.S. delegation have IEA input from the U.S. experts, largely because I forged that liaison by serving on both for several years.

14.5.3 International Organization for Standardization

14.5.3.1 Organization

The International Organization for Standardization (ISO) was founded in 1947 and, with its sister organization, the International Electrotechnical Commission (IEC), is headquartered in Geneva, Switzerland. Together, the ISO and IEC issue voluntary international consensus standards (IS) on product specifications, test methods, and practices for nearly every technical field associated with world commerce. IEC has responsibility for electrical standards; all other areas fall under the auspices of ISO. Because standards for solar heating and cooling are covered by ISO, the emphasis here is on ISO.

Whenever possible, new ISO standards are derived from the existing national standards of ISO member countries. However, methods are developed within the international manufacturing and technical communities when necessary, and new test methods usually involve interlaboratory round-robin testing prior to the development of international consensus. Documents developed by the many ISO committees are voted on by the member countries of the relevant technical committee, and finally by all "P-Member" countries (participating and observer nations are designated "P" and "O," respectively.) Negative votes are resolved to give as high degree of consensus as possible.

14.5.3.2 ISO Technical Committee on Solar Energy

Technical Committee ISO/TC 180 on Solar Energy (Thermal Applications) was organized in Sydney, Australia, in May 1981. It is one of approximately 180 currently active technical committees of the Interna-

Table 14.4
Committee program for ISO/TC 180 solar energy (as of 1989)

Group identification		Subject	Operating agent
Working Groups	WG 1	Nomenclature	United States (ANSI)
	WG 2	Materials	France (AFNOR)
Subcommittees	SC 1	Climate	F. R. Germany (DIN)
	SC 4	Systems	United States (ANSI)
	SC 5	Collectors and components	France (AFNOR)

tional Standards Organization. The secretariat of TC 180 is held by the Standards Association of Australia (SAA). The organizing meeting of TC 180 was attended by official representatives from Australia, Canada, Sweden, Germany, France, Israel, New Zealand, Japan, Kenya, the United Kingdom, and the United States. In addition to the countries attending the organization meeting, other "P" members are Denmark, Greece, the Netherlands, Saudi Arabia, South Africa, Switzerland, Tunisia, and the CIS (formerly the USSR). The committees and working groups responsible for individual programs are presented in Table 14.4.

14.5.3.3 U.S. Activities on Behalf of ISO/TC 180

Largely as a result of deliberations on the ANSI Solar Standards Steering Committee (ANSI 1979b), a U.S. technical advisory group (TAG) was formed in early 1981 to represent the technical positions of U.S. industry and the U.S. solar community interested in standards activities.

The U.S. TAG has representatives of ASHRAE, ASTM, SEIA, and SRCC. This representation has ensured that the U.S. delegation to technical committee meetings and meetings of subcommittees and working groups is properly briefed on U.S. positions as defined by the constituent organizations.

Members of the U.S. Technical Advisory Group for ISO/TC 180 represent their employers in serving on the TAG. All delegates to technical committee (TC), subcommittee (SC), and working group meetings represent the American National Standards Institute (ANSI); they do not represent themselves, their companies or employers, or the U.S. government at those meetings. In addition, members of the U.S. delegation are not themselves members of ISO, or of ISO committees. They are members of the U.S. Delegation to ISO/TC 180. In other words, because each country has only one vote on each issue, technical positions must be agreed upon prior to the TC 180 meetings and must be unified (ANSI 1986).

Although for most technologies, individual membership on U.S. TAGs is funded by the participant, or by the participant's employer, this has not been the case for participation in TC 180. Also, the funds required for ANSI and domestic standards organizations (ASTM, ASHRAE, IEEE, etc.) to manage U.S. Secretariats of either technical committees, or subcommittees, as well as for administration of the TAGs themselves, are usually provided by industry groups and trade associations.

Because this level of funding was not available from the Solar Energy Industries Association (SEIA), funding for participation in ISO/TC 180 activities, whether for administration of the Secretariat for Subcommittee SC4 on Solar Systems, for managing the U.S. TAG, or for travel to domestic and international meetings, has been wholly provided by the U.S. Department of Energy. Initially, these funds were furnished to ANSI; however, ASHRAE has administered the TAG and the SC4 secretariat for the past four years, and has been the recipient of the DOE funding provided for these activities. Over the thirteen years since TC 180 was formed, DOE has provided funds for U.S. participation in nine international meetings, including the expenses of hosting a meeting in the United States. The United States, Greece and Hungary have each hosted one international technical committee meeting, while Australia, Germany, and France have each hosted two.

By continuing to fund low-level participation in the activities of ISO/ TC 180, DOE has helped ensure that a reemergent U.S. solar industry will not be bereft of the latest standards development and testing technologies that may be required in both near-term and future international trade opportunities. It is important to note that, as a result of U.S. participation in ISO/TC 180, U.S. solar standards and testing technologies became resource documents, sometimes primary resource materials, for the developing European (largely) solar industries. The international standards and testing innovations that have matured since the 1986-demise of the U.S. solar industry will be readily available for updating and revitalizing U.S. solar standards and test procedures for use by a renewed industry.

14.6 Lessons Learned

Government stimulus for the development of laboratory accreditation programs, and to some extent of product labeling programs, was certainly not new in the case of solar energy technologies. But, it is to be hoped that the complexities and time-consuming deliberations that marked the development of the 1980s vintage solar product certification program will not be visited on a reemergent solar industry in future years. Nonetheless, one can summarize the entire process of developing the solar hot water collector rating, certification, labeling and laboratory accreditation, program as follows:

With government funding, the private sector, using experts from a mix of government, institutional and commercial organizations, successfully developed a viable solar product certification program by creating a private, nonprofit certification body (the Solar Rating and Certification Corporation).

In spite of the many early biases in various segments of the solar community, in the short span of seven years (1976 to 1983) the solar industry progressed from one having only self-regulation and essentially no standards to one with a national product certification program that had a definite and beneficial effect on the quality of the solar products available to the consumer.

Yet the sense of urgency that made it imperative that standards be quickly written around just-emerging scientific and engineering developments often overwhelmed industry. It is often said that standards should follow rather than lead technology and that the reversal of these steps often results in problems in the marketplace. As necessary as the standardization efforts were, it may be that there simply was not enough time for the industry to develop a healthy reliance on standards and the strength to survive the end of the federal tax credits—the industry lasted only days after the residential tax credits ended.

Given the difficult birthing of the solar industry and the abrupt withdrawal of the tax credits on which the industry had become hooked, it is perhaps our good fortune that a very modest, but viable, solar product certification program still remains in place. However, a future, revitalized solar industry may not be ready to accept a strong third-party product certification program in an absence of mandates from the Sunbelt states; memories of ill-conceived government initiatives are ingrained in both the remnant solar industry and in manufacturers poised for a return to a "new" solar market.

In my response to the presentation of Hoffman (1978), and others, at the 1977 Symposium on Competition in the Solar Energy Industry, sponsored by the Federal Trade Commission (FTC), I cited the rule of reason as applied to solar standards, and stated that we must be ever mindful of developing trade-offs between:

• protecting the consumer versus excluding manufacturers from the markets,

• providing a commonality of language that permits interstate and international commerce versus promoting regulations and standards that are valueless as a result of high commonality,

• providing clear choices to the consumer versus making the acceptance of cheaper and inferior products overly attractive to them, and

• protecting the buyer to such a high extent that the high costs drive him out of the decision-making process versus accepting inferior products. (Zerlaut 1978)

The critical point is that these trade-offs are as important for today's re-emerging solar industry as they were in 1977. New manufacturers must be educated to the need for, and role of, the country's still-viable solar product certification and labelling program. At the same time, the various state and public sector organizations with new solar mandates must resist the temptations to overregulate and overprotect that often marked their roles in the 1975–1985 decade of high solar energy utilization.

Acknowledgment

Although many friends and associates helped shape my views and my education in the solar standardization processes that marked the halcyon days of solar, none were more loyal, more helpful, or more persuasive than Sheldon H. Butt, president of the Solar Energy Industries Association. His perseverance and integrity in the face of an incredible array of contradictory pressures has helped steady the rest of our years for all of us who had the opportunity to work with him.

References

ANSI (American National Standards Institute). Steering Committee on Solar Energy Standards Development. 1979a. Minutes of 15 February 1979 meeting. SCSE 41. New York, 26 April.

ANSI. Steering Committee on Solar Energy Standards Development. 1979b. Minutes of 14 June 1979 meeting. SCSE 46. New York, 26 September.

ANSI. Steering Committee on Solar Energy Standards Development. 1979c. Minutes of 18 October 1979 meeting. SCSE 48. New York, 7 November.

ANSI. 1982. *Procedures for the Development and Coordination of American National Standards.* New York. 1 September.

ANSI. 1986. *Guide for U.S. Delegates to ISO Meetings.* SR14c. New York.

ANSI. 1988. *1988 Progress Report.* New York.

ARI (Air-Conditioning and Refrigeration Institute). 1981a. *ARI 910-81.* "Standard for Solar Collectors." Superseded by *ARI 910-82.* Arlington, VA.

ARI. 1981b. *ARI 920-82.* "Standard for Solar Domestic Hot Water Systems." Arlington, VA.

ARI. 1982. *ARI 930-82.* "Standard for Application of Solar Collectors." Arlington, VA.

ASHRAE (American Society of Heating, Refrigeration, and Air-Conditioning Engineers). 1977. *ASHRAE 93-1977.* "Methods of Testing to Determine the Thermal Performance of Solar Collectors." New York.

ASHRAE. 1980. *ASHRAE 96-1980.* "Method of Testing to Determine the Thermal Performance of Unglazed Solar Collectors." New York.

ASHRAE. 1981. *ASHRAE 95-1981.* "Methods of Testing to Determine the Thermal Performance of Solar Domestic Water Heating Systems." New York.

ASHRAE, 1986. *ASHRAE 93-1986.* "Methods of Testing to Determine the Thermal Performance of Solar Collectors." New York.

ASTM (American Society for Testing and Materials). 1978. Minutes, First Meeting of Committee E44 on Solar Energy, Philadelphia, May.

ASTM. 1984. *ASTM E548.* "Practice for Preparation of Criteria for Use in the Evaluation of Testing Laboratories and Inspection Bodies." Philadelphia.

ASTM. 1988. *Regulations Governing ASTM Technical Committees.* Philadelphia, September.

Bard, L. A., and M. W. Rupp. 1981. "An Operational Domestic Hot Water System Test Facility." *Proceedings of the Annual Meeting AS/ISES,* American Section, International Solar Energy Society, vol. 4.1, p. 686, Philadelphia, June 1981.

Bard, L. A. 1984. "DHW System Testing: Lessons Learned." *Proceedings of the 1984 Annual Meeting of the American Solar Energy Society,* June, Anaheim, CA.

Butt, S. H. 1977. SEREF director's letter to Ronald Scott, U.S. Department of Energy, Washington, DC, 8 November.

Butt, S. H. 1978a. "Standards for the Solar Industry." *The Solar Market: Proceedings of the Symposium on Competition in the Solar Energy Industry,* 15–16 December 1977. Washington, DC: Federal Trade Commission, Bureau of Competition, June.

Butt, S. H. 1978b. President of SEIA's position statement and status advisory presented before the ANSI Steering Committee on Solar Energy Standards Development, Solar Energy Industries Association, 21 November, Washington, DC.

Butt, S. H. 1979a. President of SEIA's letter to ANSI Subcommittee on Solar Collector Thermal Performance Standards, Solar Energy Industries Association, Washington, DC, 17 January.

Butt, S. H. 1979b. President of SEIA's memorandum to the SEIA Executive Committee, Solar Energy Industries Association, Washington, DC, 18 June.

Butt, S. H. 1979c. President of SEIA's letter to the Executive Committee, Solar Energy Industries Association, Washington, DC, 19 April.

Butt, S. H. 1979d. President of SEIA's letter to Mr. Alvin Lai, Secretary, ANSI Steering Committee on Solar Energy Standards Development, Solar Energy Industries Association, Washington, DC, 31 May.

CEC (California Energy Commission). 1980. *Test Results from Testing and Inspection Program for Solar Equipment (TIPSE)*. CEC-P500-80-056. Sacramento, CA, October.

Chandra, S., and M. Khattar. 1980. Analytical Investigation of the Relative Solar Rating Concept, Report FSEC TT80-6, Florida Solar Energy Center, Cape Kennedy, FL, June.

Clark, E. J., and W. E. Roberts. 1982. *Weathering Performance of Cover Materials for Flat Plate Solar Collectors*. NBS Technical Note 1170. Washington, DC: National Bureau of Standards.

Clark, E. J., W. E. Roberts, J. W. Grimes and E. J. Embree. 1980. *Solar Energy Systems: Standards for Cover Plates for Flat-Plate Solar Collectors*. NBS Technical Note 1132. Washington, DC: National Bureau of Standards.

DOE (U.S. Department of Energy). 1978. *Development of Performance Standards and Criteria for the Production and Installation of Solar Energy Systems*. Grant EM-78-G-01-4147 with the American Society for Testing and Materials. Washington, DC.

DOE. Office of Solar Applications. 1979. "Proposed Voluntary Framework of Federal, State and Local Agencies to Provide for Coordination and Reciprocity of Testing and Certification Programs for Solar Energy Equipment." Working Document. Washington, DC.

FEA (Federal Energy Administration). 1977. *Development of a Rating, Certification and Labeling System for Solar Collectors and Water Heating Systems*. FEA contract CB-05-70311 to the Solar Energy Research and Education Foundation. Washington, DC, 27 September.

FSEC (Florida Solar Energy Center). 1980a. *Test Methods and Minimum Standards for Certifying Solar Collectors*. FSEC-77-5-R80. Cape Canaveral, 1 December.

FSEC. 1980b. *Operation of the Collector Certification Program*. FSEC-77-6-R80. Cape Canaveral, 1 December.

FSEC. 1990. Personal communication from David L. Block.

FTC (Federal Trade Commission). 1978. *The Solar Market: Proceedings of the Symposium on Competition in the Solar Energy Industry*, 15–16 December 1977. Washington, DC: Federal Trade Commission, Bureau of Competition, June.

Geisheker, P. G., W. J. Putman, and L. A. Bard. 1981. "Development of Facilities and Methods for Thermal Performance Tests of Passive Domestic Hot Water Systems." *Proceedings of the 6th National Passive Solar Conference*, Portland, OR, September 1981.

Gilligan, J. E., J. Brzuskiewicz, and J. E. Brzuskiewicz. 1978. *Handbook of Materials for Solar Utilization*. Pts. 1 and 2. U.S. Energy Research and Development Administration Contract EY-76-C-02-0578-034.A002. Chicago: IIT Research Institute, February.

Hill, J. E., J. P. Jenkins, and D. E. Jones. 1979. *Experimental Verification of Standard Test Procedure for Solar Collectors*. NBS Building Science Series 117. Washington, DC: National Bureau of Standards.

Hill, J. E., and T. Kusuda. 1974. *NBSIR 74-635*. "Method of Testing for Rating Solar Collectors Based on Thermal Performance." Washington, DC: National Bureau of Standards.

Hill, J. E., W. B. May, and E. R. Streed. 1979. "Follow-up to Report of June 1979." NBS memorandum to the ANSI Ad Hoc Committee on Rating of Solar Collectors. National Bureau of Standards. Washington, DC, June.

Hill, J. E., E. R. Streed, G. E. Kelly, J. C. Geist, and T. Kusuda. 1979. *Development of Proposed Standards for Testing Solar Collectors and Thermal Storage Devices*. NBS Technical Note 899. Washington, DC: National Bureau of Standards, February.

Hoffman, J. E. 1978. "Antitrust Issues in Setting and Enforcing Product Standards." *The Solar Market: Proceedings of the Symposium on Competition in the Solar Industry*. 15–16

December 1977. Washington, DC: Federal Trade Commission, Bureau of Competition, June.

Hottel, H. C., and A. Whillier. 1958. *Transactions of the Conference on the Use of Solar Energy.* Vol. 2. Tucson: University of Arizona Press.

Hottel, H. C., and B. B. Woertz. 1942. "The Performance of Flat-Plate Solar Heat Collectors." *ASME Transactions* 64:91.

HUD (U.S. Department of Housing and Urban Development). 1977. *Intermediate Minimum Property Standards Supplement: Solar Heating and Domestic Hot Water Systems.* Document 4930.2. Washington, DC.

HUD. 1978. "Solar Standards: Laying the Groundwork for a Solid Industry: A Special Report." In *HUD Solar Status.* HUD-PDR-189-8. Washington, DC.

ISCC (Interstate Solar Coordinating Council). 1980a. *ISCC 80-1.* "Test Methods and Minimum Standards for Solar Collectors." Cape Canaveral, FL, 1 December.

ISCC. 1980b. *ISCC 80-2.* "Operating Guidelines for Certifying Solar Collectors." Cape Canaveral, FL, 1 December.

Johnson, N. 1984. Personal communication, International Division, National Oceanic and Atmospheric Administration, Rockville, MD.

Kirkpatrick, D. L. 1983. *Flat-Plate Solar Collector Performance Data Base and User's Manual.* SERI/STR-254-1515. SERI subcontract B-0-5254-1-M1, DOE contract EG-77-C-01-4042. Glen Moore, PA: Sun Systems, Inc., July.

Landis, J. W. 1975. Letter to participants memorializing meeting between ANSI-FEA on the need for solar standards development, American National Standards Institute, New York, 13 May.

Masters, L. W., J. F. Seiler, E. J. Embree, and W. E. Roberts. 1981. *NBSIR 81-2232.* "Solar Energy Systems: Standards for Absorber Materials." Washington, DC: National Bureau of Standards, January.

Mather, G. R., Jr. 1979. Letter from chairman, SEIA Rating Subcommittee, to S. H. Butt, president of SEIA, Solar Energy Industries Association, Washington, DC, 11 October.

May, W. B., J. E. Hill, and E. R. Streed. 1979. "Analysis of Solar Collector Rating Methods." NBS letter report prepared for the ANSI Ad Hoc Committee on Rating Solar Collectors. Washington, DC: National Bureau of Standards, June.

Neissing, W. J. 1978. *Laboratories Technically Qualified to Test Solar Collectors in Accordance with ASHRAE Standard 93-77: A Summary Report.* NBSIR 78-1535. Washington, DC: National Bureau of Standards, November.

Pytlinski, G. T. 1979. Letter to Daniel Smith, director, International Standards Activities, ANSI, New Mexico State University, Las Cruces, NM, 19 January.

SEIA (Solar Energy Industries Association). 1979. Minutes of SEIA Executive Committee meeting. Washington, DC, 11 October.

SEREF (Solar Energy Research and Education Foundation). 1977. Bylaws. Washington, DC, 28 August.

Simon, F. F. 1975. "Status of the NASA-Lewis Flat-Plate Collector Tests with a Solar Simulator." *Proceedings of the Workshop on Solar Collectors for Heating and Cooling of Buildings,* New York, 21–23 November 1974. NSF-RAN N-75-019. College Park, MD: University of Maryland, May.

SRCC (Solar Rating and Certification Corporation). 1981a. *SRCC RM-1.* "Methodology for Determining the Thermal Performance Rating for Solar Collectors." Washington, DC, June.

SRCC, 1981b. *SRCC OG-100.* "Operating Guidelines for Certifying Solar Collectors." Washington, DC, September.

SRCC. 1982. *SRCC RM-2.* "Methodology for Determining the Thermal Performance Rating for Tracking Concentrator Solar Collectors." Washington, DC.

SRCC. 1983. *SRCC 200-82.* "Test Methods and Minimum Standards for Certifying Solar Water Heating Systems." Washington, DC, April.

SRCC. 1986a. *State Solar Directory.* Washington, DC.

SRCC. 1986b. *Directory of SRCC-Certified Solar Collector Ratings.* Vol. 6, no. 2. Washington, DC.

SRCC. 1987. *Directory of SRCC-Certified Solar Water Heating System Ratings.* Vol. 5, no. 1. Washington, DC.

Streed, E. R. 1977. National Bureau of Standards. Personal communication to G. A. Zerlaut, Desert Sunshine Exposure Test Laboratories, Inc.,

Streed, E. R., W. C. Thomas, A. G. Dawson, III, B. D. Wood, and J. E. Hill. 1978. *Results and Analysis of a Round-Robin Test Program for Liquid-Heating Flat-Plate Solar Collectors.* NBS Technical Note 975. Washington, DC: National Bureau of Standards, August.

Streed, E. R., and D. Waksman. 1981. *Uncertainty in Determining Thermal Performance of Liquid-Heating Flat-Plate Solar Collectors.* NBS Technical Note 1140. Washington, DC: National Bureau of Standards, April.

Vakerics, T. V. 1977. Advisory memorandum to SEREF Board of Directors regarding "Procedural Guidelines for SEREF Solar Standards Committees," O'Connor & Hannon, Washington, DC, 9 December.

Waksman, D., J. H. Pielert, R. D. Dikkers, E. R. Streed, and W. J. Niessing. 1978. *Plan for the Development and Implementation of Standards for Solar Heating and Cooling Applications.* NBSIR 78-1143A. Washington, DC: National Bureau of Standards, June.

Waksman, D., E. R. Streed, T. W. Reichard, and L. E. Cattaneo. 1978. *NBSIR 73-1305A.* "Provisional Flat-Plate Solar Collector Testing Procedures." 1st rev. Washington, DC: National Bureau of Standards, June.

Waksman, D., E. Streed, and J. Seiler. 1981. *NBS Solar Collector Durability/Reliability Test Program Plan.* NBS Technical Note 1136. Washington, DC: National Bureau of Standards, January.

Waksman, D., W. C. Thomas, and E. R. Streed, 1984. *NBS Solar Collector Durability/ Reliability Test Program: Final Report.* NBS Technical Note 1196. Washington, DC: National Bureau of Standards, September.

Whillier, A., and S. J. Richards. 1961. "A Standard Test for Solar Water Heaters." *Proceedings of the Conference on New Sources of Energy*, Rome, 21–31 August. Paper S/97.

WMO (World Meteorological Organization). 1983. "Measurement of Radiation." In *Guide to Meteorological Instruments and Methods of Observation.* 5th ed. WMO-8. Geneva.

Wood, B. D. 1978. Arizona State University, Tempe, AZ. Personal Communication to G. A. Zerlaut, Desert Sunshine Exposure Test Laboratories, Inc.

Zerlaut, G. A. 1977. Letter project suggestion provided to the National Bureau of Standards, Desert Sunshine Exposure Tests, Inc., Phoenix, AZ, 28 March.

Zerlaut, G. A. 1978. "Remarks on Competition, Consumers, and Standards." *The Solar Market: Proceedings of the Symposium on Competition in the Solar Energy Industry*, 15–16 December 1977. Washington, DC: Federal Trade Commission, Bureau of Competition, June.

Zerlaut, G. A. 1979. Letter to Alvin Newton regarding "Joint ARI/SEIA Rating Method: A Memorandum of Understanding," Solar Energy Industries Association, Washington, DC, 17 July.

Zerlaut, G. A. 1981. "The Challenge of Solar Energy: ASTM Committee E-44." *ASTM Standardization News* April: 8–12.

Zerlaut, G. A. 1985. News conference at TÜV, Munich, Federal Republic of Germany, sponsored by Kernforschungslage-Jülich on occasion of ISO/TC180 Meeting, September.

Zerlaut, G. A. 1989. "Solar Radiation Instrumentation." In R. I. Hulstrom, ed., *Solar Resources*, 173–308. Cambridge: MIT Press.

Zerlaut, G. A., and T. E. Anderson. 1989. *Commercial Solar Materials Exposure Studies*, DSET Laboratories, Inc., report DSET-R2658-Final. Los Alamos National Laboratories subcontract 9-L34-Q8088-1. Los Alamos, NM: Los Alamos National Laboratories, 1 February.

Zerlaut, G. A., and B. L. Garner. 1983. "International Standardization and the Plastics Industry." *ASTM Standardization News* April: 8–13.

15 Consumer Assurance

Roberta W. Walsh

15.1 Federal Consumer Assurance Mechanisms

The federal government's strategy for reaching the goal of deriving 20% of the country's energy needs from the sun by the year 2000 focused considerable attention on the residential sector. President Carter's "Message on Solar Energy" (President 1979) called for residential tax credits for solar applications and financing incentives, among other initiatives to accelerate this segment of the potential solar market.

Special attention was given to the residential consumer out of recognition that consumer confidence in the market would be necessary for a technology that was generally unfamiliar to the public. Accordingly, the president's message further asked for

a strengthening of current efforts to meet the consumer's need for valid information and *assurance* that solar equipment and systems purchased under Federal incentive programs will *perform and last as expected* [and] for a study of alternative mechanisms for providing Federal assistance to development and implementation of *programs to provide solar consumer assurance.* (President 1979, 9; emphasis added)

As suggested by the emphases in the above statement, there were many issues and concerns underlying this portion of the overall message. It is the background and outcome of those and related matters that is the subject of this chapter. This section sets the foundation for an understanding of the terms, significance, and mechanisms surrounding consumer assurance.

15.1.1 Consumer Assurance, the Solar Market, and the Role of Government

What is *consumer assurance*? The term itself defied easy definition during the period under discussion. *Protection* and *fraud* were more commonplace terms to those in the consumer advocacy and consumer law enforcement communities. Similarly, those concerned with marketing could more readily identify with terms such as *satisfaction* and *acceptance* as they deliberated how solar businesses should relate to the consumer.

Moreover, models from past experience were lacking to convey the nature and scope of consumer assurance in solar energy. Although many products had been rapidly introduced and accepted in the market in the post–World War II economy, virtually none had been promoted through a strong government role, either directly or indirectly. Traditionally, government had acted only when marketplace problems arose—through regulation and enforcement of federal and state laws that prohibited deceptive business practices—not before a market was established.

The term *consumer assurance* addressed this unique situation as it confronted all groups concerned with accelerating solar commercialization—especially including consumers themselves. In the field of solar energy, consumer assurance generally means consumer confidence in market products and services as well as satisfaction in realizing the expectations generated by that market. This meaning differs from the more psychologically based interpretation that conveys optimism due to pride of purchase (i.e., being "confident" about solar and/or "satisfied" with the decision to buy). *Consumer assurance* can be thought of as an umbrella term encompassing *protection*, *acceptance* and *satisfaction*.

From an industry perspective, the "marketing concept," as distinguished from the "selling concept," probably most closely approximates the meaning of consumer assurance. Its origin in modern marketing management literature is traced to Theodore Leavitt's (1960, 49) classic article in which he states:

Selling focuses on the needs of the seller; marketing on the needs of the buyer. Selling is preoccupied with the seller's need to convert his product into cash; marketing with the idea of satisfying the needs of the customer by means of the product and the whole cluster of things associated with creating, delivering and finally consuming it.

This comprehensive approach to marketing is what underlies the goal of consumer assurance in all markets. For solar energy in particular, the goal was to form a reliable, healthy, and sustained market. The role of government is the significant distinguishing factor between Leavitt's concept and its application to solar energy. Government's rationale for supporting the principle of consumer assurance was the belief that the government would be a critical factor in increasing domestic energy supplies and maximizing return on investment in tax incentives and program expenditures.

15.1.2   Mechanisms for Providing Consumer Assurance

Consumer assurance was viewed by its proponents as fulfilling the interests of consumers, business, and government. Its role in accelerating solar

commercialization was to minimize risk to the consumer in market interactions and take steps to assure complete and truthful product claims,
competent installers, and legal and effective warranties. However, identifying the means to achieve consumer assurance and the institutions responsible for its implementation became more complicated. This section
describes some specific mechanisms and the substance of the major issues
involved in the ability of those mechanisms to achieve consumer assurance goals.

15.1.2.1 Information

For purposes of this chapter, it is necessary to differentiate between *public*
information or education about solar energy in general (addressed in
chapter 17) and *consumer* information (also addressed in chapter 16) that
is specifically related to actual consumer decision making in the pre-,
point-of-purchase, and postpurchase phases of solar energy utilization.
Consumer information includes the facts and data necessary for informed
consumer choice in areas such as (1) quality, reliability, and efficiency of
competing products; (2) competence in workmanship; (3) realistic system
performance expectations; (4) availability of product use information and
servicing; and (5) knowledge of consumer rights in dealings with contractors, retailers, and manufacturers.

The ability of consumer information to contribute to consumer assurance depends upon its accessibility, accuracy, practicality, and ease of
comprehension. In the developing solar market, however, it was not possible to assure that these criteria could be met. For example, performance
comparisons required standardized procedures, and a thorough knowledge of consumer rights required an understanding of the complexities of
contract law. In this context, the mere existence of an information source,
such as a brochure or lecture was insufficient. The information had to be
tailored in such a way that it dealt clearly with areas such as warranties,
standards, and training of installation workers, even though these measures were not fully developed in the industry.

There were two major difficulties inherent in relying on consumer
information as an assurance mechanism. First, from the consumer's perspective, there was little, if any, knowledge as a basis for what information to seek or what questions to ask. Second, from the perspective of a
developing industry (no matter how consumer-oriented), quality of information would be secondary to other business start-up matters. Relying

too heavily on information as a consumer assurance mechanism in this environment had the potential of placing the consumer at a disadvantage, leading to possible early disenchantment with the solar market and susceptibility to fraudulent or misleading claims.

A final information-related issue was concerned with who should have the major role and where should the responsibility lie for information as a consumer assurance mechanism. Identifying and coordinating the wide range of public and private sector organizations involved in developing content and/or communicating information made the feasibility of addressing this mechanism an enormous task.

15.1.2.2 Training and Related Mechanisms

To many consumers, and to those involved in the solar industry and in government, the key to consumer assurance lay in mechanisms that addressed competency of personnel installing, servicing, and inspecting solar systems. Issues surrounding these mechanisms focused on balancing the restriction of entry into the trade (thereby increasing consumer costs for services) against the degree of reliability that restriction would provide.

Mandatory state licensing represented the most restrictive approach, as contrasted with voluntary measures such as state-sanctioned certification and factory or vocational training. Regulation of training raised quality-related issues such as expertise of trainers, content of curriculum, competency testing of trainees, and means for updating trained personnel on new and improved techniques and methods affecting quality installation. Another approach relied on the enactment and enforcement of codes and solar system installation standards as an incentive to strengthen training programs. Where it arose, the potential for stifling innovation in a new industry often came into conflict with providing consumer assurance. (See chapters 14 and 19 respectively for additional information on standards and training.)

15.1.2.3 Product Testing

Product testing (also described in detail in chapter 14) is important for consumer assurance because it enables consumers to compare performance among systems. Inherent in this mechanism were difficulties in establishing a basis for tests and performance ratings and in making consumers aware of their appropriate use and limitations.

Certifying components through testing procedures raised a caution in many quarters that flexibility be built into any such program to permit technical innovation. In addition, the site-specific nature of solar systems made them a product category for which efficiency in performance could not be reasonably predicted (unlike automobiles or major household appliances).

15.1.2.4 Warranties

Although warranties had long been regarded as a means of providing consumer assurance in other product areas, could warranties have this effect in the emerging residential solar market? Would traditional consumer advice to "shop around for the best warranty terms" be practical in the new industry?

Consumers could find it difficult to make valid comparisons among warranties of competing firms if technical terminology were used. Consumers without monitoring equipment and unable to evaluate system performance would be less likely to identify problems covered in a warranty. A long-term, comprehensive warranty could be cost-prohibitive to both the consumer and the firm. Additionally, responsible dealers would be reluctant to offer long-term warranties that they might be unable to honor in later years because of the risks involved in any new technology.

Considering the complexities of both active and passive solar energy systems, consumers would best be served by total system warranties and a single point of contact in the distribution chain rather than several individual warranties from various manufacturers and dealers.

15.2   Federal Activities

Consumer assurance could not be accomplished in isolation. Rather, it had to be integrated into virtually every aspect of solar commercialization planning, implementation, and evaluation. In this section the activities of federal agencies and organizations are described as they affected or were affected by consumer assurance needs.

15.2.1   Studies of Solar Domestic Hot Water Systems

Although it is beyond the scope of this chapter to provide technical or management detail on solar demonstrations and other solar residential studies, the results of these efforts are important in examining consumer assurance because they called attention to problems in performance occurring primarily in solar domestic water heating systems.

15.2.1.1 New England Electric System Experiment
In 1975–1976 New England Electric System (NEES), a holding company composed of three utilities distributing electricity in Massachusetts, Rhode Island, and New Hampshire, became involved in a project to determine energy conservation and economic benefits of solar domestic hot water (SDHW) systems. The project consisted of installing and monitoring state-of-the-art systems in 100 residences in all three states and was the first such test undertaken in the United States on a large scale. Preliminary results (ADL 1977) revealed that (1) major malfunctions occurred in 50% of the units and (2) consumers' overall energy savings average was 17%, ranging from less than 5% for the fifteen worst systems to 37% for the fifteen best systems—in contrast to manufacturers' claimed 50% expected savings. This poor performance record was attributed to faulty design and improper installation and maintenance. Once these problems were rectified after the first year, the overall average energy contribution increased to 41% (DeFelice and Brown 1986). It was the first-year results, however, that received widespread publicity (e.g., Webster 1977; Ackerman 1977), often accompanied by calls for protection of the consumer (Boston Globe 1977).

In an independent study of the systems, a 34-step process was recommended to consumers considering a SDHW installation, including obtaining legal counsel and professional inspection assistance (Smith 1977). Brookhaven National Laboratory (BNL), which commissioned the study with the support of the Department of Energy's (DOE) Solar Division, also formed a consumer education committee, consisting of area public and private sector representatives, to publish and distribute a consumer protection brochure (BNL 1978). While these events and activities probably had little impact on the average consumer, they highlight the extent of government's and other institutions' response to problems of SDHW systems.

15.2.1.2 Solar Heating and Cooling Demonstration Act of 1974
Although DOE eventually assumed overall responsibility for federal solar demonstrations, the Department of Housing and Urban Development (HUD) carried out the provisions of the Solar Heating and Cooling Demonstration Act of 1974 (PL 93-409) (as described in chapter 9).

Among other objectives, the act was intended to obtain information on residential system performance and market acceptance in an effort to demonstrate the practical use of active and passive solar heating technology in residential buildings. Over the span of its existence (1974–1981), the program awarded 783 grants for demonstration projects that involved 12,734 dwelling units at a cost totalling $22,914,421 (Moore 1981). As reported by the program's manager during its final year: "This experience has given the HUD solar staff two impressions of the marketplace. First, the quality of many solar products, designs and installations has been dismal; second, it is possible to do the job right, but it is not easy" (Moore 1981, 454).

What is interesting about the HUD demonstration program in relation to consumer assurance is that even through did not set out to do so, it had the effect of reinforcing the severity of solar product reliability problems. Moreover, the implied endorsement of SDHW systems that accompanied grants to homeowners contributed to a loss of faith in government when systems failed and no recourse was readily available.

15.2.1.3 Florida Solar Energy Center

Under contract to DOE, the Florida Solar Energy Center (FSEC) surveyed experiences of 522 SDHW consumers who owned private sector installations. The Center reported that owners' satisfaction levels were not a reliable indicator of whether or not problems existed with their systems. Twenty-eight owners had reported being "satisfied" with systems that had frozen, as did ten owners whose units had experienced sensor failure (Yarosh 1979). The FSEC study, unlike the NEES experiment and HUD demonstration, was specifically oriented to uncover consumer protection issues inherent in systems that had been installed by the existing solar market.

15.2.2 Congressional Response

Results of the above studies and other reports of system failures led a House of Representatives subcommittee to hold public hearings on whether promoting solar energy commercialization, (particularly active systems) was in the consumer interest. Among the subcommittee's conclusions was that although solar technologies were mechanically and economically feasible if properly designed, manufactured, and installed, government agencies had been remiss in failing to develop a workable consumer protection plan. The subcommittee recommended that (1) solar standards development include paid consumer participation, (2) the Federal Trade Commission (FTC) promulgate solar consumer protection trade regulation rules, and (3) minimum short-term warranties and maintenance and operating instructions be required on all systems eligible for tax credits (U.S. House 1978).

Although it did not address consumer issues per se, a General Accounting Office (GAO) analysis, in response to reports of poor performance in federal demonstration programs, later questioned the advisability of continuing to conduct demonstrations when evidence suggested the technology as not "practical." (GAO 1979). In a later report, GAO (1981) included solar among other products advertised as energy-saving that conveyed potentially inaccurate and misleading information. The report also urged the FTC and DOE to improve their consumer protection efforts.

15.2.3   Executive Branch: Domestic Policy Review of Solar Energy

President Carter's "Message on Solar Energy" followed a Domestic Policy Review (DPR) conducted between May and September 1978. An important feature of the DPR was its emphasis on citizen participation and input. Ten regional public hearings were held for this purpose. Although it did not dominate these hearings, consumer assurance was clearly shown to be a matter of public concern and disagreement. The following comments were representative of those offered on the subject:

"[B]alanced" regulation [is needed to] protect consumers but not stifle embryonic businesses.

[T]here [needs] to be some method of comparing solar systems as to cost and efficiency, and there [needs] to be education to protect people against exaggerated claims.

We are going overboard with concern for solar standards. (DOE 1978)

Consumer assurance was addressed in the review under the general heading of "Institutional Barriers and Incentives" to increased solar use. In its response memorandum to the president, DOE cited the following efforts in describing federal solar consumer assurance operations: (1) labor and training programs, (2) the National Bureau of Standards (NBS) interim performance criteria, (3) minimum warranty requirements in HUD and DOE demonstration programs, (4) development by NBS and

DOE of solar equipment standards, and (5) identification of laboratories qualified to test and certify solar industry products (DOE 1979a).

The memorandum further raised the proposal of a federal warranty reinsurance program as a solution to the quandary of consumers' need for assurance and undercapitalized firms' inability to provide sound or long-term warranties. It concluded, however, that "it does not appear appropriate at this time for the Federal Government to enter into a full-scale warranty reinsurance program because the industry has not exhausted all possibilities for private insurance" (DOE 1979a, Attachment, p. 22).

15.2.4 Department of Energy, Assistant Secretary for Conservation and Solar Energy

In late 1977 DOE eventually assumed the lead role in federal solar programs. Responsibility within DOE for the broad range of consumer-related matters encompassing the acceleration of commercialization (e.g., standards, installer training) was primarily under the Office of the Assistant Secretary for Conservation and Solar Energy (CSE). As a program area, consumer assurance functioned separately from, but of necessity interacted with, program management in these numerous related areas. Additionally, under the assistant secretary for intergovernmental and institutional relations, the Office of Consumer Affairs engaged in solar consumer protection activities.

15.2.4.1 Florida Solar Energy Center Consumer Protection Workshop

Because of its prior experience in studying consumers' solar problems, CSE also gave the FSEC funding to "develop and recommend alternative strategies which could provide systematic means to reduce the potential risk to consumers of encounter with fraud or incompetence in their experience with the early solar market." (DOE 1979b, 1)

As a part of that project, a four-day Solar Energy Consumer Protection Workshop was held in May 1978 in Atlanta. Its objective was to bring together some 100 consumer protection and energy experts to explore and evaluate existing consumer protection mechanisms as they applied to solar applications. As this was the first such formal interaction on a large scale between industry and consumer representatives, a summation of their dialogue on the subject of self-regulation is revealing. There was support for the self-regulating activities of the Solar Energy Industries Association (SEIA) and Solar Energy Research and Education Foundation

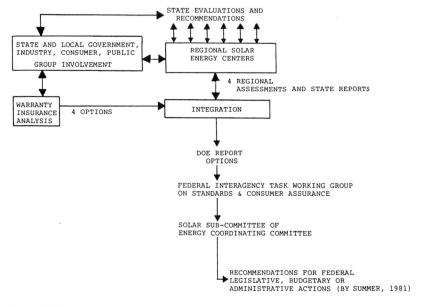

Figure 15.1
SOLCAN organization (U.S. Department of Energy, 1980, p. 6–8).

(SEREF) conducted in close coordination with state and local regulators. There was, however, an "apparent ... significant distrust of the entire concept [of solar industry] self-regulation" (DOE 1979b, 9).

15.2.4.2 Solar Consumer Assurance Program

CSE carried out the portion of the President's message requiring "alternative mechanisms ... for development and implementation of [consumer assurance] programs" in three separate but related parts: (1) state/industry/consumer assessments of mechanisms conducted by the regional solar energy centers (RSECs), (2) a federal interagency working group, and (3) warranty insurance analysis. To convey the integrated nature of the three parts, the term *network* was used, creating the acronym SOLCAN (Solar Consumer Assurance Network) to describe the total effort. An activity organization chart of SOLCAN appears in figure 15.1.

State Evaluations State-by-state assessments were administered by the RSECs, which managed and coordinated the participation of numerous groups and organizations having a stake in consumer assurance. These included solar industry representatives, state consumer protection agencies and energy offices, solar advocates, and private consumer services (such as media complaint handling services and Better Business Bureaus). The plan for conducting assessments to include interaction of these varying interests took into consideration the potential for longer-term cooperative relationships that could be developed along with the final product: a systematic, workable program for consumer assurance in each state that would protect consumers from undue risk while at the same time accelerating the solar market.

This part of the SOLCAN project was probably the most far-reaching of the three. A summary of results and follow-up activities is reserved for section 15.4 below.

Interagency Working Group This group consisted of representatives of fourteen agencies and functioned within a Solar Subcommittee of an Energy Coordinating Committee organized in response to the President's energy message. The working group's purpose was to promote a responsible, coordinated approach among federal agencies in carrying out the goals of consumer assurance, thus reducing unnecessary, duplicative, or contradictory regulation or support of the solar industry. A task description approved by the group underscored the need for interagency cooperation and recommended that a mandate and resources be provided to carry out its work. As part of the SOLCAN strategy, the group was to have made its final recommendations in early 1982, but it became dormant in 1981.

Warranty Insurance The concept of a federally supported warranty insurance program had appeal not only as a consumer assurance mechanism that would enhance information and confidence but also as a means to compensate for start-up weaknesses in the solar market. Before a plan for government investment in such a program could be developed, however, it was necessary to understand how warranty insurance would (a) be applied to different solar technologies and their respective marketing channels, (b) be consistent with or depart from existing solar and related industries' (including insurance) business practices, and (c) relate to prevailing consumer and business attitudes. Accordingly, a detailed draft analysis of the issues and complexities involved in warranty insurance was completed. The next step—distribution of the draft analysis to industry, state, and consumer representatives for comment—would have been integrated with the state assessments described above and considered as a single, comprehensive approach to consumer assurance. The step was never undertaken, however, because of federal budget reductions affecting solar commercialization programs.

15.2.4.3 Related Activities

Solar consumer assurance issues emerged in the implementation of programs functioning tangential to the government's overall solar commercialization effort. The Residential Conservation Service (RCS) Program and the Energy Conservation Program for Consumer Products (ECPCP) are discussed here by way of example, although other instances include residential solar promotions of the Tennessee Valley Authority and Bonneville Power Administration, as well as solar consumer information programs of the Energy Extension Service.

Residential Conservation Service   The RCS came into being with the passage of the National Energy Conservation Policy Act (NECPA) of 1978 (PL 95-619). Its purpose was to encourage the installation of both energy conservation and renewable energy resource measures in homes of customers of gas and electric utilities through state-administered home energy audit programs. The law required states to incorporate consumer assurance measures in their RCS program plans.

Although NECPA did not single out solar energy as being more problematic than other energy conservation industries, as a practical matter, the commonly held perception of the solar industry as being in its infancy led to greater concern for its application in the RCS than for other energy sources. (For example, although fraudulent home improvement practices might occur in the program, mechanisms for consumer redress were already in place through state consumer protection offices and in the private sector through Better Business Bureaus.) In an effort to assist states in dealing with solar consumer protection aspects of their RCS plans, the Solar Energy Research Institute (SERI), under a DOE contract, drafted a comprehensive manual describing effective consumer protection strategies already in existence, including contact persons, where applicable (Vories 1980). Although a useful document, the manual unfortunately was never issued; it is not known whether by itself it would have improved consumer assurance in the RCS. Much of the success of consumer assurance mechanisms was dependent upon effective working relationships, communication and cooperation among state energy and consumer protection agencies—a situation that varied widely from state to state.

Energy Conservation Program for Consumer Products The ECPCP called for energy efficiency labeling of products commonly considered to be household appliances (42 U.S.C. 6293 [a]). Whether solar energy products should be incorporated under the law was a matter debated within DOE, but not in terms of the merits of disclosure of energy efficiency data to consumers as an assurance measure. Instead, the focus of attention was the regulatory burden that would be imposed on the solar industry; further, implementation issues surrounding determination of efficiency ratings could not be resolved. Ultimately, it was decided to exclude solar consumer products from coverage. In so doing, the Department of Energy's Office of General Counsel compared solar products to products it maintained were intended for inclusion under the statute:

Products currently under the ECPCP serve clearly defined needs, ones for which consumer demand and expectations are firmly identified and readily expressed. Consumer confidence regarding life expectancy, maintenance requirements, reliability, and safety is relatively high.... Solar systems, on the other hand, are being presented in a market where neither consumer needs or expectations have been developed. [They] are ... being presented as energy-saving devices, per se,... as a way of meeting what is now primarily a national, rather than a consumer need. (Mussler 1979, 2, 3)

The situation confronting DOE in this instance provides a good example of conflicting objectives in solar consumer assurance. Comparative information was deemed applicable only to products for which a market was developed and for which confidence had already been established. The proverbial chicken and egg dilemma that plagued solar consumer assurance becomes readily apparent through this example.

15.2.5 Department of Energy, Office of Consumer Affairs

The Office of Consumer Affairs (OCA) was intended to coordinate consumer and other public input into DOE's policymaking and program planning and implementation; its mandate was closely aligned with the responsibility of CSE to incorporate consumer assurance in the federal solar program. OCA coordinated much of the public comment on the DPR of 1978 and, through means such as a consumer affairs advisory committee, brought consumer and solar advocates into DOE decision processes. Moreover, in three projects, described below, OCA focused specific attention to consumer protection in solar energy.

15.2.5.1   Energy Saving Devices: Fraud Prevention Project

Conducted by the Metropolitan Denver District Attorney's Consumer Office, this project consisted of a clearinghouse and monthly newsletter devoted to (1) collecting information on alleged fraudulent, deceptive and/or misleading energy and energy-related products, services, and firms, and (2) disseminating this information throughout a network of fifty state attorneys general and sixty-six district attorneys. The clearinghouse employed a consultant to conduct a technical review and analysis of advertising claims for misleading representations. This activity led to the prosecution of a number of offenders as well as to voluntary action on the part of several companies in modifying their advertising claims. Through the newsletter and clearinghouses, the project provided expertise often lacking in consumer protection law enforcement agencies to take swift action against offending firms. In one such matter, the Federal Trade Commission became informed through the newsletter that the clearing-house had conducted tests of a particular window shade. The agency requested the test results and was able to proceed with a case which, due to a lack of technical analysis, had been in abeyance for over a year (Craighill 1981). In its 1981 report the GAO acknowledged the role of the clearinghouse in assisting state and local consumer protection agencies but found the clearinghouse to be ineffective due to a lack of participation at both the state and federal levels. The GAO report also stated that a complete database of test results on energy saving products was not developed by the clearinghouse, although such a database was, in fact, available at the project's end.

15.2.5.2   Guide for Attorneys General

In an effort to assist states in their enforcement of consumer protection statutes as they pertained to energy, OCA published and distributed a consumer protection guide for attorneys general (DOE 1979c). One chapter devoted to solar cited advertising problems such as false claims of "free hot water," "no fuel bills," "never wears out"; deceptive statements about availability of grants, eligibility for tax credits, and representations of nonsolar equipment as solar. Facts about the status of mechanisms such as solar warranties, certification, product standards, building codes, and installer licensing were also included in the guide, along with appro-priate contact persons for obtaining further information.

15.2.5.3 Solar Consumer Protection Workshop

OCA sponsored a two-day workshop on solar consumer protection initiatives in May 1979 at the Solar Energy Research Institute in Golden, Colorado. Invited attendees included representatives of DOE, state consumer protection and energy offices, the four regional solar energy centers, the Department of Housing and Urban Development, and the Federal Trade Commission. An executive summary (DOE 1979b) described the nature and extent of information sharing that occurred among the participants in areas such as standards, testing, certification, warranties, training, and other activities ongoing at the state and federal levels. According to the report, many of the attendees were unaware of the variety of approaches that had been undertaken by different parties.

Workshop participants identified what they considered to be appropriate roles of the agencies they represented in order to minimize duplication of effort and to promote effective communication among the agencies. Additionally, the group unanimously passed a resolution focusing special attention on the role of the then-proposed Residential Conservation Service (RCS). Among the resolution's several statements on conservation measures, one was directed exclusively toward problems of solar energy:

Implied benefits of these proposed [RCS] regulations raise false expectations in the minds of consumers concerning the availability of well-designed, correctly sized, and properly installed and warranted equipment and systems, and the certainty of system performance. It is totally unclear that these expectations can be fulfilled by implementation of the proposed Residential Conservation Service Plan, and failure to achieve these expectations will serve to further erode consumer confidence [in solar energy]. (DOE 1979b, 7)

A set of recommendations concerning future activities of the group suggests that there was a consensus about continuing the communication that had begun at the workshop. The need was expressed to formalize contact through regular programs on selected topics at regional and national follow-up meetings to which industry representatives would also be invited; the DOE Office of Consumer Affairs was asked to coordinate and fund this effort. There is no evidence to confirm that the follow-up activity ever formally materialized, although informal interaction among the participating agencies can be assumed to have resulted from the workshop.

15.2.6   Federal Trade Commission

The Federal Trade Commission (FTC), as an independent regulatory commission having authority to promote competition and protect consumers from unfair and deceptive business practices (15 U.S.C. 41-58), looked at the solar industry from within its Bureaus of Competition, Economics, and Consumer Protection. The areas of concern of these bureaus are inextricably linked to the broad concern of the consumer interest and consumer assurance in the solar market. With regard to promoting competition, the FTC focus on standards and certification is probably most relevant. In exercising its authority to conduct investigations of, and to issue cease and desist orders against, offending firms, the agency took some direct actions to protect both consumers and the legitimate solar business community.

15.2.6.1   Standards, Certification, and Competition

Although not a standards-setting body, the FTC is involved in the process of their development. The agency's role is to provide guidance on how to prevent standards and resulting product certification from restricting competition, market entry, and innovation. Two activities of the FTC in addressing solar standards were (1) a 1977 symposium on competition in the solar energy industry and (2) an inquiry into standards and certification of all industries.

15.2.6.1.1   Solar Market: Symposium on Competition in the Solar Industry   During the 1977 symposium, FTC Chairman Michael Pertschuk stated that the agency had been "engaged in studies and investigations of structure, performance, and competitive conditions in the principal energy industries [and solar is] a logical next step" (FTC 1977, 3). With specific reference to standards, Pertschuk noted the commission's concern for the reconciliation of competition and consumer protection goals: "Are we inadvertently erecting entry barriers? Are there alternative forms of regulation equally compatible with consumer protection and competition goals?" (FTC 1977, 4).

Other speakers' presentations at the same symposium addressed the consensus process in standards and certification development as one avenue to preserve competition. Solar Energy Industries Association (SEIA) President Sheldon Butt commented on the importance of consensus and on his organization's commitment to the representation of large and small companies, design professionals, users, consumers and other interested groups on standards committees, which assured that "all points of view are ... represented and ... that no single point of view is overrepresented" (FTC 1977, 19). Susannah Lawrence of Consumer Action Now made the following observation concerning consumer representation: "With only one voice..., [a] consumer may feel very isolated and powerless. The problem is compounded by the fact that there are few consumer advocates with expertise in both solar energy and standards" (FTC 1977, 87).

Comments such as these during the symposium brought into focus the practical difficulties inherent in implementing a standards-setting procedure that would result in providing consumers with a means of gauging product quality without limiting choice in the market. However, as Sheldon Butt indicated, standards, by their very nature, have some tendency to limit innovation: "From a fundamental point of view, the path of innovation cannot be predicted and the ... group preparing a standard has no way of anticipating the nature of innovative products or designs which may be around the corner" (FTC 1977, 20).

15.2.6.1.2 Trade Regulation Rule on Standards and Certification This activity began in 1974 and ended with a proposed trade regulation rule in 1983 that was never promulgated, largely because of a change in philosophy of the FTC leadership on the broader issue of trade regulation rules in the marketplace, rather than the merits of the proposed rule itself. Chairman Pertschuk evaluates the history of this broader issue during the Carter administration in his book *Revolt against Regulation* (Pertschuk 1982).

15.2.6.2 Energy Task Force
Within the Bureau of Consumer Protection's Division of Energy and Product Information (later part of the Division of Advertising), an Energy Task Force was organized in 1979 to pursue questionable advertising and other practices in a wide range of energy firms. Although many investigations of practices in the solar industry were conducted over the years, only three matters resulted in cease and desist orders. Among the violations cited by the FTC in these cases were exaggerated performance capability claims, misleading payback period representations, safety hazards, and inoperable conditions caused by major defects. In the most flagrant of these, the FTC issued restraining orders and required the offending companies (Champion Home Builders of Dryden, Michigan,

and Solar America of Fairfield, Ohio) to give financial redress to consumers who had purchased their systems (Dershowitz 1986).

15.2.7 Department of Health, Education and Welfare

One of the earliest efforts to recognize the consumer pitfalls in the development and marketing of solar products originated in the Office of Consumer Affairs within the Department of Health, Education, and Welfare
(HEW, now the Department of Health and Human Services). A 71-page
publication titled *Buying Solar* (Dawson 1976), issued jointly by HEW
and the Federal Energy Administration (FEA), warned consumers about
the need for "self-protection" and the possibility of "more disappointed
buyers than satisfied owners" in the early solar market. Throughout, the
booklet emphasized an "informed consumer" as the key assurance mechanism, including extensive tables of such technical specifics as monthly
and annual insolation, normal total heating degree-days, life-cycle costing, and collector configurations. References to "risk" and the need to
employ "engineering counsel" suggest the sponsoring agencies recognized
that information alone would not be a sufficient preventive measure.

15.3 State-Level and Other Initiatives

Because there is only fragmentary documentation of consumer assurance
initiatives undertaken at other than the federal level, this section cites
illustrative examples of activities carried out by state governments and
other groups, rather than attempting a comprehensive coverage of such
efforts.

15.3.1 State Governments

The key roles of state governments in solar consumer assurance rested
with (1) existing consumer protection responsibilities, (2) administrators
of major federal energy conservation grant programs to which solar was
applicable, and (3) solar commercialization projects and programs origi-
nating from funding through the regional solar energy centers (RSECs).
Some state activities may have been linked, directly or indirectly, to the
federal government's role. For example, the Interstate Solar Coordinating
Council (ISCC), jointly sponsored by DOE and states, was formed to
develop common testing standards and certification procedures (Block

Table 15.1
States and programs reviewed

State	Initiative	Primary responsible agency(-ies)
California	Information outreach	California Energy Commission
	Consumer protection	California Energy Commission
	Tax credits	Franchise Tax Board California Energy Commission
Florida	Information outreach	Florida Solar Energy Center The Governor's Energy Office
	Consumer protection	Florida Solar Energy Center
Minnesota	Information outreach	Minnesota Energy Agency
	Consumer protection	Minesota Energy Agency
New Mexico	Information outreach	Energy and Minerals Department New Mexico Solar Energy Institute
	Consumer protection	Energy and Minerals Department New Mexico Solar Energy Institute
New York	Information outreach	New York State Energy Office
		New York State Research and Development Agency
	Consumer protection	New York State Energy Office

Source: Koontz Neuendorffer, and Green (1981, 2).

1980). For the particular activities described here, however, it was the respective state governments themselves that assumed responsibility.

The kinds of solar-specific consumer assurance activities in which states engaged are set forth in table 15.1, based on a study conducted by SERI (Koontz, Neuendorffer, and Green 1981) of programs in five states (California, Florida, Minnesota, New Mexico, and New York). Although this section discusses some of these states' programs as well as other efforts, the classification shown in table 15.1 provides a useful perspective of the roles the states assumed.

15.3.1.1   California's Solar and Insulation Unit

This effort represented a postpurchase/"after-the-fact" means of addressing consumer assurance, in that it was a remedial rather than preventive approach to consumer problems. California's Solar and Insulation Unit (SIU) served a complaint-handling function specifically for solar and insulation products by conducting mediation in consumer and business disputes and referral to other agencies, as appropriate. Notwithstanding this purpose, complaints about false advertising for example, often originated with competing installers and manufacturers (Koontz, Neuen-dorffer, and Green 1981).

Data from quarterly reports of the SIU's consumer services data covering the period from 1 October 1979 to 30 September 1980, revealed that solar cases outnumbered those related to insulation and conservation by a ratio of 3 to 1 (California Department of Consumer Affairs 1980), but of 220 cases in a representative quarterly report (March 1980) 107 (about 50%) were resolved by mediation, while another 66 (33%) were referred to the Contractor's Board for allegedly either (a) doing business without a license or (b) violating licensing requirements. License revoca-tion rested within the jurisdiction of the board and, according to Koontz, Neuendorffer, and Green (1981, 37), "this ha[d] yet to be much of a threat since the contractor's board ha[d] been unwilling to take such [a] step."

The SIU was unique in focusing exclusively on solar and insulation products and services; its existence probably provided consumer assur-ance, as intended, even though it had no law enforcement authority. The difficulty addressing solar consumer complaints in most states was that typical consumer protection offices had virtually no solar expertise and tended to classify these problems under the general heading of home improvement complaints. Most state energy offices were more likely to possess solar expertise but lacked consumer protection knowledge and skills. California's SIU, therefore, reduced consumers' chances of being shuffled from one agency to another when they had a solar-related com-plaint or problem. As the volume of solar consumer complaints grew smaller, the SIU was disbanded; solar complaints were then treated in the same manner as home improvement problems.

15.3.1.2   New York State's Solar Energy Products Act

In 1979 the New York State Energy Office issued proposed rules and regulations governing the state's Solar Energy Products Warranty Act. The purpose of this act was "to encourage the use of solar energy and promote the development of a viable solar industry by providing pur-chasers with access to effective, well-designed, carefully manufactured solar thermal systems; properly installed and properly serviced solar thermal systems; and warranty protection on solar thermal systems" (New York State Energy Office 1979). Directed exclusively at active SDHW systems, the proposal enumerated fifty-six different possible dis-closures required of sellers to consumers, if a warranty accompanied the sale, consistent with provisions of the federal Magnuson-Moss Warranty Act. The list included not only information such as cost and model number commonly available in the course of business transactions, but also the designation of the flat-plate collector's "SunBtu,"—a figure derived from one of several complex formulas based on *ASHRAE 93-77* or *NBSIR 74-635*. The purpose of disclosing this performance information was similar to that of disclosing energy efficiency information—to encourage the manufacture and purchase of more efficient models in the market.

As might be expected, the New York proposal never became law, largely because of objections raised during public hearings by solar industry representatives from the manufacturing and retail levels. To amass the consumer information and technical data required by the proposal and to make these available to consumers would have been nearly impossible for the industry, even if linkages among different levels of manufacturing, distributing, and selling were well established. The technical nature of the information also led to questions about practicality of enforcement.

Such a requirement probably would have had the effect of systems being sold "as is" or entirely without warranty (permissible under Magnuson-Moss) resulting in less, not more, consumer assurance. It is interesting to note, however, the assumption this proposal made about consumer assurance—that by mandating warranty disclosures, the information would be used by consumers to make good choices among competing products. This, of course, does not follow in markets other than solar, and the justification for such a mandate in these cases lies in the principle of consumers' "right to know" certain information about the products they buy.

15.3.1.3 State of Maine's Training and Certification Program
The approach to solar consumer assurance taken in Maine focused on the need for competent installers, through a state-sanctioned, voluntary training and certification program. Structured to avoid barriers-to-entry problems associated with licensing and other mandatory measures, the Maine program made no restriction on eligibility for certification, available by one of four means: (1) completion of state-administered training and examination, (2) demonstration of experience, (3) demonstration of prior training, or (4) reciprocity. A record of consumer complaints against

an applicant had the effect of withholding issuance of a certificate until the problems have been rectified, and following certification, evidence of fraud or negligence was a basis for suspension or revocation.

Generally, this approach has been regarded among policy analysts as preferable to licensing because it (a) permits an open market for entry and innovation, (b) provides a basis for information about installer competency to consumers, and (c) increases the number of trained installers (Casperson 1981).

Voluntary certification does not, of course, preclude an individual from practicing the solar installation trade if a certificate has not been granted (due to failure to apply or meet requirements for issuance, or to suspension or revocation). Such conditions only prevent an installer from being identified as certified. Success of such a program depends largely on whether consumers are aware of certification and make their choices based on this qualification, therefore providing a market incentive for noncertified installers to become certified.

15.3.2 Private Sector

The majority of private sector activity related to consumer assurance was devoted to standards development by the SEIA and the complementary role of private testing facilities. These are referred to elsewhere in this chapter and volume, leaving two other relevant initiatives for this discussion.

15.3.2.1 National Association of Solar Contractors

In its capacity as a trade association for solar installers, the National Association of Solar Contractors (NASC) developed a plan for a self-regulation mechanism in the form of a mediation system for consumer complaints, patterned in form after other industries' successes, the most notable of which being the Major Appliance Consumer Action Panel (MACAP) of the Association of Home Appliance Manufacturers. In late 1980 NASC printed and disseminated a brochure announcing the availability of the service; actual institution of procedures, however, was to await receipt of consumer complaints.

Among solar consumer assurance advocates, the plan was viewed as having the self-serving effect of making the industry appear consumer-oriented in the eyes of consumer protection enforcement agencies (Walsh 1981). That NASC had neither expertise in consumer affairs nor input from consumer groups in developing its plan further undermined the credibility of the program. Programs such as MACAP were carefully structured so that consumer representatives served in the mediation-arbitration process, allaying any suspicion of industry bias in complaint handling and resolution.

15.3.2.2 Consumers Union

One of the most widely recognized, independent organizations contributing to consumer assurance in general is Consumers Union (CU), publisher of *Consumer Reports* magazine. In its report on SDHW systems (Consumers Union 1980, 325), CU concluded that "[s]olar hot water systems are feasible from a technological point of view, but homeowners do not yet have adequate assurance that the systems they buy will be dependable or durable, or properly installed and maintained. They do not have an adequate way to compare the costs, saving, and performance of one brand against another."

In the same issue, CU reported on performance results of five off-the-shelf systems it had purchased as complete, brand name packages, specifying hot water needs for a family of three. The systems were installed at CU's headquarters in Mount Vernon, New York, and had been in operation under what the organization acknowledged were "worst-case" conditions (short, cloudy, winter days) at the time of the report. A savings range between 16% and 55% was reported for the three of the five systems which remained operational through the period.

This report was met with considerable disfavor within the active solar water heating industry, which felt CU had been remiss in publishing only preliminary data. However, in its advice to consumers, the organization stopped short of recommending against the purchase of an SDHW system. It dismissed the figures as "unimportant," stating further that "they ... indicate to us that a well-designed and well-constructed system, operating properly, could provide virtually all of a family's hot water during the summer and 50 to 75 percent of it annually" (Consumers Union 1980, 323).

15.4 SOLCAN: The Solar Consumer Assurance Network

The portion of the SOLCAN project which consisted of state-conducted assessments of consumer assurance mechanisms (introduced in section 15.2.4.2) was probably the most far-reaching aspect of the program. This section discusses both the issues involved in the program's administration and the substantive nature of the work within the four RSECs. In addition, reports to DOE completed by the Consumer Energy Council of America are summarized.

15.4.1 General Implementation Issues

The Department of Energy's Office of Conservation and Solar Energy channeled funds through the RSECs to administer the state assessments of consumer assurance mechanisms. Although this was considered appropriate in concept, some difficulties arose in the implementation of the task, largely from the prescriptive nature of the work. To meet the objectives of SOLCAN, it was essential that public and private organization representatives from the solar and consumer interest communities participate; further, RSEC management needed to become thoroughly familiar with the intricacies of consumer assurance mechanisms and the competing interests involved in order to effectively deal with sensitivities inherent in the project. As a result, DOE closely directed program implementation in a manner tailored to address a "problem" (consumer assurance) from a sound, but highly analytical perspective. RSECs, in general, saw themselves as having results-oriented roles in establishing and maintaining relationships with their clientele; they were not fully convinced that a problem existed and were therefore not enthusiastic about contributing to its resolution, particularly in a prescriptive way. SOLCAN was perceived as bringing together clientele who did not share the same opinions about how solar commercialization should be accelerated. The program's objectives were difficult to communicate, the issues it addressed were sensitive, and its results were not easily measured.

This general implementation obstacle took different forms in different RSECs, depending upon the nature of their working relationships with DOE headquarters (addressed in chapter 20 on the RSECs). For the purposes of this chapter, the obstacle is worth mentioning as a partial explanation for why each RSEC took its own approach to conducting the state assessments portion of SOLCAN.

15.4.2 Substantive Assessments

15.4.2.1 Mid-American Solar Energy Complex

MASEC conducted state assessments in two phases. Phase 1 was an identification of consumer protection statutes and other solar-related regulations in each of the twelve states in the MASEC region (MASEC

1981a). Phase 2 consisted of a report on results of each state's assessment mode—whether by a "town meeting" or "advisory panel" approach or by staff report to MASEC.

Among the conclusions in its phase 2 report (MASEC 1981b, 69), MASEC noted that the dialogue initiated by SOLCAN was useful, but that if the program were to continue in a formal way, efforts should be focused on "specific activities (such as) funding industry-sponsored, voluntary contractor/installer certification ... or developing information/ education materials and a distribution process which could be replicated in other states." As noted above, this expectation of support for specific activities was a view generally held by the RSECs; yet such support was not explicitly stated among SOLCAN objectives.

15.4.2.2 Northeast Solar Energy Center

NESEC's approach to meeting its responsibilities under SOLCAN was probably the most highly organized among all the RSECs. In each of the nine states it served, NESEC supported a state consumer coordinator, located in most cases in the state energy office, to carry out the project. To complement expertise in general consumer protection and consumer participation matters, a state consumer assurance consultant served in an adjunct capacity, reporting to the NESEC program manager. All project leaders met periodically as a group to share information on project management and substantive issues.

The NESEC report (Walsh 1981) consisted of an analysis of the states' reports, rather than a summary of their recommendations. The report concluded that "numerous deficiencies make it impractical under present conditions to rely on any single mechanism to provide consumer assurance.... Consumers, for example, often make unrealistic demands on solar technology while industry tends to place undue burdens on consumer education and information programs" (Walsh 1981, 49). NESEC's final recommendation was nonspecific, calling simply for continued government roles at local, state, regional, and federal levels to "identify and provide the appropriate resources, enabling the solar industry to make its maximum contribution to the nation's domestic energy supply" (p. 50).

15.4.2.3 Southern Solar Energy Center

SSEC's wide-ranging service region—fourteen states from Delaware to Oklahoma, as well as Puerto Rico and the Virgin Islands—made implementation of any program a management challenge. The methodology employed for its role in SOLCAN was to direct designated state solar contacts to conduct "legislative research and interviews in order to assess the perceived problems with solar energy products and businesses and how the various existing consumer protection mechanisms may or do provide prevention and/or resolution of these consumer problems in each jurisdiction" (SSEC 1981, 1).

Based on the above-referenced document and report to DOE, it is not clear why, except for logistical reasons, SSEC adopted the interview approach over the recommended interaction-of-participants format. There is some suggestion, however, that SSEC regarded the program as having the potential for imposing regulatory impediments to the solar industry and felt uncomfortable in being identified with it. For example, in its instructions to state solar contacts on conducting the interviews, SSEC cautioned that:

the purpose of this assessment is to determine the specific laws, agencies and mechanisms which exist in each state. This information will not necessarily result in change of any system but is only an attempt to investigate how the commercialization of solar energy is or could be impacted by consumer protection. This perspective should be communicated to all respondents (especially industry representatives) so that they will understand that the research is geared toward a positive rather than negative attitude with respect to solar energy. (SSEC 1981, 2)

Notwithstanding any lack of enthusiasm for the project, SSEC (1981) noted a need for increased coordination among consumer information outreach efforts, including more publicity about tax incentives, and a training program for code officials to improve the quality of system installation inspections.

15.4.2.4 Western Solar Utilization Network
Although the MASEC and SSFC reports are more descriptive and exploratory in addressing consumer assurance issues, the WSUN report (Sierra Energy Group 1982) exhibits the most purposefully analytical approach of all the RSECs in implementing WSUN's part of SOLCAN. However, WSUN did not adhere to DOE's management plan; it chose instead to develop a "model" SOLCAN program in each of four states in its region, rather than working with all thirteen states.

In developing the model programs, WSUN used a set of ten criteria to analyze consumer assurance mechanisms:

- Cost to the consumer, industry, and the public;
- Availability of resources to make the mechanisms work;
- Ability to respond to new or innovative changes in solar technology;
- Time line for implementation;
- Experience with other consumer products and services;
- Administrative requirements (e.g., data gathering);
- Barriers to enactment and implementation;
- Potential for problem resolution;
- Practical application to consumers and industry; and
- Necessary technical expertise to appraise consumer problems.

Following the drafting of the model programs, in-state meetings of individuals who would be instrumental to their implementation were held, and necessary revisions were made based on these individuals' input. An additional meeting of representatives of all thirteen states, identified as an "inter-tie conference," brought the concepts and strategies of the four models to the full spectrum of industry, consumer, and public agency participation in the WSUN region.

An example of this modeling approach is set forth in figure 15.2 which describes the organization of a proposed Arizona Solar Consumer Assurance Network. Its emphasis on a coordinated complaint-handling process is representative of other model programs developed in the WSUN project.

15.4.3 Summary Reports: Consumer Energy Council of America

The Consumer Energy Council of America (CECA) was, at the time of the SOLCAN project, a consumer interest and research arm of the Consumer Federation of America (CFA), devoted exclusively to energy issues. (It is now a separate organization and no longer affiliated with CFA.) Under contract to DOE's Office of Conservation and Solar Applications, CECA's research staff conducted two separate summary analyses of the SOLCAN state assessments.

The first was an evaluation of the approaches undertaken by the nine states served by NESEC to implement the project (CECA 1980). Referred to as a "rapid feedback evaluation," the analysis was based on information obtained from interviews with state coordinators and attendance at group meetings of project participants. It was intended to assist DOE,

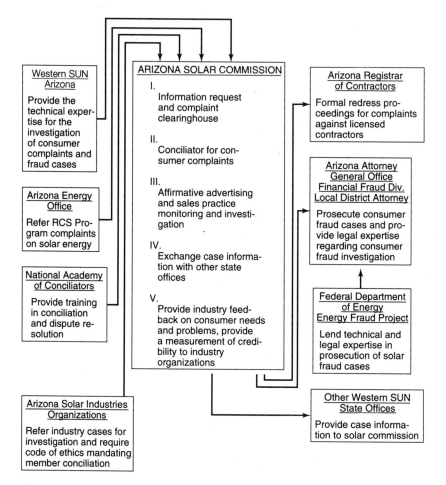

Figure 15.2
Arizona SOLCAN organization (Sierra Energy Group, 1980, p. 16).

RSECs, and states in other regions by examining how approaches to suit different states' circumstances evolved in one region. The evaluation also had an administrative purpose, according to the author of the final report, of enabling DOE and the RSECs to "specify the project steps more closely" (CECA 1980, 3) and recommend elements of the program's design that needed to be strengthened or discarded.

CECA identified four project models that revolved around a general implementation plan devised by NESEC program management. States adopted one of the four, depending upon several factors:

• State's size and diversity (e.g., urban versus rural);

• Dynamics of the administering state energy office, as affected by its energy production (versus conservation) orientation and willingness to devote staff time;

• Solar industry's size, diversity and level of organization;

• Consumer activity with regard to consumer protection legislation and enforcement and organization of nongovernmental consumer advocates. (CECA 1980, 12)

Whether other RSECs and other states made use of the CECA evaluation is not known, inasmuch as none directly referenced the organization in their reports of state assessments. For the purposes of this chapter, however, some of the recommendations affirm some of the general implementation difficulties noted earlier in this section. For example, it was suggested that each RSEC (in this case, NESEC) play a more direct role in areas such as contacting industry representatives about the nature and purpose of the project and "defin[ing] an end product on the basis of what exists in the state and what can be produced in an interactive process" (CECA 1980, 35).

The second study (CECA 1981) consisted of three parts: (1) a general assessment of the need for consumer assurance in the solar market based on survey and industry data, (2) a synthesis of the substantive work of the participants, and (3) an evaluation of the fit between needs identified in (1) and the results of (2). After distilling the states' recommendations by specific types of mechanisms, CECA noted that, in general, the states preferred to avoid the regulatory mind-set of codes and standards and to support assistance to the market by "reinforcing the quality of information available . . . and ensuring effectiveness of warranties" (p. 93). CECA concluded its evaluation by stating that, to be workable, a consumer assurance framework based on the principles developed in the state assessments would have to provide consumers prepurchase information that (a) established realistic expectations based on solar energy system performance and (b) was linked to specific criteria that could be used to trigger an enforceable warranty, where necessary.

15.5 Evaluative Evidence

To what extent was consumer assurance enhanced by the federal government's efforts in solar commercialization? The major effort, SOLCAN, was nearing completion just when budget reductions took effect; therefore, its capability to act further based on states' efforts never was tested. No comprehensive study has been conducted to identify whether states initiated any programs as a result of the SOLCAN experience. Because RSEC funding was no longer available to state energy offices to conduct solar programs, and because funding was also reduced for other energy programs directed towards the residential sector (such as general residential conservation), it is unlikely that solar consumer assurance programs would have been assigned high priority in many states.

Thus, the situation which prompted a government concern for solar consumer assurance in the first place was re-created: a government-sanctioned market incentive was available with no built-in mechanism to minimize risk. Consumers were left to free-market forces for any protections they might require, an outcome held to be desirable in the laissez-faire environment guiding policy and program decisions after the 1980 elections.

How, then, do we evaluate consumer assurance efforts in the years during and after the government's activities described in this chapter? Although a comprehensive response is beyond the scope of this volume, a review of several indicators can help provide a reasonable perspective on consumer assurance efforts: the research literature, including the popular press; at least one warranty insurance effort undertaken without government involvement; and other trends and developments.

15.5.1 Literature Reviews

For the purposes of this chapter, two literature reviews were conducted: one of research in the fields of marketing and consumer affairs from 1982 to 1986 and another of the popular press from 1978 to 1985.

15.5.1.1 Marketing and Consumer Affairs Research

A review of this literature revealed that few, if any, research results were published on the consumer assurance aspects of solar commercialization; current results appear to focus on identifying barriers to the use of solar in residences.

In one trade journal, a report of a California Public Utilities Commission (CPUC) survey of potential solar adopters received some attention (Best 1984). In addition to lack of information and high price, the CPUC identified "distrust of the solar industry—manufacturers, retailers, and installers" as a deterrent to the "high percentage of people who are interested in solar energy equipment [but] back away from actually buying it" (Best 1984, 17). Theoretically, at least, results such as these should have prompted attention by the trade to taking positive steps toward resolution.

15.5.1.2 Popular Press Coverage

In general, the solar consumer focus in the popular literature during the 1978–1985 period was oriented toward information and education rather than consumer protection. The literature review conducted for purposes of this section consisted of an enumeration of titles cited in *Consumers Index* for those years. It revealed a total of ninety solar-related entries, only six of which could be classified as consumer protection subject matter, and all of these were concentrated in the late 1970s (e.g., "The Latest in Consumer Hustles and Rip-Offs" [Harris 1977] and "New Danger: Solar Rip-Offs" [Coffee 1978]). A similar trend is evident in the category of books, with only one apparent consumer protection title (*How to Buy Solar Heating Without Getting Burnt!* [Wells and Spetgang 1978]) appearing among some twenty entries. As a library reference, *Consumer Index* does not include the numerous pamphlets, brochures, and flyers developed and distributed during the same period by state and local consumer groups and agencies (e.g., "Choosing a Solar Hot Water System/Choosing a Solar Contractor" [New Jersey Department of Energy n.d.] and "Consumer Guide to Solar Energy Systems" [Shaw and Bauer 1978]).

It is probably safe to assume that when consumer problems became publicized, consumers had a certain expectation that the government would direct some attention to these matters. Given the general support for accelerating the solar market and the existence of tax credits, this would be true whether or not they were aware of the SOLCAN project, participation in standards development, consumer protection law enforcement, or other activities. Training and product testing (explained in chapters 19 and 14, respectively, and presented in relation to consumer assurance in section 15.1.2) received some attention in the popular consumer press. Articles worthy of note appeared in *Consumer Reports* and *Changing Times*.

In 1981 *Consumer Reports* published a second but more definitive report on its testing of five solar water heaters. For readers looking for economic reasons to install a system, the headline summary of the 1981 article was not encouraging: "You can save energy," it read, [b]ut you might not save enough money to make the switch worthwhile" (Consumers Union 1981, 256). Of the five systems tested, two were recommended, another two were described as "creditable performers" (p. 261) and one was not recommended.

Among the marketplace issues addressed by Consumers Union (CU), improvements were noted in the availability of prefabricated systems and roof-mounting racks to reduce the risk of poor installations. Although CU did not employ an installer, the article noted that installers' experience "is bound to have improved over the past few years, as more and more systems have been installed around the country" (Consumers Union 1981, 259).

In an article published in March 1985, *Changing Times* (CT) was more enthusiastic in its support for solar water heating installations, especially because of the time limit then remaining on federal tax credits. CT does not test products, as does CU, so no systems were recommended. However, CT's report emphasized precautions such as contacting ASHRAE's Solar Rating and Certification Corporation for directories of approved manufacturers and equipment performance ratings. Other advice, although well-intentioned, was not necessarily practical. For example, the suggestion to "pick a licensed contractor" (*Changing Times* 1985, 71) is of little value without indicating the *type* of trade license to look for, particularly because few states license solar contractors and not all states license general contractors. Similarly, advising readers that their state consumer protection offices will "reveal any previous complaints against [a] contractor" is not helpful to consumers who reside in states where consumer agencies do not make public the names of firms under investigation unless the agencies have taken formal legal action against them. Although these kinds of suggestions are common in prepurchase information directed to consumers, they can be even more frustrating and detrimental when applied to products with which the consuming public has had little market experience.

15.5.2 Warranty Insurance

Warranty insurance was viewed as a measure that would minimize risk for both consumers and solar retailers (equipment sales, installation, and servicing) in the event actual performance fell below warranteed expectations. Initially, such insurance was thought to require government subsidy because of the high risk perceived by underwriters in the insurance industry and lack of data to conduct actuarial analysis.

There developed, however, an emerging market for warranty insurance in the case of shared-savings energy conservation programs. As a built-in protection mechanism, energy service companies sought insurance to pay the difference between the energy dollar savings they guaranteed their customers over a specific period and the actual realized savings. At least one insurance firm, Republic Hogg Robinson (RHR) of Pennsylvania, Inc., is reported to have entered this market, providing nearly $50 million of coverage in about eighty individual policies. As of late 1984, another $200 million in approved applications was waiting to be written, and this figure represented only about 50% of all companies which applied and were accepted for coverage (Connor 1984).

This development suggests a workable market mechanism for consumer assurance in the solar industry. Although not universally available due to an absence of claim history, warranty insurance has appeal because of the incentive for the solar industry to monitor and improve performance of systems, reducing the incidence of claims, and thus reducing the premiums policyholders would have to pay for the insurance coverage.

A government subsidy would, of course, overcome the inability of undercapitalized firms to obtain such coverage (much like a consumer applying for credit without a credit history). Without government support, proponents contended, only large stable firms would be able to afford the premiums or absorb the losses of warranty claims. On the other hand, an infusion of federal funds would create an artificial market, attracting marginal, less reliable firms into the industry.

15.5.3 New England Electric System's 1984–1985 Experiment
In July 1984, New England Electric System (NEES) installed 100 new SDHW systems in three states within its service area as part of a second

residential experiment titled "Solar 100/Phase II." Performance results
reported following one year showed a 64% savings for the best twenty-five
systems and 58% for the best fifty systems, with an average for of 49% for
all systems. Among the developments cited as contributing to the improved performance record compared to the 1977 experiment (see section
15.2.1.1) were new types of collectors, components and system designs
that were not available in the earlier project (e.g., vacuum tube collectors
mounted on parabolic-shaped reflectors, copper absorber plates with
highly selective surface coatings and improved glazing materials, and
computer programs for more accurate sizing). Also noted was a generally
higher level of expertise and workmanship on the part of solar contractors
as evidenced by a dramatically improved maintenance record (DeFelice
and Brown 1986).

In 1986 NEES began conducting tests of batch-type solar systems with
passive applications added, and photovoltaic systems in homes in New
Hampshire and Massachusetts, respectively (DeFelice and Brown 1986).
The company's stated objective of holding down growth in demand for
electricity through such projects has been met with less skepticism among
solar and consumer advocates more recently than during the period of the
first experiment (see, for example, Munson 1985).

15.5.4 Florida Solar Energy Center

A systematic state-by-state study would be needed to accurately report
the status of solar consumer assurance at this writing; because no such
undertaking is contemplated, an informal assessment of current activities
in one state is presented as an alternative. Specifically, Florida is examined because of its relatively heavy involvement in consumer assurance
aspects of solar energy. Unless otherwise noted, this section is based on
discussions with Colleen McCann Kettles, associate general counsel and
director of FSEC's Institutional Affairs Division (Kettles 1986).

The Florida Solar Energy Center (FSEC) had early experience in matters relating to consumer assurance. Created by the state legislature in
1974 and located in Cape Canaveral, FSEC serves as a research and
development organization of the state's university system under the administration of the University of Central Florida (FSEC n.d.).

Over the years, Florida has had several mechanisms in place to address
consumer assurance in the solar market. Collector certification labeling
and contractor licensing requirements are considered successful, but only

insofar as they are enforced at the local (city or county) level. General consumer protection statutes, when applied to the solar industry, are regarded as less effective, due to inadequacies in the law and shifting priorities of enforcement by attorneys general at the state level. The same evaluation is applied to private sector efforts, such as Better Business Bureaus, that are not solar-specific. To cite a case in point, for ten years a firm successfully skirted public and private measures aimed at halting deceptive sales tactics and continued to mislead unsuspecting consumers into purchasing solar systems with false claims and representations.

The role of consumer education in alerting the public to such practices, although worthwhile, has its limits, according to FSEC's spokesperson on this subject. Consumers cannot be expected to become familiar with technical aspects of energy savings claims made in the context of a sales presentation. Short of "making the decision for them," it is unrealistic to assume that consumers can be given adequate information to assure that they will make an appropriate choice among solar systems and/or installers.

In 1986 FSEC instituted a program to deal with "orphan systems," created when solar firms go out of business, as many did after the demise of tax credits. In a program FSEC organized in conjunction with the Florida Solar Energy Industries Association (FlaSEIA), consumers who own such systems are referred via a toll-free telephone line to a reliable solar business in their area, although neither FSEC nor FlaSEIA can promise that problems will be rectified. The predicament of these consumers was viewed as something akin to a "cruel hoax": the tax credit served as an incentive to make the solar purchase initially, yet it is consumers who were left "holding the bag" when the companies they did business with went out of existence along with the tax credit.

The center's technical staff (Tiedemann 1986 and Cromer 1986) regards Florida's method for efficiency rating and labeling as being most beneficial to consumer assurance, because it addresses the most critical aspect of a system purchase, is straightforward and relatively easy to explain and understand. According to one FSEC engineer, use of most SDHW systems still requires more time and attention on the part of the consumer than other similar products. The industry is not yet ready to serve those consumers who expect a "plug-it-in-and-forget-it" kind of appliance, although it is possible it could reach that point with mass marketing (Tiedemann 1986).

15.6   Lessons Learned

15.6.1   Ability of Government to Resolve Consumer Assurance Issues

In the field of political science/public administration, literature that addresses implementation of public policy is useful in analyzing the ability of government to deal effectively with consumer assurance issues. One landmark article (Sabatier and Mazmanian 1979) provides a set of five criteria for successful program implementation:

• The program should be based on sound theory relating changes in target group behavior to the achievement of the desired end-state [objectives] (p. 486).

• The statute (or other basic policy decision) should contain unambiguous policy directives and structure the implementation process so as to maximize the likelihood that target groups will perform as desired (p. 487).

• The leaders of the implementing agencies should possess substantial managerial skill and be committed to the statutory objectives (p. 494).

• The program should be actively supported by organized constituency groups and by a few key legislators (or the chief executive) throughout the implementation process (p. 496).

• The relative priority of statutory objectives should not be significantly undermined over time by the emergence of conflicting public policies that undermine the statute's "technical" theory or political support (p. 499).

Although it is not possible to measure the SOLCAN program and the various activities involved in every consumer assurance mechanism against these criteria in the confines of this chapter, a brief, general assessment of consumer assurance may be conducted using Sabatier and Mazmanian's five criteria.

15.6.1.1   Relationship of Theory to Changes in Target Group Behavior and Achievement of Objectives

The fundamental theory underlying consumer assurance was that risk to consumers (who were being encouraged by government policies and programs to adopt solar technologies) could be minimized. While consumers themselves could be identified as a "target group," they were, in reality, the beneficiaries of expected behavior change on the part of the solar industry (including designers, manufacturers of component parts, distributors, retailers, and installers) and various government entities (including federal, state and local consumer protection agencies, state legislative bodies, municipal code authorities, and the like).

Although it is a worthy policy objective to promote consumer assurance for the long-term benefits it brings to all of the above parties, it was probably unrealistic to expect consumer assurance to govern the short-term behavior of all parties. Rather than being distinct and predictable, the target "group" was diverse and always changing in an uncertain market. Indeed, this lack of definition of a target group defies efforts to make concrete associations between underlying theory and actual practice and therefore presents a serious obstacle to implementation.

15.6.1.2   Clarity of the Policy Directive

The policy directive can be identified as the statement in the president's message cited earlier, which called for a "strengthening of current efforts to meet the consumer's need for valid information and assurance that solar equipment and systems [would] perform and last as expected." "Current efforts," although not defined, can be taken to mean existing market mechanisms. This lack of clarity makes it difficult to structure an implementation process that will achieve the desired outcome.

Assuming that it was the intent of the directive to strengthen existing market mechanisms, there were further difficulties involved in defining areas and degrees of perceived weaknesses in those mechanisms. Left open to interpretation among target group representatives, there is (and was, as we have seen) much room for disagreement as to what exactly constituted a strengthening of the status quo.

15.6.1.3   Managerial Skill and Commitment of Leaders of Implementing Agencies

This Sabatier and Mazmanian criterion assumes a very direct relationship between the policymakers (federal government) and the implementers of the policy directive; however, such was not the case in the promotion of solar consumer assurance. The government's role was one of assistance and support to already existing institutions, without (in most cases) having direct lines of authority or reporting. As a result, the individuals upon whom consumer assurance was dependent (state energy office directors, trade association representatives, and consumer group leaders) often harbored feelings of conflict and mistrust toward the federal government in addressing the issues involved. This situation manifested itself as a lack of commitment to the objectives of consumer assurance, regardless of the degree of the federal government's management skill.

15.6.1.4   Support of Chief Executive and Organized Constituency Groups throughout the Implementation Process

Based on the context of the president's "Message on Solar Energy" cited above, support for consumer assurance existed at the chief executive level. Consumer assurance or confidence in the solar market was viewed as essential to the successful acceleration of solar adoption.

As with many presidential messages that endorse principles or render guidelines, the force of law was lacking. It was left to the executive branch agencies to carry out such principles in the course of program implementation. For example, President Kennedy's "Consumer Bill of Rights" guided consumer protection agencies' protocol, without providing consumers with any definitive recourse if such "rights" were thought to be violated. Similarly, the actual implementation of consumer assurance was not within the realm of executive oversight beyond a general statement of support. Although helpful because it acknowledged that consumer assurance needed to be addressed, President Carter's message was of only abstract value in the implementation stage.

Active support for consumer assurance among organized constituency groups can also be said to have existed, but only in principle. Agreement as to the importance of its role in achieving a strong solar market was virtually universal. Difficulties arose, however, in defining just what the mechanism should be to achieve consumer assurance objectives and whether and how these mechanisms should be changed. The lack of consensus in defining the issues involved in achieving consumer assurance made support by constituency groups very difficult to achieve.

15.6.1.5   Effect of Emerging Public Policies on Political Support

A shift in ideology at the executive level in relation to the appropriateness of a government role in solar commercialization clearly would have a negative effect on support for consumer assurance. This, of course, is exactly what happened as a result of the 1980 elections. Severe budget cuts reduced the ability of institutions participating in consumer assurance efforts (state energy offices, RSECs, consumer groups, and industry associations) to keep their focus on this function.

Accordingly, political support for consumer assurance also diminished. Budget reductions were not limited to the solar commercialization program but were all-pervasive, particularly in social programs. Elected officials who had established reputations as solar advocates found themselves having to respond to a host of other constituencies in addressing budgetary matters for many kinds of programs. In realigning their positions, they often ranked solar energy below other programs.

Another factor that should not be overlooked is the public backlash during the early 1980s against the consumer protection statutes of the 1960s and 1970s. Whether this apparent reversal of support for business regulation was a true reflection of consumers' views or the result of an organized effort among business interests is debatable (see, for example, Pertschuk 1982). Nonetheless, it is generally believed that public sentiment, which once favored strong enforcement of laws against false advertising and consumer fraud, became less enthusiastic about proposals thought to be intruding on parental judgment or consumer choice. Examples included the banning of certain kinds of advertising directed toward children and the mandating of air bags in automobiles, with the resulting increase in the purchase price of vehicles. As authors in the consumer affairs literature observed, this did not mark an end to the consumer movement, but rather another phase in its long evolution in the United States (Bloom and Greyser 1981; Metzen 1986).

15.6.2 Implications for Solar Technology Transfer: The Passive Technology Example

Many factors contributed to a dramatic change in the domestic solar water heating market after it was the focus of federal involvement as described in this chapter. Budget reductions severely reduced programmatic activity in the area of commercialization, including consumer assurance. The elimination of tax credits for installations removed government from the role of encouraging solar purchases, and hence from the mission of protecting consumers.

Under this new set of circumstances, a demand for consumer assurance had to be expressed in the market and met by private sector response. However, the decline in world oil prices dampened consumer demand for alternative energy applications, providing little incentive for industry to assume consumer assurance initiatives.

Can consumer assurance goals be achieved in this climate? An example is provided in the area of passive residential solar applications. The resolution of consumer issues was first directed toward active solar water heating because the tax credit incentive was initially only applicable to that technology and because performance and reliability problems had

been reported in early demonstration programs. Concerns about consumer expectations and ultimate performance are no less significant to
passive applications, even though they differ somewhat in nature and
scope.

Obtaining performance and occupant information about passive solar
homes was the objective of a DOE monitoring program conducted by
SERI beginning in 1980. Of three classes of evaluations developed
by SERI to gather the data, two were relevant for analyzing implications
for consumers: (1) class B, which focused on monitoring building performance at low cost through a microcomputer-based acquisition system and
(2) the noninstrumented class C methodology, which used energy audits,
fuel bill assessments, and questionnaires to obtain data about buildings,
performance, and occupants. The third level (class A) was a highly
instrumented level, primarily of use in research and design tool development. When the federal government withdrew from solar commercialization activities following budget cuts in 1981, DOE continued to promote
the use of class B results as a part of its technology transfer work. Its
strategy was to fund the National Association of Home Builders Research
Foundation (NAHB/RF) to analyze the performance results and create
information in a form that would be useful to the home building industry
and other relevant public and private sector parties. NAHB/RF formed
the Residential Passive Solar Performance Evaluation Council (RPSPEC),
consisting of representatives of home builders, lenders, utilities, product
manufacturers, architects, engineers, planners, code officials, and consumers, to assist in with this task (Frey and Zarker 1980). Although the
effort was not officially completed, the concept of a government role in
supporting the transfer of technology research is presented here because
of its potential contribution to consumer assurance.

An approach such as this has the advantage of anticipating consumer
problems before mass marketing occurs, thus reducing the need for
establishing remedial measures. For example, class B of the SERI evaluations included the following information for builders, their suppliers, and
consumers: (1) overheating is less likely to occur in passive residences with
sloped glazing (versus vertical glazing) and properly constructed window
overhangs and (2) more even heat distribution occurs if there is good
internal air circulation (Bishop and Frey 1984). Not only do specific
findings like these assist designers and the construction industry, but they
also alert consumers about what to look for in a passive solar home.

Similarly, class C results (although not formally part of the Residential Passive Solar Performance Evaluation Council's responsibilities) revealed that people who live in passive solar homes identified the following problems: (1) extreme temperature swings, (2) keeping window and glass areas clean, and (3) fading of furniture, walls, and carpeting as problematic (Sachs 1982). This information can help the building industry focus attention on needed improvements; it also can help consumers take preventive measures where possible.

Through a group such as the council, research findings can be integrated into marketing and product development by member industries; these same findings can be incorporated into consumer information channels. The result can be a healthy, stable industry and increased consumer satisfaction, with minimum distortion of competition (further specifics along these lines are developed in Walsh 1985). Groups similar to the council can be organized through already existing trade associations, making government less intrusive than in some of the attempts we have described in this chapter, where a more direct governmental role was often resisted.

15.6.3   Future of Active Solar Thermal Technologies

Is the market for solar thermal technologies on the road to decline because the federal tax incentive no longer exists? This question is explored based on informal discussions with various observers of developments in the industry. Solar proponents often argue that renewables should not be excluded from the government subsidies enjoyed by conventional sources. However, many simply accept that such subsidies are a thing of the past, and some even hold an optimistic view about the future of active solar water heating. They believe that, in the long run, the incentive may have done a disservice to both consumers and the industry by encouraging the entry of marginal firms into the market and that having to respond to the fluctuations of conventional energy prices, rather than relying on incentives, will result in a more viable, credible, and stable solar industry.

Although the period of substantial government involvement in the area of consumer assurance was not without its problems, many advances were made in the design, manufacture and installation and maintenance of SDHW systems. What one expert identified as a "copycat" effect (Tiedemann 1986) seems to have characterized these aspects of the industry, that is, there has been consistent improvement over the years.

The consuming public has grown more aware that energy prices are highly subject to ebbs and flows. As this awareness becomes reinforced over the long run, a stronger demand for energy resources that are less vulnerable to price fluctuations will be expressed in the market. Whether or not there is a federal solar program, the challenge to the solar industry remains what it has always been: to develop and maintain consumer confidence by providing products and services that do the job well, that do not break down often, and that can be quickly and inexpensively repaired when they do.

References

Ackerman, J. 1977. "Solar Heater Tests Prove Disappointing." *Boston Globe*, 15 June.

ADL (Arthur D. Little, Inc.). 1977. *Interim Report on the New England Electric Residential Solar Water Heating Experiment.* Westborough, MA: New England Electric System.

Best, D. 1984. "Why Some Consumers Don't Buy Solar." *Solar Age* June: 17.

Bishop, R. C., and D. Frey. 1984. *Lessons Learned from Monitoring Passive Solar Homes.* Draft. Washington, DC: National Association of Home Builders Research Foundation, November.

Block, D. L. 1980. *Interstate Solar Coordination Council.* Cape Canaveral, FL: Florida Solar Energy Center, September.

Bloom, P., and S. A. Greyser. 1981. "The Maturing of Consumerism." *Harvard Business Review* November–December: 130–139.

BNL (Brookhaven National Laboratory). 1978. *Solar Water Heating Consumer Guide.* Upton, NY.

Boston Globe. 1977. "Solar Protection." Editorial, 19 June.

California Department of Consumer Affairs. 1980. *Consumer Affairs Quarterly Report* March.

Casperson, P. R. 1981. *Competent Solar Installers: Certification and Licensing Issues and Solutions.* Draft. Boston, MA: Northeast Solar Energy Center, November.

CECA (Consumer Energy Council of America). 1980. *Project Models and Project Management Recommendations for the Department of Energy's Program to Evaluate State Capabilities for Solar Consumer Assurance: A Rapid Feedback Evaluation.* AX-0-9406-1. Draft. Washington, DC: U.S. Department of Energy, November.

CECA. 1981. *Assuring Consumer Information and Protection in the Solar Market: Need, Status, Strategy.* EG-77-C-01-4042. Draft. Washington, DC: U.S. Department of Energy, October.

Changing Times. 1985. "Lower Fuel Bills and a 1985 Tax Credit are Reasons for Warming up to Solar Water Heaters Now." March: 68–71.

Coffee, F. 1978. "New Danger: Solar Ripoffs." *Mechanix Illustrated* February: 70ff.

Connor, L. T. 1984. Internal Memorandum, U.S. Department of Energy, Washington, DC, 27 December.

Consumers Union. 1980. "Solar Water Heaters: What They Can Cost, What They Save." *Consumer Reports* May: 323–325.

Consumers Union. 1981. "Solar Water Heaters." *Consumer Reports* May: 256–261.

Craighill, P. 1981. Internal Memorandum, U.S. Department of Energy: Washington, DC, 20 May.

Cromer, C. 1986. Interview with author at Florida Solar Energy Center, Cape Canaveral, 30 December.

Dawson, J. 1976. *Buying Solar.* FEA/G-76/154. Washington, DC: Federal Energy Administration, June.

DeFelice, A. D., and G. Brown. 1986. *Solar 100/Phase II.* Westborough, MA: New England Electric System.

Dershowitz, M. 1986. Telephone conversation and correspondence with author, U.S. Federal Trade Commission, Washington, DC, July.

DOE (U.S. Department of Energy). 1978. *The Great Adventure.* HCP/4635401. Washington, DC, October.

DOE. 1979a. *Domestic Policy Review of Solar Energy.* TID-22834. Washington, DC, February.

DOE. Office of Consumer Affairs. 1979b. *Solar Energy Consumer Protection Workshop*, 16–18 May 1979, SERI, Golden, CO. Washington, DC, June.

DOE. 1979c. *Consumer Protection Issues in Solar Energy: A Guide for Attorneys General.* DOE/IR-0055. Washington, DC, September, November.

DOE. 1980. *The Solar Consumer Assurance Network Briefing Book.* DOE/CS-0190. Washington, DC, June.

Frey, D., and L. Zarker. 1980. *Development of an Industry-Based Residential Building Thermal Performance Evaluation Program: Revised Management Plan.* Washington, DC: National Association of Home Builders Research Foundation, May.

FSEC (Florida Solar Energy Center). N.d. "Inside the Florida Solar Center." Cape Canaveral, FL.

FTC (Federal Trade Commission). Bureau of Competition. 1977. "The Solar Market." *Proceedings of Symposium on Competition in the Solar Energy Industry.* Washington, DC, December.

GAO (U.S. General Accounting Office). 1979. *Federal Demonstrations of Solar Heating and Cooling on Private Residences: Only Limited Success.* Washington, DC. October.

GAO. 1981. *Consumer Products Advertised to Save Energy: Let the Buyer Beware.* HRD-81-85. Washington, DC.

Harris, M. 1977. "The Latest in Consumer Hustles and Ripoffs." *Money* September: 51–65.

Kettles, C. 1986. Interview with author at Florida Solar Energy Center, Cape Canaveral, 30 December.

Koontz, R., J. Neuendorffer, and B. Green. 1981. *State Solar Initiatives: A Review.* SERI/TR-722-882. Draft. Golden, CO: Solar Energy Research Institute, August.

Leavitt, T. 1960. "Marketing Myopia." *Harvard Business Review*, July–August: 45–64.

MASEC (Mid-America Solar Energy Complex). 1981a. *State Solar Consumer Assurance Network (SOLCAN) Report, Phase I.* MASEC-R-81-003. Minneapolis.

MASEC. 1981b. *State Solar Consumer Assurance Network (SOLCAN) Report, Phase II.* MASEC-R-81-004. Minneapolis.

Metzen, E. 1986. "Consumerism in the Evolving Future." In Bloom, P., and R. B. Smith, eds., *The Future of Consumerism.* Lexington, MA: Lexington Books.

Moore, D. C. 1981. "Lessons Learned from the HUD Solar Demonstration Program." Reprint from *Proceedings of the 1981 Annual Meeting*, American Section of the International Solar Energy Society. Newark: University of Delaware.

Munson, R. 1985. *The Power Makers*. Emmaus, PA: Rodale.

Mussler, R. 1979. Internal memorandum, U.S. Department of Energy, Washington, DC, 2 August.

New Jersey Department of Energy. N.d. *Choosing a Solar Hot Water System/Choosing a Solar Contractor.* Newark, NJ.

New York State Energy Office. 1979. *Summary, Proposed Rules and Regulations, Solar Thermal Systems.* Albany, NY, December.

Pertschuk, M. 1982. *Revolt against Regulation.* Berkeley: University of California Press.

President. Office of the Press Secretary. 1979. "Fact Sheet: The President's Message on Solar Energy." Washington, DC, 20 June.

Sabatier, P., and D. Mazmanian. 1979. "The Conditions of Effective Implementation: A Guide to Accomplishing Policy Objectives." *Policy Analysis* May: 481–503.

Sachs, B. 1982. "Class C Passive Solar Performance Evaluation: Summary of National Analysis and Regional Comparisons." In J. Hayes and C. B. Winn, eds., *Progress in Passive Solar Energy Systems*, vol. 7, pt. 1. Boulder, CO: American Solar Energy Society.

Shaw, N., and J. Bauer. 1978. *Consumer Guide to Solar Energy Systems.* Trenton, NJ: New Jersey Public Interest Research Group.

Sierra Energy Group. 1982. *The Solar Consumer Assurance Network.* March.

Smith, R. 1977. *Report Summary of Performance Problems of 100 Residential Solar Water Heaters Installed by New England Electric Company Subsidiaries in 1976 and 1977.* 419-919-S. Upton, NY: Brookhaven National Laboratory, October.

SSEC (Southern Solar Energy Center). 1981. *Solar Consumer Protection State Assessment Guide.* Atlanta, March.

Tiedemann, T. 1986. Interview with author at Florida Solar Energy Center, Cape Canaveral, FL., 30 December.

U.S. House. 1978. *Solar Energy and Today's Consumer.* Report by the Subcommittee on Oversight and Investigations of the Committee on Interstate and Foreign Commerce. 95th Cong., 2d Sess. Committee Print 95-75. Washington, DC.

Vories, R. 1980. *Solar Consumer Protection: A Discussion of Needs and Sample Programs.* Working draft. Golden, CO: Solar Energy Research Institute, May.

Walsh, R. 1981. *Solar Consumer Assurance: An Assessment of Mechanisms in the Northeast.* Boston, MA: Northeast Solar Energy Center.

Walsh, R. 1985. "Consumer Issues in Energy Efficient Construction: Public and Private Sector Roles and Responsibilities." *Proceedings of the American Solar Energy Society* 420–426.

Webster, B. 1977. "Test Indicates Solar Water Heating in Homes Not Competitive Yet." *The New York Times*, 15 June.

Wells, M., and I. Spetgang. 1978. *How to Buy Solar Heating Without Getting Burnt!* Emmaus, PA: Rodale.

Yarosh, M. M. 1979. *Inspections and Case Histories of Private Sector Solar System Installations in Florida: Interim Report.* FSEC-ESH-79-3. Cape Canaveral, FL: Florida Solar Energy Center.

V SOLAR THERMAL INFORMATION

16 Consumer Information

Rebecca Vories

Consumer information is a catchall category that covers a wide variety of materials and audiences. The working definition for consumer information in this chapter will be information with two primary purposes: (a) to explain federal programs to the general layperson and/or (b) to explain solar thermal technologies to the general layperson or to organizations primarily serving the "general public."

Among government agencies, consumer information or "outreach" and frequently even "technology transfer" efforts are generally directed to a broad, undifferentiated audience and cover a wide variety of formats, although print is the primary method used. Even though there is much discussion of target audiences, these are usually still seen in very broad terms—for example, homeowners, low-income energy users, and small businesses.

Marketing professionals, on the other hand, rarely attempt to address the "general consumer" but, rather, identify a number of market segments—differentiated by demographic characteristics (education, income, profession) and psychographic characteristics (beliefs, attitudes, knowledge, behavior)—that are more or less likely to adopt a particular product or idea in the most immediate time frame. In addition, there are influence groups or organizations that play a very important role in legitimizing products and ideas and that have different information needs from those of specific market segments. Various formats are used to reach the many audiences that need information with as much reliance as economically possible on electronic media and existing print media, rather than on stand-alone publications.

Most solar thermal energy consumer information programs undertaken by the federal government concentrated on providing undifferentiated, stand-alone printed materials. Although some programs did contact and work with influence organizations, they rarely prepared materials specifically aimed at helping these audiences more successfully influence the ultimate consumer. A few programs did attempt to more carefully define specific market segments but were rarely able to provide the types of materials and incentives that would motivate those segments to take action.

This chapter will focus mostly on the preparation of consumer information rather than on its distribution (covered in chapter 17); it will not address solar consumer assurance or fraud issues (covered in chapter 15).

The rare occasions on which specific programs were developed to reach target audiences, influence audiences, or use market research to prepare information will be covered in the discussion of efforts by each organization.

16.1 General History

This section will briefly describe the federal agencies involved in solar thermal consumer information and the types of programs they operated. Additional history will be provided in the discussions of each specific program.

16.1.1 Mandates for Consumer Information

There have been virtually no specific legislative mandates for the development and distribution of solar thermal consumer information. The impetus for providing this type of information comes more generally from the federal government's role in providing information on its activities. This implied mandate has been perceived in different ways over the years. In 1973, with the initial impetus for solar thermal programs, the general attitude was that the federal government should provide as much information as possible at no charge or at minimal cost recovery. Since the early 1980s, this mandate has been viewed more narrowly throughout the federal government, and fewer dollars have been committed to any type of information programs. What information is provided is generally sold at full cost recovery.

The one program that did have a specific legislative mandate for providing solar thermal consumer information was the Energy Extension Service (EES) as it was created as part of ERDA in 1977 (P.L. 95-39, Title V, Energy Research and Development Administration Appropriation Authorization). Its primary emphasis as stated in the law was "successful implementation of energy conservation and new energy technologies will require both public awareness and individual capability to use the conservation opportunities and new technology ... [which] can only be achieved on a national basis by an active outreach effort" (P.L. 95-39, Sec. 502). The law also required that a comprehensive plan and program for "Federal energy education, extension and information activities" be prepared by the ERDA administrator to coordinate the activities of all federal entities in this effort (P.L. 95-39, Sec. 508). This led to a yearly publication generally entitled *The (First) Report to Congress: Compre-*

hensive Program and Plan for Federal Energy Education, Extension, and Information Activity (DOE 1978–1984a, 1985a, 1986), which described the energy conservation and renewable technologies activities carried out by between eight and eighteen federal agencies (depending on how many agencies were involved each fiscal year). The main concerns to be addressed by the ERDA/EES plan were duplication of effort and avoidance of unfair competition with private sector efforts, although it was generally acknowledged in the reports that little was done to mitigate duplication. Much of this chapter's information on budgets, goals, and objectives is taken from this series of reports, which ran from 1978 through 1986.

16.1.2 Overview

This chapter will discuss major consumer information programs rather than individual consumer information pieces, except for a few examples. The programs to be covered are those administered by the following agencies: the National Science Foundation (NSF), the Federal Energy Administration (FEA), the Energy Research and Development Administration (ERDA), the Department of Energy (DOE, which replaced FEA and ERDA), the National Center for Appropriate Technology (NCAT), the Appropriate Technology (AT) Small Grants Program, the Solar Energy Research Institute (SERI), the American Council to Improve Our Neighborhoods (ACTION), the Department of Agriculture (USDA), the Department of Health and Human Services (HHS), the Department of Housing and Urban Development (HUD), the Tennessee Valley Authority (TVA), and the Bonneville Power Administration (BPA).

Where available, the following kinds of information will be provided on each program: general description of program, organizational location, budget history, audience(s) addressed, goals and objectives, general approach to preparing materials, method of distribution, results, evaluation activities, and lessons learned.

16.1.3 General Characteristics of Consumer Information Activities

The provision of solar thermal consumer information can be divided into three periods: the initial years (1973–1977), the growth years (1977–1981), and the lean years (1982–present). These periods basically track the changing political emphasis in how to deal with domestic energy needs.

The initial years reflect the first response to the Arab oil crises; the growth years reflect the view that solar and other renewable technologies were a major element in supplying future energy needs; and the lean years reflect a change in policy direction to a more laissez-faire approach to energy policy.

The initial years consisted primarily of activities carried out by FEA and ERDA; almost all information generated in this period consisted of brochures. During the growth years, many programs were carried out within DOE and other federal agencies; many more communications techniques were used during this era, even though the primary technique was still brochures. Other techniques used were mobile and stationary exhibits, photo displays, books, audiovisual presentations (primarily slide shows, but some films as well), press releases, magazine feature stories, and television news clips. The emphasis during this period was to use state Energy Extension Service (EES) offices and regional organizations such as the regional solar energy centers (RSECs) as the primary delivery mechanism for solar information.

Since 1982 (the lean years), there has been relatively little new consumer information generated, and once again, printed materials have been the primary communication form.

During none of these periods was there an overall, long-term communications strategy regarding what to communicate to whom, or when and how to communicate it, either for the federal effort as a whole or for any particular agency. In general, the information produced during all these periods was developed in the following manner:

• A technical program office would decide independently that information was needed. This office would provide draft copy that would be turned into an actual publication by an in-house public information office or, under contract, by a communications firm. Alternatively, an in-house public information or similar office would decide to produce a document, exhibit, and so on; interviews would be conducted with technical staff to obtain information, and production would be carried out in-house or under contract. There was rarely any outside or private sector input into this process.

• Printed materials were more likely to be developed in-house (except for actual printing), whereas exhibits and other audiovisual materials were usually contracted to outside providers with program oversight of copy and images selected. For the most part, printed materials were $8\frac{1}{2}'' \times 11''$ in format (or smaller), under 20 pages in length, and one- or two-color.

• The writing style was generally very straightforward and concentrated on technical descriptions, and payback or life-cycle economics of the technologies, accompanied by photos or line drawings. For better or worse, the solar thermal industry imitated this approach, particularly in the early years, and used much of the federally produced information as their primary information vehicles.

• Yearly budgets were incomplete, indicating only that some number of brochures, slide shows, and so on would be produced for a specific amount. Occasionally, specific items were spontaneously arrived at to meet a specific program'need.

• Most of the consumer information materials described DOE or other agency programs. Where solar thermal technologies were described, it was generally to say what DOE and others were doing to develop the technology, rather than to provide decision-making information (cash flow implications, where equipment was available, how to finance, etc.) consumers could use to actually take advantage of the equipment available.

• Materials that did address consumer decision criteria tended to focus on payback calculations that showed very favorable returns for investing in solar energy systems. Until the early 1980s it was widely assumed that utility rates would continue to escalate well above the general rate of inflation, and this assumption contributed to very optimistic payback projections.

16.1.4 General Budget History

The annual reports to Congress cited earlier (see section 16.1.1) attempted to give ballpark figures on the amount of money dedicated to energy education, extension, and information activities by all agencies. This was frequently done by taking a percentage of the total budget of an agency or particular program rather than by identifying specific budget items dedicated to communications. In some cases the total budget for a program was given with no attempt to determine the amount dedicated to communications. Usually these budgets did not reflect federal personnel costs and were based on appropriations rather than actual expenditures. The percentage of overall program dollars spent on these functions were usually well under 10%. These budgets covered information programs for all forms of energy conservation and all renewable technologies (see table below):

FY	Total budget (in millions)
1977	$79.1
1978	76.0
1979	148.0
1980	285.0
1981	335.2
1982	222.4
1983	246.0
1984	*
1985	*

* No funds earmarked for information.

No funds were provided specifically for this purpose after 1982. As stated in the Seventh Report (FY 1983), "a less promotional approach to information dissemination has permitted budget reductions and resulting efficiencies. Refinement and economy of operations have eliminated unnecessary costs for periodicals, audiovisual productions and exhibits" (DOE 1984a, 2).

16.1.5   Evaluation Activities

As stated in the first report on federal information activities, scientific evaluation was not viewed to be a cost-effective way to spend limited money available for information efforts and this continued to be the case throughout the federal effort. The majority of programs assessed their impact by counting the number of publications distributed or the number of calls or letters answered. Informal surveys and response cards were also used to determine how well people responded to the information provided. Most program managers indicated that the most useful information was obtained by sharing experiences with other program managers about what did and did not work and why. One of the main reasons advanced for not undertaking more in-depth evaluation was that it would be impossible to separate the impact of any particular federal effort, given the amount of information that was available from other sources and the lack of standards to determine what a successful program would be (DOE 1978, 28–34).

The Third Report to Congress (FY 1979; DOE 1980, 45–49) observed that

- the effort evolved "from the bottom up," by individual program response to perceived needs rather than from a central vision with guiding principles;
- activities were not viewed as part of a comprehensive, integrated program;
- insufficient information existed to determine whether the most effective and efficient activities were being pursued;
- insufficient information was available to determine whether existing duplication was desirable; and
- insufficient emphasis was placed on program evaluation to determine impact on client behavior and energy savings.

Because many years have passed since some of the programs were initiated, and the documentation of many agency information activities was sparse at best, it has not been possible to review each program in depth. What information remains that was reasonably accessible has been used to provide information on individual agency activities.

16.2   Programs Conducted by NSF, FEA, and ERDA

These efforts took place early on and were later supplanted by activities undertaken by DOE and the national laboratories.

16.2.1   National Science Foundation

NSF's primary role was in technology development and to a lesser degree in developing education programs; the foundation had no specific consumer information program for solar thermal activities.

Among its technology transfer activities, NSF supported a variety of intergovernmental programs that assisted state and local governments in using science and technology to address policy issues. Carried out primarily through the Urban Consortium, Urban Technology System, and the Community Technology Initiatives Program, these programs covered many technology areas. Their energy activities seem to have focused on local government energy management issues; at annual budgets ranging from $4 million to 5 million, most of these activities were undertaken during FY 1979–FY 1981 and seem to have disappeared after that time.

16.2.2   Federal Energy Administration

The FEA briefly preceded ERDA, but after both organizations were in existence, they divided their roles so that FEA undertook more policy and public information activities and ERDA focused more on funding and administering research, development and demonstration programs. Information was prepared for the general public. At least in some cases, the goal was to convince the public that although solar energy was in its infancy, consumers had the major responsibility to make solar a success in the economy (Dawson 1976). FEA's objective was to provide the most technically accurate information possible.

One example of the type of information prepared by the FEA was *Buying Solar* (Dawson 1976). This 70-page publication describing active water and space heating systems is typical of publications developed during that era. *Buying Solar* combined a concern for consumer protection with an attempt to provide technical details about the operation and installation of residential-scale solar thermal systems. It covered such topics as types of collectors, storage systems, system operation, system installation, and life-cycle cost analysis; it presented tables showing heating degree-days and mean radiation data for most areas of the country, as well as several case studies showing sizes and types of systems recommended for different situations. For additional insight, the publication was reviewed by people at ERDA and the Solar Energy Industries Association (SEIA).

16.2.3   Energy Research and Development Administration

ERDA's primary mission was to develop all energy sources to make the nation self-sufficient in energy while protecting health and the environment. Although its specific consumer information mandate is not known, ERDA did provide public information on the technologies it developed.

The ERDA Office of Public Affairs had much of the responsibility for providing consumer information, which was prepared for the general public. ERDA appears to have relied on materials prepared by others as its main approach to making materials available. It put together packages of reprints from such sources as *Mother Earth News, Mechanix Illustrated, Science News, International Solar Energy Society-American Section,* and *Popular Science.*

ERDA also sponsored the preparation of booklets such as *Solar Energy* (Eaton 1976). The text was prepared by an engineering professor at Yale University; it covered solar thermal, solar electric technologies, biomass conversion, and energy planning for the future, focusing on technical developments in each field.

ERDA encouraged inquirers to contact the Technical Information Center in Oak Ridge as the primary source for publications.

16.3   Energy Extension Service

Because of the perceived success of the national Cooperative Extension Service in the agricultural sector, Congress felt that a similar program should be created to provide more one-on-one information on conservation and solar energy. The National Energy Extension Service (EES) Act was passed in 1977.

16.3.1   Program Description

The EES was first established as a two-year pilot program. ERDA issued a Request for Proposals (RFP) to the states in 1977, and from their responses ten states were selected to establish pilot programs that would measurably reduce small energy users' consumption through promotional, outreach and demonstration activities.

After a year of operation, the pilot program was evaluated (DOE 1980; see section 16.3.6 for details), and all states were invited to participate in the program. Since 1980, all states and seven territories have operated such a program.

The four primary services provided by the EES have been distribution of a wide variety of brochures and reports covering many energy topics; workshops on various topics; practical demonstrations of available technology, and a centralized clearinghouse of information and one-on-one counseling (walk-in and by telephone). The main focus has been on providing personalized information and technical assistance on energy conservation and the use of renewable resources. Since 1985, the EES has been the only major consumer outreach program that is still locally available, and its budgets have been significantly reduced.

16.3.2   Organizational Location

The EES program was started under the auspices of ERDA's Conservation, Buildings, and Community Systems Program, directed by Maxine Savitz and located with the Assistant Administrator for Institutional Relations, under the direction of Judith Liersch. Later it was passed on to DOE and went through several organizational locations; Mary Fowler and Ron Kessler were long-term federal employees who played a major role in shaping the program. The EES is currently managed by the Office of Financial and Technical Assistance.

At the state level, various organizational structures have been developed or used to house this program. Most are supervised by state energy offices. In some states the program is run by the same university office that operates the agricultural extension service; in a few others it is operated by a state consumer affairs office, another university office, or a nonprofit organization. In some states the activity is centralized in one office, while in others there are "extension" offices throughout the state providing information on a local level.

16.3.3 Budget History

Between FY 1977 and FY 1987, $127.7 million in federal grants was given to states to support the EES program (DOE 1984b, 1984c, 1985b). States provided a 20% funding match for these grants. Since FY 1982, the program budget has been cut by 50% (see table below). Additional funding has been made available to this program in some states from oil overcharge monies. For example, the EXXON settlement made over $2 billion available to states, to be used in five conservation programs including the EES (Western 1986), but there was no requirement about how much money would go to each program.

FY	Total appropriations (in millions)
1977	$7.5
1978	7.5
1979	15.0
1980	25.0
1981	22.0
1982	9.6
1983	10.0
1984	10.0
1985	9.6
1986	7.5
1987	4.0*
1988	0.0

*EES 1987.

16.3.4 Audiences Addressed/Goals and Objectives

The initial audience addressed by EES offices was residential customers. Over time, however, states have developed programs targeted for many other "small energy-consuming" audiences, particularly agriculture, small businesses, local governments, and public institutions. Each state has been comparatively free to determine its own strategy and priorities.

The primary goal of the EES program has been to provide the technical assistance necessary to help people make "good" decisions regarding energy improvements for their buildings or operations. It was felt that EES offices would provide the same kind of technical credibility that agricultural extension·agents have provided in assisting the agricultural sector to adopt new farming practices. Initially, information focused on the technical intricacies of conservation products and practices; later the focus was more on economic implications. Another primary goal was to pull together as many existing resources as possible in a cooperative effort to address energy issues.

Although energy conservation information was the original focus of the program, over the years more and more solar thermal information was added. Some state programs have focused more on solar thermal than others. For example, in Georgia fifteen local governments were selected to sponsor hands-on solar demonstration projects, ranging from breadbox collectors to sunspaces installed on public buildings (EES 1987). With the decline in federal support, there has been much more emphasis by some program operators in finding other sources of funding.

16.3.5 Approaches to Preparing and Distributing Materials

Apparently, no one has ever attempted to catalog all the materials developed and used by the various EES offices, although Ron Kessler at DOE's headquarters gathered much of the material produced in each state during the early years of the program and made it available to other states. Much of the material is in the form of brochures and booklets initially prepared by various federal agencies. Many of the state and local EES programs then used their own staff to modify these materials to meet their own needs or produced entirely original materials.

In general, workshop materials were prepared at the local level and consisted of handouts gleaned from a variety of sources or specially created workshop handbooks in addition to the slides and other illustrative materials used in presentations. Total publications number well in the thousands, with hundreds of workshops having taken place as well.

Most materials have been distributed on request or through workshop handouts. Many EES centers have a regular newsletter that goes out to a mailing list that has been developed by keeping the names of people who call in for information or attend workshops. These newsletters announce the availability of new printed materials or opportunities to attend workshops. Targeted announcement brochures have been used to attract people to workshops and occasionally to tell of the availability of new publications.

16.3.6   Results and Evaluation Activities

Every state has operated an energy extension service at some time under this program. Because each state has taken a different approach to implementing the program, it is difficult to assess the results across the board. In general, the programs have been successful in providing information to residential customers, but little is known about how this information was then used by the customers in actually deciding whether to use conservation or solar technologies.

The EES is one of the few programs that has been frequently evaluated, although the sophistication of the evaluations has varied tremendously between states and from year to year. The most sophisticated evaluation took place while the EES was still a pilot program, before the service was established in all the states. This initial evaluation made two main recommendations: (1) that each state develop its own model rather than using a single approach; and (2) that communications be maintained among states so that they might learn from one another.

The evaluation of the pilot phase (DOE 1981a, A-55) concluded that:

• the cost of energy savings achieved by clients of the EES was the equivalent of spending $8.50 for a barrel of oil (during a time when imported oil cost $30–35/barrel);

• $16 of private investment was leveraged for every dollar of program funds;

• 90% of those surveyed found EES information to be as useful as, or more useful than, similar information from other sources; and

• 44% of those who had received information from an EES office planned to take additional actions.

Since the pilot program evaluation, state programs have submitted reports on the number of contacts with citizens, the amount of information distributed, and the estimated amount of energy saved based on these activities. The reliability of these reports varies greatly and it has never been corroborated or consistently compiled and analyzed for more than a one-year time period. As of December 1981, DOE concluded that there had been a favorable net return on the investment of the federal funds in this program (DOE 1978–1981b).

16.3.7 Lessons Learned

• When programs are based on the idea that information alone will result in savings, they frequently do not assess the actual savings achieved as a result of this information. When funds become short, the lack of cost-benefit data makes it hard to justify continued spending of funds on the programs.

• Audiences need to be carefully targeted and provided with information that meets their specific requirements, rather than generic information.

• Although the impact of EES activities since 1982 has not been documented, there is a feeling in some states that the EES program is probably the most effective program funded by the federal government to address consumer information needs. The ability to assist citizens one-on-one has led to the development of programs at the state level that are reasonably responsive to the marketplace and the information-seeking habits of targeted energy users in each state. By making available toll-free numbers and encouraging residents to come to them for information, some EES offices have also provided a feedback mechanism regarding abuses in the marketplace as well. Inquiries to these hot lines regarding sales practices or energy-saving claims of certain firms have frequently served as a first alert to shady practices.

16.4 Appropriate Technology Grants and National Center for Appropriate Technology

Both of these programs focused on overseeing the development and use of small-scale, locally adapted technologies for relatively small energy consumers.

16.4.1 Appropriate Technology Small Grants Program

Established at ERDA by Congress, the AT grants program began in 1977 as a pilot program in Federal Region IX (California, Arizona, Hawaii, and the Pacific Island territories); it continued as a pilot program in the Northeast and Midwest, with national funding starting in 1978. The focus was on research and technology transfer projects to develop and demonstrate small-scale energy-related technologies appropriate to local needs and skills. Three levels of grants were available: up to $10,000 for concept development; up to $50,000 for development of a concept into a practical technology; and up to $50,000 for demonstrating a technology under operating conditions (DOE n.d.).

Between 1978 and 1981, about 2,200 projects were funded, covering a wide range of conservation, solar, and other renewable energy technologies (DOE 1985c). Ideas were solicited by announcements in the *Commerce Business Daily*, newspapers, trade and technical journals, and the publications of state and local governments and other associations and groups.

16.4.1.1 Organizational Location

The AT grants program had a number of different homes. In 1977 it was part of DOE's Division of Buildings and Community Systems; in 1980 it was in the Office of Small-Scale Technology and headed by Web Otis; and in 1983 it was part of the Office of State and Local Programs.

Funds were divided on a regional basis and overseen by the DOE regional offices. State advisory committees were formed to evaluate and select the specific efforts to be supported.

In 1983 the program issued an RFP for the establishment of the National Appropriate Technology and Assistance Service (NATAS), which was awarded to NCAT (see below). The AT grants program was phased out in 1984.

16.4.1.2 Budget History

The initial pilot program provided $1.3 million (DOE n.d.).

During FY 1978 and FY 1979 very little of the overall budget was devoted to information efforts. Information efforts started in 1980, but little is known about the part of the budget devoted specifically to information activities (see table below for total budget figures).

FY	Total budget (in millions)
1978	$3.0
1979	8.0
1980	12.0*
1981	**
1982	2.9
1983	1.5
1984	-0-

* $1 million earmarked for information.
** No funds earmarked for information.

16.4.1.3 Audiences Addressed/Goals and Objectives

There were two principal audiences for the AT grants program. The first was local inventors and entrepreneurs (including nonprofit organizations) who had ideas about technical research and information transfer activities that would meet primarily specific local needs. The second was consumers within the state who could make use of the technical improvements or information approaches proposed. Specific audiences varied widely for each project.

The primary objective of this program was to tap the talents of innovators and entrepreneurs in a way that would address the need for different solutions to energy problems in different parts of the country. Technology transfer was a major goal, not just research; another goal was to provide access to DOE resources that might not be used by innovators and entrepreneurs without such a program.

16.4.1.4 Approaches to Preparing and Distributing Materials

All grantees were required to deliver final reports to the appropriate DOE regional office. Some AT grants directly funded information development; these were generated and usually distributed by the grantee. In other cases, a state agency might turn the final report into a booklet for wider distribution. For example, in 1983 the program published thirteen reports covering specific projects it had funded as a way of letting others know of the successes achieved.

Many organizations were involved in the distribution of the results of the AT grants program. DOE regional offices published lists of the grants that were awarded with brief summaries. The grantees themselves made a variety of efforts to disseminate the technologies they developed or information about the design and results of those technologies. Beyond this, it appears that there was no formal distribution plan, and much depended on the specific project, the enthusiasm of the grantee for getting the word out, and the priority placed on the specific project or the AT grants program by each DOE regional office and or state energy office.

16.4.1.5   Results and Evaluation Activities
The initial AT grants pilot program received 1,100 proposals and funded 108 of them. Between 1978 and 1981, over 2,200 projects were funded and a wide variety of products and information developed; some of the technologies found their way into the marketplace.

The funding of new projects was discontinued in 1982, when attention was turned to evaluating the impact of this program. Results of this evaluation are unknown.

16.4.1.6   Lessons Learned

• Program administrators were swamped by the response to the pilot program and later solicitations. The program tapped a deep well of creativity and interest and promoted the idea that local solutions were just as important as national solutions to energy production and conservation.

• There was very little outside corroboration of the results achieved by the AT grantees. A program which funds such a large number of projects stands to fund a number of "losers" as well as "winners". The AT grants program had a good level of tolerance for this, but it is hard to tell how many bad ideas received attention in the marketplace because of the decentralized nature of the program and the lack of in-depth technical expertise on the part of its administrators.

• Based on the enthusiastic response by inventors, the AT grants program was one of the most popular programs funded by DOE.

16.4.2   National Center for Appropriate Technology

The purpose of the NCAT program was to provide a national information source regarding low-cost, small-scale, locally differentiated technologies that met local needs. In addition to research, information activities carried out included creation and distribution of printed materials, workshops, workbooks and reports, an inquiry and referral service, and one-on-one counseling for consumers and small businesses. NCAT for some time operated a toll-free number for national queries on appropriate technology subjects (the National Appropriate Technology Assistance Service, or NATAS).

It might be said that the NCAT program was the most ideological in nature of all the federally funded programs as witnessed by an early publication of the first director, Isao Fujimoto, which stated in part,

A basic premise of appropriate technology is that it is a technology where all people can get involved—a technology more sensitive to people and the earth's resources ... a comparison between the values of appropriate technology and big technology shows real contrast—as real as the difference between self-reliance and dependence, self-determination and centralization, cooperation vs. competition and accountability as opposed to exploitation. (Fujimoto 1977)

The NCAT information program has always been multifaceted and has provided information through multimedia presentations, publications, field extension workers, newsletters, and workshops or conferences. In addition to information activities it provided grants and conducted research. It also made extensive use of regional advisory panels (NCAT annual).

The NCAT grant program was similar in nature to the DOE's AT grants program except that proposals were solicited to carry out predefined "appropriate technology" programs, and grantees were restricted primarily to community action agencies and community development corporations. According to a report on grants awarded during FY 1979, out of 174 grants funded, 6 were for solar water heaters, 100 for greenhouses, 9 for Trombe walls, and 2 for solar produce drying.

16.4.2.1 Organizational Location and Budget

NCAT began life as an independent nonprofit organization funded by the Emergency Energy Conservation Services Program of the U.S. Community Services Administration (CSA) in January 1977. The center has been located in Butte, Montana, but had between ten and fifteen regional representatives throughout the country and an office in Washington, D.C. CSA was abolished in FY 1981, and the nonprofit organization that runs NCAT has primarily continued its work through DOE contracts.

Budgets for grants averaged around $1 million per year between 1977 and 1980. Total NCAT budget for FY 1980 was $3.7 million. NATAS funding was $.5 million in 1984 and $1.5 million in 1985. Total NCAT contract funding was $2 million and $2.4 million in 1986 and 1987, respectively; as of 1988, NCAT had diversified its sources of funding considerably (NCAT 1988).

16.4.2.2 Audiences Addressed/Goals and Objectives

NCAT responded to inquiries from anyone, but the low-income consumers, as represented by community action agencies, community development corporations, appropriate technologists, and grassroots groups, were its primary focus. Since NATAS has become its outreach mechanism, NCAT's audience has been broadened to include more middle-class consumers and small businesses. With the addition of the Appropriate Technology Transfer for Rural Areas Program, an emphasis on sustainable agriculture has also been developed.

As stated at the time of NCAT's formation (NCAT annual), the goals of the effort were

• to develop viable short- and long-range AT solutions to energy and energy-related problems experienced by low-income consumers (who then numbered 5 million);

• to create an awareness of appropriate technology opportunities and promote adoption by low-income communities;

• to provide financial and technical assistance to the poor by encouraging community-based technologies;

• to provide high-quality, cost-effective options to the poor through research and development; and

• to assist in overcoming barriers to widespread use of small-scale technologies by the poor.

16.4.2.3 Approaches to Preparing and Distributing Materials

For the most part, NCAT prepared its materials in-house with a staff of technical writing and communications specialists. They put a high priority on developing "networking" materials that allowed the audiences they served to gain access to existing materials and to learn from each other's experiences. Primarily, these were bibliographies, resource lists, short technical notes and newsletters (including "ncat briefs," *A.T. Times*, and separate newsletters for each region). NCAT also published research papers, survey reports, training manuals, and how-to brochures.

NCAT conducted a number of policy studies such as *Energy and the Poor*, which analyzed social, economic, and legal barriers to adoption of appropriate technologies. These materials covered a very broad range of topics all aimed at assisting low-income communities to take more control over the production and use of their basic needs—food, housing, energy, influence on community planning. An average of twenty to thirty major publications were developed each year (NCAT 1980).

NCAT determined what information was needed by staying in close touch with and conducting surveys of its community-based clientele. Most materials were provided free of charge (until the mid-1980s) and distributed in response to individual requests to NCAT headquarters. Availability of the materials was announced through program description brochures and the newsletters. In addition, a good deal of information was provided through one-on-one discussions with NCAT's technical staff at Butte or with its regional representatives. Another frequently used technique was workshops to provide hands-on learning experiences for low-income people or those serving them.

16.4.2.4 Results and Evaluation Activities

In FY 1979 NCAT distributed over 250,000 copies of publications (DOE 1980). It is not known if any specific evaluation studies were made of the impact of NCAT programs.

16.4.2.5 Lessons Learned

• By focusing on a somewhat more defined target audience (low-income consumers) than most other programs, NCAT was able to develop a well-received program that provided a definable constituency for its ongoing programs.

• Such a targeted focus also allowed NCAT to more easily define its mission and the types of information that would be most useful to the clientele served. Working with an already established network of community development and community service agencies enabled the Center to make a quick impact on the programs of these organizations.

• Much remains to be done; low-income consumers still spend relatively more of their incomes on energy than other consumers. Loss of other funding by many of the community agencies has reduced low-income consumers' ability to engage in energy programs other than the Low-Income Energy Assistance (LIHEA) Program, which provides money to cover utility bill payments, but not more than 15% of program funds are available to improve the energy efficiency of homes.

• For this reason, NCAT is finding other niches, such as sustainable agriculture, where it can use its technical and technology transfer expertise to serve specific audiences.

16.5   Department of Energy and the Solar Energy Research Institute

After its establishment in 1976, SERI became the federal government's major arm for solar thermal information, especially technical information, although individual DOE offices, such as the Solar Energy Technology Transfer Program (directed by Lawnie Taylor), produced some of their own materials, particularly about program objectives and progress.

16.5.1   Program Description

Like the other agencies, SERI initially produced mostly general information aimed at "general consumers" or technical audiences. Because SERI's mission covered almost all of the renewable technologies, the information it produced also covered these technologies—solar thermal (active and passive), biomass (solid and liquid fuels), photovoltaics, wind and ocean thermal applications.

The most ambitious information programs were conducted by SERI's Commercialization Division through its several audience specialists (consumer, architectural, industrial, legal, etc.). In addition, substantial information was generated by SERI's Public Information Office, Solar Energy Information Data Bank, Communications Branch, and External Relations Office. Each office saw its role in a somewhat different light (see section 16.5.4).

16.5.2   Organizational Location

As in most agencies, there were no sharp organization lines in SERI dividing consumer information programs from other programs. The Consumer Specialist in the Commercialization Division (later the Planning, Applications, and Impacts Division, Community and Consumer Branch) was nominally in charge of consumer issues and information. Because there was no agreed-upon definition distinguishing consumer information from other kinds of information, many different programs addressed what they perceived to be consumer needs.

The Commercialization Division's consumer budget ranged from $100,000 to $300,000 per year from 1978 to 1981. Budgets for other organizational units are not known.

16.5.3 Audiences Addressed

None of the SERI programs was designed to deal with individual consumers on a high-volume basis. Most programs were aimed at providing guides for states, design and construction professionals, nonprofit and trade organizations, and industry on how to be more effective in communicating with and marketing to residential, commercial, and industrial consumers.

In general, responsibilities within SERI were seen to be as follows:

• *Senior Consumer Specialist*, a position held by the author, was responsible for consumer and community organization information, consumer market research, consumer interest group liaison, and consumer assurance and protection issues; information programs were devised to promote action rather than to just provide information; terminated in 1981.

• *Other Audience Specialists*, developed materials to specifically address the needs of their audiences, for example, design manuals and design tools for architects, legal journals for lawyers, use assessments for industry sectors, and so on; eliminated in 1981.

• *Community Group*, addressed local and state planning offices and advocates under the direction of Robert Odland with methods of developing community-wide planning approaches, incentives, and so on that would facilitate wide-scale use of individual solar energy systems or of community-scale solar energy systems; eliminated in 1981.

• *Public Information Office*, responded to general inquiries regarding SERI's activities and the full spectrum of solar technologies that SERI was responsible for; cut back in 1982.

• *Solar Energy Information Data Bank*, was primarily interested in meeting the needs of outside researchers and industry; program focus shifted to filling the internal needs of SERI researchers in 1982.

• *External Relations Office*, staffed by Hank Rase and Liz Moore, worked primarily with government, nonprofit, and private organizations; terminated in 1983.

• *Communications Branch*, headed by Keith Haggard, was part of the Technology Commercialization Division and worked closely with all

SERI branches to develop materials on each technology; terminated in 1982; many of the branch's responsibilities were assumed by the Solar Technical Information Program.

16.5.4 Goals and Objectives

The overall goal of the Consumer Specialist program was to encourage widespread use of solar technologies and techniques by consumers across the country. It was recognized, primarily by the Commercialization Division programs, that not all consumers would have equal interest in solar, nor would their time frame for becoming interested be similar. For that reason, this division concentrated on developing programs that would meet the needs of "early adopters." Most efforts were aimed at providing information that would help consumers choose the types of solar energy systems that made the most sense for their particular applications.

The specific goals of the program as stated just before its demise in early 1981 (Vories 1981) were (1) to gain a better understanding of what motivates consumers to use conservation and solar technologies and to speed up the adoption of these technologies through promotional activities that concentrate on specific benefits to specific target audiences; (2) to ensure the availability of consumer decision-making tools by developing credible economic information (in terms that homeowners could understand more readily than "payback" and "life-cycle costing") and by providing information on how different types of systems perform and how to identify qualified contractors; (3) to ensure that the decision to adopt solar would be a positive experience by providing an early alert to fraud in the marketplace and access to recourse for nonworking systems; and (4) to provide for consumer input into the regulation and development of community-scale solar technologies.

16.5.5 Approaches to Preparing Materials

Senior Consumer Specialist Conducted the following five main programs:

1. Production of a how-to manual for community groups to assist them in developing local outreach programs. This manual, *reaching up, reaching out: a guide to organizing local solar events* (Vories et al. 1980), was produced in-house, yet it benefited from the help of an outside consultant and from review by more than ninety people from the regional solar energy centers (RSECs), nonprofit organizations, and solar ex-

perts throughout the country. Twenty of those people attended a three-day meeting that reflected on every aspect of the draft book prior to publication.

2. Codirection of market research with other SERI branches to determine the kinds of information most important to consumers, as exemplified by the National Study of the Residential Solar Consumer. This survey of 2,000 of the nation's homeowners and 4,000 solar users resulted in *America's Solar Potential: A National Consumer Study* (Farhar-Pilgrim and Unseld 1982), which provided essential market information to industry and government agencies about which consumers were most likely to purchase solar equipment and how to best reach them over the upcoming five years. The survey was developed in-house with the help of consultants; field work was conducted by the Gallup Organization; all analysis was conducted in-house; and the final draft was published by a private publisher, with the suspension of this effort at SERI after the data were collected. It is doubtful that the conclusions of this survey were ever put to practical use by any federal organization.

Relevant conclusions of this study (Farhar-Pilgrim and Unseld 1982) were the following:

• Respondent attitudes were highly favorable to solar energy and to the use of it in their homes.

• Most homeowners were optimistic about the national energy situation over the next five years but hesitant about the state of solar technology, so saw no need to move quickly toward using it.

• Of homeowners surveyed, 46% wanted more technical information (how systems work, do-it-yourself information, expected performance, state-of-the art developments, feasibility of home retrofits); 32% wanted information on costs and savings potential (initial system and installation costs, operating costs, payoff potential, how to make financial arrangements); 5% wanted practical, day-to-day information (what it is like to actually live with a solar energy system: system durability, life expectancy, and reliability; availability of maintenance service; effects of weather on systems; safety; amount of time required to maintain system; aesthetics); and 3% wanted political and business information (who really benefits from the use of solar, what is the government doing about it, who supplies and installs systems and how reputable are they).

• Consumers preferred to turn to local sources of information such as solar energy companies, state energy offices, public libraries, and utilities to obtain the information they want, although federal sources of information such as the National Solar Heating and Cooling Information Center (NSHCIC), Government Printing Office (GPO), the Solar Energy Research Institute (SERI), the regional solar energy centers (RSECs) and the Department of Energy (DOE) were also viewed favorably. They preferred to obtain information from brochures, from the mass media, particularly the print media, from local solar contractors or solar home-owners, from solar seminars and workshops, and from advocacy organizations such as the Solar Lobby. This was further borne out by current solar users, who indicated that their best sources of information were solar companies, state energy offices, and federal offices and agencies.

• Consumer knowledgeability and readiness to consider using solar varied widely. Of those surveyed, 29% were basically ignorant of the systems, another 29% had some initial awareness, 24% were gathering information, and 18% felt they had enough information to decide whether to purchase a system. Less than 1% had actually decided to use solar and had purchased a system or undertaken a passive solar retrofit.

• Potential solar users were highly concerned about the possible risks of solar ownership, particularly system safety and reliability (availability of homeowners insurance, prohibitive local building codes, damage to the system from environmental sources or vandalism) life-style and aesthetic effects, and advisability of purchasing a system (initial cost, dependability of solar firms, expense of maintenance, warranty protection). By contrast, current solar users overwhelmingly considered their systems to be practical, cost-effective, and durable in a variety of climates, although they had had many of the same concerns, before purchasing their systems, as the potential users surveyed.

• The main motivators for potential solar users considering purchasing a system were reducing utility bills or protection from future cost increases, conservation of natural resources and protecting the environment, and the fun of experimenting with something new. Most would have been willing to pay more than their current utility bill to own solar, but there was a lot of uncertainty about whether or not using solar would result in enough utility bill savings to make it worthwhile. Current solar users had essentially these same motivations for purchasing their systems with the additional motivation of increasing self-reliance. Most felt that their investment had been economical.

• Interestingly enough, most homeowners were unaware of federal tax credits for solar energy systems, and over three-quarters of homeowners in states offering additional tax credits were unaware of those credits. Most also substantially underestimated the cost of typical solar energy systems (both passive and active).

• Most homeowners would have preferred low-interest loans to tax credits as the best way of financing systems.

• Most homeowners felt that it would be impossible to retrofit their homes with solar.

• When asked what roles they would prefer to have various organizations play, homeowners gave the following responses: the federal government should provide consumer information and financial incentives and act as warrantor of last resort; state governments should list solar contractors and also provide consumer information and financial incentives; local government should mandate the use of passive solar design and domestic hot water systems in new homes, inspect solar energy systems, list solar contractors, and provide consumer information; and utilities should conduct home energy audits, inform consumers, inspect solar energy systems, and list solar contractors.

• With regard to demographics, there were key differences among respondents due to gender, age, education, occupation, and income. Those most likely to be interested in solar were males under 45 who were more highly educated, affluent professionals or business owners and who were already more knowledgeable about solar.

• Based on the information gathered, the population as a whole was divided into eight "tiers" according to their likelihood of adopting solar over the next five years; each tier was described in terms of information-seeking habits, attitudes toward solar energy, and demographics.

3. Oversight of project funded to Consumers Union and reported in *Consumer Reports*, which evaluated the performance, aesthetics, and so on of five solar domestic hot water systems offered by different national manufacturers.

4. Collaboration with technical program officials to obtain accurate information on solar technologies and to translate this for information specialists working in the SERI library and data bank programs.

5. Collaboration with federal and state consumer protection agencies to identify abuses.

Other Audience Specialists developed a series of brochures describing design tools available for designing different types of solar energy systems, held workshops around the country to train architects in solar design, published the *Solar Design Workbook*, and developed the *Solar Law Reporter*. All of these were developed in-house, with contributions from outside experts around the country.

Other Community and Consumer Branches provided information with regard to community-scale development of solar technologies. Staff developed a wide range of documents concentrating on how to undertake solar access protection, such as *What Every Community Should Do about Solar Access* (SERI 1980a), and how to develop community energy plans; they sponsored the two national Community Renewable Energy Systems Conferences (SERI 1980b).

Public Information Office developed a number of brochures explaining solar technologies and a number of exhibits that toured both general audience and technical trade shows promoting solar thermal technologies. For the most part, these items were conceived in-house and executed by local outside contractors.

Solar Energy Information Data Bank developed a wide variety of computerized databases and provided a referral service. Most of this information was developed in-house and used technical reference specialists to answer questions. Although the databases were primarily designed to serve energy professionals, the unit received many calls from consumers and responded to these as best it could.

External Relations Office engaged in a number of joint programs with other government, private, and nonprofit organizations; held several workshops aimed at identifying the needs of low- and middle-income consumers; and worked with service organizations such as Rotary and the American Legion to channel information to a broader consumer audience. The office provided significant support to National Sun Day, which led to the visit to SERI by then President Carter on 2 May 1978 and to numerous national press releases highlighting the importance of solar technologies to the future of the country.

Communications Branch produced some general information on the technologies it was studying and promoting in each of its technical program. A typical example is *Passive Design: It's a Natural* (Snyder 1980), which was aimed at giving owners of existing homes information about the many different passive approaches to adding solar systems to their homes. Most publications were produced by staff with in-house graphics assistance. There was also some attempt to provide news clips for national television news programs. An outside contractor scripted and filmed several 1- to 5-minute items intended to be shown as part of evening news shows.

16.5.6 Approaches to Distribution

Most information was distributed free of charge either to highly targeted mailing lists (e.g., all EES offices or state energy offices) developed by each program area, through the Government Printing Office, or through SERI's Public Information Office or Document Distribution Center. Depending on the nature of the publication, it was sent directly to the mailing list, or a notice, announcement, brochure, or press release was created on the availability of documents. The regional solar energy centers (RSECs) and in some cases, outside organizations such as the National Council of Churches were provided with large quantities of publications for redistribution to their clientele at the cost of printing or for no charge.

Exhibits, mall displays, and other large visual displays were either loaned out to requesting organizations or staffed by SERI employees at major trade and technical shows or shopping centers around the country. A good deal of effort was expended in identifying appropriate display opportunities and making displays available. TV news clips intended for inclusion on prime time news shows were distributed to national network affiliates.

16.5.7 Results and Evaluation Activities

Measuring results of any of these programs was a hit-and-miss activity. About the only approach used was to identify the number of people contacted or the number who requested copies of information. What individuals or organizations did with the materials was rarely tracked or analyzed.

As for the consumer programs in the Commercialization Division, it is known that over 7,000 copies of *reaching up, reaching out* were distributed to interested organizations. In addition, based on telephone calls to SERI and the regional solar energy centers (RSECs) a minimum of 100 workshops and other events in 1980 and 1981 resulted from the availability of the handbook.

Very little scientifically valid evaluation was carried out. Most impact information was qualitative at best. It was also very hard to separate the impacts of a specific program from those of all the other activities and influences going on at the same time.

16.5.8 Lessons Learned

No formal lessons have been documented for this effort. Certainly, based on the response to the materials prepared and the workshops held, there was an appreciative audience for the more targeted information approach to "early adopters." Although it would be hard to prove that this approach influenced other federal organizations and the solar thermal industry to adopt a more targeted approach, they increasingly did so until the early 1980s, when all such efforts were cut back.

It is certain that the failure of the federal government to move officially to more of a market segment approach and the limited budget and staff resources devoted to this approach over less than four years resulted in an uphill fight that did not have a lasting impact, particularly in view of all the other policy and market factors that came into play after the early 1980s.

16.6 Other Agency Programs

In the late 1970s through early 1980s quite a few other federal agencies developed their own programs encouraging the use of energy conservation and solar thermal technologies. Many of these were carried out with at least partial funding from the Department of Energy.

16.6.1 ACTION

The American Council to Improve Our Neighbourhoods (ACTION) is a national agency that encourages volunteerism among Americans through such programs as RSVP (Retired Senior Volunteer Program),

VISTA (Volunteers in Service to America), and the Peace Corps. Its efforts concentrate on solving problems through citizen development of local resources.

After successfully demonstrating how citizens could be involved in energy planning and action activities in Fitchburg, Massachusetts, in the late 1970s, ACTION worked with over fifty-five individual communities to launch voluntary, self-help, energy conservation campaigns. In some locales, retrofit solar greenhouses were included in the measures recommended for implementation. NCAT staff were used to provide much of the technical training to community volunteers who would spearhead this program in their own towns. In addition, $5,000 minigrants were provided to cover low-cost materials for weatherizing low-income homes. The community was expected to provide any other needed materials through in-kind donations (ACTION 1981).

With DOE funding, the Community Energy Project was established in 1980 in ACTION's Office of Voluntary Citizen Participation. The program folded at the end of FY 1983 (see table below).

FY	Total budget (in millions)
1981	$2.0
1982	1.0
1983	1.2
1984	-0-

16.6.1.1 Audiences Addressed/Goals and Objectives

The primary audience for this program was the volunteer leadership of the communities that participated. Leaders were frequently from local government agencies, as well as from nonprofit organizations, businesses, utilities, schools, and universities. These people in turn planned activities that were aimed at creating a broad outreach to the communities at large.

ACTION's main objective was to show that communities could be mobilized to address energy issues on their own if small incentives (grants of $5,000) and ACTION training in how to organize such programs were provided. Specific goals were

• to mobilize significant numbers of citizens to conserve energy in their homes and community;

• to demonstrate tangible savings, increased production of energy, or both;

• to demonstrate that small amounts of funds and training from government can mobilize citizens to act on conservation and renewables;

• to involve a significant number of low- and moderate-income citizens; and

• to establish a foundation to address energy issues over the long run;

Each community identified its own goals and objectives. Many concentrated on efforts to help low-income and senior citizens reduce their energy bills.

ACTION prepared a how-to manual and provided one-on-one technical and organizational assistance and training to interested communities. There were a number of conferences sharing the communities' results. For the most part, ACTION staff shared the information with interested communities and tried to locate and encourage other communities around the country to participate.

16.6.1.2   Results and Evaluation Activities

Although energy-saving information on the fifty-five subsequent programs is unavailable, over a nine-week period in the Fitchburg pilot program, 3,000 out of 14,000 households participated, reducing fuel use by an average of 14%. Communities from around the country expressed considerable interest in ACTION's efforts and prepared proposals to receive the grants and technical assistance. In the ACTION programs, workshops were held where citizens learned to build simple solar collectors or greenhouses, door-to-door auditing of energy use took place, and community-wide energy assessments and long-range plans were developed.

A number of communities and states are still carrying out programs based on this approach, for example, the North Carolina Energy Corporation's Community Energy Campaign, Minnesota's Governor's Energy Challenge, and the Santa Monica Energy Fitness Program. Also during the Bicentennial, TVA provided grants to communities to display the results of their energy activities aboard a river barge that went from community to community.

16.6.1.3   Lessons Learned

Given appropriate assistance and training, communities can develop the enthusiasm to accomplish a great deal in the way of low-cost conservation and solar improvements in the residential sector, but it takes identifying and working closely with dedicated volunteer leadership to accomplish this. Both the communities and ACTION became more sophisticated in their approaches to developing these programs as the experience of each community added to the range of involvement techniques and issues that could be addressed.

The ACTION approach continues to be one of the most often used approaches to renewable energy information at the local level, even though there is no longer any federal support for such activities. Recently, through work sponsored by Nebraska and the Rocky Mountain Institute, this approach has expanded beyond community energy needs to address overall community economic development issues.

16.6.2 Department of Agriculture

For the most part, USDA programs were add-ons to existing cooperative extension efforts and concentrated mostly on conservation activities, although there was some information developed to promote use of solar grain dryers and homebuilt domestic hot water systems on farms. These programs are covered in greater depth in the chapter on agriculture (see Chapter 11).

16.6.3 Department of Health and Human Services

HHS's Division of Energy Policy and Programs worked with institutional health care services to consider alternative energy to replace traditional fossil fuels. This program developed several pamphlets on alternative energy resources that could be used in health facilities and held workshops on these issues at health conferences. It also sponsored eleven demonstration projects for solar space and hot water heating as well as district heating and waste-to-energy projects.

These programs were started in 1979 with annual funding of about $300,000; in the last year of the program, 1981, funding was at $1.7 million. DOE cooperated with the program by supplying about $600,000 per year for the demonstration projects until they were ended in 1983.

16.6.4 Department of Housing and Urban Development

HUD programs are covered more extensively in other chapters; as an early actor in the Solar Heating and Cooling Program along with DOE,

HUD provided information on solar for consumers. HUD-supported studies on the building, buying, and selling of solar homes were particularly important in raising interest among building industry professionals in providing new homes that were energy-efficient and made maximum use of solar thermal resources (HUD 1980).

Several HUD offices were involved in developing outreach information on solar energy and renewables. The most active was the Solar Heating and Cooling Demonstration Office eliminated in 1982 (see table below for total budget figures).

FY	Total budget (in millions)
1977	$5.0
1978	5.0
1979	5.0
1980	4.8
1981	3.0
1982	-0-

16.6.5 Tennessee Valley Authority

In the late 1970's, under a congressional mandate to become the nation's solar showcase and under the leadership of David Freeman, TVA set the model for the country in terms of its dedication of resources to developing the use of conservation and solar. It provided a wide range of information and incentive programs to encourage all classes of electricity users to conserve or acquire solar technologies.

TVA's two primary solar thermal programs were the Solar Hot Water Program and the Solar Information Service (SIS). All of its programs were conceived and developed on a pilot basis by the conservation and solar staff in TVA's Chattanooga, Tennessee, headquarters and then offered to the over 100 utilities in the authority's power grid. Seven regional TVA offices provided technical and other assistance to each utility that agreed to undertake the program(s) in its service area.

The Solar Hot Water program tested and approved individual solar hot water systems and provided loans and other incentives to encourage their adoption. The program started with a pilot program in Memphis, Tennessee, and then spread to other utilities in the area.

TVA developed Solar Homes for the Valley, the Energy Saver Home Programs, and the Solar Modular Homes Project to encourage the use of solar design in new construction and the adoption of solar standards for new homes, supporting these programs with a design portfolio and advertising campaign.

The Solar Information Service (SIS), established at TVA in 1979, served as a general public information source on all solar technologies. Each solar technology program was responsible for its own advertising and promotion, so the SIS's primary focus was on educating consumers about the different types of solar technologies available and providing a general entry point to the other programs available. This was done by publishing straightforward, factual brochures and tabletop displays on each technology, as well as general bibliographies and other resource materials. To enhance its communications efforts, every six to twelve months the central staff would choose a particular solar technology to concentrate on, produce printed materials, and then train its regional staff about the technology. Based on inquiries generated through press release and feature stories, the regional staff would then set up workshops for interested organizations on the current "campaign" or other topics.

The Solar Information Service was the largest of several programs run by the Education Section of TVA's Solar Applications Branch of the Energy Conservation and Rates Department. The other programs were part of TVA's Conservation and Energy Management or Solar Applications Branches.

As of 1986, TVA found itself in a serious excess power situation and has reduced many of its conservation and solar programs to minimum maintenance levels until such time as the need for additional generating capacity returns.

16.6.5.1 Budget History

Early on, the budget for the SIS's central staff programs averaged over $2 million per year; in 1984 the budget was $1.3 million and in 1985, $1.4 million. Each District Office prepared an additional budget for activities within the district.

Budgets for the other programs are shown on the table below.

Total Budgets

FY	Energy Saver Home	Solar Hot Water	Solar Homes for the Valley/Solar Modular Homes
1981	$630,000	*	$2,000,000
1982	1,300,000	$645,000	900,000
1983	110,000	695,000	180,000
1984	*	515,000	*
1985	*	1,500,000	*

* Unknown.

None of these programs is currently in operation.

16.6.5.2   Audience Addressed/Goals and Objectives

Homeowners were the primary audience addressed by the SIS and other TVA solar programs. Much of the material prepared by the SIS was aimed at assisting homeowners with do-it-yourself projects. There were some special materials developed for community action organizations to reach low-income consumers.

The main goal of the SIS program was to educate a large number of consumers each year about solar technologies. It was hoped that the programs would reach about 200,000 people each year directly and a total of 1.4 million through press release and other media activities. In general, the program was seen as one of several ways to promote the overall goals of reducing energy consumption and promoting economic development in the region. Goals of the Solar Hot Water and Energy Saver Home Programs were to get audiences to take advantage of the incentives provided and to accelerate usage of the specific technologies being promoted.

16.6.5.3   Approaches to Preparing and Distributing Materials

Almost all the materials were prepared in-house by TVA Education Section staff with assistance on displays from the Architecture Branch. Each campaign developed a series of general and more technical brochures that had its own color theme. Emphasis was on practical, do-it-yourself information for homeowners. The TVA Communications Branch prepared most of the other materials used in TVA technical programs.

SIS printed materials were distributed primarily through workshops and public relations activities of TVA District Office staff and participating power distributors; printed materials were also available from TVA's Public Information Offices. Usually 100,000 or more of each brochure were published. The technical programs used some of these same distribution techniques and worked closely with the participating utilities; they also used considerably more paid advertising as a means of encouraging quick response to the incentives offered.

16.6.5.4 Results and Evaluation Activities
Although the SIS program got off to a slow start because of staffing constraints, by 1980 the program was exceeding its goals for contacts with Tennessee Valley residents usually obtaining 200–300,000 direct contacts each year.

The approach to determining results was primarily to have each District Office count the number of people who attended workshops, called to ask for advice, or visited displays. Little attempt was made to determine what action was taken based on these contacts. Regular training sessions provided to district staff were used to get feedback about how well the programs were working in the field, the need for additional information, or increased interest in particular subjects. The effort to promote the planning and addition of sunspaces was considered to be the most effective.

TVA's Market Research Branch conducted a number of surveys to measure awareness of particular programs and activities among other objectives. There was, unfortunately, little attempt to tie increases in awareness to the efforts of any particular program.

One outside evaluation (Vories 1983) concluded that although the program had been very successful in providing a substantial amount of information to Tennessee Valley residents, it would benefit from more carefully targeting its activities to convey particular benefits to specific residential sectors and from working more closely with a wide range of other trade, nonprofit, and government bodies to increase the credibility of its message and to reach even more potential consumers. The study also recommended that surveys be conducted of those attending workshops to determine what follow-on actions had been taken.

Over 6,600 solar energy systems were installed under the Solar Hot Water Program since 1978. This service was provided by half of TVA's member utilities. By 1982, about 1,800 Energy Saver Homes had been certified or were under construction; in 1983, as program emphasis declined, 75 homes were certified.

16.6.5.5   Lessons Learned

• TVA demonstrated that major inroads could be made in consumer awareness through a concentrated program.

• Even so, much more could be accomplished by using more of a market sector target approach.

• Although the decentralized delivery approach had some drawbacks, particularly in terms of supervising staff, it served as an invaluable feedback mechanism for keeping in touch with consumer interests and concerns.

16.6.6   Bonneville Power Administration

BPA has had active conservation and solar programs since the early 1980s. Because it started somewhat later than others, BPA was able to avoid many of the mistakes made by other agencies.

Most of BPA's programs concentrate on energy conservation activities. Like TVA, BPA sells electricity through approximately 120 public retail utilities and 7 investor-owned utilities. Usually, BPA conducts a pilot program with several of its member utilities and then offers the program to the rest of the utilities in the region. However, none of the solar programs except the Technical Assistance efforts was deemed cost-effective enough to be offered system-wide, and so never went beyond the pilot stage.

BPA has sponsored the following solar pilot programs:

• Solar Homebuilders

• Solar Hot Water Heater Workshops

• Technical Assistance (through state and local government offices)

• Solar and Heat Pump Water Heater Market Test

16.6.6.1   Audiences Addressed/Goals and Objectives

• *Solar Homebuilders* served primarily the building industry, but potential homebuyers as well. Goals were to demonstrate to builders, designers, realtors, and the general public that passive solar homes could be attractive, affordable, functional, energy-efficient, *and* marketable. About ten homes were built in Portland, Oregon, and Spokane, Washington and their floor plans made available to interested parties.

• *Solar Hot Water Workshops* were do-it-yourself workshops designed for those willing to build and install their own systems. The workshops covered system theory and design. A $500 incentive was paid to participants who installed their own systems.

• *Technical Assistance* provided workshops to audiences similar to those addressed by typical Energy Extension Service activities. For example, part of the Technical Assistance money awarded to Oregon went to the Solar Energy Association of Oregon, which has helped in funding their annual conference and quarterly magazine. This program has also funded an initiative developed by the four states served by BPA to provide the most advanced "solar access" ordinance developed in the nation. It continues to be promoted in several of the states.

• *Solar and Heat Pump Water Heater Market Test* (Columbia Information Systems 1986a) had as its main purpose testing the effectiveness of different approaches to promoting these technologies and providing incentives to residential customers to participate in these programs. Eleven utilities participated in the test, which divided them into two groups. In one group's territory, a low level of utility-generated promotion effort was combined with both a low level of incentive (about $200 per system) and a high level of incentive (about $500 per system). In the other group's territory, a high level of professionally developed promotion effort was combined with both a low and high level of incentives.

16.6.6.2 Approaches to Preparing and Distributing Materials

• *Solar Homebuilders* provided technical information primarily through booklets covering the design of the program's seventeen pilot homes; seminars were also given on lessons learned in the construction and marketing of the homes.

• *Solar Hot Water Heater Workshops* prepared and delivered hands-on workshops.

• *Technical Assistance* unknown.

• *Solar and Heat Pump Water Heater Market Test* provided, to inquirers in all areas, the consumer booklet "Water Heater Options," a list of eligible dealers, and a decision list for making a system selection. In the high-promotion areas the following were also available: posters, counter cards, sales brochures, bill statement stuffers, newspaper ads and radio spots, media tours resulting in feature stories, and newsletters. (Up to the end of 1986, the total cost of promotion in the low-promotion area was $5,344; in the high promotion area it was $252,484.)

As mentioned earlier, most BPA programs were delivered by member utilities or by state and local energy offices. BPA provided backup materials in the form of brochures, workbooks, design materials, and so on.

16.6.6.3 Results and Evaluation Activities

• *Solar Homebuilders* distributed 6,500 copies of the design booklet "Solar Home Show II" in FY 1983.

• *Hot Water Heater Workshops* conducted 20 workshops, which resulted in 117 installed systems.

• *Technical Assistance* conducted 150 workshops on both conservation and renewable energy technologies.

• *Solar and Heat Pump Water Heater Market Test* resulted in the installation of 418 water heater systems, of which 261 were solar energy systems, at an average cost of $4,285. Surveys were conducted both of residential users and the retailers who sold the systems; it was judged that incentives had more impact than promotion, but that high promotion combined with high incentives had the most impact. Based on the consumer surveys, it was found that people who had never even thought of buying unconventional water-heating systems, especially those who purchased the heat pump systems, responded to appeals that combined high promotion with high incentive (see table below). Most frequently cited sources for awareness of the program were dealers, newspapers, friends and relatives, and utility mailings. The main motivator for purchase of the systems was reducing utility bills, followed by energy conservation, tax credit (part of the incentive in one state), and utility incentive.

	Rates of response
Low promotion–low incentive	1.9%
High promotion–low incentive	3.3%
Low promotion–high incentive	11.0%
High promotion–high incentive	41.2%

In a companion survey (Columbia Information Systems 1986a, 1986b) of solar and heat pump water heater dealers, it was found that 60% of the dealers also carried more conventional water-heating equipment. It was also found that most of these dealers did very little advertising on their own and received little assistance from their manufacturers or distributors. Those who used telemarketing and other proactive marketing techniques prior to the test program sold 650% more systems than those who did not. Of all dealers contacted, 37% said the test market program had increased their sales, while 18% said it had harmed theirs; 25% of the dealers decided not to participate in the program because of restrictions or because they had other products they were more interested in promoting.

BPA has developed perhaps the most comprehensive and sophisticated evaluation program by any agency involved in conservation or renewable energy. The administration has been dedicated to assessing both the management and implementation (process) and the savings (impact) of its programs. Most BPA programs have been planned from the beginning with this in mind so that data gathered as the program progresses will allow for "easy" evaluation at several points during the program's lifetime.

16.6.6.4 Lessons Learned

• BPA's approach to evaluation has allowed it to undertake pilot programs, assess their potential impact, and then either implement them or "mothball" them until such time as the relative cost of the measures fall and additional savings are required. In this manner, BPA hopes it will not have to reinvent the wheel when its current electricity surplus vanishes and it must once again vigorously pursue conservation and renewable energy efforts.

• BPA has also used some of the most sophisticated market research and communication techniques available to promote its programs; combination of incentives, mutlimedia promotion, and ongoing market research has, BPA believes, considerably improved its results over those achieved by other programs around the country.

16.7 Summary

Many federal agencies became involved in providing consumer information to the publics they served. Most of the information concentrated on energy conservation activities rather than solar thermal technologies. The most popular programs appear to have been those carried out on the local level through involvement of community groups or provision of walk-in or call-in technical advisory services. Based on numerous conversations with persons active in the adoption of solar thermal technologies, the author's opinion is that local advisory programs were more effective than those providing only printed materials from national distribution services, but there is little solid evaluation to back up this assertion.

Although, in retrospect, much can be identified as lacking in these consumer information programs, there seems to be little doubt that the plethora of information provided through federal government programs had a significant effect on the awareness of the general public regarding the availability of solar thermal technologies. Most public opinion surveys conducted since the mid-1970s have indicated that citizens prefer solar energy over any other energy form for use in buildings (Farhar-Pilgrim and Unseld 1982). Indeed, the federal information effort, along with the existence of tax credits, was probably the main reason there was *any* market for solar between 1975 and 1985. As federal dollars spent on information programs decreased, the market for solar energy systems leveled off (1983–1985), even when energy prices were still rising. When, in mid-1989, prices were perceived as stable and the tax credits were not renewed, the market for solar was about 10% of what it was in 1984; even with the tremendous investment in information, research, and demonstration, less than 2% of consumers were using solar thermal technologies of any kind. (In this regard, conservation efforts have been considerably more successful in reaching a larger percentage of the population, in part because of the lower cost of adopting such technologies, in part because of the efforts of the Residential Conservation Service, and later because many public utility service commissions mandated that regulated utilities provide tailored information and incentives to their customers.)

It would seem that the impact of solar consumer information created more awareness than action. In many ways, it should have been expected that a long period of awareness would probably be required before most consumers would adopt solar thermal technologies. The high hopes of the early solar programs for major penetration in five years were based on the widely held assumption that, if you build a better mousetrap, people will come knocking at your door. This led to programs more interested in developing a particular technology than in understanding and providing what it would take to get people to adopt it.

In the author's opinion, the failure of solar thermal consumer information programs can be attributed primarily to (1) the absence of a coherent adoption and/or information strategy, either in the short or long term; (2) the combined effect of a new and evolving (therefore flawed) technology, a tax credit approach to incentives that applied to only a small segment of the potential market, and a very immature and undercapitalized industry; and (3) the absence of mechanisms in place to provide ongoing evaluation and feedback to program designers and implementors. These observations are discussed in greater depth below.

16.8   Concluding Observations

• There was very little acknowledgement that the cost of solar energy systems might have more impact on adoption than anything else. Providing tax credits certainly made systems more accessible to some middle- and upper-middle income people, but tax credits were not a usable incentive for most citizens. This placed an undue burden of expectation on what could be achieved by information programs. The cost of most solar energy systems, from $3,000 (solar domestic hot water) to $20,000 (passive solar retrofit), made a solar energy system the third or fourth most expensive household purchase a consumer might ever make. There were no programs that successfully addressed this capital and disposable income issue.

• The federal government was ambivalent about its real role in promoting the use of solar thermal energy. Although many individuals throughout the federal government and affiliated agencies were sincere advocates of solar adoption, the government itself never decided that it should be primarily responsible for achieving major solar usage in the country. Its attitude might be summed up along the following lines: "We really want people to use solar energy; we think it is in the country's and citizens' best interest. We will provide them with objective information and tax credits and leave it up to industry to create a market for the systems that are available." Even this was going much further than the government normally does in promoting products. However, the government never delved into whether or not this role would be sufficient to achieve major penetration considering the size, economic resources, and business and marketing acumen of the solar thermal industry.

• There was no concerted strategy between the federal government, state governments, utilities, and industry in how to promote the use of solar energy.

• In general, there was an arm's-length relationship between the federal government and industry with regard to promotional activities, although there was cooperation on technical issues.

• There was also little thought given to consumer problems that might result from promoting equipment when very little reliability testing had been done. There appeared to be an inherent trust that industry would provide only good, reliable products and would design and install them properly. Information alone could not make up for lack of system and designer certification or installer licensing; many systems that were later inspected turned out to be defective in many minor and major ways. Although warranties were provided, these were no help to people when the companies providing them went out of business.

• There was little theoretical understanding of the role of information in consumer decision making or in the overall commercialization process. In general, there was a feeling that rational consumers would be delighted with objective information and would immediately see the value of adopting these technologies without the need for other assistance. Over time, financial incentives and more one-on-one outreach activities were provided, but there was still more emphasis on extolling technical virtues than on giving people reasons to buy in terms that would excite potential buyers, rather than technical program monitors.

• Even though most consumers prefer to get their information as close to home as possible and to talk to people who can help them with their specific situation, there was no such system, well-informed in solar thermal technologies, in place during the mid-1970s. Providing the level of training and materials needed to make this system available would have required a very substantial financial investment indeed, and there was no single entity available to see that such an investment was made. Although the Energy Extension Service might have been used in this way, it had many other missions and did not have sufficient resources to do this adequately. Nor was the solar thermal industry in a position of financial strength that would have allowed it to be the primary information source for consumers. For the most part, state and local governments did not invest any additional resources in such services; most information was created at the federal, "generic" level and made available through whatever local outlets were able to access that information. Local resources were sometimes used to modify the information for greater local applicability, but usually this was only done in informal discussions between salespeople or extension information staff. Although probably better than nothing, this arrangement could not really deal with the problem of having a centralized government effort provide information about a decentralized and highly variable resource.

• Federal programs that allow states to devise activities that best meet their needs are a double-edged sword. On the one hand, there appears to be general support for states to devise programs tailored to the energy consumption patterns and needs of their respective residents. On the other, such an approach makes it difficult to assess the overall impact of a national program and to determine which efforts are more effective than others. (In the author's opinion, the most effective way of dealing with this issue would be for the involved federal agency to develop and fund an evaluation methodology to accurately determine savings achieved and, perhaps more importantly, to provide methods of comparing the impact of similar programs conducted in different states, or different programs conducted for the same target audience so that the most effective methods and techniques could be identified.)

• There was even less understanding of marketing theories such as the diffusion of innovation. Had such theories been applied more effectively in developing programs and information, there would have been an earlier use of market segmentation and an earlier acceptance of the fact that different types of consumers would adopt solar technologies over time. However, even this understanding might have been difficult to achieve because of federal reluctance to support programs that appear to benefit one sector of the population over another (except perhaps in the case of low-income consumers). The federal approach of trying to include everyone from the outset and of measuring short-term gains rather than looking at an adoption process that might take fifteen to thirty years to achieve, makes it difficult to devise programs that are truly market segment–specific.

• As more and more technologies became "market-ready," the marketplace became very confused. It was hard for consumers to know just what solar technologies would be best for them. In fact, there was considerable federal program rivalry for consumer attention to say nothing of the marketplace rivalry. Among technical experts there was also disagreement, so that it is likely that many consumers tuned out anything that had to do with solar for their own use because it was too complicated to make a decision. The author's reading of the continuing positive attitudes to solar is that consumers, rather than delving into all the conflicting information to determine which solar energy systems best suit their needs, hope the technology will be included in any new home they buy and are relying on the builder to sort out what is best to install.

• There was no agreement in the federal program about what was most desirable for consumers to "buy." Every program pretty much conducted its own "campaign." This resulted in many conflicting messages being sent out about the importance of conservation versus solar and the value of competing solar technologies. Indeed, it seemed that the federal government expected consumers to sort out all of this conflicting information for themselves. Although this is an appropriate attitude about most things in the marketplace, in this case it served to contradict the hopes for aggressive adoption by greatly delaying decisions to buy new products, except for consumers sold on the strength of the tax credits or by high-pressure marketing organizations.

• It seems that consumer information efforts are at best an afterthought of most technically oriented programs; they do not receive the level of planning and commitment that is given to the technical efforts. Because little is done to quantify the impact of these information efforts, they are the first to be axed when budgets get tight. This leads essentially to lip service activities aimed at the general public rather than to programs to motivate consumers to take advantage of technological advances. Information alone does not produce action.

• The two federal organizations having more direct relationships with consumers (TVA and BPA) certainly were more successful in promoting specific solar installations than were the more general federal efforts. On the other hand, because these organizations still had to work through their retail networks (local utilities), the information available and its delivery were only as good as the local utility's interest and commitment to the programs, and this varied considerably. Also, as sellers of electricity, TVA and BPA were placed in the awkward position of promoting a technology that would reduce their revenues; indeed, both stopped promoting the technologies when they encountered substantial surpluses of electricity.

16.9 Recommendations

Should such an adventure be undertaken again in the future, this author would recommend that:

1. a more conscious decision about an appropriate federal role be made (How seriously does the federal government want adoption to take place? How can it ensure that all the pieces are in place to make it happen, if it decides to take on that role?);

2. delivery of information take place at the local level or through national and local influence organizations and a specific strategy be developed to see that the resources are available to make this happen;

3. market research and evaluation research be the cornerstone of the development of the strategy and frequent feedback be available to program planners and information agents;

4. market segments be identified and prioritized for action, information, incentives, and so on, designed to speak to the needs and concerns of those segments; and

5. incentives (if offered) be tied to certified performance of solar energy systems—and solar installers, repair and maintenance workers, and so on be trained and licensed.

References

ACTION (American Council to Improve Our Neighborhoods). 1981. Information from two undated press releases and article "ACTION Assists Local Communities in Energy Conservation, [February].

Columbia Information Systems. 1986a. *Solar and Heat Pump Water Heater Program: Second Interim Report.* For Bonneville Power Administration. Portland, OR, September.

Columbia Information Systems. 1986b. *Retailer Report: Solar and Heat Pump Water Heater Program Evaluation.* DE-AC79-8311645. For Bonneville Power Administration. Portland, OR, May.

Dawson, J. 1976. *Buying Solar.* FEA/G-76/154. Washington, DC: Federal Energy Administration and Department of Health, Education, and Welfare.

DOE (U.S. Department of Energy). 1978. *The First Report to Congress: Comprehensive Program and Plan for Federal Energy Education, Extension, and Information Activities.*

DOE/IR-005. Washington, DC: Assistant Secretary for Intergovernmental and Institutional Relations, February.

DOE. 1979. *The Second Report to Congress: Comprehensive Program and Plan for Federal Energy Education, Extension, and Information Activities.* DOE/CS-0071. Washington, DC: Assistant Secretary for Conservation and Solar Applications, Office of State and Local Programs, January.

DOE. 1980. *The Third Report to Congress: Comprehensive Program and Plan for Federal Energy Education, Extension, and Information Activities.* DOE/CS-0151. Washington, DC: Assistant Secretary for Conservation and Solar Applications, Office of State and Local Programs, March.

DOE. 1981a. *The Fourth Report to Congress: Comprehensive Program and Plan for Federal Energy Education, Extension, and Information Activities.* DOE/CE-0023. Washington, DC: Assistant Secretary for Conservation and Renewable Energy, May.

DOE. 1981b. *The Fifth Report to Congress: Comprehensive Program and Plan for Federal Energy Education, Extension, and information Activities—Fiscal Year 1981.* DOE/NBM-3002, DE83-002512. Washington, DC: Assistant Secretary for Conservation and Renewable Energy, December.

DOE. 1983. *The Sixth Report to Congress: Comprehensive Program and Plan for Federal Energy Education, Extension, and Information Activities.* DOE/CE-0023/2. Washington, DC: Assistant Secretary for Conservation and Renewable Energy, March.

DOE. 1984a. *The Seventh Report to Congress: Comprehensive Program and Plan for Federal Energy Education, Extension, and Information Activities.* DOE/CE-0023/3. Washington, DC: Assistant Secretary for Conservation and Renewable Energy, May.

DOE. 1984b. *Annual Operating Plan '85: Office of State and Local Assistance Programs.* Washington, DC: Office of State and Local Assistance Programs.

DOE. 1984c. *Department of Energy State and Local Assistance Programs: 1984 Report.* DOE/CE-0118. Washington, DC.

DOE. 1985a. *The Eighth Report to Congress: Comprehensive Program and Plan for Federal Energy Education, Extension, and Information Activities, Annual Revisions.* DOE/CE-0023/4. Washington, DC: Assistant Secretary for Conservation and Renewable Energy, January.

DOE. 1985b. *Program Activities: Department of Energy State and Local Assistance Programs, 1985 Report.* DOE/CE-0144. Washington, DC: Assistant Secretary for Conservation and Renewable Energy, October.

DOE. 1985c. *Single-Family Building Retrofit Research: Multi-Year Plan, FY 1986–FY1991.* Washington, DC: Oak Ridge National Laboratories for Building Energy Retrofit Research Program, Building Services Division, Office of Buildings and Community Systems, December.

DOE. 1986. *The Ninth Report to Congress: Comprehensive Program and Plan for Federal Energy Education, Extension, and Information Activities, Annual Revisions.* DOE/CE-0144. Washington, DC: Assistant Secretary for Conservation and Renewable Energy, January.

DOE. N. d. "Appropriate Technology." DOE/CS-0049. Washington, DC: Office of Buildings and Community Systems and Assistant Secretary for Conservation and Renewable Energy. Four-page program announcement brochure.

Eaton, W. W. 1976. *Solar Energy.* LCC: 74-600179. Washington, DC: Energy Research and Development Administration, Office of Public Affairs.

EES (Energy Extension Service). 1987. "Energy Extension Service." Briefing document taken from a set of viewgraph copies, January.

Farhar-Pilgrim, B., and C. T. Unseld. 1982. *America's Solar Potential: A National Consumer Study.* New York: Praeger Special Studies.

Fujimoto, I. 1977. *The Values of Appropriate Technology and Visions for a Saner World.* Pub. no. 010. Butte, MT: National Center for Appropriate Technology.

HUD (U.S. Department of Housing and Urban Development). 1980. *Selling the Solar Home '80: Market Findings for the Housing Industry.* HUD-PDR-514. Washington, DC: Real Estate Research Corporation, February.

NCAT (National Center for Appropriate Technology). Annual. *All about NCAT.* Pub. no.024. Butte, MT. Updated yearly.

NCAT. 1980. *NCAT, the National Center for Appropriate Technology: Third Annual Report, 1979–1980.* Butte, MT, September.

NCAT. 1988. *National Center for Appropriate Technology: Progress Report into the Nineties.* Butte, MT.

Public Law 95-39. 3 June 1977. Energy Research and Development Administration Appropriation Authorization. Title V: Energy Extension Service.

SERI (Solar Energy Research Institute). 1980a. *What Every Community Should do about Solar Access.* SERI/SP-744-590. Golden, CO: Community and Consumer Branch, March.

SERI. 1980b. *Community Energy Self-Reliance: Proceedings of the First Conference on Community Renewable Energy Systems,* University of Colorado, Boulder, 20–21 August. SERI/CP-354-421. Sponsored by the Department of Energy and the Solar Energy Research Institute. Golden, Co.

Snyder, R. 1980. *Passive Design: It's a Natural.* SERI/SP-432-521, GPO 061-000-00410-7. Golden Co: solar Energy Research Institute, Passive Technology Program for the Department of Energy, National Passive/Hybrid Heating and Cooling Program, April.

Vories, R., Coreen Young, Rachel Snyder, Nancy Carlisle, Susan Sczepanski, Carolyn Bartz. 1980. *reaching up, reaching out: a guide to organizing local solar events.* SERI/SP-62-326, GPO 061-000-00345-2. Golden, CO: Solar Energy Research Institute.

Vories, R. 1981. *Multi-Year Opportunities for the Consumer Program at SERI: A SERI Internal Working Paper.* Golden Co, February, pp. 5–14.

Vories, R. 1983. *Evaluation of TVA's Solar Information Service and Recommendations for Future Programs.* Denver: Infinite Energy, May.

Western (Western Interstate Energy Board). 1986. *Petroleum Violation Evaluation Report.* For the Colorado Office of Energy Conservation. Denver, 31 January, p. 1–2.

17 Public Information

Kenneth Bordner and Gerald Mara

17.1 Needs for Information on Solar Technologies

Intense curiosity and widespread enthusiasm about solar energy followed the energy shortages that began in 1973. The agencies involved in the various Federal Solar programs immediately recognized that decisions to employ solar energy heating systems would depend in part upon the availability of concrete answers to specific questions. How much energy (and money) would solar energy systems save? How much money would the systems cost? Where would the systems be purchased? Who could design and install the systems?

These questions were not, of course, limited to the general public. Builders needed to know if there were complete, easily installed systems available off the shelf. Was it as easy to include a solar energy system in a new home as it was to equip the home with an oil burner or gas furnace? How easily could a solar energy system be added to an existing building? Could solar contractors be found as easily as plumbing or electrical contractors?

Installers of heating, plumbing, and electrical systems wanted to know how easily they could acquire the skills they needed to capture some of the solar market. Financial institutions wondered how solar energy systems would affect the appraised value of a property. Local zoning and building code officials were uncertain as to how they should treat solar buildings, which often relied on designs and materials that were different from those used in standard practice.

For all these reasons, all partners in the Federal Solar technology implementation effort quickly perceived that a responsive and stimulating information program was essential to their mission.

17.2 Legislative and Administrative Mandates

Before 1974 Federal solar energy information efforts largely stemmed from administrative decisions made by the relevant agencies rather than from legislative mandate. As the solar energy programs matured, specific provisions for information were incorporated into Federal solar legislation. Explicitly recognizing the need for information if solar energy technologies were to be widely adopted, these provisions gave the agencies involved broad authority to inform the public.

17.2.1 Solar Demonstration Act

The main purposes of the National Solar Heating and Cooling Demonstration Act of 1974 (PL 93-409) were to encourage the widespread use of solar energy for residential and commercial heating and cooling and to promote general awareness of the feasibility of solar energy usage. The principal mechanism was intended to be a series of demonstrations in which builders of residential or commercial structures would be awarded grants for the inclusion of solar energy systems in their projects. Residential demonstrations (discussed in detail in chapter 9) were to be managed by the Division of Energy, Building Technology, and Standards of the U.S. Department of Housing and Urban Development (HUD).

Grants for commercial applications were originally to be overseen by the National Aeronautics and Space Administration (NASA), but the function was quickly transferred to the Solar Energy Division, Solar Heating and Cooling Branch, of the Energy Research and Development Administration (ERDA) when that agency was created in 1975. In 1977 responsibilities for the commercial demonstration program (reviewed in chapter 10) and related activities were transferred to the newly created U.S. Department of Energy (DOE), under the direction of the assistant secretary for conservation and solar energy, through the Office of Solar Applications.

Although the demonstrations were envisaged primarily as stimulating professional involvement and public interest, the initial legislation also recognized their potential for generating much-needed information. In most of the existing solar buildings, data on system cost and performance were either unavailable or anomalous. Monitoring the demonstrations would, it was believed, supply information that was both reliable (it would be generated, collected and analyzed by experienced professionals contracted by HUD and NASA/ERDA/DOE) and valid (the applications would be replicable examples rather than unique creations). Ideally, then, the demonstrations would provide the foundation for a comprehensive data bank on residential and commercial solar applications. Thus, Section 12 of PL 93-409 called for the secretary of HUD to establish a Solar Heating and Cooling Information Data Bank to collect and process solar heating and cooling information for dissemination to both the general public and interested professionals. The act specified that the data bank should utilize existing databases in the various federal agencies, adding to them as necessary as new information became available, principally from the demonstration projects.

17.2.2 Solar Energy Research, Development, and Demonstration Act

One major provision of the Solar Energy Research Development and Demonstration Act (PL 93-473) was the creation of the Solar Energy Research Institute (SERI). SERI's role in solar energy research and development was to encompass all solar energy technologies and not simply heating and cooling. All phases of solar energy research, development, and commercialization were to be included.

The legislation also mandated the creation of a Solar Energy Information Data Bank (SEIDB), which would serve as a comprehensive repository for all relevant information on the full spectrum of the solar energy technologies. Although the location of the SEIDB was not specified in the act, it eventually came to be administered by the Information Division at SERI (this problem is discussed in chapter 18).

17.3 Organizational Histories

17.3.1 FEA, ERDA, and DOE

The Federal Energy Administration (FEA) was created in 1973 to provide the Energy Resources Council with analysis and options in the formulation of national energy policy. Thus, FEA's key role in the development of solar energy was to integrate proposals for solar energy use into a national energy strategy; as a subsidiary function, it was also responsible for promoting the use of solar energy technologies by supplying information.

Under the Energy Reorganization Act of 1974 (PL 93-438), Congress created the Energy Research and Development Administration (ERDA) and awarded it "central responsibility for policy planning, coordination, support and management of all research and development programs respecting all energy sources." ERDA's responsibilities in the area of solar energy included not only basic research but also demonstrations, studies, and information dissemination. ERDA became responsible for the creation of the SEIDB and was also charged with disseminating the results of the commercial demonstration projects to a variety of technical audiences.

ERDA divided its solar demonstration and information functions between two separate offices within the Division of Solar Energy. Initially, the demonstration projects were managed by the Solar Heating and Cooling Branch under the supervision of the assistant director for direct thermal applications.

Subsequently, the organizational matrix was altered to create an Assistant Directorate for Heating and Cooling, headed by Ronald D. Scott, which managed the demonstration program through the Demonstration Program Branch. Information functions were carried out under the Assistant Director for Administration Robert Annan and through the Technology Utilization and Information Dissemination Branch, managed by Lawnie Taylor.

In 1977 ERDA's and FEA's functions were united in the U.S. Department of Energy. Initially, agency responsibilities for solar energy were divided between the assistant secretary for energy technology, whose office supported SERI and other basic research functions, and the assistant secretary for conservation and solar applications, whose office conducted the demonstration and information programs. DOE was reorganized in 1979 to coordinate all of the functions relating to a specific technology. Under that new organizational framework, all activities related to solar energy were overseen by the assistant secretary for conservation and solar energy.

Both the demonstration and information functions were the responsibility of the Market Development Division, headed by Robert Jordan. Demonstrations fell under the aegis of the Market Testing and Applications Branch, directed by Norman Lutkefedder, while the information activity—the Solar Technology Transfer Program—was managed by the Education and Commercial Branch under the direction of Lawnie Taylor.

17.3.2   Department of Housing and Urban Development

The residential component of the federal solar Demonstration program was managed by HUD's Division of Energy, Building Technology, and Standards. HUD's responsibilities within the program ranged far beyond the funding, selection and administration of grant awards. The agency was also charged with overseeing the collection of all relevant information generated by the residential demonstrations and with seeing that information was used to further the broad goals of the original legislation. Although HUD's solar budget was provided first by ERDA, then DOE, all programmatic and administrative decisions connected with the residential demonstrations were exercised autonomously by HUD.

As opposed to the continuing flux at ERDA/DOE, the organizational structure in the HUD program office remained nearly uniform for the duration of the agency's involvement. From the outset, the solar program was a separate, autonomous unit within the division. The program manager, David Moore, reported directly to the division director, Joseph Sherman. David Engel was responsible for all "market development" activities within the program. Michael Lenzi, who reported to David Engel, managed all information activities until his death in 1980. In practice, there was very close cooperation among all members of the HUD program during the demonstration years. As a result, the information activities under HUD would be related to the activities and initiative in the residential demonstrations.

17.3.3 Other Federal Agencies

The solar information activities of other Federal agencies were, for the most part, either indirect or small scale. Two participants are, however, worthy of special note.

The National Bureau of Standards (NBS), a part of the U.S. Department of Commerce, was responsible for developing and recommending reliable standards and methods that would ultimately be used to test and evaluate solar energy products and systems. NBS also served as the centralized database for most of the data collected from the demonstration projects (see chapters 9 and 10).

The Energy Extension Service (EES), operated by DOE and patterned after the USDA Extension Service, was created in 1977 to help provide energy conservation and renewable energy information on the state level (see chapters 15, 16 and 29).

17.4 Major Contractors

17.4.1 National Solar Heating and Cooling Information Center

HUD's major contractor for information transfer was the National Solar Heating and Cooling Information Center (NSHCIC). NSHCIC was designed to serve as a "one stop" source for all federal information on solar heating and cooling. The center was operated by the Franklin Research Center in Philadelphia. Franklin had been selected on the basis of its response to a competitive Request for Proposals (RFP) issued by HUD. When the RFP was released, Franklin was already operating a small solar energy information clearinghouse called SOLAPIC (Solar Applications Information Center) directed by Frank Weinstein and funded by seed money from Franklin itself. The contract to Franklin was awarded in May 1976, and Franklin began answering its first solar information requests under the HUD contract in October 1976. The center's original director was Jeremy A. Lifsey. Soon after his promotion to vice president of Franklin in 1977, the day-to-day duties were assumed by the deputy director, Kenneth Bordner.

NSHCIC was the only major actor in the federal solar program whose sole and complete mission was the dissemination of information. The broad charge of the solar information program was to encourage the widespread use of solar energy for residential and commercial heating and cooling and to promote a general awareness of the feasibility of solar energy usage.

Organizationally, NSHCIC reported for the majority of its activities directly to HUD with only indirect responsibility to ERDA/DOE. During each year of the center's operation, however, a small portion of its budget was awarded directly by ERDA/DOE for special projects. Annual budget figures for NSHCIC separated by HUD and ERDA/DOE funding are included in table 17.1. NSHCIC saw its budget increased dramatically the first year, when its main function expanded from telephone and mail

Table 17.1
HUD and ERDA/DOE funding of NSHCIC (thousands of dollars)

	Total		HUD portion		ERDA/DOE portion	
	current year	1985$	current year	1985$	current year	1985$
Year 1 (1976–77)	1,824	3,228	1,824	3,228	0	0
Year 2 (1977–78)	4,917	8,162	4,741	7,870	176	292
Year 3 (1978–79)	5,699	8,833	5,282	8,187	417	646
Year 4 (1979–80)	6,041	8,578	5,165	7,334	876	1,244
Year 5 (1980–81)	6,207	8,069	5,636	7,327	571	741

Source: Franklin Research Center 1977, 1978, 1979, 1980, 1981.

response to include exhibits, publications, and conferences (discussed in section 17.5).

In 1981 NSHCIC's contractual relationship with HUD ended as HUD's role in the Solar Demonstration Program diminished. In May of 1981 NSHCIC was reorganized as the Conservation and Renewable Energy Inquiry and Referral Services (CAREIRS), under the sole supervision of DOE.

17.4.2 Other Major Contractors

As the demonstration program expanded, other HUD and ERDA/DOE contractors played increasingly important roles in generating some of the information that was used by NSHCIC. Charged with monitoring the technical performance of the residential demonstration projects were the Boeing Company (overall administration and performance), the American Institute of Architects Research Corporation (general system and building) and Dubin Bloome Associates (specific system design and troubleshooting). This last organization's role increased dramatically as project experience was accumulated and as installations and system problems began to surface. The Real Estate Research Corporation (RERC) had principal responsibility for monitoring the market (financial, land use, code enforcement) experiences of builders and purchasers of solar homes.

All of the above organizations contributed to the demonstration data used by NSHCIC, both for its responses to questions and for the development of its publications, audiovisual products and conferences. Dubin Bloome would eventually cooperate with NSHCIC in producing a book and film on solar installation. (Dubin Bloome Associates 1978) RERC and NSHCIC cooperatively wrote a comprehensive report to builders called *Selling the Solar Home*. (RERC 1978, 1980)

NSHCIC was one of the principal distribution outlets for other products generated by more specialized contractors in the HUD program. These included a computer model on the economic feasibility of residential solar applications (the Residential Solar Viability Program— RSVP) created by Booz, Allen & Hamilton, and guidebooks concerning solar access and land use, written by Martin Jaffe and Duncan Erley of the American Planning Association (APA; Jaffe and Erley 1979; Erley and Jaffe 1979).

Table 17.2
Telephone and mail requests to NSHCIC per contract year (in thousands)

	Telephone requests	Mail requests	Total requests
Year 1 (1976–77)	57	55	112
Year 2 (1977–78)	101	82	183
Year 3 (1978–79)	96	39	135
Year 4 (1979–80)	149	72	221
Year 5 (1980–81)	155	93	248
Totals	558	341	899

Source: Franklin Research Center 1977, 1978, 1979, 1980, 1981.

Given the complex organizational relationship with DOE, NSHCIC's uses of and relations to major DOE contractors were more diffuse. In NSHCIC's third and fourth years, more use was made of performance reports on instrumented demonstration projects, compiled first by IBM and, then, by VITRO. However, the bulk of NSHCIC's contacts with DOE contractors were with the Solar Energy Research Institute and the regional solar energy centers. The relationship between NSHCIC and these other organizations is discussed in section 17.6.

17.5 NSHCIC Program Activities

In May 1976 NSHCIC came into existence; in October it established its first toll-free service. NSHCIC provided an enormous number of people with valuable information on solar energy during its five-year existence. During this period the center responded to nearly 900,000 telephone and written inquiries. The numbers of telephone and mail requests per contract year are included in table 17.2. A substantially larger number of persons were reached through publications, exhibits, and conferences (see table 17.3).

17.5.1 Dissemination Strategies

Under the direction of HUD, NSHCIC management was faced with developing a set of dissemination strategies to match the evolving federal program. Although the contract to operate NSHCIC was awarded in May 1976, concrete strategies were not firmly established until later that year. As a matter of policy, HUD, and thus NSHCIC, chose to concentrate on developing as much information as possible for the build-

Table 17.3
Exhibit attendance and bulk mailings by contract year (in thousands)

	Number of exhibits	Exhibit attendance	Bulk mailing
Year 1 (1976–77)	28	175	100
Year 2 (1977–78)	71	432	343
Year 3 (1978–79)	53	220	429[a]
Year 4 (1979–80)	16	66	503
Year 5 (1980–81)	9	54	450
Totals	177	1,353	1,825

Source: Franklin Research Center 1977, 1978, 1979, 1980, 1981.
[a] Year 3 totals do not include the 2 million items shipped for Sun Day 1978 during an eight-week period.

ing professional. The reasons behind this decision are obvious enough. Although the individual homeowner makes a single purchase decision, members of the building professions are capable of influencing many such decisions. Thus dollars spent on information for professionals active in solar technology implementation appeared to be much more effective than equal amounts spent on "general" or "public" information.

As experience with the program was gained, however, it became obvious that the general public was going to be a major audience; in fact, the general public became the largest user of NSHCIC's services. To facilitate public access to NSHCIC, it was decided to install a toll-free telephone hot line. Although this approach was not even considered until August 1976, the hot line was operational by early October and soon proved to be a significant factor in NSHCIC's effectiveness.

Thus, by the end of the first contract year, NSHCIC developed two sets of strategies to reach its audiences: a reactive or "response" strategy for the general public, and a proactive or "outreach" strategy for building professionals. In summary, the following methods were developed within both of these strategies:

Reactive or Responsive

• Toll-free hot line providing direct answers to questions and publications.

• Special mail department providing answers to written questions and publications.

• Distribution of bulk quantities of publications for use by state and local organizations.

Proactive or "Outreach"

• The Solar Van, exhibited primarily at professional society chapter meetings.

• Conferences and workshops.

• Audiovisual materials.

• Public awareness programs/media coordination.

17.5.2 Organization

NSHCIC's organizational structure was shaped by its major functions. In the first contract year (1976–1977), the center was organized into three main functional components: dissemination, data files and market development. As data became more varied (information from HUD demonstration projects as well as from the private and industry sectors), as market development strategies became more focused and as center staffs' technical expertise grew, NSHCIC's organization became more multifaceted and complex.

By the third contract year (1978–1979), separate groups existed for dissemination, data files, publications, communications and public relations, demonstration data, bibliographic data and library services, technical services, DOE liaison, and solar conferences. This general organizational structure remained until the final NSHCIC contract year.

17.5.3 User Needs

Fortunately, the center did not have to begin its assessment of user needs starting from square one. In 1976 HUD had commissioned a report by Arthur D. Little, Inc., which identified (by targeted surveys) the types of information needed by specific user groups. This preliminary study guided the center's initial decisions about the nature and format of the information it would provide. (ADL 1976)

NSHCIC soon recognized, however, that the information needs of all user groups are dynamic, growing in sophistication as users became more knowledgeable and as experiences with solar energy applications multiplied. Accordingly, the center monitored the kinds of questions it received in order to assure the adequacy of its current information and to plan the development of new materials. Assessing user needs was thus a continuous empirical process based on the kinds of inquiries actually received.

For all five years of NSHCIC's operation, consumers were the dominant user group, although members of the building professions and trades did constitute a sizable minority. The only major group within the housing industry whose interaction with NSHCIC was disappointing was the lending community. This group did not participate in significant numbers until the DOE workshops for the financial community were initiated in 1978.

Questions asked by all audiences grew rapidly in sophistication. In the first year of the center's operation, the majority of questions from both private citizens and building professionals concerned the nature of solar heating. By the third year (1978–1979), technical questions, although sometimes very general ones (e.g., "What size solar collector is best for my home?"), were more numerous than basic ones (e.g., "How does solar energy work?"). This growing sophistication cannot be attributed solely to NSHCIC's efforts (70% of inquiries in the third year were from individuals who had not contacted the center previously) but, rather, to a more general improvement in the level of public knowledge concerning solar energy.

Assessing user needs for outreach activities was less formal and involved more direct interaction with members of the different user communities. For example, agendas for the numerous state solar conferences sponsored by NSHCIC from 1977 to 1979 were developed in close cooperation with technical or outreach staff in the various state energy offices. Specialized publications were developed using the target audiences as reviewers. For example, material for builders was reviewed by the National Association of Home Builders (NAHB) and material for lenders by the U.S. League of Savings Associations (USLSA). Information needs of these audiences became more specialized as their experience with the federal demonstration program and other publicly or privately developed solar projects grew.

17.5.4 Meeting User Needs: Response

The first task facing any information center is to make its projected audiences aware of its existence. NSHCIC soon found that it could stimulate high levels of inquiry through appropriate publicity strategies. For example, during the first year of operations, the number of phone calls increased from 500 to 2,400 in one week when the center's telephone number was published in *U.S. News and World Report*. In order to encourage inquiries from members of the housing industry, publicity was often focused on the trade and professional journals read by builders, lenders, contractors, or architects. Initially, the center placed notices publicizing its own existence; later the center was listed routinely as an information source as a part of features on specific solar applications.

Inquiries were also stimulated indirectly, as a by-product of outreach activities. Literature distributed at exhibits for building professionals contained descriptions of the center's services; the same information was also provided in the materials available at the state solar conferences (see section 17.5.5.1).

Persons wishing to contact NSHCIC could do so by telephone or mail. The number of toll-free Wide-Area Telecommunication Services (WATS) lines in the telephone response center stayed constant at six for the continental United States (plus one for Pennsylvania) during the entire contract period, indicating realistic initial planning. Two additional WATS lines for Alaska and Hawaii were opened in the third year. The phone center physically located in Philadelphia responded to calls Monday through Friday from 9:00 A.M. to 8:00 P.M., EST, making the center available during the entire business day throughout the continental United States.

Letters were sent to a national post office box in Rockville, Maryland. Questions that could be answered by form letters and general publications were processed directly there; questions requiring more specialized responses were forwarded to the center's office in Philadelphia.

Whenever possible, inquiries were answered with available publications, rather than with long and detailed letters. When the center began operating, its principal publications were very general in nature. These included a short, introductory brochure on residential solar heating, "Solar Energy and Your Home", a general reading list on solar energy, and a brief description of the services provided by the center. After one or two years of operation, the publications provided to inquiries were much more specific and a great deal more technical.

17.5.4.1 Data Bank

Anticipating the more specialized needs of all its audiences, the center also began to compile more detailed information on a variety of topics. In some cases, this information took the form of specialized bibliographies. However, since many questions required current information on the state of the solar industry, the center also began to build a series of comprehensive databases. By the end of the final contract year (1981), NSHCIC had constructed data resources of over 17,500 entries in seven major categories:

Solar Manufacturers—a directory of companies manufacturing solar heating and cooling products and related equipment;

Solar Products—information about the products manufactured by companies listed in the manufacturers' file;

Solar Legislation—listings of federal, state, or local laws related to solar heating and cooling;

Solar Professionals—listings of individuals or firms active in solar design and/or construction;

Solar Installations—identification of residential or commercial applications using solar heating;

Solar Speakers—information on individuals able to speak on solar applications; and

Financial and Loan Programs—listing of grants and loan programs financing solar applications.

In the first year of the center's existence, information from the data bank was categorized in fairly general terms. In the later years, as questions became more sophisticated and as more experience accumulated, the data bank was upgraded to provide more detailed the discriminate information.

For example, lists of solar products began to be organized according to more detailed subcategories (lists of collector manufacturers were subdivided into liquid flat-plate collector manufacturers, air flat-plate collectors manufacturers, concentrating collector manufacturers, etc.). The increasing level of detail in the data bank was shown by the sophistication of the short information handouts prepared by NSHCIC's technical writers. The topics changed from the very general (e.g., "Insulation") to the very specific (e.g., "Orientation and Tilt of Solar Collectors") over the course of the center's existence. By the fifth contract year (1980–1981), NSHCIC had produced over 180 different publications.

17.5.4.2   Methods to Ensure High Quality

As interest in any technology increases, as users become more knowledgeable, and as the technology develops, a technical information center is faced with the issues of response time and quality control. Although some telephone inquiries could be answered immediately by the responding analyst, most required sending one or more publications. Indeed, analysts were encouraged to send publications instead of giving verbal answers in order to make sure that the caller received a comprehensive response, to keep telephone costs to a reasonable level, and to ensure that fewer callers would receive busy signals.

In the first year of operation, all but the most general telephone or written requests required special attention from the responding analyst. The novelty of many of the questions, together with the relative absence of specialized publications, meant that many of the responses had to be completely constructed by the analyst involved. Because most of the center's information analysts were not experts in solar energy, consultations with NSHCIC's engineering staff were often necessary.

The need to prepare specialized responses created problems for response time as the inquiries to the center increased. The goal for all responses was seven working days; in practice, however, response times often exceeded ten working days for calls and eleven working days for letters. Although upgrading the analysts' knowledge, increasing staff levels, and introducting word processors improved response time to some degree, it was nonetheless clear that the center had to make its responses much more automatic.

This was accomplished through two major innovations. First, the center's more knowledgeable technical staff (consolidated into a Technical Services Group in the center's third year) began to produce brief but detailed and specialized written materials that could be sent to everyone asking a question on a given subject. Second, this same group began to prepare "standard" paragraphs that could be mechanically assembled into customized responses.

The most effective quality control is a well-trained staff. NSHCIC was fortunate in having a relatively high staff retention rate. By the end of the fifth year, the retention rate among telephone analysts was 50%. Among the more specialized technical staff, only about 15% of the positions had turned over.

In the first years of the center's operation, staff were trained through a one-week orientation and occasional (usually monthly) "technical sessions," in which a member of the engineering staff or another technical expert would give a presentation on specific technical issues. As the center's operations became more routine, a comprehensive two-week training curriculum for new employees was developed, covering the administrative aspects of the telephone operations, the basics of solar heating and cooling technologies and general skills in dealing with callers.

The training program drew on the expertise of the more seasoned staff members and led to the creation of several resource books containing annotated copies of the center's information materials. Weekly refresher courses for all the telephone analysts also became part of the staff development program. Staff training was supplemented by monitoring randomly selected responses to gauge overall quality and to identify particular problem areas. Like staff development, this process was rather informal in the first year of the center's operation.

During the fifth year a more formal quality control plan was developed by a consulting firm in Philadelphia. As a result, random samples of fifty responses to telephone calls and letters were evaluated each month by an experienced staff member to assess the accuracy and the appropriateness of the responses. Quality control information was also obtained by telephone and mail surveys in which samples of the center's users were asked about the usefulness of the information they received. (Franklin Research Center 1981)

17.5.5 Meeting User Needs: Outreach

Consistent with its overall strategy, most of the center's outreach activities were directed toward influential groups within the housing industry. These activities fell under the broad description "market development," meaning information designed to resolve problems that otherwise might prevent or impede builders, contractors, architects, and others from integrating solar applications into the housing industry.

17.5.5.1 Outreach Strategies

In any outreach effort, decisions about the form and content of the information delivered must recognize the needs and interests of the relevant users. In dealing with each audience, HUD, DOE, and NSHCIC decided very early in the program to make use of the formats and forms that were familiar to and accepted by the groups in question. Thus, for the most part, NSHCIC presented exhibits at regular trade shows and professional conferences and placed articles in existing trade and professional journals. There were also a number of more active efforts to create special opportunities for targeted dissemination. Mostly, these took the form of special conferences, frequently cosponsored with organizations having high visibility for the targeted audience.

For example, from 1977 through 1979, the center cooperated with 27 state energy offices to host 56 conferences for local building professionals. As a part of the HUD Solar Hot Water Initiative, NSHCIC worked with 10 states to conduct 31 installation workshops. In cooperation with 4 major trade associations, DOE and NSHCIC sponsored 72 solar workshops for the financial community in 1979 and 1980.

17.5.5.2 Exhibits

To make use of these conference and workshop opportunities, NSHCIC created a number of special products. Probably the most successful of these was its series of exhibits; the most visible of these in the early years of the program was the HUD Solar Center Van. The van featured working models of solar energy system applications, videotapes, and literature and was staffed on a rotating basis by two NSHCIC staff members. From its first appearance in late 1976 to mid-1979, the van traveled over 61,000 miles and was visited by over 90,000 people. The van was retired in 1979 because of rising fuel costs and the growing inadequacy of that format as solar information became more complex and potential users more knowledgeable.

The stationary exhibits used at trade and professional shows continually improved in their appeal, sophistication, and informativeness. NSHCIC's exhibits began as fairly standard booths with pictures and literature, culminating in a "walk-through" model that demonstrated all of the functions performed within a passive solar home. In its five years of operation, NSHCIC provided stationary exhibits to over 130 trade shows and professional conferences.

17.5.5.3 Publications

One of the values of NSHCIC's exhibit program was that it provided opportunities for the large-scale dissemination of center publications to a variety of audiences. These publications also became more technical as demonstration projects and other real-world experiences accumulated.

Over its five-year existence, the center produced over thirty specialized publications for members of the housing industry. These ranged from the fairly brief *HUD Solar Status Reports*, which presented relevant information (e.g., builders' experiences, lending practices, zoning and building code problems) in a newsletter format, to large technical reports and guidelines. The most technically sophisticated of this latter group was *Installation Guidelines for Solar Domestic Hot Water Systems in One- and Two-Family Dwellings*, written by the director of the center's Technical Services, Peter Hollander. (HUD 1979a)

In addition to distributing its own specialized publications, the center also coordinated the dissemination of documents written by other contractors to the HUD demonstration program, including two editions of *Selling the Solar Home*, prepared by the Real Estate Research Corporation (RERC 1978, 1980), and two guidebooks on Planning for Solar Access, sponsored by the American Planning Association (Jaffe and Erley 1979; Erley and Jaffe 1979). Dissemination of these materials was accomplished not only through the exhibit and conference programs but also through specialized mailings to various groups on the center's mailing lists, which totaled 250,000 names by 1981.

In the later years of its operations, NSHCIC also began developing substantive articles targeted for inclusion in magazines and journals read by differing audiences. Special features were included in publications such as *Builder*, *Contractor*, and *Nation's Cities*; more than two dozen such articles appeared either written by NSHCIC staff or using information provided by the center.

17.5.5.4 Audiovisuals

To aid its outreach activities further, the center produced approximately fifteen audiovisual features, films, and slide presentations, ranging from fairly general introductions to solar energy and reports on the experiences of selected solar builders to fairly sophisticated presentations on design and installation "lessons learned." These materials compressed a substantial amount of information into relatively brief and accessible formats. In addition to using these products at center-sponsored events, NSHCIC also made them available to other groups on a purchase or rental basis. By 1981 NSHCIC had responded to approximately 3,500 such requests.

17.5.5.5 Residential Solar Viability Program

To help deal with questions about the economic feasibility of solar energy systems (questions that were necessarily application-specific), HUD staff contracted with the Washington consulting firm Booz, Allen & Hamilton to develop a computer program capable of assessing the long-term economics of specific solar heating applications. The Residential Solar

Viability Program (RSVP) was available at major trade and professional conventions and was used on a limited basis to answer questions by mail. In keeping with HUD's and NSHCIC's overall dissemination strategy, the program was largely reserved for the use of building professionals.

17.5.5.6 Staff Presentations

Center staff members also regularly participated at exhibits, conferences, and workshops, both those conducted by the center and those sponsored by other groups. Once again, the information presented by center staff became more sophisticated as information accumulated and expertise grew.

Initially, most of the appearances by center staff, even at fairly technical conferences, were confined to informing the audience about the existence of the center and the nature of its services. By 1979, however, members of the Technical Services staff were regularly giving sophisticated lectures at NSHCIC's workshops for the financial community.

17.6   Relation to Other Organizations

17.6.1   Solar Energy Research Institute

SERI's major official information responsibility was the creation and maintenance of the Solar Energy Information Data Bank (SEIDB). SERI's other major information activities were conducted as parts of other programs, for example, those managed by its Commercialization Branch and directed toward consumers, building professionals, and state and local governments. (The information activities of SERI are described in more detail in chapters 16, 18 and 22).

The most extensive area of interaction between NSHCIC and SERI concerned the SEIDB. NSHCIC would provide access to its own substantial database and, would in turn utilize SEIDB resources when appropriate. For example, NSHCIC used the SEIDB's Solar Educational Data Base to retrieve information about curricula and courses available throughout the country (see also chapter 19). Interactions with other SERI programs were generally ad hoc. Programs conducted with SERI's Commercialization Branch sometimes used information provided by NSHCIC, and NSHCIC made every effort to make its users aware of SERI's mission and resources.

17.6.2   Regional Solar Energy Centers

Four regional solar energy centers (RSECs) were created in 1977, in part to perform market development and consumer information services in each of the four major geographic regions of the United States (Northeast, Southeast, Midwest, and West) and in part to respond to political pressures. None of the RSECs engaged in public information activities comparable in scope to those of NSHCIC. Instead, the RSECs generally concentrated on developing special programs or special information packages designed to meet special needs in their own areas. (The relevant activities of each of the RSECs are discussed in detail in chapter 20.)

NSHCIC's relations with the RSECs were largely informal. NSHCIC supplied a variety of information packages to different programs conducted by the RSECs and made sure to inform people in the regions of the existence and nature of the RSECs; the RSECs in turn publicized the existence of NSHCIC.

17.7   Evaluation

17.7.1   Federal Oversight Structure

In the authors' opinion, the complex relationship between HUD, DOE, and NSHCIC created a number of recurring problems. NSHCIC was occasionally drawn into interagency disputes, for example, over the virtue of the series of workshops for the financial community, strongly desired by DOE but questioned by HUD. Moreover, the immediate ERDA/DOE project office was, for the duration of the center's existence, a relatively small component of the agency's overall solar program. One of the office's roles, then, was to serve as a conduit for requests for NSHCIC services from other ERDA/DOE areas; in practice, the project officers had little effective authority to rule on the importance of any particular request or to set priorities among competing ones. This problem intensified as NSHCIC's interactions with SERI and the RSECs, each under a different DOE project office, increased.

17.7.2   Dissemination Strategies

In its five years of operation, the center responded to the direct inquiries of nearly one million people. Another three million persons were reached through exhibits and bulk mailings. Responses to the center's quality control surveys suggest that most of the inquiries were answered to the user's satisfaction. (Franklin Research Center 1981) In terms of cost, an informal analysis by DOE suggested that costs per inquiry averaged between $10 and $11 in 1984 dollars. DOE considered this to be a very reasonable cost figure, although no exact cost comparisons were ever made.

Perhaps the most significant structural problem was that the center was prevented by its mandate from answering questions on energy conservation and on renewable energy technologies other than solar heating and cooling. However clear the reasons for this restricted scope were to federal program managers, the public found it difficult to understand why the center would not handle questions on energy conservation or wind energy.

Although this problem persisted until the end of NSHCIC's operations, the center was nevertheless allowed to compile bibliographies and fact sheets on energy conservation and other renewable energy technologies. This reduced the frustrations of callers interested in other energy-saving measures, while maintaining the center's focus on solar heating and cooling.

The center's most troubling substantive problem was that many of the most important questions about solar heating and cooling could not be answered with confidence. These questions concerned the cost effectiveness of solar heating systems and the reliability of particular solar products or manufacturers. The costs of solar energy systems varied tremendously according to manufacturer, application, and geographic area. Expected energy savings also depended upon a large number of technical and situational variables. Moreover, the degree to which energy savings were reflected in dollar savings also varied, depending on the type and price of the conventional fuel being replaced and on the rate structure of the local utility.

In the early years of the center's existence, answers to cost and performance questions were largely conjectural owing to the relative absence of reproducible solar applications. In later years, meaningful average cost figures became available, although reliable performance data continued to elude most analysts. The center's information in this area was often limited to projected performance figures (not independently verified) supplied by solar manufacturers. The unstable condition of the solar industry also meant that product reliability information was available sporadically and of limited value. Moreover, as a matter of policy, HUD and NSHCIC agreed that it would be inappropriate to issue what could be seen as endorsements of particular products.

NSHCIC addressed the need for information on product reliability by informing consumers where they might obtain this information (e.g., from state energy or consumer offices, which often accumulated collector or system test results and which sometimes commissioned tests) and by clarifying the meaning of different measures of product performance (e.g., collector efficiency). Nonetheless, the lack of reliable product evaluations was a continuing source of frustration for questioners and respondents alike.

17.7.3 Outreach Activities

By all reasonable measures, NSHCIC succeeded in reaching the professionals making up its target audiences. By 1981 NSHCIC's mailing lists in its professional categories totaled over 75,000 names. The audiences at state solar conferences, installation seminars, and workshops for the financial community were composed largely of housing or building industry professionals. The proper trade shows were attended and the pertinent publications utilized. On the basis of quality control surveys and conference responses, interest in the services of NSHCIC was high. However, in the authors' opinion, NSHCIC was not as effective as it could have been in answering some of the most significant questions, nor were the activity and visibility of the information program matched by a commensurate increase in the number of homes using solar energy.

Both of these difficulties, of course, were largely the result of factors beyond HUD's, DOE's, and NSHCIC's control: the increasing costs of and uncertain reliability of solar energy systems, the development of cheaper and more effective energy conservation technologies, and the decline of oil prices.

17.7.3.1 Problems in Outreach Efforts

In the authors' opinion, the development of specialized products for the center's target audiences was never as coherent or timely as it could have been. HUD and DOE officials, especially the more experienced ones at HUD, were often reluctant to release significant information prematurely or to use formats or language they felt would compromise their image as cautious professionals. As a result, review processes were lengthy and some products were canceled or drastically revised in midstream.

In the authors' opinion, although the impulse behind this caution and restraint was a good one and its results were often beneficial, HUD's and DOE's caution was often as damaging as it was helpful. By refusing to approve informational products that were not completely above reproach, the outreach program was effectively subordinated to the goal of damage limitation.

On more than one occasion, this resulted in the release of information materials that were seriously outdated. For example, a relatively optimistic report on builders' experiences with active solar heating systems was not released until federal and industry interest in active systems was declining (RERC 1980). Likewise, publications or reports were often delayed for very long periods or extensively revised with little corresponding improvement in the final product. For example, a long-delayed report on design and installation problems in active systems appeared after several years of review in a form and with a content similar to those in a previously rejected version (HUD 1982).

These difficulties were exacerbated by the absence of professionals with relevant information materials development expertise (as distinct from writing ability or production management skills) in NSHCIC and its Federal project offices. Thus, even though writing, reviewing and production management responsibilities were met clearly and competently, the initial design of the product, which determined the effectiveness of all subsequent steps in the process, was generally vague and impressionistic.

The market development effort was also hindered by problems in both the form and the substance of the information collected from the Federal demonstration projects. Necessarily, the form in which demonstration data were collected was developed very early in the program. Decisions on data format were made by technical program management rather than information dissemination specialists. Consequently, potential users of the data were not always given prime consideration.

For example, to ensure privacy, projects were identified for data collection purposes by number rather than by name. Project numbers were assigned by the contractor responsible for collecting the kind of data in question—design, marketing, performance, and so on. Initially, there was little care taken to see that the various numbers assigned to a single project could be adequately cross-matched. Thus every attempt to correlate technical performance, customer reaction, and utility payments had to be verified for each individual project.

In the later years of the program, a computer program for correlating project data across the different files was developed. But the absence of such a system in the earlier years made it difficult to produce any comprehensive reports on demonstration experiences.

A closely related formal problem concerned data reliability and validity, particularly involving "instrumented" projects. In order to obtain direct, objective information on system performance, a sample of systems was monitored by a diverse array of sensors recording temperatures, air and water flow, heat losses from storage systems, and so on.

Initially, the reliability of these data was highly questionable because of system idiosyncrasies and sensor or transmission malfunctions. Moreover, in the earlier years of the program, the data were reported in a highly abstract format that was too narrowly focused on the performance of the solar energy system components; for instance, the initial instrumentation reports assumed that none of the heat lost from air systems during storage and distribution contributed to heating the building (Freeborne, Mara, and Lent 1979). These and similar shortcomings compromised the reports' usefulness for an audience interested in how much of the building's heat was supplied by the solar energy system. Although these problems diminished over time, the reliability and usefulness of the instrumented data were of continuing concern throughout the program.

Of much greater concern, however, was the failure of many of the demonstration systems to live up to expectations. Experiences with active space heating systems, whether air or liquid, were particularly disappointing. (The design and installation problems that plagued these systems are discussed further in chapters 9–13 and other volumes of this series.) Here it should simply be noted that the absence of reliable and positive data on system performance was, in the authors' opinion, the most severe barrier to NSHCIC's market development.

17.7.4 Relations with Other Information Programs

In the authors' opinion, several problems developed among the various solar information programs; the most serious concerned redundancy in functions and data. Over the five years of public information efforts, considerable confusion accumulated over the division and assignment of responsibilities. In spite of the theoretically clear delineation of responsibilities among the various information programs, all of them mistakenly tried to do the same thing. Attempts at cooperation among NSHCIC, SERI, and the RSECs were often hindered by conflict; no Federal program office was able to set rules and priorities. Fortunately for NSHCIC, HUD management was able to provide both a clearly delineated set of responsibilities and a buffer against internal DOE squabbles. Moreover, difficulties arose with the SEIDB. The cost of designing and implementing this massive database became so great that DOE eventually terminated the effort in 1981. (The SEIDB is discussed in chapter 18.)

17.8   Creation of CAREIRS

The NSHCIC contract was due for renewal in May 1981; however, with the change in Federal administrations, governmental interest in renewable energy, especially commercialization efforts, diminished considerably. DOE management, therefore, planned to discontinue NSHCIC. Paradoxically, NSHCIC was receiving requests for information at rates greater than any other time in its history.

Faced with a flood of protests, including some from Congress, DOE decided to continue the NSHCIC activity with two important differences: (1) the subject matter was extended to include all renewable energy technologies and energy conservation and (2) the budget was reduced by 75% (see table 17.4).

Table 17.4
CAREIRS budgets by contract year (in thousands of dollars)

Year	Current year	1985 $
1981–82	1,500	1,782
1982–83	1,300	1,452
1983–84	1,300	1,399
1984–85	1,200	1,240
1985–86	1,000	1,000
1986–87	1,000	975
1987–88	1,000	945
1988–89	1,000	914

Source: Franklin Research Center 1977, 1978, 1979, 1980, 1981.

NSHCIC's name was changed to CAREIRS (Conservation and Renewable Energy Inquiry and Referral Service) to reflect the new scope of the center. At a meeting of all solar information program entities in early May 1981, Robert Annan of DOE outlined a concept called the "Conservation and Renewable Energy Information Network" (CAREIN), of which CAREIRS was to be the inquiry, response, and referral component, and the only entity with toll-free hot line services. SERI would have primary responsibility for publications and databases, and the RSECs would deal with users at the regional level. Unfortunately, CAREIN never materialized, databases under the SEIDB disappeared, the RSECs were discontinued and it was impossible to get approval for any new publications. Thus CAREIRS was left as a response center without a support network.

To its credit, CAREIRS managed to continue operations during this period by salvaging publications from DOE offices that were still functioning and by placing greater reliance on the computer-generated responses in its own system. The solar information program literally lived from month to month until early in 1983, when DOE, with approved funding from Congress, finally contracted CAREIRS on a long-term (five-year) basis.

17.9 Conclusions

CAREIRS continues to operate through a small mail and telephone staff in Philadelphia. Over the first seven years of its existence (1983–1989), the service handled over 1.2 million inquiries, the majority of which were from the general public. Paralleling development in the solar and energy conservation industries, the bulk of the inquiries now concern energy conservation and passive solar energy applications. In this curtailed form, information dissemination remains a part of the Federal program on conservation and renewable energy.

References

ADL (Arthur D. Little, Inc.). 1976. *Residential Solar Heating and Cooling: Constraints and Incentives.* Cambridge, MA.

Dubin Bloome Associates. 1978. *Building the Solar Home.* HUD-PDR 246-1. Washington, DC: U.S. Department of Housing and Urban Development.

Erley, D., and M. Jaffe. 1979. *Site Planning for Solar Access: A Guidebook for Residential Developers and Site Planners.* HUD-PDR 481. Washington, DC: U.S. Department of Housing and Urban Development.

Franklin Research Center. 1977. *Solar Information Program, March 20, 1976–March 19, 1977.* Philadelphia.

Franklin Research Center. 1978. *Solar Information Program, March 20, 1977–March 19, 1978.* Philadelphia.

Franklin Research Center. 1979. *Solar Information Program, March 20, 1978–March 19, 1979.* Philadelphia.

Franklin Research Center. 1980. *Solar Information Program, March 20, 1979–March 19, 1980.* Philadelphia.

Franklin Research Center. 1981. *Solar Information Program, March 20, 1980–March 19, 1981.* Philadelphia.

Freeborne, W., G. Mara, and T. Lent. 1979. "The Performance of Solar Energy Systems in the Residential Solar Demonstration Program." *Conference Proceedings: Solar Heating and Cooling Systems Operational Results.* SERI TP 245-430. Golden, CO: Solar Energy Research Institute.

Jaffe, M., and D. Erley. 1979. *Protecting Solar Access for Residential Development: A Guidebook for Planning Officials.* HUD-PDR 445. Washington, DC: U.S. Department of Housing and Urban Development.

RERC (Real Estate Research Corporation). 1978. *Selling the Solar Home.* HUD-PDR. Washington, DC: U.S. Department of Housing and Urban Development.

RERC. 1980. *Selling the Solar Home, 1980.* HUD-PDR. Washington, DC: U.S. Department of Housing and Urban Development.

HUD (U.S. Department of Housing and Urban Development). 1979a. *Installation Guidelines for Solar DHW Systems in One- or Two-Family Dwellings.* Washington, DC.

HUD. 1979b. *The Use of Active Solar Heating and DHW Systems in Single-Family Homes: Technical Findings and Lessons Learned from the HUD Solar Demonstration Program.* Washington, DC.

HUD. 1982. *Installation Guidelines for Solar DHW Systems in One- or Two-Family Dwellings.* Washington, DC.

18 Technical Information

Paul Notari

18.1 Background

18.1.1 Early Federal and Congressional Solar Technical Information Initiatives

The importance of the interchange of the latest state-of-the-art technical information on solar technologies was recognized as early as 1971 by the National Science Foundation (NSF). In that year NSF created a new program called Research Applied to National Needs (RANN). Basic research on solar energy, which had been supported earlier by NSF and the National Aeronautics and Space Administration (NASA), was assigned to the RANN program and its character was changed to encompass applied research as well as basic research. Along with this change NSF first began to issue formal reports on the results of its research. Prior to that, the agency relied almost exclusively on the publication of its scientific achievements in scholarly journals which receive very limited readership (NSF 1984).

As more and more of the new reports were generated, concern arose as to how NSF could best get them into the hands of those who would most benefit from them. To satisfy this need, the RANN Document Center (RDC) was created by the NSF and located in Washington, D.C., and was managed by a private subcontractor, Capitol Systems, Inc. The responsibility for document distribution was later transferred to the newly formed Technical Information Center (TIC) at Oak Ridge, Tennessee. At the RDC, and later at TIC, abstracts of all solar research reports were prepared and distributed throughout the scientific and technical community. Copies of the reports were then dispersed from the center as they were requested. Shortly after, a new law (PL 84-74) was passed requiring that all federal research reports be sent to the newly formed National Technical Information Service (NTIS); thereafter, the reports became available through that outlet as well.

This was a start, but it was still not enough. An NSF study issued in May of 1974 concluded: "A need exists for more interactive, user-guided information transfer. Passive access to voluminous stores of textual and numeric data does little to promote timely, selective retrieval and application of high-quality results."

Whether inspired by this study or not, a new law, PL 93-409, was passed on 11 September 1974 that establised the National Solar Heating and Cooling Information Center (NSHCIC) under the Department of Housing and Urban Development (HUD). This Center provided information on solar heating and cooling technologies to a variety of audiences including architects, engineers, builders, and contractors. Shortly after the center began operation, however, it became evident that in order to promote the use of solar energy for residential heating and domestic water heating, it would be necessary to educate consumers through various means. A solar hot line was established and widely publicized; a traveling exhibit was built and transported around the country to acquaint the general public with solar technologies. (For more details about NSHCIC, see chapter 16 of this volume.)

Also in the fall of 1974, Congress passed the Nonnuclear Energy R&D Act at 1974 (PL 93-577) that outlined specific policies and principles for the Energy Research and Development Administration (ERDA), which began operation in January 1975. The new agency was charged specifically with responsibility for federal participation in nuclear and nonnuclear research and development. Consequently, all of the solar energy research activities of NSF were transferred to ERDA and, along with them, all of the information transfer activities. During this same period, in response to the energy crises of 1973, the Federal Energy Administration (FEA) was formed to consolidate all of the federal non-research energy functions into a single organization. Under both ERDA and FEA, information activities included substantial development and dissemination of research reports. On 26 October Congress passed the Solar Energy Research, Development and Demonstration Act of 1974 (PL 93-473), which specifically promoted dissemination of solar energy information and which led to the formation of both the Solar Energy Research Institute (SERI) and the Solar Energy Information Data Bank (SEIDB).

18.1.2   Creation of the Department of Energy, SERI, and the SEIDB

A significant new era in solar technical information development and dissemination began in 1977 when the U.S. Department of Energy was formed and PL 93-473 was implemented. During the first month of the Carter administration it was decided that energy had become of such national importance that it deserved cabinet representation. As a result, in 1977 the Department of Energy (DOE) was created; all of ERDA and

FEA, as well as parts of other federal agencies devoted to energy R&D, were combined into this single energy department. Since that time, DOE has assumed the key role in all U.S. energy research and development, and has been a major player in the development and dissemination of technical information on all aspects of energy, including solar energy.

In July 1977 SERI was established in Golden Colorado under the management of Midwest Research Institute, Inc., and under the direction of Dr. Paul Rappaport, a renowned scientist from RCA. The new organization was eventually to become the foremost solar energy research facility in the world. At its start, however, SERI consisted of a few leased offices in a commercial building adjacent to the land on which the final laboratory would be built. By the middle of 1978, considerable staff was added and laboratories were constructed in leased quarters adjacent to the offices; the institute was fully functioning by the fall of 1978.

Among its many charges, SERI was responsible for managing the Solar Energy Information Data Bank (SEIDB). As the name implies, SEIDB was to be a storehouse of information on solar energy technology and was to be readily accessible to scientists and engineers all over the United States. Although the SEIDB could have been located anywhere, SERI was the logical choice because the center of all federal solar R&D activities would eventually be situated there. In August of 1978 SERI formed the Information Systems Division (ISD), under the directorship of Herbert Landau, mainly to implement and manage the SEIDB. Originally, three branches were formed within the division: the Computer Systems Branch, the Database Systems Branch, and the Information Center Branch; later, two more branches were added; the Information Dissemination Branch and the Inquiry and Referral Branch. The entire organization was staffed and fully operational by the end of 1979.

Along with SERI, four regional solar energy centers (RSECs) were also established. Located in Portland (Oregon), Minneapolis, Boston, and Atlanta, the centers were specifically charged with the responsibility for bringing solar energy technology to their regional constituencies. One of their principal functions was information dissemination; to perform this function, they were to operate as integral parts of the SEIDB network. DOE headquarters had the responsibility for coordinating the information activities of the RSECs with one another and with SERI under the overall SEIDB umbrella. (The information activities of the RSECs are further discussed in chapters 15 and 20.)

18.1.3   Other Federal Solar Energy Information Activities

To fully appreciate the importance that Congress and the federal agencies placed on the dissemination of solar technical information in earlier years, one should be aware of the information activities that were carried out in addition to the SEIDB effort. Some of these activities concerned themselves mainly with educating and motivating the general public. Although their role in technical information dissemination was subordinate to their principal mission, it was significant and should be reported here.

As mentioned earlier in this chapter, both NSHCIC and the RSECs were heavily involved in information dissemination. Both, however, concentrated more on providing general information to the public as opposed to technical information to scientists, engineers, technologists, and other professionals. Details of the activities of each of these organizations are provided elsewhere in this volume.

The Technical Information Center (TIC) at Oak Ridge, Tennessee, originally operated by the Atomic Energy Commission (AEC), became part of ERDA in 1975 and of DOE in 1977. TIC had begun to include solar energy research reports in its automated bibliographic database as early as 1972 and continued to provide this service to the solar energy technical community through January 1989, cataloguing hundreds of thousands of titles for retrieval by solar professionals. Furthermore, TIC's Energy Data Base (EDB) was made available to private commercial brokers for several years, allowing convenient access by their clients. In addition, TIC provided duplicate tapes of its EDB listings and updates to NTIS for incorporation in their own bibliographic database and catalog. Twenty-five copies of each of the reports listed are also provided to NTIS for their fulfillment of orders; when this supply is exhausted, NTIS provides microfiche "blowback" copies of the reports to its customers.

Other federally supported organizations that dealt to some extent with the dissemination of technical information were the National Center for Appropriate Technology (NCAT) and the state offices of the Energy Extension Service (EES). NCAT was formed in the 1960s by the Office of Economic Opportunity (OEO) and became involved in energy activities in the early 1970s; EES was established by the DOE in 1977.

Located in Butte, Montana, NCAT functioned mainly to educate grassroots groups and individuals on the benefits of energy conservation and renewable energy technologies. NCAT staff prepared and distributed literature directed specifically to the needs of their clients and provided direct technical assistance to inventors and those who expressed a desire to adopt some of the technologies.

EES was funded and operated though the DOE State and Local Assistance Program. Under this program, energy offices were established in all of the fifty States, plus Puerto Rico and the District of Columbia. The program was patterned after the successful USDA Extension Service and served the same end, namely, to provide hands-on assistance to people and groups at the state level who showed a likelihood and desire to adopt conservation and renewable energy technologies. In many ways the state energy offices functioned like NCAT and indeed often collaborated with that organization in many activities. (For details on NCAT and EES, see chapters 16 and 29, respectively.)

Another program that aimed to provide technical information to the solar field, the Technical Information Dissemination (TID) program, was established at SERI in late 1978 and also funded through the DOE. The TID program only continued for a little over two years but was responsible for developing and widely distributing or displaying a number of technical brochures, slide shows, and exhibits at various professional conferences and events. Administered by the Commercialization Division of SERI, the TID program was dissolved at the same time the division was discontinued in 1981. Since then, many of the TID functions have been taken over by the Solar Technical Information Program (STIP) which began in 1981.

Concurrent with the creation of FEA in 1974 and now part of DOE, the Energy Information Agency (EIA) was formed mainly to provide reliable statistical information on resources, fuel consumption, and use of energy in the United States. EIA also has provided such solar-related information as the number of direct hot water solar systems sold and the number of photovoltaic (PV) units shipped and installed. Several years ago, after the general collapse of the solar market, all of these solar statistics were dropped. As of the date of publication of this volume, the publication of this data has not yet been resumed.

The development of an alcohol fuels industry became a top priority of the Carter Administration in 1979. To aid in this development, a National Alcohol Fuels Information Center (NAFIC) was established at SERI. This center was to provide information on alcohol fuels to potential producers from the farm level to large industry. Two cornerstones of this program were the development and wide dissemination of state-of-the-art publications on the subject, and the maintenance of a hot line to answer inquiries from the field. Administration and operation of the NAFIC program was carried out by SERI; more on the subject of NAFIC will be presented later in this chapter.

18.2 Development of the Solar Energy Information Data Bank

18.2.1 Original Concept and Objectives

According to the Solar Energy Research Development and Demonstration Act of 1974 (PL 93-473), the SEIDB was conceived by Congress for the purpose of "collecting, reviewing, processing, and disseminating information and data in all solar technologies ... in a timely and objective manner...". Specifically the act directed that

"Information and data compiled in the bank shall include:

(A) technical information (including reports, journal articles, dissertations, monographs, and project descriptions) on solar energy research, development and applications;

(B) similar technical information on the design, construction, and maintenance of equipment utilizing solar energy;

(C) general information on solar energy applications to be disseminated for popular consumption;

(D) physical and chemical properties of materials required for solar energy activities and equipment; and

(E) engineering performance data on equipment and devices utilizing solar energy."

In addition, the act stated that retrieval and dissemination service for the above information was to be provided for:

"(A) Federal, State, and local government organizations that are active in the area of energy resources (and their contractors);

(B) universities and colleges in their related research and consulting activities; and

(C) the private sector upon request in appropriate cases."

Additionally, the act included these provisions:

Existing data bases of scientific and technical information in Federal agencies shall be utilized where feasible. Any information developed or accumulated under this Act that does not already reside in these existing data bases shall be added.

Special studies and research on incentives to promote broader utilization and consumer acceptance of solar energy technologies shall be performed.

Arrangements and other steps shall be entered into "as may be necessary or appropriate to provide for the effective coordination of solar energy technology utilization with all other technology utilization programs within the Federal Government."

It should be noted that one provision of PL 93-473 duplicated a provision in the legislation creating NSHCIC (PL 93-409). Both provisions called for collection of general information on solar energy applications and its dissemination for popular consumption; this caused some duplication of effort, although, close coordination between the two programs kept this to a minimum.

18.2.2   Creation of the Information Systems Division at SERI

To provide an organization and the mechanisms to build and implement the SEIDB, an Information Systems Division was formed at SERI in August of 1978. Originally the division consisted of three branches:

Computer Systems Branch, first headed by John Jones and later by Larry Harmon, was charged with developing or acquiring the computer hardware and software for the SEIDB.

Database Systems Branch, first headed by Brigitte Kenney and later by Howard Shirley, was made responsible for collecting and compiling data for the data bank and for integrating this data into interactive databases.

National Solar Energy Information Center Branch, first headed by Robert Lormand and later Jerome Maddock, was charged with establishing a central library of solar energy and related publications, to be accessible to all members of the solar professional and nonprofessional communities. It also was charged with developing a solar bibliographic database as part of the SEIDB; in addition, it was commissioned to produce a thesaurus of all terms used in the solar field.

Later in 1979 two new branches were added:

Information Dissemination Branch, headed by Paul Notari, was charged with the development of technical publications, exhibits, videos, and films and with their dissemination to appropriate audiences.

Inquiry and Referral Service Branch, headed by Stephanie Norman, was formed to provide answers to inquiries from professionals in the field as well as the general public on all subjects related to solar energy.

18.2.3 Funding and Management of the Program

In the early years, funding for the SEIDB was provided through prorated contributions from each of the several DOE solar energy programs. In 1981, for the first time, a separate line item was included in the DOE budget, designated "Solar Information" and was intended exclusively for the support of solar information development and dissemination. In the first two years that "Solar Information" was included in the budget, it supported three separate operations: NSHCIC, TIC solar energy activities, and the SEIDB. Prior to this time, NSHCIC was funded by HUD and TIC solar activities by another DOE program.

SEIDB funding (in millions of dollars) for the years of its existence was as follows:

FY	Total budget
1978	$3.4
1979	6.0
1980	8.0
1981	6.7

Because in its early years the SEIDB was funded solely from the DOE solar energy programs, there was considerable controversy between the DOE technical divisions, SERI, and the SEIDB staff. Members of the DOE technical divisions questioned whether the money diverted from their programs to the SEIDB could not better be spent on research, development, and commercialization of solar energy; they looked at the SEIDB as an unnecessary drain on their limited funds. It took DOE and SERI management several years to smooth things over and to elicit full cooperation between the members of the technical divisions at DOE and the SEIDB staff at SERI.

In the formative years of the SEIDB, Robert Annan was given responsibility for managing the program at DOE; that responsibility was passed on to Michael Pulscak in 1981. Howard Shirley managed the SEIDB program at SERI from the program's conception in 1977 until August of 1978, when Herbert Landau took charge. When the program first began, few controls were imposed; as it matured, however, frequent program status and financial reviews were carried out, and a regular reporting system was inaugurated.

18.2.4 Design and Implementation of the National Solar Energy Information Network

The first step in building the network was to establish agreements and good working relationships with the various players in the field who would eventually become members of the network. This was by no means an easy task. Most of the participating organizations by this time were either well established or in the throes of building their own information programs; they looked upon the SEIDB, at best, as a necessary evil and, at worst, as an intruder into their private affairs. It took some doing to convince each organization that it was in its best interest to cooperate.

The network participants were to include the RSECs, TIC, NSHCIC, the state energy offices of EES, and SERI. All were to be interlinked to each other by means of a central computer to which each would have online access at any time. The plan was to locate remote terminals at each of the principal offices of these organizations, permitting telephone line connection to a central computer.

The original network, first put into operation in late 1978, leased a central computer at Stanford University. The SEIDB databases being developed at SERI were the core of the system; a special software package called "SPIRES" enabled SERI and the RSECS to access and maintain the databases. Later it was decided to install two large mainframe computers at SERI to replace the Stanford computer. One of these was to be a high-speed scientific computer, which could be used by any of the network members for very large "number-crunching" chores; the other was to be a large-memory, high-capacity computer that would store a large number of databases containing information relative to different aspects of solar energy technology.

A Control Data CD7600 was purchased and installed in late 1979. This high-capacity computer was intended to eventually become the exclusive scientific computer for the network. Because of delays in procuring the database computer, however, the CD7600 served a dual purpose as both the scientific computer and the database computer for the entire life of the SEIDB. By mid-1980, the first two computer nodes were installed at the offices of the RSECs in Minneapolis and Boston; additional nodes were purchased for the other two RSECs and were in the process of being installed when the program was curtailed in January of 1981 after the new administration took power.

The ultimate plan was that each node would serve up to 100 remote terminals. These terminals would be located in various state energy offices and other strategic locations within the respective regions covered by the RSECs. From these terminals, anyone seeking information from the central data bank would merely gain access by using an assigned password and begin extracting data as needed. Another feature of the system was that it would be interactive, meaning that users could input data into the system as well as extract it. The intent was that only certain select parties in the system would be given this privilege; those who were very familiar with the workings of the system and those who could be entrusted to maintain the accuracy of the data. Anyone would be allowed to extract data from the system at any time and to use the scientific computer for computational purposes whenever it was needed.

Two organizations were connected directly into the main frame computer at SERI from the start. They were TIC and NSHCIC. Both were given limited interactive user status. TIC ultimately was to be given full discretion to input, update or extract bibliographic data in the "Solar Biblio" file at SERI. The Solar Biblio was to be a subset of TIC's Energy Data Base, for which SERI and TIC would share the responsibility for upkeep; unfortunately, the SEIDB system was terminated before this ever happened. NSHCIC, on the other hand, not only extracted data from the system but also inputted it as the occasion required. Select members of the NSHCIC staff attended training sessions at SERI to learn the intricacies of the system and NSHCIC ultimately became a full partner in maintaining some of the data in the system.

Owing mainly to the general lack of commitment to the program by the participants, the network really never jelled as originally hoped. To remedy the situation, Robert Annan, who headed the solar information program at DOE, called a summit meeting of all of the players on 6 May 1981; from that meeting, the Conservation and Renewable Energy Information Network plan was forged. This plan changed the name of NSHCIC to the Conservation and Renewable Energy Inquiry and Referral Service (CAREIRS) and attempted to define the role of all participants in the network. The plan never was given the chance to prove itself since shortly after it was issued, the RSECs were dissolved and the basic nature of the solar information program was changed.

18.2.5   Database Development

The SEIDB system was to be unique in that each database ultimately installed could be interlinked with any other database in the system and with any computational effort being carried out on the scientific computer. To do this, a very complex management system called "INQUIRE" was installed to replace the more simple SPIRES management system used with the Stanford computer. INQUIRE did provide the capability desired but required considerably more time and effort than would be required with a less complex system.

At the time of its termination, the SEIDB contained these databases:

Solar Bibliography ("Solar Biblio") contained complete bibliographic information on all literature published on the subject of solar energy, both within and outside of the federal government, including all solar entries in the TIC Energy Data Base (EDB), and all those compiled from other sources by the SERI/SEIDB staff. Several printed specialized bibliographies and reading lists were spun off of this database.

Solar Radiation provided detailed solar insolation data for all areas of the United States. A *Solar Radiation Data Manual* was produced and published from this database.

Solar Education listed every university, community college, and trade school that offered courses in solar energy, including complete details of every course offered by each institution. Three editions of the *Solar Education Directory* and the *Solar Training Directory* were produced and published from this database.

Manufacturers Directory listed all known manufacturers of equipment and components used in solar energy systems, describing the products produced by each manufacturer and identifying various contacts within the company who could be addressed for further information or other business. The directory also listed manufacturers according to the particular solar technology fields they served: photovoltaic (PV), biomass, active and passive solar buildings, wind, ocean and solar thermal; several *Manufacturers Directory* catalogs were produced from this database.

Solar Calendar listed all major solar events, such as seminars, conferences, special meetings, exhibits, and so on. The *Solar Events Calendar*, a quarterly printed version of the database, was published and widely disseminated to professionals in the field.

International listed institutions and contacts involved in energy matters, particularly solar energy, throughout the world. Several country solar information source directories were produced from this database.

At its peak a staff of approximately fifteen professionals was engaged in the development of databases for the SEIDB.

18.2.6 National Solar Energy Information Center

The largest collection of solar information in the world was assembled at SERI under the SEIDB program. It was physically located within the leased quarters of SERI but was open to anyone who wished access to the latest state-of-the-art information on solar energy. A listing of all holdings was to be made available to all interested parties as part of the Solar Biblio database, which would at first be accessible via any terminal on the solar information computer network. Ultimately, users would be able to search the database directly via regular telephone lines using a home computer and a modem; any document could then be obtained from the center by means of interlibrary loan through any local library. Unfortunately, the database was never put on-line before the SEIDB network was dissolved.

In order to achieve consistency in terminology throughout the SEIDB system, it was deemed necessary to create a single terminology standard or thesaurus. This would be particularly helpful in establishing descriptors that could be universally used in all bibliographic data files. The responsibility for producing such a thesaurus was given to the Solar Energy Information Center branch at SERI. Staff of this branch were to work very closely with the staff at TIC to ensure compatibility with the EDB. A consulting firm, expert in thesaurus preparation, was hired in 1980 to lead this effort and managed to produce a preliminary draft. Unfortunately the work was never completed because funding for this effort was terminated. Although the information center never did achieve the national status originally envisioned, it did evolve into the SERI technical library, which to this day contains one of the most complete collections of solar energy literature in the world.

18.2.7 Information Product Development and Dissemination

The principal mission of the SEIDB was to disseminate pertinent technical information related to solar energy to potential users of such information. Although the national computer network was, to a large extent, expected to serve this purpose, it was never intended to replace the print and audiovisual media; indeed, considerable effort was spent in the development of publications, slide shows, motion pictures, and exhibits within the purview of the SEIDB program.

In addition to publications derived from the various databases stored in the system, numerous brochures, manuals, handbooks, and textbooks were produced from original manuscripts written by scientists, engineers, or professional writers employed by SERI or one of the other federal laboratories, or under subcontract to SERI. These treated every solar energy subject from passive solar energy systems to photovoltaics, at every level from the average layman's to the research scientist's.

From March 1979 through February 1981 some twenty-four different documents, ranging from 4 to 400 pages long, were produced and distributed to over one-half million people who had indicated an interest in solar energy. Most of these people were registered in a Technical Audience Profile (TAP) file maintained at SERI under the SEIDB program. One of the most popular documents produced and distributed under the SEIDB program was a handbook for beginning alcohol fuel producers, *Fuel from Farms*. Distributed to almost 300,000 people, this book has been largely credited with stimulating the interest in ethanol production that resulted in the present alcohol fuels industry.

Also prepared were several solar exhibits, displayed at various meetings, and two slide shows with accompanying audio cassettes, distributed to various professional associations and community organizations. A sound motion picture was filmed but not completed before funding was terminated, and therefore was never reproduced and distributed.

In order to meet the most pressing information needs of solar energy professionals and potential users, a massive user needs study covering all solar technologies and many professional disciplines was carried out in 1978, 1979, and 1980. The results of the study determined to a very large extent what publications were produced, what methods were used to present information, and to whom the information was presented.

18.2.8   Solar Technical Inquiry and Referral Service

Although NSHCIC served the general public well by providing information on solar heating and cooling, there was no information service that provided information on the other solar technologies. Furthermore,

NSHCIC was not staffed to provide answers to highly technical inquiries, which left many scientists and technologists no place to go for more comprehensive answers to their technical questions. To meet this need, in late 1979 a technical inquiry and referral service was established at SERI and given a toll-free telephone number that was widely publicized throughout the country. At its peak, the service had six phone lines, answered over 250 telephone and letter inquiries each day, and employed eighteen information specialists. The service, carefully coordinated with NSHCIC to keep duplication of effort at a minimum, answered an estimated 100,000 inquiries over the one and a half years of its existence.

The distinct feature of the SERI Solar Technical Inquiry and Referral Service was its capability of providing custom responses to technical questions from professionals at all levels. Although "canned" and automated responses were given to callers seeking general information, an appreciable amount of effort went into researching answers for callers seeking highly technical information. In this regard, having the service at SERI, where many solar scientists and engineers were clustered, was a major advantage.

The cornerstone of the SERI Solar Technical Inquiry and Referral Service was the National Alcohol Fuels and Information Center (NAFIC). In 1980 the nation became so highly interested in the conversion of biomass into liquid fuel for automobiles that DOE was plagued with inquiries about the subject. To meet the public need for this information, DOE established NAFIC at SERI. While the center mainly responded to questions about alcohol fuels, it also displayed an alcohol fuels exhibit at various professional meetings and accumulated the world's largest collection of alcohol fuels literature.

18.3 Assessment of Federal Solar Information Activities, 1971–1981

Although intentions were universally of the highest order, many of the information programs dealing with solar energy in the 1970s were not as successful as had been expected. Some dealt only in generalities, rarely providing comprehensive information to those who desperately needed it; others were too technical, publishing highly complex information in scholarly journals that failed to reach the audiences that could have best applied it; and still others were just too expensive to operate for the benefit they provided.

18.3.1 Energy Information Agency

EIA provided much more useful information to the oil, gas, coal, and nuclear industries than to the solar industry; because solar was so minuscule in comparison to the others, this DOE statistical service could not justify large expenditures of funds to conduct exhaustive surveys of solar equipment producers and users. EIA failed to collect and publish needed statistical information in many areas of importance to the solar field; the few solar statistics the agency originally compiled and published were dropped from its reports in the mid-1980s.

18.3.2 National Center for Appropriate Technology

Situated in Butte, Montana, NCAT carried out the largest share of its direct assistance programs in surrounding areas, which benefited many citizens in the western states but not most of the population in the United States. As a result, many of the demonstration projects and other hands-on activities carried out by NCAT went largely unnoticed by the rest of the country.

NCAT became far more national in scope in 1984, when it was given a contract to manage a new DOE outreach program called the "National Alternative Technology Assistance Program" (NATAS). Under this program, NCAT installed a toll-free telephone number and solicited inquiries from all fifty states; hundreds of calls were received daily, and answers or referrals were provided to every caller.

In its early years, NCAT concentrated its efforts on education of the general public, community organizations, and activist groups, introducing hundred of thousands of people to the basic concepts and importance of solar energy. However, few engineers, scientists, or technologists were ever provided meaningful technical information or assistance through NCAT. Although many home builders and construction workers were instructed on how to build new homes or renovate old structures to make them energy-efficient, too large a portion of this direct assistance was confined to the West and did not benefit those outside of the region.

In 1984, when NCAT took on the responsibility for the NATAS program, its character changed and it began to serve more technical clients. Although it offered little aid to scientists and engineers, it began to provide direct assistance to builders, manufacturers, engineers, entrepreneurs, potential investors, inventors, developers, technologists, and others engaged in the application of conservation and solar energy technologies.

18.3.3 National Solar Heating and Cooling Information Center

If any organization was successful in fulfilling its mission, it was NSHCIC. It was formed primarily to inform professionals in the building industry and the general public about solar heating and cooling and to encourage its adoption. (Full details of this operation are provided in chapter 17.) Suffice it to say that millions of Americans first learned about solar energy through the efforts of NSHCIC. Although the center did dispense some technical information already on hand, the normal procedure was to refer inquirers to more knowledgeable sources such as SERI.

When, in 1981, the name of the National Solar Heating and Cooling Information Center was changed to the Conservation and Renewable Energy Inquiry and Referral Service (CAREIRS), the mission of the organization was changed as well. The essential difference was that the new service was commissioned to answer questions on energy conservation and other renewable energy technologies, as well as those on solar heating and cooling. Although the scope of CAREIRS services was increased, no additional funding was provided to expand the resources of the organization; indeed, overall funding for the organization was reduced 25%. The result was the elimination of most custom responses to specific questions posed by inquirers; much more reliance was placed on providing automated "canned" replies. Thus many serious inquirers did not receive the in-depth answers they would have received under NSHCIC.

18.3.4 Technical Information Center

TIC was designated by DOE as its central depository of energy information. There, copies of all solar research reports were abstracted and stocked for targeted distribution and fulfillment of requests from the federal government or its subcontractors. In addition, titles and descriptive data about the reports were entered into the Energy Data Base which was made available to various federal libraries and private information brokers for searches by their clients.

The service provided by TIC was very helpful to technical users insofar as it went. Its drawback was that TIC rarely generated information itself, depending mostly on information supplied by others to fill its inventory. Furthermore, federal restrictions prevented TIC from disseminating its documents freely to companies and individuals outside of the government network. Scientists, engineers, and others who were not affiliated with the government were referred to other agencies such as NTIS for information. This severely restricted TIC's ability to serve the private sector. In 1986 TIC's name was changed to the Office of Scientific and Technical Information (OSTI); it is still basically the same organization and performs the same functions.

18.3.5 National Technical Information Service

NTIS was established by Congress in the 1960s to stock copies of all publications generated under government funds and to make them available for purchase by the public (NSF 1984). All solar documents included in the TIC's Energy Data Base were automatically included in the NTIS inventory; because of limited storage space, however, NTIS only maintained an inventory of twenty-five copies. After these were sold, inferior-quality microfiche blowbacks were provided to customers; in many cases data and photos shown in color or in small detail in the original document were unclear or unreadable. Furthermore, because of the expense in generating and storing thousands of microfiche copies of government reports, the blowbacks were priced relatively high in order for the government to recover its cost.

18.3.6 Superintendent of Documents, U.S. Government Printing Office

The government does provide a service whereby the public may buy government documents in their originally published form at very reasonable prices. This service is maintained by the Superintendent of Documents, U.S. Government Printing Office (SUPDOCS). The only difficulty is that SUPDOCS is very selective as to what titles it will stock. The general rule is that SUPDOCS will only inventory publications that are salable in large quantities; this excludes such documents as research reports, bibliographies, reading lists, catalogs, programmatic documents, and flyers. A good number of solar technical documents are carried in SUPDOC's inventory but many more important solar publications are not.

18.3.7 Energy Extension Service

Beginning in 1978, DOE dispensed a limited amount of conservation and solar technical information through state offices of the Energy Extension Service (EES). In its early period, the principal aim of the program was to curb the unnecessary waste of energy at the grassroots level. Most efforts centered on direct technical assistance to various energy users who could not afford to improve the energy efficiency of their homes or businesses on their own. Although some technical information on solar was dispensed through this channel, it was not substantial; most EES representatives concentrated on the implementation of conservation measures such as the weatherization of buildings.

18.3.8 Regional Solar Energy Centers

The regional centers (RSECs) did dispense a good deal of solar technical information, although, with the exception of the Northeast Solar Energy Center (NESEC), they did not produce much of it on their own. Two centers, NESEC and the Mid-American Solar Complex (MASEC) disseminated much solar information during the years of their existence. The other two, the Southern Solar Energy Center (SSEC) and the Western Solar Utilization Network (WSUN), were established much later than the first two and therefore did only a modest amount of technical information transfer before they were discontinued. The regional centers as a whole originated little technical information on solar energy, mainly passing on what they received from other sources.

18.3.9 Technical Information Dissemination Program

The TID program was in existence for only a short time. Funded through DOE and administered by SERI from late 1978 through early 1981, TID produced several solar energy brochures, slide shows, and exhibits and participated in many meetings and exhibits, but a large part of its efforts went toward carrying out special projects and producing programmatic reports and documents for the various technology divisions at DOE headquarters. This may have been the program's ultimate downfall. When funds became scarce in 1981, the TID program was evaluated according to the amount of technical information produced and distributed for the dollars spent. Because much time and expense had been consumed in DOE-directed projects that did not relate directly to technical information dissemination, the program was not found to be cost-effective and was canceled.

18.3.10 Solar Energy Information Data Bank

The SEIDB program was built around PL 93-473, enacted in 1974. While the law's vague and indistinct language left much room for interpretation, it was construed as a mandate to create a national information network interlinking all major metropolitan areas of the country with a single data bank of solar information in one central location. On that basis, a national computer network was conceived in 1979 and implemented in various stages through 1981. In parallel, an information development and dissemination effort was also undertaken.

The principal failure of the SEIDB program was that it was too far ahead of its time. The hardware and software used to implement the system were prohibitively expensive and complex, especially by today's standards; the telephone line communication links, however expensive, were not yet of a caliber that would allow transmission of complex data without frequent error; and the staff required to build and support such a system was necessarily large and high-salaried. Another major shortcoming of the system was that it was far too expensive and sophisticated for the amount of data available at the time and for the few people around the country who had need for such data. As an analogy, the SEIDB effort could be likened to building a mammoth water distribution system for a reservoir that held only a few hundred gallons of water and served only a handful of people. In all fairness, while members of the database development branch at SERI did their best to input data that would be of value to users, they found very little.

The bottom line was that the SEIDB system was not at all cost-effective and had little prospect of becoming so in the near future. This realization lead to its final termination in 1981. The CD7600 computer was eventually sold off by SERI, and the nodes installed at the RSECs were removed from service as the centers themselves were closed. It is interesting to speculate whether the time is approaching when sufficient data will be available and the cost of personal computers, communication links, and peripherals will be low enough to make such a system practical.

18.4   Redirection of Solar Information Activities

With the change in administrations that took place in 1981, the goal of the solar information program changed radically. "Commercialization," which was the major thrust of the program up to that point, was no longer to be part of its mission. The transfer of technology from government research programs to the private sector was to become the program's main and only mission. With this change in mission, funding for information transfer activities was reduced sharply. Consequently, several information programs were dropped entirely, those duplicating services provided by others were restructured to avoid redundancy, and other programs were curtailed substantially (GAO 1981).

With the shift in administration philosophy, all solar information programs were radically changed. The main audience to be addressed was no longer the general public. Most general interest exhibits, demonstrations, and audiovisual presentations were eliminated. Publications that were intended to educate and motivate the general public were discontinued. The national data network, which to a large extent served public needs, was dismantled. Public service announcements on television were discontinued. Hot line inquiry services for the public were either eliminated or severely reduced (see chapter 17). And article placement in the print media was restricted to scientific and technical journals.

Specifically discontinued in 1981 were the TID program, the SEIDB program, the NAFIC program, and various information programs conducted by the four RSECs. CAREIRS (formerly NSHCIC) took over responsibility for all public communications within the fields of conservation and renewable energy, essentially absorbing all of the public inquiry service functions performed by each of the discontinued programs. It should be noted, however, that the breadth of CAREIRS's public service activities was reduced considerably. While the service would still be allowed to answer inquiries from anyone who called or wrote to them, it no longer could promote the availability of the service to the public; which, over several years, reduced the volume of inquiries substantially.

As part of the consolidation of solar information activities in 1981, TIC took over the Solar Biblio database from the SEIDB and merged it into its EDB database. TIC also was given the responsibility for final development of the solar energy thesaurus, a task it later abandoned in favor of expanding its overall energy thesaurus to include just the most often used solar terms. In addition, many of the technical assistance functions performed by the RSECs were taken over by the state offices of the EES and by a new organization formed at NCAT in 1983, the National Appropriate Technology Assistance Service (NATAS).

The single solar information activity that the new administration supported was technical information development and dissemination, with the emphasis on "technical." In addition, the administration took the position that information of value to users should be paid for by users.

Free distribution was frowned upon. As a result, funds remaining at SERI originally earmarked for the SEIDB were redirected to fund a new program embodying these new principles, the Solar Technical Information Program (STIP).

18.5 Solar Technical Information Program

Inaugurated in 1981, STIP was founded upon the premise that "the Program shall contribute to the advancement of all solar technologies by the development and transfer of clearly understandable technical information to concerned professionals from all areas of the country in the most cost-effective manner possible."

The program continued uninterrupted from 1981 until the beginning of FY 1991, when the word "solar" was dropped from its name and its scope was expanded to include conservation and energy efficiency technologies. Emphasis was also changed from the development of technical documents to more programmatic and general audience publications.

From its inception to its termination, the Solar Technical Information Program was managed by Paul Notari of SERI. From 1981 through 1984 the program was funded as part of the "Solar Information" line item in the renewable energy budget; since the beginning of FY 1985, its funding has come from the "Solar Technology Transfer" line item. Funding and spending levels for the program from FY 1982 through FY 1990 were as follows:

FY	Funded	Spent
1982	$4,800,000	$2,440,000
1983	1,205,000	2,840,000
1984	2,315,000	2,740,000
1985	3,115,000	2,960,000
1986	2,627,000	2,530,000
1987	1,710,000	2,113,000
1988	1,400.000	1,672,000
1989	1,325,000	1,485,831
1990	1,259,000	1,176,341

Over the years, STIP fulfilled its mission in several ways. Foremost, it generated and published new technical publications dealing with every aspect of solar energy. Some of these publications treated highly technical subjects such as the latest state-of-the-art advances emanating from federal research. Others were less technical but dealt with information of value to technical audiences; an example would be *Technical Information Guide*, which identified research centers and other creditable sources where technical information on a given technology could be obtained. Categorized as technical communication publications with particular emphasis on communications, STIP publications are characterized by their conciseness and understandability, regardless of the complexity of their content.

A large number of STIP technical communication publications were sold through SUPDOCS because they appealed to wide audiences, were well written, and were advertised by the program. Normally, SUPDOCS declines to stock any technical documents because of their limited sales potential. Over the years, SUPDOCS has carried over seventy STIP titles and has sold some 75,000 copies.

STIP also published some documents that were distributed free to targeted technical audiences. As part of the program, a special mailing list of professionals interested in solar technologies, the National Solar Technical Audience File (NSTAF), was maintained and updated regularly. Fact sheets, technology briefs, announcements of the availability of STIP publications through SUPDOCS, and notices of special solar conferences or meetings were typical of the types of information distributed using the NSTAF mailing list.

Two special reference series were also produced under the STIP. These were distributed to approximately 500 university and private technical libraries around the country. The first, Solar Research Report Desktop Library, was a compendium of federal research report summaries indexed and bound in a special binder. The compendium was widely used by library patrons to identify what reports were available on a given subject and what they contained; an interested patron could then purchase any one or several of the reports from the sources identified in the volume. The second reference work, the Solar Techbook series, consisted of a set of loose-leaf binders of technical information pertaining to each of the five principal solar technologies: photovoltaics, biofuels, solar buildings, solar thermal, and wind energy. The information was contained in small booklets known as information modules that were inserted into the binders as they were received. The Techbooks enabled library patrons to stay abreast of the latest state-of-the-art developments in the various solar technologies. The libraries received regular updates for both the Desktop Library and Techbook series as new information became available.

STIP also published many publications in collaboration with private publishers, trade associations, and professional societies. STIP-originated textbooks were published by MIT Press, Van Nostrand Reinhold, John Wiley & Sons, Elsevier, and Hemisphere Press. Other STIP documents were published by the American Solar Energy Society, the Solar Energy Industries Association, the Passive Solar Industries Council, the National Association of Home Builders, the American Society of Mechanical Engineers, the American Society of Heating, Air-Conditioning, and Refrigerating Engineers, the Institute of Electrical and Electronic Engineers, the Electric Power Research Institute, and the American Wind Energy Association.

In addition to its larger publications, STIP prepared and placed technical articles in the more popular technical and business press. The articles were often authored by top authorities in the field but were almost always restructured and edited by STIP communication specialists experienced in translating highly technical ideas into readily understandable terms. Although discouraged by the DOE during most of the 1980s, occasional exhibits and audiovisual presentations were also produced under the STIP program for display at technical conferences.

To provide comprehensive answers to technical questions received from scientists and other professionals in the field, STIP maintained the Technical Inquiry Service (TIS) at SERI. Many inquiries were referred to TIS from the other DOE information services, such as NATAS, CAREIRS, and TIC. Although the service had no toll-free telephone line, TIS responded to an average of 2,500 technical inquiries a year.

Over the years of STIP's existence, 1981 through 1990, it is estimated that over one-half million professionals purchased or otherwise obtained a STIP document for their enlightenment on a solar energy topic. An equal number likely used the Desktop Library or a Solar Techbook in their local technical library to obtain needed information. And there is no accounting of how many people have read the STIP reference books and other publications published by private publishers, professional societies, and trade associations. In addition, hundreds of thousands have also read articles generated through the STIP program in various technical and business magazines. And about 25,000 inquirers have received in-depth answers to their technical questions using the STIP Technical Inquiry Service.

18.6 Summary Assessment and Future Directions

Almost $45 million of federal appropriations have gone into solar information programs from 1978 through 1990. Some have returned solid dividends; others have not. In retrospect, the SEIDB was too far ahead of its time and too expensive for the value it could provide. Other information programs, such as those conducted by the RSECs, EES, NAFIC, and NCAT, either were too ambitious for the funds available or concentrated too heavily on the wrong audiences. And some programs were simply ineffective for a wide variety of reasons.

Overall, however, federal solar technical information programs disseminated much information, and many people benefited. Although no quantitative studies have been conducted, it is this author's opinion that many of the advances made in solar energy over the last decades are directly attributable to the quality and quantity of information dispensed.

As we approach the turn of the century, the nature of solar technical information is very likely to change. Throughout the 1980s we concentrated on educating the scientist, engineer, and technologist in state-of-the-art solar technologies. In the years ahead I suspect more emphasis will be placed on educating the business executive, the entrepreneur, the corporate decision maker, and the utility executive on the practical advantage of introducing solar technologies into the marketplace. In this respect, more economic analysis results will be disseminated, and more business-oriented publications will be published. Although scientific and technical documents will still be produced, their relative numbers will diminish appreciably.

Technology transfer programs will shift from passive technical publication production and dissemination to more direct technical interaction activities. Information exchange events such as industry forums, specialty seminars, and the like will be held more frequently. One-on-one meetings between scientists from the federal laboratories, industry, and universities will increase; research and development collaborations between these three groups will be commonplace. In common practice, videos will replace publications, computer compact disks containing reams of reference data will become readily available, and computer networks will be widely used to exchange information between sites all over the world. Publications will still be used to record and convey technical information but they will serve primarily to augment other communications media and active technology transfer activities. While the character of future technology transfer programs may differ from what they are today, their importance will not diminish with the change; on the contrary, their importance will increase appreciably.

References

GAO, (U.S. Government Accounting Office). 1981. *Improvement Needed in DOE's Efforts to Disseminate Solar Information.* Report No. EMB-81-101. Washington, DC, June.

NSF (National Science Foundation). 1984. *Scientific and Technical Information Transfer: Issues and Options.* Report No. NSF/PRA-84015. Washington, DC, March.

19 Training and Education

Kevin O'Connor

19.1   Scope and History of Federal Support for Solar Energy Education

19.1.1   Scope

This chapter examines the history of federal involvement in the field of solar energy education and training, where the federal government has provided both direct program support and indirect stimulation. Previous chapters in this volume have covered the topics of consumer information, public information dissemination, and technical information, activities that might, in an informal sense, be viewed as educational. This chapter will concentrate on the historical development of programs and activities (courses, curricula, and materials) in the more formal solar education arena.

This section describes the scope and history of federal energy agency educational activities. Section 19.2 describes the solar energy education programs that have been supported by the Department of Energy (DOE) and its predecessors. Section 19.3 details the solar educational activities of non–energy-related federal departments. Section 19.4 lists and describes major solar conferences and workshops supported by the federal government during the period of significant solar funding for such activities in the early years (1977–1980) while the detailed education activities of the Solar Energy Research Institute (SERI) are related in section 19.5. Section 19.6 provides a prospectus of the growth and emphasis of solar education and training, tracking information and educational institutional involvement and section 19.7 describes the detailed activities of solar vocational education development. And section 19.8 gives the author's perspective of what can be gained from this historical experience.

19.1.2   Historical Development of Education Programs within Energy Agencies

Duggan (1983) has summarized the beginning of energy education activities in the United States. The earliest activities began with the Atomic Energy Commission (AEC) in 1947. The first energy-related training programs were formalized beginning in 1964 (DOE 1978a) and focused on the peaceful uses of atomic energy. Graduate fellowships, trainee-ships, and faculty institutes and workshops for both college faculty and secondary school teachers were initiated. This pattern of educational support has continued to grow across the many disciplines represented by the growth of the federal energy agencies.

The transformation of the federal energy agencies (discussed in more detail in chapters 1 and 2, but reiterated here for clarity of the following sections) is well summarized by Duggan (1983, 447):

In response to the Arab-imposed oil embargo of 1973–1974, new Federal agencies were established to deal with the "energy crisis." First was the Federal Energy Office (FEO), which became the Federal Energy Administration (FEA). This agency was primarily a policy and regulatory agency, although it did conduct some R&D as well. Shortly after the FEA was founded, an attempt was made to link all the Federal energy R&D efforts together; the AEC was absorbed into the new Energy Research and Development Administration (ERDA), which existed for a few years in tandem with (and sometimes in competition with) the FEA. In 1977, both agencies, as well as the Federal Power Commission, were merged into a new Cabinet-level agency, the Department of Energy.

19.2   Solar Education Activities Supported by DOE and Its Predecessors

This section describes the various solar educational activities of DOE and its predecessor agencies. An attempt is made to give examples of funding levels and number of participants in programs of the late 1970s and early 1980s, although complete data are not readily available. According to Duggan (1983), the federal government's role in energy education can be divided into two areas: training and materials development.

19.2.1   Training

University/Laboratory Cooperative Program   Formalized in 1964 to "provide the opportunity for faculty and students to become involved in the cutting edge of energy research" (Duggan 1983, 448), this program included research participation, thesis research, student/faculty seminars and workshops, and so on, and covered all energy disciplines, with the exposure being determined by each individual participant. "The primary function of this program is to give college students and faculty the opportunity to enter into collaborative research and training activities over the summer at one of 30 or more DOE Laboratories, Contractors, or Energy Research Centers" (DOE 1978a). According to Duggan (1983), the research-related program activities attracted 1,000 students and faculty per year, with an additional 1,000 students and faculty involved in the several instructional programs supported annually under the same umbrella. Funding for this program totaled $3.21 million in 1977.

Graduate Traineeships Started under the AEC in 1965, this program expanded its role as it became absorbed into ERDA, and then into DOE. According to Duggan (1983, 448), "the fields of graduate traineeships expanded to include ... geothermal, solar, conservation." From 1971 through 1978 over $5.5 million was spent for over 850 fellowships.

Faculty Development Program Initiated under the AEC in 1971, this program had its mission broadened under ERDA in 1977 "to include a wide variety of energy topics, such as solar, geothermal, fossil, nuclear, conservation, and energy economics." The program continued through 1982. "During the life of the program, $9 million was spent in hundreds of our Nation's colleges and universities, and the program reached over 16,000 teachers" (Duggan 1983, 449).

Other DOE-Supported Training Programs Duggan (1983, 449) goes on to state that there were several other training programs supported by DOE: "In 1977, eight universities were selected to house solar meteorological training centers.... Training programs for solar installers were created for contractors, correctional institutions, and low income communities. A number of training programs were also created to train farmers and others in techniques of small-scale alcohol fuel production."

One of the more notable examples of a specialized DOE-supported program, Solar Energy Utilization, Economic Development, and Employment in Low-Income Communities, was a joint effort between DOE and the Department of Labor (DOL). As part of the Comprehensive Employment Training Act (CETA), this popular program was begun in 1979 and ended in 1980. Initially, the "project began in fiscal year 1979 and funded 15 state and local CETA prime sponsors to train 320 unskilled participants to install active and passive solar energy systems on low-income homes" (DOE 1980a, 14).

19.2.2 Teaching Materials Development

The DOE and its predecessor agencies have continually provided materials to teachers of all levels to assist in their correct understanding of energy matters. Two organizations which specifically impacted the solar field are

• *National Science Teachers Association*   From 1975 to 1980 the NSTA was involved in one of the two major curriculum infusion projects sponsored by the DOE (initiated by ERDA). NSTA (1979) developed the Project for an Energy Enriched Curriculum (PEEC), which, by the end of FY 1982, had distributed nearly 2 million teacher fact sheets on energy technologies, and 1.5 million packets of 15 lesson plans to teachers in grades K–12. This project "was an attempt to infuse energy-related information into already existing courses through an interdisciplinary approach" (Duggan 1983, 449).

• *Oak Ridge Associated Universities*   In 1977 ORAU developed the Science Activities in Energy program, which took a hands-on, disciplinary approach. "In this program, discovery-type activities were developed for junior and senior high school science classes, to demonstrate the scientific principles of energy production, distribution, conservation and consumption" (Duggan 1983, 449–450). *Solar Energy*, *Solar Energy II*, *Wind Energy*, and *Conservation*, were a series of experiments aimed at a variety of student levels (ORAU 1978).

Several other solar curriculum projects were also initiated by ERDA and continued with the DOE's Office of Solar Programs in 1979. The New York State Education Department's "Solar Energy Project," in cooperation with the State University of New York (SUNY-Albany), was involved in the development of a teaching packet for secondary education (NYSED 1981). The SUNY project is summarized by Langford and La Hart (1981), who notes that during the 1979 school year, some 3,000 teachers were introduced to this curriculum by Northeast Solar Energy Center (NESEC) workshops. The University of Southern California (USC) developed a packet for elementary teachers (Lampert et al. 1978, 1980). The dissemination of USC's curriculum is described by Lampert and Yanow (1981).

19.2.3 Solar Energy Research Institute

Part of the original charter of SERI was "to Perform Research, Development, and Related Functions in support of the National Solar Energy Program" (PL 93-473). SERI's broadly defined programmatic functions in support of the above mission included the development and dissemination of information related to solar energy and the performance and support of solar energy research, development and demonstration activities. With that mandate, SERI initiated educational and research support for academic programs. SERI's activities in solar education and training are detailed in sections 19.5 and 19.6.

19.2.4 Regional Solar Energy Centers

The RSECs were involved in a variety of educational activities, including the conduct of seminars and workshops and the publication and distribution of solar energy educational materials. (Chapter 20 of this volume covers the activities of the RSECs in detail.)

19.2.5 Residential Conservation Service: Auditor Training

Mandated as part of the National Energy Conservation Policy Act (PL 95-619) and implemented by the DOE Technical Assistance Program, RCS was intended to reduce national energy consumption by having public utility companies offer energy audits to homeowners. The audits not only included energy conservation measures but also offered the use of alternative energy resources for the homeowner's consideration. SERI and Oak Ridge National Laboratory developed a primary manual for training RCS home auditors (DOE 1980b), as well as a manual for homeowners entitled *Your Home Energy Audit: The First Step to More Energy-Efficient Living.*

19.2.6 Education Commission of the States

In 1978 ECS was awarded a grant from the U.S. Department of Energy to determine the status of state energy education policies and to develop policy recommendations. Governors' offices, state legislatures, state education agencies, and energy offices were surveyed, and the findings reported in *The Status of State Energy Education Policy* (Petrock 1979). Using this information, a national task force, chaired by Governor Richard D. Lamm of Colorado, worked with the ECS staff in preparing *Policy Issues in K–12 Energy Education* (Petrock 1980). Both the status report and the policy recommendations were widely distributed, and the project generated considerable interest nationwide. With renewed funding in 1981 from DOE for a program designed to help states develop and implement energy education policies and programs, ECS also produced

Energy Education: What, Why, and How (Bauman 1981) and *A Policy Development Handbook for Energy and Education* (Petrock 1981).

19.2.7   Florida Solar Energy Center

As part of DOE's Solar Technology Transfer Program the Florida Solar Energy Center (FSEC) produced the *Solar Water and Pool Heating: Installation and Operation Manual* (FSEC 1979), designed to be used as a textbook for those attending the Solar Installation Short Course, as encouraged by the Florida Solar Industry Association.

19.3   Solar Education Activities Supported by Non–Energy-Related Federal Departments

19.3.1   U.S. Department of Commerce: CSU Course Manuals

The Department of Commerce funded the Colorado State University's Solar Energy Applications Laboratory for the development of two of the most comprehensive manuals oriented to training the solar instructor and practitioner: *Solar Heating and Cooling of Residential Buildings: Sizing, Installation* (CSU 1979b) and *Operational Guidelines, Solar Heating and Cooling of Residential Buildings: Design of Systems* (CSU 1979a). This course was offered many times in one-week modules, but some schools used the manuals, along with supplemental materials, as the basis for their own training programs. Each manual is over 700 pages long.

19.3.2   U.S. Department of Labor: CETA Programs

In a joint venture with DOE (described earlier), the Department of Labor sponsored a program to encourage employment and training in renewable energy resources. The program was in partial fulfillment of the Comprehensive Employment Training Act (CETA), which was intended to provide training to improve skills of the unemployed and underemployed.

19.3.3   U.S. Department of Housing and Urban Development

Although many solar publications have been produced by HUD, probably one of the most instructive, in an educational sense, was *Installation Guidelines for Solar Domestic Hot Water Systems* (HUD 1979), which was written for the professional contractor or skilled homeowner but which had application in the solar technical training arena.

19.3.4   U.S. Office of Education

The Center for Renewable Resources (CRR) played a very active role in the generation of materials and ideas that influenced and assisted teachers, students, policymakers, and others in the realm of solar energy education. Under funding from the U.S. Office of Education, CRR produced *A Solar Energy Education Packet for Elementary and Secondary Students* (CRR 1980b) as well as the *Solar Energy Education Bibliography* (CRR 1979a) and *Solar Energy Education Bibliography: Books, Films, and Slides* (CRR 1980a).

The U.S. Office of Education's Energy and Education Action Center made its contribution to solar energy education in two major ways: (1) by hosting a national conference in February 1980, entitled "Meeting Energy Workforce Needs: Determining Education and Training Requirements" (OOE 1980a), reported in detail in section 19.4.4; and (2) by producing *A Selected Guide to Federal Energy and Education Assistance* (OOE 1980b), a detailed funding roadmap to areas such as curriculum materials development and employee/teacher training.

19.3.5   National Science Foundation

The NSF funded the Solar Energy Technician Training Program at Navarro College in Corsicana, Texas, and three associated junior colleges. The main purpose of the program was to develop curriculum and course materials for training solar installers. The story of this project was reported by Meyers (1981).

19.4   Federally Sponsored Education-Related Conferences and Workshops

19.4.1   Solar Energy Task Force Report: Technical Training Guidelines

SERI convened a workshop on technical training guidelines in September 1978; the results are reported in section 19.7.1.

19.4.2   National Energy Education, Labor, and Business Conference

Attended by approximately 1,000 persons and sponsored by the DOE's Office of Education, Business, and Labor Affairs in January 1979, this conference addressed the question of whether the nation had sufficient trained and skilled manpower to meet the requirements of renewable energy resource implementation. Discussions centered on such issues as

the educational and technical skills required for energy-related employment and sources of trained manpower. Perspective and positions regarding the ability of the current labor workface and training facilities
to meet energy-related employment needs varied immensely.

19.4.3 Open Workshop on Solar Technologies

The DOE's Office of Solar, Geothermal, Electric, and Storage Subsystems, with SERI in the role of coordinator, held the Second Open
Workshop on Solar Technologies on 23–24 October 1979 (SERI 1980b).
The workshop was divided into six panels, each addressing a separate
topic; each panel prepared a report to DOE for consideration as part of
its policy determination and implementation (SERI 1980a). The "Solar
Job Training" panel made a number of major recommendations:

• Job training programs should focus on providing reliable, durable, and
economically viable systems.

• Training the disadvantaged should be emphasized.

• Both new training and retraining programs should be funded and
should be responsive to available local solar jobs.

• An examination and assessment should be made of the reasonable
expectations of employment in solar.

• DOE should be aware of the upcoming need for training people to
perform Residential Conservation Service (RCS) solar and energy conservation audits.

• Community awareness programs oriented to giving the general population a feeling of confidence about solar and general knowledge of how
systems are installed, maintained, and operated should be considered.

• Technical assistance programs at the regional level should be funded,
utilizing existing educational institutions.

• Training for specific solar jobs should be periodically evaluated to
address the need for continued and/or revamped training.

• The Solar Utilization for Economic Development and Employment
(SUEDE) program should be reinstituted, and the solar portion of the
Comprehensive Employment and Training Act (CETA) should be imple-
mented.

• DOE should do its part to promote tax investment credits for the solar
industry.

• A central, coordinated effort of information dissemination to the solar job-training community is needed.

19.4.4 Meeting Energy Workforce Needs: Determining Education and Training Requirements

This workshop was sponsored by the U.S. Office of Education's Energy and Education Action Center and a host of other agencies, including DOE, the Department of Labor, the American Association of Community and Junior Colleges, and the Education Commission of the States. Held 26–28 February 1980, the conference focused on

• assessments of job opportunities in energy-related occupations;

• programs and curricula available for occupational training where needs are projected in energy-related career opportunities;

• capabilities of the nation's schools, colleges, and other training facilities to meet anticipated needs;

• improved linkages between educators and industry, and between government and labor, to assure best use of education and training resources.

19.4.5 Solar Energy Education Workshop

In what proved to be the waning days of federal support for energy education programs, an enthusiastic group of people was brought together by SERI for the Solar Energy Education Workshop in January of 1981. Representatives assembled from the regional solar energy centers (RSECs), SERI, DOE, the Department of Education, the Education Commission of the States, the Center for Renewable Resources, and a select group of additional organizations involved in curriculum and materials development and evaluation. The purpose of the workshop was to discuss the state of the art of renewable energy education and the direction renewable energy education should take. Workshop tasks included envisioning where we wanted to be five years from now; identifying the barriers to reaching the vision, and planning strategies and programs which would overcome the barriers.

One outcome of the workshop was to form an organization known as the "Friends of Renewable Energy Education" (FREE). FREE's overall purpose was to identify, coordinate, stimulate, and promote quality programs in renewable energy information and education activities that would create a stronger national posture with regard to security, decentralized energy systems, and individual and community self-reliance. FREE provided a united front on renewable energy education issues and an opportunity to communicate and work together, share and implement ideas, and avoid duplication of effort.

In discussing this workshop, the author wishes to emphasize the need to implement a network of important organizations and people as soon as renewable energy comes to the fore again. The Center for Renewable Resources conducted a similar workshop on 15–16 November 1979, noting in its appraisal that

renewable energy education has been well launched toward cohesive, comprehensive development. A dedicated, interested advisory group now exists to lend support and encouragement to the Center and other groups that are willing to undertake renewable energy education projects. Valuable ideas emerged from these sessions and they will serve as guideposts and sources for further ideas for educational projects. (CRR 1979b)

19.5 Solar Energy Research Institute

Educational activities at SERI were concentrated in its first five years (1977–1981) under the Academic and University Programs Branch.

19.5.1 National Solar Energy Education Data Base and Associated Publications

In order to answer public requests for information on educational activities at the postsecondary level, a computerized solar energy education database was established in the fall of 1978. This effort continued through the spring of 1981. Information for the database was collected annually from a survey of postsecondary educational institutions. The survey had support from the office of U.S. Congressman George E. Brown, Jr., and the Congressional Solar Coalition. The collected information became part of SERI's Solar Energy Information Data Bank (SEIDB), in SERI's Information Systems Division. By the spring of 1980, the data bank contained information collected from 892 postsecondary educational institutions, offering 2,308 courses and 367 programs.

Publications from the database included

• *National Solar Energy Education Directory* Three editions of this publication were produced (Corcoleotes, Kramer, and O'Connor 1979a,

1980a, 1981a). The information contained in this directory reflected the information collected and stored in the national database described above; detailed information was presented on all solar courses, programs, and curricula as submitted by the educational institutions on the survey instruments.

• *Solar Energy Technical Training Directory*   Three editions of this document were produced to provide information about institutions offering solar technical training programs, usually vocational-technical schools and community or junior colleges (Corcoleotes, Kramer, and O'Connor 1979b, 1980b, 1981b).

Hundreds of searches were run on the database to provide detailed information in response to individual requests from the public and private sectors. For example, many students inquired as to whether there were any solar degree programs offered in their state, where they were offered, and what types of degrees were available.

19.5.2   University Research Program

Established by DOE in March 1979 to encourage colleges and universities to perform advanced research in solar energy, the University Research Program was intended to

• stimulate new ideas that might or might not be associated with existing solar technology;

• support feasibility or "proof of concept" studies of these ideas;

• add to the store of fundamental knowledge for advancing the use of solar energy; and

• stimulate the growth of a solar-related, intellectual base in the academic community.

DOE contracted with SERI to provide proposal review and subsequent technical monitoring services for this program, which continued for three years.

19.5.3   Summer Intern Program

For three years SERI also offered a nationally advertised summer intern program to expose promising college juniors and seniors to the broad problems associated with the practical, widespread use of solar energy, including the technical, environmental, economic, legal, and social problems. Students selected for the program were invited to spend ten weeks at SERI assigned to a specific project on which they would work half of the time, attending special courses, seminars, and discussion sessions and going on field trips for the other half. Approximately sixty students were afforded this research and educational opportunity.

19.5.4   Sabbatical Leave Program

Through its Sabbatical Leave Program, advertised to the nation's colleges and universities, SERI provided a 4–12 month residency program for distinguished faculty members in mathematics, engineering, and the physical, life, and social sciences.

19.5.5   International Science and Engineering Fair

From 1978 through 1992 SERI supported the ISEF by providing judges for this DOE-funded project. In 1986 SERI instituted a solar award to be presented at the fair for the outstanding solar science project; SERI has also twice hosted the top ten ISEF energy award winners and their teachers for a one-week stay at the SERI laboratory with visits to surrounding energy projects.

19.6   Postsecondary Education and Training: Growth and Emphasis

This section presents an analysis of the wealth of data accumulated in the national Solar Energy Education Data Base at SERI.

19.6.1   Sources of Information and Definitions

Solar energy educational information was important in the analysis of the solar technology which was emphasized during the middle to last part of the 1970s and the early 1980s. A number of organizations published educational tracking materials. Probably the largest effort was the establishment of the Solar Energy Education Data Base (SEEDB) at SERI. Additionally, the American Association of Community and Junior Colleges (AACJC) put together a compendium of institutions offering energy related programs in *Energy-Related Activities in Community, Junior, and Technical Colleges: A Directory* (AACJC 1981).

Some definitions and notes of caution are in order. First, here are the definitions used for entries in the SEEDB:

• *Solar Course*   A course in which one-third or more of the course content was devoted specifically to teaching solar topics.

• *Solar Program*   A set of courses leading to a solar or solar-related degree. Solar programs were divided into three categories:

1. *Solar Curriculum*   A program in which the student receives a degree or diploma in a solar field, such as a master's in solar engineering or an associate degree in solar installation.

2. *Curriculum with Solar Study*   A program in which the student has the opportunity to receive solar education while working on a related degree or diploma, for example, a Ph.D. in physics with solar emphasis or a B.S. in architecture with solar design experience.

3. *Solar Technical Training*   A nonacademic degree program, such as for solar technicians, in which the student receives a certificate for study in solar energy or a solar-related field.

Some caution should be used in drawing conclusions from the data analysis. While follow-up efforts were made to obtain responses from all educational institutions, there are certainly some who did not respond, even though the institution may have had solar offerings. On the other hand, it is certain that some of the courses conformed only marginally to the one-third course content criterion. In addition, it was often difficult to determine whether a program was actually a solar-related program. Therefore, the reader is encouraged to exercise caution in interpreting the data presented.

Note that references to years means the academic school year, for example, 1980 means the 1980–1981 school year. The survey was conducted in the spring of each of the school years 1978, 1979, and 1980. The appendix contains a copy of the survey instrument used.

19.6.2   Postsecondary Educational Institutional Involvement

The effect that DOE and SERI, among many other organizations, had on the growth of solar education and training activities is dramatically demonstrated by the information analyzed from the Solar Energy Education Data Base. The dramatic growth in the field of solar-related educational offerings is indicated in table 19.1.

Table 19.2 shows student enrollment in solar-related courses from 1973 to 1980. Note that this information represents historical course enrollment figures from estimates supplied by the responding institutions. The

Table 19.1
Summary of postsecondary educational institutions offering solar-related programs and courses, 1978–1980.

	Organizations		Courses		Programs	
	Number	% increase	Number	% increase	Number	% increase
1978	679	—	1,307	—	174	—
1979	760	11.9%	1,740	33.1%	243	39.7%
1980	892	17.4%	2,308	32.6%	367	51.0%

Source: Solar Energy Research Institute.

Table 19.2
Solar-related course participation, 1973–1980

	Enrollment[a]		Average enrollment	Courses[b]		Actual number of courses
Year	Total	% increase		Number	% increase	
1973	5,043	—	58.0	86	—	c
1974	6,266	24.2	54.0	116	34.9	c
1975	9,541	52.2	45.7	207	78.4	c
1976	12,899	35.2	43.7	295	42.5	c
1977	30,508	136.5	41.1	742	151.5	c
1978	44,192	44.9	41.2	1,072	44.5	1,307
1979	55,570	25.7	41.0	1,357	26.6	1,740
1980	62,813	13.0	45.0	1,395	2.8	2,308

Source: Solar Energy Research Institute.
[a] These figures do not represent individual students, rather the total enrollments in all solar-related courses where information on enrollment is reported.
[b] Number of courses for which enrollment statistics are reported.
[c] Based on information supplied in 1978, 1979, 1980 annual surveys of postsecondary institutions. Information not available for 1973–1977.

most dramatic piece of data is that, from the oil embargo in 1973 to just before the advent of cheaper oil and the national de-emphasis of renewable energy in 1980, the number of student enrollments in solar courses increased over twelvefold. The largest spurts in enrollment coincided with the increased national resolve to pursue and implement renewable energy technologies. Naturally paralleling the enrollment increase was the sixteenfold rise in the number of course offerings over the same time period. The average class sizes for solar courses were large, averaging forty-five students.

Table 19.3
Solar-related program participation, 1973–1980

Year	Students completing programs[a]		Average participation	Programs[b]		Actual number of programs
	Number	% increase		Number	% increase	
1973	785	—	34.1	23	—	c
1974	911	16.1	39.6	23	0.0	c
1975	1,372	50.6	45.7	30	30.4	c
1976	1,863	35.8	50.4	37	23.3	c
1977	5,483	194.3	43.2	127	243.2	c
1978	8,161	48.8	49.2	166	30.7	174
1979	10,524	29.0	48.5	217	30.7	243
1980	12,955	23.1	50.8	255	17.5	367

Source: Solar Energy Research Institute.
[a] Although these figures are estimates of actual numbers of students completing solar-related programs, they do not imply that these students are solar experts. It means students have taken as a minimum one solar-related course offered as part of a solar-related program.
[b] Number of programs for which participation statistics are reported.
[c] Based on information supplied in 1978, 1979, 1980 annual surveys of postsecondary educational institutions. Information not available for 1973–1977.

The increase in student involvement in programs is shown in table 19.3. From 1973 to 1980 there was a sixteenfold increase in the number of students completing programs with at least some solar content. The number of programs offered over the same time period (for those institutions reporting on the question of involvement) increased more than elevenfold. Based on the information gathered, approximately fifty students participated in each program for the years 1978–1980.

Information on the types of degree offered is presented in table 19.4. About one-third of the degrees offered at colleges and universities are bachelor's degrees, one-third master's, and one-sixth doctorates. Of the associate degrees offered, about three out of four were offered at junior and community colleges, with one of six offered in vocational/technical schools. Three-fourths of all certificate programs were offered at either junior and community colleges or vocational/technical schools. Overall, the degrees being offered were about equally split among bachelor's and advanced degrees and associate and certificate offerings.

Table 19.5 categorizes the type of curriculum by the degree offered. Most of the bachelor's and advanced degrees were offered as part of a curriculum with solar study in which students received a variety of solar

Table 19.4
Organization type by degree offered, 1980

Degree	Colleges and universities		Junior and community colleges		Vocational/ technical schools		Other		Total	
	No.	%	No.	%	No.	%	No.	%	No.	%
Doctoral	43	16.9	0	—	0	—	0	—	43	9.1
Master's	87	34.1	0	—	0	—	0	—	87	18.4
Bachelor's	91	35.7	0	—	0	—	0	—	91	19.2
Associate	8	3.1	67	49.6	16	23.5	2	13.3	93	19.7
Certificate	22	8.6	68	50.4	44	64.7	13	86.7	147	31.1
Other	4	1.6	0	—	8	11.8	0	—	12	2.5
Totals[a]	255	100.0	135	100.0	68	100.0	15	100.0	473	100.0

Source: Solar Energy Research Institute.
[a] Totals will be greater than the number of programs because more than one degree may be offered in a program; based on 97% question response rate.

Table 19.5
Solar curriculum type by degree offered in solar programs, 1980

Degree	Solar curriculum		Curriculum with solar study		Solar technical training		Total	
	No.	%	No.	%	No.	%	No.	%
Doctoral	3	7.5	40	13.3	0	—	43	9.1
Master's	7	17.5	80	26.6	0	—	87	18.4
Bachelor's	6	15.0	83	27.6	2	1.5	91	19.2
Associate	21	52.5	59	19.6	13	9.8	93	19.7
Certificate	3	7.5	30	9.9	114	86.4	147	31.1
Other	0	—	9	3.0	3	2.3	12	2.5
Totals[a]	40	100.0	301	100.0	132	100.0	473	100.0
Percentage		8.5		63.6		27.9		100.0

Source: Solar Energy Research Institute.
[a] Totals will be greater than the number of programs because more than one degree may be offered in a program.

Table 19.6
Solar training specialization by program type, 1980

Solar training specialization	Solar curriculum No.	%	Curriculum with solar study No.	%	Solar technical training No.	%	Total No.	%
Architecture—solar specialization	12	7.9	45	6.7	30	5.8	87	6.5
Solar energy education	10	6.6	52	7.7	40	7.7	102	7.6
Solar administration/policy	6	4.0	19	2.8	6	1.2	31	2.3
Scientific R&D in solar energy	9	6.0	74	11.0	6	1.2	89	6.6
Engineering—solar specialization	9	6.0	79	11.7	15	2.9	103	7.7
Mechanical/electrical contracting	6	4.0	27	4.0	14	2.7	47	3.5
General contracting—solar design	8	5.3	34	5.0	40	7.7	82	6.1
HVAC—solar specialization	14	9.3	77	11.4	54	10.5	145	10.8
Residential solar installation	15	9.9	59	8.8	80	15.5	154	11.5
Industrial/commercial installation	10	6.6	30	4.4	37	7.2	77	5.7
Solar technology—controls/ design/maintenance	18	11.9	62	9.2	43	8.3	123	9.2
Electricity–solar specialization	8	5.3	35	5.2	26	5.0	69	5.1
Plumbing—solar specialization	9	6.0	21	3.1	37	7.2	67	5.0
Sheet metal—solar specialization	4	2.6	10	1.5	18	3.5	32	2.4
DIY—home installation	9	6.0	42	6.2	67	13.0	118	8.8
Other Training	4	2.6	9	1.3	3	0.6	16	1.2
Totals[a]	151	100.0	675	100.0	516	100.0	1,342	100.0
Percentage		11.3		50.3		38.4		100.0

Source: Solar Energy Research Institute.
[a] Totals will be greater than the number of programs (367) because more than one solar specialization may be included in the program.

Table 19.7
Institution type by level of course offering, 1980

| Academic levels | Institution type | | | | | | | | | |
| | Colleges and universities | | Junior and community colleges | | Vocational/ technical schools | | Other | | Total | |
	No.	%	No.	%	No.	%	No.	%	No.	%
All levels	217	16.0	242	37.8	31	31.6	19	63.3	509	24.0
College graduate	363	26.8	3	0.5	3	3.1	5	16.7	374	17.6
Junior/senior	645	47.6	37	5.8	2	2.0	0	—	684	32.2
Freshman/ sophomore	131	9.6	358	55.9	62	63.3	6	20.0	557	26.2
Totals	1,356	100.0	640	100.0	98	100.0	30	100.0	2,124	100.0
Non-academic levels										
	No.	%	No.	%	No.	%	No.	%	No.	%
Managerial	40	14.1	99	13.7	9	7.2	14	14.9	162	13.2
Professional	104	36.6	156	21.6	25	20.0	25	26.6	310	25.3
Skilled labor	31	10.9	189	26.1	50	40.0	21	22.3	291	23.7
Layperson	109	38.4	279	38.6	41	32.8	34	36.2	463	37.8
Totals	284	100.0	723	100.0	125	100.0	94	100.0	1,226	100.0
Grand totals[a]	1,640		1,363		223		124		3,350	
Percentage		49.0		40.7		6.6		3.7		100.0

Source: Solar Energy Research Institute.
[a] Because courses may be offered at several levels, the totals may be greater than the individual number of courses offered; based on 94% question completion by respondents.

courses as part of their educational experience. This approach seems to make sense in that students are more broadly trained in a solar-related field, rather than concentrating solely on solar. In fact, only 8.5% of the degrees offered specifically trained students to become pure solar specialists. About two-thirds of the programs were curricula with solar study, and one-fourth offered technical training. It appears from tables 19.4 and 19.5 that a reasonable mix of both technical and professional training was being offered by our nation's educational institutions.

Information on training specializations offered in solar programs is presented in table 19.6. Information for this table was obtained by asking respondents in what fields students would be trained as a result of completing the program offered. The most important point demonstrated by this table is the variety of solar programs offered, especially considering the short time in which the study of solar energy had become popular; most, of course, were oriented to solar thermal.

Table 19.7 depicts the level of solar courses being offered. For each course, respondents were asked to identify the academic or nonacademic level. At the academic level, nearly one-quarter of the offerings were of a general nature, offered at all levels; one-sixth of the courses were at the college graduate level, one-third at the junior/senior level, and one-quarter at the sophomore level. At the nonacademic level, one out of eight courses was available at the managerial level, about one out of four at the professional level, and one out of four at the skilled labor level. In other words, it appears that solar offerings were fairly well distributed in both the academic and nonacademic communities.

19.7 Vocational Education

As the need for technical training in the emerging solar industry was recognized by the nation's educational institutions, many organizations became involved in the development of program, course, and curricular materials. This section will note some of the more prominent of these and will also summarize the recommendations of the Solar Energy Task Force on Technical Training Guidelines (O'Connor 1979), which stemmed from a meeting held in September 1978 at SERI and were presented at the National Education, Labor, and Business Conference in January 1979.

19.7.1 Solar Energy Task Force Report: Technical Training Guidelines

In the fall of 1978 leaders in education, labor, and business who were involved with solar technical training programs assembled at SERI to study solar training problems. The task force focused on solar space and water heating applications to provide guidelines for technical training courses and curricula for vocational and technical schools and community and junior colleges. Participants dealt with curriculum development; jobs, tasks, and skills that required training; and equipment recommendations in establishing a solar training center. In their report they recommended procedures for the establishment of institutions offering solar technical training; they identified the following jobs as necessary for solar training programs: sales, designer, installer, electrician (not necessarily trained in the solar program, but needed in some solar installations), supervisor, and service-maintenance technician. The task force report also suggested sample curricula and programs for students and instructors (for information on student training programs, see Corcoleotes, Kramer, and O'Connor 1981a, 1981b and AACJC 1981).

The task force's cautions and recommendations, probably the most significant contributions to be made to initiators of future solar training programs, included the following:

• Solar workers should not be strictly trained as solar technicians, mechanics, installers, and so on. Because the demand for purely solar jobs may be uncertain, training a person for job entry as a "solar" anything might not ensure employment. Follow-up studies on continued employment in the solar field and assessment of the current job market were suggested before any solar program be undertaken at any educational institution.

• When creating a solar program, an institution should understand the contents of the curriculum and coursework, the required tasks and competency levels for the personnel, and the equipment necessary for the solar laboratory.

• A solar advisory council should be formed from members of local industry and labor organizations, solar installers and contractors, building contractors, and so on. Such a council could advise solar programs on technical training needs, maintain awareness of manpower requirements in the solar market to alleviate mistraining or overtraining, provide speakers for programs, and create links with business and industry to facilitate job placement.

• Solar students should be well rounded in heating, ventilating, air-conditioning (HVAC), and plumbing fundamentals. Schools and institutions already offering HVAC and plumbing programs could most easily expand their programs to offer solar training as the need arose.

• Solar engineers should continue to be trained in traditional disciplines with specialization in solar energy, so as to not preclude entrance into other career patterns.

19.7.2 Solar Energy Employment and Labor Force Requirements

Although chapter 31 addresses the issues of law, environment and labor, the topic of solar education and training would not be complete without addressing some of the work done during the brief time when the solar employment market burgeoned during the second half of the 1970s and into 1980–1981. Several investigations reached widely differing conclusions as to the needs for specialized solar training and the numbers of jobs that might be created by a developing solar industry. In the conference report *Training Programs and Advanced Education in Solar Energy* (O'Connor 1980, p. 258), I wrote that "A quantification of this (training) demand is not clear. However, a middle of the road figure, derived from the Solar Energy Domestic Policy Review (DOE 1978b), suggests that about 150,000 to 200,000 jobs of all types per year will be created from 1985 to the year 2000 in solar hot water heating and space heating and cooling applications if the President's 20% goal of energy needs supplied by solar are to be met by the end of the century. While there will be a demand for jobs created by the emerging solar technologies, the actual number of jobs is a much disputed topic."

Analyses of solar energy employment and labor force requirements or projections can be found in several references (Burns and Blair 1979; DOE 1980c; Grossman, Danaker and Wasserman 1979; Mason 1978; Schacter 1979). Although most studies attempt to estimate the total number of jobs that will emerge as a result of a traditional, large-scale approach to energy systems, Grossman and Danaker (1977, p. 8) argue that "if the nation decides to pin its hopes on inefficient, large-scale energy systems, such a vast quantity of resources and money will be consumed and so much havoc will be generated through all levels of society that energy and job options for the future will be choked off."

19.7.3 Vocational Curriculum/Materials Development

DOE's Office of Conservation and Solar Applications sponsored a Community College Instructor Training Program. Under contract with DOE's Solar Technology Transfer Program, the League for Innovation in the Community College, a national consortium representing fifty colleges, developed curriculum materials and procedures for training community college instructors who must teach solar installation skills. Specifically, this model program was designed to

• train a minimum of sixty community college vocational/technical faculty members from nine districts (thirty-eight colleges) in the western United States to teach skills necessary for the installation of solar systems;

• train two-member teams from the nine districts to develop and implement training programs for solar system installers;

• duplicate for eastern and midwestern community college districts the results of the Western Region Program;

• explore training by television for solar energy system installers;

• develop a training program for installers; and

• establish a network for the community colleges to share information about solar systems (see also CSU 1979a, 1979b).

19.8 Where Are We Now?

19.8.1 Lessons Learned

The national solar energy education and training efforts suffered for a variety of reasons. Most notable was the lack of a coordinated, centralized effort in one department of the federal government. Consequently, there were many fragmented efforts with no central focus for organizations competing for energy program dollars. Solar program funding spread across many agencies, resulting in much duplication of programs, courses, and curricula developed independently by DOE, DOC, DOL, HUD, the Office of Education, and NSF, not to mention the independent efforts of the 892 postsecondary educational institutions (offering 2,308 courses and 367 programs in 1980).

I believe that more attention should have been paid (and should be paid if renewable energy education has a resurgence) to increasing awareness of what is available, rather than developing materials and curricula that duplicate already existing materials. Certainly, materials may need to be evaluated and updated, but prevention of the NIH (not-invented-here) syndrome should be emphasized. A new covering on an old book does not change the book. An example of the need *not* to generate additional volumes of information can be seen in the 75-page solar bibliography published by the Office of Education's Center for Renewable Resources (CRR 1980a).

19.8.2 What about the Future?

Although I can only guess what the future needs of solar education will be, I can strongly recommend that we attempt to learn from past experience. If there is an additional need for renewable energy manpower, and thus for education and training materials, I would recommend that a group of experts representing all facets of the educational process (education, labor, and business as a minimum) be assembled to act as the coordinating, policy-recommending body. This group of experts should at a minimum:

- attempt to avoid duplication of effort, evaluating what exists first;
- determine that there is a real need for additional training (i.e., analyze manpower needs);
- develop a program that is fair in its allocation of federal support;
- make information dissemination an important goal;
- provide a mechanism for materials and curriculum evaluation;
- monitor projects closely to ensure that they are spending research dollars wisely; and
- offer CETA-like training and programs for the disadvantaged.

19.8.3   Final Note

Within the Department of Energy during the 1980s, at least two programs endured the downturn in renewable and solar energy education activities: the Solar Technical Information Program (STIP) and the Conservation and Renewable Energy Inquiry and Referral Service (CAREIRS). In addition to these programs, the *Ninth Report to Congress: Comprehensive Program and Plan for Federal Energy Education, Extension, and Information Activities: Annual Revisions* (EES 1986) lists twenty-one DOE programs/activities in the area of energy education in FY 1985, some of which may also be related to solar and renewables. The *Twelfth Report to Congress* (EES, 1988) lists eighteen such DOE programs/activities for FY 1988. Both of these reports provide informative snapshots of recent emphasis in all aspects of the federal government's involvement in energy education. Anyone seeking the current status of solar education and training programs and activities would be well advised to start by perusing the latest report to Congress on these activities.

References

AACJC (American Association of Community and Junior Colleges). 1981. *Energy-Related Activities in Community, Junior, and Technical Colleges: A Directory.* Compiled by S. T. Duncan and J. R. Mahoney.

Bauman, P. 1981. *Energy Education: What, Why, and How*. Denver: Education Commission of the States.

Burns, B. and L. Blair. 1979. *Beyond the Body Count: The Qualitative Aspects of Solar Energy Employment*. SERI/RR-345-395. Golden, CO: Solar Energy Research Institute.

Corcoleotes, G., K. Kramer, and K. O'Connor. 1979a. *National Solar Energy Education Directory*. Vol. 1. Golden, CO: Solar Energy Research Institute.

Corcoleotes, G., K. Kramer, and K. O'Connor. 1979b. *Solar Energy Technical Training Directory*. Vol. 1. Golden, CO: Solar Energy Research Institute.

Corcoleotes, G., K. Kramer, and K. O'Connor. 1980a. *National Solar Energy Education Directory*. Vol. 2. Golden, CO: Solar Energy Research Institute.

Corcoleotes, G., K. Kramer, and K. O'Connor. 1980b. *Solar Energy Technical Training Directory*. Vol. 2. Golden, CO: Solar Energy Research Institute.

Corcoleotes, G., K. Kramer, and K. O'Connor. 1981a. *National Solar Energy Education Directory*. Vol. 3. Golden, CO: Solar Energy Research Institute.

Corcoleotes, G., K. Kramer, and K. O'Connor. 1981b. *Solar Energy Technical Training Directory*. Vol. 3. Golden, CO: Solar Energy Research Institute.

CRR (Center for Renewable Resources). 1979a. *Solar Energy Education Bibliography (1979)*. Washington, DC.

CRR. 1979b. *A Summary of the Renewable Energy Education Planning Workshop*, 15–16 November 1979. Washington, DC.

CRR. 1980a. *Solar Energy Education Bibliography: Books, Films, and Slides (Expanded)*. Washington, DC.

CRR. 1980b. *Solar Energy Education Packet for Elementary and Secondary Students (1978)*. Washington, DC.

CSU (Colorado State Universtiy). 1979a. *Solar Heating and Cooling of Residential Buildings: Design of Systems*. Fort Collins.

CSU. 1979b. *Solar Heating and Cooling of Residential Buildings: Sizing, Installation*. Fort Collins.

DOE (U.S. Department of Energy). 1978a. *Activities of the Department of Energy in Energy Education*. DOE/IR-0008. Springfield, VA: National Technical Information Service.

DOE. 1978b. *Solar Energy Domestic Review, Appendix: Labor Impacts of Solar Energy Development*. Reports to the Impacts Panel by the Employment Impact Task Force, Washington, DC.

DOE. Office of Consumer Affairs. 1980a. *Activities of the U.S. Department of Energy in Education*. DOE/CA-0002. Annual status report FY 1979. Washington, DC.

DOE. Office of Conservation and Solar. 1980b. *Residential Conservation Service: Auditors Training Manual*. In cooperation with Oak Ridge Associated Universities, University of Massachusetts Cooperative Extension Service, and Solar Energy Research Institute. August.

DOE. 1980c. Office of Education, Business, and Labor Affairs and Office of Solar Applications. *Solar Energy Employment and Requirements 1978–1985*. DOE/TIC-11154. Washington, DC.

Duggan, D. D. 1983. *The Federal Role in Energy Education in the U.S.A. Alternative Energy Sources V*. Pt. F, *Energy Economics/Planning/Education*. Amsterdam: Elsevier.

EES (Energy Extension Service). 1986. *The Ninth Report to Congress: Comprehensive Program and Plan for Federal Energy Education, Extension, and Information Activities, Annual Revisions*. DOE/CE-0144. 42 U.S.C. 7007, Section 508, National Energy Extension Service Act and 42 U.S.C. 7373, Section 404, Energy Security Act. Washington, DC: U.S. Department of Energy.

EES. 1988. *The Twelfth Report to Congress: Comprehensive Program and Plan for Federal Energy Education, Extension, and Information Activities, Annual Revisions.* DOE/CE-0241. Section 508(c), National Energy Extension Service Act, PL 95-39 (42 U.S.C. 7007(c)), and Section 404(2), Energy Security Act, PL 96-294 (42 U.S.C. 7373). Washington, DC: U.S. Department of Energy.

FSEC (Florida Solar Energy Center). 1979. *Solar Water and Pool Heating: Installation and Operation Manual.* Cape Canaveral.

Grossman, R., and G. Danaker. 1977. "Jobs and Energy." *Solar Age* September:8.

Grossman, R., G. Danaker, and H. Wasserman. 1979. *Energy, Jobs and the Economy.* Boston: Alyson Publications.

HUD (U.S. Department of Housing and Urban Development). 1979. *Installation Guidelines for Solar Domestic Hot Water Systems.* In cooperation with U.S. Department of Energy. GPO626-716/1659. Washington, DC.

Lampert, S., K. M. Wulf, and G. Yanow. 1981. "The Dissemination of Solar Energy Curriculum for Elementary Schools." *AS/ISES 1981: Proceedings of the 1981 Annual Meeting.* Newark: University of Delaware, pp. 1284–1287.

Lampert, S., K. M. Wulf, J. Grishaver, G. Yanow. 1978. "A Solar Energy Program for Elementary Schools." *AS/ISES 1978: International Solar Energy Society, Proceedings of 1978 Annual Meeting.* Denver, CO, August, pp. 521–525.

Lampert, S., K. M. Wulf, G. Yanow 1980. *A Solar Energy Curriculum for Elementary Schools.* Washington, DC: U.S. Department of Energy, Office of Solar Applications, February.

Langford, D. D. and D. E. La Hart. 1981. "Solar Energy in the Secondary School Curriculum." *AS/ISES 1981; Proceedings of the 1981 Annual Meeting.* Newark, DE: University of Delaware, pp. 1282–1283.

Mason, B. 1978. *Labor Manpower and Training Requirements: Technical Progress Report.* SERI/PR-53-073. Golden, CO: Solar Energy Research Institute.

Meyers, A. C., III. 1981. "Solar Energy Technical Curriculum: Report on Phase II Activities." *AS/ISES 1981: Proceedings of the 1981 Annual Meeting.* Newark: University of Delaware, pp. 1268–1272.

NSTA (National Science Teachers Association). 1979. *Energy Education Workshop Handbook: A Guide to Materials by the Project for an Energy-Enriched Curriculum (PEEC).* DOE/TID/3841-11. Prepared for the U.S. Department of Energy, 1978–1979. Distributed by the DOE's Technical Information Center. Washington, DC: The Guide references the 15 PEEC lesson plans developed by NSTA.

NYSED (New York State Education Department). 1981. *Solar Energy Education. Reader Part I: Energy, Society, and the Sun; Reader Part II: Sun Story; Reader Part III: Solar Solutions; Reader Part IV: Sun Schooling.* Prepared in cooperation with SUNY Atmospheric Sciences Research Center for the U.S. Department of Energy, Office of Conservation and Solar Energy, Office of Solar Applications for Buildings.

O'Connor, K. 1979. *Solar Energy Task Force Report: Technical Training Guidelines.* SERI/SP-42-345. Golden, CO: Solar Energy Research Institute.

O'Connor, K. 1980. *Training Programs and Advanced Education in Solar Energy.* Report of a conference on Meeting Energy Workforce Needs. Silver Spring, MD: Information Dynamics, Inc., pp. 258–279.

OOE (U.S. Office of Education). 1980a. *The National Conference on Meeting Energy Workforce Needs: Determining Education and Training Requirements,* 26–28 February. Sponsored by the Energy and Education and Action Center, Washington, DC.

OOE. 1980b. *A Selected Guide to Federal Energy and Education Assistance.* Washington, DC: Energy and Education Action Center.

ORAU (Oak Ridge Associated Universities). 1978. *Science Activities in Energy: Solar Energy and Solar Energy II* (HCP/U00033-01, May 1978); *Wind Energy* (DOE/IR-0037); and *Conservation* (HCP/U0033-02). Oak Ridge, TN: U.S. Department of Energy, Technical Information Center.

Petrock, E. 1979. *The Status of State Energy Education Policy.* Denver, CO: Education Commission of the States.

Petrock, E. 1980. *Policy Issues in K–12 Energy Education.* Denver, CO: Education Commission of the States.

Petrock, E. M. 1981. *A Policy Development Handbook for Energy and Education.* Denver, CO: Education Commission of the States.

Schacter, M. 1979. "The Job Creation Potential of Solar and Conservation: A Critical Evaluation". U.S. Department of Energy, Policy and Evaluation Office, Washington, DC, 6 November Mimeo.

SERI (Solar Energy Research Institute), 1980a. *Department of Energy Responses to Panel Recommendations from the Open Workshop on Solar Technologies*, 23–24 October 1979. SERI/SP-742-732. Sponsored by the U.S. Department of Energy. Golden, CO.

SERI, 1980b. *Proceedings of the Open Workshop on Solar Technologies*, 23–24 October 1979. SERI/CP-741-683. Sponsored by the U.S. Department of Energy. Golden, CO.

Appendix: Solar Energy Educational Survey

Institution Information (1980–1981)

Please type or print legibly

Institution Name _____

Address _____

City _____ State _____ Zip _____

Important

We request the following information of the person filling out or coordinating the response to this survey.

Name _____ Position _____

Department _____

Phone ( ) _____ Ext _____

1. Institution Type (Circle single most appropriate response):
   - U University/College/Graduate School
   - J Junior/Community College
   - V Vocational/Technical School
   - R Research Institute/Laboratory
   - T Trade Association
   - L Labor Organization
   - I Industrial Organization
   - G Government Organization
   - E Adult/Community Education
   - A Political Action Group
   - P Professional Association
   - O Other _____

2. What is your institution's total educational enrollment? _____

3. Control or affiliation of your institution (Circle one)
   - 1 Public   2 Private   3 Combination—Public/Private

4. What is the minimum admission level for students attending your institution? (Circle one)
   - 1 Only the ability to profit from attendance
   - 2 High school graduate or equivalent
   - 3 High school graduate and superior academic aptitude
   - 4 Two-year college graduate
   - 5 Four-year college graduate
   - 6 Other (Specify) _____
   _____

5. How is your school year divided? (Circle one)
   - S Semester   Q Quarter   T Trimester   N Not Applicable   O Other (Specify) _____

6. What are the tuition and fees for a full-time student for one term: $ _____ Resident
   $ _____ Non-resident

7. If you have no solar-related programs or courses, circle the response, stop here, and return this part of the survey. Thank you   N No solar programs or courses

If you have solar-related programs or courses, please go to Part I

I Educational Data Base—Educational Institutions

Solar Energy Program Information (1980–1981)

Note: If more than one solar program is offered during 1980 or 1981, please make copies of this part for each

1. Which of the followings does your institution offer or plan to offer during 1980 or 1981? NOTE: Please see Program Definitions on instruction page

   SC Solar Curriculum. Awarding a Degree or Diploma in a Solar Field

   SS Curriculum with Solar Study. Awarding a Degree or Diploma in a Solar-Related Field

   ST Solar Technical Training. Awarding a Certificate for Study in Solar Energy or a Solar-Related Field

If you have not circled one of the above program types, go to Part II

2. Program Name: _____

3. Department Offering Program: _____

4. Head of Program: _____

5. Phone: (____) _____ Ext. _____

6. What degree does the student receive? (Circle appropriate responses)

   D Doctorate   B Bachelor's   C Certificate

   M Master's   A Associate   O Other (Specify) ____

7. What is the discipline or subject area of the above degree type? _____

8. Estimate the number of students completing the above program for the years indicated:

   1977: _____   1978: _____   1979: _____   1980: _____   1981: _____   1982: _____

9. As a result of having completed the above program, students will be trained in the following area(s) (Circle as many as apply)

   ARC Architecture—Solar Specialization

   HOM Do-It-Yourself Home Installation

   ELE Electricity—Solar Specialization

   ENG Engineering—Solar Specialization

   CON General Contracting—Specialization in Solar Design Installation

   HVC Heating, Ventilation, Air Conditioning—Solar Specialization

   MEC Mechanical Electrical Contracting Solar Specialization

   PLB Plumbing—Solar Specialization

   RES Scientific Research and Development in Solar Energy

   SHM Sheet Metal—Solar Specialization

   ADM Solar Energy Administration/Policy

   EDU Solar Energy Education

   INI Solar System Installation—Industrial/Commercial

   INR Solar System Installation—Residential

   SOT Solar Technology—Training in Instrumentation, Controls, Design, Maintenance, Etc

   OTH Other Solar Specialization (Specify)

   _____

10. Does your institution offer a job placement service?   Yes   No

11. Does your institution have any procedures for following student employment after program completion?   Yes   No

12. If yes, what is the estimate of the percentage of students who find employment in the field for which they are specifically trained as a result of having completed the above program: _____ %

II Educational Data Base—Course Information

Solar Energy Course Information (1980–1981)

Note: If more than one solar program is offered during 1980 or 1981, <u>please make copies of this part</u> for each

ORG _____

CRS _____

PRG _____

1. Course Title: _____

2. Course No.: _____

3. Instructor or Contact Person: _____

4. Phone: (____) _____ Ext. _____

5. Department Offering Course: _____

6. Is course also taught in conjuction with other departments/ schools? ☐Yes ☐No

7. Is course part of a program(s) described in Part I of this survey? ☐Yes ☐No

8. If yes, give program name(s): _____

_____

ACTV	BIOM
CHEM	HYBR
LOHY	OTHR
PASS	PHVC
PROC	SATS
SEAS	SOLG
STOR	THER
WIND	

9. Number of times course taught to date: _____

10. Average enrollment per class: _____

11. Estimate the number of students completing the course for the years indicated:

1977: _____ 1978: _____ 1979: _____ 1980: _____ 1981: _____ 1982: _____

12. Is the course offered for academic credit? ☐Yes ☐No

13. If yes, how many credits? _____

14. Are continuing education units offered for this course? ☐Yes ☐No

15. If yes, how many units? _____

16. Level for which course is offered:

Academic Level	Non-Academic Level
☐AL All Levels	☐MA Managerial
☐CG College Graduate	☐PR Professional
☐JS College Junior/Senior	☐SK Skilled labor
☐FS College Freshman Sophomore	☐LA Layperson
☐OA Other (Specify) _____	☐ON Other (Specify) _____

17. Duration of Course:

No. of weeks _____		No. of Days _____
Hours per week _____	OR	Hours Per Day _____
Total Course Contact Hours _____		Total Course Contact Hours _____

18. Detailed Course Contact Hours for Duration of Course

Lecture/Discussion _____ Hrs.		Workshop _____ Hrs.
Research/Independent Study _____ Hrs.		Laboratory _____ Hrs.
On-the-Job Training _____ Hrs.		Fieldtrip _____ Hrs.
Seminar _____ Hrs.		

Other (Specify) _____ Hrs.

19. When is this course offered? (Circle as may as apply)

☐D Day ☐E Evening ☐W Weekend

II Educational Data Base—Course Information (Continued)

CRS _____

20. Tuition and fees for a student taking just this course: $ _____ Resident

    $ _____ Non-resident

21. What are the principal texts used in this course?

    AUTHOR   TITLE

    1. _____   _____

    2. _____   _____

    3. _____   _____

22. Are there any prerequisites for taking this course?   [Yes]   [No]

23. To what extent are the following topics covered in this course? Circle a 1 or 2 using the following code definitions:

    1-Topic Covered Extensively (Major course topic or concept)

    2-Topic Covered In Some Detail (Minor course topic or concept)

    Do not code topics covered superficially.

[1] [2] ALCH—Alcohol/Gasohol Fuels	[1] [2] SSPS—Satellite Solar Power Systems
[1] [2] ALTE—Alternate Energy Sources	[1] [2] SHMT—Sheet Metal Techniques
[1] [2] APRT—Appropriate Technology	[1] [2] ARCH—Solar Architecture
[1] [2] BIOM—Bioconversion	[1] [2] SHAC—Solar Cooling
[1] [2] CENT—Centralized Solar Power Systems	[1] [2] HOTW—Solar Domestic Hot Water
[1] [2] COMP—Components—Solar	[1] [2] AUDT—Solar Energy Audit
[1] [2] DIST—Distributed Solar Power Systems	[1] [2] POLD—Solar Energy Policy Development
[1] [2] CNSV—Energy Conservation	[1] [2] SHAK—Solar Heating
[1] [2] STOR—Energy Storage Systems	[1] [2] CNST—Solar Home Construction
[1] [2] GRHT—Greenhouse Techniques	[1] [2] LAWL—Solar Law/Legislation
[1] [2] HETR—Heat and Energy Transfer	[1] [2] MRKT—Solar Marketing/Economic Analysis
[1] [2] HYBR—Hybrid Systems	[1] [2] SWPL—Solar Swimming Pool Heating
[1] [2] INTR—Introduction to Solar Energy	[1] [2] SSYD—Solar Systems Design
[1] [2] MATR—Materials	[1] [2] INST—Solar Systems Install Maintenance
[1] [2] SEAS—Ocean Systems	[1] [2] TEST—Solar Systems Testing/ Evaluation
[1] [2] PASS—Passive Solar Systems	[1] [2] CHEM—Thermochemical Conversion
[1] [2] PHOT—Photoconversion	[1] [2] WIND—Wind Energy Conversion Systems
[1] [2] PHVC—Photovoltaics and Solar Cells	[1] [2] OTHR—Other (Specify) _____
[1] [2] PLMB—Plumbing Techniques	
[1] [2] PROC—Process Heat	

20 Regional Solar Energy Centers

Donald E. Anderson

20.1 Mandates

On 24 March 1977 the Energy Research and Development Agency (ERDA) announced that the new Solar Energy Research Institute (SERI) would be established in Golden, Colorado, and would be managed and operated by the Midwest Research Institute (MRI). At the same time, it announced that the regional solar energy centers (RSECs) would be established to work with SERI in accelerating the commercialization of solar energy applications.

Representatives of the state governors (usually from the state energy agencies) were invited to attend a one-day meeting in Denver on 26 April 1977. This conference was hosted by MRI for ERDA (MRI 1977). At this meeting, ERDA representatives invited the states to organize themselves into four distinct regions. The states of each region were invited to designate an entity that would, under a grant from ERDA, develop a plan for joint regional activities that could be effectively provided through a single entity in their region.

20.2 Planning Phase

In the midwestern region two organizations that had been incorporated to present competitive proposals to manage and operate the Solar Energy Research Institute offered to form a joint venture to provide the focus for the planning phase. One of these was the Central Solar Energy Research Institute (CSERI), incorporated in Minnesota; the other was the Central Solar Energy Research Corporation (CSERC), established in Michigan.

At a meeting in Chicago with representatives of the state energy offices, these two organizations agreed to form a single entity for planning purposes, and had formally done so by the time they submitted a grant proposal to ERDA on 22 June 1977. The joint venture, named the Midwest Regional Solar Energy Planning Venture (MRSEPV), operated under a board of advisors that included CSERI, CSERC, and MRI. Although planning roles and staffing assignments were balanced between CSERI and CSERC, the chief executive officer of CSERI (Donald Anderson) was named CEO and principal investigator of the planning grant.

In the other three regions, comparable planning grants were awarded to entities that had been formed to present competitive proposals to manage and operate SERI. In the southeastern region, that entity was the Southern Solar Energy Center (SSEC), under Barry Graves; their planning activities were headquartered in Atlanta. In the northeastern region, the Northern Energy Corporation directed by Lawrence Levy in Boston headed the planning activities. Finally, the western region acted through a planning grant to a Compact of States office headed by Ray Gilbert in Denver. (The center that resulted—the WSUN—operated from Portland under Donald Aitken.)

The planning grants were received from ERDA in July 1977. The objectives of the four planning entities were similar; all included planning activities throughout their regions. The objectives of the grants were to

• develop and submit to ERDA (later to become the Department of Energy) a plan (and an offer to contractually implement it) for the accelerated introduction and commercial utilization of solar energy systems in the region in a manner supportive of national energy policy goals as they developed;

• cooperatively develop a regional profile of resources, needs, and program capabilities;

• involve, at policy levels, each of the participating states in the process of plan development and coordination;

• assist qualified individuals from each state in identifying the needs and resources extant in the region and in developing a matrix of programs, with prioritization, supportive of regional solar energy objectives;

• identify regionally supportive state and multistate programs for implementation within the framework of a five-year process (with identification of a first year's program); and

• initialize, for implementation during the five-year period in coordination with SERI, the other regions, and other DOE solar-related organizations, a client-oriented database supportive of consumer and industrial solar energy component and systems installation.

Each of the four regions performed similar studies under these grants; I will describe the work of the joint venture in the north central states as an example.

The CSERI team was established at a site in Eagan, Minnesota, that had been offered as the Minnesota interim location for SERI; initial staffing was twelve at CSERI. A comparable staff was in place in the State Executive Office Building in Detroit under the planning grant, a number of organizational steps and processes were established as the basis for the regional center MASEC—the Mid-American Solar Energy Complex; these included the following:

• Subgrants to each of the twelve states in the north central region, under which they developed plans for the operation of State Solar Offices (SSOs) under DOE funding in the operational phase.

• Restructuring and renaming one of the two corporations of the joint venture (CSERI) to establish the MASEC Corporation as a free-standing, not-for-profit, 501c(3) entity, reporting to a board of directors nominated by the twelve governors and intended solely to manage and operate the regional solar energy center (RSEC) for those states under contract to DOE.

• Establishing an Advisory Council of States (ACS) to focus on programmatic issues, to oversee operating programs, and to recommend and plan new programs. These ACS members were in most cases representatives of the state energy offices. They were particularly effective in planning for, and later implementing, linked SSOs that could handle requests and assignments from DOE, through MASEC, while also acting in and for a given state.

• Establishing, during the planning phase, a balanced panel of private sector advisers nominated by the ACS members. These advisers were to meet and confer as a source of expert and informed opinion on the many technical, economic, and institutional issues and barriers facing the different solar technologies.

• Identifying a number of technical issues in the research and development phases of manufacturable solar collectors that were of particular interest to the other participant in the joint planning venture, CSERC (for details on these issues, see MRSEPV 1977, vol. 5). This led to a series of subcontracts where CSERC reported to SERI under the SERI/MRI operating contract to DOE.

In developing the regional plan, the private sector advisers participated through a structured advisory and polling technique involving topical conferences and questionnaires; this technique was formalized as the Knossos process, an extension of the Delphic technique often invoked in working toward consensus. The Knossos process (see MRSEPV 1977, vol. 2) involved the participation of a subset of the state-designated experts in topical conferences, where the general issues of importance to each given group were explored. Questions were developed that could be used to poll all identified "experts" in the same area in order to determine whether a particular issue was of high relevance and what the appropriate role of an RSEC might be.

As an example, figure 20.1 shows the bivariate analysis of the responses of all participants regarding the appropriate role of an RSEC for specific activity items of relevance to the general mission of "accelerating the commercialization process for solar energy utilization in the region." This figure, taken with the accompanying table 20.1, provided a clear and unambiguous focus for the regional plan.

Of equal significance, it became obvious in working through these sub-groups of expert, informed opinion on a given issue that the removal of institutional barriers (such as local codes, financial guidelines, organized labor) could most effectively be provided by the continued involvement of the groups—in a distributed but interactive network. As a consequence, the planners proposed to continue this involvement in the operational phase. The group ultimately grew from the 852 nominated by ACS designees during the planning phase in 1977 to over 6,300 members of Solar Resource Advisory Panels (SRAPs) in the twelve states by 1981.

Planning activities in the other three regions led DOE to establish four regional solar energy centers at the end of the planning period. The four RSECs were managed and operated by four free-standing, not-for-profit corporations under contracts from DOE; the regions served by the four RSECs are shown in figure 20.2.

20.3   Operational History of MASEC

Based upon the submission of the five planning volumes by MRSEPV on 31 December 1977, a contract was negotiated between DOE and the MASEC Corporation as of 1 May 1978; this contract was later renegotiated into a five-year operating contract effective 1 April 1979. Under this contract:

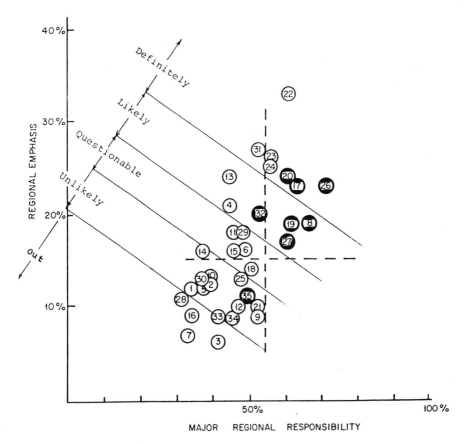

Figure 20.1
Graph of regional activity items by regional emphasis and major regional responsibility, showing likely mission divisions (MRSEPV, 1977). For item codes, see table 20.1. Darkened circles indicate that the regional level was given a higher percentage of "major" or "all" ratings than any other level.

• The corporation was renamed, adopted new bylaws, and seated a new regional board of directors at a first annual meeting in July 1978.

• Each of the twelve states nominated a member for the board of directors, the Advisory Council of States, and the State Solar Energy Resource Advisory Councils. All of these groups attended annual meetings from 1978 through 1981.

• All twelve states established State Solar Offices (SSOs) to disseminate information and to handle data flow to and from the RSEC in Minnesota (and hence to the other RSECs and to SERI).

Table 20.1
Ranked products of regional emphasis and responsibility, general activities

Rank	Item	Product	Activity
1	22	0.20	Collecting, disseminating, and exchanging solar energy information regarding basic solar energy research
2	26	0.16	Collecting, diseminating, and exchanging solar energy information regarding climate issues
3	23	0.15	Collecting, disseminating, and exchanging solar energy information regarding technologies (systems, hardware)
6	17	0.14	Doing solar energy-related research and/or policy analysis on climate issues
6	20	0.14	Collecting, disseminating, and exchanging solar energy information regarding economic/financial issues
"Definitely"			
6	31	0.14	Providing solar energy-related educational programs for legislative/regulatory bodies
7	24	0.13	Collecting, disseminating and exchanging solar energy information regarding educational/instructional issues
8	8	0.12	Assigning responsibility for solar energy–related research and/or policy analysis on information about climate issues
9	19	0.12	Collecting, disseminating, and exchanging solar energy information regarding legislative/legal issues
10	32	0.11	Providing solar energy–related educational programs for professional groups (architects/engineers/lawyers)
11	13	0.10	Doing solar energy–related research and/or policy analysis on basic solar energy
"Likely"			
13	27	0.10	Collecting, disseminating, and exchanging solar energy information regarding environmental impact issues
13	4	0.10	Assigning responsibility for solar energy–related research and/or policy analysis on basic solar energy
15	11	0.08	Doing solar energy–related research and/or policy analysis on economic/financial issues
"Questionable"			
15	29	0.08	Providing solar energy–related educational programs for colleges
18	6	0.07	Assigning responsibility for solar energy–related research and/or policy analysis on educational/instructional issues
18	15	0.07	Doing solar energy–related research and/or policy analysis on education/instructional issues
18	18	0.07	Doing solar energy–related research and/or policy analysis on environmental issues

Table 20.1 (continued)

Rank	Item	Product	Activity
19.5	14	0.06	Doing solar energy–related research and/or policy analysis on technologies (system, hardware)
19.5	25	0.06	Collecting, disseminating, and exchanging solar energy information regarding manufacturing/distribution issues
21	21	0.06	Collecting, disseminating, and exchanging solar energy information regarding sociological/psychological issues
27	2	0.05	Assigning responsibility for solar energy–related research and/or policy analysis on economic/financial issues
27	9	0.05	Assigning responsibility for solar energy–related research and/or policy analysis on environmental impact issues
"Unlikely"			
27	10	0.05	Doing solar energy–related research and/or policy analysis on legislative/legal issues
27	12	0.05	Doing solar energy–related research and/or policy analysis on sociological/psychological issues
27	30	0.05	Providing solar energy–related education programs for vocational/technical schools
27	35	0.05	Providing solar energy–related education programs for energy suppliers (utilities/oil companies)
29.5	1	0.04	Assigning responsibility for solar energy–related research and/or policy analysis on legislative/legal issues
29.5	5	0.04	Assigning resposibility for solar energy–related research and/or policy analysis on technologies (systems/hardware)
29.5	33	0.04	Providing solar energy–related education programs for trades
29.5	34	0.04	Providing solar energy–related education programs for manufacturers
33	3	0.03	Assigning responsibility for solar energy–related research and/or policy analysis on sociological/psychological issues
33 "Out"	16	0.03	Doing solar energy–related research and/or policy analysis on manufacturing/distribution issues
33	28	0.03	Providing solar energy–related educational programs for elementary/secondary schools
35	7	0.02	Assigning responsibility for solar energy–related research and/or policy analysis on manufacturing/distribution issues

Source: MRSEPV 1977.

Figure 20.2
Division of the regions for solar commercialization. The Mid-American Region (MASEC) is
highlighted. Western SUN Includes Hawaii and Alaska. SESEC includes Washington, D.C.,
and Puerto Rico.

MASEC initially continued to operate in the leased office space of a
general manufacturing facility in Eagan, Minnesota. In November 1979 a
move was made to a building in Bloomington, Minnesota, that was more
appropriate to facilitating information flow, with conference space and
open offices. Total staffing at the center reached the planned level of 89 in
the fall of 1980; in addition, forty-three people in the twelve SSOs were
directly linked to MASEC through grants to the states or by employment
contracts from MASEC.

It should be emphasized that the role of the center of the complex was
viewed as that of coordinating regional activities and disseminating
information between and among DOE, the other RSECs, SERI, state
agencies, and the private sector. As such, no development or demonstration facilities were planned or provided at the center. Some 142 subcontracts were awarded for regional solar projects, totaling over $1.8
million; these used existing resources to the greatest extent possible. In
comparable fashion, existing facilities in the twelve states were used for
SSO operations and to link the SRAP groups; these entities were provided
with over $3 million for their operations.

Table 20.2
Estimated cumulative budget for MASEC under operating contract from DOE (thousands of dollars)

Income		
1 May 1978–1 April 1979	$1,130	
1 April 1979–30 September 1979	1,798	
1 October 1979–30 September 1980	4,961	
1 October 1980–30 September 1981	5,800	
Total	$13,689	
Distribution		
Center operations	$8,009	58.5%
SSO operations = SRAP	3,000	21.9
Subcontracts	1,863	13.6
Earned fees (to SSOs)	817	6.0
Total	$13,689	100

Throughout the operational history of all of the RSECs, initial and incremented funding elements tended to be late in authorization by DOE and subject to midyear adjustments. Program cutbacks and cancellations began in late 1980 as it became clear that federal support for solar commercialization activities and the level of support for all solar energy activities would be changed by the election. Thus, at the time of the MASEC annual meeting in July 1981 the Center staff had been cut to 53; the twelve SSOs had been informed of DOE plans to discontinue their funding and in fact were funded from earned fees of MASEC (by vote of the MASEC Board of Directors) in the last half of FY 1981. DOE had by that time established guidelines for future work that would retain the support of high-risk, long-term research projects in solar energy but that would de-emphasize information dissemination and the support of commercialization of near-term applications, at SERI as well as through the RSECs.

The approximate breakdown of the total performance budgets and staffing for MASEC are summarized in table 20.2. This covers the period from 1 May 1978 through December 1981, when the operating contract for MASEC was terminated by DOE.

20.4 Information Dissemination

A significant part of the effort of all four RSECs, and of SERI, focused on providing for the efficient and prompt access of many different audiences to current information on solar energy alternatives. This process was facilitated by the establishment of a national Solar Energy Information Data Bank (SEIDB). In the midwestern region, MASEC provided the access point for nationally available information. In a number of cases, regionally specific data were provided as input to the national data bank through the SSOs; this effort is documented in the final report for this project (Mortison 1981a).

As specific activities provided at the center through this program, 23 solar models and 17 solar databases of the national network were made available in the region. The overall flow of information into the region throughout the operational period from 1978 until termination in 1981 is documented in the final report of MASEC (1982). Acting joinly, center staff, the SSOs, and the SRAPs

• conducted 530 workshops and conferences with over 20,000 attendees;

• handled more than 210,000 public inquiries;

• distributed 2,482,000 items of solar information;

• provided 7,000 technical assistance consultations;

• participated in 870 energy fairs reaching over 840,000 individuals;

• made 935 speaking appearances reaching 18,720 individuals; and

• made 85 major radio and TV appearances, reaching an estimated 950,000 individuals.

Taking into account all ways in which the network provided access to decision makers in the region, it is estimated that 5.3 million people, or 9.1% of the total regional population of 58 million, were provided with some information on solar energy systems and applications.

20.5   Passive Solar Programs: Solar 80

In the mid-American region, a plan for creating generic single-family structures was developed in 1978; it focused on providing broad marketing support through design, development, and demonstration teams selected in all twelve states. This program had three basic products: (1) house plans, (2) workshops, and (3) homes constructed, demonstrated, and sold.

Table 20.3
MASEC Solar 80 home designs

Design team	Design name, location	Finish floor area (ft²)	Estimated cost with lot (1980$)	Heating degree-days	EER (Btu/ft²-degree-day)
Architectural Alliance/ Marv Anderson Const.	Solar Ridgeway, Plymouth, MN	1,640	115,000	8,329	1.8
District One Technical Inst.	Eau Claire, WI	1,800	80,000	8,609	2.5
Gordon Clark Associates/Mar-Kel Construction	Indianapolis, IN	2,600	190,000+	5,699	1.1
Kirkwood Community College	Kirkwood Energy Saver, Cedar Rapids, IA	1,200	60,000	6,614	1.0
LaPeer County Vo-Tech Center	Solar Salt Box, Attica, MI	1,337	60,000	6,991	1.5
Londe-Parker-Michels	St. Louis, MO	1,400	62,500	4,661	1.1
University of Nebraska	Sol-Deck House, Omaha, NE	1,380	79,000	6,601	2.0
Steven Winters Assoc./ Unibuilt Industries	Unisol I, Vandalia, OH	1,740	65,000	5,628	1.8
Waterford School District	Solar Wedge, Waterford, MI	1,800	97,001	6,783	2.0

Source: Pogany and Kraft 1980.

The teams who were competitively selected to design and build these homes, with supporting plans and documentation and workshops, were invited to design to a net energy consumption that constituted an 80% reduction in the use of fossil fuels, compared to the average conventional home. The performance design goal for these teams was posed in terms of an energy efficiency ratio of better than 2.5 Btu/ft²/degree-day, compared to a regionwide average for single-family residences of 12.5; thus one rationale for the "80" in Solar 80: an 80% reduction in fuel costs. In the first nine homes designed and built, the average EER was 1.65. A summary of the pertinent features of these homes is given in table 20.3. An eleven-page special article on these passive solar homes was made available through *Independent Banker* (1980) magazine to all subscribers.

Plans for the nineteen houses meeting the Solar 80 criteria were originally sold through MASEC and continue to be available through Garlinghouse Publications, Wichita, Kansas, and through *Better Homes and*

Gardens, which featured several of the designs in a special energy efficiency issue in 1981.

In total, over 1,500 sets of the plans for one of the houses meeting the Solar 80 design criteria were known to have been sold as of 1982. Approximately 13,000 building professionals and members of the general public attended one of 150 passive solar/Solar 80 workshops sponsored by MASEC. This program was being expanded to cover retrofit and multifamily housing, with similar plans to develop training aids with design and construction workshops, when the RSEC program was terminated.

20.6   Active Solar Programs

The MASEC active solar program was designed to focus on providing education, training, and information to retailers, installers, and homeowners. The program also involved a series of solar energy conferences for labor leaders, the financial community, building officials, and educators. Economic analysis and decision-making support were provided for 54 sites in the region for pilot test marketing, plus 16 specific cities for business applications.

With the other RSECs and SERI, MASEC had five major tasks in the integrated Residential Conservation Service (RCS) program: utility solar background briefings, training support, industry outreach, financing, and retrofit documentation. A short course entitled "Solar Energy and Conservation for Buildings" was held 17–21 August 1981, in Saint Paul, Minnesota. This course was attended by approximately 500 design and construction professionals from private firms and government agencies throughout the nation. The course was cosponsored by MASEC and SERI and included presentations on the state-of-the-art in conservation and solar applications for buildings with emphasis on commercial and government buildings.

A project was initiated in FY 1981 to develop active solar energy systems for new homes. The strategy was to develop energy-efficient designs, build prototypes using design/build teams, and accompany this with concentrated market development activities.

A contract with the National Association of Home Builders (NAHB) was initiated in the fall of 1981. NAHB was to select a large-volume home builder willing to adapt the design of a standard model, build it, and instrument both the solar home and an identical nonsolar home. Both were to be monitored for a year in order to analyze cost and operating differences and to demonstrate the applicability of active/passive integrated systems. This contract was transferred to the DOE for continuation after MASEC's termination (see chapter 9 for further details).

A market analysis was completed during FY 1981 for active residential solar systems in 54 different cities in the mid-American region. Economic analyses were completed using the MASEC-1 life cycle costing computer model (Laulainen 1979). The data and conclusions for all 54 sites were contained in *State-level Market Analysis for Active Solar Systems* (Mortison 1981b); this document identifies primary geographic areas in the region for pilot test marketing of active solar water heating systems for single-family new and retrofit applications.

Another market analysis completed in FY 1981 addressed commercial solar space and hot water systems. This study included 16 cities in the region and identifies incentives and barriers relating to the commercial market (MASEC 1981). A related study focused on solar industrial process heat; 4 cases selected from 30 potential solar industrial installations that had been carried through the design stage in the twelve midwestern states were examined for internal rate of return (Western Reserve 1981).

20.7   Other Solar Technologies

MASEC planning demonstrated significant regionally specific interest in biomass and wind energy systems; this provided an impetus for specific programs in these areas.

In the case of biomass, the twelve state region produces 82% of the nation's corn and 53% of the nation's wheat while consuming 47% of the nation's total agricultural energy. MASEC hosted a major biomass energy workshop at Purdue University in 1979 that helped stimulate regionwide interest in alcohol fuels, anaerobic digestion, gasification, and direct combustion (MASEC 1979). Subsequent workshops and conferences on wood combustion and residential safety, nonwoody agricultural biomass, liquid fuels, and gasification were also conducted in the region (see Hohmann 1980; MASEC 1980).

MASEC awarded subcontracts for studying the potential of small wind energy conversion systems as an integral part of the generating mix for a regional utility. MASEC conducted a Small Wind Energy

Conversion System Information Development and Demonstration Program, which included the development and dissemination of information concerning the Public Utility Regulatory Policies Act (PURPA). Regional conferences and seminars were conducted for consumers, alternative energy planners, and utilities.

20.8 Lessons Learned

The development and operation, under federal contract, of regional centers in order to accelerate the commercialization of alternative energy sources were difficult but worthwhile processes. In particular, it should be clearly recognized that federal involvement in the advanced research and development stages of new technology is quite different from federal involvement in the later steps leading to commercialization, where the many parties involved in governmental, regulatory, financial, business, and consumer affairs must participate in the decision-making processes. The regional centers were thus challenged to perform the anatomically difficult task of standing with both feet firmly in all three (federal, state, and private) sectors.

In recognition of this difficult role, the planners for MASEC selected the word "complex," not "corporation" or "center," as the last word in their corporate acronym. The word "complex" was chosen because it "suggests the *unavoidable* result of a *necessary* combining or folding, and does not imply a fault or failure" (MRSPV 1977).

In the case of the RSECs, each of the four regions selected a different model for planning and for operation. The Northeast Solar Energy Center (NESEC) model was that of a private-sector board-managed corporation. The Southern Solar Energy Center (SSEC) model involved state energy officials much more directly in management. WSUN (Western Solar Utilization Network), operated as a subsidiary of a Compact of States instrument chartered by the Western states for all energy policy coordination.

MASEC chose the early involvement of a large number of regional sources of expert opinion on the institutional issues of concern, ranging from technical experts to those in industry, labor, finance, legislative, educational, and consumer affairs. This involvement, using the Delphic-Knossos polling technique, provided a source of prioritization for programs and operating procedures attempting to gain access to, and elicit the participation of, a diverse set of interest groups.

MASEC was becoming quite successful in information dissemination, in providing a "friend of the court" presence for local institutional barrier removal, and in similar activities. On the other hand, the active involvement of interest groups made it more difficult for MASEC staff to respond to nationally directed activities arising routinely in the normal operations required by the Department of Energy. Thus, a differentiation between MASEC's role as participant in regional programs (including prioritization, evaluation, and budgetary detailing) and its role as captive contractor providing all requested involvement in national programs should have been more clearly established and maintained. If this differentiation had been made during the planning process, it would have helped to make the expectations of those invited to help in planning more realistic.

On balance, it must be remembered that the states were invited to establish through the RSECs a mechanism for accelerating the process of commercialization. This process clearly must involve dynamic interaction between the many participants—hopefully, all acting as experts, informed and sensitive to all aspects of the process. I believe that MASEC and the other RSECs, by structuring regionwide activities in a comprehensive network, did provide the potential for that mechanism.

References

Hohmann, M. A. 1980. *On-Farm Production of Fuel Alcohol in Mid-America: Technical and Economic Potential.* MASEC-TP-80-0099. Minneapolis, MASEC March.

Independent Banker. 1980. "Passive Solar Homes: Making More Sense in Today's Housing Market." August.

Laulainen, L. A. 1979. *F-Chart Evaluation: The Economic Analysis.* MASEC-R-79-045. Minneapolis, MASEC December.

MASEC (Mid-American Solar Energy Complex). 1979. *Conference Proceedings: Mid-American Biomass Energy Workshop at Purdue University,* 21–23 May. MASEC-R-79-034. Minneapolis.

MASEC. 1980. *Conference Proceedings: Mid-American Wood Combustion Conference at Northwestern Michigan College,* 8–9 November. MASEC-CF-80-005. Minneapolis.

MASEC. 1982. *The MASEC Corporation: Mid-American Solar Energy Complex, Final Report.* MASEC-R-82-002. Minneapolis March.

Mortison, J. 1981a. Solar Energy Information Data Bank (SEIDB) Program FY 1981: Final Project Report. MASEC-R-81-074. Minneapolis, MASEC.

Mortison, J. 1981b. *State Level Market Analysis for Active Solar Systems (54 sites)* MASEC-R-81-046. Minneapolis, MASEC.

MRI (Midwest Research Institute). 1977. Report of National Governor's Conference on Regional Solar Energy Centers. Kansas City. 26 April.

MRSEPV (Midwest Regional Solar Energy Planning Venture). 1977. *Final Report.* 5 vols. Planning Grant EG-77-G-01-4103. Minneapolis. 31 December.

Pogany, D. Z., and D. E. Kraft. 1980. "Mid-America's Passive Homes." *Solar Age* April.

Western Reserve Associates. 1981. *Financial Barriers to the Use of Solar Industrial Process Heat.* MASEC-SCR-81-011. Minneapolis, March.

VI SOLAR THERMAL TECHNOLOGY TRANSFER

21 Liaison with Industry

Daniel Halacy

21.1 Technology Transfer

The transfer of an obviously useful technology to the market would seem to be easily accomplished and very likely to succeed. In practice, however, the process is often difficult and sometimes seems impossible. Part of the reason is the great variety of technologies there are to be transferred: design, engineering expertise, computer software, testing facilities, materials, products, and production techniques, for example.

Because of this complexity, some observers have considered the technology transfer process more art than science. This is so for a number of reasons, including human factors as well as institutional and social barriers of many kinds. The initial complexity is also compounded by the technology producer's lack of awareness of consumer needs and/or methods of marketing technology. Similarly, industry and commerce may be unaware of new developments available in the laboratory and how to make use of them. Furthermore, both the inventors and the users of a new technology are faced with problems of patents and licensing.

There are barriers of many kinds to the transfer of solar heat technology in particular. These barriers range from consumer skepticism to vested interests in competing energy sources; from unfamiliarity with solar technologies, legal restraints, and other barriers to the difficulties of innovation. A major problem is that while potential users appreciate that solar devices reduce the need for fuel, they are also aware that they cost more initially than do conventional devices.

21.1.1 Definition and Methodology

A congressional report (CRS 1988), provides the following definition and explanation:

Technology transfer is the process by which technology, knowledge, and/or information developed in one organization, one area, or for one purpose is applied and utilized in another organization, in another area, or for another purpose. Various technologies, techniques, and expertise resulting from the Federal Government's sizable investment in R&D may be amenable to transfer to the private sector. There they may be further developed to meet market demands for new and different products, processes, and services. In addition, through the transfer of technologically oriented problems in both the public and private sectors. It should be noted that the concept of technology transfer can have different meaning in different situations. In some instances it refers to the transfer of legal rights, such as the assignment of title to a patent to a contractor, or the licensing of a Government-owned patent to a company. In other cases the technology endeavor involves the informal movement of information, knowledge, and skills from the Federal laboratories to the private sector through person-to-person interaction.

The value of technology transfer becomes most effective when it results in the commercialization of a new product or process, or the improvement of existing technologies or techniques. Technology development is an important factor in economic growth. It provides for new products and processes to be sold in the marketplace, as well as for new processes which can improve productivity and the quality of goods. Technological advance can be the basis for new companies which form to take advantage of these ideas. It can also help to make traditional industries (particularly manufacturing) more efficient. While the Federal Government directly funds basic research—and applied research and development to meet the mission requirements of the Federal departments and agencies—commercialization is the responsibility of the private sector.

There are two basic methods of transferring a technology from laboratory to consumer. The more obvious is for the technologist to sell the idea to the user: the "push" method. Pushing a technology from laboratory to marketplace can be done by conducting workshops to demonstrate the technology, publishing information about it, and otherwise marketing it. The second method is to "pull" the technology into the commercial marketplace through demand, using advertising, funding of prototypes, media attention, and government tax incentives. For example, from 1974 through 1985 the federal government spent appreciable amounts of money on various solar tax credit programs. Individual states also offered solar tax credits of their own (Frankel 1984).

The federal laboratory system, which has extensive science and technology resources developed as a consequence of meeting the mission requirements of federal departments and agencies, is a potential source of technology, technical expertise, information, and state-of-the-art facilities that can be utilized by the business community and other government entities. In particular, a portion of the laboratories' information, technologies, and techniques may have commercial application. Because, however, the federal government does not have the authority or capability to develop, refine, adapt, and market the results of this research and development beyond legitimate government mission objectives, there is expanding interest in transferring technology to the private sector, which has the resources to undertake commercialization activities.

21.1.1.1 Commercialization

Successful technology transfer results in the commercialization of that technology. In the federal solar energy program, such commercialization was facilitated by organizations including the four regional solar energy centers (RSECs), the National Center for Appropriate Technology (NCAT), the Conservation and Renewable Energy Information and Referral Service (CAREIRS), and the Solar Technical Information Program (STIP). In addition, solar energy demonstration programs were carried out by agencies including the Department of Housing and Urban Development (HUD), the National Science Foundation (NSF), and the National Aeronautical and Space Administration (NASA). Commercialization was also facilitated for some time by federal and state tax credit programs that subsidized the costs of solar equipment. The various methods of commercializing solar energy in the United States are described in detail in other chapters of this volume.

21.1.1.2 Technology Transfer

Interpreted broadly, technology transfer includes a great variety of activities involving the dissemination of research or technical information. These activities include the flow of such information from laboratory to industry, from one industry to another, and from industrialized nations to developing nations. Technology transfer involves direct approaches such as applying laboratory research findings for similar commercial applications, as well as indirect approaches that spin off commercial technologies developed for one purpose for use in other, often quite different, areas.

This chapter describes the transfer of solar thermal technology from the Department of Energy (DOE) national laboratories to university and industry researchers and from there to the marketplace. The three chapters that follow expand on this general description with specific examples of technology transfer from several such laboratories.

21.1.2 History of Technology Transfer

Evidence of the reluctance of the community to adopt solar technologies can be seen in their lengthy history. The real pioneers of solar heat technology were Hero and Archimedes in ancient Greece. Scientist Antoine Lavoisier built sophisticated solar thermal furnaces and did important research with them before he was executed in the French Revolution. American engineer Charles Wilson built a large solar still in Chile in 1872, and Frank Shuman, also an American, built a a 70-horsepower solar heat engine that pumped irrigation water from the Nile until World War I ended its operation (Halacy 1963). Yet few American scientists and engineers were aware of this background in solar heat technology when the oil crisis of 1973 prompted the U.S. government to initiate its solar energy program.

Although *technology transfer* is a relatively new term and the process itself generally perceived as a recent development, the Extension Service of the U.S. Department of Agriculture (USDA) long ago successfully transferred technology developed in USDA laboratories to the farm and forest product industries. The Bureau of Mines similarly transferred extractive metallurgy techniques and processes to the mining industry. The National Institutes of Health (NIH) and the National Bureau of Standards (NBS) likewise transferred medical technology and standards to the private sector. The National Advisory Committee on Aeronautics (NACA) and its successor, the National Aeronautics and Space Administration (NASA), have since the 1920s aided in developing the highly sophisticated aircraft and airlines industries through technology transfer.

Even more successful was the transfer of nuclear energy technology developed in the national laboratories. The private U.S. nuclear energy industry developed almost entirely from the innovative and highly secret work in federal laboratories that led to the atomic bomb. Nuclear electric power plants and other commercial uses of nuclear technologies resulting from government research are clear examples of successful technology transfer.

Nuclear fission and fusion were developed first in university and federal research laboratories and were thus well understood by government scientists and engineers. In general, however, this was not the case with solar energy technologies. Most of these originated in the private sector and were little known to federal researchers. With hindsight, one might argue against involving federal researchers in a technology intended at the outset for the commercial market, particularly if those researchers are not expert in the technology. A better approach would have been to assist appropriate private industries to develop solar energy equipment; DOE did so, although only a small private solar industry evolved.

21.1.3 Modern Technology Transfer

A major transfer of energy technologies from federal laboratories to the public sector came in the mid-1970s with the passage of several solar energy bills following the oil shortage. Argonne, Lawrence Berkeley and Lawrence Livermore, Oak Ridge, Pacific Northwest, Sandia, the Jet Propulsion, and other national laboratories worked on a variety of solar technologies intended for direct transfer to the commercial marketplace. So did a number of large aviation and industrial firms, including Grumman Aerospace, Boeing, General Electric, McDonnell Douglas, and Honeywell. However, for reasons covered in section 21.7, the transfer of solar thermal technology has not been as successful as the earlier transfer of nuclear technologies.

21.2 Mandates for Solar Energy Technology Transfer

The federal government is responsible for the nation's welfare, including an adequate response to energy needs; thus it was logical that federal legislation be offered in the area of alternative energy development when problems arose with conventional energy resources (Halacy 1976). Such federal legislation dates back to 1951, when H.R. 4286, a wind energy bill, was introduced by Representative John Murdock of Arizona. This measure did not pass, however, nor did a long series of other renewable energy bills including S2318 introduced by Senator Alan Bible of Nevada in 1959 and S2849 introduced by Senator Hubert Humphrey in 1962 and again in 1964 and 1965. None of these bills passed because conventional fuels were then cheap and plentiful.

Not until the 93d Congress was solar legislation enacted, and then only because of the sudden and disruptive shutoff of Mideast oil to the United States. In 1974, four key pieces of legislation became law in quick succession:

The Solar Heating and Cooling Demonstration Act (PL 93-409), which authorized $60 million for solar demonstration projects through 1979;

The Energy Reorganization Act (PL 93-438), which created the Energy Research and Development Administration to take over all energy sources not delegated to the Nuclear Regulatory Agency; and

The Solar Energy Research, Development and Demonstration Act (PL 93-473), which called for a solar resources inventory, the demonstration of eight solar technologies, establishment of a Solar Energy Information Data Bank (SEIDB), and a Solar Energy Research Institute;

The Federal Non-Nuclear Energy Research and Development Act (PL 93-577), which strongly recommended spending in the range of $20 billion on solar energy research during the next decade.

21.2.1 Implementation of Solar Programs

With passage of these bills (described in detail in chapter 2), solar energy R&D seemed to be off and running. Congress and the administration were also concerned with environmental problems, and what was termed the need for a level playing field for producers of solar energy equipment. One result of this concern was passage of PL 95-617, the Public Utilities Regulatory Policies Act (PURPA). PURPA mandated that electric power utilities buy power at avoided costs from small producers (up to 30 megawatts) of wind, water, and direct solar power; this resulted in a number of small solar thermal power plants feeding electricity into utility grids or providing industrial process heat for a variety of applications (these solar programs are described in section 21.8).

NSF, NASA, and HUD were the first federal agencies charged with developing practical solar energy technologies with a series of solar heating and cooling demonstration projects. The Solar Heating and Cooling Demonstration Act of 1974 was a five-year program, originally budgeted at $60 million, to provide a major demonstration of solar heating and cooling technology in residential and commercial buildings. Well-intended but less well carried out, the program ultimately funded 287 commercial and 497 residential demonstration projects (with an additional 446 awards granted for designs only). These projects totaled 10,098 living units and 1,255 solar energy system demonstrations.

In 1975, under the Energy Reorganization Act of 1974 (PL 93-438), solar program responsibilities were transferred to the newly created Energy Research and Development Administration (ERDA). ERDA's responsibility was much broader than heating and cooling demonstrations and in addition to those categories included industrial solar thermal projects, wind energy, biomass fuels, ocean thermal energy, and photovoltaics. This expanded research and development effort to commercialize solar equipment cost an estimated $3.5 billion over a ten-year period.

21.3 Implementation of Effective Technology Transfer

One of ERDA's major responsibilities was the expeditious transfer of solar thermal technology from federal laboratories to the commercial solar market and to that end ERDA established the Solar Technology Transfer Program (STTP). Five federal laboratories were designated to serve as regional field agents for STTP: Brookhaven National Laboratory (northeastern states), Oak Ridge National Laboratory (southeastern states), Sandia National Laboratory (southwest and Rocky Mountain states), Battelle Pacific Northwest Laboratory (northwest/upper midwest states), and Lawrence Livermore Laboratory (western states).

This was the beginning of a successful technology transfer program, passed on to DOE in 1977. In 1980 the Technology Innovation Act (PL 96-480) mandated that federal laboratories carry out aggressive technology transfer programs. To implement this responsibility, each laboratory was to set up and operate an Office of Research and Technology Applications (ORTA) to cooperate with the private sector in making new technologies developed by the laboratory widely available. PL 96-480 also mandated that "each Federal agency which operates one or more Federal laboratories shall make available not less than 0.5% of the agency's research and development budget to support technology transfer...." However, the law allowed the agency head to waive this requirement if it could be established that alternative plans for technology transfer met the goal of the legislation. DOE chose such a waiver and did not enforce the 0.5% requirement on its laboratories or programs.

Another important development was the passage in 1986 of the Federal Technology Transfer Act (PL 99-502), which formalized the Federal Laboratory Consortium for Technology Transfer (FLC), and included provisions for cooperative research and development between the federal government and the private sector. In 1988 the FLC had more than 300 member laboratories representing more than 85% of the laboratories in ten federal agencies. The bill mandated that each agency set aside 0.005% of its total R&D funds yearly to finance the FLC, which thus received about $500,000 annually for operational expenses. Mandated funding was increased to 0.008% in 1988 and continued through FY 1992, after which the agencies were expected to continue to subscribe the amounts voluntarily in support of the FLC.

Technology transfer was further encouraged in "Small Business Incubators." By 1988 several such innovation centers had been established, teamed with nearby federal laboratories. These included the Tennessee Innovation Center adjacent to Oak Ridge National Laboratory (ORNL), the Los Alamos Small Business Center at Los Alamos National Laboratory

(LANL), the Business and Innovation Center of Jefferson County, Colorado, adjoining the National Renewable Energy Laboratory (NREL; formerly the Solar Energy Research Institute) in Golden, Colorado, and a similar facility near the Idaho National Engineering Laboratory (INEL). Although none of these innovation centers receive direct federal funding, all have in kind support from the local laboratory, with DOE sharing facility costs for the Idaho center.

In 1988, Lawrence Berkeley Laboratory (LBL) gave special awards for "Excellence in Technology Transfer" to five of its scientists for the development of the energy-saving *Solar Window 2* computer program and its subsequent transfer to the solar industry. The LBL scientists provided about 100 major window manufacturers with the program formated for microcomputers, and arranged for its distribution to another 500 smaller firms.

In addition to its large and continuing program of interactive technical conferences, seminars, workshops, and demonstration project, the federal government has initiated a number of "research associate" programs aimed at the effective transfer of renewable energy technologies. These include the programs of the National Bureau of Standards (NBS) and NREL, as well as other federal laboratories. Not only has there been spin-off of technologies and products but of personnel as well. Just as federal scientists and engineers from nuclear and aviation/space research often entered the private sector to commercially develop their inventions, many DOE and federal laboratory scientists and engineers are now directly involved in the solar energy industry. As an example, more than a dozen solar-related companies have been started by former NREL employees (SERI 1988). By giving more visibility to technology transfer at the federal level, federal legislation (and the programs resulting from it) has increased the flow of government-developed technology into the private sector and thus yield a greater return to taxpayers for their investment in federal laboratories.

21.4 Special Federal Programs with Scientific/Trade Organizations

In addition to these formal solar thermal technology transfer programs, the federal government also worked in close coordination with many existing organizations in the solar energy and related fields. These include

- American Solar Energy Society (ASES)
- Solar Energy Industries Association (SEIA)
- National Association of Home Builders (NAHB)
- Passive Solar Industries Council (PSIC)
- Certification and testing groups, such as the trade associations Solar Rating and Certification Corporation (SRCC) and Desert Sunshine Exposure Testing (DSET), plus many state testing and certification laboratories
- American Federation of Labor/Congress of Industrial Organizations (AFL/CIO) Solar Program
- Sheet Metal Workers International Association (SMWIA)
- Sheet Metal and Air-Conditioning National Association (SMACNA)
- Real estate and lending organizations

The American Solar Energy Society had been in existence for two decades when DOE became active in the solar energy field; thus it was natural for interactive federal solar programs to involve ASES. Solar equipment trade exhibits were organized and state, regional, national, and international scientific conferences were held. Similar involvement with SEIA, NAHB, and the PSIC accomplished important solar thermal technology transfer between the DOE laboratories and industry through scientific conferences, the exchange of technical data, and the testing and improvement of solar devices. For example, PSIC in cooperation with local NAHB chapters assisted in evaluating and promulgating solar design guidelines developed at Los Alamos National Laboratory and at NREL.

Certification of solar equipment is vital for product reliability and efficient performance, and DOE worked closely with the established testing agencies such as DSET and SRCC and with numerous state certification agencies. In most cases, state tax credits for solar equipment were allowed only for devices meeting performance and other standards prescribed by these agencies.

To reach installers of solar energy equipment and ensure that they were using the latest and best procedures, DOE participated in the AFL/CIO solar program and also those of SMWIA and SMACNA. These efforts contributed appreciably to proper installation of well-designed and maintained equipment.

21.5   National Renewable Energy Laboratory

The Solar Energy Research Development Act of 1974 (PL 93-473) established the Solar Energy Research Institute (which became the National Renewable Energy Laboratory in 1991). This facility is DOE's single-mission laboratory dedicated entirely to the development of renewable energy technologies.

In the mid 1990s NREL staff was approximately 700, including about 50 visiting researchers and graduate students. Two-thirds of NREL personnel are research professionals involved in the Solar Electric Conversion, Solar Fuels, and Solar Heat Divisions; 45% hold doctorates in the sciences. These three divisions support ten DOE program areas:

Solar thermal energy

Solar energy storage

Solar buildings

Building energy R&D

Ocean energy

Photovoltaics

Wind

Biofuels

Resource assessment

Solar technical information.

NREL researchers participate in about 125 collaborative research studies each year, involving 45 to 50 industrial organizations and 60 to 65 universities, government laboratories, and nonprofit institutions.

As of 1992, NREL researchers have won twelve prestigious *Research and Development* magazine IR-100 awards, given annually for the best 100 industrial inventions. One of the first such awards was given to the NREL Biomass Gasifier in 1982 and patent rights were subsequently transferred to the private SynGas Corporation, which produced a commercial version of the gasifier. Other IR-100 awards received by NREL have been in the fields of alternative fuels, industrial technologies, photovoltaics, solar buildings, and wind energy. Between 1977 and 1992, NREL had been issued 79 patents, with 26 more pending. Recent patents include wood-to-ethanol conversion technology and vacuum and electro-chromic windows.

An active member of the Federal Laboratories Consortium for Technology Transfer (FLC), NREL encourages the formation of new businesses to market its solar technology. One of these firms is Industrial Solar Technology, which manufactures and installs large distributed receiver solar thermal hot water systems. NREL has also assisted existing private industries in producing cost-effective materials and components for solar applications. An example is the stretched-membrane heliostat mirror for solar thermal power plants manufactured by Solar Kinetics, Inc., and Science Applications International Corporation (SAIC). Other NREL solar technologies introduced to industry include heat exchanger designs, for use in solar ponds and the ocean, that are potentially useful in many industrial applications where low-grade heat is presently being wasted.

21.6 Invention and Patent Policy

The information in this section comes from a review by Brown, Franchuk, and Wilson (1991). Early in the federal solar energy program lawmakers became aware of considerable frustration on the part of solar inventors and entrepreneurs who were unable to participate in what they considered their own program. At that time, ERDA's Solar Division received about 5,000 requests for evaluations of solar inventions; the National Bureau of Standards (NBS) Energy-Related Inventions Program received more than 600 similar requests. Because of inadequate staffing timely action was not taken on many such requests.

The Federal Non-Nuclear Energy Research and Development Act (PL 93-577) had established a comprehensive national program for research and development of all potentially beneficial energy sources and utilization technologies. In October 1977 DOE assumed the responsibilities and programs of ERDA. One of the new department's early acts was to direct that the NBS evaluate all promising energy-related inventions other than nuclear; DOE would then fund further development of those found most promising.

The resulting Office of Energy-Related Inventions assisted small inventors in developing their ideas. Upon evaluation as worthwhile, inventions are referred to DOE, which determines if they should be supported for further development. In addition to monetary grants, DOE assistance at times includes testing of the invention, marketing assistance, invitation to the inventor to participate in competitive solicitations, or direct negotiation of a contract with the inventor. DOE also normally waives its title to the patent and licensing in favor of the small inventor. Between 1974 and 1982 NBS received more than 18,000 such inventions and completed evaluations of about half of them; DOE supported 165 of these, with average grants in the $70,000 range.

Between 1980 and 1990, 486 inventions were recommended to DOE by the National Institute for Standards and Technology (NIST), which screened all submitted inventions in terms of technical merit, potential for commercial success, and potential energy impact. By the end of 1990 at least 109 of these inventions had entered the market, generating total cumulative sales of more than $500 million. With $25.7 million in grants awarded from 1975 through 1990, and $63.1 million in program appropriations over the same period, DOE's Energy-Related Inventions Program (ERIP) generated a 20 to 1 return in terms of sales values to grants, and an 8 to 1 return in sales versus program appropriations. An estimated 25% of all ERIP inventions had achieved sale by the end of 1990.

Although these programs were of benefit in making use of the solar patents of private inventors, it was evident that problems still remained in the transfer of solar technology from federal laboratories to the infant solar industry. This concern increased with a finding by Senator Harrison Schmitt of New Mexico (a leading solar energy state) that of 30,000 federal patents in all fields, only about 5% had been commercialized. This lack of patent use resulted from several factors, not the least of which was that researchers in federal laboratories frequently were not primarily concerned with end uses for their inventions, traditionally leaving it up to market pull to discover applicable inventions and develop them.

Another major barrier to technology transfer resulted from the understandably stringent patent and licensing practices of the DOE, which always held title to research in its laboratories and the patents resulting therefrom. Although generally willing to grant nonexclusive licenses to its patents, the federal government has been unwilling until recently to grant exclusivity to work paid for by the taxpayers in general. Few large industrial firms are willing to risk large amounts of money in development costs for an invention or process without holding an exclusive license for its commercial use, and few are willing to share developmental costs of an invention when the public is already paying for such work.

The result of the patent and licensing situation was a stalemate, with only a limited number of government research projects ever seeing the light of day in the commercial marketplace. As this situation became obvious, Senator Schmitt and other legislators worked to speed movement of important technologies from federal laboratories to industry; the 96th and 97th Congresses passed several important laws germane to the topic of technology transfer.

In 1980 the Small Business Patent Procedure Act (PL 96-517) granted universities and small businesses the right to patents resulting from federally sponsored research. The Trademark Clarification Act (PL 98-620) in 1986 extended to nonprofit, government-owned, contractor-operated federal laboratories the rights granted under PL 96-517; most of the thirty-nine laboratories chartered by DOE may now retain invention rights and license them to private industry, retain a portion of royalty income to support further R&D, and make invention awards to staff members. This is of significance to industry because several hundred new inventions will be available for license each year.

In 1987 President Reagan signed Executive Order 12591 relating to technology transfer. Citing the provisions of the Federal Technology Transfer Act of 1986 (PL 99-502), the Trademark Clarification Act of 1984 (PL 98-620), and the Small Business Patent Procedure Act of 1980 (PL 96-517), this order called for facilitating access by university and industry researchers to technology developed in the national laboratories. The president appended the following personal statement to the order:

I believe a vigorous science and technology enterprise involving the private sector is essential to our economic and national security as we approach the 21st century. Accordingly, I have today issued an Executive Order "Facilitating Access to Science and Technology."

It is important not only to ensure that we maintain American preeminence in generating new knowledge and know-how in advanced technologies but also that we encourage the swiftest possible transfer of federally developed science and technology to the private sector. All of the provisions of this Executive Order are designed to keep the United States on the leading edge of international competition. (President 1987)

21.7   Lessons Learned

The federal solar energy program to date has been an expensive but worthwhile learning process. A most important lesson was that it is very difficult to transfer a new technology in a very short time, particularly with the number of management changes and policies and the different political climates under which NSF, NASA, HUD, ERDA, and DOE operated.

The time required for a new technology to be diffused into the economy has been estimated from as short as five years, under the best of circumstances and with a small impact, to fifty years or longer for a technology with a market potential as large as that for solar energy. Even the lower limit suggests that the government had insufficient time to implement so comprehensive a solar thermal program as seemed necessary in the 1970s.

The resulting crisis-oriented nature of the solar energy program worked strongly against success in the short time frame. For example, instead of the budgeted $60 million, the Solar Heating and Cooling Demonstration Act cost about $500 million through 1981. The overrun resulted from overly optimistic estimates of solar energy system costs from $4 to $8 per square foot of collector, when actual costs reached $77 per square foot in the first cycle of demonstrations and averaged $38 per square foot even in later, hot-water-only demonstrations (Frankel 1984). In the NSF and NASA programs, overruns resulted in part from government overspecification and unnecessary technical complexity. However, even in the HUD program, where grants were given directly to contractors to purchase solar components available from manufacturers, system costs ranged from $25 to $40 per square foot.

Additional problems included excessive time required to complete solar projects, faulty installations, and poor solar equipment performance. For example, a GAO investigation in 1979 found that of 91 HUD-financed residential solar energy systems, 46 were either not working or were experiencing operational problems. Summing up, solar cooling was uneconomical because of high cost and low efficiency; active solar space heating was marginally economical, unreliable, and inefficient; and only solar domestic hot water and passive solar space heating systems were cost-effective and provided high performance.

A criticism of some programs was that they used the gold-plated approach of large military contractors, leading to much higher developmental and production costs than might have resulted from using solar manufacturers and giving them more autonomy. Unfortunately, a mature solar equipment technology did not then exist and there was no course other than to contract with large firms. Furthermore, multimegawatt central receiver solar thermal power plants and ocean thermal power plants could be managed only by large, experienced contracting firms.

Another important lesson learned was that the technology to be transferred must be compatible with the existing infrastructure, in this case the established energy utilities, fuel delivery systems, and so on. The crash attempt to commercialize active solar heating and cooling was therefore only a limited success. Likewise, programs for large-scale solar thermal industrial process heating were not cost-effective—again because of failure to consider the well-established existing infrastructure of conventional fuels and because fuel shortages and prices did not increase as predicted.

Some observers feel that the technology of simpler devices such as solar water heaters and passive solar space heating was more effectively transferred to the market by small firms. Such a decentralized approach produced a variety of different designs and made use of those technologies most effective for particular applications. Linked with tax credits at the federal and state levels, this approach apparently provided a strong market pull in the technology transfer of solar water heating and similar applications.

21.8 Conclusions

While not as costly as the nuclear research and development program, federal and state solar energy programs have spent billions on research, development, and demonstration programs. However, solar thermal technology as yet makes only a small contribution to U.S. energy needs.

Nevertheless, an estimated one million U.S. homes use solar water heaters; many others are passively heated with solar energy systems that operate quietly and efficiently. At the outset of U.S. solar programs in 1973, no solar industry existed as such. By 1980, 133 firms were manufacturing solar collectors, 161 were installing them, 195 were doing system design and consulting, and 239 wholesale and 182 retail firms were distributing solar products (Frankel 1984). National business was reportedly in the $1 billion range. Solar equipment testing and certification facilities aided by government programs worked to guarantee efficient and reliable hardware. And although the ranks of solar businesses have thinned greatly since then (in large part because of the absence of tax credits), building them back to strength should be easier next time for the lessons learned in the process.

At the public and private utility levels, the high-temperature solar thermal demonstration program has been encouraging. A consortium of utilities is working with DOE in mid-1992 to modify and restart the 10-megawatt demonstration solar thermal power plant near Barstow, California, as the prototype for a new family of 100-megawatt "power towers." Nearby, privately built solar thermal power plants generate several hundred megawatts for the Southern California Edison grid, using simpler, single-axis trough concentrators (NREL 1992).

The federal solar thermal energy program has yet to achieve its goal of appreciably supplementing conventional energy sources and we are still far short of President Carter's admittedly optimistic goal of meeting 20% of our energy needs with solar power. But the lessons learned in the government's program will be of great benefit when we begin to make major use of solar energy.

References

Brown, M. A., C. A. Franchuk, and C. R. Wilson. 1991. *The Energy-Related Inventions Program: A Decade of Commercial Progress*. ORNL/CON-339. Oak Ridge, TN: Oak Ridge National Laboratory, December.

CRS (Congressional Research Service). 1988. *Commercialization of Federally Funded R&D: A Guide to Technology Transfer from Federal Laboratories*. Report prepared for the Subcommittee on Science, Space, and Technology. U.S. House of Representatives. 100th Cong., 2d sess. Washington, DC, September.

Frankel, E. 1984. *Technology, Politics, and Ideology: The Vicissitudes of Federal Solar Energy Policy, 1973-1983*. Report prepared for the Committee on Science and Technology. U.S. House of Representatives. 98th Cong., 2d sess. Washington, DC, January.

Halacy, D. S., Jr. 1963. *The Coming Age of Solar Energy*. New York: Harper & Row.

Halacy, D. S., Jr. 1976. "The 94th Congress and Solar Energy." *Solar Engineering* November: pp. 7, 8.

NREL (National Renewable Energy Laboratory). 1992. *Solar Thermal Electric and Biomass Power*. DOE/CH 10093-148. Washington, DC: U.S. Department of Energy, June.

President. 1987. Executive Order 12591. "Facilitating Access to Science and Technology." The White House, Washington, DC, 10 April.

SERI (Solar Energy Research Institute). 1988. *R&D Technology Transfer: An Overview*. SERI/MRI-320-3279. Golden, CO, January.

22 Solar Energy Research Institute

Barry L. Butler

22.1 SERI Materials Technology Transfer

This chapter will review several examples of work conducted at, or under contract to, the Solar Energy Research Institute (SERI) in its first decade, 1977–1987. Much of this work focused on quantifying how solar heat collector systems capture solar energy and match it to end uses. The goal was to collect solar energy less expensively and to use it more effectively to meet end-use applications—in other words, to reduce the cost of solar energy while increasing the value of the energy or process output. Starting in 1980, SERI experienced reductions in staff as the Reagan administration de-emphasized solar technologies. Over 150 of the 220 staff members who worked in SERI's Solar Heat Technology Division between 1978 and 1986 are now in industry, government, and universities, with perhaps 10% utilizing their experience directly or indirectly to aid the development and acceptance of solar heat technologies. SERI is now the National Renewable Energy Laboratory.

The federal government's solar thermal program began in 1972 at Sandia National Laboratories in Albuquerque (SNLA), and much of the planning for SERI's solar thermal technology work was done over the course of the next five years at Sandia and other national labs (Butler and Claassen 1980). Once it was established, SERI took the initiative to define a national solar materials agenda. With the support of the Department of Energy (DOE), SERI chartered a working group, the National Solar Optical Materials Planning Committee (SOMPC) consisting of solar scientists from SERI (Barry L. Butler, Patrick J. Call, Keith D. Masterson), SNL (John Vitko, Richard B. Pettit), Los Alamos National Laboratory (Stan W. Moore), the Jet Propulsion Laboratory (William F. Carroll), Battelle Pacific Northwest Laboratory (Michael A. Lind), and the National Bureau of Standards (Joseph C. Richmond). The SOMPC documented the state-of-the-art and materials needs in the areas of solar reflectors, solar transmitters, and solar absorbers (Butler 1979; Call 1979b; Masterson 1979; Call, Jorgensen, and Pitts 1981; Carroll and Schissel 1980).

Cooperation between SERI and Sandia was limited, and in many cases SERI was funded to complete or augment work Sandia had identified as crucial. In the areas of solar thermal materials and membrane collectors, SERI built upon Sandia-developed concepts and research. SERI enhanced and moved membrane heliostat collector technology forward, while Sandia Research and Development was limited to evaluating existing solar thermal central receiver and trough systems. This unnatural division needed to be forced by DOE because Sandia and SERI both had charters to support the development of solar thermal systems. Sandia already had a strong relationship with industry, whereas in the beginning SERI was either unknown or distrusted by industry.

22.1.1   SERI Solar Heat Research Directions

The solar collector manufacturing industry benefited significantly from work done between 1978 and 1986 at SERI and other national laboratories in support of the SOMPC plans. The industry received data on system performance and the optical properties of materials. SERI's research, in concert with the other national laboratories, helped the suppliers of critical optical materials, such as silvered mirrors and high-temperature selective absorbers, to improve and evaluate their products for use in solar energy systems; such advanced optical materials are the basis for today's lower-cost, higher-quality solar collectors.

The solar heat collector industry, including makers of flat-plate and concentrating solar collectors, benefited from improved materials and improved system design data. Improved materials included high solar transmittance glasses; high-reflectance and longer-life silvered glass mirrors; silvered polymer mirrors; and high solar absorptance, low thermal emittance selective absorber coatings (Call 1979a; Butler and Livingston 1980). Materials suppliers such as Corning Glass Co., Flacht Glass (a German company), the Environmental Control Products Division of 3M, and Telec Corporation developed products for use by solar collector manufacturers. SERI also produced collector design handbooks to aid solar collector manufacturers in developing high-efficiency solar energy delivery systems, including collectors, pumps, valves, sensors, and controllers (Farrington 1984; SERI 1988). Many of these improvements in materials and system design are evident in solar energy systems that have survived low oil prices and the post–solar tax credit economy. During this same time period, Sandia's results from the Modular Industrial Solar Retrofit (MISR) and the Central Receiver Test Facility

(CRTF) at Albuquerque, New Mexico, and Solar One at Barstow, California, were also providing valuable data to industry, end users, and utilities.

22.1.2   SERI/Industry Technology Transfer

The diffusion of the technological advances achieved by SERI using public funds occurred via two major mechanisms: first, through conferences, reports, and handbooks; and second, through the migration of SERI's research staff to universities, government, businesses, and industries. The migration was a result of SERI's massive layoffs of nearly 50% of its technical staff. The solar industry itself was devastated by the low oil prices ($12–18 per barrel) of the 1986–1990 period and by the abandonment of both federal and state tax credits to stimulate solar use. The industry decreased from over 200 to less than 30 companies between 1981 and 1986. The solar industry capitalized on the SERI staff reductions to the extent it could, and many of the remaining solar industries have ex-SERI staff in key positions. The scientists, engineers, and analysts who came from SERI were committed to a clean solar energy future and privileged to have been brought together to develop the ideas, concepts, technologies, designs, and plans needed to make it happen. A book entitled *A New Prosperity* (Irwin 1981) was based on the DOE-sponsored Domestic Policy Review report, to which SERI Solar Heat Division staff had contributed many sections.

Jointly funded research contracts will become the major technology transfer mechanism for SERI in the future. As research results become useful to industry, jointly funded and conducted research programs will give industry access to the SERI staff and SERI's state-of-the-art capital equipment needed to accelerate technology advances. This method of technology transfer needs the close involvement of management and strong incentives to make it work effectively. Training and then laying off people has proven to be an extremely effective method of technology transfer, albeit very disruptive to careers and lives.

The following sections highlight different stages of technology diffusion processes and are snapshots of the continuing evolution of solar heat technology. Examples are arranged to illustrate the challenge, the SERI pioneering research, the methods of technology transfer, and the resulting commercial products, processes, or designs.

22.2 Thermal Collector Systems Research

22.2.1 Solar Reflector Materials

Silver bathroom and "looking glass" mirrors for inside use are unstable in outdoor environments. Bathroom mirrors become corroded and lose their reflecting silver layer after a few months outside. Specially prepared weatherable bathroom mirrors last only a few years. Although aluminum and chromium are much more stable than silver in outdoor environments and tarnish more slowly, the solar specular reflectance of silver (92%–94%) is significantly higher than that of aluminum (84%–87%) or chromium (about 54%). A solar energy system using a silver reflector versus an aluminum reflector would thus collect 8% more energy per year or, alternatively, could be about 8% smaller and provide the same annual energy. It is therefore desirable to create weatherable silver reflecting layers that can be applied to transparent glass and plastic sheets or plastic films to form solar mirrors.

The researchers' task was to find out why the silver corroded and to create ways to prevent it from happening (Schissel and Czanderna 1987). The goal was to develop silvered glass and silvered polymer mirrors with thirty-year life expectancies in outdoor environments. This problem required a team of solid-state physicists, electrochemists, and surface scientists using the most advanced high-vacuum surface probes, electron microscopy, and analytical techniques. The equipment dedicated to this project at SERI and Sandia cost in excess of $1.5 million, an investment much greater than fledgling solar companies could afford for mirror research. SERI's research team led the national effort to understand silver mirror corrosion.

In addition to the SOMPC effort, several major companies joined the research team, including Corning Glass Co., the Environmental Control Products Division of 3M, Falconer Glass Industries, Pittsburgh Plate Glass, Donnolly Mirror, Gardener Mirror, and Carolina Mirror. Each of these companies saw the potential sale of reflectors to the growing solar energy collector market as an extension of their existing product lines. All but 3M were makers of "bathroom" or automobile mirrors. Reflectors for the fluorescent and high-intensity lighting industry included polished aluminum reflectors made by Alcoa (Alzak) and Kingston Industries (Kinglux) and aluminized acrylic films made by 3M (Scotch Cal 5400).

These glass, plastic, and aluminum reflector materials were the only mirrors available for solar energy systems when the outdoor weatherable silver reflector research began in 1979.

The research conducted by SERI and others concluded that small amounts of chloride ions greatly accelerated silver corrosion (Schissel, Neidlinger, and Czanderna 1987). The chloride ions were either left on the glass surface by tin chloride sensitization, on the polymer surface from the polymer preparation, or deposited from atmospheric pollutants. The exact mechanism is still somewhat open to question, but simplistically, a chlorine molecule reacts with the metallic silver to form the compound silver chloride (nonreflecting); a water molecule then reacts with silver chloride to form silver oxide (also nonreflecting), and the chloride ion joins the hydrogen and moves off to find another silver atom. Keeping both water and chloride ions away from the silver is critical.

Technical details of the silver corrosion process are described by Schissel, Neidlinger, and Czanderna (1987). Industry has responded to the SERI/SOMPC results reported in conferences and technical papers in several ways. The "bathroom" mirror industry relies on a wet process that plates silver onto clean and sensitized glass from a complex silver solution; tin chloride is the preferred sensitizer, because it is effective and inexpensive. New sensitizers have been developed, however, that provide uniform coating thickness but do not use chloride ions. Mirrors made by these advanced processes are more resistant to outdoor environmental corrosion. Some companies now laminate standard wet process silver (put down with tin chloride sensitizer) between two pieces of glass or between glass and a metal sheet. The chloride is still there, but no moisture can get to the reflecting layer to activate the degradation process (Czanderna, Masterson, and Thomas 1985).

For silver coatings on transparent polymers, three major modifications need to be made. First, photocatalyzed oxidation of silver needs to be inhibited. Second, migration of chloride, oxygen, and hydrogen ions must be minimized in the transparent polymer layer adjacent to the silver. Third, chloride ions in the transparent polymer must be minimized. The silver is applied to the polymer using a vacuum evaporation or sputtering technique; thus no sensitizers are used. Cleaning and activation of the polymer surface may be accomplished prior to silver deposition.

The 3M Company embarked on a joint research venture with SERI in 1985 to produce and evaluate advanced, durable silvered acrylic films

(Benson 1987). 3M directed their efforts toward process development, while SERI conducted research on the durability of 3M-produced test films. SERI researchers used an Atlas Corporation Weather-O-Meter™, which subjected reflector film samples to intense ultraviolet radiation, dry heat, and moisture alternately. Outdoor exposure testing was also conducted by SERI and Desert Sunshine Exposure Testing Inc. of Black Canyon Stage, Arizona. The 3M-Factory Experimental Kit 300 (FEK-300), which was a UV-stabilized acrylic film silvered and then backed by proprietary polymer layers, was tested by SERI. Energy Control Products 300 (ECP-300), the commercial product based on FEK-300, had improved weatherability; further research resulted in a more durable version that inhibited moisture diffusion. The more durable silvered polymer reflector, with expected solar reflectance loss of less than 5% over seven years, was introduced in 1990 as 3M-ECP-305 for the solar collector market. However, this film is susceptible to delamination at the silver-polymer interface under conditions of standing water or high humidity. In 1994 ECP-305 + was introduced with a thin copper layer behind the silver to add galvanic silver protection. Current research is aimed at mitigating this problem and increasing the acrylic surface abrasion resistance, so that wet wash contact cleaning techniques will not scratch the surface (Benson 1989).

Some glass mirror companies have taken the SERI data and unilaterally altered their processes and techniques in proprietary ways to improve mirror life. Glass-silver-glass mirrors that exhibit about 93% solar specular reflectance with less than 5% solar reflectance loss over 25 + years are now available from Flacht Glass. Other mirror manufacturers may also be using the more stable silver coating in their "bathroom" and automobile mirror products.

22.2.2 Solar Absorber Materials

Solar selective absorber materials absorb all wavelengths of the solar spectrum and convert them into heat, usually at greater than 95% conversion efficiency. They have a low thermal emittance, usually less than 10–35% at the receiver operating temperature. Thus maximum solar heat is absorbed and minimum heat reradiated, resulting in more useful energy collected. Common black paint has high solar absorptance (80%–90%) and high thermal emittance (95%). A parabolic trough or flat-plate solar heat system that uses a selective absorber coating on its receiver tube or plate can achieve 10%–20% efficiency improvement over nonselective black paint coatings. Selective coatings are produced by putting a thin solar absorbing coating over a low-emittance metal surface. If the absorbing coating is too thick, its emittance is raised; if it is too thin, the solar absorptance is reduced. Hence precise coating thicknesses are required. Electroplated coatings of zinc sulfide on nickel, developed by Harry Tabor of Israel, have been widely used in flat-plate collectors (Solar Energy Society 1958), but they proved to be thermally and environmentally unstable above 250°F (120°C). Electroplated black chrome coatings over nickel were widely researched by SERI and Sandia and have proven stable up to 570°F (300°C), which is acceptable for flat-plate collectors but not for high-temperature parabolic troughs and dishes.

The SOMPC identified a need to develop high-temperature selective coatings. Two approaches were taken: improvement of black chrome coatings and identification and development of new, advanced coatings. Black chrome coatings were spalling (flaking) off the mild steel receiver tubes used in parabolic trough collectors; this was a result of the combined effects of thermal cycling and atmospheric oxidation of the mild steel. These black chrome receiver tubes were made by cleaning the mild steel and acid-etching the surface, then electroplating nickel over the steel to provide a low-emittance surface, and finally electroplating a mixed chrome–chrome oxide layer to absorb the solar radiation. The mixed chrome–chrome oxide layer had to be a precise thickness, thick enough to absorb most of the solar radiation but not overshadowing the low emittance of the nickel below. SERI's research was aimed at altering the interface between the mild steel and nickel coating to prevent corrosion.

Cermet coatings, which are combined metal and ceramic materials, were postulated to be manufacturable as possible selective absorbers. The cermet selective absorber approach yielded both improved mild steel substrate corrosion resistance and higher-temperature operation. SERI pursued several cermet coatings based on vacuum deposition and sputtering. This in-house and subcontracted research led to the identification of cosputtered platinum-aluminum and aluminum oxide cermet coatings. Radio frequency sputtering was the only practical way to electrically deposit insulating materials such as aluminum oxide. The small amount of fine platinum dispersed in the aluminum–aluminum oxide cermet (3%) provides the solar absorptance, while aluminum provides the low emittance.

SERI researchers worked closely with coatings researchers at DOE's Rocky Flats Plant in Broomfield, Colorado, who made and evaluated cermet coatings during the early studies to determine optimum thickness and composition. SERI assembled equipment to measure solar absorptance and emittance of surfaces heated up to 1,470°F (800°C). As research progressed, research contracts were placed with Telec Corporation of Santa Monica, California, to develop the sputtering process by creating techniques to coat round tubes and flat surfaces. After three years, the Telec coating proved to be highly selective and thermally stable. SERI test data verified its high selectivity and long life at 930°F (500°C). DOE incorrectly declared the project a success too early and discontinued the SERI research and Telec's subcontract in 1985 because of solar budget cuts.

Industrial Solar Technologies Inc. (IST) troughs are used for process heat and operate up to 480°F (25°C). IST used carefully cleaned and etched mild steel tubes coated with nickel and electroplated black chrome at Tehachapi, California; these worked well for two years, but concerns about long-term durability still exist. The principals of IST are both ex-SERI engineers.

Luz Industries developed large fields of parabolic troughs for electric power generation (figure 22.1). As their engineers improve their concentrating reflector accuracy, they could reduce the receiver tube diameter and increase its temperature to 750°F (400°C) from 555°F (290°C), which would increase system performance by 10%.

In the words of an engineer from Luz, "The purpose of the cermet selective coating on the heat collection pipe is to increase absorption of solar radiation and help retain collected energy as it travels to the heat exchanger. The black chrome coating used in earlier systems was applied by a conventional coating production line, but this coating proved unstable at the higher operating temperatures needed for future generations of Luz systems" (Kearney 1992).

In 1987 Luz undertook an eighteen-month effort with private funds to develop cermet selective absorber receiver tubes for their parabolic troughs. They employed David W. Kearney, a key ex-SERI expert familiar with the cermet project. Luz introduced a new production line at its Jerusalem facilities that used a high-tech sputtering process to coat the heat collection pipes, a key element in the company's solar heat technology. The highly complex machine was designed over a period of eighteen

Figure 22.1
Luz Solar Electric Generating System (SEGS), Barstow, California Luz Installed 364 MWe
by 1992.

months by a team of Luz engineers in cooperation with Vac-Tec, a U.S. sputtering technology company. The continuous sputtering system was designed to achieve better coating properties than those produced by the SERI/Telec batch-sputtering equipment.

The sputtering machine is composed of a series of twelve chambers stretching over 98 feet (30 meters). The metal pipe or heat receiver tube is washed and automatically loaded into the entrance chamber of the machine, where it moves along a conveyor system through the process chamber. Inside this chamber, the pipes are coated with four successive layers.

The fully automated sputtering line is the only machine of its kind in the world, applying recent developments in microelectronics sputtering to much larger, rounded surfaces. The development of the machine and the sputtered coating technology has enabled Luz to proceed with the development of their third-generation solar collector, which raises system temperatures to 750°F (400°C). Luz has built solar power plants that, to date, can produce a total of 364 MWe annually. The new LS-3 and LS-4 collectors will all use advanced cermet selective absorbers.

22.2.3   Low-Cost Solar Concentrators

Reflecting optical concentrator elements (mirrors, supports, trackers) represent approximately half of a solar system's cost (Mavis 1987). System cost reductions required lower-cost troughs, heliostats, and dishes; this was recognized by SERI in 1980, and a low-cost concentrator committee was set up to develop new approaches to reduce concentrator costs for troughs, heliostats and dishes. The members were all from SERI but came from many different branches: Barry L. Butler, materials; Frank Kreith, heat transfer; R. K. Collier, solar cooling; Frances H. Arnold, systems; Randy Gee, mechanical design; Charles F. Kutcher, systems; Roger L. Davenport, low-cost flat plates; David Benson, optically switchable windows; Cecile M. Leboeuf, solar ponds; and Steven Sargent, solar ponds. The committee developed and brainstormed ideas. Denis Hayes, then director of SERI, provided both seed money and prototype concentrator funding. Funding the low-cost collector committee was one of Hayes's wisest decisions, based on later industrial use of this committee's output.

Four research projects generated by the low-cost collector committee have been commercialized by industry. These are the parabolic trough project, the membrane heliostat and dish project, the desiccant cooling project, and the solar pond project. The results of the first two projects have been used directly by industry for product development and are described below. The last two projects have helped industry in an indirect way and are not reported in detail here. The desiccant cooling product development work was used by American Solar King of Waco, Texas, to improve their developmental desiccant cooling systems, which are now commercial as gas-fired cooling systems. The U.S. Bureau of Reclamation used SERI's solar pond data to support the design and testing of their solar pond in El Paso, Texas.

The low-cost trough project included the design, fabrication, and testing of a low-cost trough concentrator system. The project, led by Randy Gee, resulted in the collector shown in figure 22.2. This collector featured a paper honeycomb reflector and a multiple row tracker drive system. The cable drive tracker worked well; the paper honeycomb also worked well but was considered too expensive. Paper honeycomb had low materials cost but high fabrication costs, and moisture degradation was possible. Shortly after the successful completion of the project in 1983, Randy Gee left SERI to help start Industrial Solar Technology, a solar trough manufacturing and installation company.

Figure 22.2
Low-cost parabolic trough system at SERI using a single motor to drive multiple rows.

I headed the low-cost membrane collector project team, which started by reviewing the research on low-cost collectors that had been done by both Boeing and McDonnell Douglas (Ginz and Gillett 1977; Steinmeyer 1977). They had used stressed polymer membranes inside plastic bubbles. SERI developed a patent for DOE on the use of structural membranes not needing a bubble for protection from wind and hail loads (Butler 1985).

The membrane collector project proceeded with the design and fabrication and testing of a 9.8-ft (3-m) diameter prototype. The frame and membrane assembly was a trampoline tensioned at 30 lb/in (5.36 kg/cm) along the perimeter. The stiff springs allowed only 1 in (2.54 cm) center deflection with a center load of 200 lb (90 kg) on a 1 ft² (0.0946 m²) area. One foot square (0.3 m × 0.3 m) mirror tiles were bonded to the polyethylene mesh trampoline using silicone rubber. The prototype is shown in figure 22.3. The high optical quality and low structure weight to reflective area ratio of 2 lb/ft² (9.60 kg/m²) versus 10 lb/ft² (47.00 kg/m²) for Solar One glass metal heliostats at Barstow, California, convinced both Gerald Braun, the DOE solar thermal program manager, and

Figure 22.3
SERI stressed membrane heliostat proof-of-concept unit. Trampoline membrane with
0.0946 m² (1 ft²) glass mirrors attached (1980).

Sandia Labs that membrane heliostats without plastic bubbles were feasible. In a DOE-funded effort, aluminum and stainless steel membrane
heliostats were analyzed, and optical properties of 3.3-ft (1-m) diameter
heliostat mirror modules were measured (Murphy, Simms, and Sallis
1986; Murphy and Sallis 1986).

In 1984 I left SERI to join Science Applications International Corporation (SAIC) of San Diego and Lawrence M. Murphy took charge of the
project. The SERI structural analysis and testing resulted in several vali-
dated membrane dish and heliostat membrane design tools. These were
used to compute membrane stresses and deformed shapes under differential pressure loading (Murphy 1984, 1985). In late 1984 Sandia awarded
two contracts, one to Solar Kinetics, Inc. (SKI), of Dallas and one to
SAIC for the design and fabrication of a 23-ft (7-m) membrane heliostat. SAIC used a 0.003 in (0.0762 mm) thick stainless steel membrane,
and SKI chose a 0.010 in (0.254 mm) thick aluminum membrane.
SERI aided both contractors with design assistance and testing of 9.8-ft
(3-m) diameter proof-of-concept membrane modules (figure 22.4). The
optical quality of the SAIC 23-ft (7-m) prototype heliostat is shown in
figure 22.5.

Figure 22.4
SERI test unit from SAIC. A 3-m diameter membrane heliostat prototype with 0.0762 mm
(0.003 in) thick 304 stainless steel membranes front and back.

SERI also funded a separate contract with SAIC for the development
of a 9.8-ft (3-m) diameter polymer membrane dish formed on a stretched
stainless steel membrane tool for optical evaluation by SERI (figure 22.6).
The composite membrane dish had poor optical accuracy because it did
not replicate the stretched-membrane tool. SKI successfully built 9.8-ft
and 23-ft (3-m and 7-m) diameter stretched membrane dish concentrators
with focal length to diameter ratios of 0.6 and have pursued commercial
concepts based on these units. SERI's results helped SAIC and SKI to
develop viable designs, which helped Sandia to manage successful membrane projects.

Low-cost trough technology has flourished, and IST has become a
viable trough supplier, using an upgraded drive system (figure 22.7) partially patterned after the original SERI system. IST has developed an all-
aluminum sheet metal trough with a welded aluminum frame and thin
tempered aluminum reflectors coated with 3M reflecting film. The company has sold several systems for heating hot water for prisons and has
achieved very low cost process heat troughs.

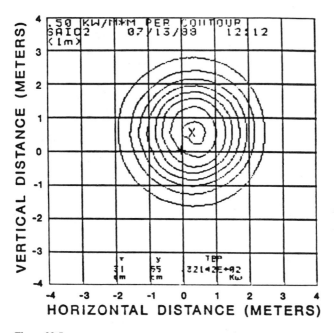

Figure 22.5
Measured flux contours for SAIC's second mirror module from 13 July 1988, at 12:12 MDT.
Direct normal isolation was 949 W/m². Winds were calm. Heliostat was 461.5 m (1500 ft)
from the target.

The heliostat program has continued through three design cycles. SAIC
and SKI have both produced and installed at Sandia first- and second-
generation membrane heliostats (Alpert, Mancini, Houser, Grossman
1990; Alpert, Houser, Hecker, and Erdman 1990; Alpert, Houser, Heckes,
Erdman, Beninga, and Konnerth 1990; SAIC 1987, 1989; SKI 1987,
1990; Alpert and Houser 1988, 1989; see figures 22.8 and 22.9.) SAIC
has patented an advanced low-cost focus control system, and SKI has
improved their fan-driven focus control system (Butler and Beninga
1989). Both second-generation heliostats have shown optical quality as
good as glass-metal heliostats, and costs are projected to be 50% lower
than those of glass-metal heliostats in mass production quantities. Central
receiver system designers, such as Bechtel Corporation, now consider
membrane heliostats as the most cost-effective choice for central receiver
plants that are under consideration, but no large purchase of membrane

Figure 22.6
SERI glass-reinforced 3-m diameter polyester composite membrane dish developed for proof-of-concept and optical evaluation.

Figure 22.7
IST trough system deployed at Adams County Jail.

Figure 22.8
SAIC 100-m^2 stainless steel membrane heliostat collector prototype (1990).

Figure 22.9
SKI 50 m² aluminum membrane commercial prototype.

heliostats has taken place. In the membrane dish area, a commercial prototype (23-ft or 7-m diameter) has been built by SKI and tested by Sandia. This prototype is shown in figure 22.10. Faceted dish collectors using around twelve membrane facets have been produced by SKI, SAIC, and Cummins/LaJet to power 25kWe Stirling engine systems.

22.3 Conclusions

SERI and Sandia, whose solar thermal technology development roles created a dynamic competition, have both played a significant part in

Figure 22.10
Test at Sandia of SKI 7-m diameter proof-of-concept prototype membrane dish. Pre-deformed stainless steel membranes are used.

the development of solar thermal technologies. With Sandia, DOE, and major industry involvement, the technologies championed by SERI have come to the marketplace. The wise allocation of research dollars by SERI has produced innovative and cost-effective materials and components needed to make solar thermal process heat and electricity applications practical in the 1990s.

Among the major lessons the experience of SERI has taught us are the following:

1. Industry-driven needs for research results must be present during program planning. (SOMPC plans included the industry.)

2. Technology transfer by layoff is extremely effective.

3. If future work at SERI is to be transferred to industry, a high incentive will be needed as well as cooperative SERI/industry planning. The Cooperative Research and Development Agreement partially satisfies this need.

4. SERI patents help staff achieve recognition, but exclusive licensing must be done to motivate industry. (Nonexclusive licenses do not protect industrial investment.)

5. Industry supports Sandia and SERI both. Competition has kept the research performed at SERI from being too esoteric for use by industry, while providing incentives to excel on a national level.

6. It is sound industrial policy to have industries and national laboratories work together to make the highest-quality, most reliable products in the world.

References

Alpert, D. J., and R. M. Houser. 1988. *Optical Performance of the First Prototype Stretched-Membrane Mirror Modules.* SAND88-2620. Albuquerque, NM: Sandia National Laboratories.

Alpert, D. J., and R. M. Houser. 1989. "Evaluation of the Optical Performance of a Prototype Stretched-Membrane Mirror Module for Solar Central Receivers." *Journal of Solar Energy Engineering* 111: 37.

Alpert, D. J., R. M. Houser, A. A. Heckes, and W. W. Erdman. 1990. *An Assessment of Second-Generation Stretched-Membrane Mirror Modules.* SAND90-0183. Albuquerque, NM: Sandia National Laboratories.

Alpert, D. J., R. M. Houser, A. A. Heckes, W. W. Erdman, K. Beninga, and J. Konnerth. 1990. *The Development of Stretched-Membrane Heliostats in the United States.* SAND90-0273. Albuquerque, NM: Sandia National Laboratories.

Alpert, D. J., T. R. Mancini, R. M. Houser, and J. W. Grossman. 1990. "Solar Concentrator Development in the United States." *In International Energy Agency, 5th Symposium on Solar-High Temperature Technologies.* Davos, Switzerland: International Energy Agency.

Benson, B. A. 1987. "Silver-Polymer Film Development." *Energy* vol. 12, No. 3–4, pp. 203–207.

Benson, B. A. 1989. *Industrial Support on Silver/Polymer R&D.* St. Paul 3M Company.

Butler, B. L. 1985. *Lightweight Diaphragm Mirror Module System for Solar Collectors.* Patent 4,511,215. Golden, CO: Solar Energy Research Institute/Department of Energy.

Butler, B. L., ed. 1979. *Solar Glass Mirror Program: A Planning Report on Near-Term Mirror Development Activities.* SERI/RR-31-145. Golden, CO: Solar Energy Research Institute.

Butler, B. L., and K. Beninga. 1989. *Focus Control System for Stretched-Membrane Mirror Module.* Patent 5,016,998. San Diego: Science Applications International Corporation.

Butler, B. L., and R. S. Claassen. 1980. "Survey of Solar Materials." *Journal of Solar Energy Engineering* 102: 175.

Butler B. L., and R. Livingston. 1980. *The Accelerated Commercialization Program for Materials and Components: Solar Sheet Glass*. SERI/TR-733-603. Golden, CO: Solar Energy Research Institute.

Call, P. J. 1979a. "Applications of Passive Thin Films." In L. L. Kazmerski, ed., *Properties of Polycrystalline and Amorphous Thin Films and Devices*. New York: Academic Press.

Call, P. J. 1979b. *National Program Plan for Absorbed Surfaces R&D*. SERI/TR-31-103. Golden, CO: Solar Energy Research Institute.

Call, P. J., G. L. Jorgensen, and J. R. Pitts. 1981. "The Effect of Receiver Optical Properties on Solar Thermal Electric Systems." (SERI/TP-641-796.) *Journal of Solar Energy Engineering* 103:3 pp. 207–212.

Carroll, W. F., and P. Schissel. 1980. *Polymers in Solar Technologies: An R&D Strategy*. SERI/TR-334-601. Golden, CO: Solar Energy Research Institute.

Czanderna, A., K. Masterson, and T. M. Thomas. 1985. *Silver/Glass Mirrors for Solar Thermal Applications*. SERI/SP-271-2293. Golden, CO: Solar Energy Research Institute.

Farrington, R. 1984. *Failure Testing of Active Solar Energy System Components*. SERI/TR-253-2187. Golden, CO: Solar Energy Research Institute.

Ginz, L., and R. Gillett. 1977. *Polymer Bubble Projected Membrane Heliostats: Final Report, Advanced Heliostat Program*. Seattle, Boeing Company.

Irwin, N., ed. 1981. *A New Prosperity: Building a Sustainable Energy Future. The SERI Solar/Conservation Study*. Anderson, MA: Brick House.

Kearney, D. W. 1992. *Tomorrow's Energy Today*. Luz Annual Report. Westwood, CA: Luz International Limited.

Masterson, K. D., ed. 1979. *Solar Optical Materials Planning Committee: Progress Report for Period July 1, 1978–January 31, 1979*. SERI/PR-31-137. Golden, CO: Solar Energy Research Institute.

Mavis, C. 1987. *A Description and Assessment of Heliostat Technology*. SAND-87-8025. Livermore, CA: Sandia National Laboratories.

Murphy, L. M. 1984. "Technical and Cost Potential for Lightweight Stretched–Membrane Heliostat Technology." *Solar Engineering—1984: Proceedings of the ASME Solar Energy Division Sixth Annual Conference*; Las Vegas, Nevada; 8–12 April 1984, Goswami, D. Y., ed. New Yok: The American Society of Mechanical Engineers; pp. 220–227. SERT/TP-253-2079.

Murphy, L. M. 1985. *A Variational Approach for Predicting the Load Deformation Response of a Double Stretched-Membrane Reflector Module*. SERI/TR-253-2626. Golden, CO: Solar Energy Research Institute.

Murphy, L. M., and D. V. Sallis. 1986. *Analytical Modeling and Structural Response of a Stretched-Membrane Reflective Module*. SERI/TR-253-2101. Golden, CO: Solar Energy Research Institute.

Murphy, L. M., D. Simms, and D. V. Sallis. 1986. *Structural Design Considerations for Stretched-Membrane Heliostat Reflector Modules with Stability and Initial Imperfection Considerations*. SERI/TR-253-2338. Golden, CO: Solar Energy Research Institute.

SAIC (Science Applications Iternational Corporation). 1987. *Development of the Stressed-Membrane Heliostat Mirror Module: Final Report*. SAND87-8179. Livermore, CA: Sandia National Laboratories.

SAIC. 1989. *An Improved Design for Stretched-Membrane Heliostats*. SAND89-7027. Albuquerque, NM: Sandia National Laboratories.

Schissel, P., and A. W. Czanderna. 1987. *Stability of Reflectors with Polymer Coatings.* SERI/TP-255-1497. Golden, CO: Solar Energy Research Institute.

Schissel, P., H. H. Neidlinger, and A. W. Czanderna. 1987. "Silvered Polymer Reflectors." *Energy* 12:197.

SERI (Solar Energy Research Institute). 1988. *Engineering Principles and Concepts for Active Solar Systems.* New York, NY: Hemisphere.

SKI (Solar Kinetics, Inc.). 1987. *Development of the Stressed-Membrane Heliostat.* SAND87-8180. Livermore, CA: Sandia National Laboratories.

SKI. 1990. *Design and Demonstration of an Improved Stretched-Membrane Heliostat.* SAND89-7028 Albuquerque, NM: Sandia National Laboratories.

Solar Energy Society. 1958. *Conference on the Use of Solar Energy, 1955.* Tucson, AZ.

Steinmeyer, D. 1977. *Inflatable Polymer Dome for Heliostat Wind Protection: Final Report, Advanced Heliostat Program.* Huntington Beach, CA: McDonnell-Douglas Astronautics Company-West.

23 Los Alamos National Laboratory

J. Douglas Balcomb and W. Henry Lambright

23.1 Introduction

Passive utilization of solar energy is achieved by building components that admit, transport, and store the energy within the building itself with little or no mechanical intervention. By 1984 there were more than 200,000 residential and 15,000 commercial passive solar buildings in the United States. That number has increased greatly and continues to increase. Thus passive solar is a proven and accepted technology. (More information about the techniques for passive use of solar energy can be found in volumes 7, 8, and 9 of this series.)

The building industry is diverse and decentralized and includes many different people. For this reason, one strategy for commercializing passive solar energy was to develop and disseminate readily usable design methods. The Los Alamos National Laboratory (LANL) experience in doing so is reviewed in this chapter as an example of the transfer of design technology from a national laboratory into mainstream practice. (Chapter 5 in this volume deals with commercialization within the overall federal passive solar program.)

Much of this chapter is based on an evaluation of passive solar energy technology transfer by Lambright and Sheehan (1985), sponsored by the Department of Energy (DOE). The Lambright and Sheehan report is one of the very few evaluations of any of the federal commercialization efforts, and covers much more than the LANL work.

A conceptual model for the technology transfer process is presented in section 23.2, followed by a summary of the LANL technology transfer activities and a comparison of those actions with the model; analysis and conclusions; and guidelines for future actions.

23.2 Conceptual Framework

23.2.1 Life Cycle of a New Technology

It is useful to conceive of the transfer of passive solar energy technology from R&D organizations (national laboratories and universities) to the building industry as involving the following five steps:

1. *R&D planning* is the front end of the technology transfer process, yet often technology transfer is not considered in R&D planning.

2. *R&D performance* follows from the plan; it may lead to new products and processes.

3. *Initial application* is a proof of concept or utility of the new technology; the R&D result is adapted into a form that is more acceptable in a real-world environment.

4. *Replication* of the technology is made at a small number of additional sites, perhaps not as "friendly" as the initial ones.

5. *Diffusion* of the new technology, having been tested and made ready, occurs on a large scale.

The above linear model is believed to represent key steps in the technology transfer processes, but reality is not as linear and straightforward as this model implies. The R&D system produces a technology, and then the question is, How to transfer it? Strategies for getting a new technology applied frequently entail a "push" from the developer; that is, the developer takes the initiative to find and exploit uses. A "pull" from potential users is also desirable; however, where new technology is at issue, users do not necessarily know what is available, much less what could be developed in R&D.

An alternative to this linear model is an interactive model, wherein users and R&D personnel are involved at the front end of the process. User input is provided in R&D planning, and there is interaction over time, even during the R&D period. The key to the successful implementation of this model is to establish a dynamic equilibrium of mutual understanding of user needs by R&D personnel and of technological capabilities by users. The interactive model, which is governed by principles of exchange—of knowledge and support—is an ideal. The linear model is the dominant mode of laboratory technology transfer, and variations on this theme can be seen in the case described in the next section.

23.2.2 Decision Making

Within each one of the above noted stages of technology transfer, there is a decision-making process with six decision phases:

1. *Awareness*: A decision maker (a laboratory, a user) becomes aware of a problem or opportunity.

2. *Trigger*: Some event, usually external to the decision maker, catalyzes an attempt to do something about this problem or opportunity.

3. *Search/Planning*: A concerted effort to find an appropriate response is undertaken. Initially, this is a fairly wide-open search, but as options are narrowed, gradually it resembles more a planning process.

4. *Adoption*: A decision is made to use resources (money, people) toward putting a particular response policy into effect.

5. *Implementation*: The policy is carried out.

6. *Transition*: The effort set in motion by the adopted policy is completed and passed on to the next step of technology transfer.

The foregoing model assumes continued forward motion, but termination is possible at any point. Options are filtered out, preferred choices are selected, and what is adopted may be canceled prior to full implementation. The barriers to innovation far exceed the forces that facilitate it. Our expectation is that the most important and difficult issues in technology transfer lie at the points of transition.

23.2.3 Actors

The actors in the technology transfer process are many. Who takes the leading role differs, depending upon where the technology is positioned along the course from R&D to diffusion:

Researchers are involved (e.g., scientists and engineers) in the R&D stages at laboratories.

Users are those who receive the technology and put it into operational practice (e.g., builders and developers).

Beneficiaries are those who benefit from deployment of the technology (e.g., home buyers).

Suppliers are those who reduce the technology to operational practice so that it can be acquired by users (e.g., architects and engineers who design buildings, and builders and manufacturers who produce them).

Intermediaries are those who help facilitate linkages among and between others (e.g., professional organizations).

Sponsors are those who pay for activities in the process (e.g., federal agencies).

Policymakers are those who establish the policies and rules within which other actors operate (e.g., elected and appointed public officials and corporate management).

Entrepreneurs are those who take the initiative to push or pull a technology through the steps of the transfer. This is a general role that can be played by any of the above, and that must be played if technology transfer is to take place.

A key to rapid technology transfer lies with the interaction among the actors, particularly at the transition points.

23.3   Background: Case Description

The following is a condensed description of key steps in the Los Alamos National Laboratory (LANL) solar design tool technology transfer. Detailed presentations of this and other cases are found in Lambright and Sheehan (1985).

23.3.1   R&D: Los Alamos

The "Los Alamos design tool" is the product of an awareness that began as early as 1975 and grew continuously thereafter. One entrepreneurial scientist at LANL began diversifying into solar energy and soon had DOE support, which grew from $150,000 to $1 million. There is no evidence available to indicate any institutional planning for the creation and transfer of the Los Alamos design tool. In the mind of the key scientist and research entrepreneur, however, there may well have been such planning.

R. D. McFarland, J. D. Balcomb, S. W. Moore, J. C. Hedstrom, and coworkers at LANL began research by testing and simulating simple passive solar designs. For several years they continued this work with more complex designs and in an increasing number of geographical areas. They made great advances by insightful interpretation and correlation of the many results and reduced these results into passive solar design methods and guidelines.

The information generated was encapsulated in three volumes of a *Passive Solar Design Handbook*. Volume 1 was largely a qualitative introduction to the field written by Total Environmental Action and volume 2 was quantitative (Balcomb, Barley, et al. 1980a, 1980b). Volume 3 was a guide for engineers who sought simpler tools than simulation analysis (Balcomb, Jones, et al. 1982); this volume discussed 94 different passive design systems at 219 locations across North America. Volume 3 constitutes a complex version of the "LANL design tool," which is based on the solar load ratio (SLR) method developed at LANL. The tool is a computational technique that enables builders to determine the performance of numerous passive designs at locations throughout the country.

Program management in Washington was apparently interested in working with LANL in moving its design tool to an articulated mission, culminating in the three-volume handbook. Users were not a part of this process, except indirectly. One way a user perspective came indirectly into play occurred when LANL served as a judge in a solar design competition sponsored by HUD and had to relate to HUD as a sponsor. Another occurred when a researcher became a user by purchasing a solar home. DOE/LANL's involvement in this effort was that of a sponsor: providing the R&D function. This support included the research itself and extended to funding the handbook's preparation.

23.3.2   Initial Application: New Mexico Showcase

The New Mexico Showcase of Solar Homes was a public promotion of passive solar homes in the Santa Fe–Albuquerque area. It was a highly publicized example of what could be done with passive solar energy. A complete decision-making process took place (awareness, trigger, search/planning, adoption, implementation, and transition).

The Los Alamos design tool and the New Mexico Showcase of Solar Homes are closely linked. The Showcase of Solar Homes presented an opportunity for Los Alamos. The developer, Amrep, wanted to promote new subdivisions in Santa Fe and Albuquerque. A factor that finally triggered these groups into action was the falling of interest rates in late 1982. Project planning began with considerable publicity and careful financing. A committee was formed to oversee the project, representing the developer, the builders, Public Service Company of New Mexico, and the New Mexico state government (*Solar Energy Intelligence Report* 1984). The committee wanted to promote good passive solar design but needed technical backup, so it approached the scientists at Los Alamos for assistance. The result, put together in a joint effort between the committee and the scientists, was a simple analysis procedure or tool for evaluating prospective designs. The tool estimated performance and comfort, based on the more complex procedures in the LANL volumes, using a set of three worksheets with tables developed specifically for each locality. There were two versions, one for Santa Fe and one for Albuquerque. The idea worked; many builders with no previous knowledge of the analysis concepts were able to successfully complete the worksheets, and the committee was satisfied. The quality of passive solar design was definitely improved and many of the Amrep builders continued to use the worksheets to assist in the design of their ongoing projects.

Ultimately, twelve firms were selected for one construction site and eight for another (Rational Alternatives 1983, 3). Carefully calculated designs meeting detailed specifications were constructed during the summer of 1983 (*New Mexico Solar Review* 1983). In September the showcase opened in concert with the eighth National Conference on Passive Solar Energy (*Solar Engineering and Contracting* 1983). One of the subdivisions, Eldorado in Santa Fe, has had more than 300 passive solar homes built (Conkling 1991).

The guidelines that were developed for the New Mexico Showcase of Solar Homes are very short. Primarily they consist of only the three worksheets. Their main purpose was to assure the oversight committee appointed to advise on the project that the houses followed good passive solar design practice.

The key problem that existed in 1982, when the showcase was conceived, was that most solar houses were not well balanced. Although passive houses were being constructed in great numbers throughout northern New Mexico, they were frequently underinsulated and often did not have enough thermal mass to store the large amount of solar gain admitted through their oversized south windows. The resulting houses used little backup heat but were often uncomfortable. The guidelines explicitly addressed both problems, and houses that passed the criteria in the guidelines had good performance and were comfortable.

Prior to the actual transfer, this technology existed only as a very sophisticated and technical way of choosing optimum designs. In this mode, it was quickly deemed inappropriate for use by local builders and hence underwent extensive translation by a subcommittee of the project's advisory board to make it readable, comprehensible, and amenable to use by builders.

The Public Service Company of New Mexico (PNM) has promoted use of the LANL tool to interested parties in the area it serves. It has adapted the tool itself and in some cases has put its own name on the adaptation.

23.3.3 Replication: Wichita, Kansas

In a letter to a LANL technologist, a real estate developer from Global Alternatives in Wichita, Kansas, expressed interest in conducting a program in Kansas similar to the New Mexico showcase. The developer was impressed by the credibility of a program under the technologists' guidance and by the impact such a showcase could have in a mature market. The LANL technologist responded by preparing a "quick and dirty" reference worksheet for the area. A workshop, conducted in December 1982 by a passive solar consultant, put flesh on the developer's thinking about solar housing. Although the solar market in Wichita was young, the developer speculated that the geographic suitability of the region, coupled with a peculiar energy cost situation, would make the innovation successful. The developer dealt with the dearth of local designers by buying five plans from Ambassador Homes in Kansas City. Six builders were involved and, between mid-1982 and mid-1983, thirteen homes were constructed. In June 1983 the Kansas Showcase of Solar Homes was formally announced. The homes were largely sold on aesthetic rather than solar merits. For a number of reasons—not the least of which was that technology transfer was not the priority goal—the Wichita replication appears to have lost some of the original innovative qualities in the transfer from New Mexico to Kansas in spite of its apparent success as a real estate venture.

The Wichita replication differed in fairly substantial ways from New Mexico. The level of interest and acceptance of solar technology in Kansas, unlike New Mexico, was very low. This problem was compounded by higher land costs and competition from other housing developments having conventional construction. The Wichita replication was thus less successful, as a passive technology transfer, than the New Mexico program because the emphasis was less on the homes' solar features.

23.3.4 Diffusion: Software Case

The LANL design tool guidelines for passive solar housing were regarded by their creators and initial users as a potentially significant advance in the industry. Various software companies who cater to the solar housing industry became aware of the LANL work in the late 1970s. They saw the guidelines as capable of being converted into computer programs, which they in turn could market. They could do this by creating user-friendly programs that appealed to architects and builders by making the process easily understood and manageable. At least five companies converted volume 3 of the LANL handbook techniques to microcomputer programs. The computer software case added a new twist. Here the LANL design tool technology served as a basis for the development of a new tool—computer software for passive solar design. This software provided the LANL tool with a mechanism for diffusion in that software is more readily applied than technical manuals; a relatively complex technology was thus made operational. Because LANL made the basic technology available to the software companies free of charge, the financial costs were low. Opportunity costs, however, were significant for the small firms that undertook translating the LANL design tool as one of their main lines. Computer software firms have effectively distributed the LANL design tool to builders who would not have waded through the complex manuals. The user-friendly approach to solar design has served to expand the concept's feasibility.

23.3.5 Diffusion: ASHRAE

The American Society of Heating, Refrigerating, and Air-Conditioning Engineers (ASHRAE) has become a major transfer mechanism for the LANL passive solar design tool. In 1982 it was felt by LANL and DOE that although volume 3 of the LANL *Passive Solar Design Handbook* was available as a government document, something more was needed to bring the Los Alamos method to more widespread use. A DOE manager suggested that the LANL solar energy group do something jointly with ASHRAE to lend its work more credibility with engineers. Funds provided by DOE in 1982 inaugurated the effort, which culminated in *Passive Solar Heating Analysis: A Design Manual*, published in June 1984 (Balcomb, Jones, et al. 1984).

ASHRAE's decision to become another transfer mechanism for the LANL design tool was not as momentous as those of the software companies. The ASHRAE expansion and endorsement of the LANL design tool required the labor of only a few people—a small proportion of their resources—and involved little risk. The additional research and compilation were conducted in cooperation with LANL and were readily feasible, separable, and reversible; the minimal effort required was well invested. The positive impact on the industry was manifested primarily by the new passive solar design manual. The manual was technically accurate (acceptable to LANL), functionally feasible, and professionally endorsed.

Los Alamos scientific personnel played active roles in translating their research into user-relevant needs. A successful transfer channel was created through the cooperation of sponsor, researcher, and supplier.

23.3.6   Diffusion: PSIC Guidelines

In the few years after the 1983 New Mexico Showcase of Solar Homes, interest in passive solar decreased as the public perception changed. Energy concerns were put on the back burner, and government-funded research and commercialization efforts were cut back. The Los Alamos solar program was phased out, and J. D. Balcomb transferred to the Solar Energy Research Institute (now the National Renewable Energy Laboratory NREL). The guidelines work remained dormant for more than two years.

Although the resemblance seems faint, today's residential guidelines project is an evolution of the New Mexico experience. The 58-page package, *Passive Solar Design Strategies: Guidelines for Home Builders* produced by the Passive Solar Industies Council (PSIC 1989), is more comprehensive and much more professional in appearance. A cooling analysis has been added, based on a simplified analysis technique developed by Robert McFarland at Los Alamos (McFarland and Lazarus 1989), and the process of generating the performance tables has been automated. The most important modification, however, is a complete shift of emphasis. The entire first section of the guidelines is devoted to design guidance. The worksheets are still included but appear toward the end. The key feature that distinguishes both the original New Mexico guidelines and the current evolution is the focus on a single locality.

23.3.6.1   PSIC Guidelines

The basic approach in the guidelines is to reference everything to a base-case house design. This provides a logical starting point for the designer. The base case is a standard 1,500-ft^2 house built in conformance with typical insulation levels used in the location. The guidelines present the annual heating and cooling loads of the base-case house in the location and clearly indicate the effectiveness of various conservation, passive solar, and natural cooling strategies. Several example house designs are described that will save 20%, 40%, and 60% compared with the base-case energy requirement.

A second key part of the guidelines are the four one-page, fill-in-the-blanks worksheets. These enable designers to quickly evaluate the energy and comfort characteristics of a proposed design in their location. Annual heating and cooling energy and conservation and comfort indices are calculated with the aid of a set of location-specific tables. The values obtained can be compared with the base-case house presented in the guidelines; the design can then be adjusted as necessary to meet design goals. These worksheets take the guidelines beyond the realm of the typical write-up that gives only general advice into that of a useful tool by providing a design procedure capable of quantitative evaluation. This is essential if the designer is serious about energy efficiency.

An example house evaluation included in the guidelines book illustrates how to use the worksheets. This example employs improved insulation, a sunspace, shading, some added mass, and a ceiling fan to reduce heating and cooling loads. The annual energy reduction is typically more than 50%, depending on location. Filled-out worksheets included in the guidelines are intended to be used as a case study during the workshop presentation.

A parallel effort at NREL has resulted in *BuilderGuide*, a computerized version of the guideline worksheets, which allows designers to complete the worksheet calculations in a fraction of the time needed for the hand-written version. PSIC disseminates *BuilderGuide* as an integral part of its overall guidelines program.

An important innovation in the *BuilderGuide* program is the automatic generation of a base-case house. The user has only to specify the location, the size of the house, the number of stories, and the floor type (slab-on-grade, crawl space, or basement). The program then automatically generates a complete building description, fills out the worksheets, and performs the energy calculations, a process that takes only seconds. The insulation values used are typical of new construction practice in the locality. This house serves as a base case, a reference point against which to compare the actual design. The user then modifies the worksheets using values for the actual proposed design. The program completes the analysis and automatically compares the results with the base-case numbers.

23.3.6.2 Single-Locality Approach
A difficult problem facing authors of passive solar guidelines is how to deal effectively with a wide range of weather situations, a particular problem in the United States. Climates range from nearly arctic to nearly tropical. Most authors either present a general method that can be used with monthly weather data or they attempt to regionalize the weather.

The first approach invariably leads to an unacceptable loss of accuracy and specificity or is far too unwieldy to present and too complicated to apply, thereby losing its desired audience (as with the ASHRAE volume). The second approach does not work well because there are just too many climatic variations to be considered. For example, there are seven important climate zones in Arizona and nine in California.

Passive Solar Design Strategies: Guidelines for Home Builders takes a different approach. Each version of the booklet is based on weather data for one specific place, and thus it addresses only that locality. This complicates the production and distribution of the booklets but makes the user's life much easier.

23.3.6.3 Producing the Guidelines

The text of the guidelines book is generated by merging two computer files. The first is a word processor file containing the basic text that does not change from location to location; the second contains numbers and text that are location-dependent. This second file is produced by running a computer program that calculates performance numbers based on long-term monthly weather and solar data compiled by the National Oceanic and Atmospheric Administration (NOAA) for a particular location. The merge operation places the numbers and text in the second file into their correct locations in the first file. The merged file is then laser-printed to produce the camera-ready manuscript.

In this way, it is possible to produce individual guideline booklets as they are requested. The packages can be generated for 205 U.S. locations, or they can be modified and customized to apply to adjacent sites, as long as long-term monthly weather data are available. Thus guidelines for more than 2,000 sites can potentially be created. These booklets are available from the Passive Solar Industries Council (PSIC), which also presents seminars based on the material; by 1994, booklets had been generated for a total of 180 locations, encompassing nearly every state in the United States.

23.3.6.4 Oversight

The entire evolution of the guidelines was overseen and reviewed by leaders in the home-building industry. The purpose of this broad oversight was not to seek lip-service endorsement of the product but to assure that the result would be acceptable and useful. In fact, the form of the guidelines was shaped and reshaped by the participating organizations.

More than one midcourse correction was made, and although somewhat grueling, this process produced a credible product that is well received.

The current guidelines were developed over a four-year period by a unique and creative partnership that included NREL, the Los Alamos National Laboratory, Charles Eley Associates, the member associations and companies that form the Passive Solar Industries Council, and the Standing Committee on Energy of the National Association of Home Builders. Funding was provided by the U.S. Department of Energy.

23.3.6.5 Dissemination

PSIC has launched an aggressive technology transfer and information dissemination campaign to introduce the guidelines to all the key groups within the U.S. housing industry through presentations to national organizations, technical and promotional articles, and development and support of local workshops. Workshops have been presented in several locations throughout the United States and interest is increasing rapidly. As of 1994, more than 50 one-day workshops had been presented, reaching more than 3,100 builders and designers, and the program had received substantial national attention throughout the building industry.

23.4   Analysis and Conclusions

The Los Alamos case reveals that technology transfer in passive solar energy is indeed possible—but also very difficult. What it takes are linkages between those who develop technology and those who can use it. Such linkages can be deliberately furthered by government policy. Or they can happen in spite of government policy when key technologists reach out to users and work with them to translate new scientific information into appropriate tools they can employ. Small innovative companies frequently are the ideal entities to take technology from federally funded sources and move it forward. What is clear is that innovation takes time and that federal policy can only accelerate the process somewhat.

In the Los Alamos design tool case, there was little R&D planning, but there was R&D performance and a transfer from R&D to initial applications and beyond. The demonstrated worth of the design tool eased its conversion to computer language via software companies and to manuals via ASHRAE and PSIC.

There was technology transfer from Los Alamos to the New Mexico showcase. Here the key mechanism or entrepreneur was an ad hoc committee. This committee included various interests, most notably technical and user. It bridged a gap between R&D and end use, and did so actively and energetically, literally rewriting R&D language into simplified language that users could understand. This committee filled a gap in the system.

In moving from this initial application in the New Mexico showcase to a replication in Wichita, Kansas, the developer/user in Wichita played the entrepreneurial role. The technical translation was not consummated, however. There was no advisory committee linking technical information to the project. The replication appears to have been a partial one in terms of the design tool. Some builders used it, but most did not, and there was little push or pull on its behalf, although, from a solar real estate perspective, the Wichita development was a clear success.

Design guidelines based on targeting a single user constituency and a single locality have proved to be effective. This concept originated with Los Alamos for the New Mexico Showcase of Solar Homes and was carried forward at the National Renewable Energy Laboratory. The logistical problems of producing such guidelines have been solved; the residential guideline packages for both new residences and the remodeling of existing residences are being disseminated. Similar packages for small commercial buildings are under development.

The design tool was diffused through the mechanisms of software companies, ASHRAE, and PSIC; this tool is now available to designers and builder. The degree to which this information was incorporated in buildings is unknown, but certainly diffusion of the design tool took place.

An important factor in this example is continuity. The fact that one of the design-tool originators carried through the process from inception to conclusion not only kept the project going but assured the technical integrity of the final product.

The Los Alamos case reveals a range of mechanisms: national laboratories, a local ad hoc committee linking technical and user interests in a specific project, a federal agency (DOE), private companies, a national professional association, and an industry association. Getting a technology to the point where industrial policies can take control from public policies is a considerable task. What we know is that there are certain critical factors in technology transfer success. There must be an entrepreneurial push or pull. The Los Alamos case is an example of a successful push and pull. DOE itself is seldom in a position to be the entrepreneur; it is a sponsor and influences the process indirectly, most in the early stages and least when a technology shifts to either replication or diffusion.

23.5 Guidelines for Action

Future federal programs should identify the innovators in the private sector and help them to play a strong role in the transfer process. Those within government and national laboratories also need a favorable policy environment to facilitate their roles in transfer. Technology transfer is a delicate process of coalition building across the public and private sectors. Such coalitions and the linkages that make them possible are difficult to assemble and easy to break; they must be constructed over many years. More often than not, it takes a public push to get a private sector pull for technology transfer. The public push may be more appropriately and successfully guided by public-private linkages established up front in the process. The need for DOE's experimenting with better mechanisms for doing so is clear.

23.6 Recommendations

General

• Be sensitive to and aware of the needs and interests of other actors; R&D professionals, architects, and builders have different perspectives that must be bridged.

• Attempt to secure a stable managerial/administrative environment.

• Provide for extensive monitoring, evaluation, and dissemination of project information.

• Get the technology to be responsive to user needs.

• Keep the technology transfer process a primary goal.

• Be sure that the target population—here the building industry—*leaders* are solicited to participate in the program.

R&D Performance

• Develop the technology fully before attempting its transfer.

Initial Application

• Use competition to solicit a broad talent base.

• Solicit input from a range of actors—technologists and users at a minimum.

• Use utilities to enhance initial applications and further developments.

• Translate the technology to make it more adoptable by the users.

Replication

• Be cognizant of the environmental differences from one setting to another, and the resultant need to adapt the technology.

• Involve all key actors in the process; absence of certain actors, such as builders and developers, can inhibit replication and diffusion of even successful initial applications.

Diffusion

• Secure professional/trade associations as disseminators of information.

• Be careful to accurately assess the market for the technology before selecting a specific dissemination strategy.

References

Balcomb, J. D., D. Barley, R. McFarland, J. Perry, W. Wray, and S. Noll. 1980b. *Passive Solar Design Handbook*. Vol. 2, *Passive Solar Design Analysis*. DOE/CS-0127/2. Washington, DC: U.S. Department of Energy.

Balcomb, J. D., R. W. Jones, C. E. Kosiewicz, G. S. Lazarus, R. D. McFarland, and W. O. Wray. 1982. *Passive Solar Design Handbook*. Vol. 3, *Passive Solar Design Analysis*, ed. R. W. Jones. DOE/CS-0127/3. Washington, DC: U.S. Department of Energy; Boulder, CO: American Solar Energy Society, 1983.

Balcomb, J. D., R. W. Jones, R. D. McFarland, and W. O. Wray. 1984. *Passive Solar Heating Analysis: A Design Manual*. Atlanta: American Society of Heating, Refrigerating, and Air-Conditioning Engineers.

Conkling, M. 1991. "A Solar Community: Eldorado at Santa Fe." *Solar Today*, November–December.

Lambright, W. H., and S. E. Sheehan. 1985. *Improving the Transfer of Passive Solar Energy from DOE National Laboratories: Linkages and Decision Processes*. Final Report for DOE contract DE-AC03-83-SF11966. San Francisco: U.S. Department of Energy Operations Office, November.

McFarland, R. D. and G. Lazarus. 1989. *Monthly Auxiliary Cooling Estimation for Residential Buildings*. LA-11394-MS. Los Alamos National Laboratory.

New Mexico Solar Review. 1983. "Lt. Governor Runnels Dedicates New Mexico Solar Showcase." Vol 4, June.

PSIC (Passive Solar Industries Council). 1989. *Passive Solar Design Strategies: Guidelines for Homebuilders*. Washington, DC.

Rational Alternatives, Inc. 1983. "New Mexico Showcase of Solar Homes Builder Fact Sheet."

Solar Energy Intelligence Report. 1984. 23 January, p. 28.

Solar Engineering and Contracting. 1983. "Showcase to Open during Passive 83." Vol. 2, September–October, p. 20.

Total Environmental Action. 1980a. *Passive Solar Design Handbook*. Vol. 1, *Passive Solar Design Concepts*. DOE/CS-0127/1. Washington, DC: U.S. Department of Energy.

24 Argonne National Laboratory

William W. Schertz

24.1 A Nonimaging Collector

The original concept of using nonimaging optics for the concentration of light was developed by Roland Winston of the University of Chicago for use in high energy physics experiments. In 1974 Robert Sachs, director of Argonne National Laboratory (ANL), asked Roland Winston whether the same technique could be used for the collection of solar energy, and if so, would it have any advantages as compared to imaging concentrators. After a short investigation it became apparent that this method could be used to concentrate solar energy and that it did offer a potential advantage. Concentrators using nonimaging optics could effectively concentrate solar energy with little movement, whereas other concentrators require continuous movement of some part of the equipment in order to be effective. Thus was born the compound parabolic concentrator (CPC); for more information on the CPC and nonimaging optics, see Winston 1990, 1991; O'Gallagher et al. 1980; and McIntire 1980).

The development of the CPC for solar energy collection was done by ANL in a joint program with the University of Chicago. ANL worked closely with private industry in a joint development and commercialization program (Allen et al. 1977). This example of successful technology transfer is described in this chapter.

24.2 CPC Development

Development began in 1974 with initial funding from the National Science Foundation, under the Research Applied to National Needs program (NSF/RANN). The CPC program stressed the theoretical and practical understanding of the optics of the concept, and the results indicated that nonimaging optics did offer a significant optical advantage for the collection of solar energy. The initial collectors were modifications of flat-plate collectors, using the concentrator elements to reduce the net area available for reradiation and convection heat losses. The early prototype collectors performed well optically but were not particularly good performers thermally.

Early in the program it was recognized that significant industry interaction would have to be incorporated into a successful development program and that some independent assessments of the concept by industrial firms would be useful in setting goals for the future. In early 1975 several study contracts were placed, along with two hardware fabrication contracts; these early contracts had a significant impact on the program. Two studies (ADL 1975; Bechtel 1975) evaluated the concept for solar thermal applications; their key findings were

1. The compound parabolic concentrator should be used for applications that require temperatures higher than ordinary flat-plate collectors can easily obtain, and lower than the higher-concentration (30 times) parabolic collectors. The temperature range of 180°–300°F represented an area where the CPC was likely to have a unique advantage in performance.

2. The chief competition for the CPC in this temperature range was represented by the selectively coated evacuated tubes that were being developed by Owens-Illinois and Corning Glass at the time. However, it was noted that coupling the CPC to evacuated tubes offered a good possibility for reducing of the overall cost of delivered energy.

3. Consideration should be given to the use of the CPC as a terminal coupling element in focusing optical systems.

Another study was commissioned to investigate the possibilities of using the CPC with photovoltaic cells as the receiver rather than a thermal receiver. Mobil-Tyco investigated using the CPC as a primary concentrator onto a ribbon-type photovoltaic cell, and Spectrolab investigated the possibility of using the CPC as a secondary concentrator onto photovoltaic cells for a primary line-focus system. The study showed that significant economic benefits could be derived from the low value of concentration available from the CPC because of the high cost of the photovoltaic cells (Spectrolab 1975).

Two fabrication contracts were initiated in early 1975 at Chamberlain Manufacturing and American Science and Engineering (AS&E) to make nonevacuated collectors based on the optical designs that had been developed at ANL and the University of Chicago. The study contracts and fabrication contracts were the first substantial interaction with industry on the project.

The results of these contracts and the internal investigations of performance at ANL and the University of Chicago led to the following major conclusions:

1. The optics worked as theory had indicated that they should.

2. The thermal performance of the collectors was lower than expected because of losses from the nonevacuated receivers.

3. A serious evaluation of the use of evacuated receivers coupled to the CPC optics should be started.

4. The feedback from the industrial fabrication contracts on the problems they had faced in making the prototypes was very valuable, and the Industrial Participants Program was initiated.

In 1975, based upon the results from the first round of contracts, a series of competitive procurements was initiated for reflectors and receivers (evacuated). The philosophy behind these procurements was to decouple the need for each participant to make both glass and metal components of the collector (which had been a problem in the previous fabrication contracts)—to let glass-oriented companies work on the evacuated tubes and metal working companies work on fabricating structure and mirror assemblies. ANL would then interface the best elements from each of the industrial firms into a complete collector. This concept allowed great flexibility, and the components from these collectors were used in several different configurations for test and evaluation; these included

- plastic reflectors (thermoformed)
- metal reflectors, flat horizontal absorbers in evacuated tubes
- vertical flat absorbers in evacuated tubes
- cylindrical absorbers

During this time, the Industrial Participants Program was active, and a representative from one of the potential manufacturers was in permanent, year-long residence at ANL. This representative worked with the solar energy group and participated in the successes and failures of each design as it was fabricated and tested.

Photovoltaic applications (from the first round of studies) were also under development in late 1975 and early 1976. A dielectric version of the CPC was developed for photovoltaic applications and was a winner of an Industrial Research IR-100 award in 1976. Much industrial interest in the overall concept was generated by the publicity that went with the award.

In 1977 development of the thermal collectors had progressed to the point that a working prototype of a stationary, evacuated tubular receiver CPC with a 1.5 concentration ratio had been built by ANL using the components supplied by industry subcontractors, and another IR-100 award for the thermal collector was received. This again resulted in a fresh flurry of industrial interest in the concept. A standard packet of information was assembled that would be mailed out whenever a query came in.

One of the photos that had been taken for use in the 1977 IR-100 award display was subsequently used in an article in *Solar Engineering* Magazine (Schertz 1977). The optical characteristics of the reflector design were clearly evident (figure 24.1), and this article led to one of the key interactions with the industry. Sunmaster, Inc., had been exploring the use of reflectors to augment the performance of evacuated tubes in a collector of their own design, and when they saw this article, they realized that significant help might be obtained from ANL.

Initial contact was made by D. Michael Platt of Sunmaster, which was followed up by a visit by W. Schertz to the Sunmaster facility in Corning, New York. In an all-day design and discussion meeting, a first cut of an "optimum" design was established, incorporating the system design features unique to Sunmaster with the optics and reflector design of the ANL and formed the basis of the final product design. This initial meeting was followed by many subsequent meetings both at Sunmaster's facilities and at ANL, as part of a deliberate attempt to transfer knowledge and expertise that had been developed in-house at ANL and the University of Chicago to firms that could use the technology.

A flood of other requests for information and help in the design and evaluation of alternate designs followed the publicity generated by the IR-100 awards and the numerous trade and energy shows in which ANL participated. As a result, a specific task on interactions with industry was added to the program. A key technical staff member (W. R. McIntire) was assigned responsibility for interacting with and assisting any and all industrial firms that requested information and assistance. The basic ground rules were

• To help anyone with their design problems as long as it was related to CPC-type collectors on which DOE was funding the work.

Figure 24.1
Solar thermal collector designed at Argonne National Laboratory, 1977.

• To provide as much assistance as requested as long as it did not prevent some other requester from being helped. If the workload became too great, the level of assistance to all requesters was to be scaled back equally until a balance was achieved. (In practice, this turned out to be a non-problem. The most that had to be done was delay some requests for a few days to balance the level of assistance.)

During the next two years, McIntire provided significant help to a large number of companies, both small businesses and larger corporations. This assistance included the following:

• Design and analysis of the performance of mirror shapes a manufacturer felt would be more economical to manufacture relative to the ideal performance of a perfectly shaped mirror, taking into account such variables as manufacturing tolerance on assembly.

• Examination and analysis of finished collectors to determine why they might not perform as expected.

• Design of techniques to be used by manufacturers/developers to ensure that the product they were making was within optical tolerances (for example, a quick technique was developed to allow a production line to determine whether mirror assemblies were correctly shaped or were deformed).

Table 24.1 displays the pattern of funding to support the CPC solar collector work at ANL. Funds were supplied by NSF/RANN in 1974 and by ERDA, and then DOE, thereafter. The last year of funding was 1986.

24.3 Evaluative Evidence

Numerous manufacturers requested help. Some companies manufactured prototype panels and then progressed no further; others went on to begin production of the CPC-based collectors as a product line. Among those companies was Sunmaster Corporation, which built a CPC collector with an evacuated tube absorber. Sunmaster sold some 25,000 collector modules between 1978 and 1987 but ceased manufacturing CPC collectors in 1987. Although no CPCs have been manufactured in the United States since 1987, they are still being built in South Korea, Japan, and Israel.

Table 24.1
CPC funding history at Argonne National Laboratory (thousands of current year dollars)

FY	1974	1975	1976	1977	1978	1979	1980	1981	1982	1983	1984	1985	1986
Totals	100	220	883	850	424	500	645	35	440	200	222	175	90

24.4 Lessons Learned

The successful transfer of a technology from government-funded work into a product requires several ingredients:

1. Dedication on the part of the research staff to work closely with industry over the life of the product.

2. Working directly with industry technical personnel (not sales people or managers). Simply publishing results does not achieve much in the way of utilization of the technology.

3. Integration of the development being transferred into the designs and systems the manufacturer is developing on its own, even if this means some compromise in the basic design developed in the laboratory. The manufacturer must have a proprietary right to and interest in its portion of the design to give it a marketing edge over the competition.

4. Follow-through cooperation on the part of industry. Smaller businesses seem to be more responsive to working with a laboratory in the development of the product than large corporations. Equal help was given to both, but the larger corporations did not follow through on the designs and the eventual manufacture of collectors.

5. Two-way information flow. ANL learned as much or more from their close association with the manufacturers as did the manufacturers learn from ANL. The practical considerations that a company making the device must face cannot be ignored by the research and development team in the laboratory. But effective consideration of these factors can only be achieved if good two-way communication is established. The recipient and laboratory must establish a mutual level of trust and confidence in each other's competence and integrity.

6. Ongoing, long-term dedication of significant resources. A short course or series of publications does not do much, because the feedback from industry to the researchers is missing.

7. Current, updated information. Old publications can sometimes mislead. Periodically, we were put in contact with a new entry into the field that had picked up an old publication on the concept and had proceeded to try to develop the concept form that point, without the benefit of subsequent, published research. By having a person whose job was dedicated to the process of interacting with the industry developers, we were able to help some of these people leapfrog over the old technology and catch up quickly.

References

ADL (Arthur D. Little, Inc.). 1975. *Goals Study for Technical Development and Economic Evaluation of the Compound Parabolic Concentrator Concept for Solar Energy Collector Applications.* ANL-K-75-3192-1. Argonne, IL: Argonne National Laboratory, June.

Allen, J. W., N. M. Levitz, A. Rabl, K. A. Reed, W. W. Schertz, G. Thodos, and R. Winston. 1977. *Development and Demonstration of Compound Parabolic Concentrators for Solar Thermal Power Generation and Heating and Cooling Applications.* ANL-76-71. Argonne, IL: Argonne National Laboratory, January.

Bechtel Corporation. 1975. *Phase-Zero-Goal Study for the Technical and Economic Evaluation of the Compound Parabolic Concentrator (CPC) Concept Applied to Solar, Thermal, and Photovoltaic Collectors.* ANL-K-75-3192-1. Argonne, IL: Argonne National Laboratory, June.

McIntire, W. R. 1980. "Optimization of Stationary Nonimaging Reflectors for Tubular Evacuated Receivers Aligned North-South." *Solar Energy* 24: 167–175.

O'Gallagher, J. J., A. Rabl, R. Winston, and W. R. McIntire. 1980. *Solar Energy* 24: 323–326.

Schertz, W. W. 1977. "High-Temperature Collectors." *Solar Engineering* July: 28–29.

Spectrolab. 1975. *Preliminary Evaluation of Two-Element Optical Concentrators for Use in Solar Photovoltaic Systems.* ANL-K-75-3191-1. Argonne, IL: Argonne National Laboratory, June.

Winston, R. 1990. "Optical Research and Development." In F. de Winter, ed., *Solar Collectors, Energy Storage, and Materials* Cambridge: MIT Press.

Winston, R. 1991. "Nonimaging Optics." *Scientific American* March: 76–81.

VII SOLAR THERMAL INCENTIVES

25 Tax Credits

Daniel Rich and J. David Roessner

The U.S. government has often used tax subsidies to promote the production of energy and to foster the commercialization of new technologies. Subsidies have included tax credits, depreciation allowances, and other deductions that reduce tax liability. Cone (Cone et al. 1980; Cone 1982) estimates that between 1918 and 1978 the federal government provided $202 billion in 1977 dollars ($335 billion in 1985 dollars) in direct subsidies to the nuclear, coal, oil, natural gas, and electricity industries. Between 1950 and 1976 alone, the United States spent $870 per capita (1976 dollars; $1,540 in 1985 dollars) on incentives for energy production (Cole et al. 1981).

Despite extensive tax subsidies for nonsolar energy industries, solar technologies received virtually no favorable tax treatment under federal law before 1978. The solar energy tax credit was intended to change this situation by creating incentives to accelerate commercial investment in, and consumer adoption of, renewable energy technologies. In October 1978 Congress passed five bills that together constituted the National Energy Act. The Energy Tax Act provided for an income tax credit of 30% of the first $2,000 and 20% of the next $8,000 for a maximum homeowner credit of $2,200 on expenditures of $10,000 for installation of residential solar devices. The act also created a tax credit of 10% for commercial solar investments. The 1979 *Internal Revenue Code* defined eligible residential renewable energy source property as property

which, when installed in connection with a dwelling, transmits or uses (i) solar energy, energy derived from geothermal deposits ... or any other form of renewable energy which the Secretary specifies by regulations, for the purpose of heating or cooling such dwelling or providing hot water for use within such dwelling, or (ii) wind energy for nonbusiness residential purposes. (IRS 1979, 33)

Later legislation provided additional incentives for solar investments. The Windfall Profit Tax Act of 1980 increased the residential credit to 40% or a maximum of $4,000 on an expenditure of $10,000, and the business credit was increased to 15%. The 1980 legislation also extended eligibility to a wide array of renewable technologies including hydropower, wind, geothermal, and biomass applications. The business energy credit was expanded to include nine categories of energy property including solar or wind energy property. Solar or wind energy property was defined as property that "uses energy from the sun or wind to (1) heat or cool a structure, (2) provide hot water for use in a structure, (3) generate electricity, and (4) provide solar process heat" (GAO 1985, 28); the property had to remain in operation for at least five years and had to meet any performance and quality standards prescribed by the Secretary of Energy.

Even before the federal government adopted solar tax credits, a number of states—such as California, New Mexico, Hawaii, Arizona, and Montana—enacted legislation creating solar financial incentives. In 1976 California enacted a 55% tax credit, the largest among all the states. Claims under the California law for the tax years 1977 and 1978 totaled $22 million; no other state's expenditures for solar financial incentives was more than a tiny faction of this amount and in some cases the credits were so small that their role can, at best, be viewed as symbolic (Roessner 1982, 6–7). Nonetheless, by mid-1985 forty-two states offered tax incentives for renewable energy equipment purchases, and twenty-seven states provided tax credits specifically for residential solar purchases. In addition, twenty states enacted business tax credits for commercial installations (Sawyer and Lancaster 1985, 172). For the most part, state tax credits were intended to complement and amplify the inducements provided by the federal credits, and in this regard, some states made their credit programs contingent on the federal tax credit.

The magnitude of the financial subsidy provided by the federal tax credits has been substantial. The Internal Revenue Service has not released complete data on claims since 1980 (see Thompson and Hillelson 1982). Estimates are available, however, based on the IRS's annual sampling of individual returns for the years between 1978 and 1984. The IRS estimates that during this period over one million individual returns claimed the residential renewable energy tax credit (see table 25.1). The total value of the subsidy provided by the tax credits between 1978 and 1984 (equivalent to public expenditures in the form of revenues forgone) was $1.5 billion ($1.7 billion in 1985 dollars). The average amount of each claim increased from $464 in 1978 to $1,745 in 1984; the average amount per claim over the entire period was $1,294 (table 25.1). In addition, a DOE study using data from the the Treasury Department's Office of Tax Analysis estimates that between 1978 and 1984 the value of claims for the business tax credit was $645 million in 1985 dollars (DOE 1985, 4). Based on these estimates, the total value of the residential and business tax

Table 25.1
Estimates of residential renewable energy tax credit claims, 1978–1984

Year	Number of claims (thousands)	Amount of claims (millions of current $)	Amount of claims (millions of 1985 $)	Amount per claim (current $)
1978[a]	69	32	49.5	463.79
1979[a]	77	44	62.5	571.49
1980[a]	155	166	216.4	1,070.79
1981	225	263	312.5	1,168.89
1982	229	322	359.8	1,406.11
1983	193	300	322.9	1,554.40
1984	220	384	396.2	1,745.45
Totals	1,168	1,511	1,719.8	1,293.66

Source: U.S. Internal Revenue Service, Statistics on Income Division, unpublished data, 1987.
[a] Data on years 1978–1980 are consistent with data reported in Richard Thompson and Rich Hillelson 1982, 4.

credits was in excess of $2.3 billion (1985 dollars) for the seven-year period. On a per capita basis, this level of subsidy amounted to about $10 a person over the seven-year period.

Evaluation of the effectiveness and impacts of the tax credits is an essential part of a comprehensive assessment of federal solar commercialization strategies and overall federal solar policy. The purposes of this chapter are to review available evaluation studies and to analyze and synthesize their results. On the basis of these results and in light of policy experience, we identify many of the strengths and weaknesses of the tax credits and assess their role as part of a solar commercialization strategy.

25.1 Policy Expectations and the Rationale for the Tax Credits

From the outset, government commercialization programs for solar technologies were justified by pointing to limitations of existing energy markets. Market failures were seen as inhibiting and perhaps even precluding a socially desirable level of investment in solar options. The following were cited as reasons for a substantial federal government role in solar commercialization: (1) the existing subsidies for conventional fuels (e.g., the oil depletion allowance, regulation of natural gas prices); (2) the high risks of investing in new technologies with uncertain markets or without rapid payoffs; (3) the existence of a variety of institutional "barriers" to the acceptance of solar technologies (e.g., inadequate consumer information, limited availability of financing for initial costs); and (4) the inability of markets to reflect the social and environmental benefits of renewable energy use (e.g., reduced environmental pollution, increased fuel diversity, decreased vulnerability to energy supply disruptions and price increases, encouragement of energy-efficient attitudes). Even so, it did not necessarily follow that the federal role was most appropriately expressed by providing tax credits. The logic behind this policy choice reflected a number of general expectations concerning the features of solar thermal technologies, the markets into which they were to be introduced, and the political appeal of tax credits as a policy instrument.

In the case of solar thermal applications for residential use in heating and cooling, the government's role was shaped by the assumption that the technology was, for the most part, market-ready, requiring mainly financial incentives to prospective purchasers (to overcome the inertia created by decades of artificially low conventional energy prices) and a variety of public demonstration and information programs to reduce institutional barriers to acceptance. Indeed, as early as 1974, advocates for substantial government support of solar heating applications in residential buildings claimed that the technology was already developed. In the apparent absence of opposing information, Congress authorized and appropriated money to produce and disseminate information about residential solar energy systems and to demonstrate the technology to the construction industry. The policy judgment that residential solar heating technology was commercially ready also led to a decision to focus commercialization efforts on the stimulation of market demand through subsidies to purchasers. Programs oriented toward strengthening supply (e.g., through loans and loan guarantees to small business) and infrastructure (e.g., through training programs for installers and consumer protection) received relatively little federal attention and support (HUD 1980, 5–6).

With few exceptions, the question that dominated analyses of government policies to accelerate the spread of "market-ready" technologies such as solar water and space heating systems was, What is the appropriate size of the financial incentive? This was, in part, a response to several efforts to identify the "barriers" that were inhibiting the rapid spread of solar technologies. These analyses, conducted prior to government enactment of incentives for solar energy, consistently identified the high initial capital cost of solar systems as one key factor slowing their acceptance in the market (ADL 1976; Booz, Allen & Hamilton 1976; Bezdek et al. 1977). This and the artificially low prices for conventional energy sources induced by federal energy policies were seen as the major impediments to adoption and diffusion of solar technology.

The residential and commercial solar tax credits represented an extension of the federal government's already established emphasis on a demand-oriented commercialization strategy for solar hot water and space heating systems. Because they were based on the assumption that technological uncertainties and operating performance problems had been reduced to a very low level, the solar tax credits were, in important respects, unlike tax subsidies for other energy sources. In contrast to other energy sources that enjoyed continuing, and in some cases virtually permanent, federal subsidies, the renewable energy tax credits were established with a specific "sunset date" for expiration (31 December 1985). The sunset provision was intended to limit potential revenue loss and may well have been politically important for achieving passage of the legislation. Sawyer and Lancaster (1985, 175) point out, however, that the sunset provision also reflected "the assumption of the 1970s and early 1980s that rising traditional fuel prices and decreasing renewable energy equipment costs (as a result of technical improvements and economics of scale) would eliminate the need" for subsidies.

From the outset then, the credits were viewed as a temporary policy initiative. The expectation was that the tax credits would provide "support of an infant industry" within the context of a largely nonfree energy marketplace (OTA 1985, 9) but that this support would not be needed beyond an initial boost. Thus, while it was assumed that economic incentives were necessary to "offset subsidies given to conventional energy sources and help the diffusion of innovation," it was expected that such incentives would follow the traditional innovation pattern; "that is, the problem is to get over the sluggish first phase, and then allow the industry to proceed on its own momentum" (Maidique 1983, 251).

Although most conventional energy subsidies have been directed to energy producers, the renewable energy tax credits made subsidies available to consumers. This feature is consistent with general policy expectations that many solar technologies were market-ready and that only sufficient financial incentives were needed to stimulate initial demand. It also helps to explain why the tax credits had widespread political appeal.

The tax credits were viewed by government, industry, and solar advocates as a particularly suitable instrument to promote solar thermal applications in residential use because these applications are capital-intensive and because underinvestment was expected without capital assistance. Thus it was expected that the credits would help to offset the high front-end capital cost of solar options until consumers and financial institutions became accustomed to evaluating the life-cycle advantages of renewable energy options. In addition, the tax credits might help to offset the general ignorance of potential buyers or their prejudice against new, unfamiliar, and unproven technologies. Both of these conditions appeared to be particularly significant obstacles to the emergence of the industry.

In addition to other rationales, there were a number of practical operational advantages to relying on tax credits as a policy vehicle. First, tax credits were easily adapted to work with the wide variety of renewable energy technologies. Second, by providing assistance to consumers who made purchase decisions rather than earmarking particular designs or suppliers, tax credits would foster the competition, experimentation, and flexibility needed for a still-emerging technology. Third, tax credits were relatively easy to design and administer, and they could be integrated with existing income tax processes, thereby avoiding the costs of establishing new administrative arrangements. The tax credits offered other practical benefits, although some of these were never exploited. Thus, for example, the credits potentially provided a mechanism for consumer protection. Suppliers could be required to provide minimum warranties and performance labeling to qualify for the commercial tax credit (Sawyer 1986). Congress, however, refused to approve these protections as part of the credit program.

In addition to their practical advantages, tax credits also offered a number of general political advantages. Tax credits were highly visible and touched a broad constituency. In addition, the costs of tax credit programs were concealed because they were contained in lost revenue rather than requiring direct appropriations. Added to these advantages, in the late 1970s, the solar tax credits performed important symbolic functions. When the federal government was under public pressure to do something about the energy crisis, the tax credits demonstrated the federal government's support for popular technologies and showed that the government was taking concrete policy actions to alter energy conditions. Thus the tax credits met the criteria of political viability. According to Whittington (1985, 117):

[there was] relatively little opposition to the enactment of the federal solar tax credit provisions ... when they were passed and amended. Expert testimony was virtually unanimous in the opinion that under certain applications, the "new" solar technology would lead to measurable energy savings.

In part because of their widespread support, the tax credits were used as a political lever in a broader controversy over national energy policy. Frankel (1986, 75) points out that "the solar tax credits became bogged down in the wrangling over natural gas pricing and other controversial provisions of the [National Energy Act]" and that President Carter "refused to allow them to be passed separately, in part in order to force solar advocates to support his entire package." Frankel concludes that the delay in enacting the credits "significantly hurt the solar industry, since consumers withheld purchases of solar equipment awaiting a better deal after passage of the tax credit." The delay was perhaps also a sign of the political uncertainties that would surround the tax credits in subsequent years leading up to their sunset dates. Indeed, in arguing before the Senate Finance Committee for a continuation and expansion of the tax credits in 1983, Jack Conway, chairman of the Renewable Energy Institute (REI), pointed out the cost of such uncertainty (Conway 1983, 3):

If there is anything that the financial market requires, it is stability, or at least predictability. The energy tax credits have been wracked with policy instability during their short life.... The items lending a sense of instability include attempts that have occurred to repeal the credits, weakening of the credits through provisions in last year's Tax Equity and Fiscal Responsibility Act (TEFRA) and, now, the scheduled termination of the credits.

When the tax credits were approved in 1978, there was virtually unanimous expert testimony that they would "lead to measurable energy savings," but there was "no clear agreement on the magnitude of these savings" or the costs that would be incurred by the federal treasury (Whittington 1985, 117). Thus, as Whittington points out, a study published in 1978 by the Energy Information Administration concluded that by 1985 "the incremental energy savings resulting from the solar tax credit program alone would be approximately 0.085 quads per year," but it "did not report either the number of projected claimants of the federal credits or the cost of the tax credits to the federal treasury" (1985, 117).

25.2 Tax Credit Evaluation Research

Are tax credits effective? This is a complex question, despite its straight-forward appearance. Simplistically, any program is effective if it accomplishes its goal. In this context, federal tax credits, which are a subsidy to purchasers of a product, may be judged "effective" if they increase the number of purchasers, over and above the number that would have purchased the product in the absence of the subsidy. But public programs nearly always have multiple goals, and the solar tax credit program was no exception. As we have demonstrated, the objectives of the solar tax credits were political as well as programmatic; they incorporated expectations about continuing and long-term impacts on the development of the solar industry, on the environment, and on the security of energy supplies, as well as immediate effects on solar purchasing decisions. Moreover, the effectiveness of any public program, however measured, should be weighed against its costs.

Generally, when we refer to the effectiveness of the tax credits, we mean *cost effectiveness*: the overall balance between the total benefits of the tax credits and their costs. In this context, a comprehensive evaluation would seek to measure and compare the full range of social benefits and costs of the program. The benefits would include not only obvious items such as reduced levels of energy consumption, conserved barrels of oil, or the number of new solar installations attributable to the credits, but also indirect social benefits that, as we noted earlier, were important rationales for the tax credits (e.g., increased consumer confidence in solar technology, reduced pollution, increased fuel diversity, and stimulation of other conservation actions). An obvious cost is tax revenue forgone; less obvious costs include the distributional inequities of tax credits, induced price increases in solar energy systems, and reduced economic efficiency in consumer choices involving energy consumption. A full evaluation of the tax credits would also compare their cost effectiveness with other policies for achieving the same results using criteria such as cost per Btu saved, cost per barrel of oil displaced, or cost per additional solar energy system installed.

Although several studies of the federal solar tax credits have been done since they were initiated in 1978, none approaches the comprehensiveness of a full social benefit-cost accounting. Moreover, there have been no systematic comparisons of the tax credits with other policy options to promote solar commercialization. Most studies are either narrower analyses of the performance of the tax credits measured against a single criterion or surveys of consumers that seek to identify attitude or behavioral changes than can be attributed to the tax credits (such as accelerated decisions to purchase solar systems, or increased awareness of, or preference for, solar energy sources). This limited scope of evaluation studies is probably the result of a combination of factors: the lack of reliable data, the high costs of systematic cost effectiveness analyses, and the difficulties of isolating the influence of the tax credits given the complexity of the processes relating economic subsidies to changes in purchasing behavior. With these limitations, a reasonably complete assessment of what is currently known about the tax credits must be constructed through a review and synthesis of the fragmentary evidence that exists in the general literature.

25.2.1 Evaluation Methods

Evaluations of the federal solar tax credits have been conducted by federal and state agencies, the solar industry and its representatives, and academic and professional policy analysts. Consistent with this diversity, evaluations incorporate diverse methodological approaches. A review of evaluation research on the tax credits must therefore take into account the strengths and weaknesses of the methodologies that have been used.

Five different methods for evaluating the effectiveness of solar tax credits can be identified in the literature: (1) process engineering analysis, (2) econometric analysis, (3) modeling, (4) surveys, and (5) quasi-empiricism (Lazzari 1982). Many evaluations use some combination of these.

Process engineering analysis typically is used to project the energy savings expected from the installation of various energy-saving technologies. Engineering estimates of the energy savings associated with particular types of solar energy systems or conservation technologies can be combined with data on climate, residential housing characteristics, fuel use, and estimates of the number of expected installations of that type of system to yield projections of total energy savings. Such analyses can be helpful in policy planning, but their use in evaluating the actual results of policies and programs is limited. Estimates of energy use often deviate widely from actual energy use observed in the field because of numerous nontechnical factors (Stern 1986).

Econometric analysis involves the use of multivariate statistical techniques, usually regression analysis, to assess the relative contribution that a number of possible causal factors make to changes in a dependent variable of interest (e.g., average residential energy consumption or proportion of a population purchasing solar energy systems). Explicitness of theory, ease of replication, and the apparent rigor of quantitative analysis are among the strengths of this approach. These benefits are counterbalanced by difficulties in obtaining the most appropriate data and in interpreting the results, omission from the analysis of factors that cannot easily be quantified, lack of validity of some of the measurements used, attribution of causal relationships where only association is indicated, and problems with the theory (for example, some analysts have challenged certain assumptions made by economists about the elasticity effects of changing fuel prices; see Kouris 1981; Byrne and Rich 1984).

Modeling studies combine a process engineering approach with economic or econometric analysis to forecast the anticipated market response to alternative values of a variety of input variables, including the existence and levels of solar tax credits. Usually an explicit market penetration model is incorporated into the study's larger model to estimate the number of solar installations expected to result from a given combination of energy prices, solar system prices, climatic conditions, and incentive levels. The strengths of the modeling studies derive from their explicitness and their ability to be adapted easily to a wide range of conditional "what if" questions. These features are often very valuable for policy planning and program management, but they are of little help to evaluators seeking evidence from actual experience. The weaknesses of modeling studies also frequently derive from limitations of their underlying assumptions (e.g., that consumers act solely on the basis of economic rationality and purchasing decisions are based solely on price comparisons) as well as from limitations created by the absence of sufficient reliable data to document a particular model. As a method for evaluation, modeling studies are of limited value because they do not yield experiential information on the causal connections between the existence of solar tax credits and such variables as household energy consumption.

Surveys of actual or potential purchasers of solar systems yield data on the levels of information about solar energy systems and/or the tax credits, the proportion of consumers of a given type who purchased solar systems, and the importance of the tax credits relative to other factors that influence decisions to purchase solar systems. Data from carefully done surveys reveal much about the factors that affect energy-related behavior, but validating the data has been extremely difficult and costly. In addition, surveys often suffer from problems of low response rates, non-representative samples, and bias or ambiguity of results due to limitations of the survey instruments.

Quasi-empiricism involves the analysis and interpretation of secondary data in the absence of an explicit theoretical model or research design. Researchers use data on trends and levels of activity to draw limited conclusions about the processes involved. One example of quasi-empiricism is the use of tax return data on the number of persons claiming credits, the amounts claimed, and the types of systems purchased to draw inferences about the effects of the tax credits. Although inexpensive, such studies reveal little or nothing about causal relationships between the existence of the tax credits and the number of additional solar energy systems that were purchased.

Few of the evaluations conducted to date restrict their attention to the effectiveness of the federal solar tax credits alone. Within the larger evaluation literature, there are studies of the combined federal solar and conservation credits embodied in the 1978 act, amended in 1980. There are also evaluations of the various state incentives intended to conserve energy and/or to stimulate the purchase of solar energy systems. Most studies have focused on residential credits alone, excluding business tax credits.

25.2.2 Modeling, Process Engineering, and Econometric Studies

Studies using the modeling approach, or a combination of modeling, process engineering, and econometrics, were the first contributions to the evaluation literature on the solar tax credits. Many of these studies (Bezdek et al. 1977; Bezdek, Hirshberg, and Babcock 1979) actually predated passage of the credits and were used to estimate the cost of the tax credits to the federal government and their impact on energy use. It is useful to review a few of the major modeling studies to document evolving expectations of performance and to compare their findings with those of evaluations based on actual experience.

Two studies commissioned by DOE's Office of Conservation and Renewable Energy used market penetration modeling and economic analysis to estimate the economic value of the energy savings induced by tax credits and the effects on the treasury of different levels of tax credits over the period 1980 to 1990. A study by Arthur D. Little, Inc. (ADL 1981) examined the federal tax revenue consequences of five different levels of tax credits for a variety of renewable energy equipment (solar ponds, flat-plate collectors, evacuated tubes, parabolic troughs, wind turbines) intended for industrial markets. The study suggested that the economic value of the estimated energy savings would exceed the net cost to the treasury up to a tax credit level of 70%. Part of the study involved interviews with a small number of firms that had purchased solar equipment. Purchasers indicated that the payback period was too long to justify solar purchases on economic grounds, and they cited such factors as fear of fuel shortages, pioneering spirit, and public relations benefits as instrumental in their decisions to invest.

The second study, by Urban Systems Research and Engineering (1981), also estimated the net federal revenue consequences of various levels of tax incentives for solar thermal and wind electric industrial equipment. Using an internal rate of return criterion to determine the profitability of each solar technology, the study concluded that the competitiveness of solar equipment was sensitive to the level of the tax credits, but the dominant influence would probably be the rate of increase in energy prices. In the direct heat market, the study concluded, net federal tax revenues would increase over the life of all types of equipment taken together, but in the case of process steam they would decrease over the life of the equipment if the level of subsidy was high. The most serious shortcoming of this and the Arthur D. Little study is, of course, that their conclusions regarding the tax credits were not based on direct empirical evidence.

Charles River Associates (1981) developed an econometric model of household investments in insulation, storm windows and doors, and solar water heaters over the period from 1977 to 1985. Using process engineering methods combined with economic analysis, the study found that net social benefits (defined as energy savings per dollar of tax revenue lost) from energy tax credits were greatest for insulation and storm windows, while solar water heaters resulted in much lower values. Several, possibly unrealistic, assumptions were made, however, that limit the confidence one may have in the results: for example, that a household's energy consumption behavior does not change after the investment is made and that all households have the same discount rate and attitudes in analyzing decisions. In addition, noneconomic factors likely to influence the purchasing behavior of households were excluded from the analysis (Lazzari 1982, 32; GAO 1982, 12–13).

In 1982 Lazzari reported the results of an econometric analysis of the federal solar and conservation tax credits, using as his criterion for effectiveness variations in levels of household energy use for the period 1960–1979. The analytical question posed was, what amount of reduction in household energy use could be attributed to introduction of the tax credits in 1978 relative to alternative explanations, such as the price of energy, the price of energy conserving items, and climate? Forty regressions were run under different assumptions. The results showed that the coefficient of the tax credit variable usually was insignificant and had the wrong sign. Lazzari (1982, 56) concluded that "there is at this point little, if any, evidence that they [the tax credits] have proven to be an effective tool of energy conservation." His explanations for this result are severalfold:

• The rise in energy prices dwarfed the effect of the tax credits;

• The rate of credit for most conservation devices was low (15%);

• Householders lacked information about the credits;

• The initial cost of conservation equipment was too high; and

• Landlords and tenants had little incentive to purchase conservation equipment.

Lazzari's analysis had several weaknesses in addition to those inherent in the econometric method. As he observes, the price of energy conservation devices was omitted because of the lack of data; the measure used for the tax credit was not its effective rate; the analysis probably was conducted before some of the effects of the tax credits materialized and therefore it was too early to draw conclusions (Lazarri 1982, 57–58).

Sawyer and Wirtshafter (1984, 1017) recently observed that "despite the cost and widespread commitment to tax incentives, their effectiveness has not been rigorously tested in a systematic, nationally based analysis." This was attributed in part to methodological and data problems. Because the credits are universally available, there is no concurrent control group.

In addition, the use of time series analysis is limited by the number of years of available data and the inability to discriminate among the many factors that influence solar and conservation equipment sales (Sawyer and Wirtshafter 1984, 1018).

Sawyer and Wirtshafter sought to overcome these constraints by analyzing the effect of state tax incentives on the number of residential solar installations in 1980 and 1981. They first tested for significant relationships between the size of a state tax incentive and the state's solar adoption rate (the proportion of attached and detached single-family dwellings with solar domestic hot water installations). They found no significant relationship regionally or nationally, for either year, for any combination of incentives (income tax credits, sales tax, property tax, etc.). A multiple regression analysis that included incentive values, energy prices, energy expenditures per household, climate, and several other variables showed that state incentives (of all types) explain about 9% of the variation in state adoption patterns. The authors concluded that, while no statistically significant relationships could be identified, the solar adoption process is too complex and poorly understood to dismiss the incentives altogether (Sawyer and Wirtshafter 1984, 1017). They list several limitations of their analysis: reliance on overall average system costs and homeowner income levels, the unknown effect of the 40% federal credit, and the omission of possible regional preferences for space heating and passive designs over domestic hot water systems.

Using the same DOE data on 1980 and 1981 solar installations by state, Lancaster and Berndt (1984) performed an analysis similar to that of Sawyer and Wirtshafter but drew somewhat more positive conclusions about the effectiveness of the tax credits in stimulating solar purchases. Lancaster and Berndt employed a slightly different measure of solar adoption rate (square feet of collectors per capita). Their data contained an implicit "solar attribution rate"—that is, the percentage of solar installations actually attributable to the tax credits—which was determined to be 33.4%. They also found that a 10% difference in the magnitude of tax credits was associated statistically with a 1%–3% difference in the solar collector installation rate.

These latter two studies of state tax credits do not address directly the effectiveness of the federal credits, but they are useful in assessing the influence that tax credits and other financial incentives have on decisions to purchase solar systems. From a methodological standpoint, their underlying model, which predicts the behavior of solar purchasers state by state, is of greater validity than a model that predicts the effects of the federal credits alone, because the size of the credits and the dependent variable selected can exhibit much greater variation. Still, it is problematic to draw inferences about the effectiveness of federal credits from experience with state credits alone. One reason is the role that variable factors such as levels of information about the credits play in explaining solar adoption rates. A second, noted by Sawyer and Wirtshafter (1984), involves interaction effects between state and federal tax credits; the 40% federal credit may dilute the influence of state credits.

In a final example of evaluations of this type, Sav (1986) combined economic and engineering process models to create a dynamic model of successive investment decisions in capital stock—conventional, solar, or mixed—that would produce hot water for residential and commercial buildings. Sav's program computed the optimal time to invest in conventional and solar hot water systems in the Northeast (Boston) and Southeast (Atlanta) under different assumptions about the price of solar systems, the price of conventional fuels, the cost of capital, the discount rate, and the existence of tax credits and other financial subsidies to purchasers of solar systems. Under base-case assumptions (no tax credits, 20% discount rate, solar energy system prices at the level prevailing in 1979), Sav found that neither the level of tax incentives reflected in the 1978 tax act nor a "full acceleration" level (reflecting both the federal tax credits and state incentives exempting solar systems from sales and property taxes) would induce consumers to substitute solar for oil-fired systems at the reference (1979) system price of $28 per square foot ($300 per square meter). For electrical systems, Sav predicted substitution to occur around 1990 at best, but generally well beyond 2000.

Sav's results, which were highly sensitive to the discount rate assumed, are interesting because they suggest that the level of subsidy required to induce substitution within a short time is very high. Because it is well known that consumers discount the future heavily when purchasing equipment such as appliances and hot water heaters, Sav concludes that producers, rather than consumers, should be the target of subsidies. Sav's results suggest that, for market penetration to have occurred in the electrical sector in 1979, two-thirds of the price of solar systems would have to have been subsidized (Sav 1986, 65). As is the case with most of the other studies described in this section, Sav's analysis is more useful for planning than for evaluation. Like many other market penetration models, Sav's assumes that consumers make decisions on financial grounds alone. Nevertheless, had this type of dynamic investment model been employed during the early stages of solar policy planning, it might have alerted policymakers to the size of subsidies likely to be required and to the importance of accurate estimates of the consumer discount rate.

25.2.3 Survey Research

A substantial number of studies of solar tax credits are based on survey research results that indicate the reasons people decide to purchase solar energy systems and also offer information about the significance of tax credits as an incentive. Unseld and Crews (1979) completed a review of eleven studies that relied on questionnaires to gather data from solar users. Most of the samples were small and unrepresentative of all solar users, and few directly addressed the question of financial incentives. The findings of these surveys showed that respondents claimed that both financial and nonfinancial factors played a role in their purchase decisions. Results were inconclusive on the question of the relative importance of tax credits and other financial incentives on decisions to purchase solar energy systems.

Sawyer and Feldman (1981) reported the results of a survey of researchers that sought to assess their judgment of the barriers to residential solar commercialization and the incentives necessary to overcome these barriers. High initial cost was perceived as the most significant barrier and, accordingly, income tax credits were identified as the most important incentive required. Sawyer and Feldman also reported the findings of homeowner assessments of the barriers to solar energy use based on two surveys of homeowners (with samples of 177 and 179) who had purchased solar domestic water and space heating devices. In both surveys high initial cost was seen as the dominant barrier.

A comprehensive study of solar purchasers was part of the National Study of the Residential Solar Consumer conducted by the Solar Energy Research Institute (*Solar Age* 1981, 29–31). SERI surveyed 3,800 solar homeowners; the results showed that initial cost was one of the most important concerns of purchasers *before* they made their decisions. In retrospect, purchasers identified many advantages of solar energy systems as important to their decisions to purchase; among them, three were related to cost and five to noncost factors. The before and after wording of the questions and the noncomparable lists of factors in the two sets of survey questions complicated the process of drawing conclusions about the relative significance of financial incentives.

Petersen (1982) surveyed consumer attitudes and practices toward conservation and solar equipment in eleven states. Most of his respondents were aware that tax credits for solar and conservation expenditures were available, but relatively few (about 10%) said they made decisions to purchase in response to the credits. This response, combined with information on the respondents' tax burden and the size of the solar purchase made, led Petersen to conclude that the tax credits stimulated a 12.3% increase in conservation and solar expenditures and that they largely resulted in a windfall for taxpayers with above-average incomes. The Petersen study has been criticized, however, on the grounds of biased sampling of states, unexplained differences between the data used and IRS tax return data in six of the eleven states, and low validity of Petersen's model relating respondents' demographic characteristics to incentive effects of the tax credits (Lazzari 1982, 34–35).

Carpenter and Chester (1984) collected data on the use of the federal energy tax credit from a survey of 8,369 households in ten western states in 1981. Although much of the reported analysis deals with the tax credit for conservation, some data are provided on the solar tax credit. Carpenter and Chester report that 63% of those households who had installed solar hot water or space heating indicated that they would have made improvements without tax credits. They compare this figure with the responses of households who had *not* made solar investments but had invested in conservation activities; fully 95% of these households reported they would have made improvements without tax credits. They conclude that the tax credits may be more important for larger, more costly installations (such as solar hot water and space heating) than for less costly conservation measures.

Petersen (1985) has criticized the Carpenter and Chester (1984) study, claiming that serious problems with the data cast doubt on the findings. In particular, he argued that the study was characterized by ambiguous questions and that the assignment scheme used to classify responses may have resulted in underestimation of the actual importance of the tax credit to solar investors. Petersen (1985, 133–135) reports the results of a later survey of 5,820 households in eight western states in 1983 using a survey instrument similar to that used in the 1981 survey by Carpenter and Chester:

> Only about 9 percent of those reporting solar expenditures indicated they would have spent the same amount without a tax credit. Over one-third stated that no money would have been spent on solar if the credits had not been available ... [T]he renewable energy credit probably provides a substantial stimulus for investing in qualifying systems ... [T]he data imply that the federal tax credit for purchase of renewable energy systems has significantly increased the demand for solar energy systems. The cost of a solar space or water heating system can range from $2,000 to $15,000. Often it is impossible to justify such systems using conventional cost-benefit analysis. But with an energy tax credit of 40 percent of system cost, the tax saving may be sufficient to make the system cost-effective.

25.2.4 Quasi-Empirical Studies

Quasi-empirical studies of solar tax credits typically have presented descriptive information such as the proportion of taxpayers claiming the credit, their household income, the average size of the claim, and a breakdown of claims by type. Little about the effectiveness of the tax credits, other than that they have not been ignored and that the claimants tend to have above-average incomes, can be inferred from such data.

Typical of these kinds of analyses is one for the Solar Energy Industries Association (SEIA) by Robert R. Nathan Associates (1985). Nathan used tax credit claims data to show that claims for the federal credit increased threefold over the period 1978–1982. Also, a sharp increase in claims between 1979 and 1980 was attributed largely to the increases in the size of the federal credit and in the ceiling on claims that went into effect in April of 1980. Nathan classified the claims data by states, grouped according to whether the state had its own solar incentives, to show that relatively more claims were filed in states that had additional incentives than in states that did not. Of course, trend data on claims alone are at best a weak basis for drawing inference about the effectiveness of the tax credits. Oil prices rose dramatically in the middle of the period covered by Nathan's analysis and, as noted in the previous section, many nonfinancial factors influence consumer purchasing decisions.

One quasi-empirical analysis that extends this line of thinking is Roessner's (1982) examination of the relative effectiveness of financial and nonfinancial incentives on decisions to purchase residential solar energy systems. Roessner combined findings from research on the diffusion of innovations, studies of solar purchasing decisions, and analyses of state solar tax credits. He concluded that nonfinancial factors—such as system reliability, warranty protection, and confidence in system suppliers and installers—are at least as important as initial system costs to purchasers during the early stages of market penetration. This conclusion rests on inferences drawn from weak data, however, most of which are only indirectly related to the effectiveness of the tax credits. Roessner also argues that federal and state solar tax credits provided the solar industry with a useful marketing tool, a conclusion supported by industry surveys.

25.3 Research Conclusions and Policy Experience

In composite, what do the various studies indicate about the cost effectiveness of federal tax credits? Earlier we referred to several evaluation reviews of solar tax incentives that appeared in the early 1980s (GAO 1982; Lazzari 1982; Rodberg and Schachter 1980). These reviews are consistent in concluding that, at the time they were conducted, evidence was insufficient to enable firm judgments about the cost effectiveness of tax credits. Thus, with respect to state solar and conservation tax incentives, Rodberg and Schacter (1980, 35) state:

There is an important and legitimate role for state tax incentives in promoting conservation and renewable energy. It is not possible, however, on the basis of the present evidence, to conclude that the incentives that now exist have stimulated new investment.

The GAO's (1982, 11) review of twelve studies of the federal conservation and alternative energy tax credits was similarly equivocal:

These findings provide some insight into the tax effectiveness issue, but because in many cases they are preliminary and limited to a few residential energy conservation and renewable tax incentives, they do not provide a solid foundation for formulating policy. No current reliable estimates are available on the energy savings or the production effects for the majority of the tax incentives available.

Lazzari's previously cited 1982 evaluation reviewed seven studies of the effectiveness of the federal residential solar and conservation tax credits. He concluded that "none of the studies conclusively determines the overall effectiveness of the energy tax credits" (Lazarri 1982, 28).

More recently, Sawyer and Lancaster (1985) reviewed the available studies of state renewable energy tax incentives. They, too, were unable to reach a firm conclusion about the cost effectiveness of state tax credits as a tool for saving energy or stimulating the spread of solar equipment.

On the basis of available empirical studies, we can be no more conclusive than earlier analysts concerning the documented overall cost effectiveness of the federal credits in solar commercialization. Funds for substantial evaluation efforts in the alternative energy field declined drastically with the budget constraints and changed priorities of the Reagan Administration. With dramatic declines in the resources devoted to evaluation studies, little new empirical analysis has become available to enrich or alter the conclusions reached by previous reviews. Indeed, given the available research, virtually nothing can be said about the performance of the business tax credits. The results of studies of the residential tax credits generally indicate a positive influence on the rate of solar commercialization, but no study offers a definitive assessment of the magnitude of that influence. On the more complex issue of cost effectiveness, the evidence is mixed and incomplete.

But even though the various studies do not provide conclusive evidence for or against the cost effectiveness of the tax credits, they do provide insights into the strengths and weaknesses of the solar tax credit program as part of federal energy policy. Financial cost figures for public revenues forgone or attribution rates for solar installations, even if complete and reliable data were consistently available for analysis, provide only one vantage point for evaluating the tax credits. A comprehensive policy evaluation should consider a large number of additional factors, many of which are difficult to measure with accuracy and precision. Many of the strengths and weaknesses of the solar tax credits are listed in table 25.2, and these should be taken into account in assessing the overall advantages and disadvantages of the federal tax credit program.

25.3.1   Strengths of the Solar Tax Credits

As noted in section 25.1 anticipated social benefits—such as reduced environmental pollution, increased fuel diversity, decreased vulnerability to energy supply interruptions and price increases, and stimulation of other conservation actions—loomed as important rationales for federal efforts to promote solar commercialization. The tax credits should be assessed at least in part on the basis of the social benefits derived from renewable energy use. Moreover, as Sawyer and Lancaster (1985, 188) point out, "revenue losses caused by the credits are one-time expenses, while social

Table 25.2
Summary of strengths and weaknesses of solar tax credits

Strengths
- Promote the social benefits of renewable energy use.
- Revenue losses are a one-time expense, while benefits continue through the operating life of the technology.
- Expenditures are small compared to subsidies for convertional energy sources.
- Help to overcome the institutional barriers to solar commercialization.
- Promote wider choice among energy options.
- Stimulate growth in the renewable energy industry.
- Easily adapt to different renewable energy technologies.
- Relatively easy and inexpensive to design and administer.

Weaknesses
- May divert resources from more cost-effective energy-saving alternatives.
- May encourage installation of inferior technology that would not survive under fully competitive market conditions.
- Produce windfalls for those who would have purchased the new technology without subsidy.
- Encourage abuse as a tax shelter.
- Provide financial benefits to the relatively affluent rather than to lower income groups, who suffer the greatest burden of high energy prices.
- Unresponsive to variations in cost effectiveness of solar technology across different geographic regions.

benefits continue through the operating life of the technology." But social benefits are difficult to measure and typically have been undervalued in assessments of the tax credits. Moreover, it is difficult to isolate the social effects of any single policy factor such as the tax credits. As a result, some of the continuing advantages of federal efforts to promote renewable energy use have been neglected in evaluations that focus heavily on short-term, quantifiable variables.

Although the total revenue loss resulting from the tax credits may appear substantial, it was modest compared to the overall cost of federal energy subsidies during the same period. According to Morgan (1985), for example, the estimated value of the federal income tax provisions to assist energy industries totaled between $26 and $28 billion in 1984 alone ($27 to $29 billion in 1985 dollars, cited in Sawyer and Lancaster, 1985, 171–172). Other estimates are even higher. The Center for Renewable Resources (CRR), for example, calculates that in 1984, the federal government expended $44 billion ($45.5 in 1985 dollars) in energy subsidies. Even with the tax credits and other supports provided for renewable energy and conservation alternatives, "the bulk of federal subsidies—more than $41 billion [$42 billion in 1985 dollars] ... [went] to mature energy technologies which long ago reached commercial status" (Heede et al. 1985, 4–5). Moreover, subsidies would seem most defensible in cases like the solar tax credits that direct public resources to new technologies, expanding the range of energy choice, that promise greater energy or economic efficiency, and that face institutional barriers to commercialization. In this context, the costs of the tax credit program may be viewed as a small and necessary counterbalance to the market distortions and institutionalized inertia created and sustained by decades of subsidizing conventional energy options.

The rate of growth of solar energy use when tax credits were in force has been significant and much greater than other energy sources receiving far larger subsidies during the same period (Flavin 1985). The Department of Energy's Office of Renewable Energy (DOE 1985, 9) estimates that between 1975 and 1984 the incremental additional contribution of renewable energy to the energy supply base of the United States was equivalent to $39 billion in 1982 dollars ($43.5 billion in 1985 dollars). Although it is not clear precisely what role the credits played in this growth, it is clear that the tax credits have worked together with additional factors—specifically higher prices for energy from conventional sources and declining prices for some renewable options—to generate substantial increases in the use of renewable energy. Further, Heede et al. (1985) have conducted an analysis of the incremental additions of energy produced per federal tax subsidy dollar; they conclude that even though the benefits of investment have been greater for conservation than for renewables, the benefits of investment in renewables are still many times the benefits of investments in fossil and nuclear energy.

However unclear the cost effectiveness of the tax credits may be, there appear to have been a number of real market stimulation effects. Among these, the most important may have been the support that the credits offered for the growth of the solar industry. The Office of Technology Assessment (OTA 1985, 9) claims that "the current system of Renewable Energy Tax Credits has been an important contributor to the Federal policy of supporting the infant renewable energy industry." Strengthening the solar energy industry means that some of the short-term costs of the tax credits in forgone public revenues may be offset by long-term gains in economic development that result in new investment, employment, technological innovation, and public revenues. There is evidence of substantial growth in the industry during the period the tax credits were in force, and industry representatives and advocates argued that earlier growth would not have been as rapid without the credits and that the industry would continue to be severely handicapped if the tax credits were discontinued.

In addition to the strengths of the tax credits noted above, there are the other advantages described earlier: the tax credits were relatively easy and inexpensive to administer; they promoted competition, experimentation, and flexibility in consumer choice of solar technologies; and, at least for a time, they had substantial political appeal.

25.3.2 Weaknesses of the Solar Tax Credits

Solar tax credits exhibit a number of weaknesses as an energy policy instrument. One of the most significant of these is the potential they create for diverting resources from least-cost energy alternatives. For example, the credits can encourage installation of high-cost solar energy systems compared with equivalent or greater benefits from conservation programs at perhaps much lower costs to government and the consumer. There is no evidence that, during the period in which tax credits were in effect, consumers substituted subsidized solar purchases for improvements in energy efficiency or that the tax credit program was responsible for reducing federal commitments to energy efficiency improvements. Nonetheless, Sawyer and Lancaster (1985, 178) question the wisdom of spending public resources on tax incentives for renewable energy development "when many of the same objectives may be achieved more cost-effectively through programs to increase energy efficiency."

A potentially serious criticism of the tax credits is that they may discourage least-cost alternatives within solar energy markets. By supporting some solar technologies and excluding others, they have the potential of leading markets to higher-cost designs and away from less expensive options (like passive solar design). The Office of Technology Assessment (OTA 1985, 6, 16) points out that there are instances where the tax credits have "prompted installation of inferior technology that has little possibility of commercial success" and that if a subsidy is to be provided perhaps it "should be awarded on the basis of energy produced regardless of the technology employed, thereby reducing the possibility of artificially supporting inferior technology."

Tax credits have been criticized for promoting other inefficiencies in solar energy markets. When solar energy systems are subsidized by the government, producers and installers may be less conscious of costs and consumers less sensitive to price. The tax credits allowed suppliers to inflate initial costs because the subsidy provided a substantial guaranteed discount to the purchaser and because the maximum allowable expenditure was higher than some, including representatives of the industry, believed was appropriate to encourage prudent purchases. The inflated prices of solar energy systems, in turn, as well as the lack of quality assurance through certification requirements, made them difficult to sell without government assistance and contributed to a perception of solar systems as too expensive and unreliable. The tax credits have been further criticized for subsidizing purchases that would have been made in the absence of the credits. In this case, the tax credits would be " 'windfalls,' payments to people who would have pursued renewable energy regardless . . . " (Sawyer and Lancaster 1985, 178).

The Office of Technology Assessment (OTA 1985, 16) has pointed out that the business tax credits were abused and that in some cases "investors with considerable income have used projects primarily as tax shelters, benefiting even if the projects fail to operate properly and provide any gain for the technology." Similarly, the General Accounting Office's review of business tax credits also indicated cases of abuse. The study pointed to Internal Revenue Service reports of a large number of "inappropriate tax benefit claims and the use of the credit in alleged abusive tax shelter schemes" (GAO 1985, 4). The IRS examination of claimed business energy credits has resulted in recommendations for substantial disallowances, although most of the recommendations have been disputed by the taxpayers. The recommended disallowances for corporations generally involved questions of when a property was acquired or placed in service or the eligibility of the property. "The recommended disallowances on the partners or other individual returns were based, in all cases, on the revenue agent's determination that the investments involved 'abusive tax shelters' " (GAO 1985, 20).

The tax credits may also be criticized as inequitable because they largely assist the more affluent segments of the population rather than low-income households that experience the greatest burden of higher energy prices. Individuals and firms must generally have sizable taxable income to offset in order to benefit fully from such subsidies. Further, credits only help those who have, or can obtain, sufficient capital to make the investment even after the subsidy has covered part of the cost. Tax credits, therefore, appeal to higher-income groups, and there is evidence that the greatest proportion of claims comes from middle- and upper-income homeowners (Roessner 1982; Koontz and Neuendorffer 1981).

Finally, the solar tax credit legislation makes no provision for regional differences in the cost effectiveness of solar energy systems. The tax credit could be claimed whether or not claimants' systems were cost-effective without a subsidy, or whether their payback periods were longer than the life of the system. In this sense, Congress implicitly defined the potential solar market as all homeowners in the United States, even though active solar heating, for example, is less cost-effective in some parts of the United States where winter insolation levels are relatively low (Roessner 1982, 9).

25.4 Conclusions and Lessons Learned

The available evidence suggests that changes in energy prices are far more important than tax incentives in influencing consumers to purchase residential solar energy systems. High and stable prices for conventional energy options encourage both energy efficiency and the development of alternative energy sources, including solar. This, however, hardly resolves the question of the viability of the tax credits as part of a federal commercialization policy for solar energy. The credits were instituted in response to energy prices that were not only high but volatile. In addition, these higher energy prices still did not incorporate the social and environmental costs of energy use. Moreover, the impact of the tax credits was expected to result from the confluence of many factors, prices included, rather than from the isolated influence of tax credits alone. The key policy issue is not whether tax credits are more influential than other factors in influencing the future of solar markets but, rather, whether tax credits constitute an appropriate, viable, and cost-effective component of federal government policy aimed at promoting the commercialization of solar technologies.

25.4.1 Tax Credits and Commercialization Policies

The viability of tax credits as policy instruments to promote commercialization needs to be assessed in the context of the expectations that accompanied their enactment and that provided the rationale for their role in federal solar policy. When so assessed, it is not at all certain that tax credits were the most appropriate policy vehicle to facilitate growth in solar markets, given the state of the solar industry and solar technologies in the mid- and late 1970s. The tax credits were created on the basis of assumptions that technologies for solar hot water and space heating were market-ready and therefore that commercialization policies should focus on stimulating demand. There is, however, reason to question whether the state of the industry and the technology delivery system justified the assessment of market-ready status (Roessner 1984).

Since the mid-1970s, both the entry rate and exit rate have been high for businesses in the solar heating industry. Technical and investment barriers to entry have been low relative to many other manufacturing industries, encouraging entry, but the exit rate has been high. One observer of the solar industry in Florida estimated the turnover rate in that state at between 25% and 30% annually, while the authors of a study of solar manufacturing activity in the San Francisco Bay area estimated that approximately one-third of the solar businesses exited the market each year (Rogers et al. 1980). Observers have attributed the high exit rate to the often inadequate technical and business skills of entering solar entrepreneurs. Few companies reported an adequate return on investment and many suffered net losses; observers of the industry argued that sales were insufficient to sustain the existing number of firms, though a "thinning out" process had not yet occurred (Roessner 1984, 239).

The solar heating industry clearly was in its early stages when the demand-side commercialization strategy was formulated; it was a new industry, based on what appeared superficially to be an "old" technology. The industry exhibited a weak technical base, poorly developed infrastructure, and, with some notable exceptions, inadequate business management and marketing skills; it did not possess the resources to train installers properly, correct technical problems with system components, or conduct significant R&D. Residential and commercial demonstration programs have been the subject of several evaluations supporting the judgment that solar heating technology was, in the mid- to late 1970s, still characterized by substantial technical uncertainties (HUD 1980; GAO 1980; Koontz, Genest, and Bryant 1980).

Because substantial uncertainties about the technology's performance and reliability remained, Roessner argues that the "commercially ready" label applied by the government was inappropriate at the outset.

Demand-oriented commercialization strategies (such as financial incentives to potential purchasers) are appropriate when a technology works and works well and when the key elements of the technology delivery system are in place and functioning well. Roessner concluded that, in the case of solar heating, these assumptions did not hold. A strategy suited to a competitive or mature industry was applied to an emerging one. A more appropriate strategy, he argues, would have recognized the systemic weaknesses in the solar heating industry and allocated resources to supply and infrastructure elements of the delivery system in amounts that balanced those devoted to stimulating demand (Roessner 1984, 243–244).

Despite apparent mismatches between the commercialization strategy represented by the tax credits and the state of solar industry and technology, it seems clear that during the period in which the federal and state tax credits were in effect, there was significant growth in solar energy markets. The Center for Renewable Resources (Heede et al. 1985, 16–17) reports that:

In 1974 there were only 45 companies engaged in the manufacture of solar collectors in the U.S. By 1984 there were more than 200. The volume of their production has increased from 1.2 million square feet in 1974 to 16.8 million square feet in 1983. And the number of people employed in production, sales, and installation by some 2,600 firms is estimated at 30,000.... Nationwide, the number of active solar installations was estimated at 1.1 million at the end of 1983.

It is not clear how much of the growth in solar markets may be directly attributable to the availability of the credits, but the solar energy industry undoubtedly benefited from them. Moreover, some analysts have argued that the tax credits, despite a number of liabilities, were more appropriate and effective than other facets of the federal commercialization policy for solar energy and that their flexibility contrasted favorably with policies that concentrated on specific technologies and designs. After reviewing policy experience with a range of solar commercialization policies since 1974, Eugene Frankel (1986, 82) concluded that tax credits, "which allow the marketplace to select the technology and the application but guide the market in directions favored by overall national policy ... appear to be effective in aiding the diffusion of new technologies...."

25.4.2 Political Aspects of the Solar Tax Credits

Energy commercialization policies need to be understood as outcomes of the prevailing politics of technology policy. During the late 1970s tax

credits played important symbolic roles as evidence to consumers that
government had confidence in solar technology and as evidence to taxpayers that policymakers were acting to help solve the energy crisis and
reduce energy costs. The political appeal of tax credits and the desire
to show that government was taking some action to ameliorate energy
problems probably also accounts for the underestimation of the obstacles
to successful commercialization of solar technologies. Frankel (1986, 83)
has warned, that "probably the most serious problem posed by politics
for technology development is the balancing of a desire to demonstrate a
commitment or achieve an early result with the actual time it takes to
accomplish significant technological change."

As the sunset date for the credits approached, much of the initial
political capital of the tax credits had been exhausted; the solar industry
lobbied hard for extension. Eventually, business energy investment tax
credits for solar systems were restored retroactively as part of the 1986 tax
reform bill. The business credits were set at "15% in 1986, 12% in 1987,
and 10% in 1988 for solar thermal and photovoltaic systems; 15% for
three years for ocean thermal energy conversion (a credit that has not yet
been claimed); 15%, 10%, and 10%, respectively, in the three years for
geothermal energy; and 15% in 1986, 10% in 1987, and zero in 1988 for
bioenergy conversion" (*Solar Energy Intelligence Report* 1986, 303). Subsequently, the business credit was extended at the 10% level through 31
December 1989. The credits for wind farms and hydroelectric plants were
not extended.

Efforts to extend the federal credits for residential use, however, were
unsuccessful. In addition, solar tax credits in a number of states have been
eliminated. In large part this may be interpreted, not as a sign of the
ineffectiveness of the credits in promoting solar commercialization, but as
a reflection of the reduced salience of energy issues on the policy agenda
of the 1980s. It is also consistent with general federal government cutbacks in all funding for solar technologies since 1981.

The declining political fortunes of the tax credits have significantly
impacted the solar industry. As Frankel (1986, 79) indicates, the industry
has faced the loss of the credits "before it has had a chance to establish a
solid industrial base." Before the elimination of the tax credits, some
analysts warned that elimination "at a time when renewable energy technologies are making critical headway will push their commercialization
back five to ten years" and that "it will delay mass production and the

growth of the market indefinitely" (Heede et al. 1985, 15, 17). Advocates also pointed out that elimination of the tax credits would return energy markets to a situation where conventional sources receive government subsidies, while no subsidies are available to support renewable options. The Office of Technology Assessment argued that the favorable tax treatment afforded by the credits had proved important to the development and deployment of commercial applications for a number of renewable technologies, and warned that, if favorable tax treatment ceased, such development might be delayed significantly as firms lost access to sources of capital and distribution networks (OTA 1985, 36). In fact, there had been some contraction of the industry even before the tax credits were eliminated. The contraction of the industry since the tax credits were eliminated, however, has been extraordinary. The Energy Information Administration's (EIA) survey of solar thermal collector manufacturing demonstrates a dramatic decrease in industry activity between 1984 and 1986: total shipments of solar thermal collectors declined by more than 70%—from 16.4 million square feet in 1984 to below 4.6 million square feet in 1986—and the total number of firms shipping solar thermal collectors decreased from 224 in 1984 to 97 in 1986 (EIA 1987, 3). How much of this contraction is the result of the loss of the tax credits and how much the result of changes in the prices of other energy sources is not clear. Both factors were mentioned by industry respondents to the EIA's survey as "the key events that led to the decline of the industry" (EIA 1987, 4). At the very least, the elimination of the credits added to the industry's burden at a time when it was already difficult for it to sustain a stable, much less growing, market position. Measured in terms of numbers of firms and total sales, the solar thermal collector industry in 1986 was actually smaller than it had been a decade earlier, before the initiation of the tax credits (EIA 1987, 4).

25.4.3 Summary Assessment

Evidence suggests that the federal tax credits have benefited the solar industry and, in combination with rising conventional energy prices, have increased the pace of solar commercialization. Nonetheless, the overall cost effectiveness of the federal solar tax credits has not been conclusively documented by existing evaluation research. Furthermore, given the state of the industry and the characteristics of the technology delivery system in the late 1970s, it is questionable whether the tax credits were the most suitable instrument to promote solar commercialization at that time.

Despite apparent mismatches between the commercialization strategy represented by the tax credits and the state of the solar industry and technology, it seems clear that during the period when the tax credits were in effect, there was significant growth in solar energy markets. Indeed, evidence suggest that the federal tax credits benefited the solar industry and contributed to its development. The credits played an important role in supporting the multiple objectives embodied in solar commercialization. Beyond their particular strengths and weaknesses as a policy instrument, the federal tax credits performed a valuable political function: they gave practical expression to an active federal role in solar commercialization at a time when expansion of the range of energy choice was recognized as having important social, environmental, and national security benefits. Given the nature of existing energy markets, the history of federal energy subsidization policies, and the anticipated social and environmental benefits of greater renewable energy use, there remains ample political justification for an active federal role in solar commercialization.

With appropriate attention to timing and to the state of the technology and its supporting infrastructure, a well-designed tax credit program— one with realistic ceilings for allowable expenditures, certification requirements for systems, performance-based incentives, and effective monitoring to prevent abuse—can be a viable part of government technology commercialization strategies. What is ironic, however, is that political support for solar tax credits, was exhausted by the mid-1980s, just when the conditions in energy markets may have enhanced the credits' impact and when financial incentives might have dominated other factors in the eyes of prospective purchasers. The elimination of the residential tax credits has contributed to the contraction of the solar industry, and the federal government has adopted no other policy to facilitate solar commercialization.

References

ADL (Arthur D. Little, Inc). 1976. *Residential Solar Heating and Cooling Constraints and Incentives: A Review of the Literature*. Cambridge, MA.

ADL 1981. *The Cost of Federal Tax Programs to Develop the Market for Industrial Solar and Wind Energy Techniques*. Cambridge, MA.

Daniel Rich and J. David Roessner787

Bezdek, R. H., et al. 1977. *Analysis of Policy Options for Accelerating Commercialization of Solar Heating and Cooling Systems*. Washington, DC: George Washington University.

Bezdek, R. H., A. S. Hirshberg, and W. H. Babcock. 1979. "Economic Feasibility of Solar Water and Space Heating." *Science* 203:1214–1220.

Booz, Allen & Hamilton. 1976. *The Effectiveness of Solar Energy Incentives at the State and Local Level*. Bethesda, MD.

Byrne, J. and D. Rich. 1984. "Deregulation and Energy Conservation: A Reappraisal," *Policy Studies Journal* 13:331–344.

Carpenter, E. H., and S. T. Chester, Jr. 1984. "Are Federal Energy Tax Credits Effective? A Western United States Survey." *Energy Journal* 5:139–149.

CEC (California Energy Commission). 1983. *California's Solar, Wind and Conservation Tax Credits*. Sacramento, CA.

Charles River Associates, Inc. 1981. *An Analysis of the Residential Energy Conservation Tax Credits: Concepts and Numerical Estimates*. Boston.

Cole, R. J., B. W. Somers, P. Eschback, C. Sheppard, W. J. Lenerz, D. E. Huelshoff, and A. A. Marcus. 1981. *A Comparison of the Incentives Used to Stimulate Energy Production in Japan, France, West Germany, and the United States*. Richland, WA: Pacific Northwest Laboratory.

Cone, B. W., et al. 1980. *An Analysis of Federal Incentives to Stimulate Energy Production*. Richland, WA: Pacific Northwest Laboratory.

Cone, B. W. 1982. "A Historical Perspective on Federal Incentives to Stimulate Energy Production." *Energy*. 7:51–60.

Conway, J. 1983. *Statement before the Subcommittee on Energy and Agricultural Taxation, Senate Finance Committee*. Washington, DC: Renewable Energy Institute.

DOE. Office of Renewable Energy. 1985. "Renewable Energy Profile 1975–1984." Working paper. Washington, DC.

EIA (Energy Information Adminstration). 1987. *Solar Collector Manufacturing Activity 1986*. Washington, DC.

Flavin, C. 1985. *Renewable Energy at the Crossroads*. Washington, DC: Center for Renewable Resources.

Frankel, E. 1986. "Technology, Politics and Ideology: The Vicissitudes of Federal Solar Energy Policy, 1973–1983." In J. Byrne and D. Rich, eds., *The Politics of Energy Research and Development*. New Brunswick, NJ: Transaction Books.

GAO (U.S. General Accounting Office). 1980. *Federal Demonstrations of Solar Heating and Cooling on Commercial Buildings Have Not Been Very Effective*. Report to Congress. Washington, D.C.

GAO. 1982. *Studies on Effectiveness of Energy Tax Incentives are Inconclusive*. Washington, DC.

GAO. 1985. *Tax Policy: Business Energy Investment Credit*. Report to the Joint Committee on Taxation, Congress of the United States. Washington, DC.

Heede, H. R., R. E. Morgan, and S. Ridley. 1985. *The Hidden Costs of Energy: How Taxpayers Subsidize Energy Development*. Washington, DC: Center for Renewable Resources.

HUD (U.S. Department of Housing and Urban Development). 1980. *Interim Report on Performance Data from the Residential Solar Demonstration Program*. Washington, DC.

IRS (U.S. Internal Revenue Service) *Internal Revenue Code, 1979*. St. Paul, MN: West.

IRS. Statistics on Income Division. 1987. "Individual Income Tax Returns, Analytic Table A." Unpublished data. Washington, DC.

Koontz, R., M. Genest, and B. Bryant. 1980. *The National Solar Heating and Cooling Demonstration: Purposes, Program Activities, and Implications for Future Programs.* Golden, CO: Solar Energy Research Institute.

Koontz, R., and J. Neuendorffer. 1981. *State Solar Initiatives: A Review.* SERI TR-722-882. Golden, CO: Solar Energy Research Institute.

Kouris, G. 1981. "Elasticities—Science or Fiction?" *Energy Economics* 3:66–70.

Lancaster, R. R. and M. J. Berndt. 1984. "Alternative Energy Development in the U.S.A.: The Effectiveness of State Government Incentives." *Energy Policy* 12:170–179.

Lazzari, S. 1982. *An Economic Evaluation of Federal Tax Credits for Residential Energy Conservation.* Report 82-204 E. Washington, DC: Congressional Research Service.

Lazzari, S. 1988. *A History of Federal Energy Tax Policy: Conventional as Compared to Renewable and Nonconventional Energy Resources.* Report 88-455 E. Washington, DC: Congressional Research Service.

Maidique, M. A. 1983. "Solar America." In R. Stobaugh and D. Yergin, eds., *Energy Future: Report of the Energy Project at the Harvard Business School.* 3d ed. New York: Random House.

Morgan, R. E. 1985. *Federal Energy Tax Policy and the Environment.* Washington, DC: Environmental Action Foundation.

OTA (Office of Technology Assessment). 1985. *Commercialization of New Electric Power Technologies: Experience with Renewable Energy Tax Credits.* Staff memorandum. Washington, DC, October.

Petersen, H. C. 1982. "Survey Analysis of the Impact of Conservation and Solar Tax Credits." Draft paper prepared for the National Science Foundation.

Petersen, H. C. 1985. "The Impact of State and Federal Solar Tax Credits." Cited in Sawyer and Lancaster 1985, 181–182.

Robert R. Nathan Associates, Inc. 1985. *The Development of Solar Energy and Federal Income Tax Credits.* Prepared for the Solar Energy Industry Association. Washington, DC.

Rodberg, L., and M. Schachter. 1980. *State Conservation and Solar Energy Tax Programs: Incentives or Windfalls?* Washington, DC: Council of State Planning Agencies.

Roessner, J. D. 1982. "U.S. Government Solar Policy: Appropriate Roles for Nonfinancial Incentives," *Policy Sciences* 5:3–21.

Roessner, J. D. 1984. "Commercializing Solar Technology: The Government Role." *Research Policy* 13:235–246.

Rogers, E. M., V. C. Walling, D. Leonard-Barton, and D. W. Gibson. 1980. *Modeling Technological Innovation in Private Firms: The Solar and Microprocessor Industries in Northern California.* Stanford, CA: Institute for Communication Research.

Sav, T. 1986. "The Failure of Solar Tax Incentives: A Dynamic Analysis." *Energy Journal* 7:51–66.

Sawyer, S. W. 1982. "Leaders in Change, Solar Owners and the Implications for Future Adoptions Rates." *Technological Forecasting and Social Change* 21:201–211.

Sawyer, S. W. 1984. "State Energy Conditions and Policy Development." *Public Administration Review*, May–June, 205–214.

Sawyer, S. W. 1985. "State Renewable Energy Policy: Program Characteristics, Projections, Needs," *State and Local Government*, 17(1):147–154.

Sawyer, S. W. 1986. "Outline for Assessment of the Federal Solar Tax Credits." Unpublished memorandum. Golden, CO: Solar Energy Research Institute.

Sawyer, S. W., and S. L. Feldman. 1981. "Technocracy versus Reality: Perceptions in Solar Policy." *Policy Sciences* 13:459–472.

Sawyer, S. W., and R. R. Lancaster. 1985. "Renewable Energy Tax Incentives: Status, Evaluation Attempts, Continuing Issues." In S. W. Sawyer and J. R. Armstrong, eds., *State Energy Policy: Current Issues, Future Directions.*" Boulder, CO: Westview Press.

Sawyer, S. W., and R. M. Wirtshafter. 1984. "Market Stimulation by Renewable Energy Tax Incentives." *Energy* 9:1017–1022.

Solar Age. 1981. September.

Solar Energy Intelligence Report. 1980. June.

Solar Energy Intelligence Report. 1986. 30 September.

Stern, P. C. 1986. "Blind Spots in Policy Anaysis: What Economics Doesn't Say About Energy Use." *Journal of Policy Analysis and Management* 5:200–227.

Thompson, R., and R. Hillelson,. 1982. "Residential Energy Credit, 1978–1980." *SDI Bulletin*, fall, pp. 1–8.

Unseld, C. T., and R. Crews. 1979. *A Review of Empirical Studies of Residential Solar Energy Users.* Golden, CO: Solar Energy Research Institute.

Urban Systems Research and Engineering, Inc. 1981. *Analysis of the Impact of Federal Tax Incentives on the Market Diffusion for Solar Thermal/WEGS Technologies.* Washington, DC: Urban Systems Research and Engineering.

Whittington, D. 1985. "An Examination of Tax Credits for Solar Hot Water Heating Systems." *Energy Systems and Policy.* 9:115–140.

26 Financing

Steven Ferrey

26.1 The Solar Bank

Federal financing efforts for solar energy systems were primarily embodied in federal tax credits and the Solar Energy and Energy Conservation Bank (the "bank"). This chapter will focus on the Solar Energy and Energy Conservation Bank in the context of all financing programs, highlighting the relationship of the bank to federal tax credits.

The following sections concern shortfalls in tax incentives and limitations in direct federal financing programs, the impact of President's Carter's Domestic Policy Review on financing issues, the legislative history of the Solar Energy and Energy Conservation Bank as a legislative response, the organizational history, budgets and performance of the Solar and Energy Conservation Bank, and the lessons gleaned from operation of the program.

The Solar Energy and Energy Conservation Bank originated as a mechanism to finance renewable energy and energy conservation improvements for low- and moderate-income households, as well as commercial and agricultural interests. The purpose of the bank was to promote the technology, and especially to bridge financial gaps for access to credit to make solar energy investments. The early executive and legislative branch efforts, while serving well certain segments of the population, left significant pockets unserved.

26.2 Limitations of Other Financing Programs

26.2.1 Tax Incentive Shortfalls

The Energy Tax Act of 1978 created tax credits for conservation and renewable energy resource investments in principal residences.[1] Absentee landlords therefore could not utilize these credits. The credits were nonrefundable and had to be taken against current tax liability,[2] thus low-income families with no taxable income or with minimal tax liability could not take advantage of the residential energy tax credits.

Solar,[3] geothermal, and wind systems and the requisite labor, assembly, and installation expenses were eligible for the credit. The original credit of 30% of the first $2,000 and 20% of the next $8,000, to a maximum of $2,200, was revised to a flat 40% credit on the first $10,000.[4]

Tide II of the Energy Tax Act provided a 10% business credit,[5] later amended to 15%,[6] for alternative energy investments. The credit was in addition to the normal 10% investment tax credit for which renewable energy resource improvements to business property are eligible; this created an effective 25% credit for solar investments in multifamily dwellings. Furthermore, this business credit was refundable, entitling businesses to a rebate on the credit amount in excess of annual tax liability. These credits were later dropped as part of tax reform legislation.

Tax credit mechanisms performed poorly in catalyzing solar energy improvements across all income groups in the population. Internal Revenue Service (IRS) income tax return data showed a maldistribution in the usage of the residential energy tax credit. In 1978, 88% of the residential credits claimed were taken by those with adjusted gross incomes in excess of $15,000, with fully 80% of the credits claimed by those earning more than $20,000 annually.[7] Those with adjusted income in excess of $15,000 constituted only 38% of the returns filed, while those with incomes in excess of $20,000 filed only 25% of all tax returns.[8]

Table 26.1 shows renewable energy credits indexed by income level. Income groups earning less than $10,000 and $5,000 annually, representing 47% and 25% of all returns filed, accounted for only 6% and 0.01% of the renewable credits claimed, respectively. For 1980, a family of four taking the standard deductions and exemptions at the median income level of $19,500 would have a tax liability of $2,165; at 75% of the median

Table 26.1
Total individual residential renewable resource credits, 1978 (before limitations)

Adjusted gross income ($)	Number of returns claiming credits	Percentage	Total credits claimed ($1,000s)	Percentage
0–5,000	8	0.01	70	0.06
5,000–10,000	4,361	6.00	1,612	1.00
10,000–15,000	4,982	7.00	12,129	10.00
15,000–20,000	6,415	10.00	9,405	8.00
20,000+	52,336	77.00	92,324	80.00
Totals	68,102	100.00	115,540	100.00

Source: U.S. Internal Revenue Service, *1978 Statistics of Income*, table 8 (1978).

income level, or $15,000, this tax liability would fall to $1,247; and at half the median income level, or $10,000, a federal income tax liability of only $378 would be incurred.[9] Thus, even spreading the credit over the allowable two years, a family earning $15,000 would not be able to realize the maximum $4,000 solar tax credit; a family earning $10,000 could cumulatively realize less than an aggregate of $800 in energy credits. One consultant to HUD calculated that a tax credit of this magnitude would be unavailable in full to 78% of U.S. taxpayers, and 65% of all homeowners, because of insufficient tax liability to offset the full credit.[10] (The tax credit program is more fully discussed in chapter 25.)

26.2.2   Early Federal Direct Financing Programs

Prior to the creation of the bank, direct financing for solar energy assistance was provided principally by three major programs and a number of smaller programs. First, HUD administered the Residential Solar Heating and Cooling Demonstration Program.[11] Twenty-seven million dollars in grants were distributed in five award cycles to stimulate active and passive solar applications in approximately 23,000 homes. Second, HUD administered a $400/unit domestic hot water initiative program in eleven East Coast states. (These two HUD programs are further described in chapters 9 and 27.) The hot water program resulted in nearly 7,000 units being installed. Third, the Residential Conservation Service (RCS) program required all qualifying regulated and nonregulated utilities,[12] pursuant to state plans, to assist residential utility customers with conservation and solar improvements. Each utility was required to (1) provide information to customers regarding methods of conserving energy, (2) provide a residential energy audit, (3) suggest cost-effective conservation or solar measures as determined by audits, and (4) assist interested customers in arranging financing and installation of selected measures.[13] (The RCS program is described in chapter 29.)

   In addition, a number of federal agencies were empowered to make, underwrite, or purchase in the secondary market home loans that could include the cost of solar improvement or to fund these improvements directly:

• Farmers Home Administration made direct or insured loans in rural areas;

• Rural Electrification Administration made direct or guaranteed loans through rural electric cooperatives;

• Economic Development Administration in the Department of Commerce provided direct or guaranteed loans or grants to projects in redevelopment area;

• Small Business Administration provided loans or loan guarantees to small businesses unable to obtain conventional credit;

• Community Services Administration provided grants for renewable energy projects benefiting low-income individuals or economically depressed areas;

• Department of Housing and Urban Development, through its urban development action grants and community development bloc grants, funneled through local communities, provided funding for renewable energy purposes; and

• Veterans Administration provided direct loans or loan guarantees for veterans who installed renewable energy technologies.

Because none of these programs had as its explicit purpose the promotion of renewable energy technologies, however, their impact was sporadic, uncoordinated, and indirect; they did not provide a consistent financing mechanism for the many solar energy applications they funded. The citizenry was not generally aware of these financing sources, many of which did not provide funding directly to individuals, nor was this financing generally available in all geographic areas.

26.2.3   State and Local Financing Programs

State and local solar financing initiatives were of three types: (1) direct assistance, (2) subsidized loans, and (3) tax inducements. However, these programs were fiscally small, limited in their scope, and inconsistent in their coverage. All of the direct assistance and loan programs predominantly benefited owners rather then renters, and provided minimal assistance for low-income owners. In California, which provided the most ambitious solar tax credit, only 6% of the credits, comprising less than 2% of the total dollar value, were claimed by that half of the population below median family income; 31% of the credits were utilized by those few households with 1977 adjusted income in excess of $50,000.[14] In many cases, the incentives merely facilitated investment of by those already predisposed and economically able to undertake residential energy improvements. (Chapter 29 further describes these state and local efforts, many of which were funded through the regional solar energy centers or RSECs, discussed in chapter 20.)

26.2.4 Domestic Policy Review

In 1978 President Carter convened a Domestic Policy Review (DPR) on solar energy. Its purpose was to examine the state of solar technologies, institutional barriers and financing opportunities, and make policy recommendations to the president. This process drew primarily upon personnel from the relevant government agencies, with some input from members of the public and outside consultants.

On 25 August 1978 the DPR issued a status report for public review on a variety of options. This report highlighted eight "choices" to "provide relief for the first-cost and cash flow problems purchasers (of solar equipment) now face. In essence, the choices facilitate decisions that benefit both society and the individual decision maker."[15] The first option listed was to "provide assistance through new or existing programs of relevant Federal agencies or through a solar Development Bank."

After conducting nationwide public hearings, the final Domestic Policy Review on Solar Energy was submitted to the president. The document recommended the creation of a Solar Bank to ensure adequate financing for solar energy technologies in the residential and commercial sectors. Citing the need for $700 to $800 billion in solar investments under its base-case scenario, a Solar Bank was envisioned to encourage the necessary capital flows and to assist borrowers with financing costs.[16] Assessing the other financing and credit mechanisms available with the federal government, the report concluded that "other credit mechanisms now in place within the Federal Government do not appear to be sufficiently broad to address the solar energy financing problem."[17]

The DPR envisioned the Solar Bank as a government corporation which would operate principally in the secondary market to purchase bank loans. It was envisioned that the Solar Bank would reduce the interest rate and extend the terms on solar loans. This problem was to be large in scope: financing by the year 2000 an estimated 12 million incremental solar heating or hot water systems,[18] which were to save more than 0.6 quads of energy in the residential sector at a cost of approximately 40¢ per million Btu. The total cost of the bank over five years was estimated to be approximately $3 billion.[19] This set the stage for legislative action.

26.3 Legislative History

26.3.1 Beginning

President Carter proposed to create a Solar Bank only if it was tied to the passage of his Energy Security Act. The president proposed an initial funding level of $100 million annually for the bank. The Carter proposal would have the board of directors of the bank set the interest rate on Solar Bank loans. Numerous bills proposing a solar development bank eventually reduced to three vehicles in the House (which became the primary actor among the two chambers): H.R. 2974 (the administration's proposal), H.R. 605 (sponsored by Representative Neal, D-NC), and H.R. 2343 (sponsored by Representative McKinney, R-CT).

Prior to the DPR, Representative Stephen Neal (D-NC) filed his H.R. 605, to create a solar energy development bank. Neal's bill was strongly backed by solar advocates. The primary differences between Representative Neal's bill (H.R. 605) and the Carter administration's bill were:

• Fixed interest subsidies (Neal) versus flexible interest subsidies or subsidies of principal (Carter).

• Higher (Neal) versus lower (Carter) levels of funding.

• Independent funding (Neal) versus funding dependent on the creation and fulfillment of an Energy Security Trust Fund generated from the Windfall Profits Tax (Carter).

• Having bank operate indirectly, through the secondary loan market (Carter), versus directly, through loan subsidies (Neal).

• Creation of advisory boards to the bank (Neal).

• Establishment of specific set-aside for low-income persons (Neal).

• Establishment of bank lending rates at 6% below the prevailing Federal Housing Administration lending rates (Neal).

Representative McKinney's bill would have created an independent solar bank for a period of forty years. It would be governed by a seven-member board of directors comprising three cabinet secretaries and four presidential appointees representing the public and the scientific community.

The legislative impetus for passage of the Solar Bank came from four sources:

• Members of Congress (particularly including Representatives Stephen Neal, Stuart McKinney, and Robert Drinan, and Senators Edward Kennedy and Mark Hatfield)

• Industry groups representing solar manufacturers and installers

• Research organizations such as the Congressional Office of Technology Assessment and the Solar Energy Research Institute

• Public interest organizations

Public interest groups clearly were the dominant element among the four groups pressing for creation of a solar and conservation bank. The bank's enactment was the key legislative priority for several public interest organizations and was a significant goal for a united front of environmental, public interest, and low-income advocacy organizations. As is typical in public interest politics, a few organizations took the lead for the enactment of the Solar Bank. The work of two organizations was preeminent. The Solar Lobby, an outgrowth of the organizers of the Sun Day event in May 1978, applied the resources of their congressional lobbyists Herb Epstein and Suzette Tapper to the issue; their membership of 30,000 persons in all fifty states mounted a grassroots campaign to influence key legislators regarding the need for a Solar Bank. In addition, the National Consumer Law Center, a Boston-based public interest law firm which represented low-income and other citizen interests around the country, devoted some time of Steven Ferrey, their senior energy counsel, to affecting passage.

Meetings were held during 1979 and the early part of 1980 on the various proposals. The basic concept as proposed by President Carter became the prime vehicle for legislative markup. This concept included a government corporation, governed by five cabinet secretaries, with broad discretion to use various means to subsidize financing for solar energy improvements.

26.3.2 Critical Alterations

Two dramatic alterations occurred that strengthened the chances for authorization of the Solar and Conservation Bank. The first major alteration occurred when Senator Edward Kennedy (D-MA), through his staff representatives on the Senate Joint Economic Committee, successfully added an amendment in the Banking Committee creating conservation grants for low- and moderate-income persons. This amendment, embodied in S932 (the Senate version of the Energy Security Act), fundamentally altered the bank and harmonized this provision with the original Neal bill. It provided a direct mechanism to include those who would not otherwise be able to obtain bank loans. And by allowing grants, it included a much more direct role not only for nonprofit corporations but also for state and local government agencies, which might not be interested in making loans but which could now extend grants and target these grants to special, locally determined needs. The constituency for creation of the bank was thus broadened to include states and cities.

The second major alteration occurred after hearings on 16 and 17 October 1979 before the House Subcommittee on the City (Committee on Banking, Finance, and Urban Affairs) and the House Subcommittee on Oversight and Investigations (Committee on Interstate and Foreign Commerce), ostensibly to examine renewable energy and the city. Banking Committee Chairman-to-be Henry Reuss (D-WI) drafted amendments later sponsored successfully by Congressman Thomas Ashley (D-OH), chairman of the Housing Subcommittee on Banking, Finance, and Urban Affairs. The Reuss-Ashley amendments (1) provided subsidy levels sensitive to recipient incomes, (2) provided set-asides for low- and moderate-income recipients, (3) included cooperatives and other joint ventures as eligible recipients for funding, (4) included planning and technical assistance as eligible funding items, and (5) specifically included nonprofit groups, local governments, tenants, and community-based organizations as eligible recipients.

These amendments dramatically altered the bank, which now became a specialized financing mechanism to reach those not served by the conventional tax credit programs, rather than a general financing mechanism. Because of the grant provisions sponsored by Senator Kennedy, the program became an important substitute for the Community Services Administration solar grant programs that were soon to expire with the demise of that agency (see chapter 29 for more information). And the legislation became an important priority for groups serving the low-income community, thereby adding an additional group pressing for enactment.

The bank became part of the Energy Security Act; the final legislation as passed by Congress was a hybrid of the complementary approaches. That hybrid form was forged largely to meet the specific recommendations of the prosolar energy and public interest communities. As enacted, the bank was governed by five cabinet secretaries, included advisory committees, provided a set-aside from the bank for low-income individuals, allowed flexible subsidies, had authority to operate in the secondary loan market, and contained significant multiyear funding, as further discussed in section 26.5.

26.4 Organizational History

The board of directors of the bank consisted of the Secretaries of the U.S. Departments of Housing and Urban Development, Energy, Commerce, Treasury, and Agriculture. Therefore, the board changed as cabinet secretaries changed. This did provide for some discontinuity in program administration. However, the bank was operated day-to-day by a president, two vice presidents and subordinate staff.

In 1980 President Carter appointed Joseph Bracewell as president of the bank, but the nomination was not confirmed by the Senate before the transition to the Reagan Administration. The president also appointed Nancy Carson Naismith as vice president for conservation and Joseph LeVangie as vice president for solar energy; these officers left upon the inauguration of President Reagan.

President Reagan did not appoint any officers or subordinate staff to the bank during the first seventeen months of his administration. On 23 June 1982 Robert W. Karpe, president of the Government National Mortgage Association (GNMA), was appointed to serve simultaneously as manager of the bank.[20] The comptroller of the bank later was designated as an acting executive vice president. In 1983 President Reagan nominated Richard Francis as president of the bank. After the initial hearings,[21] the Senate did not act to confirm Francis's nomination, although he was subsequently confirmed in August 1986, following a second hearing. No other vice presidents were designated as officers of the bank by President Reagan.

The statute created two five-member advisory committees to the bank—one for conservation and one for solar energy. Members of the advisory committees were appointed by the president. One member of each advisory committee was to represent consumers, lenders, builders, industry, and architectural/engineering interests. For the Conservation Advisory Committee, President Carter appointed Steven Ferrey, Joseph Honick, Elizabeth Bogosian, Aldon McDonald, and Jimmie Wortman.

For the Solar Energy Advisory Committee, President Carter appointed Harry Schwartz, Paul Sullivan, Carol Hoover, Lee Cohen, and David Chavez.

In mid-1983 President Reagan reconstituted the Conservation Advisory Committee, retaining Steven Ferrey and Aldon McDonald, but replacing the other members with Rick Fore, Pasquale Alibrandi, and James Hughes. On the Solar Energy Advisory Committee, the president retained Harry Schwartz and Carol Hoover, but replaced the other members with Peter Wold, Robert Gardner, and Reed Robbins. The advisory committees were required to issue an annual report and their views were to be included in the annual report of the bank,[22] from 1982 through 1984, because of minimal bank activity. The bank found no need to convene meetings of the advisory committees or to have them issue a report.

Although not contemplated in the statutes, the states became major actors in the delivery of bank services. The evolutionary role of the states became a significant factor in the organizational history of the bank.

26.5 Solar Energy and Energy Conservation Bank as a Financing Mechanism

On 30 June 1980, upon signing into law the Energy Security Act of 1980,[23] President Carter stated: "Its scope, in fact, is so great that it will dwarf the combined efforts extended to put Americans on the Moon and to build the entire Interstate Highway System of our country."[24] The bank was one component of this act, along with provisions to create the Synfuels Corporation and to extend energy audits to multifamily buildings.

The Solar Energy and Energy Conservation Bank was neither a bank nor a regulatory program. The bank made no loans, guaranteed no credit, insured no financing, and imposed no regulatory sanctions on established banks; it did not open branch offices, hire tellers, or accept deposits. What the bank did do was direct an arsenal of flexible subsidies to lending institutions to overcome barriers that had prevented access to adequate credit for making energy conserving and solar investments in residential, agricultural, and commercial property.

The bank was physically and organizationally situated in HUD and was delegated the same powers as the Government National Mortgage Association.[25] The bank could transact business with the national network of lending institutions by two methods. First, the bank could make lump-sum payments to financial institutions; these payments in turn could be used to reduce the principal or interest on solar energy loans made to eligible recipients.[26] Second, the bank was empowered to engage in secondary market loan financing by purchasing any loans made in whole or in part for residential solar energy improvements.[27] Significantly, this secondary market authority was limited to single- and multifamily residential properties and could be employed to purchase loans made to residential customers by utilities.

Tenants were specifically made eligible for conservation loans and grants, although not for solar energy loans.[28] Builders of new residences could receive subsidies for the purchase and installation of a solar energy system if the bank board made three findings: (1) a direct builder subsidy was necessary to encourage a greater number of single-family solar dwellings; (2) a direct builder subsidy was a more effective expenditure of bank funds; and (3) a system was established to prevent both builders and purchasers from claiming a bank subsidy and residential energy tax credit for the same expenditure.[29]

The solar component of the bank was structured flexibly to transcend income barriers using (1) subsidies that were inversely a function of recipients' income; (2) tailored subsidies that were delivered through loan interest subsidies or loan principal reductions; and (3) maximum income limitations and prohibitions against "double subsidization."

As table 26.2 illustrates, maximum solar energy subsidy levels were 40% of costs for more affluent owners of single-family dwellings and all other multifamily, commercial, or agricultural borrowers,[30] or for owners of multifamily dwellings whose buildings were primarily occupied by low-income tenants,[31] up to 40% MAI (median area income). For solar loans made after 31 December 1985, no owner whose income exceeded 250% MAI was eligible.

Although the bank generally disallowed "double dipping" into the bank and other financing programs, it did allow simultaneous use of (1) utility financing of solar energy through rates;[32] (2) indirect federal tax assistance, such as federally tax-exempt state or municipal revenue or general obligation bonds;[33] and (3) state and local direct grant, loan, or tax subsidies.[34]

The bank exempted any individual receiving financial assistance from the bank from reporting such assistance as gross income or as an increase

Table 26.2
Maximum loan subsidies for solar energy systems (active/passive systems only; building owners/purchasers only)

	Assistance as percentage of investment	Maximum assistance[a] (dollars)			
Residential (1–4 units)		1 unit	2 units	3 units	4 units
Space heating[b]		5,000	7,500	10,000	10,000
Domestic hot water	40	1,000/dwelling unit			
Passive space cooling (new construction only)		5,000			
Multifamily buildings					
Space heating[c]		2,500/dwelling unit			
Domestic hot water	40	12,500			
Agricultural and commercial buildings					
Space heating[c]		100,000			
Domestic hot water	40	12,500			

Source: U.S. Department of Housing and Urban Development, *Solar Energy and Energy Conservation Bank Program Summary*, Washington, DC, 26 March 1984.
[a] Space heating and space cooling subsidies are limited based on energy savings.
[b] Includes any combination of space heating with passive space cooling and domestic hot water systems. All income levels are eligible except owners of existing 1- to 4-family residential buildings with income exceeding 250% of median area income after 31 December 1985.
[c] Includes any combination of space heating and domestic hot water systems; no income limitation.

in the basis of any real property.[35] This prevented bank assistance from being used to disqualify a recipient from eligibility for welfare, Supplemental Security Income (SSI), or other assistance programs. In those states that adopted utility-financing programs for solar energy improvements, individuals could effectively "triple dip" into utility rate-financed subsidies, state tax subsidies, and either federal tax or bank subsidies (but not both) for residential renewable energy investments.[36]

One of the most important contributions of the bank was the subsidization of a broader scope of solar energy measures than those included under conventional tax credits. For solar energy technologies, the measures made eligible by the statute were "purposely broad," including active, passive, and photovoltaic systems; wood-burning stoves or appliances; solar process heat devices; and earth-sheltered homes.[37] It is of contrasting note that although the bank was encouraged by statute to increase the percentage subsidy of individual conservation expenditures in direct relation to the amount of money *invested* in conservation by a consumer,[38] for solar projects it was encouraged by statute to increase the percentage subsidy provided in direct relation to the amount of energy potentially *saved* by the solar investment. Passive solar energy improvements were required to be subsidized in relation to energy saved, to the extent practicable,[39] while active solar energy improvements could be subsidized on this savings basis, if practicable, after 1 January 1983.[40]

Eligible bank lenders included state and local governments, neighborhood housing services, credit unions, mortgage companies, and charitable organizations and foundations. Many of these unconventional lenders could make bank credit available to individuals denied access to conventional credit markets. In addition, utilities financing solar energy improvements were eligible as lenders of bank-subsidized funds.[41] A maximum of 10% (20% if the board so authorized) of solar energy appropriations could be passed through utility lenders.[42] The bank could utilize its secondary market financing authority to purchase loans made by utilities for solar energy purposes.[43]

Warranties were required for all financed program measures. For solar improvements, written warranties specifying a minimum of three years on manufacture and a minimum of one year on supply and installation were required;[44] the board retained discretion to require longer warranties.[45]

26.6 Budgets

As originally authorized, the Solar Bank contemplated hundreds of millions of dollars expended annually to help finance solar energy improvements. The authorization levels were established originally for the first four years as depicted in table 26.3.

For FY 1981, on 15 December 1980 Congress appropriated $125 million for the bank. This was subsequently reduced to $121.5 million by an across-the-board budget reduction. President Carter, on 15 January 1981, submitted an FY 1982 budget to Congress that suggested an appropriation equal to that of FY 1981 for the bank. On 5 June 1981, in anticipation of an upcoming FY 1982 appropriation, the Congress rescinded all unspent FY 1981 appropriations for the bank except for $250,000, which was already expended for administrative purposes.

Even though bank funding had already been authorized, in 1981, under the new administration, it was reauthorized at lower levels. In 1981 Congress approved a concurrent budget resolution[46] intended to reduce

Table 26.3
Solar energy and energy conservation bank authorizations and appropriations (millions of dollars)

	1981	1982	1983	1984	1985	1986	1987–88
Authorizations							
Energy conservation	200/10	625/7.5	800/7.5	875/7.5	—	Prior	—
Solar energy	100/10	200/7.5	225/7.5	—	—	recaptured	—
Appropriations		21.85	20	25	15	funds	1.5

Source: 12 U.S.C.A. 3620 (1980).
Note: The second figure represents the maximum amount of the appropriation that could be expended for promotion. (For example, of the $200 million authorized for energy conservation in 1981, only $10 million could be spent for promotion.)

federal spending significantly.[47] As required by the Congressional Budget Act,[48] the Senate Committee on Energy and Natural Resources recommended a new authorization ceiling for the act of $50 million.[49] Congress ultimately incorporated this ceiling in Section 1071 of Title X, Subtitle G of the Omnibus Budget Reconciliation Act (OBRA) of 1981, PL 97-35, 95 Stat. 357, 622 (13 August 1981). The reduced authorization ceiling in OBRA applied to appropriations for the fiscal years 1982 through 1984 of the bank's operation.

Shortly before the beginning of FY 1982, the House Committee on Conference of the House and Senate Appropriations Committees approved a $25 million FY 1982 appropriation for the bank. The House Committee on Conference, in reporting the appropriation bill to Congress, unanimously voted to "direct the Secretary of HUD to expedite all Bank implementation activities by moving rapidly to publish regulations, secure an agent, staff the Bank, and disburse loans and subsidies at the earliest possible date."[50]

Congress issued the bank interim funding even during the budget reduction process. Pending the passage of the FY 1982 appropriation bill, Congress enacted an interim continuing resolution that temporarily appropriated $3.5 million for the bank, corresponding to a $25 million annual rate of bank spending.[51] Senator Hatfield (R-OR), with the unanimous consent of the full Senate, inserted language in the continuing resolution specifically mandating the bank's directors to operate the program and expend funds prior to the bank's regular appropriation;[52] on 23 December 1981 the Congress enacted the regular FY 1982 appropriation for the bank, in the amount of $23 million.[53]

Shortly after the passage of the continuing resolution, the president proposed the deferral of the bank's interim budget authority.[54] The comptroller general reported to Congress, in an opinion required by statute, that the president's action was illegal.[55]

One and one-half months after the passage of the regular FY 1982 appropriation for the bank, the president proposed a rescission of this appropriation.[56] Although the president did not propose to defer any of the bank's regular budget authority, the Office of Management and Budget (OMB) refused to provide any of this budget authority to HUD.

In each case except for FY 1984, the Senate accepted the president's recommendation of zero funding for the bank; the House of Representatives rejected the recommendation and appropriated at least $25 million each year. The final appropriations figure represents a compromise worked out in the Joint Conference Committee, with Representative Boland (D-MA) and Senator Garn (R-UT) heading the conference delegations from each house. In FY 1984, the Senate voted to appropriate $15 million; the compromise with the House resulted in an appropriation of $25 million.

On several occasions in 1987, Congress took action to extend the authorization for the bank.[57] Ultimately, the bank was extended until 15 March 1988.[58] On the same day, 21 December 1987, Congress agreed to a compromise of the housing bill that did not contain an extension of bank authorization.[59] Also on the same day, Congress enacted the last continuing fiscal resolution for FY 1988.[60] This contained a $1.5 million appropriation for the bank for FY 1989, to remain available until 30 September 1989.

In both FY 1986 and FY 1987, the bank received no direct appropriation; rather, funds appropriated previous years were rolled forward into subsequent years. In FY 1988, the final year of bank funding, $1.5 million was appropriated for the bank, which represented a Conference Committee compromise between $3 million sought by the House and zero funding provided by the Senate.

26.7 Operation

When originally created, with an initial annual appropriation of more than $100 million for solar energy financing, the bank had a broad and ambitious agenda. The Carter Administration cabinet secretaries staffed the bank, hired consultants, and drafted operating regulations. In January 1981 President Reagan's cabinet secretaries assumed management of the bank. Shortly after January 1981, these new directors withdrew bank regulations drafted by the bank's previous board, dismissed bank staff, terminated all consulting contracts, and allowed bank managerial positions vacated by resignation to remain unfilled. OMB impounded the bank's FY 1982 appropriation, pending unsuccessful budget rescission attempts directed at the bank, which the President proposed, but the Congress did not act upon.

In the second half of FY 1982 a suit was initiated in federal district court in New York to restore bank functions and funding. Plaintiffs included four low-income individuals, two businesses engaged in the construction of solar homes, a credit union, the cities of St. Paul and Philadelphia, five members of Congress representing both major political parties, six public interest organizations whose members, in all fifty states and territories, were eligible for financial assistance under the act, three members of the two Presidentially appointed five-member advisory committees to the bank, and the State of New York. Defendants to the action were President Ronald Reagan, OMB Director David A. Stockman, the five cabinet secretaries designated by the act as the bank's board of directors, and HUD.

Plaintiffs' suit sought judicial review of the OMB impoundment of the bank's FY 1982 appropriation, the failure to promulgate regulations as required by the act, the failure to staff the bank or to operate the bank program, and the failure to convene the advisory committees to the bank and to issue the 1981 annual report required by statute.

Immediately after plaintiffs filed their 8 April 1982 legal complaint in federal court, the board took action to restore the Bank to the operational status mandated by the Congress. On 22 April HUD Secretary Pierce designated President Robert W. Karpe as manager of the bank, and allocated six full-time permanent HUD positions to Karpe to operate the bank program. On the same date, the bank for the first time requested that OMB pass along the budget appropriation provided by the FY 1982 Appropriations Act to the bank; on 26 April OMB provided the requested budget authority.

On 18 May plaintiffs asked the Court to issue preliminary injunctive and declaratory relief; the federal court, after a hearing on June 16, issued a decision in favor of the plaintiffs, which required prompt action from the board to expend bank funds, and asserted continuing court supervision to ensure that the statute would be executed as Congress intended. The court determined that the bank's authorizing statute mandated the expenditure of funds by defendants:[61]

> [P]laintiffs are entitled to an order at this time which will place the defendants responsible for the implementation of the Bank under a court-supervised obligation to implement the Act, and make the appropriated funds available for expenditure to qualified applicants as expeditiously as the good faith efforts of those defendants may permit. Such an order will be consistent with the Court's hold that, contrary to defendants' contention, Congress intended that the fiscal year 1982 appropriation be made available for distribution within that year.[62]

On 19 October the federal court entered a further order mandating "continued judicial supervision" of the board's "long-term compliance with the court's June 29, 1982 Order."[63]

By 27 August 1982 the bank board had substantially completed selection of a full-time bank staff. The Board further decided to draft interim regulations under which bank funds would be distributed to recipients indirectly through the states, as administrative agents for the board. By 1 January 1983, the bank board made a "tentative allocation" of bank funds totaling approximately $30.4 million to applicant states. Delaware, Mississippi, and Wyoming returned their "tentative allocations" to the bank, citing the cumbersome administrative structure established by the bank to distribute its funds. Despite more than $150 million in state requests for bank funding, the January "tentative allocation" by the board awarded only $30.4 million, excluding $11 million of the bank's FY 1983 appropriation.

The board's treatment of the FY 1983 funds prompted sharp Congressional reaction:

> Almost two and one-half years have passed and a single loan has yet to be awarded. [The House] Committee [on Banking, Housing, and Urban Affairs] is extremely distressed by the reluctance of the present Administration to implement a program that has strong Congressional support and by the failure of the Bank to propose regulations that comply in major respects with the basic intent of the original program. In addition, the Administration has attempted to diminish the impact of this program by refusing to award $11 million of the funds appropriated for fiscal year 1983 and by proposing to reprogram these funds for the Low-Income Energy Assistance Program operated by the Department of Health and Human Services.

Most of the amendments to the authorizing legislation are necessary, not because the basic statute itself is flawed, but because the department has failed to use its existing authority to assure the creation of an effective program.[64]

To attempt remedial action, the Congress accompanied this report with legislation requiring that the bank, within 90 days of 30 November 1983, promulgate regulations to

• include active solar energy systems for funding;

• permit metropolitan cities and counties to distribute financial assistance for the bank;

• permit the use of tax-exempt financing to deepen the subsidies;

• raise the administrative portion of funds from 2% to 10% and raise the promotional ceiling from 1% to 2% on funds distributed by the states; and

• require that any bank funds recaptured from any state be reallocated to other eligible financial institutions for distribution.[65]

On 18 September 1984 Richard Francis, manager of the bank, announced that the bank would "recapture" any FY 1982 or FY 1983 bank budget authority not obligated by the states by 1 March 1985. At the close of 1984, bank staff predicted that $7.9 million of the FY 1982–1983 funds would revert to the treasury because of this first recapture on 1 March 1985. On 3 January 1985 the litigating plaintiffs from the 1982 court suit requested a new court order to block the board's attempt to "recapture" bank funds and return them to the treasury.

On 20 March 1985 the Federal Court granted injunctive and declaratory relief in favor of plaintiffs.[66] The court held that the recently amended Section 3618(b) (6) of the bank statute overrode any discretion the board possessed to return to the treasury previously obligated but recaptured FY 1982–1983 bank budget authority:

The enactment was a direct message from Congress that the Bank would not be allowed to undercut the program by recapturing and returning appropriated funds to the Treasury. Congress underscored its intention that all appropriated funds be given to eligible recipients by specifically requiring reallocation of recaptured funds.[67]

After this 1985 court action, the board proceeded to distribute bank funds. Between September 1983 and March 1985, the Bank delegated to the states for distribution to eligible financing recipients FY 1982–1984 bank appropriations totaling $66.85 million; the FY 1985 appropriation was distributed in August 1985. Some of these delegations were made on the basis of a weighted formula using population, income, and weather data, while others were made on the basis of competitive applications from the states.

In 1984 the administration of the Bank returned approximately $263,000 of appropriated FY 1982 monies to the treasury, principally funds returned by the State of Alaska, which elected not to continue its participation in the bank, as well as approximately $12,000 in administrative expenses returned by the bank board to the U.S. Treasury.[68] This action, which would appear to violate the Federal Court's order in *Dabney v. Reagan*, was only uncovered years later during a routine financial audit of the bank by the GAO.[69]

The bank appeared poised in September 1986 to return to the U.S. Treasury funds recaptured at the end of FY 1986 from the Bank. This issue arose because the Congress had not yet adopted an appropriation for FY 1987. Although both houses of Congress had adopted language directing the bank board to recycle and reappropriate in FY 1987 any funds unspent or recaptured at the conclusion of FY 1986, these bills were not yet law.

The Solar Bank was a small program inside the large HUD-Independent Agencies appropriation. While larger forces stalled the entire legislation, the bank was exposed to a potential Constitutional dilemma. Could the bank board use the time between 1 October 1986 and the point of passage of the HUD FY 1987 appropriation to unilaterally return recaptured and unspent bank funds that Congress was on the verge of reobligating for bank purposes?

Plaintiffs in the still-pending *Dabney v. Reagan* litigation were confronted with the choice of seeking an injunction to restrain the bank board from using this legislative window to preempt the not yet fully expressed intent of Congress. A significant confrontation was avoided at the eleventh hour, when, on 30 September 1986, attorneys for the bank board and for the plaintiffs in the litigation reached a negotiated accord. The parties agreed that the bank board could not and would not return FY 1982–1984 funds recaptured at the end of FY 1986, as such funds were clearly covered by the court's orders enjoining impoundments by the bank board. Moreover, the bank board agreed not to return any FY 1985 recaptured funds to the treasury during this legislative window, provided Congress in a continuing resolution or FY 1987 appropriation ultimately directed that funds recaptured at the conclusion of FY 1986 were to be reobligated for FY 1987.

In return, the plaintiffs in the lawsuit agreed not to seek an additional emergency temporary restraining order removing control from the bank board over these financial issues. The parties adhered to this agreement. The language in the House and Senate bills, directing the bank board to reobligate unspent and recaptured funds, was embodied in the FY 1987 appropriation and was signed by President Reagan; the bank reobligated these recaptured funds for FY 1987.

In January 1988 OMB apportioned $1.5 million of FY 1989 appropriations to the bank; the bank did not spend the funds. In July 1988 OMB withdrew the budget authority for this $1.5 million from HUD, arguing that the bank's authorization had expired.

In fact, the authorization for the bank had naturally expired on 15 March 1988. In August 1988 Congress enacted and the president signed the HUD-Independent Agencies Appropriations Act for FY 1989, with a specific provision rolling over any recaptured funds from this $1.5 million appropriation into FY 1989. OMB refused to reapportion any of these funds, submitting that the funds had been returned to the treasury in July of that year, when bank authority temporarily lapsed, and that no funds remained to be recaptured in FY 1989.

On 10 August 1988, the chairman of the House Subcommittee on HUD-Independent Agencies, Representative Boland, asked the comptroller general to review the propriety of these actions by OMB. The comptroller general concluded that the unclear legislative history of the appropriation for the Bank in its last year could be construed to mean the appropriation applied only until the bank's 15 March 1988 expiration, as there was no specific language in the appropriation mandating that it continue to any particular date. Therefore, by electing not to expend any of the FY 1988 bank funds prior to 15 March 1988, the board was able to take advantage of the legislative window between March and August 1989 to return these funds to the U.S. Treasury.

Had the bank desired, it could either have expended or obligated the funds prior to 15 March 1988 or not returned these funds to OMB prior to August 1988, when the bank was reauthorized. Either action would have preserved the funds for the bank. During FY 1989, $1.5 million of funds were estimated to be recaptured by the bank from prior years pursuant to the decision in *Dabney v. Reagan*.

26.8   Evaluative Evidence and Lessons Learned

The history of the Solar Energy and Energy Conservation Bank is marked by conflict between Congress and the executive branch. Under the Reagan administration, the officers and board of the bank apparently believed that deregulation of the prices of other energy sources, notably oil and natural gas, would alone provide sufficient incentive for the financing of solar energy systems where appropriate. Moreover, they apparently believed that the bank provided an unfair and unnecessary economic subsidy to solar energy systems.

Sponsors of the bank legislation (from both political parties) took exception to what they often regarded as the arbitrary and inefficient management of the bank. This conflict is evident in the *Congressional Record* and in the transcripts of congressional committee correspondence with the bank. In September 1981 Senator Hatfield, through a variety of other correspondence, urged special, expedited funding for the bank; on 22 March 1982, Chairman Richard Ottinger of the House Subcommittee on Energy Conservation and Power took issue in writing with the slow development of regulations for the bank.[70] Representative Ottinger again took issue in September 1982 with the bank's exclusion of cities as direct program participants and also with its prohibition of local use of tax-exempt bond funding to augment bank resources.[71]

Other key members of Congress expressed discontent with the administration of the bank. On 3 February 1983, Representative Steven Neal expressed to Chairman Henry Gonzalez, of the House Banking Committee, his concern that bank regulations excluded commercial buildings (the exclusion was removed for FY 1985 funding cycles), allowed no measures to be funded with paybacks greater than seven years, restricted eligibility for owners of 1- to 4-unit rental buildings, and proscribed use of tax-exempt bond proceeds at the local level.[72] On 16 December 1982 ranking Republican Representative Stanton and Chairman St. Germain, of the House Committee on Banking, Finance, and Urban Affairs, wrote Solar Bank Chairman of the Board Samuel Pierce regarding changes they requested in the operating regulations of the bank.[73]

On 12 April 1984 a bipartisan coalition of representatives, including the chairman of the Committee on Energy and Commerce, the chairman of the Committee on Banking, Housing and Urban Affairs, the chairman of the Subcommittee on Energy Conservation and Power, the chairman of the Subcommittee on Housing, as well as the key Republican and Democratic sponsor of the initial solar bank legislation, wrote the administrators of the bank objecting to the distribution mechanism of funds to the states, which excluded city applications and in some funding cycles distributed funds on a per capita basis rather than with regard to need or capability; the delay in distributing approximately 50% of appropriated funds to the states; the use of bank funds to finance less efficient air-conditioners; the restriction of eligible measures to the Residential Conservation Service (RCS) program measures list; restrictions on rental housing eligibility; and the failure for more than three years to convene the advisory committees as required by statute (after three years, the president was able to entirely reconstitute both advisory committees with members of his choosing, and thereafter the advisory committees did meet).[74] Finally, Housing Subcommittee Chairman Gonzalez wrote Secretary Pierce on 7 February 1985, objecting to the recapture and return of bank funds to the treasury.[75]

The result of this tension between Congress and the administration was a continuing process: administrative regulation, congressional complaint, administrative perseverance, congressional amendment of statute to require as a matter of law the requested changes, administrative compliance. Eventually, by requiring through statutory amendment the regulatory changes sought, Congress was able to fine-tune and shape the bank program to its intentions. The lawsuit in *Dabney v. Reagan* also dramatically altered the operation of the bank. But the necessity for Congress to legislate changes, rather than obtain voluntary regulatory changes by the bank, affected the smooth operation of the program. Consequently, there was a series of regulatory changes, delays, and uncertainty injected into the program.

Forty-eight states plus the District of Columbia, the North Mariana Islands, and Puerto Rico acted as distributors of bank funds. Official bank data (internal) through FY 1984 on bank fund recipients provide the following information regarding the income of beneficiaries:

- 59.3% had incomes below 80% of median area income;
- 27.1% had incomes below 80%–150% of median area income; and
- 13.6% had incomes greater than 150% of median area income.

The types of buildings and applications receiving assistance varied:

- 65.9% of funds benefited 1- to 4-family buildings for conservation investments;
- 15.9% of the funds benefited multifamily buildings (more than four families) for conservation investments;
- 11.4% of the funds benefited single-family residences employing passive solar design;
- 1.5% were multifamily solar applications; and
- 1.5% were commercial or agricultural solar applications.

The type of financial assistance provided by the bank also varied:

- 59.2% of funds were used for loan subsidies; and
- 34.5% of the funds were used for conservation grants to low-income families;
- 6.3% of the funds were used for state and local administration and promotion.

From these profiles several trends are clear. First, only a limited fraction of bank funds were devoted to solar rather than conservation technologies. This was a result of a preference in the regulations and among states to fund conservation and of the disparity between solar energy residential tax credits (40%) and residential conservation tax credits (15%). Second, only eleven states elected to use any bank funds for multifamily housing, through FY 1984. Third, fully twenty-one states or jurisdictions did not elect to offer conservation grants from bank funds to low-income recipients.

Although the Solar Bank appears to have reached the target income population not served by tax credits, it did not make a dramatic penetration into the multifamily and rental sectors where the tax credits did not apply. With subsequent funding cycles, penetration increased as states became more sophisticated in outreach and marketing to these sectors, but again not dramatically. The primary problem reported by the states was a lack of sufficient funds for administrative and promotional expenses; for a small state with a small grant, the original 3% limit on the amount of funds that could be devoted to these purposes was clearly inadequate. Congress subsequently amended the statute to require the bank to allow states to expend 12%, or quadruple the original 3% program limitation on administration and promotion. Sufficient funding for administration and promotion is key to the success of the bank as a financing mechanism.

Discussions with state and local officials reveals a continuing need for solar energy financing, whether from the bank or from other sources. The difficult administrative history of the bank is testament to both the pitfalls and benefits of the American system of checks and balance. When the executive and legislative branches reached a standoff on the legal requirements to operate the bank, only the judicial branch could resolve this conflict; the courts were required to intervene on two occasions to exercise supervision over the bank board's administrative decision making. These problems seemed to stem not so much from the Congress's decision to vest administrative management of the bank with HUD as from fundamental policy disagreements about the bank between the new administration and members of Congress.

The Solar Energy and Energy Conservation Bank underwrote a wide variety of energy conservation and, to a lesser degree, solar energy improvements. States and communities used the funds a variety of innovative and inventive manners (see, in this regards, chapter 29). Particularly with conservation technologies, the bank financed a number of programs in cold-climate states that delivered previously unavailable energy efficiency technologies to low-income dwellings. Some retail utility companies are continuing programs to promote energy efficiency that were seeded with bank funds; in this sense, the bank "spun off" private sector efforts for energy efficiency.

It is clear that even under judicial command, the administration of any sophisticated energy program requires good faith cooperation of the administrative agency to operate smoothly. Reports from the states indicate significant demand for bank funds but insufficient promotional budgets to advertise the program. For a thinly funded program, operating at a shadow of its originally designed scope, the bank demonstrates that who participates in utilizing solar energy and energy conservation technologies, at least with regard to a large segment of the population, is a function of how those technologies are financed.

Notes

1. Residences qualify if "substantially completed" by 20 April 1977. 26 U.S.C. 44C(c) (1) (Supp. III 1979).

2. Conservation credits of less than $10 annually, representing a $70 expenditure, cannot be taken pursuant to IRS computation procedures.

3. Eligible solar improvements do not include systems serving a "significant structural function" in a dwelling. S. Rept. 1324, 95th Cong., 2d sess. 44 (1978).

4. Sec. 202 of Crude Oil Windfall Profit Tax of 1980 (PL 96-223), 94 Stat. 258 (1980), codified at 26 U.S.C. 44C (b).

5. 26 U.S.C. 46(a) (2) (C) (Supp. III 1979).

6. PL 96-223, 94 Stat. 230, codified at 26 U.S.C. 46(a) (2) (1980).

7. IRS, *1978 Statistics of Income*, table 8 (1978). Dollar amounts of total residential energy credits claimed were parallel to the incidence of returns, showing 82% and 65% of expenditures falling to income groups with over $15,000 and over $20,000 annually adjusted income, respectively.

8. Ibid.

9. This assumes a standard deduction applied against adjusted gross income from wages, with no other credits. 26 H.S.C. 1 (1980).

10. Regional Urban Planning Institute, *Federal Incentives for Solar Homes: An Assessment of Program Options*, pt. 3, at 23–24 (prepared for the U.S. Department of Housing and Urban Development; 1977).

11. 42 U.S.C. 5501 *et seq.* (Supp. III 1979).

12. All utilities with annual retail residential natural gas sales in excess of 10 billion cubic feet or annual retail residential electric sales exceeding 750 million kilowatt hours are covered. 42 U.S.C. 8212 (Supp. III 1979).

13. 42 U.S.C. 8216 (Supp. III 1979).

14. California Energy Commission, *California's Solar Energy Tax Credit: An Analysis of Tax Returns for 1977*, Doc. 500-79-012.

15. U.S. Department of Energy, "Status Report on Solar Energy Domestic Policy Review," 25 August 1978, at VI-3.

16. *Domestic Policy Review of Solar Energy*, attachment, at 23.

17. Ibid.

18. Ibid., at 25.

19. Ibid., at 26.

20. Karpe was simultaneously president of the Government National Mortgage Association and did not serve full-time at the bank.

21. Senate Hearing 98-870, 23 February 1984.

22. 12 U.S.C. 3606.

23. Energy Security Act of 1980 (PL 96-294), 94 Stat. 611 (1980), codified in scattered sections of 2, 5, 7,10, 12, 15, 16, 18, 26, 30, 42, 49, 50 U.S.C.

24. Energy Security Act, President Carter's remarks on signing S. 932 into Law, 16 *Weekly Compilation of Presidential Documents* 1253 (30 June 1980).

25. 12 U.S.C.A. 3606 (a) (1980) states that the bank shall have the same powers as those given to the Government National Mortgage Association (GNMA) by 12 U.S.C.A. 1723a (a) (1980). The powers of the GNMA include the authority to enter into and perform contracts, leases, or other transactions on terms deemed appropriate; to sue and be sued; to avoid attachment of its property; to lease, purchase, or acquire property that can be operated and maintained or improved; to sell or otherwise dispose of any property it holds; and to make rules necessary to govern its conduct and affairs.

26. Eligible recipients of assistance for loan-financed solar improvements are owners of
existing buildings for the purchase and installation of solar energy systems, builders of newly
constructed or substantially rehabilitated residential buildings that will contain solar energy
systems (pursuant to board determination), and purchasers of new or substantially rehabilitated buildings containing solar energy systems. 12 U.S.C.A. 3607 (a) (1) (B) (1980).

27. 12 U.S.C.A. 1723g (1980). This authority is vested in the bank unless the board determines that such authority is unnecessary. 12 U.S.C.A. 1723g (a). The loans purchased in the
secondary market must have terms between five and fifteen years with no prepayment pen-
alties and a face amount less than $15,000 at an interest rate and security acceptable to the
board. 12 U.S.C.A. 1723g (g). Moreover, the Federal Home Loan Mortgage Corporation
(FHLMC) Act has been amended to permit the FHLMC to make commitments to purchase
residential mortgages carried by any public utility if the mortgage is approved for participation in any mortgage insurance program under the National Housing Act. 12 U.S.C. 1454
(a) (1) (Supp. III 1979). Sec. 534 of Energy Security Act of 1980 (PL 96-294). Under 12
U.S.C.A. 1717 (b) (3) (1980) the Federal National Mortgage Association has similar powers.

28. 12 U.S.C.A. 3607 (a) (1)–3607 (a) (3). Tenants must receive written permission from
their landlord before receiving subsidies for conservation improvements. 12 U.S.C.A. 3612
(a) (4) (1980).

29. 12 U.S.C.A. 3612 (b) (1) (1980). The builder is required to inform purchasers of the
amount of credit claimed and to provide such information to the bank as necessary to prevent double subsidization. The board has not elected to provide builder subsidies to date.

30. 12 U.S.C.A. 3610 (1980).

31. "Primarily occupied by low-income tenants" is defined to mean that a majority of tenant
householders must have incomes below 80% of median area income. 12 U.S.C.A. 3612 (b)
(2) (1980). There is no indication in the section that an owner/recipient of such a solar subsidy must maintain the low-income character of his or her building after receipt of the subsidy.

32. Energy Security Act; Conference Report, S. Rept. 824, 96th Cong., 2d sess. 294 (1980).

33. 12 U.S.C.A. 3612, 3613 (1980).

34. 12 U.S.C.A. 3612, 3613 (1980).

35. 12 U.S.C.A. 3607 (c).

36. Utility-financed subsidies had to be generated from utility revenues, and could not be
mere pass-throughs of bank or other federal or state tax subsidies, in order to qualify as
exempt from double-dipping exemptions. Conference report, 96-824, 96th Cong., 2d sess. (to
accompany S. 932), at 294.

37. 12 U.S.C.A. 3602 (8) (1980). Passive technologies are not eligible for the federal tax
credit. See 26 H.S.C. 44C (1980).

> The definition of "solar energy system" is purposefully broad in order to include any
> solar technology likely to be commercially available during the life of the Bank....
> The Conferees expect the Bank, during its first years of operation, to focus on sub-
> sidizing commercially viable solar technologies and to specify the circumstances
> under which products presently under development could be considered commercially
> viable and eligible for subsidy. The criteria developed by the Bank are not to dis-
> criminate against simple passive or hybrid solar energy systems. (Conference Report
> 96-824 at 279)

38. Conference Report 96-824 at 293. Sliding scale subsidies "should not apply" to low-
income families, who should receive "[t]he maximum percentage subsidy level set by the
Bank."

39. 12 U.S.C.A. 3610 (d) (2) (1980).

40. 12 U.S.C.A. 3610 (d) (1) (1980).

> The determination of whether it is practicable to apply the energy efficiency principle to active solar energy systems should be based, among other factors, upon whether a reliable industry-wide energy efficiency rating system is in place and whether the Bank can develop an uncomplicated method of applying this concept to active solar systems. It is not the intent of the Conferees that the energy-efficiency test act as an impediment to the acceptance of solar systems, rather it is the intent of the Conferees that persons purchasing solar energy systems should be encouraged to purchase those systems that would save the greatest amount of energy.
>
> It is the intent of Conferees that energy savings are to be estimated using the difference on an annual basis between (1) the amount of oil, gas, electricity or other conventional fuels required to meet the heating/cooling load of the solar equipped buildings, and (2) the amount of such fuels required to meet the heating/cooling load of a reference residential building in a similar location. No assistance in excess of the amounts specified in subsection (a) of this section shall be provided. Additionally, the Conferees intend that the installation of a solar energy system in an existing building does not constitute *per se* a substantial rehabilitation of the building. (Conference Report 96-824 at 284)

41. 12 U.S.C.A. 3602 (a) (1980). Eligible utilities must participate pursuant to Title II of the National Energy Conservation Policy Act or meet qualifications designated by the board. Conference Report 96-824 at 279.

42. 12 U.S.C.A. 3615 (a) (1) (1980).

43. Sec. 531 of the Energy Security Act of 1980 (PL 96-294), amending 12 U.S.C.A. 1723 h (1980).

44. 12 U.S.C.A. 3610 (a) (2) (A) (1980).

45. Conference Report, 96-824 at 287.

46. 2 U.S.C. 632, codifying Sec. 301 of the Congressional Budget Act, Titles I–IX of PL 93-344, 88 Stat. 297, 306 (1974).

47. House Concurrent Resolution 115, 97th Cong., 1st sess. (21 May 1981).

48. 2 U.S.C. 641 (c–f).

49. Omnibus Budget Reconciliation Act of 1981, S. Rept. 97-139 97th Cong., 1st. sess. (17 June 1981), reprinted in 1981 *U.S. Code, Congressional and Administrative News* 396, 635. For FY 1984, this amount was subsequently lowered to $35 million.

50. House Conference Report 97-222, 97th Cong., 1st sess., at 7 (11 September 1981).

51. PL 97-51, 95 Stat. 958, 961 (1 October 1981).

52. *Congressional Record*, 24 September 1981, at S10378–87.

53. PL 97-101, 95 Stat. 1417, 1420 and 1438–1441. Under the authority conferred by Sec. 501 of PL 97-101, 95 Stat. 1417, 1438-1441, OMB subsequently reduced the amount of the appropriation to $21.85 million.

54. Deferral D82-184 (29 October 1981).

55. Opinion of the Comptroller General, B-205053(7) (31 December 1981).

56. Rescission R82-22, H. Doc. 97-140, at 27 (8 December 1981).

57. PL 100-122 (extending Solar Bank until 31 October 1987); PL 100-154, 101 Stat. 890 (extension to 15 November 1987); PL 100-170, 101 Stat. 914 (extension to 2 December 1987); PL 100-179, 101 Stat. 1018 (extension to 16 December 1987).

58. PL 100-200.

59. PL 100-242, 101 Stat. 1964.

60. PL 100-202, 101 Stat. 1329.

61. *Dabney et al. v Reagan et al.*, 542 F. Supp. 756, 765.

62. 542 F. Supp. 756, 768.

63. 556 F. Supp. 861, 866.

64. House Rept. 98-123, 98th Cong., 1st sess., at 79 (13 May 1983).

65. Ibid., at 256; Sec. 464 (e) of Title IV of the Supplemental Appropriation Act of 1984 (PL 98-181), 97 Stat. 1153, 1235 (30 November 1983), codified at 12 U.S.C. 3618 (b).

66. *Dabney v. Reagan*, Civ. 82-2231 C.S.H., Memorandum Opinion and Order.

67. Memorandum Opinion and Order, at 23.

68. U.S. General Accounting Office, *Financial Audit: Solar Energy and Energy Conservation Bank's Financial Statements for 1981 through 1985*, GAO/AFMD-87-7, November 1986.

69. Ibid.

70. Letter from Representative Richard L. Ottinger to Chairman of the Board Samuel R. Pierce, Jr., 22 March 1982.

71. Letter from Representative Richard L. Ottinger to Chairman of the Board Samuel R. Pierce, Jr., 30 September 1982.

72. Letter from Representative Steven Neal to Representative Henry B. Gonzalez, 3 February 1983.

73. Letter from Representatives Stanton and St Germain to Chairman of the Board Samuel R. Pierce, Jr., 16 December 1982.

74. Letter from John D. Dingell, Fernand J. St. Germain, and other representatives to Richard Francis, Manager, Solar Energy and Energy Conservation Bank, 12 April 1984.

75. Letter from Representative Henry Gonzalez to Chairman of the Board Samuel R. Pierce, Jr., 7 February 1985.

27 Grants

Seymour Warkov and T. P. Schwartz

In the 1970s energy policy analysts were of two minds concerning the role of government in facilitating the adoption of residential solar technologies. "Free-market" advocates opposed public investments in solar energy, despite the historic pattern of subsidization of conventional energy; in contrast, most solar advocates promoted the use of various incentives to foster the use of solar alternatives and, in turn, reduce national dependence on imported fuels. Suggested incentives included income tax credits or deductions, sales or use tax exemptions, and property tax exemptions (Rodberg and Schacter 1980; see also chapter 25). One incentive developed by the federal government was a direct subsidy in the form of a grant to pay for part of the initial expenditure for solar equipment. As noted in ERDA-23, *National Plan for Solar Heating and Cooling*, such an incentive had the advantage "of being easier to terminate after having outlived its usefulness because it is subjected to continuing budget authorization. On the other hand, tax incentives are more politically acceptable, but have the disadvantage of not being easily terminated and not being subject to annual review" (ERDA 1975, 108–109). This chapter focuses on the grant mechanism for promoting solar and specifically reports on the U.S. Department of Housing and Urban Development (HUD) Solar Hot Water Grant Program, the single (and highly visible) example of this type of incentive.

The HUD Solar Domestic Hot Water Initiative Program, an effort by the federal government to help bring solar heating systems into the marketplace, had its origins in several laws passed in 1974 by the second session of the 93d Congress (see chapter 2). The program was initially designed to cover a time frame of some thirty months (March 1977 through September 1979), but delays in implementation extended its operation to fall 1980. Significant cooperative effort by diverse federal and state agencies, businesses, and interest groups facilitated this relatively quick delivery. Our chapter offers some insight concerning the translation of government policy into concrete programs and the management of interorganizational relationships.

What follows is an account based on our effort to reconstruct an "interorganizational memory" based on retrospective interviews with individuals who had major roles in initiating and implementing the program. We have tried to cross-validate facts and figures provided by individuals on the basis of reports, documents, surveys, and memoranda on file in various federal and state agencies.

In the sections that follow we sketch the legislative and organizational origins of the HUD grants program. Attention is given to agencies implementing the program and to critical decisions and events occurring during its evolution. We present a case study of program implementation and evaluation in the state of Connecticut. Finally, we review evaluative evidence concerning the program's strengths and weaknesses and conclude with observations on some significant issues arising from this review.

27.1 Origins

The oil and gasoline shortages of 1973–1974 prompted many legislative responses at the federal level, including several laws passed by the second session of the 93d Congress in the fall of 1974. These laws served as the basis for many federal energy programs (further discussed in other chapters; see especially chapter 2) during the remainder of the 1970s, including the HUD grants program that is the subject of this chapter. The four landmark laws are as follows:

• The Solar Heating and Cooling Demonstration Act (PL 93-409), enacted on 3 September 1974, is probably the most frequently cited source legislation for the grants program. Its stated purpose was "to provide for the early development and commercial demonstration of the technology of solar heating and combined solar heating and cooling systems." This law authorized $5 million to HUD for the fiscal year ending 30 June 1975, and $50 million, "in the aggregate," for 1976, 1977, 1978, and 1979, "to carry out the programs established by this Act" (PL 93-409, 1079). It is noteworthy that the law did not specify how HUD was to use that authorization.

• The Energy Reorganization Act of 1974 (PL 93-438) resulted in frequent changes in structural arrangements within and among various federal agencies, particularly ERDA and its successor, DOE. The number and frequency of the changes may well have contributed to some of the delays that occurred in 1977 and 1978 in reaching consensus concerning the HUD grants program and its implementation.

• The Solar Energy Research, Development, and Demonstration Act of 1974 (PL 93-473) authorized the establishment of a "Solar Energy Coordination and Management Project" and the vesting of power in the chairman of this project, including responsibility for managing and coordinating HUD's efforts to foster the use of solar energy for the heating and cooling of buildings.

• The Federal Non-Nuclear Energy Research and Development Act of 1974 (PL 93-577) enacted at the very end of 1974, was in our view crucial in conceptualizing the HUD grants program in 1977 as an effort to "commercialize" rather than to "demonstrate" a particular solar heating technology. This was the concept that some key actors and agencies held of the grants program by 1979, a point that will be discussed later in this chapter. This law specified ERDA's responsibilities in considerable detail and authorized separate acts to provide ERDA "amounts as are required for demonstration projects for which the total federal contribution to construction costs exceeds $50 million" (PL 93-577, 1895).

At least three observations can be made at this point concerning the legislative antecedents of the HUD grants program. First, no single piece of legislation served as its origin. Rather, a complex but interrelated set of laws was passed in short order. These laws repeatedly reinforced the notion that Congress was intent on having virtually all of the relevant federal agencies directly involved in a massive, well-funded effort to bring solar home heating and cooling system into the marketplace and into the homes of U.S. citizens.

Second, these laws were very specific in naming federal agencies to be involved in these efforts and in assigning responsibilities and appropriations of funds to implement agency tasks. At the same time, however, the laws did not instruct the agencies on how to fulfill their responsibilities; rather, the laws set policy and permitted the agencies to spell out the relevant programs. Furthermore, the laws did not specify how designated agencies were to interact other than to establish a "Solar Energy Coordination and Management Project" (PL 93-473, 1432).

Third, the laws made available substantial amounts of money to the agencies, more than $50 million in any one year, for a period (1975–1979) that was neither so short that it required the agencies to leap before they looked nor so long that the agencies would spend all of their time simply "looking" or putting into place new self-perpetuating bureaucracies.

Nonetheless, some delays and uncertainties occurred as the agencies responded to the their congressional mandate, as we shall see in the sections that follow.

27.2   National Level

Some of the key events in the evolution of the HUD grants program are listed chronologically in figure 27.1. Each of these will be discussed in turn along with some relationships between the events and the processes that were involved along the way.

1972–1973	Solar energy background hearings before panel of the House Committee on Science and Astronautics.
Fall 1973	Solar demonstration legislation introduced.
1973–1974	"Energy Crisis" in the United States.
1/74	93rd Congress, 2nd Session, convenes.
9/3/74 to 12/31/74	Congress enacts four laws concerning the use of solar energy: PL 93-409, PL 93-438, PL 93–473, and PL 93-577.
3/75	"National Plan for Solar Heating and Cooling (ERDA-23)" is promulgated.
Spring 1976	Congressional Committee on Science and Technology holds hearings and encourages HUD to develop more aggressively a residential solar heating program.
5/76–6/76	Grants program takes shape at HUD.
7/76	HUD discusses planned grants program with ERDA (assistant secretary for conservation and solar) and with the energy offices of various states.
Fall 1976	HUD continues to discuss the planned grants program with states, agencies, and interest groups.
3/77	HUD announces the grants program, and a formal announcement is sent by HUD Secretary Harris to the first ten states selected for the program.

Figure 27.1
Key events in the history of the program

6/77	States complete plans and arrange for technical evaluation of proposed systems. "Notice to Solar Energy Industrial Community" sent by Polytechnic Institute of New York (PINY) and Florida Solar Energy Center (two principal evaluation centers) to nine participating states.
Summer 1977	DOE and RSEC created.
7/77	States announce the program. HUD completes publication of New Intermediate Minimum Property Standards. Delay in publication means that systems cannot be tested by initial target date of 9/77.
8/77	Participating states requested by HUD to delay system announcements until 1/78 to permit testing (original deadline for the installation of 10,867 systems is 8/31/78).
9/77	Many states complete acceptance of applications and selection of households. Some continue to advertise. Pennsylvania significantly below target. (Revised deadline for completion of all grant awards: 9/31/77.)
Fall 1977	Delays occur in awarding grants while technical standards are being established and certification of suppliers and installers takes place.
12/77	Technical data for qualification of systems to be received by Polytechnic Institute of New York, to be followed by announcements during 1/78. Extended cloudiness undermines scheduled completion of testing.
1/78	Formal announcement of qualified systems made by states but additional announcements forthcoming as testing is completed.
Fall 1980	Final Reports on the grants program sent to HUD by some state energy offices.
12/24/81	DOE notifies NESEC to terminate operations.

Figure 27.1 (continued)

Virtually all of the documents and interviews indicate, at least implicitly, that the 2d session of the 93d Congress laid the foundations for the
development of the HUD solar grants program (see chapter references,
especially ERDA 1975; SEAC 1979). Several sources indicate that the
Committee on Science and Technology's Energy Subcommittee chaired by
Representative Mike McCormack (D-WA) was particularly instrumental
in securing passage of the legislation and ensuring its implementation.
This subcommittee also played a key role during spring 1976 in prompting
HUD to develop the grants program and to "get it into the marketplace"
by 1977. Even so, these laws were not passed until approximately one
year after the energy crisis became a major economic, political, and social
issue (*Encyclopaedia Brittanica* 1974, 189–294). Then, in rapid fashion, the
legislation was sent to five different Senate committees. HUD was added
to the legislation by the Senate; this had to do in part with jurisdictional
issues in the House Science Committee. The House committee with oversight of HUD never held hearings, precisely so that legislation could be
speeded through. Indeed, it could be argued that critical to its quick
passage was the fact that the original legislation had been written prior to
the energy crisis. (See chapter 2 for more background to the legislation.)

Passage of the four landmark laws generated a quick response in the
agencies assigned responsibilities, with much of their solar staff coming
from the National Science Foundation. By mid-January 1975, only two
weeks after the passage of PL 93-577, ERDA was assigned lead responsibility in coordinating interagency efforts to comply with the laws
(ERDA 1975). An interagency task force was established on 10 February
1975. In March 1975 it published the *National Plan for Solar Heating and
Cooling* (*Residential and Commercial Applications*). This document, also
identified as "Interim Report" and "ERDA-23," essentially serves as a
bridge between the most significant efforts that had been taken by federal
agencies before 1975 to deal with solar energy applications in the United
States and actions that they would take during the remainder of the
decade. The document lays out the decisions that were made by the
interagency task force and the staffs' interpretation of Congress's intent
in the four landmark laws.

The plan's goals explicitly focus on stimulating and developing a market for domestic and commercial solar heating systems. Its overall goal
was to "stimulate the creation of a viable industrial and commercial
capability to produce and distribute solar heating and cooling systems

and thereby reduce the demand on present fuel supplies through wide-spread applications" (ERDA 1975, 15). No other goals or objectives related to market development are identified.

The HUD (1977) news release details many of the key aspects of the grants program as it was conceptualized in mid-1976. Apparently relatively few changes occurred between the time the grants program was conceived in May 1976 and its official inception, 28 March 1977, even though HUD exposed the program concept to review and discussion by other federal agencies, state energy offices, solar system manufacturers and installers, and public interest groups during the interim period (SEAC 1979). For instance, HUD held extensive discussions of the proposed grants program with ERDA in June and July 1976, principally the office of the assistant secretary for conservation and solar. Evidently HUD engaged in a consensus-building process that contributed to the implementation and eventual impact of the program.

The initial formulation developed by the interagency task force (and especially by HUD and ERDA) entailed the selection of several states in four regions earmarked for regional solar energy centers (RSECs), with the choice of specific states to be guided by factors such as insolation levels, energy costs, and housing characteristics. A preference for local/state utility management of the program was set aside once it was determined that not all utilities were willing to administer the program or foster public awareness of incentives. Furthermore, significant public opposition to utility "control" of solar was anticipated. Also set aside was an option involving program management through state agencies, preferably states with an established energy office and solar program. Another phase in the selection process focused on the cost of competing fuels and the calculation of electric rates that would permit a satisfactory payback of a solar investment. In sum, even though the selection process could not explicitly link states to data concerning market potential or to the number of grants required to foster a viable market (no such data were available to planning personnel), the program plan did consider alternative approaches before settling on a scheme to create a solar market that would be sustained following termination of incentives.

27.3 Program Structure

On 28 March 1977, HUD (1977) officially announced the grants program (see table 27.1), whose key features it described as follows:

Table 27.1
Participating states and their allocations (1977 dollars)

	Units	Solar grants	Administrative grants	Total
New England				
Connecticut	750	300,000	47,500	347,500
Massachusetts	1,375	550,000	78,750	628,750
New Hampshire	200	80,000	20,000	100,000
Rhode Island	250	100,000	22,500	122,500
Vermont	150	60,000	17,500	77,500
Middle Atlantic				
Delaware	150	60,000	17,500	77,500
Maryland	950	380,000	57,500	437,500
New Jersey	1,725	690,000	96,250	786,250
Pennsylvania	2,800	1,120,000	150,000	1,270,000
South Atlantic				
Florida	1,650	660,000	92,500	752,500
Totals	10,000	4,000,000	600,000	4,600,000

Source: HUD 1977.

• $4.6 million was allocated by HUD to accomplish two objectives: "to put solar-heated hot water systems in 10,000 homes," and "to step-up sales of the equipment."

• The grants program was seen by HUD as "a major expansion of the Solar Heating and Cooling Demonstration program."

• Ten states were to receive the $4.6 million and "distribute the funds to homeowners and builder-developers who want to install the solar-heated hot water systems."

• The ten states were identified, and the allocations they were to receive were designated for very specific purposes. Connecticut, for example, was to use $300,000 of its $347,500 grant to provide grants for the installation of 750 units.

• ERDA was identified as the lead agency in providing a separate program for commercial users (see chapter 10).

• The states were authorized to pay $400 per unit, and this amount was estimated to constitute "about half the cost, to homeowners and builder-developers, for the solar-heated hot water hardware." Installation costs were not included.

• Criteria for selecting the ten states were made explicit: high electric energy rates and a demonstrated "interest in encouraging the use of solar energy in residences."

• The grants program was explicitly intended to give the solar market "its initial boost."

• Finally, HUD seemed to grant the states complete authority to administer the grants as they saw fit and to use business organizations for this purpose as well.

This widely disseminated announcement created very specific expectations concerning involvement in the program and benefits to be derived from participation. It should be noted that New York was subsequently added to the roster of states, with participation limited to counties (all downstate) with high electric rates (Wener et al., 1979). We have examined some of the key events that preceded HUD's announcement; at this point we turn to a unique organizational element as the program moved into the state agencies and the marketplace, namely, the regional solar energy centers (RSECs), focusing in particular on the facilitative role of the Northeast Solar Energy Center (NESEC) in fostering public support for the program.

27.4   Regional Level

The northeast region of the United States includes ten of the eleven states listed in HUD's announcement. Although the states in this region did not deal exclusively with HUD in Washington, it is evident that the linkage between the individual states, usually their energy offices, and HUD (principally David C. Moore of the Building Technology Division) was the primary relationship. The grants program involved minimal contact (if any) with the regional offices of DOE, HUD, and other agencies. However, other linkage of the state agencies involved a relatively new and unique quasi-governmental organization, the Northeast Solar Energy Center (NESEC; see Connecticut 1980).

The origins and evolution of the RSECs is treated elsewhere (see chapter 20); here we limit our discussion to the relationships between state agencies and NESEC in the context of the HUD grants program. In March 1977 the Northern Energy Corporation received a contract from

ERDA to prepare a feasibility study for a regional center to help "commercialize" solar heating and cooling for domestic and commercial
buildings in the northeast region, followed six months later by an ERDA
contract establishing NESEC to facilitate these goals in the region's ten
states. Some funding was assigned by NESEC to activities intended to
bolster the states' participation in the HUD grants program.

   Among other activities, NESEC provided participating states with
access to technical staff members who had substantial, direct experience
with solar-heated hot water systems funded in the early 1970s by other
government programs, by private businesses, and by utilities. NESEC
also offered a speakers bureau, a solar home heating library, and a variety
of educational, public relations, and outreach services that state agencies
and citizens could call upon in deciding whether and how to participate
in the grants program.

   By virtually all accounts (SEAC 1979; Merrigan and Gleman 1980;
Connecticut 1980) the initial response of citizens, interest groups, state
agencies, and manufacturers and installers of solar systems to the announcement of the HUD grants program was favorable. Enthusiasm
created by the announcement and by subsequent public information
efforts on the part of the state agencies and NESEC fostered rising expectations. Those expectations were dampened appreciably as 1977 turned
into 1978 and as long delays occurred in establishing performance standards for the solar hot water systems and in certifying qualified manufacturers and installers of the solar heating systems. Also, prospective
participants came to realize that the costs of the systems would far exceed
twice the $400 homeowner grants (the "50%" contribution to the pur-
chase of solar hardware that was identified by HUD in the original announcement of the program). Finally, prices escalated as manufacturers
and installers rushed into the marketplace. To understand these events,
we shift to an analysis at the level of the state agency and end user, with
special attention to the case of Connecticut.

27.5 Connecticut: A Case Study

The State of Connecticut's Department of Planning and Energy Policy
(DPEP) submitted a proposal in Spring 1977 to issue 750 grants in the
amount of $400 to qualified owners of single-family dwellings or to pro-
spective buyers of new homes incorporating solar domestic hot water

heating systems (Connecticut 1980). A plan developed by the department specified an award to 600 recipients of qualified systems installed in owner-occupied dwellings and to 150 builders/developers. The plan estimated costs to recipients of at least $1,000 in excess of the award. The formal proposal submitted to the HUD Office of Policy Development and Research was signed on 3 May 1977 by Governor Ella T. Grasso. DPEP publicized the program in the mass media and received some 2,750 requests for the application form by 18 July 1977. Other sources disseminating information about the program included trade fairs, exhibits, staff contacts, and the solar industry. The sheer volume of response to the announcements imposed special demands on DPEP and its successor unit, the Energy Division of the Office of Policy and Management. Accustomed to dealing with other administrative units or agencies, the staff was now required to acknowledge several thousand individual requests for information and applications forms. Nor did demands on the staff abate after the first wave of inquiries. Prior to a final grant award, applicants were to submit documentation from the installer that the system conformed to the requirements and guidelines of *NBSIR 77-1272* if installed after 30 June 1978 (see chapter 14); a copy of the building permit and a copy of the final building approval; a copy of the manufacturer's and installer's warranty; a system certificate of compliance; and a statement by the installer of 100% backup availability. Furthermore, applicants were forwarded notices by the state agency concerning extension requests, extension renewals, acknowledgment of receipt of documentation, notice of unacceptable documentation, notice of missing documentation, and program deadline requirements. From the perspective of administrative staff in Connecticut, then, HUD requirements imposed a substantial commitment of staff resources to manage the flow of documents.

27.5.1 Public Response

Public interest in innovations such as solar technology can be gauged through public opinion polls, content analysis of the mass media, responsiveness to public policy initiatives, and the like. One analysis of Connecticut's interest in household solar was indexed by responsiveness to the announcement of the HUD grants program; it used the names and addresses of respondents to map the geographic and social variation in requests for application forms and other information on the part of homeowners and builders/developers (Warkov 1979; Warkov and Meyer

Table 27.2
HUD grants awarded and solar energy systems installed in the Northeast, by state

	Initial allocation	Systems installed, as of 2-1-79	Systems installed, as of 12-31-79	Final number of systems installed	Number installed, as % of allocation
Connecticut	750	189	542	889	118
Massachusetts	1,375	190	679	1,288	93
New Hampshire	200	15	210	244	122
New Jersey	1,725	275	622	1,270	73
New York	867	300	524	956	110
Pennsylvania	2,800	400	1,113	1,646	59
Rhode Island	250	45	245	283	113
Vermont	150	20	104	162	108
Total	8,117	1,434	4,039	6,738	83

Source: Northeast Solar Energy Center, Update, Vol. 1, No. 3, 1979, p. 3; Northeast Solar Energy Center, Update, Vol. 2, No. 7, 1980, p. 2; and unpublished reports on file, state energy offices.

1982). When names and addresses were sorted by town (the state comprises 169 "towns"), the list yielded 2,254 unique household addresses; 153 business, institutional, and government addresses; and 433 duplicate household and 42 duplicate business and professional addresses. By September 1977 a total of 893 applications had been submitted for the 750 grants initially allocated to the state. Certain characteristics of the population were related to variations in the rate of public interest found in the 169 towns. Overall, interest was higher in towns with high-status populations (indexed by occupation, education, and income) and in smaller towns with low-density populations. A measure of proximity to solar dealers/installers did not predict interest in the program because the data were gathered in the very earliest stages of the technology diffusion process.

Not all states participating in the program experienced the level of interest and effective public demand for the grants evidenced by the state of Connecticut. Program data are displayed in table 27.2. New Hampshire and Connecticut experienced effective demand well in excess of the initial allocation of grants, while New Jersey and Pennsylvania installed fewer systems than expected on the basis of solar energy systems installed per hundred grants awarded at the outset of the program (see column 5). As a consequence, HUD program managers reallocated underutilized resources; in August 1979 Connecticut received an additional 150 grants, which resulted in a total of 900 potential recipients.

27.5.2 Implementation Delayed

A major barrier to program implementation was the requirement for a system certificate of compliance. On the one hand, grants could be awarded in Connecticut only for systems approved by the Connecticut State Board of Materials Review. (Under the state building code, only systems/collectors that had such approval could obtain a local building permit.) On the other hand, HUD stipulated that grantees would qualify for the award only if the solar energy system conformed to the HUD *Intermediate Standards for Solar Heating and Domestic Hot Water Systems*, including a 30-day stagnation test (as per *NBSIR 77-1272*). Participating states could meet this requirement by restricting the purchase of a solar energy system to the list of acceptable suppliers assembled by the New York Solar Energy Applications Center (SEAC) at the Polytechnic Institute of New York. Although nine states entered into a contract with SEAC, Connecticut elected an alternative to centralized coordination of technical system evaluation. (Parenthetically, DPEP's HUD solar grant program planning document indicated that the SEAC list of acceptable suppliers would be available "on or about" 1 January 1978.) By late spring 1977 HUD and state officials arrived at a compromise. Recipients of a HUD grant could receive a system certificate of compliance by (1) securing a system certificate from the Solar Energy Evaluation Center at the University of Connecticut, (2) purchasing a system from a supplier on the SEAC list, or (3) obtaining certification by a professional engineer licensed in the state of Connecticut. Whichever method the applicant elected, the system would be certified to provide one-half of the energy requirements for the annual hot water needs of a family of two adults and two minor children (50% of 70 gallons per day raised to 90°F).

Program participants were required to provide other documentation, as were applicants in other states, prior to notification of the grant award. These included (1) a copy of the building permit and a copy of the "inspection record" or certification of final approval by the local building official; (2) a one-year warranty provided by the installer on parts and labor against failure of any component or assembly due to manufacturing or installation defects, and a five-year warranty on collectors, storage tank, and heat exchanger; and (3) a statement from the installer that a

100% backup was available from a conventional or auxiliary system. In brief, recipients of the grant would be expected to clear a number of hurdles before they would see a tangible $400 benefit. Some applicants found the task daunting.

27.5.3 Warranties and Technical Evaluations

The final report of the State of Connecticut (1980) to the program manager, HUD Solar Heating and Cooling Demonstration Program, also addressed the problem of warranties and staff review of documentation to ensure "conformance" to program requirements. It noted that "difficulties experienced in obtaining required warranties arose most often under the following conditions: (1) installation of a comprehensive solar space and domestic hot water system; (2) installation of collectors and components purchased from several dealers; (3) certain instances of self-installation; and (4) system manufacturers offering several different warranties. Constant interaction between this office and local dealers and manufacturers alleviated this problem."

The report also reviewed the work of the Solar Energy Evaluation Center, which sent teams of inspectors to conduct 153 on-site technical evaluations of systems installed by first-round applicants who submitted all required documentation. All systems were inspected for 15 "critical" and 24 "serious" deficiencies. A system found to have any critical deficiency was rejected. The team prepared an eight-page report on each system with recommendations for improvement, and all critical deficiencies required correction prior to approval and payment. Only 11 systems (7%) received unconditional approval on the first inspection; 88 (58%) received conditional approval; and the remaining 54 systems (35%) were rejected on this first round of inspections. It was noted that in many cases the critical deficiencies affecting system performance were relatively easy to remedy. The report gives no indication of the proportion of do-it-yourself solar enthusiasts among early program participants, but it is clear that pervasive problems occurred in this phase of program implementation, although "[a] significant improvement in the quality of installations by professional installers was noted as a result of receiving their inspection reports." Reinspections after owners or installers made the corrections recommended by the team resulted in a reduction of rejected systems from 54 to 26 by 1 July 1979. Further reinspections eventually resulted in 21 systems receiving unconditional approval and 4 systems being rejected.

(The status of one system is unknown.) These results are instructive in that they highlight the need for an independent and competent technical evaluation to ensure adequate standards of installation in the early stages of technology diffusion.

27.6   Evaluation Studies

Given an appropriation of $4.6 million to install some ten thousand solar hot water heaters, it is appropriate to ask: Were the program's goals achieved? Did it make a difference in establishing a solar presence in the U.S. energy mix? These are issues of consequence. An evaluation might address the following four questions. First, were program services delivered to the public with reasonable efficiency? Second, did the hardware work? Third, were the systems cost-effective? And fourth, did specialists and the public (users and nonusers) respond favorably or not in evaluating one or more of the above questions?

To understand the relative paucity of systematic evidence, we turn again to the larger sociopolitical context of this and similar programs. We have emphasized congressional pressures to "do-something-about-solar" and have offered an account of a number of key developments antedating the program's inception in spring 1977. In retrospect, the political process that fostered relatively rapid expansion of public investments in solar fostered with equal fervor its demise. Specifically, J. Glen Moore (see chapter 2) and others note that the budget process was employed with surgical skill to scale down the federal solar commitment following the departure of the Carter Administration. Major technoeconomic evaluations were curtailed, and research designs premised on longitudinal studies of technical performance and economic benefits were discarded. Thus evaluative data are largely unavailable for the period after 1981.

However, bearing on the four questions above, program implementation was hampered by problems associated with the availability of system standards and testing laboratories; improper installation of systems; and warranty requirements and their enforcement. (SEAC 1979) Issues bearing on materials compatability, system sizing, toxicity of heat transfer fluids, collector testing, manuals, control systems, and freeze protection are raised (Merrigan and Gleman 1980). Documentation of national incidence and prevalence as well as regional variation in the salience of these problems among households installing solar hot water systems under the auspices of the HUD program are required but not available. (See chapter 9, however, for a summary of technical issues that arose during other parts of the residential demonstrations.)

27.6.1   Program Effectiveness: Perspectives of Participants

To get a sense of how the grant program was viewed by program participants, we draw on the results of a telephone survey conducted in spring 1979 comparing program applicants who had installed a HUD grant-assisted solar hot water system with counterpart applicants (Warkov 1979; Warkov and Meyer 1982), who completed an application form for the grant, but failed to install a solar collector. We refer to this "high-risk" solar candidate population as "near-adopters." Although adopters and near-adopters were comparable based on social class indicators such as educational attainment and household income (both groups were wealthier and more likely to be college-educated than the state's population at large), the most important characteristics differentiating adopters from near-adopters in this study population concerned their perceptions of private benefits to be derived from solar technology (e.g., solar adds to the resale value of the house); their communications behavior (e.g., spoke to and were encouraged by solar dealers, engineers, etc.); and certain dwelling attributes (no shade, compatible roof design, and southern exposure). On the other hand, the two groups exhibited no significant differences in knowledge about available financial incentives (e.g., tax credits for solar, state sales tax exemption on solar collectors, duration of exemption if town adopts a solar ordinance).

Of special interest here is the evaluation of program effectiveness overall as well as ratings on a 3-point scale of five program requirements. As seen in table 27.3, the general evaluation was lower than rated effectiveness of any one feature. Some 29% of the solar adopters rated the program overall as "not at all effective," as did 23% of the near-adopters. At the other end of the scale, nearly two in five (38%) of those using the grant to install solar found the program "very effective," as compared with only one in four (23%) of those not installing solar. Grant recipients were more negative about specific features of the program than were near-adopters in five out of six comparisons: offering a grant of $400: 31% versus 19% rating the program "not at all effective"; processing your application and required documents: 36% versus 18%; having a list of manufacturers when you needed it: 32% versus 23%; requiring a system meeting 50% of

Table 27.3
Rated effectiveness of the Connecticut HUD solar grant program, by adopter status

	Solar users		Near-adopters	
Program feature	% very effective	% not at all	% very effective	% not at all
Offering a grant of $400	55	31	53	19
Processing your application and required documents	44	36	59	18
Requiring an inspection of the system before approving award	64	19	45	20
Having a list of manufacturers when you needed it	48	32	49	23
Requiring a system meeting 50% of hot water needs	57	30	43	17
Providing information on solar energy systems in general	40	35	34	23
In general, the program, was ...	38	29	24	23

Note: Rated effectiveness is based on a 3-point scale: "very effective," "somewhat effective," and "not at all effective." The intermediate category is not displayed in the table.

hot water needs: 30% versus 17%; and providing information on solar energy systems in general: 35% versus 23%. Because grant recipients probably had more extensive contact with the program, their tendency to rate these features "not at all effective" is based on direct experience. By the same token, grant recipients rated the program more favorably than near-adopters with respect to the requirement of an inspection before approving the award (64% versus 45% very effective); and requiring a system meeting 50% of hot water needs (57% versus 43%). About half of each group rated the program "very effective" in offering a $400 grant and in having a list of manufacturers when needed.

Near-adopters also were asked to identify a list of twenty-five perceived barriers as a "major," "minor," or "no reason at all" for dropping out of the program. The three reasons mentioned by a majority of near-adopters were in the "psychoeconomic" realm: size of grant (66%); price of solar higher than expected (65%); and not convinced solar would pay for itself (59%). A multivariate analysis of the twenty-five reasons produced several factors (Warkov 1979; Warkov and Meyer 1982), of which the three most prominent were "Psychoeconomics" (incorporating the three items above as well as "could get the same benefits from conservation"; "Technology Delivery System" ("worried about manufacturer staying in business long enough to back up warranty" and "no competent distributor around to install solar"); and "Program Organization," which consisted of three items mentioned by program dropouts: delays in securing an approved list (30%); program required too much paperwork (22%); and insufficient information about solar from the program (25%). One feature of the program figured prominently in the decision to drop out. The grant incentive was simply too small in light of the price of a solar energy system, coupled with a judgment that the investment would not be recovered via savings accrued from solar-heated hot water. A secondary factor concerned the volume of documentation required to secure grant approval as well as the delay in providing prospective users with an approved list of suppliers and solar energy systems. To be sure, these were HUD-designed program requirements that state officials were obliged to enforce, although program applicants and eventual adopters would not be expected to differentiate among federal and state-level operations, however, in offering these evaluations.

27.6.2   Related Findings

Indirect evidence of the contribution of the HUD program to the enhancement of the solar industry and the residential use of domestic solar-heated water may be gleaned from a national, nonrandom sample of some 3,800 solar owners contacted by mail questionnaire during October through December 1980 (Farhar-Pilgrim and Unseld 1982). Although generalizability is limited by a bias toward self-selection, it is nevertheless the largest extant database of solar owners (and users). Fully eight in ten own flat-plate collectors, and seven in ten systems provide domestic hot water only. About 27% (1,013 cases) were recipients of government or industrial assistance other than tax credits; of these, two-thirds (608 cases) were awarded an HUD program grant. Clearly, the HUD grants program was associated with the acquisition of a significant number of systems, although the findings permit no assertion that solar adoption was contingent upon a grant award.

The dearth of systematic evaluative work on the HUD grants program can be tied directly to the abrupt shift in energy policy emphasis on the part of the federal government in early 1981, although other factors, such as competing conceptions of criteria of satisfactory evaluation, may have operated as well.

27.7 Conclusions

The authors cited earlier in this chapter (HUD 1977; ERDA 1975) were prescient in their appraisal of the political and administrative convenience of initiating a grants program as compared with other incentives to encourage solar. The very nature of grants permitted easy termination of the HUD program once the states received their respective allocations. Our account confirms the principle that a grants program can be self-limiting if policymakers are intent on placing boundaries on program development in any given policy domain.

The argument was made in some interviews that Congress pushed for an expanded federal solar effort, pressing DOE and HUD to move at a faster pace than these agencies were able to sustain. Delays in implementation are endemic to most, if not all, large-scale public programs, and energy-related programs are hardly exceptional in this respect. The grants program and others initiated in the mid-1970s were developed in response to congressional pressure to produce results quickly. Implemented in haste with insufficient staff for proper planning and assessment, decisions concerning contracts were hurried, followed by late hiring, tardy evaluations of technical defects, and additional delays in certifying solar energy systems. Because of the abrupt termination of federal support for solar programs, we do not know whether problems having to do with the performance of materials, improper installation and maintenance, and the like were significant for participating households. It is ironic that the grants incentive and other solar programs were dismantled just as solid data began to be produced in 1980–1981.

The social (distributional) impact of the grants program deserves comment. There is an equity issue involved in offering government assistance to those least in need of an economic incentive, and most solar adopters and near-adopters hardly suffered from "fuel poverty." Yet, even though an income limit would have fostered greater social equity, it also could have excluded some households unwilling to pay for a system absent the grant.

Bearing in mind that the program's explicit goal was the development of a viable industry stimulated by demand for solar products, critics nevertheless might ask, Should homeowners be investing in solar hot water before they have maximized energy savings in the form of insulation, efficient furnaces, double and triple glazing, and the like? Public and private programs encouraging these forms of energy conservation were in high gear at the time the grants program was announced, yet some owners were installing solar energy systems before the less expensive standard residential conservation technologies had been implemented (Warkov 1981). This point had been raised in the interviews as well.

The solar hot water grants program conforms to the classical top-down model of policy and program development and implementation, in which a number of policy goals such as "energy independence," "reduced consumption of imported oil," and "energy conservation" frame the activities of federal policymakers. A different conception of the grants program comes into focus if we ask, Where does a solar grants program fit into a comprehensive policy of energy conservation, alternate energy development, and environmental protection? This conception combines top-down and bottom-up processes in which federally initiated consumer energy incentives are integrated into local community programs that make greater provision for consumer involvement in design (Olsen and Joerges 1981; Gaskell and Joerges 1987). It is likely that flexible program design would have avoided the pattern of investment in relatively expensive solar hot water heating without due regard to lower-cost energy efficiency measures. Linking solar grants to comprehensive residential energy conservation could have achieved greater energy savings in the household sector and more effective use of the HUD program funds than was feasible given the relatively uncoordinated system of household incentives in place during the energy crisis. As noted in a multinational study of local energy conservation programs, conditions fostering broad consumer participation and maximal energy savings involve "an active and prudent channelling of central-level programmes, a high commitment of key persons or a core group of citizens, an imaginative and extensive use of local communication channels, a combination of various instruments and services into 'packages', and an active involvement of local business and utilities" (Gaskell and Joerges 1987, 26).

A national evaluation of all solar (and conservation) incentives would have expanded our knowledge base of the U.S. experience and possibly encouraged the inclusion of solar grants in a package of solar and conservation incentives. However, the politics of energy policymaking during the early 1980s dictated otherwise. Given current energy and environmental conditions, we can anticipate a resurgence of public demand for environmentally compatible energy technologies and, heeding Santayana's dictum, perhaps learn appropriate lessons from the recent past.

References

Connecticut. Office of Policy and Management. Energy Division. 1980. *Solar Domestic Hot Water Initiative: Final Report.* H8001. December. Hartford.

Encyclopaedia Brittanica. 1974. *Book of the Year.* Chicago.

ERDA (U.S. Energy Research and Development Administration). Division of Solar Energy. 1975. *National Plan for Solar Heating and Cooling (Residential and Commercial Applications): Interim Report.* ERDA-23. Washington, DC.

Farhar-Pilgrim, B., and C. T. Unseld. 1982. *America's Solar Potential: A National Consumer Study.* New York: Praeger.

Gaskell, G., and B. Joerges. 1987. *Public Policies and Private Actions: A Multinational Study of Local Energy Conservation Schemes.* London: Gower.

HUD (U.S. Department of Housing and Urban Development). 1977. HUD-N. 77-87. News release concerning the solar domestic hot water initiative grant program. 28 March.

Merrigan, T., and S. Gleman. 1980. *HUD/Florida Solar Hot Water Initiative Program: Final Report.* Pt. A, *Narrative Summary.* Cape Canaveral: Florida Solar Energy Center.

NESEC (Northeast Solar Energy Center). 1979. Update. Vol. 1, no. 3. Cambridge.

NESEC. 1980. Update. Vol. 2, no. 7. Cambridge.

Olsen, M., and B. Joerges. 1981. *The Process of Consumer Energy Conservation: An International Perspective.* Berlin: IIUG, Wissenschaftszentrum.

Public Law 93-409. The Solar Heating and Cooling Demonstration Act of 1974. 3 September.

Public Law 93-438. The Energy Reorganization Act of 1974. 11 October.

Public Law 93-473. The Solar Energy Research, Development, and Demonstration Act of 1974. 26 October.

Public Law 93-577. The Federal Non-Nuclear Energy Research and Development Act of 1974. 31 December.

Rodberg, L., and M. Schacter. 1980. *Solar Conservation and Solar Energy Tax Programs: Incentives or Windfalls?* Washington, DC: Council of State Planning Agencies.

SEAC (Solar Energy Applications Center). 1979. *HUD Solar Hot Water Initiative: Centralized Coordination of Technical Tasks and System Evaluation, Final Report.* Poly-M/AE 79-64. Brooklyn: Polytechnic Institute of New York.

Warkov, S. 1979. *Solar Adopters and Near-Adopters: A Study of the HUD Solar Hot Water Grant Program.* Report to the Northeast Solar Energy Center. Cambridge, MA, August.

Warkov, S. 1981. "Energy Conservation and Adoption of Household Solar." In Claxton, J. D., Anderson, C. D., B. Ritchie, J. R. and G. H. G. McDougall (eds.) *Consumers and Energy Conservation.* New York: Praeger.

Warkov, S., and J. W. Meyer. 1982. *Solar Diffusion and Public Incentives.* Lexington and Toronto: Lexington Books, D.C. Heath.

Wener, R. E., Mulcahy, F. D., McHugh, D. and R. Sviedrys. 1979. *Factors Influencing Adoption of Solar Hot Water Heaters.* Poly-M/AE 79-26. Solar Energy Applications Center, Department of Mechanical and Aerospace Engineering. Brooklyn: Polytechnic Institute of New York.

VIII SOLAR THERMAL ORGANIZATIONAL SUPPORT

28 International Activities

Murrey D. Goldberg

28.1 Program Goals

The international solar program of the United States government originated in 1973 and was carried out under the National Science Foundation, the Energy Research and Development Administration, and the Department of Energy, successively. It was organized around three primary goals:

1. Enhancement of activities carried out under the domestic research and development program,

2. Assistance to the export programs of domestic solar industries, and

3. Use of solar technologies in Third World development assistance programs.

Through bilateral and multilateral cooperative programs, much was accomplished toward realization of the first goal. This work is described below in some detail. An extensive program to address the second goal was begun in 1978, but results at the time of termination of the effort in 1981 were very hard to measure; the content of the export assistance effort will be presented. As concerns the third goal, some projects of the Agency for International Development involved renewable energy technologies (e.g., improved wood-burning cook stoves), but solar thermal technologies played no major role in these activities; Third World development efforts will not be described further here.

The chapter includes sections on multilateral work under the North Atlantic Treaty Organization (28.2) and the International Energy Agency (28.3) and on bilateral cooperation with Italy, Saudi Arabia, Spain, and the United Kingdom (28.4). These are followed by a section on the solar commercialization effort (28.5) and a summary section (28.6).

28.2 Committee on the Challenges of Modern Society

This section is based on notes by Frederick H. Morse (1977). The North Atlantic Treaty Organization (NATO) was created by treaty in 1949 as a mutual defense pact between the United States and Canada and ten western European nations. Current NATO membership includes Belgium,

Canada, Denmark, France, Germany, Greece, Iceland, Italy, Luxembourg, Netherlands, Norway, Portugal, Spain, Turkey, United Kingdom, and the United States.

The NATO Committee on the Challenges of Modern Society (CCMS) was founded in 1969 to provide a "social dimension" to the North Atlantic Alliance through cooperative responses to the complex problems facing industrialized nations. The pilot study method was used in which a lead country, in association with any other interested countries, conducted a project in an area of common interest. Such projects were open to non-NATO countries and aroused wide interest and participation.

In 1973 CCMS initiated a Solar Energy Pilot Study. The "energy crisis" was a recent memory, and many countries had expanded their national research and development programs to develop both conventional and nonconventional energy sources; CCMS pilot studies were begun in solar energy, geothermal energy, and energy conservation.

The United States was the pilot or lead country for the Solar Energy Pilot Study (NATO/CCMS 1978), with Denmark and France as copilot countries. A large fraction of energy use in industrialized countries was consumed in heating, cooling, and providing hot water in buildings, and solar energy seemed to have the potential for significant market penetration and fossil fuel displacement in this economic sector in many countries. It was therefore agreed to restrict the pilot study to space heating and cooling and hot water heating.

The main objectives of the study were (1) exchange of information on solar heating and cooling system programs and projects in each participating country and (2) promotion of cost-effective and practical applications of solar energy to heating and cooling in residential, commercial, industrial, agricultural, and public buildings. Key information exchange elements included

• preparation and distribution of special reports, in an agreed format, on certain projects in each country,

• distribution of relevant and publicly available reports on solar heating and cooling systems for buildings, and

• participation in meetings for review of research and development programs to exchange information and ideas.

Guidelines for participation in the pilot study were specified in a "Memorandum of Understanding." The memorandum embodied the belief that international cooperation in solar heating and cooling of buildings would accelerate introduction of cost-effective applications in participating countries. A comprehensive exchange of knowledge was to be instituted regarding the design, technical and economic characteristics, and performance of solar energy systems and subsystems developed and demonstrated in each country participating. Non-NATO countries were urged to join.

Signatories to the memorandum, generally representatives of government agencies, designated a contact person through whom all reports and other documents were to be exchanged. Fifteen countries signed the memorandum: Australia, Belgium, Canada, Denmark, Federal Republic of Germany, France, Greece, Israel, Italy, Jamaica, the Netherlands, New Zealand, Spain, the United Kingdom, and the United States—with additional countries, plus the European Commission, acting as observers.

Participation in the pilot study was intended to meet and satisfy the programmatic goals of participating countries. In addition to the obvious benefits of a general information exchange and periodic reviews of one national program against another, it was hoped that individual national programs would be broadened and accelerated through this cooperation. Because the study involved information exchange only, it was essential that the exchange be in sufficient detail for each country to be able to adopt the reported project as one of its own, where relevant.

Solar heating and/or cooling system performance can only be thoroughly understood if appropriate descriptions of the environment, system components and thermal characteristics, and economic factors are provided. A special reporting format was developed to effectively provide standardized information on the performance of a system or subsystem (Morse and Rose 1975). Adoption of the format was encouraged so that reports would contain all information central to a thorough understanding of system performance. The format was also envisioned as establishing an international standard for system testing because it was to provide all measurements required to fully specify system performance.

During early meetings of study participants, it became apparent that several countries had similar climatic conditions and were stressing similar applications. By drawing together experts who were working on common tasks, effective information exchanges could be developed to supplement the more formal CCMS report exchanges. Special study groups were created on zero-energy houses and on Mediterranean applications.

The United States solar energy program benefited in a number of ways from participation in the CCMS:

• The CCMS reporting format provided the basis for the technical reporting format adopted for the national data collection program for heating and cooling systems.

• Exchange of international data significantly expanded the amount of data available on system performance to the program.

• Program officials were able to identify useful programmatic strategies developed in other countries and areas that might benefit from future international cooperation.

• The specialized study groups provided access to work and data that were more advanced than work done in the United States.

• System performance became available in a wider variety of climatic conditions than were under study in the domestic program.

• Contacts with governments, research institutes, and industries proved highly useful to U.S. program participants.

The CCMS programs were gradually absorbed into the International Energy Agency (IEA) effort described below. The IEA provided a broader forum for involving countries not directly involved in mutual defense activities of the Western nations.

28.3 International Energy Agency

The steep increase in oil prices in the early 1970s severely disrupted the economies of most oil-importing countries and strained many political, strategic, and economic relationships. Some member countries of the Organization for Economic Cooperation and Development (OECD) undertook to achieve a coordinated approach to decrease their dependence on imported oil and to reduce the strategic and economic vulnerabilities such dependence caused.

The OECD was established in 1961 to promote economic and social welfare in member countries and to stimulate and harmonize efforts on behalf of developing nations. Nearly all industrialized "free-market" countries belong, with Yugoslavia as an associate member. OECD is headquartered in Paris and primarily collects, analyzes, and disseminates economic and environmental information. Present membership includes

Australia, Austria, Belgium, Canada, Denmark, Finland, France, Germany, Greece, Iceland, Ireland, Italy, Japan, Luxembourg, Netherlands, New Zealand, Norway, Portugal, Spain, Sweden, Switzerland, Turkey, the United Kingdom, the United States, and Yugoslavia.

On 18 November 1974 interested countries adopted an International Energy Program (IEP) for cooperation in energy research and development. The International Energy Agency (IEA) was established within the OECD to administer, execute, and monitor the IEP. OECD members participating in the IEP include Austria, Belgium, Canada, Denmark, Germany, Ireland, Italy, Japan, Luxembourg, Netherlands, Spain, Sweden, Switzerland, Turkey, the United Kingdom, and the United States. Later, New Zealand became a participating member, and Norway and Greece became observers. The Commission on Economic Communities also participated under special arrangements.

One committee of the IEA, the Committee on Energy R&D, was mandated to carry out cooperative research, development and demonstration programs of interest to two or more participating countries (Morse 1977). Nine expert groups were created by the Committee on Energy R&D, including the Subgroup on Solar Energy; five initial projects were selected by this subgroup in the areas of solar heating and cooling of buildings and solar radiation measurement and analysis. Each project (organized as a "Task") had a lead country, responsible for maintaining the pace and quality of the development of project details; the five tasks and lead countries were:

• Task I: Investigation of the Performance of Solar Heating and Cooling Systems—Denmark

• Task II: Coordination of R&D on Solar Heating and Cooling Components and Systems—Japan

• Task III: Performance Testing of Solar Collectors—Federal Republic of Germany

• Task IV: Development of an Insolation Handbook and Instrument Package—United States

• Task V: Use of Existing Meteorological Information for Solar Energy Application—Sweden.

(The first decade of the IEA Solar Heating and Cooling Program is described in IEA 1988.)

As of 1990, eleven additional tasks were developed and begun:

• Task VI: Performance of Solar Heating, Cooling, and Hot Water Systems Using Evacuated Collectors—United States

• Task VII: Central Solar Heating Plants with Seasonal Storage—Sweden

• Task VIII: Passive and Hybrid Solar Low-Energy Buildings—United States

• Task IX: Solar Radiation and Pyranometry Studies—Canada

• Task X: Solar Materials R&D—Japan

• Task XI: Passive and Hybrid Solar Commercial Buildings—Switzerland

• Task XII: Building Energy Analysis Tools for Solar Applications—United States

• Task XIII: Advanced Solar Low-Energy Buildings—Norway

• Task XIV: Active Solar Systems—Canada

• Task XV: Advanced Central Heating Plants with Seasonal Storage—Sweden

• Task XVI: Photovoltaics Building Applications—Germany.

United States participation in the IEP is administered directly by the Department of Energy. These multinational solar energy projects of the IEA continue to be a very fruitful source of research information for the United States solar program and an excellent mechanism for tapping foreign research capabilities for the benefit of this program.

28.4   Bilateral Cooperative Research Programs

The United States solar energy program began during the early 1970s under the auspices of the National Science Foundation (NSF). The NSF solar staff (who soon moved to the Energy Research and Development Administration and then into the Department of Energy) realized that tapping worldwide expertise in solar energy research and development could provide substantial benefits to the young domestic effort. Contacts abroad with solar energy workers and governments indicated much mutual interest. Formal bilateral discussions were initiated with several countries, usually resulting in a visit by a governmental and scientific delegation to that country, with sometimes a return visit by a similar delegation. These visits usually involved joint seminars and trips to relevant research facilities. Early exchanges included Australia, Israel (Goldberg 1980), Japan, Saudi Arabia, the Soviet Union (Goldberg 1979), and Spain.

As a result of political concerns, budgetary constraints, proprietary realities, and lack of technical interest, some of these exchanges never resulted in formal bilateral cooperative programs in areas relevant to this volume. However, U.S. scientists did learn much of value to apply to the domestic program, and useful scientist-to-scientist information exchanges were created or strengthened. Interesting bilateral projects in solar thermal technologies did result with Italy, Saudi Arabia, Spain, and the United Kingdom. These are detailed in the following sections.

28.4.1   Italy

Late in 1980, a Memorandum of Understanding (MOU) was signed by the Department of Energy and the Italian Ministry of Industry, Commerce, and Handicraft for a cooperative program in solar energy research and development. Seven projects were adopted, including work in photovoltaics, power systems, passive design, and information exchange. The MOU, with detailed program descriptions for the seven projects appended, was prepared by Matthew Sandor of the International Programs Branch in the Solar Energy Research Institute. The institute was vested with overall program coordination, serving as the secretariat to the Executive Committee for Solar Projects created in the MOU. The program was to last three years and had an initial proposed budget of $4,535,000, about 60% provided by Italy and 40% provided by the United States.

A project on Design and Testing of Passive Solar Systems and Components was initiated with a Joint Technical Meeting in Boulder, Colorado, in October 1979. The U.S. team was headed by Michael Holtz of SERI, and the Italian team was headed by Sergio Los of the University of Venice. Several multifamily and/or commercial buildings incorporating passive solar systems were to be designed, built, instrumented, and monitored in each country using common design and simulation test methods, instrumentation, and performance data analysis. Passive solar components incorporated into buildings or test structures were to be compared for differing climates, sociocultural contexts, and building types.

Considerable work was completed before the project was unilaterally terminated by the United States in 1982. Several multifamily housing projects were designed, built, and monitored, and a multifamily passive design handbook was written by the Energy Office of Massachusetts. Several passive solar commercial buildings were also built, and certain technology transfer activities were completed, peripheral to but relevant to the bilateral agreement.

28.4.2 Saudi Arabia

On 25 May 1977 Saudi Crown Prince Fahd ibn Abdulaziz al-Saud, in an address at Reston, Virginia, at dedication ceremonies for the Terraset Elementary School, proposed a joint program with the United States in solar energy research and development. After much discussion and negotiation, a Project Agreement for Cooperation in the Field of Solar Energy was signed on 30 October 1977. Principal parties included the Saudi Arabian National Center for Science and Technology (SANCT), the Saudi Arabian Ministry of Finance and National Economy, and the United States Departments of Treasury and Energy. The agreement called for the expenditure of $100 million over five years, with each government providing 50% of the total. Management responsibility for the full program, which came to be called "SOLERAS," was assigned to SERI's International Programs Branch. Eventually, a new branch was created in SERI to administer this large program, and this group was moved in 1982 to the Kansas City headquarters of the Midwest Research Institute. SERI's initial activities included preparation of a detailed management plan, leading exchange visits by groups of Saudi and U.S. solar scientists through each other's facilities, and creation of extensive research proposals for SANCT and DOE approval.

As part of its mission (Williamson 1991), SOLERAS funded the design and construction of state-of-the-art photovoltaic and solar thermal research facilities for the King Abdulaziz City for Science and Technology in Saudi Arabia. These were placed adjacent to a SOLERAS-funded photovoltaic village system and included laboratories for conducting photovoltaic cell and array experiments, testing solar collectors in both outdoor and solar simulation setups, and measuring solar radiation.

During its lifetime, SOLERAS funded a number of research projects at both Saudi and U.S. universities. Research topics included solar cooling, artificial photosynthesis, satellite measurements of solar radiation, and solar ponds. Projects were selected from competitive proposals and were funded for several years; the largest of these were thermal projects for solar cooling at all four major Saudi research universities. Two large-scale projects under the SOLERAS program involved technologies of interest here, one on solar desalination and one on solar cooling; these are described in the following sections.

28.4.2.1 Solar Desalination

Saudi Arabia was very interested in the use of solar energy to desalinate water, and SOLERAS placed a high priority on this technology. A number of systems were designed and compared for both seawater and brackish water sources. A seawater system, using an innovative design involving direct freezing, was selected, constructed, and tested adjacent to a large Saudi desalination system on the Red Sea coast near the city of Yanbu.

This Chicago Bridge & Iron Company system was powered by an array of large dish solar collectors. The energy advantages of the freezing concept were proven and showed great promise for future development. The solar collectors suffered substantial performance degradation due to inadequate mirror edge protection and the hostile environment of a coastal industrial site. After two years of testing, the project was disassembled because of continued collector problems.

28.4.2.2 Solar Cooling

SOLERAS funded several projects on the use of solar thermal technologies to power building air cooling systems. The largest of these projects was the design and testing of four innovative systems in Phoenix, Arizona.

The United Technologies Research Center (UTRC) designed an advanced Rankine cycle chiller powered by an array of solar troughs to cool a small office building. UTRC designed and built the chiller, and Suntech supplied the collectors. After two years of successful operation, the system was modified to also provide solar heating in the winter months.

Carrier Corporation designed and tested an absorption air-conditioner powered by Acurex trough collectors to cool an office/warehouse building. Chiller operation did not fully meet design specifications due to coil corrosion and fouling, but the collectors performed extremely well. The test was terminated after two years when the building was vacated and not reoccupied; the system was reassembled at Arizona State University, where solar cooling research was continued.

Honeywell Corporation designed and built a Rankine cycle chiller and electric generator powered by evacuated-tube, fixed collectors provided by Energy Systems. The system was sited at a new Arizona Public Service (APS) facility and was monitored and maintained by APS technicians. The collectors suffered considerable breakage and did not meet design requirements; the chiller design worked very well and continued in operation for some time.

Carrier Corporation designed a second absorption chiller, powered by Entech trough Fresnel lens collectors, for another office/warehouse. The project had many design problems and was never able to meet adequate operating standards; it was ultimately disassembled and abandoned.

SOLERAS cooling projects made important research advances for solar energy systems and provided a base for continued research in both the United States and Saudi Arabia.

28.4.3 Spain

As a result of a provision of the 1976 Treaty of Friendship and Cooperation between the United States and Spain, both countries expressed a strong interest in pursuing a collaborative program in solar energy research and development. An extensive screening process was applied by the Spanish government to the over one hundred Spanish proposals submitted as part of a larger Spanish solar energy effort, and five projects were selected for funding under the treaty. In addition, an U.S. specialist team visited Spain at Spain's request to review a proposal for a Spanish Solar Energy Research Center. A Memorandum of Understanding (MOU) was completed and signed in 1978 between DOE and the Spanish Ministry of Industry to cover the five selected projects for a five-year period. Spain was to provide a majority of the funds, 50% to 90% depending on the project, with most of the substantial hardware procurement to be from American sources. SERI's International Programs Branch was given full management and oversight responsibility for the MOU projects in the spring of 1978. The five projects, each with an American consultant partner, included

• a 1-MW "power tower" prototype to be built in Almeria

• a 100-kW wind energy conversion system to be built on the Spanish south coast

- demonstration of the use of low-temperature solar energy systems in Palma de Mallorca
- development of concentrating two-sided photovoltaic cells in Madrid
- demonstration of methane from urban waste outside Madrid.

The demonstration of low-temperature solar energy systems was patterned after aspects of the ongoing solar demonstration program in the United States. The Spanish principal investigator was Feliciano Fuster Jaume of Gas y Electricidad, S.A., and the U.S. support was provided by a group from the Franklin Research Center headed by Harold Lorsch. The Spanish group collected meteorological, demographic, and economic data for Spain; analyzed technical and economic trade-offs for various types of solar energy systems in different Spanish climatic regions; evaluated various water and space heating applications; performed cost-performance trade-off studies and market potential and barrier studies; and prepared a solar energy development and demonstration plan for Spain. The Franklin Research Center team provided technical consultation on all the above tasks; supplied technical information services, including reports and machine-readable databases; and trained Spanish solar technical personnel in U.S. facilities.

28.4.4 United Kingdom

In May 1984 the United Kingdom Secretary of State for Energy Peter Walker announced a new initiative in international cooperation in energy research and development through increased cooperation with the United States. Agreement was reached with Secretary Donald Hodel of the U.S. Department of Energy, a Memorandum of Understanding was signed in October, 1984, and a series of exchange meetings was arranged in four selected technology areas. The aims of these meetings included (1) development of a network of contacts among research and development practitioners, (2) preparation of a summary of relevant work in the two countries, (3) dissemination of relevant information to users in each country, and (4) identification of technical areas where future collaboration could be of mutual benefit.

A Workshop on Passive Solar Design of Buildings was held in October 1984, and further meetings during the next two years explored the following areas for cooperation: (1) design tools, (2) residential buildings, (3) nonresidential buildings, and (4) market information (Jenior 1991).

Twenty-five possible projects were identified and defined; of these, two were implemented, in daylighting and in performance evaluation. A third project, in advanced glazing materials, was recently begun and is continuing, with a new cooperative agreement signed in 1990, to continue until 2000.

Technical collaboration in daylighting involves testing a performance evaluation method developed in the United States with data gathered from short-term tests conducted on jointly selected buildings in the United Kingdom. Data requirements were specified by the United States, and data were gathered by the United Kingdom. A seminar in daylighting performance evaluation was held in March 1988, including a critical review of ongoing evaluation work in both countries.

A glazing materials assessment has been defined and is being funded by both governments and by LOF, an U.S. glass manufacturer. The major focus is on verification of short-term monitoring of optical switching glazings.

28.5 Solar Commercialization

One method for reducing the costs of solar technologies in the domestic market is to expand the market for solar hardware worldwide. Increased production should lead to economies of scale and to increased domestic employment through expansion of existing plants and construction of new ones. Increasing the flow of solar goods and services into the international marketplace eventually became an important part of the Department of Energy's international solar program (Corcoran 1978).

As the results of DOE research and development efforts flowed into the marketplace in the late 1970s, it became clear that the export potential for solar technologies by U.S. solar manufacturing and service industries was dependent on many social, political, and market factors. Most domestic solar manufacturers were small and had no experience in dealing with the complexities of the export market nor in dealing with the many cultural and political aspects of doing business in a foreign land. Some countries, thanks primarily to a history of high fuel costs and supply uncertainties, already had a mature solar industry with substantial experience in the export business; Israel and Japan were two prime examples. Nevertheless, it was felt that the quality and diversity of U.S. products, combined with the many strengths of the domestic R&D program, would provide the basis for a strong future showing in international markets.

In April of 1978, DOE's Office of Solar Applications established the International Solar Commercialization Working Group (ISCWG) to develop and oversee an international commercialization component of the domestic commercialization program. Under its mandate, the ISCWG

• met with other federal agencies and solar industry representatives interested in exploiting export markets

• identified and reviewed other current and proposed international solar commercialization activities

• conducted fact-finding visits to both industrialized and developing countries with potential solar markets

• instituted studies of the structure of export opportunities and the market potential of various solar technologies

• developed plans for an extensive information resource to provide domestic industries with detailed information on social, political, and economic factors of target countries

• conducted export-related workshops for domestic industrial participants.

In addition to the ISCWG, the assistant secretary for international affairs in the Department of Energy assembled a Solar/Renewable Task Force in 1979 to promote the use of solar/renewable technologies abroad. Participating were representatives from the offices of the Assistant Secretaries for Conservation and Solar Applications, Energy Technology, and Resource Applications.

Responsibility for the major activities of the ISCWG program was given to the International Programs Branch (IPB) of the Solar Energy Research Institute. The IPB commissioned market penetration studies (e.g., Jacobius and Chingari 1981) to define country energy needs, assess desirable joint-venture possibilities, evaluate demonstration opportunities with sales follow-on, analyze such infrastructure questions as operation and maintenance site services and distributor networks, identify major-purchase decision makers, and assess barriers to market penetration.

An extensive information resource was created at SERI to serve the needs of industry participants for country-specific data on the many aspects of energy requirements, solar resources, currency restrictions, indigenous industrial capabilities, and so on. During the course of the programs' life (up to 1981), a series of highly detailed country profiles was created; four were published: Argentina, Australia, Italy, and Mexico (Hawkins 1981; Case 1980; Shea 1980; Hawkins 1980). In addition, a series of shorter country-specific solar market condition and potential reports was commissioned from Systems Consultants, Inc. of Washington, D.C., with twenty-six published. These seven-page reports contained sections on country infrastructure, market definition and penetration, practical barriers and incentives, and the domestic solar industry.

Finally, the IPB conducted a series of one-day workshops around the country for interested industry participants. Experts on all aspects of export sales were present, and participation by those attending was strongly encouraged. Although the participants were enthusiastic in their responses, the program was terminated before it was possible to properly assess whether attendance had led to export activity (Spongberg and Corcoran 1979). Also working in conjunction with DOE and SERI, the U.S. Department of Commerce conducted trade shows thoughout the world that, in whole or part, promoted U.S. solar technology.

U.S. and international solar efforts culminated in the 1981 United Nations Conference on New and Renewable Sources of Energy (DOS/ DOE 1981), held in Nairobi, Kenya. Exhibits sponsored by DOE, SERI, and private enterprise brought U.S. products to world attention, although the transition between Carter and Reagan administrations meant that this conference did not have a major impact on U.S. solar commercialization. The World Environmental Conference held in Rio de Janeiro, Brazil, in June 1992 also had only a small solar component.

In 1985, by authority of the Renewable Energy Industries Development Act of 1983 (PL 98-370), another U.S. agency international solar energy program was established, the Committee on Renewable Energy Commerce and Trade (CORECT), composed of representatives from more than a dozen federal agencies, including the Departments of Energy, State, Interior, Commerce, Treasury, and Defense. The committee's annual reports provide data on the growing export market for the solar thermal technologies of this volume (CORECT 1985, 1986, 1987).

28.6 In Summary

International activities made a significant contribution to early efforts by the U.S. government to make solar energy a viable option for the country following the oil embargo and the energy crisis of the early 1970s. As we entered the 1980s, the domestic research and development program, in cooperation with U.S. industry and with an able assist from many skilled foreign researchers, was developing products and services that were superior to any others in the world in most solar technologies.

With the change of administration in 1981, government efforts in solar energy research and development and government assistance to solar industry were substantially curtailed. In particular, international cooperative research activities were hard hit, and essentially all export assistance programs for industry in the Department of Energy ceased. Although the large cooperative program with strategically important Saudi Arabia and joint projects under the International Energy Agency were retained, no new cooperative initiatives were undertaken for several years, and ongoing projects were abruptly terminated or allowed to wind down.

During its active years, the international solar program of the Department of Energy was well planned to achieve useful results for domestic needs and was effectively managed and carried out. Its twin focus on utilizing the skills of foreign scientists and engineers and on providing assistance to a small but growing U.S. solar industry showed strong signs of bearing substantial fruit for the country's energy position in the worldwide economy. Many opportunities for cooperative international research of major benefit to American companies remain and could be effectively exploited today.

References

Case, G. L. 1980. *Solar Energy in Australia.* SERI/SP-763-717. Golden, CO: Solar Energy Research Institute, August.

Corcoran, W. L. 1978. *Commercialization Activities: International Solar Commercialization Working Group.* Washington DC: U.S. Department of Energy.

CORECT (Committee on Renewable Energy, Commerce, and Trade). 1985. *CORECT's First Year: September 1984—October 1985.* Washington, DC, 2 October.

CORECT. 1986. *Federal Export Assistance Programs Applicable to the U.S. Renewable Energy Industry.* Meridian Corporation. Washington, DC, June.

CORECT. 1987. *1987 Annual Report.* DOE/CE-0215. Washington, DC: U.S. Department of Energy.

DOS/DOE (U.S. Department of State and U.S. Department of Energy). 1981. "New and Renewable Energy in the United States of America." *The United States National Paper for the 1981 United Nations Conference on New and Renewable Sources of Energy.* Washington, DC, June.

Goldberg, M. D. 1979. *Report on Visit of U.S. Solar Energy Specialist Team to the Soviet Union, 3–16 September 1977*. SERI/SP-43-310. Golden, CO: Solar Energy Research Institute, April.

Goldberg, M. D. 1980. *Report on Visit of U.S. Solar Energy Specialist Team to Israel, 9–14 September 1979*. SERI/SP-411-582. Golden, CO: Solar Energy Research Institute, March.

Hawkins, D. H. 1980. *Energy in Mexico*. SERI/SP-763-595. Golden, CO: Solar Energy Research Institute, April.

Hawkins, D. H. 1981. *Solar Energy in Argentina*. SERI/SP-763-1002. Golden, CO: Solar Energy Research Institute, January.

IEA (International Energy Agency). 1988. *A Decade of Advances: 1987 Annual Report of the IEA Solar Heating and Cooling Programme*. IEA/SHC/AR-87. Beltsville, MD: International Planning Associates, Inc., February.

Jacobius, T. M., and G. Chingari. 1981. *Renewable Energy Systems Market and Requirements Analysis in Italy, Spain, and Greece*. SERI Subcontract AF-1-1023-1. Chicago: IIT Research Institute, April.

Jenior, M.-M. 1991. Private communication, 9 November.

Morse, F. H. 1977. *Notes Prepared for the 31st Meeting of the Interagency Panel on Terrestrial Applications of Solar Energy*. Washington, DC, January.

Morse, F. H. and I. B. Rose. 1975. *CCMS Solar Energy Pilot Study: Report of the Annual Meeting*. UMD-4908-5. College Park: University of Maryland, Department of Mechanical Engineering, August.

NATO/CCMS (North Atlantic Treaty Organization, Committee on the Challenges of Modern Society). 1978. *Solar Energy Pilot Study*. CCMS 83/UMD-4908-13. University of Maryland, Department of Engineering, CCMS Projects Office, October.

Shea, C. A. 1980. *Solar Energy in Italy*. SERI/SP-763-718. Golden, CO: Solar Energy Research Institute, December.

Spongberg, R. C. and W. L. Corcoran. 1979. *DOE/Solar Export Opportunities Workshop: Atlanta, January 1979*. Golden CO: Solar Energy Research Institute. SERI/TP-49-186 (April 79).

Williamson, J. 1991. Private communication, 5 November.

29 State and Local Programs

Peggy Wrenn and Michael DeAngelis

This chapter describes a range of state and local solar energy programs funded by or affected by the federal government. The basic framework and tenor of state energy programs were established with the passage of the Energy Policy and Conservation Act (EPCA) in 1975. The act provided federal funds for states to establish State Energy Conservation Plans. Funding provided by the federal government under this act helped institutionalize state energy offices, which in turn both funded and stimulated local energy offices through outreach programs aimed at local governments and agricultural extension services.

As of February 1983, there were 29 state energy offices (from a high of 50 in 1980) still actively working on solar and energy efficiency programs according to a Department of Energy survey on their activities and publications (San Martin 1986). A 1985 survey of state energy offices by an independent researcher (Sawyer 1985a) identified seven federally funded state energy activities: the State Energy Conservation Programs, the Energy Extension Service, the Institutional Conservation Program, the Weatherization Assistance Program, the Residential Conservation Service, the Solar Energy and Energy Conservation Bank, and the Commercial and Apartment Conservation Service. Six of these, and other earlier programs, are described in this chapter as they relate to solar thermal applications and commercialization; the Commercial and Apartment Conservation Service was never fully implemented and did not affect solar energy applications enough to warrant treatment here.

Section 29.1 focuses on federal support for state and local solar energy programs from the passage of EPCA in 1975 through the mid-1980s. EPCA was quickly augmented by the passage of the Energy Conservation and Production Act of 1976 (ECPA). Together, these two acts provided a total of approximately $330 million between 1976 and 1983 (see table 29.1), much of which was distributed to state energy offices, and some of which was distributed through the states via the Energy Extension Service to local energy offices.

State and local governments augmented federal programs with locally appropriate information programs and many state tax incentives to encourage solar energy applications in homes and businesses, as well as in local government buildings. These diverse experiments in solar energy gave added proof to the hypothesis that energy efficiency improvements

Table 29.1
Funding (in thousands) for the state energy conservation programs under EPCA and ECPA
(DOE 1982)

Fiscal year	EPCA base	ECPA supplemental	Total
1976	5,000	—	5,000
1977	23,000	12,000	35,000
1978	47,800	23,740	71,540
1979	47,800	10,000	57,800
1980	37,800	10,000	47,800
1981	30,400	10,000	40,400
1982	24,000	—	24,000
1983	24,000	—	24,000
1984	24,000	—	24,000
Total	263,800	65,740	329,540

are the most cost-effective precursor to renewable energy technologies. These state and local efforts also contributed significantly to the advancement of solar thermal technologies for buildings by stimulating and reporting a plethora of solar experiments at all scales. The state and local efforts are covered in sections 29.2, 29.3, and 29.4.

The benefits of federal funding to state and local solar programs are difficult to quantify because state and local efforts are so dispersed, decentralized, and geared to varying climatic and political conditions. As Sawyer 1985a points out, "supporters point to econometric models that suggest significant national benefits, while opponents argue that methodological weaknesses, errors, and attribution uncertainties make any savings figures suspect." Green et al. 1982 and Hirst, Fulkerson, and Carlsmith 1982 estimate the effectiveness of government actions on total energy consumption. None of these assessments is conclusive in establishing verifiable, quantitative energy savings or solar energy production. Many state programs lacked clear goals and objectives. Often they also lacked criteria and resources for evaluation of their success.

The authors of this chapter, who worked in the states of California and Colorado, believe that state and local efforts, stimulated by federal funding in the late 1970s and early 1980s, were significant in creating at least a temporary shift in U.S. energy policy toward solar and renewable energy sources and, perhaps more importantly, toward maximizing energy efficiency. It appears to the authors that these effects resulted from both information programs and economic factors. Information programs are hardest to evaluate. Probably the most salient factors were escalating fuel prices and economic incentives such as tax credits.

29.1   Federal Programs for State and Local Governments

Federally funded programs to encourage the use of solar energy are described below in three categories: mandates, direct incentives, and other programs. This section also gives summaries of federal budgets for these programs. Section 29.5.2 describes how various state and local institutions were affected by federal solar programs.

29.1.1   Mandated Federal Programs

Three mandated programs initiated by the federal government include State Energy Conservation Programs (required through EPCA and ECPA), the Public Utilities Regulatory Policy Act (PURPA), and the Residential Conservation Service (RCS). Although these initiatives also sometimes included funding, they are treated as mandates because they required the states to take certain actions.

29.1.1.1   State Energy Conservation Programs
The first state energy conservation programs of substantial importance were based on the Energy Policy and Conservation Act (EPCA) of 1975 (PL 94-163, 42 U.S.C. 6325). This was the first significant initiative by Congress to promote energy conservation and included an EPCA authorization for each of three years (FY 1976–FY 1978). The EPCA legislation mandated that states develop a plan to achieve a 5% reduction of energy consumption in the baseline year of 1980 (DOE 1983). The Federal Energy Administration (FEA) was authorized to help the states develop and implement energy conservation plans. To be eligible for funding under EPCA, a state plan was required to include the first five of the six programs listed below.

1. mandatory lighting efficiency standards for public buildings;

2. programs to promote the availability and use of car pools, van pools, and public transportation;

3. mandatory standards and policies relating to energy efficiency to govern the procurement practices of a state and its political subdivisions;

4. mandatory thermal efficiency standards and insulation requirements for new and renovated buildings;

5. traffic law allowing right turn on red;

6. state-initiated programs (optional).

Although state plans were not limited to conservation measures only, the federal programs had no direct mandates for solar thermal technologies. However, many states included significant information on residential solar technologies in their consumer information and energy audit programs. Solar heating of buildings was presented as the next step after energy conservation measures were installed, and solar domestic hot water systems were widely encouraged, especially in Florida and California.

Congress established a second major state energy conservation program by passing the Energy Conservation and Production Act (ECPA) of 1976. This program added funds and amended the state energy conservation program under EPCA. To be eligible for these funds, a state was required to add to its state energy conservation plan the following activities:

1. public education about energy saving measures;

2. coordination among local, state and federal energy conservation programs;

3. encouragement and implementation of residential energy audits.

State energy programs pursuant to ECPA included public information programs about solar energy, and in many cases, legislative measures at the state level to implement state solar legislation and state energy tax credits. For example, the Colorado State Energy Office, like many others funded through EPCA and ECPA, worked on several successful state laws later copied by other states. These included state solar tax credits (which allowed passive solar as well as active solar technologies), state authorizing legislation for local governments to pass solar access regulations, legislation defining solar easements (copied by 22 states), and laws encouraging gasohol and other renewable fuels.

These state initiatives, as well as those described in section 29.2.1, are examples of the hard-to-quantify spin-off effects of EPCA and ECPA funding. The regional solar energy centers (RSECs), covered in chapter 20 of this volume, supported state and local government efforts. It is difficult to separate solar efforts from general energy efficiency/renewable programs of state and local governments.

By the late 1970s all states had approved State Energy Conservation Plans and were thereby eligible for federal funds through EPCA and ECPA. The federal funding for these programs peaked during FY 1978 and FY 1979, and fell to less than half these levels during the 1980s. Funding of these programs through FY 1984 is listed in table 29.1.

29.1.1.2 Public Utilities Regulatory Policy Act

As one of the five National Energy Acts of 1978, PURPA was designed to mitigate the two major constraints affecting expanded use of electricity produced by small, dispersed, renewable or cogenerating sources, including solar thermal electricity–generating installations. The first constraint addressed by PURPA was the reluctance of utilities to purchase the power from small power producers at reasonable rates, commensurate with their potentially avoidable costs for building new fossil fuel–fired power plants. The second constraint was the possibility that these small power producers might be considered electric utilities and thereby be subject to state and federal public utility regulations. The expense of being so regulated was essentially eliminated for small power producers by the language of PURPA.

The Federal Energy Regulatory Commission (FERC) issued regulations implementing PURPA and requiring states to establish "avoided cost" rates that utilities would be mandated to pay to small power producers (Berger, Royce, and Farley 1980; and FERC 1984). Each state's utility regulatory agency proceeded to conduct rate hearings to establish "avoided cost" rates pursuant to PURPA, mostly during 1980–1983. They set rates at which each individual utility had to buy power from small power producers, and they established interconnection rules and in some cases "wheeling" charges for transmitting power across utility distribution systems. In the case of New Hampshire, the legislature set an avoided cost rate by law. Local governments became involved in state hearings on PURPA both as municipal utilities and as small power producers. A 1984 review of state PURPA-rate-setting procedures showed that at least forty states had established rules or guidelines (*Energy User's News* 1984).

The documented effects of PURPA on solar thermal technologies appear to be relatively limited, except in California, where Luz International began construction of a series of large parabolic trough solar electric generation stations (SEGS) in 1984, and LaJet Energy constructed a single 4.5 MWe parabolic dish plant in the same year. The Luz plants, described in chapter 8, were organized to sell power to Southern California Edison under PURPA contracts. The Luz SEGS construction program under PURPA continued through 1991 to the completion of the SEGS XII plant, for a total of 594 MWe.

Luz's experience with PURPA is similar to that of the wind turbine industry. The first SEGS plants were small and expensive at a time when the PURPA contracts and federal and state tax credits were most generous. As the contracts available under PURPA became less generous and tax credits eliminated, the scale of the SEGS was increased, and manufacturing, construction, and O&M costs were reduced.

Although many smaller photovoltaic solar installations benefited from PURPA, the largest photovoltaic plant in the world, ARCO's Carissa Plains 6.5 MWe facility, sold its power to PG&E under PURPA.

In addition to direct effects of PURPA, there were spin-off effects because the PURPA proceedings raised the awareness of state utility regulatory agencies and state legislators and governors about the potential of small-scale renewable energy power production. A few states considered, and, in some cases implemented, rates for customers who could use off-peak energy to charge their solar thermal storage. For example, Public Service Company of Colorado filed such a rate with the Colorado PUC allowing off-peak power at less-than-residential-retail rates for solar customers who could prove they could charge their solar storage with off-peak electricity.

29.1.1.3 Residential Conservation Service

The National Energy Conservation Policy Act (NECPA) of 1978 and the subsequent amendments of the Energy Security Act (1980) mandated that large electric and gas utilities (annual sales of 10 billion cubic feet of natural gas or 750 million kWh of electricity) provide energy conservation services to their customers. The main purpose of this program, popularly known as the "Residential Conservation Service" (RCS), was to require on-site inspections of customers' homes by qualified energy auditors. The auditors had to identify specific energy conservation and solar opportunities, as well as to present costs and expected energy savings and payback periods.

Most state RCS programs included auditor assessment of solar energy opportunities and paybacks, with widespread emphasis on solar hot water heating. Ancillary services included

• identifying state-approved installers, suppliers, and lenders to help consumers undertake recommended measures, including solar installations;

• arranging for installation and/or financing of such measures;

• inspecting completed measures in certain instances; and

• providing conciliation services for consumer complaints.

The federal RCS guidelines allowed a maximum direct customer charge of $15 for the residential audit. The remainder of the program costs were generally rate-based as part of the cost of providing utility service. Although federal guidelines did not require attention to solar thermal technologies, states that failed to implement state RCS plans by a deadline date were required to force utilities to implement a federal "default" plan, which did address solar domestic hot water systems. However, the main focus of the program was on low-cost, high-energy-saving measures such as insulation, caulking, weather stripping, low-flow showerheads, and water heater blankets.

RCS funding was provided by utilities, which estimated the average audit costs at $150 and up, compared to the $15 customer contribution allowed by the regulations. No federal funds, other than the administration costs of DOE, were involved.

State energy offices played a major role in designing the specifics of the program in most cases because the states were required to implement state RCS plans or else default to a mandatory DOE RCS plan. The default plan included a requirement to assess solar thermal opportunities in the RCS audit. As in many other audit programs, solar heat was generally found to be less cost-effective than energy conservation. The utilities regulated by RCS were generally reluctant to implement its plans because of excess energy capacity they wished to sell (Sawyer 1985a). Although it was initially forecast that 90% of the existing residences would receive RCS audits within five years, during the first two years of the program only 4–6% of eligible customers requested audits (Sawyer 1985a).

29.1.1.4 Institutional Conservation Service
The ICS was initially contemplated as a mandatory program but was not mandated. Instead, it was offered as a model to states, following the RCS

federal initiative; its purpose was to encourage conservation for schools, hospitals, local government buildings, and nursing homes. This program, also known as the "Schools and Hospitals" grant program, provided up to 50% of the funding for school or hospital energy-saving measures recommended in a technical audit that conformed to the guidelines of the program. The technical audit could also be funded through the program, for schools, hospitals, and local government buildings. Although audit guidelines required an analysis of solar thermal energy opportunities, they placed so much emphasis on short payback that solar measures were rarely stimulated by this program.

The name of the Institutional Conservation Service was later changed to Institutional Conservation Program (ICP). ICP funding began in 1979 with a national budget of $100 million, rising to $199 million and 158 million in 1980 and 1981, respectively, before dropping to $46 million in 1982, $40 million in 1983 (supplemented by $50 million from the Jobs Act), and $8 million in both 1984 and 1985 (Sawyer 1985a).

The effects of the ICP program on solar applications, although hard to measure, were in all likelihood fairly small. The technical audits required an evaluation of solar opportunities, but institutional buildings generally offered much more cost-effective energy conservation opportunities. Because the equipment grants were based on paybacks demonstrated through technical audits, solar applications were rarely, if ever, funded by these grants.

29.1.1.5 Weatherization Assistance Program
Established by DOE to reduce energy consumption in low-income dwelling units, the Weatherization Assistance Program was used extensively by public housing authorities at both state and local levels to weatherize public housing. It was generally administered through state housing departments and local community action programs, rather than state energy offices.

By 1985, approximately 1.5 million of the 13.1 million eligible dwelling units had been weatherized through the program, at an average cost of $1,000–$1,500 (Sawyer 1985a). Since 1982, states have been able to direct up to 15% of their Low-Income Energy Assistance Program (LIEAP) funds to supplement weatherization assistance. This allowed for redirection of $171 million in 1982 and $194 million in 1983 for weatherization efforts. The direct appropriations for the Weatherization Assistance Program were $199 million in 1979, $175 million for both of 1980 and 1981, $144 million in 1982, $245 million in 1983 (including $100 million from the Jobs Act), and $190 million for both 1984 and 1985 (Sawyer 1985a).

29.1.2 Federal Incentives for States and Localities

Federal incentives are difficult to separate from federal mandates because congressional budgets for each year represented a mix of mandates, "guidelines," and incentives. For purposes of this chapter, federal incentives are discussed under two main categories, both funded through the Department of Housing and Urban Development (HUD) and/or the Department of Energy (DOE). These are the Comprehensive Community Energy Management Plan (CCEMP) program, funded in sixteen cities, and the Appropriate Technology Small Grants Program, for which all states were eligible. Also briefly discussed are other federal incentive programs that are treated in more detail in chapters 25–27.

29.1.2.1 Comprehensive Community Energy Management Plan

The Department of Energy (DOE), in cooperation with the Department of Housing and Urban Development (HUD), established the CCEMP program, which funded sixteen localities (cities and counties) to develop community energy management programs (Tschanz 1987). The program was managed for DOE and HUD by Argonne National Laboratory. The cities reported their results, and their experiences were offered as models in various published documents, including *Comprehensive Community Energy Planning: A Workbook* (Hittman Associates 1978).

The work of the cities and counties was to evaluate community energy usage, establish an energy management structure, develop an energy conservation action plan, and, if appropriate, increase the use of alternative energy such as solar heating. The communities all followed a federal planning method in order to receive federal funding. Under the method, communities were required to

1. establish an organization structure and work plan;

2. estimate or "audit" the community's energy resources and demands;

3. identify dominant energy issues and establish objectives for dealing with them;

4. identify, evaluate, and choose alternative strategies and actions to meet local energy objectives; and

5. prepare and adopt the community energy management plan.

According to the CCEMP Workbook (Hittman Associates 1978), the program's objectives were

• to develop, test, and demonstrate organizational arrangements for accomplishing a comprehensive community energy management program;

• to test, evaluate, and improve planning/implementation approaches and methodologies for conducting a CCEMP;

• to expand the available community energy database;

• to facilitate comprehensive community energy management through dissemination of methodologies; and

• to develop information for future federal policy related to comprehensive community energy management.

Argonne selected 16 communities from 116 respondents to the RFP, in the following categories: small cities, intermediate cities, large cities (Los Angeles and Philadelphia), counties, and metropolitan areas including both cities and counties (Moore et al. 1979).

Through detailed audits of energy uses in their jurisdictions, these communities assessed conservation and renewable energy opportunities and implemented policies to encourage them. Although most of the programs and policies produced through CCEMP were oriented toward energy efficiency, some notable solar thermal policies emerged from CCEMP cities, the most significant of which were probably the local solar access ordinances designed to protect existing sunlight striking residential and in some cases commercial buildings. The purpose of these ordinances was to protect existing and potential solar heating systems that could contribute to building heating loads.

One example well known to author Peggy Wrenn is Boulder, Colorado, which was funded at $250,000 in 1979 to establish a CCEMP according to Argonne management guidelines. A community energy audit identified the percentages of community energy consumed in residential, commercial, transportation, and government sectors. A citizen group appointed by the city council established an Energy Action Plan and a Zero Energy Growth goal, which were incorporated in the local comprehensive plan. The goal stated that Boulder would aim toward consuming the same amount of energy in 1990 as it did in 1980, regardless of community population growth. Zero Energy Growth was coupled with a growth management ordinance, and buildings with low energy consumption or solar thermal energy sources were given extra points in competing for building permits under the growth management policy. The city also adopted a solar access ordinance that required a shading analysis of any development within the city limits to ensure it would not shade the south wall or roof of any adjacent property. The solar access ordinance provided protection of access to sunshine depending on the zoning and defined by a hypothetical solar fence on the property line of the adjoining property. Applicants for building permits had to demonstrate that they would not shade their neighbors between the hours of 10 a.m. and 2 p.m. on December 21, the winter solstice. These significant solar initiatives at the local level are examples of measures initiated largely because of CCEMP grant funding through HUD.

Nationally, the CCEMP program was funded at $4.5 million for Phase I (FY 1979), which ran approximately three years. Phase II was funded in FY 1981 at $840,000 (Tschanz 1987).

29.1.2.2 Appropriate Technology Small Grants Program
Congress authorized the Energy Research and Development Administration (ERDA; now the DOE) in 1977 to undertake a grants program in small-scale, energy-related technologies using renewable energy resources, including solar heat. These technologies are considered "appropriate" technologies because of their high suitability to local needs and skills. During the years 1978–1981, the Department of Energy awarded more than 2,200 small grants worth more than $25 million to individuals, organizations, and small businesses across the nation for the purpose of researching, developing, and demonstrating appropriate technologies (NCAT 1983). The program provided for small grants (often under $5,000, and never more than $50,000) to help develop innovative projects for the following purposes:

1. idea development, for concept-demonstrating potential;

2. concept testing, for projects that have gone beyond the idea development phase and are ready for testing; and

3. demonstration, for projects that having been tested, now must be proven through actual use.

Appropriate technologies were defined as small-scale, nonpolluting, renewable, environmentally safe, decentralized, and appropriate to local cultural, economic, and social conditions.

The Appropriate Technology Small Grants Program (hereafter "AT Small Grants") continued through FY 1982, though no small grants were funded after 1981. During 1982 the Department of Energy funded the National Center for Appropriate Technology (NCAT) to develop a computer database called the "National Appropriate Technology Management Information System," containing abstracts of 1,473 projects (two-thirds of the total) funded by AT Small Grants (Sesso 1987). Projects abstracted had been completed and had submitted final reports at the time of publication (NCAT 1983).

For a total of $25,441,900 over the years 1978–1981, the program funded 2,221 projects in conservation, solar, biomass, geothermal, wind hydro, integrated systems, and miscellaneous categories. Overall, solar projects constituted 32.8% of the projects funded (NCAT 1984). Approximately one-third of the projects were also funded by nonfederal sources, with a total contribution from other funding sources of over $11 million. These projects numbered 94 in 1978, 542 in 1979, 776 in 1980, and 626 in 1981 (NCAT 1983).

Funding was allocated by DOE Region and by population; average grants were in the $12,000 range, although grants under $500 per year and up to $50,000 per year were also awarded. The nine regional DOE offices managed AT Small Grants, often using state energy offices to organize and select grant applications. Projects eligible for funding included those proposed by state and local agencies, small businesses, nonprofit organizations, individuals, and Indian tribes. Funded projects ranged from community greenhouses for Indian tribes, to feasibility studies for hydroelectric power on municipal water systems, to neighborhood demonstrations of low-cost passive solar techniques in private homes. The projects are summarized both in the National Appropriate Technology Management Information System and more briefly in *Appropriate Technology at Work: Outstanding Projects* (NCAT 1983).

The states were involved in publicizing, recommending selection, and documenting AT Small Grants. The appropriate technology projects were often important demonstrations publicized by state and local energy offices. Many states set up public review committees to assist in the selection of programs for grants. This involved local people in both selection and follow-up.

29.1.2.3   Energy Extension Service

The EES program, modeled after the agricultural extension service, offered great flexibility to states. States generally disbursed the funds through existing outreach organizations like agricultural extension services, non-profits, or local governments. States and the local outreach offices they funded through EES devised hundreds of information efforts, workshops, and energy audit programs.

Small-scale energy users account for about 40% of the U.S. energy use, so EES outreach efforts aimed at builders, farmers, hotels, landlords, tenants, homeowners, small businesses, and a great variety of other target audiences selected because of their share of state and local energy usage (Sawyer 1985a). Energy auditor trainings and on-site, targeted workshops were offered to develop expertise in auditing various types of buildings. Energy information hot lines were also frequently established through EES centers. The flexibility of the program and the emphasis on person-alized conservation advice targeted to specific needs made the program popular (Sawyer 1985a). The program was well received and contributed to widespread education on energy efficiency and solar heating. Its effects are impossible to quantify, but a few EES agencies still exist in the 1990s, years after federal funding was withdrawn. The authors believe that tar-geted information efforts geared to specific groups were more effective than programs geared to the general public.

The federal funding for EES, administered by DOE, was $58 million in 1979, dropping to $48 million for 1980 and 1981. It then dropped by half, to $24 million annually from 1982 through 1985 (Sawyer 1985a).

29.1.2.4   Solar and Energy Conservation Bank

The Solar and Energy Conservation Bank (hereafter "Solar Bank") was authorized by Congress in 1979, although it was not implemented until 1982, after a lawsuit brought against DOE by the Solar Lobby forced expenditure of the funds. The Solar Bank regulations required consid-erably more administrative accounting for loans than for grants, both of which were authorized by the program. The regulations only allowed loans for solar applications, while also allowing grants for energy conser-vation measures. The regulations limited administrative expenditures to a very small percentage of total costs. Many states chose to offer energy conservation grants (up to 50% of eligible costs determined by an audit) as well as loans. Very few Solar Bank funds were used for solar applications;

most were used instead for matching grants or loans to low-income eligible grant recipients for energy conservation. A 1986 report indicated that "nearly two-thirds of currently obligated funds [were] for loan subsidies, where the average subsidy is equal to 38% of the measure cost" (O'Hara 1986).

Funding for the Solar Bank started with $22 million in 1982 and 1983, $25 million in 1984, and ended with $15 million in 1985 (Sawyer 1985a). State energy offices applied for the funds and administered a central wire transfer accountability for each state's funds, but the funds were often directly disbursed through local agencies in various parts of each state. Only state governments were allowed to apply for the funds. Solar energy loans were included in 23 states' 1984 grant cycles, but 90% of the grants were applied to conservation activities (Sawyer 1985a). As of 1985, 38 states had received $18 million, with an additional $42 million obligated (O'Hara 1986; see chapter 26 for more on the Solar Bank).

29.1.3   Other Federal Programs: State and Local Efforts

Many states participated actively in a variety of other federal solar programs described in more detail elsewhere in this volume. Several of these programs also had specific local government components, usually delivered through the state solar energy office. These programs include state income tax credits for solar applications, a few local property tax or sales tax incentives, participation in the regional solar energy center (RSEC) programs by both state and local governments, participation in the National Solar Demonstration Programs funded through HUD, and participation in quality assurance programs. States were involved in the National Solar Demonstration Programs as recipients; many state buildings, especially educational buildings, received demonstration grants. States also publicized the demonstrations through state energy office literature and public speaking.

"Oil overcharge" funds were also distributed through the states. The largest settlement in the time period covered in this chapter, $200 million, was distributed in 1983 to governors for use in five federal-state conservation programs; the courts have also directed that the Exxon settlement (now over $2 billion) be allocated to the five programs. Eventual settlements in the Petroleum Violation Escrow Account may exceed $5 billion nationally (Sawyer 1985a).

"Oil overcharge" funds have been administered by states under oversight by DOE regional support offices, which have documented the uses of these funds and required evaluation of the funded programs. Increasingly, the funds have been expended to encourage more market research on the front end and more evaluation after program completion than was present in the earlier state and local programs discussed in this chapter, perhaps in response to lessons learned from the earlier lack of such efforts.

29.2   State and Local Incentives and Regulatory Programs

This section summarizes state and local incentives and regulatory programs, and discusses third party financing used by states and localities to augment local incentives. As previously, this section focuses on those state and local programs which directly affected solar energy. A general summary of state and local programs affecting solar applications is contained in San Martin 1986, *Overview of State Renewable Energy Programs*.

29.2.1   State Financial Incentives and Tax Credits

The number of states with financial incentives to encourage the installation of solar devices increased rapidly in the mid to late 1970s. These financial incentives were either income tax, property tax, sales tax, or other financial incentives such as grants or loan programs. Four tax incentive programs were common (Sawyer 1985b):

1. tax credits that allow a percentage of the system costs to be credited against state income taxes (29 states);

2. tax deductions that allow costs to be deducted from income in computing taxes (1 state);

3. exemption of the system's value from property tax assessments (27 states, plus 6 at local option); and

4. exemption or refund of sales tax (10 states).

The specific design of tax incentives varied tremendously from state to state. In states that provided large tax credits the value of the tax incentives for solar devices was substantial. For example, the total incentive provided to a homeowner for purchase of a $3,500 solar domestic hot water heating system exceeded $1,000 in twelve states when combined with the federal solar tax credits (Sawyer 1985b). The magnitude of state incentives offered through tax credits has been studied in Arizona, California, Colorado, and Michigan, where 1982 tax credits cost $13.1 million, $84.1 million, $11.1 million, and $1.2 million, respectively (Sawyer 1985b). These figures include energy conservation and solar tax credits; the figures for solar energy alone are not available.

Much of the state tax credit activity came on the heels of the federal solar tax credits and perhaps was stimulated by the federal initiative. Some state credits added to the federal tax credits; some broadened the federal guidelines for state incentives (see chapter 25 for more details on state solar tax credits).

The DOE assisted in development and growth of state solar financial incentives through information activities and research programs. The National Solar Heating and Cooling Information Center (NSHCIC) compiled and distributed information on the number, type, and magnitude of incentives offered by various states. Most of the NSHCIC efforts were funded by DOE. The Solar Energy Research Institute (SERI) evaluated state incentive efforts and recommended improvements (Ashworth, Green, et al. 1979). SERI completed a detailed set of logical criteria and lists of advantages and disadvantages to help guide states to develop and implement incentives that best suited their needs (Ashworth, DeAngelis, et al. 1979).

SERI also produced, with DOE funding, the quarterly publication *Solar Law Reporter*, which provided a central source of information on state and local incentive and regulatory initiatives. *Solar Law Reporter* was published between June 1979 and February 1982 and contained some of the best published synopses of state and local regulations, ordinances, laws, and incentive programs.

For example, "State Approaches to Solar Legislation: A Survey" (Johnson 1979) and "Common Problems in Drafting State Solar Legislation" (Warren, 1979) were published in the first issue of *Solar Law Reporter*. The journal also reported extensively on results of various federal solar programs, local legal issues such as solar access law, and utility programs mandated by the federal government.

Although evaluative data on state and local incentive programs are limited and often anecdotal, some data are provided in a study by F. M. O'Hara and others (O'Hara 1986). The study concerns itself primarily with energy conservation, as opposed to solar heat, but many of the findings are applicable to both. The following quote from O'Hara 1986, *Energy Efficiency in Buildings: Progress and Promise*, provides information salient to state and local energy programs in general:

California spent $38 million in 1983 on their 40% conservation tax credit ... [and] estimated that 43% of this amount was returned to the state through other tax revenues, both directly from the conservation activity and indirectly through the increase in disposable income resulting from lower energy bills. ...

Some states instituted their own loan programs. For example, California established a $20 million loan pool to help schools, hospitals, local governments, public-care facilities, and special districts undertake energy audits and conservation measures. The Minneapolis Community Development Agency issued revenue bonds to subsidize financing [10% interest, 10-year loans in 1982] for efficiency improvements in multifamily housing. The state of Minnesota also [encouraged] private-sector loans to rental housing through a loan insurance program for lenders. ... New York's Energy Investment Loan Program is targeted at small and medium-size manufacturing firms, multifamily housing owners, and not-for-profit organizations. Other states with loan programs included Alaska, Oregon, Maine, and Maryland. ...

States also [used] seed money to help local government and private organizations launch self-sustaining energy conservation activities. For example, the North Carolina Alternative Energy Corporation's Local Energy Officer program [guaranteed] the first two years of a local government's energy manager's salary in the event that not enough energy savings were produced to cover staff costs. ... Missouri [provided] grants and technical assistance to support full-time energy management offices in local governments for the first year. Program costs for subsequent years [were] then funded out of previous-year savings. ... Few (if any) evaluations of the incremental energy savings and cost effectiveness of state loan and incentive programs have been performed. ...

29.2.2   State Regulatory Efforts

The major state regulatory efforts affected by federal funding were the State Energy Conservation Programs funded through EPCA and ECPA, as they affected state building codes. State energy offices significantly affected local government buildings codes and provided model solar codes for local adoption, sometimes required by state legislation. The model codes, popularized, taught, and in some cases mandated by state energy offices, are summarized in section 29.2.3.

State Energy Conservation Programs through EPCA and ECPA funding have had significant influence on the use of solar thermal systems. Although this review does not include a survey of the status of building energy conservation programs and regulations, some examples of regulations incorporating both solar thermal and energy conservation are given below.

In California, residential energy conservation regulations were modified from a prescriptive approach (mandated levels of ceiling and wall insulation, mandated U (thermal conductance) values, prescribing double-paned windows based on climate zone) to an "energy budget" or performance approach allowing flexible design trade-offs. The energy budget approach included construction and design options for homebuilders to achieve the prescribed budget based on climate zone. Passive solar design and solar water heaters were often selected by the homebuilder to achieve the prescribed energy budget.

Other states undertook significant efforts (often federally funded) in regulating solar applications through model building codes for local officials. Some states mandated local adoption of a state energy conservation code as a minimum allowable energy code. For example, Colorado mandated that local governments adopt the *Colorado Model Energy Efficiency Construction and Renovation Standards* (Colorado 1980). This model code was developed by the state energy office using EPCA and ECPA funds, and training programs were offered to local governments to help them adopt the code. Based on consensus standards established by *ASHRAE 90-75*, the code set forth performance options, including passive solar heating and lighting.

29.2.3 Local Incentives and Regulatory Efforts: Building Code and Land Use

Local financial incentives for solar applications, although sometimes offered through sales tax and property tax rebates, were not widespread. Most direct financial incentives offered by local governments were exclusively funded by federal AT grants, Solar Bank funds, CCEMP funds, and EES funds, as described above. Local incentives were more often created through building codes and local land use regulations that provided for solar applications.

Local building code and land use regulations designed to promote solar applications fell generally into two categories: (1) adoption of model energy/solar codes; and (2) adoption of solar access regulations through land use and zoning ordinances. Many local governments also offered solar incentives through education and training programs funded through EPCA, ECPA, and EES (these are discussed in chapter 19).

Model energy/solar codes were published and adopted by reference in many local building codes during the late 1970s and early 1980s. One notable result of federal funding was the *Performance Criteria for Solar Heating and Cooling Systems in Residential Buildings* (NBS 1982). This document was published pursuant to the Solar Heating and Cooling Demonstration Act of 1974 (PL 93-409), which authorized a five-year program for research, demonstration, and market development of solar heating and cooling systems in residential and commercial buildings. Many state and local governments used the NBS criteria to qualify solar systems for their tax credits and other incentives (see chapter 14 for further details).

A second notable influence on local solar codes was the Uniform Building Code (UBC) update for 1975. This national code, adopted by many local jurisdictions in regular updates, included a section on solar thermal for the first time in 1975. The code addressed fire retardancy for roof-mounted solar collectors, atriums, and solar plumbing and valving requirements. Another widely used document was the *Recommended Requirements to Code Officials for Solar Heating, Cooling and Hot Water Systems* (DOE 1980).

Local land use regulations, zoning regulations, and related regulatory mechanisms were also used to promote solar thermal applications. Solar access legislation was the most widespread of these local initiatives. Federal funding contributed significantly to the dissemination of information about solar access law through two notable publications: *Solar Access Law: Protecting Access to Sunlight for Solar Energy Systems*, published by the Environmental Law Institute and funded by HUD and DOE (Hayes 1979); and *Solar Access Ordinances: A Guide for Local Governments*, published by the SolarCal Local Government Commission for the California Energy Commission (Mott-Smith 1981).

Federal funding also contributed to solar access legislation through the publication and wide dissemination of a book called *Protecting Solar Access for Residential Development: A Guidebook for Planning Officials*, published by the American Planning Association with funding from HUD and DOE (Jaffe and Erley 1980). This document was used by many local governments to draft legislation on zoning for solar access, on subdivision regulations and landscaping regulations for solar access, and on site-planning considerations and private agreements such as covenants and easements to protect solar access. In many cases, local governments required private solar access agreements as a condition of zoning, annexation, or building permit.

29.2.4 Third-Party Financing

Third-party financing deals were established during 1980–1982 in response to favorable tax incentives for private investors to invest in renewable energy ventures. Wind and hydroelectric opportunities benefited more from third-party financing than solar applications, in general. However, many state solar and conservation loan programs were established based on sales of tax-exempt bonds, usually mortgage revenue bonds (MRBs). For example, Boulder, Colorado, sponsored an energy loan program from 1983 to 1985 that offered homeowners loans at three-quarters of prime interest, through a tax-exempt line of credit with the local bank (tax-exempt through federal MRB enabling legislation).

The Boulder Housing Authority, for another example, installed a solar hot water system on a senior housing project using private funds through a third-party financing deal. The local solar collector company installed solar collectors on the roof of a senior housing project called "Canyon Pointe," and contracted with the Boulder Housing Authority to provide solar energy at a rate discounted from what they would otherwise pay for natural gas. The investors who provided the solar collectors were compensated through federal energy tax credits. This type of third-party financing for solar energy systems was popularized under the name "Municipal Solar Utilities."

Municipal Solar Utilities were set up to match private capital with various public programs to finance solar systems, especially domestic hot water systems (Robertson, Besal, and Strange 1981). This concept of third-party financing for solar energy systems was also used widely in California, with results published by the California Energy Commission (Saitman 1981).

Third-party financing contributed significantly to hydroelectric and wind energy applications during the early 1980s. The advantages of private tax incentives combined with PURPA-regulated power sales rates to small power producers created a vast opportunity for public-private ventures. But solar heat applications generally did not benefit from PURPA, which was directed primarily at electrical energy technologies.

29.3   State and Local Information and Training Programs Using Federal Funds

State and local information and training programs are treated in the chapters of this volume dealing with regional solar energy centers and consumer programs. A brief summary suffices for here.

In the area of passive solar technology DOE funded the work of Los Alamos National Laboratory that resulted in publication of the *Passive Solar Design Handbook* (Jones et al. 1982). A vital resource used by many state and local energy offices in outreach training programs aimed at architects, engineers, and builders, and considered by many the bible of passive solar design, this document addresses the need to balance conservation and solar energy and presents the solar load ratio (SLR) method for evaluating design trade-offs in passive solar buildings.

State and local information and training programs on solar technologies, such as those of Arizona, California, Colorado, Florida, New York, and Oregon, were often modeled on federal programs. These state programs are detailed in publications of the National Solar Heating and Cooling Information Center (NSHCIC; see chapters 17, 19), where they are supplemented by state information on subjects such as solar radiation data, directories of the state solar industry, passive solar design guides matched to state climate regimes, and state consumer guides. Workshops and seminars funded at the federal level often emphasized state and local information, such as the HVAC and homebuilder industries workshops sponsored by SERI in 1979–1981.

The National Solar Residential and Commercial Buildings Demonstration Program held regional conferences to discuss program results in the late 1970s. Substantial benefits were derived from state and local solar education and training efforts, and the authors refer the reader to other chapters in this volume for details.

29.4   State and Local Quality Assurance and Testing Programs for Solar Heating

State and local quality assurance programs for solar heat technologies were often supported directly by federal funding for standards development, and indirectly by widespread use of federally developed guidelines, standards and codes.

Federal and state activities in consumer protection began as early as 1975. Early attempts were aimed at accelerating the national standards-setting process of the American Society of Testing Materials (ASTM), the American Society of Heating, Refrigeration, and Air-Conditioning Engineers (ASHRAE), and the American National Standards Institute (ANSI). Participating in these standards programs were the Department of Energy (DOE), the Department of Housing and Urban Development (HUD), the National Bureau of Standards (NBS), the Florida Solar Energy Center (FSEC), and the California Energy Commission (CEC). The Interstate Solar Coordination Council (ISCC; see below) had a position on the ANSI Solar Standards Committee.

Florida constructed and operated the first state-owned testing facility for flat-plate collectors; the Florida certification program for solar collectors in that state began in 1978. California developed a testing and certification program that included the accreditation of private laboratories to complete the testing and that also began in 1978. Both state programs used ASHRAE testing methods but developed their own procedures and certification methods as well. Both state programs also exchanged information and participated in federal efforts to accelerate the development of national standards for the solar industry.

DOE funded the Florida Solar Energy Center (FSEC) to convene the Interstate Solar Coordination Council (ISCC) to provide a forum for state energy officials to exchange standards information and to create reciprocal state solar collector certification programs. ISCC, representing 38 states, cooperated with the Solar Energy Industries Association (SEIA) to form the national Solar Rating and Certification Committee (SRCC). That group published industry standards and solar collector performance results from some half-dozen licensed testing laboratories including those set up by California and Florida, as well as private and federally funding testing labs. ISCC was funded by the Department of Energy at $51,200 in 1980 and an additional $191,124 in 1981 through the Florida Solar Energy Center (Block 1986). After 1981, support for ISCC has come only through paid memberships, and over half of those states have been paid members every year since then.

An important federal solar quality assurance document, funded by HUD and published by the National Bureau of Standards, was *Performance Criteria for Solar Heating and Cooling Systems in Residential Buildings* (NBS 1982). The document established "minimum criteria" for solar heating systems and addressed mechanical, safety and health, durability and reliability, and operation and servicing criteria. Widely used by state and local governments to qualify solar systems for state tax credits and for other incentives, it was incorporated by reference into some statewide codes, for example, by Minnesota in 1977.

Other consumer protection efforts were developed by state and local governments. The Colorado Office of Energy Conservation published a brochure called "Solar is Hot: Don't Get Burned," which referenced national standards and testing certification programs. This brochure was later reprinted for national distribution through state energy offices by SERI and the RSECs. The brochure recommended procedures for consumers to use in evaluating solar heating systems.

Examination and licensing of solar contractors began in some states; income tax incentive programs in New Mexico, Arizona, Colorado, California, and others all established warranty and other threshold requirements for solar system eligibility. The federal government and the State of California funded the Municipal Solar Utilities (MSU) program in California, which assisted local governments to establish baseline quality requirements for solar manufacturers to participate in locally designed solar programs. The Western Solar Utilization Network, the DOE-funded RSEC for the western states, published and distributed a book called *Renewable Energy: Opportunities for Local Governments* (WSUN 1981). Tailored for many of the western states, this document included a chapter on solar energy that covered incentives, solar access, and removing legal barriers to solar thermal technologies, as well as an extensive chapter on model codes and ordinances.

29.5 Evaluative Evidence and Lessons Learned

29.5.1 Evaluative Summary of Programs in this Chapter

The limited evaluative data available on the programs in this chapter are summarized below. Section 29.5.2 is a qualitative discussion of the state and local institutions affected by federal solar programs, and of some of the ways in which they were affected.

Energy Extension Service   Although widely appreciated by state energy offices, the DOE EES program was not very well documented, partly because the funding was disbursed through state energy offices to many

small, local organizations. The DOE did an evaluation of the Energy
Extension Service (DOE 1981), which concluded that EES activities for
local government and institutions produced an average of $16 in addi-
tional investment per federal dollar expended.

These data suggest that federal funds spent through local governments
may have been one of the better public investments; local governments
tended to invest their own money in energy efficiency and solar measures
partly due to stimulation of federally funded programs aimed at local
governments. The EES program, especially combined with the Institutional Conservation Service (ICS) and the Comprehensive Community
Energy Management Plan (CCEMP) program, contributed significantly
to local government awareness of energy conservation and renewable
energy opportunities.

Although there is justifiable controversy about the effectiveness of ESS
offices because they were not effectively evaluated, the authors of this
chapter can attest to the widespread awareness of EES centers and their
programs as consumer information and assurance sources. In the case of
Boulder, Colorado, even when EES funding ran out in 1984, the city
continued to fund their local EES office from general municipal revenues
at $25,000 per year (Wrenn 1987). This office serves as an energy infor-
mation center, provides audits and technical support to city energy loan
and grant programs (including one supported by the Solar Bank and
one supported by the mortgage revenue bond authority of the city), and
administers the rental rehab program funded by HUD. It is one example
of long-term spin-off effects from the federally funded EES program.
Similar examples abound throughout the country, although they are for
the most part undocumented, to the best of the authors' knowledge. The
contribution of these federally funded programs to solar thermal technologies cannot be separated from the energy conservation aspect, but
certainly solar thermal technologies were stimulated by these programs
and the awareness they raised.

Solar Bank   The impact of the Solar Bank program on solar thermal
technologies is probably negligible, although significant education about
solar heating occurred through the required audits and significant energy
conservation has been accomplished with these funds. A notable example
of creative use of these funds was to augment existing state and local
low-income programs.

Institutional Conservation Service   The state energy offices rated the ICS (later, the Institutional Conservation Program, or ICP) among the most cost-effective of federal-state energy programs in terms of energy saving per dollar of government funding; they estimated that fewer than 25% of the capital improvements would have been made independently and that paybacks averaged 3.2 years in the 1983 grant awards (Sawyer 1985a). A DOE evaluation shows that recipients of the ICP grants reduced energy consumption by 13% at a cost six times less than purchasing oil. Although the DOE evaluation also shows that ICP recipients were much more aware of other conservation and solar opportunities (DOE 1983), state energy office evaluations of the ICP program complained of erratic and unpredictable funding levels and the exclusion of local government buildings and nursing homes from the conservation equipment grants (Sawyer 1985a). The authors of this chapter conclude that, while the ICP program has been very important in accomplishing energy conservation, and may have contributed to general institutional awareness of solar technologies through its technical audits, it has been relatively insignificant with respect to actual demonstration or use of solar technologies.

29.5.2   State and Local Institutions Affected by Federal Solar Efforts

Many state and local institutions, both in the public and private sectors, were significantly influenced by federal solar and energy conservation programs. These include state governors and legislatures, state and local housing authorities, grassroots solar organizations, building code officials and homebuilders' associations, city government officials, agricultural extension services, and utilities.

State governors were impacted by federal solar programs primarily in the implementation of ECPA and EPCA and the programs of the regional solar energy centers (RSECs). Governors set up state energy offices to implement ECPA and EPCA, and often appointed their directors to serve on the Boards of Directors of the RSECs. For example, the Western Solar Utilization Network (WSUN) was set up to maximize state and local participation; this RSEC, serving the thirteen western states, was governed by a board of directors appointed by the governors of those states. In most cases, the WSUN board of directors served also as state energy office directors for their respective states, and in all cases were involved in the program implementation of ECPA and EPCA.

Another example of state participation in an RSEC is the Mid-American Solar Energy Center (MASEC), where each state was represented on the MASEC board of directors, each state established a Solar Office under MASEC, and each had a Solar Resource Advisory Panel made up of renewable energy community members to advise the state and MASEC. All the RSECs implemented solar programs by providing funding for state and local solar programs or for documents and workshops to benefit state and local public and private sector programs (see chapter 20 on RSECs). State legislatures were influenced by federal tax credit programs and often authorized state tax credits in addition to federal credits.

In the authors' opinion, one hard and bitter lesson learned from the state and federal tax programs is that the tax credits, quite unintentionally, provided a vehicle for unscrupulous solar dealers to market sometimes inferior solar water heaters at prices (from $2,600 to $6,000) that would—and eventually did—make these systems impossible to sell without government assistance. After the tax credits disappeared, so did the solar entrepreneurs. Such business practices were not, regrettably, the exception and occurred with increasing frequency as the federal tax credit renewal deadline (31 December 1985) approached. The legacy these activities left in their wake is the near elimination of the solar water heating industry and the stigmatization of solar energy as something both too expensive and unreliable. This is a very unfortunate side effect of the tax credit programs, which otherwise were among the most effective solar programs of both federal and state governments (see also Sav 1986, "The Failure of Solar Tax Incentives: A Dynamic Analysis").

State and local housing authorities were also influenced by federal solar programs, through the Weatherization Assistance Program and the Solar Energy and Energy Conservation Bank (Sawyer 1985a). The Weatherization Assistance Program was established to reduce energy consumption in low-income dwelling units, and was generally administered through state housing authorities or social services departments and local Community Action Programs. Although seldom including solar thermal applications per se, because of the low per unit funding and high paybacks required, weatherization assistance was used in rural areas for attached solar greenhouses, which provided solar heating to the residence.

Many states used housing authorities to administer the Solar Energy and Energy Conservation Bank. Federal regulations favored energy conservation improvements over solar applications. Other solar loan programs were initiated by states with such mechanisms as tax-exempt mortgage revenue bonds. These notably included New Mexico's Energy-Related Innovations for New Development program (which funded renewable energy companies at about $700,000 annually) and California's Alternative Energy Source Financing Authority, which provided initial loans to small businesses that were leveraged to obtain larger loans guaranteed by the Small Business Administration (Sawyer 1985a).

Grassroots solar energy organizations were also supported by federal funding, directly (often through funding of conferences) and indirectly (e.g., through small contracts to publish educational materials from EES centers). Exceptional examples were state chapters of the American Solar Energy Society (ASES) such as the New Mexico Solar Energy Association and Northeastern Solar Energy Association. Both these, and many other state and local energy societies, received federal funds through the American Solar Energy Society, the Center for Renewable Resources (CRR; a nonprofit arm of the Solar Lobby in Washington, D.C.), and occasionally through the state energy offices, usually in the form of contracts for services.

For example, the American Solar Energy Society and the Center for Renewable Resources sponsored the Solar Action Project in 1981, using services of grassroots organizations in twenty-seven states. The services generally included homeowner and builder education programs, which specifically taught passive and active domestic hot water solar thermal applications (AS/ISES 1981).

Several international organizations were also supported by federal funds to implement energy and solar codes, and these international organizations affected state and local energy codes. The International Conference of Building Officials (ICBO) was funded by HUD to establish a model solar heating and plumbing code which was adopted by many local and state jurisdictions as an amendment to the Uniform Building Code (UBC).

The ISCC detailed above also contributed significantly to standards and testing for solar heat applications, and continued to do so even ten years after the seed money provided by the federal government was expended. Collectively, the federally supported, solar-related efforts of many private and public building code groups permeated the building community of the United States in relatively short order during the late 1970s. By the mid-1980s, many homebuilders' associations were working with local and state jurisdictions to implement home energy-rating systems that gave credit to houses built with passive and active solar thermal features.

City officials, such as city councils, city managers, and city energy officials, were also influenced by various federal solar-related funding programs. The HUD CCEMP program discussed above resulted in solar thermal applications and ordinances, as well as energy conservation and other renewable energy programs. The National League of Cities (NLC) established a Local Energy Official network with an annual conference in conjunction with the NLC annual conference.

Agricultural extension services, traditional agricultural outreach institutions funded through agricultural land grant colleges since the early 1900s, also benefited from federal solar funds. These institutions were used widely by state energy offices to implement local energy outreach programs through the Energy Extension Service program funded by EPCA and ECPA during the late 1970s. Once having gained a foothold as local energy outreach agencies, the agricultural extension programs later benefited from court-ordered regulations concerning distribution of "oil overcharge" funds during the mid 1980s. These generally rural-oriented agencies provided farmers and homeowners with extremely practical information about attached solar greenhouses, sunspaces, and passive solar additions. The thousands of solar experiments that used this federally funded information, combined with local, private investments, yielded substantial information about net energy benefits of low-cost passive solar residential retrofits. Much of this "collective wisdom" was published in popular "energy extension service bulletins," which supplanted earlier decade's bulletins on home canning techniques and technical farming tips.

Utility companies, both public and private, as well as state utility regulatory agencies, were substantially affected by federal solar energy programs. The Residential Conservation Service (RCS) and the Public Utilities Regulatory Policy Act (PURPA), both described elsewhere in this chapter, mandated policy directions which steered electric and gas utilities toward energy efficiency and renewable energy options.

The Electric Power Research Institute (EPRI) and other utility-industry efforts significantly contributed to a new understanding of the economic value of renewable energy. Although few solar thermal electrical plants can be counted toward practical U.S. energy production, much solar thermal research and development was accomplished through U.S. electrical utilities during the late 1970s. Federal funds and federal interest in solar energy certainly contributed to this general research direction (see chapter 30 for additional information on utility programs).

29.6   Conclusions

The authors of this chapter conclude that federally funded programs aimed at state and local governments and institutions have been generally effective. Because solar thermal technologies are inherently decentralized, dispersed, and customized to local climates, federal energy programs are well advised to take advantage of existing state and local institutions.

The Energy Extension Service, the Comprehensive Community Energy Management Program, the Institutional Conservation Service, and the Appropriate Technology Small Grants Program were perhaps the most successful federally funded "state and local" programs. Each could be improved, but the common underlying principle of widely disbursing small amounts of funds is a good one. It appears to have been a successful strategy to disburse federal funding for solar technology through state and local entities.

However, the downside of this decentralized approach as practiced in the late 1970s and early 1980s is the scarcity of documentation. There is very little central access to information about the thousands of state and local efforts that resulted from federal funding. The research for this chapter indicates that perhaps the best example of a well-documented federal solar program is the Appropriate Technology Small Grants Program. The grants were funded for four years, followed by funding for the National Center of Appropriate Technology to create a computer database called "Appropriate Technology Management Information System." Perhaps in the next phase of federal funding for solar technologies, the computer age will provide the means to organize and easily share the wealth of information stimulated by federal funding of state and local solar programs.

References

AS/ISES (American Section of the International Solar Energy Society). 1981. *Solar Action: 27 Communities Boost Renewable Energy Use*. ISBN: 0-937446-03-3. Killeen, TX, May.

Ashworth, J., M. DeAngelis, B. Green, S. Parker, and D. Roessner. 1979. *State Solar Incentives Primer: A Guide to Selection and Design.* SERI/SP-433-470. Golden, CO: Solar Energy Research Institute.

Ashworth, J., B. Green, P. Pollack, R. Odland, R. Saltonstall, and L. Perlman. 1979. *The Implementation of State Solar Incentives: A Preliminary Assessment.* SERI/TR-51-159. Golden, CO: Solar Energy Research Institute.

Berger, G. J., R. Boyce, and R. Farley. 1980. *A Handbook on the Sale of Excess Electricity by Industrial and Individual Power Producers under the Public Utilities Regulatory Policy Act.* SERI/TR-98125-1. Golden, CO: Solar Energy Research Institute.

Block, D. 1986. Telephone communication with director of Florida Solar Energy Center, 10 November.

Colorado. 1980. *Colorado Model Energy Efficiency and Renovation Standards.* 3d ed. Denver, CO: Colorado Office of Energy Conservation.

DOE (U.S. Department of Energy). 1980. *Recommended Requirements to Code Officials for Solar Heating, Cooling, and Hot Water Systems.* DOE/CS/34281-01. Washington, DC: Council of American Building Officials, June.

DOE. 1981. *The Fourth Report to Congress: Comprehensive Program and Plan for Federal Energy Education, Extension, and Information Activities.* DOE/CE-0023. Washington, DC: Assistant Secretary for Conservation and Solar Applications, May.

DOE. 1982. *Annual Report to the President and the Congress on the State Energy Conservation Program for Calendar Year 1982.* DOE/CE-0016-2, DE-84002554. Springfield, VA: National Technical Information Service, U.S. Department of Commerce.

DOE. 1983. *An Evaluation of the Institutional Conservation Program: Results of On-Site Analysis.* Washington, DC.

Energy User's News. 1984. "States' Cogeneration Rate Setting under PURPA." Pts. 1, 2, 3, and 4. Vol. 9, nos. 40 (1 October), 41 (8 October), 42 (15 October), and 43 (22 October).

FERC (Federal Energy Regulatory Commission). 1984. *Annual Report.* Washington, DC: Government Printing Office.

Green, D. L., E. Hirst, J. Soderstron, and J. Trimble. 1982. *Estimating the Total Impact on Energy Consumption of Department of Energy Programs.* Oak Ridge, TN: Oak Ridge National Laboratory.

Hayes, G. B. 1979. *Solar Access Law: Protecting Access to Sunlight for Solar Energy Systems.* 1st ed. Cambridge, MA: Ballinger.

Hirst, E., W. Fulkerson, and R. Carlsmith. 1982. "Improving Energy Efficiency: The Effectiveness of Government Actions." *Energy Policy* 10 (1): 131–142.

Hittman Associates, Inc. 1978. *Comprehensive Community Energy Planning: A Workbook.* HIT-703-3. Columbia, MD, April.

Jaffe, M., and D. Erley. 1980. *Protecting Solar Access: A Guidebook for Planning Officials.* HUD-PDR-445(2), H-2573. Washington, DC: Government Printing Office, February.

Johnson, S. B. 1979. "State Approaches to Solar Legislation: A Survey." *Solar Law Reporter* 1 (1): 139–155, May/June.

Jones, R. W., D. Balcomb, C. Kosiewicz, G. Lazarus, R. McFarland, and W. Wray. 1982. *Passive Solar Design Handbook.* Vol 3: *Passive Solar Design Analysis.* DOE/CS-0127/3, UC-59. Washington, DC: Government Printing Office.

Moore, J. L., D. Berger, C. B. Rubin, and P. A. Hutchinson, Sr. 1979. *Organizing for Comprehensive Community Energy Management Planning: Some Preliminary Observations.* ANL/CNSV-TM-27. Argonne, IL: Argonne National Laboratory.

Mott-Smith, J. 1981. *Solar Access Ordinances: A Guide for Local Governments.* SolarCal-CEC contract no. 400-80-021. Sacramento, CA: SolarCal Local Government Commission.

NBS (National Bureau of Standards). 1982. *Performance Criteria for Solar Healing and Cooling Systems in Residential Buildings*. NBS Building Science Series 147. Washington, DC: Government Printing Office.

NCAT (National Center for Appropriate Technology). 1983. *Appropriate Technology at Work: Outstanding Projects*. DE-AC01-82CE159095. Butte, MT.

O'Hara, F. M., Jr., ed. 1986. *Energy Efficiency in Buildings: Progress and Promise*. Washington, DC: American Council for an Energy-Efficient Economy Series on Energy Conservation.

Robertson, C., M. Besal, and L. D. Strange. 1981. *Municipal Solar Utility: A Model for Carbondale, Illinois*. Chicago: State of Illinois Institute of National Resources.

Saitman, B., ed. 1981. *American Section of the International Solar Energy Society Annual Proceedings*. Newark, DE: American Section of the International Solar Energy Society.

San Martin, R. 1986. *Overview of State Renewable Energy Programs*. Washington, DC: U.S. Department of Energy.

Sav, G. T. 1986. "The Failure of Solar Tax Incentives: A Dynamic Analysis." *Energy Journal* 7 (3): 51–66, July.

Sawyer, S. W. 1985a. "Federal-State Conservation Programs: The States' Assessment." *Energy Policy* 13(3): 156–168, April.

Sawyer, S. W. 1985b. "State Renewable Energy Policy: Program Characteristics, Projections, Needs." *State and Local Government Review* 17 (1): 147–154, winter.

Sesso, J. 1987. Telephone communication with program manager, National Appropriate Technology Assistance Services, National Center for Appropriate Technology, Butte, MT, 17 June.

Tschanz, J. 1987. Telephone communication, Urban Planner/Urban Methods, Energy Planning, Argonne National Laboratory, Argonne, IL, 17 June.

Warren, M. 1979. "Common Problems in Drafting State Solar Legislation." *Solar Law Reporter* 1(1): 157–191, May/June.

Wrenn, P. 1987. Personal knowledge as former energy director for Boulder, Colorado (1982–1985).

WSUN (Western Solar Utilization Network). 1981. *Renewable Resources in Colorado: Opportunities for Local Governments*. UC-58a. Portland, OR, November.

30 Public Utilities

Stephen L. Feldman* and Patricia Weis Taylor

Utility response to the decentralized solar option has been diverse. Reaction has depended on the particular generation inventory, demand growth projections, and regulatory environment faced by a utility. To help clarify the costs and benefits of solar technology to utilities, the federal government has funded studies and shared the expense of demonstration projects. As a result, between 1974 and 1988, solar thermal systems have evolved from a novelty to one of several elements routinely considered in the utility generation mix.

This chapter outlines the history of federal activities to promote decentralized solar thermal applications to utilities. This history reveals an evolution of federal involvement from policy studies to addressing technical issues. The initial studies defining the interface between utilities and solar energy systems are described; the administrative framework and the role of the various actors are discussed in the context of these early policy studies. In addition, the increasing role of state regulators (see also chapter 29) and utilities in promoting solar and conservation is described. (Federal activities to promote centralized applications of solar thermal energy are discussed in chapter 7).

30.1 Early Federal Government Efforts

The interface between users of decentralized solar thermal (for heating and cooling of buildings) and utilities emerged as a federal concern early in the efforts to commercialize this technology. This was not part of the original federal program design but evolved in response to the clash in interests among federal solar policy, solar customers, and electric and gas utilities. High-temperature thermal applications for centralized electric generation were administratively convenient and less controversial to utilities, especially when highly subsidized by federal funds. Decentralized solar thermal applications included heating and cooling of buildings throughout the utility distribution system; industrial process heat or cogeneration at widely dispersed sites were also coming into use. And, eventually, small power producers set up facilities on available land and offered electricity to the utilities for sale. Due to the greater numbers of

* Deceased

facilities involved and due to technical issues, the interface between decentralized users and utilities was a controversial subject fraught with social and economic complexity.

Federal involvement began when utility and regulatory policies were identified in early policy studies as significant barriers to the commercialization of solar thermal applications. These early studies and their conclusions played an important role in shaping federal, state, and local policies regarding solar energy.

30.1.1   Initial Mandates and Their Failure to Address Utilities

The Solar Heating and Cooling Act of 1974 (see chapter 2) and its demonstration programs were the primary vehicles for demonstrating the effectiveness of solar heating and cooling technologies. The act itself was driven by the underlying assumption that technical problems were resolvable and that the widespread demonstration of the technology would result in its rapid commercialization; the legislation overlooked a formidable list of barriers that would have a dampening effect on the commercialization of the technology. Although many of these barriers are addressed elsewhere in other chapters, this chapter focuses upon the role of utilities in resisting or promoting decentralized solar thermal applications.

Utilities are not mentioned in the 1974 act, except for a suggestion that solar heating and cooling systems for buildings could use storage to reduce building demand during utility peaks. The implication of this legislation was that utilities would not have to provide backup energy to solar buildings during peaks; in fact, this was one of the major concerns utilities had in the early days—that solar thermal systems would contribute to peak demand and reduce off-peak usage.

30.1.2   Federal Research Effort Begins

In 1974 there was little evidence of utility activity or federal efforts to commercialize solar energy. Research on technical problems of off-peak storage and solar interface began with some initial National Science Foundation (NSF) grants. The largest of these grants established the National Center of Energy Management and Power in 1971 at the University of Pennsylvania (the nation's first academic energy center). Here the technical problems associated with the interface between solar energy systems and the utility were addressed in pioneering reports. Additional federal involvement in interaction of utilities and users of solar energy was not considered necessary until 1975. Even the landmark TRW, Westinghouse, and General Electric "Phase Zero" reports devoted little attention to the role of the utilities despite the prognosis of the enormous market potential for solar heating and cooling. (TRW 1974; Westinghouse 1974; GE 1974)

It is therefore appropriate to start with the NSF's efforts in discussing the role of the federal government in utility/customer relations when solar thermal systems were involved. The NSF/RANN program funded at least four modest efforts to study the effects of solar market penetration on utility loads, tariffs, customer cost effectiveness, and regulation; these studies were the first attempt at defining utility interface problems. The $1.2 million program at NSF that initially defined these utility barriers to the widespread use of solar energy was under the directorship of Lawrence Rosenberg. In these early studies, it was assumed that any solar-heated building would probably require backup energy from the utility during adverse weather conditions. It was also assumed that electricity would most likely be chosen to power these backup systems; in fact, initial solar user surveys reported that electricity *was* favored over gas and oil for backup systems (Sawyer and Feldman 1978; Feldman and Wirtschafter et al. 1979; Feldman, et. al. 1979).

Another important assumption at this time was that investor-owned utilities subject to rate of return regulation were pricing their electricity and gas at "average cost" rather than at "marginal cost" (Dickson and Feldman 1977). Because average costs did not reflect opportunity costs of electricity and gas, solar systems were competing with an artificially low price of energy; the regulated market of conventional fuel was competing with the nonregulated market of new energy sources. This environment, it was posited, would discourage investment in solar systems. It would also encourage customers to undersize the capacity of solar thermal systems (Feldman and Anderson 1975b, 1976a).

At about this time, the Federal Energy Administration's Office of Utilities was established. The Office of Utilities began a $10 million program to evaluate peak-load pricing and its impact upon utilities; pricing utility services in the peak and off-peak fashion of European countries would have an impact on the economics of solar energy systems and on the value of off-peak storage (Feldman and Anderson 1975a). The first major federal solar/utility project at this period was Project Sage, supported by NSF at Southern California Edison with cooperation from Jet

Propulsion Laboratory, which investigated gas-assisted solar heating in residential complexes (Davis and Bartera 1976).

During this period, there was a paucity of data on the actual performance of solar buildings. Researchers did not know the load factor of solar buildings or the actual effect the widespread adoption of the technology would have on utility loads. Analyses were based on simulation modeling (Abrash et al. 1978). The shortage of system performance data was exacerbated by the lack of site-specific and regional meteorological data; therefore, generalizations about system performance within specific utilities was difficult. Given the lack of resources and limited objectives, the first studies were highly qualitative and only moderately useful for generalizing results.

In 1976 the Energy Research and Development Administration (ERDA) was organized out of other agencies and took over the efforts of those other agencies in the energy area. After various reorganizations, Roger Bezdek was directed to examine barriers to implementation and incentives for adoption of solar energy; these ERDA studies would identify appropriate legislative options for the National Energy Act proposed by President Carter in 1977 (Bezdek 1977; see chapter 2 for details of legislation). The solar/utility interface problems needed to be defined in a comprehensive manner at that time.

On the technological front, ERDA also funded studies on load management in cooperation with utilities. Projects that used a number of solar-assisted technologies and on-site storage were examined. In addition, a relatively small and quick study of policy options at the solar/ utility interface was conducted by a multidisciplinary team headed by Stephen Feldman at Clark University (Feldman and Anderson 1976b). These early efforts aided in the development of a research and development agenda for ERDA. Other contracts with George Washington University and Booz, Allen & Hamilton, under the direction of Alan Hirshberg, produced a detailed study of market barriers and incentives to solar heating and cooling of buildings (SHACOB); Booz, Allen & Hamilton 1976). This was further abetted by a contract to Thomas Sparrow at the University of Houston to convene a cadre of individuals representing the solar industry, utilities, academia, government, the legal profession, and others to cite which barriers and incentives were important for policy purposes. The cadre's effort produced the first comprehensive document on the barriers and incentives to the diffusion of solar heating and cooling technologies in the marketplace (Bezdek, Cambel, and Hauer 1977); the document described the following major issues raised during the discussions of the solar/utility interface:

• Utilities predicted detrimental impacts on their loads due to bad weather.

• Utilities would not expand centralized storage to integrate solar into load management, even though on-site thermal storage and timing devices were available.

• Each utility would examine its own marginal costs, weather patterns, type of load, and so on, at the margin to select the most economic technologies.

30.1.3   Department of Energy Is Formed

In 1977 the creation of the U.S. Department of Energy (DOE) consolidated the functions of identifying solar/utility interface problems and providing analytical and policy analysis. By 1978 the early federally funded studies were augmented by more detailed policy work on the solar/utility interface funded by DOE. The earlier efforts, particularly those of NSF and ERDA, helped define DOE courses of action; DOE contractor awards followed those suggested action plans and addressed areas of concern.

At this time the founding of the Solar Energy Research Institute (SERI), in Golden, Colorado, further demonstrated federal interest in research and policy issues surrounding the development of solar energy. By 1979 many reports written by SERI analysts, as well as by contractors, addressed a multitude of questions involving solar energy development (SERI 1979, 1980, 1981).

Following up the barriers and incentives workshop of 1976, Roger Bezdek's office at DOE funded a series of workshops and conferences in 1978. Conducted by the Technology and Engineering Program at George Washington University, under Ali Cambel, these were offered as a "conflict-resolution experiment" that allowed many parties to express opinions in an anonymous fashion (Bezdek and Sparrow 1981). Workshop participants identified technical, economic/marketing, rate base, legal, and regulatory financial, consumer, public interest, and utility-specific issues. There was a workshop for utility executives, another for representatives of the solar industry, and a third for persons from public interest organizations and public utility commission representatives; each workshop was lead by a rapporteur. Then three more workshops were held, each consisting of a mixture of the previous groups and a plenary conference was held where all attendees were brought together.

Forty-five basic questions in seven areas were discussed—many that could only be answered by experience, for example (paraphrasing):

• Do dispersed solar energy systems present load management problems that are more difficult to solve than the load management problems caused, for example, by heat pump systems?

• What technical problems need to be resolved in creating a utility system consisting of a mixture of large and small energy facilities?

• How will different rate structures affect the design, sizing, and cost competitiveness of solar energy systems?

• Should solar/utility interface policies be formulated at the local, state, or national level?

In 1978 ICF, Inc., under a DOE contract, performed a case study analysis of the design of electric rates for residential customers requiring auxiliary electric service for SHACOB (ICF 1978). The study focused on five utilities in five states and found few legal or regulatory constraints on the consideration and implementation of explicit rates for solar auxiliary service. The primary reasons for the low level of rate design activity were (1) solar rate design was not perceived to be of critical importance; and (2) there were limited data and resources available to conduct cost of service and rate design analyses for solar customers.

Many of the federal efforts in the 1980s were aimed at the technical issues of performance of solar energy systems and how they might affect utility loads and costs (Feldman and Anderson 1977; Bright and Davitian 1978, 1979). Although thousands of solar buildings had been constructed by 1980, there were scant comprehensive data with which to assess solar's impact on utilities. The effects of solar heating and cooling systems on utility loads, production costs, capacity expansion, and revenues were still poorly understood. By 1980, some utilities were considering placing solar houses with electric backup on demand-charge rate schedules. A utility and regulatory concern was that the increased cost of service generated by solar customers not be subsidized by nonsolar customers; Public Service of Colorado placed demand charges on solar customers for just this reason.

It had been argued, however, that a subsidy to solar and conservation technologies equivalent to the difference between average cost and marginal cost to utilities was justified. There is much discussion of this issue in Feldman and Anderson (1975a, 1975b) and the Federal Trade Commission report's (1978) chapters by Habicht and Noll (see also Feldman, Wirtshafter et al. 1979; Feldman and Wirtschafter 1980, 1981). This logic, although not legislated at the federal level, was the rationale for establishing the large-scale, decentralized solar programs at the Tennessee Valley Authority.

DOE funded a number of studies at Lawrence Berkeley Laboratories to investigate the implications of such programs as TVA's on private utilities. The studies, led by Edward Kahn, examined the issue of financing solar energy systems (Kahn and Schute 1978; Kahn 1979); his analysis describes various utility-solar arrangements and assesses the generic impacts of utility solar financing on both the customer and the corporation (Kahn 1979). A solar financing program administered by regulated utilities would, in principle, address larger markets than the use of tax credits.

An analysis conducted at the Solar Energy Research Institute (Laitos and Feuerstein 1979) discussed the many legal issues involved when utilities own, sell, lease, finance, or service solar devices. There was concern that utilities would compete with private suppliers of solar equipment. The potential for solar-powered utilities to compete with already existing utilities was also examined; the impact of the National Energy Act statutes was seen as prohibiting utility domination of the solar market. A more subtle, but equally important effect of utility solar financing, is the potential for more optimal solar design. Early studies showed that optimization criteria for solar design on buildings will differ according to the viewpoint of the users, society as a whole, and the utility. The economic impact on utilities as well as participants and non-participants of a solar financing program must be assessed. (Bezdek and Cambel 1981)

At about this time the portions of the Public Utilities Regulatory Policies Act (PURPA) of 1978 regarding independent energy producers were being implemented. PURPA implicitly identified the "average cost" barrier as significant and required qualifying facilities and cogenerators to receive buyback rates equal to state defined "avoided cost." The PURPA requirement for utilities to interconnect with qualifying small power producers provided opportunities for solar thermal electric technologies to be developed and demonstrated in the private/utility sector even though DOE support was waning.

30.1.4 Solar Thermal Technologies Are Divided

By 1981 the Reagan administration reoriented DOE away from "commercialization" activities to the support of high-risk, long-range research and development. DOE budgets in the solar program plunged to 1975 levels, and policy studies came to a halt. The programs most affected by the Reagan administration's free-market approach were those, like solar thermal for buildings, which supported near-term technologies. The active solar heating and cooling of buildings and passive solar programs were consolidated into "solar buildings" and moved to the energy conservation branch of DOE by 1985.

A major focus of DOE's energy conservation activities for utilities (including passive and active solar heating systems) became known as demand-side management or DSM. DSM—also called "demand planning" or "energy management"—is a deliberate intervention by the utility to alter its loads in the marketplace. Using incentives and promotions, the utility influences choices on the "customer side of the meter." A utility can choose to promote energy use, encourage energy conservation, or alter usage patterns to shift peak demand, depending on its planning needs. Programs that are carefully designed and managed can benefit both customer and utility by using resources more efficiently (Kirsch and Deevy 1985). The adversarial nature of relations between utilities and users of solar energy lessened somewhat as passive solar features, solar hot water, and conservation measures began to be incorporated into DSM programs (Synergic Resources 1987).

As experience with DSM programs increased, their impact on utility supply-side planning began to be felt. Integrated resource planning (IRP) in some utilities began to incorporate activities promoted on the customer side of the meter. By 1985 a variation of integrated resource planning called "least-cost utility planning" was promoted to minimize the cost to society of electricity. In 1986 the DOE Office of Buildings and Community Systems (where the solar buildings program resided) published an agenda for research on least-cost utility planning to be conducted in cooperation with the National Association of Regulatory Utility Commissioners, EPRI, Edison Electric Institute, the American Public Power

Association, and others (DOE 1986); state regulatory agencies increasingly required this approach by 1987.

30.1.5   Rise of Decentralized Solar Thermal Electric Technologies

After the formation of DOE, high-temperature solar thermal research activities (previously directed mainly at central station projects) began to address decentralized applications. The initial markets were seen as total energy systems for small communities, industrial load centers, and military bases. Based on previous experience in the central receiver program, the Solar Thermal Technology Program made utilities an integral part of the solar thermal electric demonstration activities.

In 1982 the first large-scale demonstration of parabolic dish technology was installed with Georgia Power Company in Shenandoah, Georgia. The objective of this Solar Total Energy Project was to build and operate the system in a utility/industrial user environment; the system provided electric power, process steam, and air-conditioning to a knitwear factory. Georgia Power reported extensively on its experiences to the utility community, and the parabolic dish and trough research programs benefited from lessons learned in Georgia (see chapter 8).

Continuing the successful strategy of cooperating with utilities, DOE and Southern California Edison (SCE) demonstrated a parabolic dish module called "Vanguard," coupled with a Stirling heat engine to generate electricity at the Santa Rosa substation. SCE's interest in solar thermal electric continued and the company eventually purchased rights to the promising dish-Stirling technology in 1985.

Congress authorized the Small Community Solar Experiments in 1984. These two projects, totaling over $4 million, included the design, construction, and testing of competitively selected solar thermal dish electric systems; they were connected to utility grids to help supply community electrical needs in Osage City, Kansas, and in the service territory of Molakai Electric Company, Hawaii. Important technical issues of interconnected systems such as power quality and safety concerns were identified and addressed in these projects. (SERI 1987)

Decentralized activities such as these kept utilities informed about the technology and maintained a group of people experienced in the design and implementation of such projects. During this period, many experienced professionals undoubtedly moved to other fields due to the greatly reduced level of funding and activities.

30.2 Utility Initiatives in Decentralized Solar Thermal

Each utility has its own level of interest in or resistance to solar thermal systems. And the level of interest within a particular utility varies over time as economic and regulatory conditions change. Because virtually every person in the U.S. interacts with a utility, at home and at work, these organizations can be effective vehicles for introducing new technology. Utilities have amassed a large experience base in solar energy through their internal research and demonstration efforts, through cooperative research with government, and, increasingly, through purchasing agreements with PURPA small power producers.

30.2.1 Electric and Gas Utility Activities

Independent of the federal departmental efforts beginning in 1975 and 1976, a number of utilities engaged in solar projects of their own. The first national energy survey performed by EPRI in 1976 (EPRI 1977) showed that at least 168 solar projects were being undertaken by some 116 utilities. Nearly three-quarters of these projects were primarily single solar buildings and were for demonstration and data-gathering purposes. Unfortunately, few of these utilities monitored their buildings on time-of-day metering; little or no information of the effect of the building upon utility coincident demands was generated. Discussion with utility executives by Feldman and Anderson (1976b) indicated that one of the major concerns at this time was the impact of solar space and hot water systems on their peak loads. Therefore, in subsequent years many demonstration project designs were changed to gather time-of-day data and correlate it with overall system demands and weather patterns.

Utility interest increased and by 1981 the EPRI survey showed that 236 utilities participated in a total of 943 solar projects. Of these projects 545, or 58%, were devoted to solar heating and cooling of buildings (SHACOB). Utility spending on these projects for 1981 was $26 million, and a total of $140 million had been spent since 1975. In 1981 about 2% of EPRI's total budget, some $3.4 million, was spent on its Solar Power Systems Program (EPRI 1985). As part of this program, EPRI funded a feasibility study at Aerospace Corp. to simulate the impact of solar systems in ten utilities (Aerospace 1978). Arthur D. Little, Inc., was also supported to assess the usefulness of a demonstration project in the southwest and northeast United States for residential SHACOB (ADL

1977). The latter project was a multiyear, multimillion dollar effort but had limited application because it was conducted in only two utilities using ten residential buildings. However, instrumentation of these ten buildings did represent the most technically competent data-gathering effort to date by the utility industry.

During this period many utilities, especially municipal utilities in California, had programs to finance and install solar hot water systems. These activities, along with vigorous energy conservation programs, had direct benefits for these nongenerating utilities. In addition, the regulatory and tax environment encouraged innovative approaches such as Municipal Solar Utilities (MSU), energy service contracts, and leasing arrangements (ACEEE 1986). In 1980 the California Public Utilities Commission ordered five major utilities to finance 175,000 solar domestic water heaters, and they developed a variety of programs.

In 1980 Wisconsin Power and Light Co. (WPL) launched plans for a solar subsidiary that sold solar hot water systems. It was not unusual for a utility to sell national brand appliances, but WPL, even in the face of uncertain supplier markets for solar products, broke ground among utilities by creating a solar subsidiary. Within a few years several hundred hot water systems were sold and many other utilities promoted installation of solar hot water systems.

Most of the projects of gas utilities, in some cases performed by utilities that provided both gas and electric service, were similar to the electric utilities' demonstration projects. The Gas Research Institute was, however, more interested in testing advanced gas/solar-assisted technologies. For example, whereas electric heat pumps with solar assist were tested on a routine basis, the gas-assisted heat pump was a relatively novel innovation in itself, made more distinct by solar assist.

A drastic slowing in the growth of solar projects within many electric utilities occurred with the change in administration in Washington, the ensuing recession, and the fall in energy prices. The escalating prices of the late 1970s that were used to project the value of solar energy in future years no longer applied. By 1984 EPRI discontinued its annual surveys of solar energy, although utility interest and EPRI's programs were far from disappearing. Even with falling energy prices, utilities with growing peak demand and limited prospects for generation expansion were looking to the small power producers authorized by the PURPA legislation for relief. In 1985 Southern California Edison signed an energy purchase agreement for solar thermal electricity generated using a parabolic trough system; San Diego Gas and Electric bought power generated by a parabolic dish system.

One producer of electricity for sale to utilities, Luz International, Ltd., installed solar thermal electric power plants in southern California from 1983 to 1987; these projects were developed as third-party independent energy production facilities under provisions of PURPA. State tax credits, the remaining federal tax credits, and accelerated depreciation rates helped make these projects work financially; often utility companies themselves invested in the projects (Becker 1992). The Luz project was particularly appealing to utilities supplying both electricity and gas. The systems were actually natural gas/solar thermal hybrid systems. This assured that the facility could produce electricity whenever the utility needed peak power and provided the utility with a gas customer, while maximizing the revenue stream to the project for sale of electricity. Utility dispatching was also easier because the project was always available.

30.2.2   Federally Owned Utilities Experiment with Dispersed Solar Thermal Systems

In addition to efforts in federal energy departments, federally owned power-generating and marketing organizations were becoming interested in solar energy. In 1977 the federally owned utilities Tennessee Valley Authority and Bonneville Power Administration were beginning to think about solar policy in their service territories or for their affiliated distribution companies.

The appointment of S. David Freeman, the dynamic leader of an aggressive Ford Foundation-funded study, as chairman of the board of TVA, was to change the posture of TVA quite rapidly. In 1978 Freeman created an office of conservation and included plans for the largest solar program undertaken by any single entity in the country; the commercialization effort included solar buildings, hot water heating systems, passive design, and energy conservation. A highly leveraged, zero-interest loan program for participants was established based on the power credit methodology and having its roots in the marginal cost logic presented above. Within two years over 2,000 water heaters were installed, which represented the largest program of its kind in the nation.

The TVA effort set the penultimate example for other utilities, including its sister, the Bonneville Power Administration (BPA). TVA built an impressive staff of some 150 professionals and spent more than $40 million on solar activities that included experimentation as well as the loan program. Eventually, because of nuclear power plant overcapacity and the political demise of the chairman who led the effort, the program was reduced dramatically. Other federal marketers such as the Western Area Power Administration (WAPA), continued a steady information and assistance campaign to promote the use of solar energy and conservation among its customers (WAPA/DOE 1988).

30.3 Conclusion

The cutbacks in the federal budget for solar activities beginning in 1980–1981 immediately affected the study of the solar/utility interface, although utilities, industry, and the DOE technical research programs continued work to address these issues. Legislation such as PURPA and state regulatory interest in solar assured that some systems would be built and experience accumulated. Many types of systems were placed at a competitive disadvantage when oil and gas prices slumped, but others survived on their economic merits alone.

Two important lessons were learned from the early period of "commercialization" efforts. First, the design for any type of demonstration or prototype program in energy technology must include the utility interface as a specific consideration. Utilities have the potential to make or break an energy technology. The earlier utility concerns are incorporated into technical developments *and* institutional considerations, the better the chance for utilities to make a positive contribution to implementing new energy technologies. Second, any solar tax incentive program must also recognize the importance of utilities and their interactions with the purchaser of technology. In the end, programs that encouraged utility investment in small power producer projects helped advance the implementation of solar thermal electric systems.

In addition, utility electric and gas backup rates and buyback rates have been shown to have a direct impact not only on the economics of a solar project, but also on the system design, size, and technology configuration. With the advent of DSM and least-cost utility planning tools, the impact of rates on solar energy systems can be estimated more accurately. Although reaction among utilities to new technology options will continue to vary from extreme resistance to active promotion, any future government demonstration and commercialization program must not fail to include these important actors in the design and implementation of such efforts.

References

Abrash, M., R. Wirtshafter, P. Sullivan, and J. Kohler. 1978. "Modeling Passive Buildings Using TRNSYS." *Proceedings of Second Passive Solar Conference*. Philadelphia: Boulder, CO: American Solar Energy Society.

ACEEE (American Council for an Energy-Efficient Economy). 1986. *Financing Energy Conservation*. Washington, DC.

ADL (Arthur D. Little, Inc.). 1977. *System Definition Study—Phase 1: Individual Load Center, Solar Heating and Cooling Residential Project*. Palo Alto, CA: Electric Power Research Institute.

Aerospace Corporation. 1978. *Solar Heating and Cooling of Buildings (SHACOB) Requirements and Impact Analysis*. Palo Alto, CA: Electric Power Research Institute.

Becker, N. 1992. "The Demise of Luz: A Case Study." *Solar Today* January–February.

Bezdek, R. 1977. *Analysis of Policy Options for Accelerating Commercialization of Solar Heating and Cooling Systems*. Programs of Policy Studies in Science and Technology. Washington, DC: George Washington University.

Bezdek, R., and A. B. Cambel. 1981. "The Solar Energy/Utility Interface." *Energy* 6: 479–484.

Bezdek, R., and F. T. Sparrow. 1981. "Solar Subsidies and Economic Efficiency." *Energy Policy* December: 289–300.

Bezdek, R. H., A. B. Cambel, C. R. Hauer. 1977. *Summary of Proceedings of Solar Heating and Cooling Commercialization Workshop*. DSE/4017-1. Washington, DC: National Technical Information Service.

Booz, Allen & Hamilton. 1976. *Effectiveness of Solar Energy Incentives at the State and Local Level: An Overview*. Washington, DC: U.S. Federal Energy Administration.

Bright, R., and H. Davitian. 1978. *Electric Utilities and Residential Solar Heating and Hot Water Systems*. EY-76-C-02-0016. Upton, NY: Brookhaven National Laboratory.

Bright, R., and H. Davitian. 1979. *The Marginal Cost of Electricity Used as a Backup for Solar Hot Water Systems: A Case Study*. BNL-25501. Upton, NY: Brookhaven National Laboratory.

Davis, E. S., and R. E. Bartera. 1976. "Design and Evaluation of Solar-Assisted Gas Energy Water Heating Systems for New Apartments: SAGE Phase II Report." Technical memorandum 5030-15. Pasadena: Jet Propulsion Laboratory, California Institute of Technology.

Dickson, C., M. Eichen, and S. Feldman. 1977. "Solar Energy and U.S. Public Utilities." *Energy Policy* 5:3.

DOE (U.S. Department of Energy). Assistant Secretary for Conservation and Renewable Energy. 1986. *Department of Energy Least-Cost Utility Planning Research Agenda*. Washington, DC.

EPRI (Electric Power Research Institute). 1977. *National Energy Survey*. AP-2850-SR. Palo Alto, CA.

EPRI. 1985. *National Energy Survey*. AP-3665-SR. Palo Alto, CA.

Feldman, S. L., and B. Anderson. 1975a. "Financial Incentives for the Adoption of Solar Energy Design." *Solar Energy* 17:339–343.

Feldman, S. L., and B. Anderson. 1975b. *Nonconventional Incentives for the Adoption of Solar Energy Design: Interim Report 2.* APR-75-18006. Prepared for NSF/RANN. Worcester, MA: Clark University.

Feldman, S. L., and B. Anderson. 1976a. *Utility Pricing and Solar Energy Design.* APR-75-18006. Prepared for NSF/RANN. Worcester, MA: Clark University.

Feldman, S. L., and B. Anderson. 1976b. *The Public Utility and Solar Energy Interface: An Assessment of Policy Options.* ERDA E(49-18)-2523. Prepared for U.S. Energy Research and Development Administration. Worcester, MA: Clark University.

Feldman, S. L., and B. Anderson. 1977. *The Impact of Active and Passive Solar Building Designs on Utility Peak Loads: Interim Report.* Prepared for U.S. Department of Energy under contract EG-77-G-01-4029. Worcester, MA: Clark University.

Feldman, S. L. 1979. *The Impact of Solar Energy in Buildings upon California Electric Utilities.* Sacramento: State of California, Energy Resources Conservation and Development Commission.

Feldman, S. L., R. M. Wirtshafter, M. Abrash, B. Anderson, P. Sullivan, and J. Kohler. 1979. *The Impact of Federal Tax Policy and Electric Utility Rate Schedules upon the Solar Building/Electric Utility Interface.* Prepared for U.S. Department of Energy under contract EG-77-G-01-4029. Worcester, MA: Clark University.

Feldman, S. L., and R. M. Wirtshafter. 1981. *On the Economics of Solar Energy.* Rev. ed. Lexington, MA: D. C. Heath, 1980.

FTC (Federal Trade Commission). 1978. *The Solar Market: Proceedings.* Washington, DC.

GE (General Electric Corporation). 1974. *Solar Heating and Cooling of Buildings: Phase 0 Feasibility and Planning Study, Final Report.* 3 vols. Valley Forge, PA.

ICF, Inc. 1978. *Technical, Institutional and Economic Analysis of Alternative Electric Rate Design and Related Regulatory Issues in Support of DOE Utility Conservation Programs and Policy.* Vol. 3 of *Economic Analysis.* Washington, DC: ICG, Inc.

Kahn, E. 1979. "The Compatibility of Wind and Solar Technology with Conventional Energy Systems." *Annual Review of Energy* 4:

Kahn, E. and S. Schutz. 1978. *Utility Investment in On-Site Solar: Risk and Return Analysis for Capitalization and Financing.* LBL-7876. Berkeley, CA: Lawrence Berkeley Laboratory.

Kirsch, F. W. and L. M. Deevy. 1985. *Energy Conservation Opportunities Implemented by Small and Medium-Size Manufacturers.* DOE/CS40091-T14. Washington, DC: U.S. Department of Energy.

Laitos, J., and R. J. Feuerstein. 1979. *Regulated Utilities and Solar Energy: A Legal-Economic Analysis of the Major Issues Affecting the Solar Commercialization Effort.* SERI/TR-62-255. Golden, CO: Solar Energy Research Institute.

Sawyer, S., and S. L. Feldman. 1978. "The Barriers and Incentives to the Commercialization of Solar Energy for Residential Domestic Hot Water and Space Heating: An In-Depth Assessment by Solar Consumers." *Proceedings of the 1978 Annual Meeting of the American Section of International Solar Energy Society.* Boulder, CO: American Solar Energy Society; Denver, CO. pp. 382–386.

SERI (Solar Energy Research Institute). 1979. *Solar Energy Legal Bibliography.* SERI/TR-62-069. Golden, CO.

SERI. 1979–1980. *Solar Law Reporter.* Golden, CO.

SERI. 1980. *Solar Energy Legal Bibliography Update.* SERI/TR-733-741. Golden, CO.

SERI. 1981. *Solar Energy Legal Bibliography Update.* SERI/TR-744-876. Golden, CO.

SERI. 1987. *Solar Thermal Power.* SERI/SP-273-3047. Golden, CO.

Synergic Resources Corporation. 1987. *Proceedings: Third National Conference on Utility Demand-Side Management Programs.* Houston.

TRW Systems Group. 1974. *Solar Heating and Cooling of Buildings: Phase 0, Final Report.* Redondo Beach, CA.

WAPA/DOE (Western Area Power Administration and U.S. Department of Energy). 1988. *Energy Conservation Technical Information Guide.* Vol. 2, *Utilities.* SERI/SP-320-3341. Golden, CO: Solar Energy Research Institute.

Westinghouse Electric Corporation. 1974. *Solar Heating and Cooling of Buildings: Phase 0, Final Report.* 4 vols. Baltimore.

31 Legal, Environmental, and Labor Issues

Alan S. Miller

31.1 Federal Role

This chapter reviews federal activities related to legal, environmental, and labor issues arising from the use of solar heating technologies. The three topics cover a wide range of problems and federal activities, as indicated in table 31.1.

The federal approach to these issues reflected wide-ranging strategies and objectives. Legal research was undertaken to ascertain what changes in laws might be necessary to facilitate solar energy development. This effort was complicated by the early recognition that many of the most important issues—for example, building codes and solar access legislation—are primarily governed by state and local authorities. Some of the earliest environmental analyses were done to advance broad social arguments sympathetic to particular energy sources, while more technical evaluations came later. Both had a modest influence on the evolution of federal energy and environmental policies. Labor studies were done partly to document the merits of particular energy strategies, but federal support was also given to technical studies of solar manpower needs and of the implications of union jurisdiction for solar development. (Federal funding for solar training programs is discussed separately in chapter 19.)

Although legal, environmental, and labor issues were generally not addressed together by federal programs or contractor studies, several shared characteristics justify a unified treatment. First, a primary federal objective in all three areas was to identify potential obstacles and ancillary benefits that might hinder or accelerate the acceptance of solar technologies; the activities were therefore often similar. Federal programs used studies, conferences, and related educational efforts to identify and evaluate the extent to which legal, environmental, and labor issues might impede or facilitate the widespread use of solar technologies. Much of this chapter is devoted to a review of these efforts.

Second, all three issues cut across the jurisdiction of federal agencies as illustrated in table 31.2, causing some management problems. The interest in using solar energy became a national issue in a matter of months after the oil embargo of 1973 and led to numerous federal laws involving almost all agencies of the federal government. The Solar Energy Research Institute (SERI) was also created as a federal laboratory dedicated to

Table 31.1
Major legal, environmental, and labor issues addressed by federal solar heating programs

Legal	Environmental	Labor
Access to sunlight	Materials analysis	Union conflict
Building codes	Comparative impacts assessment needs	Manpower needs
Land use controls	Land use and urban form evaluation	Training
Taxation	Solar energy in support of environmental goals	Net job creation or loss
Utility regulation		
Insurance/warranties		
Antitrust/competition		

Table 31.2
Federal agencies involved in solar legal, environmental, and labor issues

Agency	Legal	Environmental	Labor
CEQ/EPA		×	
DOE/ERDA/FEA	×	×	×
Economic Development Administration			×
FTC	×		
HUD	×	×	
Justice	×		
Labor/OSHA		×	×
National Arts Endowment		×	
NSF	×	×	×
SBA	×		
Commerce			×

Note: Acronyms and specific agency programs are discussed in the text.

research on renewable energy. Bureaucratic coordination and efficiency were difficult to maintain in these circumstances; some duplication of effort and wasted funding occurred, but substantial expertise and resources were redirected to a new priority in only a few years.

Third, the federal role in all three issues was to varying degrees limited by or dependent upon state and local authorities and private organizations. This review of disparate federal activities is therefore in part a common story of efforts to build linkages among the federal government, other levels of government, and the private sector.

31.1.1 Rationale for Federal Role

Federal activities related to legal, environmental, and labor issues were premised on a broad consensus in the early 1970s that many solar energy technologies were nearly ready for commercialization without substantial further basic research, but that many nontechnical barriers had to be overcome. The Senate committee report supporting the Solar Heating and Cooling Demonstration Act of 1974 states that the production of reliable, durable and economic systems requires "taking into account not only the technological factors, but the problems of economic feasibility, mortgageability, and public confidence. These factors are sufficiently complex to require coordinating and encouraging the efforts in this regard of both the public and private sector" (Senate 1974).

The National Science Foundation (NSF) played a key role in defining and advancing this rationale. A Solar Energy Panel report prepared jointly by NSF and the National Aeronautics and Space Administration (NASA) in 1973 emphasized the need for demonstration projects (NSF/ NASA 1972). Several of its principal authors subsequently testified at congressional hearings on solar energy during the 93d Congress (House 1974). The Solar Energy Panel report of Project Independence, prepared under NSF direction, states that "[i]n general, the applications of solar energy do not require establishing scientific feasibilities." The major problem, the task force concludes, is to develop systems that are "economically acceptable," which requires in part dealing with social, legal, political, regulatory, and economic factors (Project Independence 1974).

An NSF program director, Arthur Ezra, expounded on these issues in *Science* (Ezra 1975). Ezra argued that "many a time, even in the presence of both a clearly perceived market and public need, industry alone cannot put the R&D results to use for the benefit of the public." While recognizing that some may see an answer in the "free workings of the marketplace," he argued for an active governmental role in creating incentives appropriate to all of the institutions involved in the introduction and use of a technology. Similar arguments appear in articles by government contractors at about the same time (see Hirshberg and Schoen 1974).

This philosophy is reflected in a 1975 program planning report of the Energy Research and Development Administration (ERDA). The report (ERDA 1975) concluded that solar technologies could supply as much as 25% of the nation's energy by 2020, although federal demonstration projects, market research, and other measures to identify and counter institutional problems would be necessary to achieve this goal.

The first two National Energy Plans (NEPs) also describe a strategy linking demonstration programs to the identification of nontechnical barriers as the basis for accelerating commercialization of solar energy. NEP I in 1977 noted:

The results of the solar demonstration programs being carried out by [ERDA and HUD] and the equipment standards being developed by HUD should help provide a basis for warranties, insurance, and mortgage valuations.... The States are also encouraged to enact legislation to protect access to the sun and to promote consumer education in the solar field. (DOE 1977)

In 1979 NEP II similarly noted that "a number of institutional barriers at the Federal, State, and local levels can affect the choices between solar and conventional systems" (DOE 1979a). The plan cites as examples building code issues and the impact of utility rates on solar economics.

31.1.2 Overview of Studies

The initial studies of legal, environmental, and labor issues served to identify the major problems. Thereafter, the sponsoring agencies shifted their priorities toward resolving problems and more directly accelerating the use of solar energy. This tended to mean different types of activities for each issue. In response to legal questions, federal programs supported more detailed analysis of problems, particularly access to sunlight, and drafting of model statutes. Results of these efforts were publicized through professional meetings and publications, including the *Solar Law Reporter*, a SERI project discussed below. Labor issues were addressed through training programs as well as more detailed analysis. The conclusions of environmental studies were incorporated in energy policy documents, including a comparative assessment of energy technologies required by Congress. These activities are reviewed in the sections that follow.

Several relevant activities are also described in other chapters, including Federal Trade Commission (FTC) review of consumer fraud issues (chapter 15), Department of Energy (DOE) involvement in state utility regulatory proceedings affecting solar users (chapter 30), federal technical assistance to state and local governments (chapter 29), and DOE and Department of Labor (DOL) training programs (chapter 19).

31.2   Statutory Guidance

31.2.1   General Mandate for Solar Development

Congress adopted several statutes in 1974 to promote solar energy (see chapter 2 for a general discussion of the legislative history). These statutes gave several agencies but primarily the Department of Energy authority to undertake research and demonstration activities in order to accelerate the introduction and use of solar energy systems, reflecting the philosophy outlined above. Several statutes specifically addressed legal, environmental, and labor issues as discussed below. However, federal activities went beyond these specific mandates, presumably justified by the broad purposes of the operative statutes.

In the Solar Heating and Cooling Demonstration Act of 1974 (42 U.S.C. 5501), Congress mandated a five-year demonstration of solar heating and cooling intended to establish the "practical use" of the technology. The act's stated premises were that demonstration would expedite commercial use and that innovation could be fostered through the experience that would result. Congress directed development of performance standards for solar equipment, monitoring of actual performance, and several activities designed to maximize dissemination of results, including creation of a central data bank and liaison activities with professional groups.

The Solar Energy Research, Development, and Demonstration Act of 1974 (42 U.S.C. 5559) similarly required studies of "incentives to promote broader utilization and consumer acceptance of solar energy technologies." Several broad provisions directed the Department of Energy to conduct research to assess the environmental impacts of energy sources. For example, the Energy Research and Development Administration Act of 1974 (42 U.S.C. 5820) authorizes research programs coordinated with the Environmental Protection Agency to minimize the adverse environmental effects of energy projects.

The Federal Nonnuclear Energy Research and Development Act of 1974 (42 U.S.C 5905) directed preparation of an annual comprehensive energy plan to include analysis of environmental problems over the near and long term. Section 5910 requires a separate annual analysis by the Environmental Protection Agency (EPA) of environmental consequences of trends in energy technologies. Following these specific mandates, one purpose of the Department of Energy Organization Act of 1977 (42 U.S.C. 7112[13]) is "to assure incorporation of national environmental protection goals in the formulation and implementation of energy programs, and to advance the goals of restoring, protecting, and enhancing environmental quality...."

The statutory basis for creation of the Solar Energy Research Institute (SERI) is surprisingly brief. SERI played a major role in the duty of legal, labor, and institutional aspects of solar energy in the late 1970s and early 1980s. Section 10 of the Solar Energy Research, Development, and Demonstration Act of 1974 (42 U.S.C. 5559) authorized SERI's establishment. The functions of the Institute were left to be defined by an interagency task force.

ERDA contracted for outside reports and proposals as part of the process defining SERI's structure and selecting a site (Blisset 1979). The National Academy of Sciences prepared the most influential report (NAS 1975), lending its general support for SERI activities related to legal, environmental, and labor issues. The report notes that decisions regarding solar energy "must combine technical, economic, sociological, and environmental analyses of changing energy resources and societal demands," and urges that SERI undertake a correspondingly broad analytical and assessment function. In particular, the report recommends that SERI "analyze incentives for communities and industry that will make the introduction of solar energy more attractive when it is economically sound," that SERI develop environmentally sensitive models for assessing candidate systems, and that SERI publish and disseminate results of the institute's work in policy-oriented journals as well as standard scientific publications.

The Energy Conservation Standards for Buildings Act of 1976 (42 U.S.C. 6831 *et seq.*) authorized federal development of building energy performance standards to limit energy consumption by new buildings. The Department of Energy proposed to encourage solar energy use in the process of implementing this program by omitting the energy provided by solar equipment from the calculation of building energy consumption. However, the issue became less consequential when Congress made the use of these standards voluntary in Section 1041 of The Energy Conservation Standards for New Buildings Act (42 U.S.C. 6831) of 1981. The Department of Housing and Urban Development (HUD) was separately directed to promote improved energy efficiency in federally insured housing through requirements of minimum property standards (12 U.S.C. 1735f-4).

31.2.2 Research on Solar Legal Issues

Section 5510(b) of the Solar Heating and Cooling Demonstration Act (42 U.S.C. 5501 et. seq.) requires studies of "the effect of building codes, zoning ordinances, tax regulations, and other laws, codes, ordinances, and practices upon the practical use of solar energy for the heating and cooling of buildings." The act further mandates analysis of "the extent to which such laws, codes, ordinances, and practices should be changed to permit or facilitate such use, and the methods by which any such changes may best be brought about, and [to] study the necessity of the program of incentives to accelerate the commercial application of solar heating and cooling technology." Numerous studies of these issues were commissioned (as described in section 31.4.) but the author was unable to identify any formally submitted response to the congressional request.

Section 8232 of the National Energy Conservation Policy Act of 1978 (42 U.S.C. 8201 et. seq.) directed HUD to study the feasibility of mandatory federal energy efficiency standards for residential buildings. (The report that resulted is discussed in section 31.4.3.) This statute also established a residential energy audit program administered by electric and gas utility companies. The audits must be low-cost to the homeowner and assess the economic benefit of enumerated "residential conservation measures," including solar devices.

Congress also addressed the employment implications of alternative energy sources. Section 829a of the Comprehensive Employment and

Training Programs Act (29 U.S.C. 801 et. seq.) directed the secretary of labor to "develop methods to ascertain ... energy development and conservation employment impact data by type and scale of energy technologies used." This law was repealed in 1982 by PL 97-300, as part of a major overhaul of the Comprehensive Employment and Training Act (CETA) program (29 U.S.C. 801 et. seq.). The legislative history does not explain the reasons for repeal or reveal whether a formal methodology was ever approved; interest in the issue may have waned by 1982.

31.2.3 Related Statutes

In addition to statutes specifically addressing solar energy, federal agencies became involved in legal, environmental, and labor issues as a result of broader mandates. The Department of Labor, for example, supported some solar training programs as discussed below. The Federal Trade Commission, the Antitrust Division of the Justice Department, and the Small Business Administration conducted studies of competition issues under their purview.

The National Environmental Policy Act (NEPA) of 1969 (42 U.S.C. 4331) requires all government agencies to prepare environmental impact statements on major federal activities that must include a comparison of the environmental consequences of alternatives. The courts, in *SIPI v. Atomic Energy Commission* (481F. 2d 1070 [D.C. Cir. 1973]), have imposed this requirement on government research supporting specific energy technologies, such as the liquid metal fast breeder reactor program. The discussion of solar energy and energy conservation alternatives included in the environmental impact statement (EIS) process tended to be pessimistic, sometimes leading to litigation, as in *Vermont Yankee Nuclear Power Corp. v. NRDC* (435 U.S. 519 [1978]).

31.3 Historical Perspective on Legal Environmental and Labor Studies

Federal programs on nontechnical aspects of solar energy, including legal, labor, and environmental issues, grew rapidly from negligible amounts in 1975 to about several million dollars a year in 1979–1980, after the formation of DOE. This section presents an overview of this critical period, and summarizes the activities that preceded and followed the centralization of resources and responsibilities in DOE; parallel nonfederal developments that significantly affected these programs are also described to provide some historical context.

31.3.1 Nonfederal Activities before the Formation of DOE

Federal programs on nontechnical aspects of solar energy were initially greatly influenced by external political arguments concerning the merits of alternative energy sources. The work of Amory Levins was singularly important. His *Foreign Affairs* article (Lovins 1976) was the subject of a 1976 Senate hearing, where Lovins argued that an energy policy based on the "soft path"—a combination of energy efficiency and renewable energy sources—was both superior and mutually inconsistent with existing energy policies. Lovins explained that the major reason his preferred investments had yet to win out in the marketplace was

a wide array of institutional barriers, including more than 3,000 conflicting and often obsolete building codes, an innovation-resistant building industry, lack of mechanisms to ease the transition from kinds of work that we no longer need to kinds we do need, opposition by strong unions to schemes that would transfer jobs from their members to larger numbers of less "skilled" worker.... (Senate 1976)

The hearing includes critiques of the Lovins article by ERDA and the Congressional Research Service.

DOE subsequently initiated more detailed review of the article. For example, James Liverman of DOE wrote a preface to a report describing it as "undertaken largely as a response to the Amory Lovins article ... " (Craig and Levine 1978). Similarly, a 1980 report of the DOE Office of the Assistant Secretary for Environment begins with the explanation that it was prepared as a response to past articles and reports "extolling the advantages of the 'soft path' " (DOE 1980a).

Several other popular publications of the period exerted considerable influence on solar energy programs, including E. F. Schumacher, *Small Is Beautiful* (1973); Barry Commoner, *The Closing Circle* (1972); Denis Hayes, *Rays of Hope* (1977); and *Energy Future: The Report of the Energy Project at the Harvard Business School* (1979), edited by Stobaugh and Yergin. One distinguishing feature of these works was an emphasis on the social consequences of energy policy, including environmental, employment, and equity considerations. For example, Denis Hayes argued that energy sources are *not* neutral and interchangeable. Some energy sources are necessarily centralized; others are necessarily dispersed. Some are exceedingly vulnerable; others will reduce the number of people employed. Some will tend to diminish the gap between rich and poor; others will accentuate it. Some inherently dangerous sources can be permitted unchecked growth only under totalitarian regimes; others can lead to nothing more dangerous than a leaky roof. Some sources can be comprehended only by the world's most elite technicians; others can be assembled in remote villages using local labor and indigenous materials. In the long run, such considerations are likely to prove more important than the financial criteria that dominate and limit current energy thinking. (D. Hayes 1977, 25–26)

The environmental advantages of solar energy relative to conventional energy were of interest from the start of the major expansion in federal energy programs in 1974. The first government analysis of this issue may have been a report of a National Academy of Sciences Committee on Nuclear and Alternative Energy Systems (CONAES) done between 1976 and 1980. The study expressed tentative support for the relative environmental advantages of solar energy, noting the absence of data and, in some cases, understanding for several important issues (NRC 1979).

One study that expressed a critical view of the environmental consequences of renewable energy also received considerable attention. In a study originally prepared for the Atomic Energy Board of Canada, Harold Inhaber (1978) concluded that the health risks of renewables are greater than those of nuclear power and comparable to those of coal and oil; this report generated considerable critical response (see Holdren, Morris, and Mintzer 1980).

The argumentative and sometimes ideological presentation of environmental and labor issues of studies such as that by Inhaber greatly influenced subsequent public and private analysis. The substance of important analytical issues sometimes became inextricably associated with particular political viewpoints supportive of and opposed to solar energy.

31.3.2 Evolution of Federal Programs before DOE

Federal activities addressed to legal, environmental, and labor aspects of solar energy were premised on the very broadly stated rationale that nontechnical factors were important to the acceptance of the technology. The responsible agencies—particularly the Department of Energy and its predecessors—were left with a mandate to explore the seriousness of these issues and report on any needed reforms. In contrast, legislation addressed to several other issues discussed in other chapters was much more specific. For example, Congress established very specific limits on utility involvement in the distribution of solar energy systems to homeowners in section 216(a) of the National Energy Conservation Policy Act of 1978 (42 U.S.C. 8217[a]), subsequently amended by section 546(a)(2) of the Energy Security Act of 1980 (PL 96-294; see chapters 2 and 30.) Congress also directed the development of criteria and standards for residential solar heating equipment in the Solar Heating and Cooling Demonstration Act of 1974 (42 U.S.C. 5503[a]; see chapters 2, 9 and 14.)

Because congressional interest was so generally stated, much of the initial federal effort was oriented towards identifying issues and evaluating reform proposals. Few clear directions emerged initially; an Office of Technology Assessment (OTA) critique of early federal solar programs found a lack of adequate direction concerning commercialization policies and procedures (OTA 1975).

31.3.2.1 National Science Foundation Studies

The National Science Foundation was the principal supporter of solar energy research prior to the creation of ERDA in 1975 (Blisset 1979, 87). NSF studies played a valuable role in the early process of sorting out issues deserving more attention. The research was funded from broadly solicited proposals and accordingly reflects diverse objectives, rather than some clearly focused mission. The structure of the program actively promoted the involvement of new audiences in thinking about these issues, and interdisciplinary research was encouraged.

31.3.2.2 ERDA/FEA Studies

The Energy Research and Development Administration (ERDA) came in existence in 1975 and assumed responsibility for the solar programs developed by NSF. ERDA, in a program managed by Roger Bezdek, and the Federal Energy Administration (FEA), in a program managed by Norman Lutkefedder, undertook or funded many important solar legal, environmental, and labor studies, building on the efforts of NSF. FEA also began its exploration of these issues during the Project Independence report noted above. These agencies were in turn merged into the Department of Energy in 1977.

31.3.3   Nonfederal Activities after the Formation of DOE

Federal expenditures on solar energy grew rapidly in the mid-1970s and became a dominant influence on solar research efforts. However, several external developments also helped shape the federal program.

Sun Day, 3 May 1978, marked perhaps a high point in public enthusiasm about solar energy. President Carter and many other political officials made speeches, and thousands of public events took place across the United States and in thirty other countries (SL/CRR 1982). The interest in "going solar" was, in part, a reaction to its perceived environmental advantages, which became of even greater public concern after the Three Mile Island accident in 1979.

The oil price shock in 1979 also was a boost to solar energy. A focus on accelerating solar energy naturally led in part to questions about legal issues that might help or hinder the process.

The election of President Reagan and the opposition to solar energy programs he expressed was associated with an almost overnight turnaround in attitudes toward these activities. Cutbacks at DOE and SERI quickly had a major impact on activities at other levels of government and in the private sector.

Finally, the decline in oil prices in the early 1980s, accompanied by the expiration of federal tax credits in 1985, inflicted serious financial damage on many solar companies.

31.3.4   Federal Activities after the Formation of DOE

After 1978, analysis of legal, environmental, and labor issues became more in-depth but no less political. The importance of these issues was highlighted in the Domestic Policy Review (DPR) of Solar Energy Response Memorandum (DOE), prepared by an interagency task force in response to President Carter's 3 May 1978 Sun Day speech outlining the steps needed to accelerate solar. The environmental benefits of solar were noted and specifically calculated by the Environmental Protection Agency (EPA) in terms of reduced air pollution. The DPR also estimated large employment benefits on the basis of "limited data" and "crude approximations."

The creation of the Solar Energy Research Institute (SERI) also greatly increased the level of attention to nontechnical issues. Shortly after it began operations in 1977, SERI developed several major programs rele-

Table 31.3
SERI projects on nontechnical barriers to solar energy, 1980

Solar Law Reporter
National Solar Energy Education Directory
Solar manpower needs studies
Market penetration workshop
Solar warranties workshop
Implementation of state solar energy laws
Solar cost data bank
Public attitudes survey
Land use barriers and incentives
Impact on antitrust laws
Utility rates and service policies
Guide to organizing local solar events
Social values study
Use of the Internal Revenue Code to foster solar energy
Solar access pamphlet for communities
State information outreach programs

Source: Hayes 1980.

vant to nontechnical aspects of solar development. Denis Hayes, the institute's second director, summarized SERI's work in this field in a May 1980 presentation (see table 31.3).

31.4 Legal Studies

31.4.1 NSF-Funded Studies

Several NSF studies managed by Arthur Knopka and Laurance Rosenberg were among the earliest federal efforts to examine the likelihood of legal barriers to solar energy. In one project the American Bar Foundation conducted a workshop and subsequently prepared a handbook containing model statutes (Thomas 1975; Thomas, Miller, and Robbins 1978). The NSF studies very early established most of the issues that became the focus for more detailed research: regulation of buildings; financing and marketing arrangements, including warranty issues; public utility regulation; land use controls; and access to sunlight (Thomas 1976). The studies also identified much of the legal history relevant to these issues. Historical analysis is important because of the deference given by courts to past cases dealing with similar problems; for example, judicial review of litigation involving obstruction of access to light and air offers an analogy for potential cases involving interference with the use of solar collectors.

A 1975 NSF study illustrates another approach toward identifying the consequences of new solar technologies, based on detailed study of one community, in this case Colorado Springs (Phillips 1975). The study used aerial photography to analyze the extent of shading on suburban rooftops and concluded that setback and side yard requirements were adequate to avoid shading of rooftops under most circumstances. A second 1975 study, done in Long Island, New York, apparently with local funding, also used aerial photography and concluded that, in general, shading was not a problem for new buildings (Dubin 1975).

A study by the Environmental Law Institute (Hillhouse, Miller, et al. 1976; Hillhouse, Kohler, et al. 1977) explored in detail the implications of solar energy for Colorado; the authors reviewed legal and environmental issues, including access to sunlight, utility rates for solar users, and the possible incentive effect of environmental regulation for solar energy. Innovative aspects of the study included a look at the use of solar energy on public lands and an analysis of changes in urban form and land use planning that might follow from solar design.

NSF also supported what was probably the first extensive survey and analysis of state solar legislation, prepared by the National Conference of State Legislatures (NCSL 1975). Many other efforts to address state and local issues followed, often involving organizations like NCSL that represent state and local governments (see chapter 29).

Another Environmental Law Institute study funded by NSF looked at the potential role for state legislation promoting energy conservation, in many cases addressing issues such as building codes that overlapped with solar energy problems. The study generated several books on state strategies for promoting energy conservation in different sectors of the economy: building (Thompson 1980), agriculture (Friedrich 1978), industry (Dean 1979), land use (Harwood 1977), government procurement (Tether 1978); another book addressed the use of economic incentives to promote conservation (Russel 1979).

A Dow Chemical Company study, prepared with NSF support, examined the implications of energy industrial centers, where power generation and use might be centralized to achieve significant savings. The study included an extensive analysis of the effect of utility regulation on industries generating their own power, a subject of considerable interest to solar energy users (Dow Chemical et al. 1975).

Although NSF's leadership role on these issues was assumed by ERDA and FEA in 1975–1976, as noted below, NSF did support some relevant studies in later years, for example, a study by the Council of State Planning Agencies on renewable resource policy that examined the changes necessary in political and economic institutions to permit a sustainable economy. Another NSF study looked at the potential and limitations of state tax incentives for encouraging energy conservation and the use of renewable resources (Rodberg and Schacter 1980).

31.4.2 ERDA/FEA Studies

In 1976 ERDA commissioned state-of-the-art reviews of several aspects of nontechnical aspects of solar development. The Environmental Law Institute reviewed legal aspects of solar energy (Miller and Thompson 1977); the chapter headings in that report outline the legal issues that guided much subsequent research: solar access and land use, building codes, home financing, utility issues, mandatory installation, patent policy, antitrust issues, labor law applicable to union conflicts, property taxation, mobile home regulation, tort liability, insurance, and warranty issues.

These early legal studies emphasized the importance of understanding uncertainty created by the different legal systems in many areas. For example, the administration of building codes creates uncertainty for any new technology because of the discretion exercised by inspectors. Builders tend to minimize the risks of added paperwork and delay by waiting until informal rules have crystallized through repeated reviews (Miller and Thompson 1977). The application of some land use regulations, such as height controls and aesthetic reviews, creates similar disincentives. One response to this uncertainty was the development of uniform code provisions for solar energy systems for adoption by state and local authorities (see chapter 29).

Another type of uncertainty results from lack of operating experience. Mortgage lenders, for example, may insist that solar homes have an expensive backup heating system because they lack information concerning the long-term performance of solar energy systems (Barrett, Epstein, and Haar 1976). One answer is to reallocate the risks away from the consumer through warranties or third-party insurance, with or without government involvement.

ERDA, like NSF, supported legal research on state and local issues. A joint project with the City of Santa Clara examined the concept of a municipal utility as an organization to support distribution and maintenance of solar energy systems (Wilson, Jones, Morton & Lynch 1976).

ERDA and HUD jointly supported the National Solar Heating and Cooling Information Center (NSHCIC). The center distributed information on the status of state solar legislation and published some additional material on solar legal issues taken from other sources, such as the Environmental Law Institute review of legal barriers to solar energy and a New York forum on solar access (NSHCIC 1977). (See chapter 17 for a more detailed review of NSHCIC activities.)

The Federal Energy Administration conducted several early studies of legal and institutional obstacles to solar energy (e.g., FEA 1976). There was a period during 1975–1976 when FEA and ERDA both asserted jurisdiction over these issues, resulting in some duplication of effort.

31.4.3 Other Public and Private Legal Studies

The Department of Housing and Urban Development assessed the significance of legal and institutional issues through the experience of the demonstration program. A HUD program managed by David Engel issued numerous reports based on survey information, such as *Selling the Solar Home* (HUD 1978), and cooperated with DOE on several urban planning studies noted below.

HUD contracted with the Urban Institute to fulfill the congressional mandate for assessing the feasibility of federal mandatory energy efficiency standards applicable to existing residential buildings. The report focused on time-of-sale requirements and concluded that the benefits could be substantial but that the program was likely to be difficult and time-consuming to administer. Other approaches, such as utility audits and disclosure of energy costs, were adjudged more feasible but less effective (Lap, Carlson, and Dubinsky 1980).

The Federal Trade Commission (FTC) expressed interest at an early stage in the antitrust and competition issues related to solar commercialization. The commission conducted a symposium on these issues in 1977 that brought together numerous national authorities on legal and economic aspects of competition to discuss the solar market; the proceedings of this meeting provide an excellent overview of these issues (FTC 1978).

31.4.4 DOE-Sponsored Studies

The federal role in addressing legal issues after the formulation of the Department of Energy in 1978 became much more detailed and focused, although not without some duplication of prior efforts. The solar access problem was addressed in two book-length reports jointly funded by DOE and HUD. Each was prepared by organizations having previous federal support for research on solar issues.

The first report, *Solar Access Law* (G. Hayes 1979) prepared by the Environmental Law Institute, focused on the problem of solar access in existing communities. The report provides practical advice to communities on how to assess the seriousness of the problem, including model language for an ordinance to protect solar access through a notice, permit, and recordation scheme.

In the second report (Erley and Mosena 1980), the American Society of Planning Officials addressed the parallel problem of protecting solar access in new residential development, outlining choices for planners, consistent with other priorities, to assure access to sunlight as one criterion in development. The authors describe means to accomplish this objective through traditional zoning, subdivision controls, and vegetation restrictions.

The feasibility of solar access protection was tested in several government-supported studies. SERI supported an analysis of the feasibility of a concept known as "solar envelope zoning," developed by architect Ralph Knowles. SERI contracted with the City of Los Angeles to apply the method to city neighborhoods. The city concluded that the system was effective and fair but complex to administer (Los Angeles 1980).

A study of solar access and planning issues in Philadelphia (Charles Burnette 1980), done with support from DOE and the National Endowment for the Arts, found that it was practical to protect rooftop access to sunlight in large parts of that city. A Boston study supported by DOE (Shapiro 1980) similarly concluded that up to half of older urban areas were suitable for solar heating applications.

Another DOE approach involved meetings and studies directed to the concerns of specific audiences. For example, workshops and a background manual entitled *Technical, Economic, and Legal Considerations for Evaluating Solar Heated Buildings* targeted lenders, appraisers, insurers, and tax consultants (DOE 1979b).

DOE supported a guide to county energy planning (Woolson 1981) prepared by the National Association of Counties that also addressed renewable energy legal and financial issues. The report describes the economic development benefit of county-based energy planning emphasizing conservation and renewables, a concept promoted by former DOE official James Benson (Okagaki and Benson 1979).

DOE also supported some analyses of nontechnical barriers to solar energy in an international setting. An international symposium was held on the subject in Brussels in 1980, cosponsored by the Commission of the European Communities and the Canadian Department of Energy (Strub and Steemers 1980).

SERI reviewed solar legal issues within its Analysis and Assessment Division (SERI 1978). Probably the most substantial undertaking, and arguably, also the most effective, was the bimonthly publication *Solar Law Reporter*, beginning in May–June 1979 and ending after three volumes in 1982. Publication of the *Reporter* served three important functions. First, a great deal of information was made accessible to the general legal community and local level officials. Although many of the initial articles were surveys or summaries of government reports, this information had not been readily accessible. Some state court decisions related to solar energy were also difficult to track down prior to the publication of the *Reporter*. Second, the existence of the *Reporter* as a forum for discussion of solar legal issues encouraged more thoughtful analysis. Later issues showed a trend away from reporting in favor of more analytical articles, such as problems associated with drafting a solar access ordinance (Danielson 1981), a discussion of the relative merits of loans versus rebates (Gardels 1981), and an analysis of methods for calculating payments to be made to independent power producers (Locke 1981). And third, the *Reporter* had begun to emerge as an important institution at the time of its demise. Students and professors interested in solar legal issues were able to affiliate with the *Reporter*, a source of some credibility as well as resources for relevant research.

The regional solar energy centers (RSECs) were created with an explicit emphasis on more geographically focused commercialization activities and therefore tended to include building codes and other legislative issues in their agendas (Blisset 1979). For example, the Northeast Solar Energy

Center (NESEC) prepared a lengthy review of barriers and incentives to solar energy, reviewing specific laws in the states it served (Wallenstein 1978).

31.4.5 Other Federal Facilitative Efforts

Apart from the studies, meetings, and other outreach efforts already described, the federal role included some activities addressed to legal issues with more direct impact on the development of the solar market. As noted above, the Federal Trade Commission (FTC) reviewed solar competition and antitrust issues, including the implications of oil company investment in solar energy business (FTC 1978).

The Department of Justice intervened on behalf of a solar homeowner in *Prah v. Maretti* (108 Wis. 2d 223[1982]), a Wisconsin Supreme Court case that addressed the right of a solar energy user to bring a nuisance suit against a neighbor for interfering with his sunlight. The suit was decided in favor of the solar user.

The Small Business Administration (SBA) also actively reviewed competition issues. Unlike FTC, SBA focused specifically on the importance of preserving the innovative role of small firms (Stambler and Stambler 1982). SBA has continued to express related concerns about utility energy conservation programs (SBA 1984).

31.5 Labor Issues

Few detailed studies were undertaken on labor aspects of solar energy development prior to 1978. One early concern of solar nontechnical studies was that union conflicts might be a barrier to solar energy because the installation requirement cut across the union job definitions dividing plumbers, sheet metal workers, and so on (Miller and Thompson 1977). A related concern was an interest in licensing installers of solar equipment by some local and state governments (Blisset 1979). This was promoted as a means of encouraging quality and promoting consumer confidence, but there was also the risk of supporting one craft over another and adding costs if unnecessary skills were required.

The Sheet Metal Workers Union showed an early interest in solar energy as a source of jobs, commissioning the Mitre Corporation to conduct an analysis of employment opportunities (Cohen et al. 1975). The union subsequently undertook solar training programs in cooperation with state and federal agencies (Blisset 1979).

SERI produced numerous reports on solar employment requirements, union acceptance, and solar training programs. SERI's analysis of labor requirements was directed toward improving available models and methodologies to improve on the "piecemeal and preliminary" nature of much prior work (Ferris and Mason 1979). SERI was not itself engaged in training programs but did support them through evaluation efforts, including a California effort to train CETA workers for jobs in the solar industry (Burns, Mason, and Mikasa 1980). SERI also sponsored activities to encourage unions' interest in solar energy and in turn to understand their concerns (Livingston and Moran 1979).

Environmentalists for Full Employment (EFFE), formed in 1975, made analysis of the employment benefits of solar energy a major focus (Grossman and Daneker 1979). EFFE's expressed mission was to show unions the benefits of supporting renewable energy over fossil fuels and nuclear power and thereby to forge an effective political alliance.

Much of the analysis of solar labor issues had political overtones insofar as it was used to promote one energy source over another to justify government funding of a particular program (Keegan 1979). An EFFE (1980) report for DOE argued that solar energy systems create more jobs than conventional fuels, that job creation should be an explicit criterion for energy policy, and that the Department of Commerce and the Department of Labor were overstating the methodological obstacles to energy employment studies (see also Rodberg 1980).

Employment issues were addressed in a 1983 congressional hearing on the low-income weatherization program. A DOE-funded analysis presented at the hearing estimated the jobs created directly by proposed government expenditure but did not assess other energy sources (House 1983, 144).

The debate extended outside DOE to other agencies and forums. A report for the Coalition of Northeastern Governors, funded by the Economic Development Administration, argued that the northeastern states could boost their economies and create tens of thousands of jobs by redirecting funds from nonindigenous energy sources—coal, oil, and natural gas—to locally obtained renewable fuels and conservation (Scales and Popkin 1983).

A similar perspective is apparent in a privately funded report of the Council on Economic Priorities, *Jobs and Energy* (Buschbaum and Benson 1979). This study compared the jobs per dollar of investment from conservation and solar with the jobs from a proposed nuclear power plant on Long Island and concluded that twice as much employment would result from the former.

The public record and informal interviews by the author disclosed no formal conclusions by DOE on the relative effects of solar energy. This may reflect difficulties in coordinating DOE and DOL. The methodological issues fell outside DOE's jurisdiction and were complicated; the assumed cost of solar energy systems tended to have a major impact on the results of some studies because much of the potential employment benefits came as a result of the fuel savings assumed to be spent locally on other commodities (Keegan 1979). There also may have been political resistance to a governmental statement that solar energy is preferable on employment grounds.

Some solar labor issues were less controversial. For example, a DOE-funded Battelle survey of solar research firms and businesses in 1978 (Levy and Field 1980) projected that solar employment would probably triple by 1983. DOE also provided substantial support to Navarro College for a solar manpower assessment and for development of a curriculum for solar training (Orsak et al. 1978; see also Barker 1978; Rodberg 1980).

DOE also supported conferences and manuals on the benefits of solar energy for "community building" and local job creation (see, for example, Totten et al. 1980). Some projects incorporating employment and environmental issues were framed around city or regional planning perspectives (e.g., Mara 1984).

31.6   Environmental Studies

31.6.1   ERDA-Sponsored Studies

Although the environmental implications of solar energy were noted in some NSF studies (e.g., Hillhouse, Miller, et al. 1976), they became a major focus of ERDA. Technical reviews were done to estimate requirements for materials, land, and other resource associated with solar energy systems (see, for example, ERDA 1976; Brannon et al. 1977). These reviews reflected the considerable uncertainty associated with possible technical developments in solar equipment but in general identified solar heating systems as very attractive from an environmental standpoint.

ERDA also supported environmental analysis in the context of comparative assessments of the social implications of different energy policies. A draft study of these issues by the Stanford Research Institute (SRI 1977) attracted considerable attention, including a congressional hearing (Senate 1976). The SRI report was critical of government energy policy as expressed in a national energy plan, *ERDA-49*, in part on environmental grounds. Solar energy was described as a practical alternative, more expensive but capable of cost reduction and vastly superior on environmental grounds. Accelerated development of more conventional fuels, the draft report stated, would be "extremely costly" both economically and environmentally.

The report's emphasis on the importance of government intervention echoed the arguments of Lovins and Hayes noted at the outset of this chapter and assured some controversy. Energy policy, the report stated, "cannot be left solely to normal economic processes ... society's decision with regard to the future role of solar energy is best considered as part of a broad societal choice involving much more than selection of an energy technology." (SRI 1977, VI-6).

31.6.2 DOE-Sponsored Studies

A 1980 DOE survey reported more than 150 projects and $25 million of federal government expenditure directed to environmental and safety research on solar energy (DOE 1980b). However, few studies can be found at a broad policy level; most addressed narrow technical questions, such as the toxicity of certain substances. For example, Brookhaven National Laboratory undertook a major effort over several years to identify any major technical or environmental barriers to photovoltaic energy systems.

A two-year SERI study based on a large macroeconomic input-output model concluded that "rapid deployment of solar energy would make a significant net contribution to environmental quality during the 1975–2000 period" (Yokell 1980). Yokell noted several significant sources of uncertainty, notably the need to assign costs to environmental damages.

Another major study (Holdren, Morris, and Mintzer 1980) reviewed all extant literature (the bibliography includes 198 references) and applied several different approaches in order to assess the relative environmental impact of renewable energy. The authors concluded that

Notwithstanding the gaps, uncertainties, and apples-and-oranges problems ... [we tentatively conclude that] some of the renewables have particular promise for reducing environmental costs, per unit of energy delivered, to well below the levels that have been associated with the use of oil land coal. (Holdren et al. 1980, 283)

One caveat of this study was the need for much more analysis of certain environmental risks, such as the fire hazards of rooftop collectors.

Another program on environmental issues supported by DOE considered likely land use changes associated with solar energy and the possible environmental consequences. For example, one concern was that solar access might require lower density residential development, which might in turn lead to environmentally undesirable urban sprawl. One DOE study (Twiss 1980) found that typical land use densities could be maintained and were consistent with the maximum solar penetration goals stated in the Domestic Policy Review of Solar Energy.

An article in the *Solar Law Reporter* by SERI staff (O'Brien and Euser 1980) proposed the use of solar energy to offset air pollution from conventional fuels. Under regulations of the Environmental Protection Agency (EPA), this might allow a solar-powered plant to be built or expanded in an area not in compliance with Clean Air Act requirements and therefore not otherwise permitted to allow development.

As noted above, environmental critiques of energy sources were politically sensitive because they could be construed to support one energy strategy over another. Sensitivity to these issues is apparent in a 1980 DOE report that argues that "on inspection, solar and conservation strategies may not be environmentally benign" (DOE 1980a). The report cites the indoor air pollution possible with tightly sealed solar houses, the pollution released by wood stoves, and the risk of erosion from increased use of biomass. The same concern was raised earlier in the Domestic Policy Review. Four years later the secretary of energy's *Annual Report to Congress* (DOE 1984, 107) would note the general conclusion of a solar technology assessment: an accelerated solar and biomass policy would cause "major stresses on national capital and finished materials resources as well as the possibility of significant air pollution and safety problems."

Peter House of the Office of Environmental Assessments took another approach (House et al. 1981), arguing that "for technical reasons, analysis can provide no definitive or rationally credible answers to the question of

overall safety." Because of inherent—and insurmountable—methodological problems, "comparative analysis cannot form the basis of credible,
viable energy policy." The House report advocates an energy policy that
leaves environmental questions to be addressed as needed for each energy
source, without broader comparison.

31.7   Results of Federal Studies

31.7.1   Pre-DOE Studies

Pre-1978 studies of nontechnical solar issues were speculative by design—
an effort to anticipate problems in advance of widespread experience.
Nevertheless, the conclusions and recommendations advanced appear in
many cases surprisingly detailed and pertinent a decade or more later.
Despite their speculative nature, early analyses were also of some practical significance, particularly on legal issues. For example, an American
Bar Foundation report (Thomas, Miller, and Robbins 1978) was one of
the authorities cited by the Wisconsin Supreme Court in a decision up-
holding a homeowner's right to bring a nuisance suit against a neighbor
blocking his access to sunlight (*Prah v. Maretti*, 108 Wis. 2d 223 [1982]).

   On the other hand, federally funded research was not the sole source of
research on these issues. Many articles appeared by experts and students
interested in the exciting new field of solar energy (see *Solar Law Reporter*
bibliography, November–December 1981). The solar access issue alone
generated many analyses even before 1977, although most of these works
were much less ambitious and often not widely available.

   Careful legal analysis by the American Bar Foundation (ABF), the
Environmental Law Institute, and other recognized institutions helped
channel enthusiasm for solar energy toward practical proposals and away
from hasty and ill-thought ideas. As the ABF report states, "Legislators,
anxious to prove their understanding of the problems and their ability to
'get things done' may be tempted to support any legislation that appears
reasonable (Thomas, Miller, and Robbins 1978). Thus one important
contribution of these reports may be as much what they helped to avoid
as what they achieved. The author believes that, for better or worse,
criticism of a 1977 New Mexico solar access protection statute developed
and communicated in federally supported reports (see G. Hayes 1979) no
doubt helped discourage other states from adopting similar measures
proposed in several states.

These early reports also led to more refined efforts by federal agencies, state and local governments, and other organizations. Subsequent work was in general much more carefully defined, although there was some repetition of effort. Another benefit was the creation of a large group of experts, many of whom later did valuable work without governmental support.

31.7.2 DOE Studies

Government programs after 1978 made substantial progress toward defining problems and solutions and building relationships with states, local governments, and the private sector. Some fairly well defined answers had been reached for many legal issues, although the process of implementation in many areas remained for other levels of government. Solar technologies were also almost universally rated environmentally preferable; the exception was the Inhaber (1978) analysis, which assumed the risks of solar energy were additive to those of conventional fuels due to the need for backup systems.

Quantifying the environmental benefit, however, proved to be a deceptively complex problem. Labor issues proved more difficult to resolve, but they were narrowed, and work plans for devising needed methodologies were developed or begun. The difficulty of separating analysis from ideology caused some labor and environmental research to founder, although there were also impressive efforts to distinguish the two, such as the analysis of environmental issues by Holdren, Morris, and Mintzer (1980).

Research on the potential for promoting community development with solar energy offered some hope for combining technical, environmental, and labor analysis with efforts to achieve broader social policy goals (Mara 1984). An Urban Institute analysis (Bechhoefer et al. 1980) concluded that neighborhood organizations were responsible for a relatively high proportion of successful residential conservation programs. However, federal support for research on this set of issues no longer exists.

31.8 Lessons Learned

Unlike demonstration projects or research programs, nontechnical activities for the most part lack hard, quantifiable goals. These activities can only be judged in the larger context of efforts to support accelerated solar development through financial incentives, demonstration, and federal procurement. Nonetheless, the underlying problems the government sought to address were real and important, as reflected in the substantial amount of independent activity by other levels of government and private groups.

Much of the federal effort might be characterized as a search for problems. "There is a danger ... of becoming so obsessed with barriers that we lose sight of, and then lose faith in, the real purpose that set us off" (D. Hayes 1980). Surprisingly, however, this concern was largely unjustified. Most legal analyses noted the absence of problems for many areas and carefully restricted their recommendations to particular locales.

Even though the federal government lost interest in the mandate that supported most of the programs described in this chapter, there is little evidence that the original concept was in error. DOE's (1982) "sunset review" of the solar program concluded that the change in federal solar priorities was justified by changing circumstance; the rise in oil prices and initial demonstration efforts had provided enough of a boost to get the industry moving. The DOE report also suggests that the experience to that time was sufficient to resolve basic questions about legal, labor, and environmental issues. In contrast, a report the same year by the agency's Energy Research Advisory Board (ERAB 1982) concluded that assessments of environmental issues and training of personnel in solar technologies should remain important federal responsibilities.

In any case, the issues have not gone away. For example, a National Academy of Sciences report (Stern and Aronson 1984) reiterated the importance of noneconomic determinants of decisions about energy, urging increased attention to understanding the range of factors that influence consumers. The issue of frameworks for making environmental comparisons of alternative energy choices continues to be actively debated (Weil 1991), comparative job creation is still promoted by solar advocates (Renner 1991), and solar ordinances are still being adopted (Kale 1989).

The *Solar Law Reporter* project deserves some specific discussion because of the large commitment of federal funds to a largely academic journal. That no private sponsorship emerged after three volumes may be viewed as an indication of its failure by market standards; however, there are several countervailing arguments. Some initial period of federal sponsorship may have been warranted to help accelerate the flow of information during the critical, formative period for the solar industry, when uncertainty was greatest. Some features of the *Reporter* were also picked up by other publications; for example, *Solar Age* began publishing detailed information on state solar tax policies.

Taking a broader view, the *Reporter* also provided benefits that might not economically justify private publication but that were nevertheless substantial. The publication was responsive to the high information cost problem, which was widely perceived as a major barrier to solar development. There may not be a market for better solar laws and policies, despite their effectiveness.

31.8.1 Conclusions

There are three major lessons that should be considered in any future federal effort to address legal, environmental, and labor issues:

1. Federal efforts prior to the formation of the Department of Energy suffered from a lack of coordination, focus, and leadership. Labor issues never became the mandate of a single agency and were never handled well.

2. Federal efforts properly emphasized outreach to important professional associations, trade groups, unions, and other relevant organizations. An impressive network was built in a very short period of time, and a wide range of skills and experience were brought together in support of a national goal.

3. Federally supported efforts indirectly encouraged parallel efforts by other levels of government and the private sector; as federal efforts increased after 1975, so did others. Federal activities probably encouraged some of these projects by providing necessary background information and analytical models.

The efforts reviewed in this chapter demonstrate that some important social issues, such as the environmental and employment implications of energy sources, cannot be divorced from policy questions. This unwelcome fact will no doubt continue to be a source of discomfort for governments to come.

References

Barker, H. 1978. *Investigation of Labor, Manpower, and Training Requirements for Selected Solar Applications, Technical Progress Report.* Golden, CO: Solar Energy Research Institute.

Barrett, D., P. Epstein, and C. Haar. 1976. *Financing the Solar Home: Understanding and Improving Mortgage Market Receptivity to Energy Conservation and Housing Innovation.* Cambridge, MA: Regional and Urban Planning Institute.

Bechhoefer, A., R. Kirby, C. Maurer, and G. Miller. 1980. *Institutional Response to Energy Alternatives.* Washington, DC: Urban Institute.

Blisset, M. 1979. *Toward a Solar America: An Institutional Assessment of On-Site Solar Technologies.* Austin, TX: Lyndon B. Johnson School of Public Affairs.

Brannon, P., H. Church, R. Luna, and W. Thomas. 1977. *The Environmental Issues Associated with Solar Heating and Cooling Residential Dwellings.* Albuquerque, NM: Sandia National Laboratories.

Burns, B., B. Mason, and G. Mikasa. 1980. *Assessment of the Labor Market Experiences of CETA-Trained Solar Workers.* Golden, CO: Solar Energy Research Institute.

Buschbaum, S., and J. Benson. 1979. *Jobs and Energy: The Employment and Economic Impacts of Nuclear Power, Conservation, and Other Energy Options.* New York: Council on Economic Priorities .

Charles Burnette & Associates. 1980. *Philadelphia Solar Planning, Working Papers.* Philadelphia.

Cohen, A., M. Harlow, A. Johnson, and P. Spewak. 1975. *Impact of Energy Developments on the Sheet Metal Industry.* Washington, DC: Mitre Corporation.

Commoner, B. 1972. *The Closing Circle: Nature, Man & Technology.* New York: Alfred A. Knopf.

Craig, P., and M. Levine. 1978. *Distributed Energy Systems and California's Future: Issues in Transition.* Berkeley: University of California.

Danielson, L. 1981. "Drafting a Solar Access Ordinance: One City's Experience." Solar Law Reporter 3:911–965.

Dean, N. 1979. *Energy Efficiency in Industry.* Cambridge, MA: Ballinger.

DOE (U.S. Department of Energy). 1977. *National Energy Policy Plan.* Washington, DC.

DOE. 1979a. *National Energy Policy Plan.* Washington, DC.

DOE. 1979b. *Technical, Economic, and Legal Considerations for Evaluating Solar-Heated Buildings.* Washington, DC.

DOE. 1980a. *Solar Technologies: Are They Environmentally Benign?* Washington, DC.

DOE. 1980b. *Inventory of Federal Energy-Related Environment and Safety Research for FY 1979.* Oak Ridge, TN: DOE Technical Information Center.

DOE. 1982. *Report to the Congress. Department of Energy Organization Act, Title X: Sunset Review Program-by-Program Analysis.* Vol. 2. Washington, DC.

DOE. 1984. Secretary of Energy. *Annual Report to Congress.* Washington, DC.

DOE. Domestic Policy Review of Solar Energy: A Response Memorandum to the President of the United States. February 1979. Washington, DC. Available from National Technical Information Service, U.S. Department of Commerce.

Dow Chemical Company, Environmental Research Institute of Michigan, Townsend-Greenspan & Co., and Cravath, Swaine & Moore. June 1975. *Energy Industrial Center Study.* Report to the National Science Foundation. No place of publication available. Prepared for the Office of Energy Research & Development Policy, National Science Foundation.

Dubin, F. 1975. *Analysis of Energy Usage on Long Island from 1975 to 1995: The Opportunities to Reduce Peak Electrical Demands and Energy Consumption by Energy Conservation, Solar Energy. and Total Energy Systems.* Suffolk County, NY: Suffolk County Department of Environment Control.

EFFE (Environmentalists for Full Employment). 1980. *An Evaluation of National Energy and Employment Policy*. Washington, DC.

ERAB (Energy Research Advisory Board). 1982. *Solar Energy Research and Development: Federal and Private Sector Roles*. Washington, DC: U.S. Department of Energy.

ERDA (Energy Research and Development Administration). 1975. *A National Plan for Energy Research, Development and Demonstration: Creating Energy Choices for the Future*. Vol. 1. *The Plan*. ERDA-48. Washington, DC.

ERDA. 1976. *Summary Report: Solar Energy Environmental and Resource Assessment Program*. ERDA 76-138. Washington, DC.

ERDA. 1977. *The Need for Deployment of Inexhaustible Energy Resource Technologies*. Report of Inexhaustible Energy Resources Planning Study (IERPS). Washington, DC.

Erley D., and D. Mosena. 1980. *Energy Conserving Development Regulations: Current Practice*. Chicago: American Planning Association.

Ezra, A. 1975. "Technology Utilization: Incentives and Solar Energy." *Science* 187:707–713.

FEA (Federal Energy Administration). Region IX. 1976. *The Legal and Institutional Barriers to Solar Development: A Summary of the Issues*. San Francisco.

Ferris, B., and B. Mason. 1979. *A Review of Regional Economic Models with Special Reference to Labor Impact Assessment*. Golden, CO: Solar Energy Research Institute.

Friedrich, R. 1978. *Energy Conservation for American Agriculture*. Cambridge, MA: Ballinger.

FTC (Federal Trade Commission). 1978. *The Solar Market: Proceedings of a Symposium on Competition in the Solar Energy Industry*. Washington, DC.

Gardels, C. 1981. "Utility Financing: Loans or Rebates?" *Solar Law Reporter* 3:477–486.

Grossman, R., and G. Daneker. 1979. *Energy, Jobs, and the Economy*. Boston: Alyson.

Harwood, C. 1977. *Using Land to Save Energy*. Cambridge, MA: Ballinger.

Hayes, D. 1977. *Rays of Hope*. New York: Norton.

Hayes, D. 1980. "Overcoming Solar Barriers: Progress in the United States." In *Non-Technical Obstacles to the Use of Solar Energy*, ed. A. Strub and T. Steemers, 2–12. New York: Harwood Academic.

Hayes, G. 1979. *Solar Access Law*. Cambridge, MA: Ballinger.

Hillhouse, K., E. Kohler, R. Liroff, and A. Miller. 1977. *Legal and Institutional Perspectives on Solar Energy in Colorado: A Case Study of Land Use and Energy Decision Making*. Washington, DC: Environmental Law Institute.

Hillhouse, K., A. Miller, R. Liroff, E. Kohler, and G. Thompson. 1976. *Solar Energy and Land Use in Colorado*. Interim report to the National Science Foundation. Washington, DC: Environmental Law Institute.

Hirshberg, A., and R. Schoen. 1974. "Barriers to the Widespread Utilization of Residential Solar Energy: The Prospects for Solar Energy in the U.S. Housing Industry." *Policy Sciences* 5:453–468.

Holdren, J., G. Morris, and I. Mintzer. 1980. "Environmental Aspects of Renewable Energy Sources." *Annual Review of Energy* 5:241–291.

House, P., J. Coleman, R. Shull, R. Matheny, and J. Hock. 1981. *Comparing Energy Technology Alternatives from an Environmental Perspective*. Washington, DC: U.S. Department of Energy.

HUD (U.S. Department of Housing and Urban Development). 1978. *Selling the Solar Home*. Washington, DC.

Inhaber, H., 1978. "Risks with Energy from Conventional and Nonconventional Sources." *Science* 203:718–723.

Kale, S. 1989. "Resource Assessment of Solar Access Ordinances in the Pacific Northwest." Paper prepared for the Bonneville Power Administration. Corvallis: Department of Geography, University of Oregon.

Keegan, K. 1979. "Employment is the Name of the Game as Solar Advocates Press Their Case." *National Journal* 11:2100–2103.

Lap, L., D. Carlson, and R. Dubinsky. 1980. *Potential Effects and Administrative Feasibility of Energy Conservation Standards for Existing Homes.* Washington, DC: Urban Institute.

Levy, G., and J. Field. 1980. *Solar Energy Employment and Requirements, 1978–1983.* Columbus, OH: Battelle Columbus Laboratories.

Livingston, R., and D. Moran. 1979. *Organized Labor and Solar Energy.* Golden CO: Solar Energy Research Institute.

Locke, R. 1981. "Statewide Purchase Rates under Section 210 of PURPA." *Solar Law Reporter* 3:419–452.

Los Angeles. 1980. *Solar Envelope Zoning: Application to the City Planning Process, Los Angeles Case Study.* Los Angeles: Mayor's Office, Department of City Planning .

Lovins, A. 1976. "Energy Strategy: The Road Not Taken?" *Foreign Affairs* 55:65–96.

Mara, G. 1984. *Renewable Energy in Cities.* New York: Van Nostrand Rheinhold.

Miller, A., and G. Thompson. 1977. *Legal Barriers to Solar Heating and Cooling of Buildings.* Washington, DC: Environmental Law Institute.

NAS (National Academy of Sciences). 1975. *Establishment of a Solar Energy Research Institute.* Washington, DC.

NCSL (National Conference of State Legislatures). 1975. *Turning toward the Sun.* Renewable Energy Project. Denver.

NRC (U.S. Nuclear Regulatory Commission). 1979. *Energy in Transition: 1985–2010.* San Francisco: W. H. Freeman.

NSF/NASA (National Science Foundation and National Aeronautics and Space Administration). 1972. *An Assessment of Solar Energy as a National Energy Resource.* Washington, DC: National Science Foundation.

NSHCIC (National Solar Heating and Cooling Information Center). 1977. *A Forum on Solar Access.* Rockville, MD.

O'Brien, K., and B. Euser. 1980. "Solar Energy and the Search for Emission Offsets: A Proposal." *Solar Law Reporter* 1:1076–1093.

Okagaki, A., and J. Benson. 1979. *County Energy Plan Guidebook: Creating a Renewable Energy Future.* Fairfax, VA: Institute for Ecological Policies.

Orsak, C., R. Barnstone, H. Gibson, Y. Jani, and J. Morehouse. 1978. *An Assessment of Need for Developing and Implementing Technical and Skilled Worker Training for the Solar Energy Industry.* Corsicana, TX: Navarro Junior College.

OTA (Office of Technology Assessment). 1975. *Analysis of the ERDA Plan and Program.* Washington, DC.

Phillips, J. 1976. *Assessment of a Single Family Residence Solar Heating System in A Suburban Development Setting.* Annual research report to the National Science Foundation. Colorado Springs: Department of Public Utilities.

Project Independence. 1974. Final Report of the Solar Energy Panel. Washington, DC: Federal Energy Administration, November.

Renner, M. 1991. *Jobs In a Sustainable Economy.* Washington, DC: Worldwatch Institute.

Rodberg, L. 1980. *Energy and Employment: A Review and Commentary*. Background paper prepared for the National Panel on Energy and Employment Policy. Washington, DC: Environmentalists for Full Employment.

Rodberg, L., and M. Schacter. 1980. *State Conservation and Solar Energy Tax Programs: Incentives or Windfalls?* Washington, DC: Council of State Planning Agencies.

Russel, J. 1979. *Economic Incentives for Energy Conservation*. Cambridge, Mass: Ballinger.

SBA (Small Business Administration). Office of Advocacy. 1984 . *Utility Competition with Small Business: Recommendations for States on Utility Energy-Related Programs and the Commercial and Apartment Conservation Service Program*. Washington, DC.

Scales, J., and J. Popkin. 1983. *Energy and Jobs: Employment Implications of Alternate Energy Development and Conservation Strategies in the Northeast*. Council of Northeastern Governors CONEG Policy Research Center, Inc.

Schumacher, E. F. 1973. *Small is Beautiful: Economics as if People Mattered.* New York: Harper & Row.

SERI (Solar Energy Research Institute). 1978. *Law and Solar Energy: A Meeting of Solar Energy Research Institute and the American Bar Association's Special Committee on Energy Law*. Golden, CO.

Shapiro, M. 1980. *Boston Solar Retrofits: Studies of Solar Access and Economics*. Cambridge: Energy and Environmental Policy Center, Harvard University.

SL/CRR (Solar Lobby and Center for Renewable Resources). 1982. *The Solar Agenda: Progress and Prospects*. Washington, DC: Solar Lobby.

SRI (Stanford Research Institute). 1977. *Solar Energy in America's Future: A Preliminary Assessment*. Palo Alto, CA.

Stambler, B., and L. Stambler. 1982. *Competition in the Photovoltaics Industry: A Question of Balance*. Washington, DC: Center for Renewable Resources.

Stern, P., and E. Aronson, eds. 1984. *Energy Use: The Human Dimension*. New York: W. H. Freeman.

Stobaugh, R., and D. Yergin, eds. 1979. *Energy Future: Report of the Energy Project at the Harvard Business School*. New York: Random House.

Strub, A., and Steemers, T. 1980. *Non-Technical Obstacles to the Use of Solar Energy*. New York: Harwood.

Thomas, W. 1975. *Proceedings of the Workshop on Solar Energy and the Law*. Chicago: American Bar Foundation.

Thomas, W. 1976. *Legal Issues Related to the Use of Solar Energy Systems*. Chicago: American Bar Foundation.

Thomas, W., A. Miller, and R. Robbins, 1978. *Overcoming Legal Uncertainties about the Use of Solar Energy Systems*. Chicago: American Bar Foundation.

Thompson, G. 1980. *Building to Save Energy*. Cambridge, MA: Ballinger.

Totten, M. N., B. Glass, M. Freedberg, and L. Webb, 1980. *Local Alternative Energy Futures*. Austin, TX: Conference on Alternative State and Local Policies.

Twiss, R. 1980. *Land Use and Environmental Impacts of Decentralized Solar Energy Use*. Berkeley: University of California.

U.S. House. Committee on Science and Aeronautics. 1974. *Solar Heating and Cooling Demonstration Act of 1973*. H. Rpt. 93-769.

U.S. House. Subcommittee on Energy Conservation and Power. 1983. *Hearings on Energy Conservation and Jobs*. Serial 98-15.

U.S. Senate. Committee on Banking, Housing, and Urban Affairs and Senate Committee on Labor and Public Welfare. 1974. *Solar Heating and Cooling Demonstration Act of 1974.* S. Rpt. 93-847.

U.S. Senate. Committee on Small Business and Senate Committee on Interior and Insular Affairs. 1976. *Joint Hearings on Alternative Long-Range Energy Strategies.*

Wallenstein, A. 1978. *Barriers and Incentives to Solar Energy Development: An Analysis of Legal Institutional Issues in the Northeast.* Cambridge, MA: Northeast Solar Energy Center.

Weil, S. 1991. "The New Environmental Accounting: A Status Report." *Electricity Journal* 4:46–54.

Wilson, Jones, Morton & Lynch. 1976. *The Sun: A Municipal Utility Energy Source.* Santa Clara, CA.

Woolson, A. 1981. *The County Energy Production Handbook.* Washington, DC: National Association of Counties Research Foundation.

Yap, L., D. Carlson, and R. Dubinsky. 1980. *Potential Effects and Administrative Feasibility of Energy Conservation Standards for Homes.* Washington, DC: Urban Institute.

Yokell, M. 1980. *Environmental Benefits and Costs of Solar Energy.* Lexington, MA: Lexington Books.

Contributors

Donald E. Anderson

Donald E. Anderson is President and CEO of Rho Delta Inc., Mankato, Minnesota, and is on the board of directors of several corporations. He is also professor emeritus of electrical engineering at Mankato State University. He has electrical engineering degrees through the Ph.D. from the University of Minnesota, Minneapolis, and was on the faculty there. Dr. Anderson has thirteen patents in solar and electronics technology.

From 1978 through 1981, he was President and Executive Director of the Mid-America Solar Energy Complex (MASEC), one of the four regional solar energy centers established under contracts with the U.S. Department of Energy.

J. Douglas Balcomb

Principal engineer at the National Renewable Energy Laboratory, Dr. J. Douglas Balcomb has been working on the quantification and evaluation of passive solar heating techniques since 1975. He organized the solar research team at the Los Alamos National Laboratory and has been very active in the leadership of both the International Solar Energy Society and the American Solar Energy Society. He has monitored numerous buildings, conducted original research on natural convective air-flow, published 120 technical papers and several books, lectured extensively including fifty-four seminars in twenty-eight countries, consulted on the design of about sixty buildings, participated on International Energy Agency and United Nations projects, and received five awards for his contributions to the field. He did his under graduate work at the University of New Mexico and received his Ph.D. in nuclear engineering from the Massachusetts Institute of Technology.

 **Kenneth R. Bordner**

Ken Bordner is currently a Principal Associate in KRB Associates, an information technology consulting group. He was formerly the Director of the National Solar Heating and Cooling Information Center (NSHCIC). He received a bachelor of science degree in commerce and engineering from Drexel University and has been involved with the design, development, and implementation of information systems for over thirty years. NSHCIC was a federally funded program, operated at the Franklin Research Center in Philadelphia, to collect, organize, and disseminate information concerning the status of and uses for solar energy research and development during the 1970s and 1980s. Mr. Bordner was instrumental in developing a computer-based inquiry and response system to handle the thousands of requests for solar energy information handled by the Center each year. He was a speaker and participant at many solar energy-related conferences and was selected to testify before Congress on the role in information systems in government energy research and development.

Barry Butler

Dr. Barry Butler is a material scientist with training in the structure property relationships of ceramics (earning his bachelor's degree in ceramics from Alfred University), metals (having received a master's degree in metallurgy from Rensselaer Polytechnic Institute), and polymers (receiving a Ph.D. in materials science from RPI). He managed the Solar Materials Program at Sandia National Laboratories, Albuquerque, then went to the Solar Energy Research Institute (SERI), now called National Renewable Energy Laboratory (NREL), to guide the development of the solar thermal technology. He currently manages the Materials and Structures Division for Science Applications International Corporation, which includes management of solar projects and the NASA Solid Propulsion Integrity Program (SPIP) Bondline work package, and is the Vice President of the Solar Energy Industries Association (SEIA).

Michael DeAngelis

Michael DeAngelis is currently the Assistant Deputy Director for Energy Technology Development at the California Energy Commission. This division completes research and commercialization activities for transportation, electricity generation, and end use efficiency technologies for California, and includes over 100 staff and over $30 million in annual contract funds. He formerly was the Manager of Research and Development at the Commission for three years, which includes the Energy Technologies Advancement Program (ETAP), Geothermal Program, Small Business Program, Wind Performance Reporting, Solar Tax Credits Support, and Energy Technologies Status Report activities. Before working at the Commission, he was employed as a senior scientist and manager from 1978 to 1982 at the Solar Energy Research Institute (recently renamed National Renewable Energy Laboratory) in Golden, Colorado. Mr. DeAngelis has authored or co-authored twenty-two published papers or reports in the alternative energy fields since 1978. He has both undergraduate and graduate science degrees in environmental sciences at California State University (Sacramento) and at the University of British Columbia.

Stephen L. Feldman

Stephen L. Feldman was Director of the Center for Energy and the Environment and Chairman of the City and Regional Planning Department at the University of Pennsylvania until his untimely death in 1990. His Ph.D. was in geography from the Hebrew University in Jerusalem, Israel, and he had an M.A. from Johns Hopkins University. He had worked on national and international energy policy for a number of years. He began working on utility/solar-thermal-energy issues in 1975, and published numerous monographs and papers on this subject. In 1980 he authored, with Robert M. Wirtshafter, *On the Economics of Solar Energy: The Public Utility Interface.*

Steven Ferrey

Steven Ferrey is professor of law at Suffolk University School of Law in Boston, and also teaches at Boston University School of Law. He is the author of more than fifty articles on energy and environmental subjects, and is the author of *The Law of Independent Power* (New York: Clark Boardman Callaghan, 5th ed., 1994), the standard treatise on the electric utility industry's evolution to independent power. Professor Ferrey consults widely on energy matters in North America, as well as working overseas for the World Bank in developing countries. He holds three academic degrees in economics, environmental planning, and law, and was a postdoctoral Fulbright Fellow in London.

Murrey D. Goldberg

Murrey D. Goldberg served from 1975 to 1979 as coordinator of the Interagency Panel on Terrestrial Applications of Solar Energy. This Washington-based panel was responsible for interagency coordination of all aspects of the U.S. nonspace solar program. In particular, the panel provided the main link between the congressionally mandated roles of the Energy Research and Development Administration (later the Department of Energy) and the Department of Housing and Urban Development in the residential demonstration program. Goldberg was a staff physicist at the Brookhaven National Laboratory when he began this work, moving in 1977 to the Solar Energy Research Institute were he was chief of the International Programs Branch until 1981. He is now retired.

Charles Grosskreutz

Charlie Grosskreutz, now retired, was active in the development of solar heat technologies beginning in late 1972. He was Manager, Solar Energy Programs, for Black & Veatch Engineers during the period 1972–77 in which he participated in the design and construction of the Central Receiver Test Facility at Sandia Laboratories in Albuquerque, New Mexico. He also contributed to the development of the solar thermal research program at the Electric Power Research Institute. In July 1977, Dr. Grosskreutz became the first Director of Research at the newly formed Solar Energy Research Institute (SERI) in Golden, Colorado. In that position, he was responsible for building and directing the research and engineering effort at SERI.

Dr. Grosskreutz returned to Black & Veatch in 1979 as Manager of Advanced Technology Projects. He organized the Solar Thermal Review Panel for the U.S. Department of Energy (DOE) in 1980 and served as its chairman until 1984. He also served as chairman of the Technical Committee for the Molten Salt Electric Experiment conducted at the Sandia Albuquerque solar test facility. At the time of his retirement from Black & Veatch at the end of 1988, he had completed a major economic evaluation of Integrated Coal Gasification Combined Cycle Power Plants.

His degrees include a B.S. in mathematics from Drury College and an M.S. and Ph.D. in physics from Washington University in St. Louis. He is a Fellow of the American Physical Society and ASTM.

Dan Halacy

Dan Halacy has been seriously involved in solar energy work since the 1955 inaugural meeting of the Association for Applied Solar Energy in Phoenix. He is the author of eight books and numerous papers on solar energy, and he holds two solar patents. He served on the Arizona Solar Energy Commission and also as State Solar Officer for the Western Solar Utilization Network. Solar energy adviser to U.S. Senator Paul Fannin for two years, Mr. Halacy also served as a board member and vice chairman of the American Solar Energy Society and general chairman of the ASES Twenty-fifth Anniversary Meeting in 1980. He spent five years at the Solar Energy Research Institute (now the National Renewable Energy Laboratory) in public affairs and technical writing and editing.

Marvin Hall

Marvin Hall retired in 1989 from the University of Illinois where he had spent twenty-seven years as an Extension Service Agricultural Engineer. Hall has agriculture degrees from the University of Missouri and the University of Southern Illinois. He was named to the Farm Building Hall of Fame in 1982.

He began working on solar energy applications for agriculture in 1965. Between 1978 and 1983 he was the principal investigator on "solar field studies" conducted by the University of Illinois with funding from the Illinois State Energy Office.

Oscar R. Hillig

Oscar R. Hillig is retired after forty-one years with the Rockwell International Corporation and Argonne National Laboratories. He received his bachelor's degree in physics from the Illinois Institute of Technology. After thirty-one years in the nuclear field, Mr. Hillig transferred to the Department of Energy (DOE) sponsored Energy Technology Engineering Center operated by Rockwell International. Here he managed several solar programs for the DOE. He was Program Manager of the Solar in Federal Buildings Program from 1981 to the completion of the program in 1992.

Mary-Margaret Jenior

Mary-Margaret Jenior has been with the U.S. Department of Energy since 1975, serving as an Architect/Program Manager for the Office of Solar Heat Technology and Office of Building Technologies. She manages research on solar and energy efficiency building technologies. She has also managed research on technology transfer and home energy rating systems. She also coordinates solar building international research programs for DOE. She has led expansion of the passive solar program into research required to integrate passive and hybrid heating, cooling, and daylighting systems, advanced glazing and storage materials, and thermal transport components with a building's system and aesthetic requirements. Ms. Jenior has a bachelor's of architecture degree from Case-Western Reserve, a master's in city planning from University of Pennsylvania, and experience as a planner for Cincinnati, Ohio, and for the U.S. Navy and Air Force.

David W. Kearney

David W. Kearney is President of Kearney & Associates, an energy consulting firm active in solar thermal technology evaluation and development. He has over fifteen years' experience working with solar thermal systems. Starting in this field with the Solar Energy Research Institute in the late 1970s, he also spent six years as a Vice President of Luz in the late 1980s during the development of the SEGS plants, which supply 354 MW to the electricity grid in southern California. Educated in mechanical engineering, he received his Ph.D. from Stanford in 1970. At SERI Dr. Kearney was responsible for the solar industrial process heat program and manager of the solar thermal development activities. He has extensive experience with international projects associated with solar thermal electric siting and development. He has been active in ASES and served on its board of directors. His work of recent years has focused on consulting with utilities, institutions, the operating SEGS plants, and industry on the assessment, evaluation, and development planning of solar thermal systems. He has many publications and presentations in these areas, and was a co-author of the chapter on solar thermal electric technology contained in *Renewable Energy—Sources for Fuels and Electricity*, edited by Johansson et al. and published as input to the 1992 UN Conference on Environment and Development in Rio de Janeiro.

W. Henry Lambright

W. Henry Lambright is a professor of political science and public administration, The Maxwell School, Syracuse University; and Director of the Science and Technology Policy Center, Syracuse Research Corporation. He has also been an adjunct professor at the SUNY College of Environmental Science and Forestry. He received his A.B. in political science at Johns Hopkins, and M.A. and Ph.D. from Columbia. He has had a longstanding interest in the relation between government and science and technology. He has written numerous articles and the following books: *Governing Science and Technology* (New York: Oxford, 1976); *Presidential Management of Science and Technology: The Johnson Presidency* (Austin, TX: University of Texas Press, 1985); *Technology Transfer to Cities* (Boulder, CO: Westview Press, 1979); and (with others) *Educating the Innovative Public Manager* (Cambridge: Oelgeschlager, Gunn & Hain, 1981). Most recently, he has co-edited a book entitled: *Technology and U.S. Competitiveness: An Institutional Focus* (Westport: Greenwood, 1992).

Carlo La Porta

Carlo La Porta is an analyst in the renewable energy field and a participant in the solar thermal energy industry. He has an M.A. in international relations from the Johns Hopkins University School of Advanced

International Studies (1972), a B.A. in history from Indiana University (1967) and a Diplome in Etudes Francaises from the Universite de Strasbourg, France (1966). Since 1976, he has worked in the solar energy field, providing analysis, contract management, and consulting to a wide range of public and private sector clients. While Director of Research for the Solar Energy Industries Association from 1982 to 1986, he edited and authored a two-volume history of the renewable energy industry, completed in 1985. La Porta served as a member of an International Energy Agency Advisory Review Board that oversaw publication of the IEA's book, *Renewable Energy Resources*, for which he provided additional research writing and editing. He also contributed to the *Greenpeace Report on Global Warming* (Oxford University Press) as a result of having managed the Forum on Renewable Energy and Climate Change in 1990. La Porta has served as Executive Director of the Maryland-D.C.-Virginia Solar Energy Industries Association (1988 to 1992) and was appointed to Governor Schaefer's Energy Task Force for Maryland. More recently he has provided management support to the Department of Energy Office of Energy Efficiency and Renewable Energy and worked on developing commercial projects for solar thermal industrial process heat systems.

Ronal W. Larson

Dr. Larson, as President of Larson Consulting, is active in solar energy policy and economic analysis—both in the United States and internationally. A Congressional Fellow in 1973–74 (while on sabbatical from Georgia Tech), he worked on the first solar legislation to pass the U.S. Congress. He later was one of the first staff at the Solar Energy Research Institute, largely working on national solar program evaluation and budgeting as SERI's first Principle Scientist. He left SERI in 1982 to head a USAID solar program in Sudan. Since 1984 on a part-time basis and since 1992 on a full-time basis, he has devoted his efforts to consulting on a range of solar topics.

Robert T. Lorand

Robert T. Lorand manages the Energy Services Division of Science Applications International Corporation (SAIC). In this capacity he is responsible for overseeing work in the areas of energy efficiency and renewable energy, demand-side management, and alternative fuels. Over the past eighteen years, he has performed numerous studies and technology assements of various solar and energy efficiency technologies for government and private sector clients. Mr. Lorand has provided technical support services to the Department of Energy's (DOE) Active and Passive Solar Programs since 1979. This has included developing plans and strategies for advancing the technology and hastening its commercialization. He has taught solar design workshops for architects/engineers. Prior to joining SAIC, Mr. Lorand was employed by Insolar, Inc., a small solar energy equipment and energy consulting firm.

Gerald M. Mara

Gerald M. Mara worked in a variety of capacities for the National Solar Heating and Cooling Information Center from 1976 through 1979. He was Policy Analyst for the Cities Project at the Center for Renewable Resources from 1980–1982. He was Research Director and principal author of *Putting Renewable Energy to Work in Cities*, a three-volume report published by the U.S. Department of Energy in 1982. A condensed version of this study was also published by Van Nostrand in the same year. He is the author (joint or sole) of a number of reports on the institutional aspects of solar energy use. Since 1982, he has worked in the Graduate School of Georgetown University. He was named Associate Dean for Research in 1983. He has also taught graduate and undergraduate courses for the department of government since 1983–84. His principal research and teaching interests are in political philosophy. He is joint editor of *Liberalism and the Good* (Routledge, 1990) and the author of a number of articles appearing in refereed journals and edited collections.

Alan S. Miller

Alan S. Miller is Executive Director of the Center for Global Change at the University of Maryland, in College Park, Maryland. Mr. Miller is a lawyer and policy analyst who has worked on environmental and energy issues for more than fifteen years at research institutions in the United States and abroad, including the World Resources Institute, the Natural Resources Defense Council, the American Bar Association, and the Environmental Law Institute. He is the co-author of *Environmental Regulation: Law, Science, and Policy*.

Mr. Miller earned his J.D. and M.P.P. from the University of Michigan and his bachelor's degree from Cornell University. He has taught energy and environmental law courses at numerous universities including Maryland, Duke, George Washington, Widener, and Iowa.

J. Glen Moore

J. Glen Moore is a research specialist with the Congressional Research Service. He became involved in the congressional oversight of solar energy issues in the early 1970s, and remained involved until congressional interest peaked in the 1980s. He provided nonpartisan committee and member staff support on solar matters throughout this period, assisting in the development of solar legislation and contributing to numerous committee publications on heating and cooling and other solar applications and issues. Mr. Moore holds a bachelor of science degree from the University of Maryland at Salisbury.

Myron L. (Mike) Myers

Mike Myers is a graduate of Washington State University and a registered professional engineer. Mr. Myers retired from NASA in 1980. In the 1960s, Mr. Myers was responsible for the acceptance testing (static firing) of all of the first stage propulsion systems (S-IC's) for the Saturn/Apollo ("moon rocket") space vehicle which, among other missions, took our astronauts to the moon in July 1969.

Prior to retirement, he managed the Department of Energy's solar energy Commercial Demonstration Program using engineers employed at NASA's George C. Marshall Space Flight Center located at Huntsville, Alabama. Some 118 projects were involved including Hotel/Motel Hot

Water Initiative projects. Other than these, there were active solar energy systems that provided solar heated transport fluid to heat water, provide space heat, and provide air conditioning.

One of Mr. Myers's major accomplishments while managing the DOE program was the formation of a team of well-equipped engineers who visited sites where solar energy systems were not performing as expected. These teams were composed of engineers with unique capabilities. These teams used various types of intrusion devises for taking measurements as well as using originally installed hardware.

Toward the end of his career, Mr. Myers headed a development program for various solar energy system components including data acquisition, collectors, etc.

Paul Notari

Paul Notari is founder and president of SciTech Communications, Inc., a firm specializing in the communication of scientific and technical ideas. Mr. Notari was formerly Manager of Technical Information Programs at the Solar Energy Research Institute in Golden, Colorado (later renamed the National Renewable Energy Laboratory). In this capacity he was responsible for all of the information programs conducted by the Institute on behalf of the U.S. Department of Energy. Mr. Notari was with the Institute from 1979, shortly after its founding, until December 31, 1991.

Mr. Notari has been a member of the Board of Directors of the American Solar Energy Society since 1982. He was elected Chairman in 1990 and 1991. Prior to his employment at SERI, Mr. Notari served as Director of Publications for the American Water Works Association, Director of Communications for the Computer and Business Equipment Manufacturers Association, President of the Association of Computer Programmers and Analysts, and Manager of Publications and Training for Motorola Inc.

J. Kevin O'Connor

Mr. O'Connor has a broad base of employment experience in private industry, government, and academic/research environments. With a B.A. from Franklin and Marshall College, and an M.A. from the University of Delaware, Mr. O'Connor commenced his alternative energy and energy conservation experience at the University of Delaware's Institute of Energy Conversion in 1973 as Program Manager of Solar One, the University of Delaware's experimental solar house project. Here he was responsible for program development geared to raising energy utilization awareness and the insurance of successful solar installations. He co-authored solar design and system installation and maintenance manuals, and taught courses in these fields at both the University of Delaware and later at the Community College of Denver—Red Rocks Campus

Mr. O'Connor moved on to the Solar Energy Research Institute (SERI) in 1978 to lead the development of a national solar energy educational database of all postsecondary institutions offering solar energy courses and programs. National and state directories of solar educational offerings were published for vocational schools and colleges and universities. During this time, Mr. O'Connor also chaired a national task force on solar energy technical training assessment.

After a hiatus while working with AT&T for eight years on the development of private business exchange telephone switching systems, Mr. O'Connor's most recent experience in the energy field has taken him back to the National Renewable Energy Laboratory, formerly SERI, where he has been involved with the development and growth of the national Alternative Fuels Data Center since 1990. In addition to AFDC involvement, he acts as technical monitor for the national Alternative Fuels Hotline, where callers from around the country can have their questions answered on any aspect of alternative fuels.

Daniel Rich

Daniel Rich is Dean of the College of Urban Affairs and Public Policy and professor of urban affairs, public policy, and political science at the University of Delaware. He is a Senior Research Associate in the Center for Energy and Environmental Policy and holds a concurrent appointment as visiting professor in the Centre for Planning at the University of Strathclyde, Scotland. Dr. Rich is co-editor of the Energy Policy Studies book series. His recent publications include *Energy and Environment: The Policy Challenge, Planning for Changing Energy Conditions*, and *The Politics of Energy Research and Development*.

J. David Roessner

J. David Roessner is professor of public policy at Georgia Institute of Technology and Co-Director of the Technology Policy and Assessment Center. Prior to joining the Georgia Tech faculty in 1980, he was Principal Scientist and Group Manager for Industrial Policy and Planning at the Solar Energy Research Institute in Golden, Colorado. He served as Policy Analyst with the National Science Foundation's R&D Assessment Program and, subsequently, as Acting Leader of the Working Group on Innovation Processes and their Management in the Division of Policy Research and Analysis at NSF. Previous to this he was Research Associate at the Bureau of Social Science Research, Inc. His first professional position was as a development engineer for Hewlett-Packard Co. in Palo Alto, California.

Dr. Roessner received B.S. and M.S. degrees in electrical engineering from Brown University and Stanford University, respectively. He returned to graduate school after working at Hewlett-Packard to receive the master's degree in science, technology, and public policy from Case Western Reserve University in 1967, and the Ph.D. in the same field in 1970.

William W. Schertz

Mr. Schertz was employed from 1966 to 1969 as an Assistant Project Engineer at Pratt and Whitney Aircraft, where he worked on the development of both base electrolyte (KOH) and acid electrolyte (H_3PO4) fuel cells. From 1969–1970 he worked on liquid sodium technology at Los Alamos Scientific Laboratories. He joined Texas Instruments in 1970 and worked on the development of thermal printers for calculators and computer terminals and on the development of liquid crystal displays for electronic watches. He joined Argonne National Laboratory in 1972 and has worked on the development of high energy density batteries, the Compound Parabolic Concentrator Solar Collector, thermal energy storage systems, and energy conservation systems.

In 1983, he was named Associate Division Director of the Energy and Environmental Systems Division responsible for Systems Engineering and Technology. In 1986, he was named Senior Chemical Engineer, and in January 1986, he became responsible for the program management of all Energy Efficiency and Renewable Energy Research Programs at ANL. From 1989–1991 he also served as Deputy Director of Energy Systems Division at ANL. He has eight patents, and has published forty-six papers and reports.

William B. Scholten

William Scholten has a Ph.D. in mechanical engineering from the University of Wisconsin—Madison. He was on the faculty of the energetics department of the College of Applied Science and Engineering of the University of Wisconsin—Milwaukee for seven years, where he became active in renewable energy research. He has been employed by Science Applications International Corporation (SAIC) since 1980. Since 1975 he has been heavily involved in the U.S. Department of Energy's (DOE) Renewable Energy Development Program. With InterTechnology Solar Corporation from 1975 to 1978 he participated in the development of Phase I of the Solar Heating and Cooling Demonstration Program, and several system designs for the "PON" cycles. With Planning Research Corporation (1978–79), Planning and Management Associates (1979–80), and SAIC he provided technical support to DOE's Active Solar Heating and Cooling Program.

T. P. Schwartz

T. P. Schwartz, Ph.D., University of North Carolina, is a sociologist who teaches at the University of Rhode Island and Providence College and conducts studies of long-term, large-scale social problems including environmental destruction and the overconsumption of energy. He has served as Executive Director of the Center for Energy Policy, Boston, and the Arson Task Force, Rhode Island.

Robert Shibley

Robert Shibley is a professor of planning and architecture at the State University of New York at Buffalo, where he served as chair of the department of architecture from 1982 to 1990 and currently serves as Director of Urban Design. He was initially the project manager and then the Branch Chief of the Department of Energy's Passive and Hybrid Solar Commercial Buildings Program from 1980 to 1982. He is also a partner of The Caucus Partnership: Consultants on Environmental and Organizational Change, currently based in Buffalo, New York. He is the author of *Urban Excellence* (Van Nostrand Reinhold, 1989) and co-author of *Placemaking: The Art and Practice of Building Community* (John Wiley and Sons, 1994).

William A. Tolbert

William A. Tolbert spent over a decade developing and managing military renewable energy programs. In 1974, he served as the lead engineer for the design and construction of the award-winning USAF Academy solar home—the first residential solar heated home in the Department of Defense. After his assignment at the Academy, Mr. Tolbert went on to manage the Air Force renewable energy R&D program, and serve as the first military Research & Technology Liaison Officer (RTLO) assigned to the U.S. Department of Energy. In 1981, he was sent to the Solar Energy

Research Institute (SERI) on an interagency assignment, and served as Manager, Building Systems Research Branch, and Task Leader for the DOE National Passive Solar Program. In 1984, he was assigned to the Air Force Space Command, where he managed a division of the military space shuttle program until he retired from active duty. Mr. Tolbert then returned to SERI as Director, Development and Communications Office. He left SERI in 1989 to pursue an executive career in industry with the Raytheon Company, where he was a national product line manager for international infrastructure programs. In 1993, he formed his own Denver-based international consulting firm, The Meneren Group (TMG). Bill has a master's degree in architectural engineering from the University of Texas and is a registered professional engineer in the state of Texas. Bill has been a member of the American Solar Energy Society (ASES) since 1974.

Rebecca Vories

Rebecca Vories is the owner and principal staff of Infinite Energy. Her primary experience is in market and evaluation research, market strategy planning, development and fielding of public information, advertising and public relations programs. Prior to founding Infinite Energy in 1981, she developed consumer information and market research programs for the U.S. Solar Energy Research Institute (now the National Renewable Energy Laboratory) and the Colorado Energy Research Institute from 1974–1981. Since 1981, she has worked for more than 180 clients on programs aimed at encouraging greater use of energy-efficient, renewable energy and/or environmentally sound technologies, evaluating the effectiveness of such programs, or in assisting policymakers and the general public to understand the issues involved in such efforts.

   She has also been a very active volunteer in the solar field beginning as founding member of the Colorado Solar Energy Association in 1976 and since 1978 serving five terms on the board of directors of the American Solar Energy Society. She is currently the Treasurer of the Society and a member of the Red Rocks Community College Solar Program Advisory

Board. She is also active in the Professional Association of Consumer Energy Educators.

She received a B.A. from the University of Colorado at Boulder and a B.I.M. from the American Graduate School for International Management in Glendale, Arizona.

Roberta W. Walsh

Roberta W. Walsh is associate professor of consumer studies in the College of Agriculture and Life Sciences at the University of Vermont. She holds the B.S. degree in education from Framingham State College in Massachusetts, M.S. in consumer economics from Cornell University, and Ph.D. in social policy from the Heller Graduate School, Brandeis University.

Before assuming her present position she was Consumer Affairs Manager at the Northeast Solar Energy Center in Boston. She also held consumer affairs positions with the Federal Energy Administration, U.S. Department of Energy, and Federal Trade Commission.

Currently Dr. Walsh teaches courses in consumer policy, energy, and housing. She is associated with the University of Vermont's Center for Rural Studies where she has served as principal investigator for evaluations of state and federal energy programs. She has been an energy program consultant in both the United States and United Kingdom, and has served on the planning committee of the International Energy Program Evaluation Conference throughout its ten-year history. Her publications include numerous articles on the design, implementation, and evaluation of energy programs. She is the co-editor of *Energizing the Energy Policy Process: The Impact of Evaluation* (Quorum Books, 1994).

Seymour Warkov

Seymour Warkov serves as professor of sociology, University of Connecticut. His research has focused on behavioral and institutional aspects of energy and environmental policy issues, including *Energy Policy in the United States: Social and Behavioral Dimensions* (editor); *Solar Diffusion and Public Incentives* (with J. W. Meyer); and refereed articles, chapters, and other papers on these topics. His most recent work concerns hazardous materials transport issues and public acceptance of resource recovery technology.

Patricia Weis-Taylor

Patricia Weis-Taylor consults on the effectiveness of technical communications for solar energy, energy conservation, space weather, and human health and safety. Ms. Weis-Taylor was among the original 125 employees of the Solar Energy Research Institute (now the National Renewable Energy Laboratory) in Golden, Colorado in 1977. Working as a Policy Analyst at SERI during the early years she observed the personalities and trends of this period from inside the DOE organization.

Ms. Weis-Taylor holds a master's of public policy degree from the University of California at Berkeley. Her publications, written under contract, include many annual overviews of U.S. Department of Energy solar programs, as well as guides to technical literature for energy conservation and wind energy technologies.

Ronald E. West

Ronald E. West is professor of chemical engineering at the University of Colorado, Boulder. West first worked on solar energy in the early 1960s and returned to the subject in 1978 as a consultant and sometime visiting professional at the Solar Energy Research Institute. Efficient utilization of energy and materials continue to be his chief professional interests. He holds a Ph.D. in chemical engineering from the University of Michigan.

Peggy Wrenn

Peggy Wrenn is working as a consultant to state and local government and writing a book on public management. She recently resigned as Assistant City Manager of Boulder (1985–1987) and previously worked as Energy Director for the City of Boulder (1982–1985). She was Renewable Energy Director for the State of Colorado Governor's Office of Energy Conservation from 1978–1981. She has built her own passive solar homes and served on the boards of directors of Solar Lobby, Western SUN, American Solar Energy Society, Colorado Council of Local Energy Officals, and dozens of local and regional professional energy associations. She is author of many publications on solar energy and energy conservation.

Robert G. Yeck

Robert G. Yeck, Ph.D., P. E. is an agricultural engineer specializing in farm building, rural housing, environmental control systems, and solar energy. He has served as a research scientist with the Agricultural Research Service, U.S.D.A., from 1948–1980 and as a visiting professor for research, teaching, and extension at the University of Maryland from 1980 to 1988. Dr. Yeck managed and/or participated in research and demonstration projects on solar houses, wind energy, solar grain storage, and solar heating of livestock shelters. He represented the U.S.D.A. on President Jimmy Carter's Domestic Policy Review of Solar Energy during its six months of deliberations and worked with the AID program on Solar Energy Applications in North Africa

Gene Zerlaut

Gene Zerlaut has been involved with active solar thermal and solar photovoltaic technologies since 1973. Mr. Zerlaut was the president of DSET Laboratories Inc. from 1973 until its sale in 1991, at which time he formed SC-International Inc., a technical and business consulting company.

Mr. Zerlaut was instrumental in the development of the original SRCC solar collector testing, rating, and labeling program, and is currently involved in developing a similar program for photovoltaic module certification. He was the founding chairman of both ASTM Committee E-44 on Solar Energy and the U.S. Technical Advisory Group (TAG) for ISO/ TC 180 Committee on international standards for solar energy. He currently serves as a U.S. delegate to ISO/TC 180 and chairs the subcommittee on materials. Mr. Zerlaut is also currently chairman of ASTM Subcommittee G3.09 dealing with standards for solar and ultraviolet radiometry. Credited with the development of numerous methods for solar device testing, for optical and durability testing of solar materials, and standards for calibrating solar radiometers, Mr. Zerlaut holds patents for solar tracking devices and solar simulators. He has more than twenty publications dealing with solar energy and numerous publications, reports, and seminars in allied fields.

Name Index

Subject Index

Acceptance testing, 333–335
Acceptance Test Procedure (ATP), 334–335
Active Heating and Cooling Division, 194, 361
Active Solar Heating Systems Design Manual, 371
Active systems, 5. *See also* Active systems development
codes, standards, and performance criteria for, 190–191
costs of, 23–24, 194
data collection, evaluation, and dissemination for, 188–190
demonstrations for, 181–187
historical background, 177–179
legislation for, 179–180
lessons learned from, 194–196
market development for, 191–194
Mid-American Solar Energy Complex programs for, 686–687
National Program for Solar Heating and Cooling of Buildings for, 180–181
R&D programs for, 187–188
Active systems development, 119
air vs. hydronic collectors in, 129–131
applications in, 129–131
geographic locations in, 131–134, 145–146
government involvement in, 139–140, 146–147
imports and exports, 143–144
market size, 144
private sector sales, 140–143
sellers in, 132, 135–139
for swimming pool heating, 126–129
technology in, 145
Acurex Solar Corp. (ASC)
cooling systems by, 851–852
military demonstration program by, 436–437
parabolic trough systems by, 155, 159
Aden agricultural project, 397–398
Advanced Energy Utilization Test Bed (AEUTB), 424
Advanco Corp., 158, 275
Advisory Council of States (ACS), 677, 679
Aerospace Corp.
for high-temperature technologies, 266
study by, 900
Agricultural and Industrial Process Heat (AIPH) Program, 244–245. *See also* Industrial process heat
Agricultural demonstration programs, 38–39

benefits and information dissemination in, 403–405
evaluation procedures in, 392
grain drying, 390–391, 393–395
greenhouse activities in, 405–407
Illinois projects, 395–403
livestock program, 389–390, 392–393
organization agreements for, 388–389
rationale for, 387–388
recommendations for, 407–408
system erection and monitoring, 391–392
Agricultural process heat, 244–245
Agricultural Research Service, 388, 405–406
Agriculture, Department of
for consumer information, 575
technology transfer in, 696
AIA Research Corp., 237, 599
Air collectors vs. hydronic, 129–131
Air-Conditioning and Refrigeration Institute (ARI), 190–191, 449–450, 454, 466, 468
Air Force
demonstration programs by, 426, 430, 432–433
energy organization, 416–417
Air Force Academy, 422–424
Air Force and Army Exchange Service (AFAES), 422, 433–436
Alternate Energy Source Financing Authority, 885
American Association of Community and Junior Colleges (AACJC), 656
American Bar Foundation (ABF) studies, 919–920, 930
American Council to Improve Our Neighbourhoods (ACTION), 572–573
evaluation of, 574
goals and objectives of, 573–574
lessons learned from, 574–575
American Institute of Architects (AIA), 165, 235
American Institute of Architects Research Corp., 237, 599
American National Standards Institute (ANSI)
organization of, 473–474
standards by, 451
Steering Committee on Solar Energy Standards Development, 474–478
American Planning Association, 599, 877
American Science and Engineering (AS&E), 748
American Society for Testing and Materials (ASTM)
Committee E-44, 481–484

enactment of, 90–92
purpose of, 180
for testing and certification, 448
Public Law 95–39 (Title V, Energy
  Research and Development Administra-
  tion Appropriation Authorization), 546
Public Law 95–88 (International
  Development and Food Assistance Act of
  1977), 97
Public Law 95–91 (Department of Energy
  Organization Act), 93–94, 180
Public Law 95–113 (Food and Agricultural
  Act of 1977), 96–97
Public Law 95–148 (Foreign Assistance and
  Related Programs Appropriation Act of
  1978), 97
Public Law 95–242 (Nuclear Non-
  Proliferation Act of 1978), 97
Public Law 95–315 (Small Business Energy
  Loan Act of 1978), 97
Public Law 95–356 (Military Construction
  Authorization Act, FY 1979), 98–99
Public Law 95–424 (International
  Development and Food Assistance Act of
  1978), 98
Public Law 95–426 (Foreign Relations
  Authorization Act, FY 1979), 98
Public Law 95–617 (Public Utilities
  Regulatory Policies Act (PURPA))
  provisions of, 95
  for public utilities, 897–898
  for SEGS systems, 155–156, 285
  for state and local programs, 863–864
  for technology transfer, 698
Public Law 95–618 (Energy Tax Act), 757,
  791–792
  enactment of, 12
  provisions of, 95
  for testing and certification, 448
Public Law 95–619 (National Energy
  Conservation Policy Act), 180, 182, 510
  for auditor training, 649
  for Federal buildings demonstration
    programs, 359–360
  for legal issues, 913
  provisions of, 95, 98–99
Public Law 95–620 (Powerplant and
  Industrial Fuel Use Act), 96
Public Law 95–621 (Natural Gas Policy
  Act), 96
Public Law 96–223 (Crude Oil Windfall
  Profits Tax Act of 1980), 102–103, 757
Public Law 96–294 (Energy Security Act of
  1980), 101–103, 800, 917

Public Law 96–480 (Technology Innovation
  Act), 699
Public Law 96–517 (Small Business Patent
  Procedure Act of 1980), 705
Public Law 97–35 (Omnibus Budget
  Reconciliation Act), 110
Public Law 97–88 (Energy and Water
  Development Appropriations Act), 111
Public Law 97–101 (HUD Appropriations
  Act), 111
Public Law 97–214 (Military Construction
  Codification Act), 422, 436
Public Law 97–418 (Military Construction
  Act of 1981), 437–438
Public Law 97–436 (Military Construction
  Appropriations Act of 1981), 437
Public Law 98–370 (Renewable Energy
  Industries Development Act of 1983), 856
Public Law 98–620 (Trademark
  Clarification Act of 1986), 705
Public Law 99–502 (Federal Technology
  Transfer Act of 1986), 699, 705
Public Service Company of New Mexico
  (PNM), 150, 271–272, 736
Public utilities, 891–892
  conclusions about, 903–904
  consumption data for, 298
  decentralized technologies for, 899–903
  demand side management for, 898–899
  and Department of Energy, 895–898
  Federal mandates for, 892
  Federal research for, 892–895
  incentives for, 900–903
  repowering designs for, 271–272
  support from, 48–49
Public Utilities Regulatory Policies Act
  (PURPA), 21
  provisions of, 95
  for public utilities, 897–898
  for SEGS systems, 155–156, 285
  for state and local programs, 863–864
  for technology transfer, 698
Purchase programs, evaluations of, 58–59
Purchasers
  interviews with, 317–319
  Solar Energy Research Institute study of,
    772–773
Push vs. pull methods, 63–64, 147, 694, 732

Quality assurance
  codes, standards, and certification for,
    39–40
  in commercialization, 19
  for consumer assurance, 40